Scattering Characteristics of Aerial and Ground Radar Objects

This book presents computations for various types of aerial and ground objects. It contains a brief explanation of the theoretical calculation methods used for obtaining the scattering characteristics of these objects. It provides working examples for the analysis of electromagnetic wave scattering processes by different objects.

Scattering Characteristics of Aerial and Ground Radar Objects is divided into two sections. The first section includes scattering characteristics for different aerial objects: aircraft, helicopters, transport and passenger airplanes, unmanned aerial vehicles, and missiles. The second section contains data about scattering for many ground objects such as tanks, surface-to-air missile systems, ground radars, and other military objects. In total, this book contains actual data for 63 aerial objects (fighters, attack aircraft, bombers, long-range radar detection aircraft, transport aircraft, helicopters, unmanned aerial vehicles, and cruise missiles) and 18 ground objects, among which are anti-aircraft missile systems and tanks. This book contains data obtained by computations such as circular diagrams of radar backscattering; mean and median radar cross section values of various objects; probability distributions of echo signal amplitude given various parameters of illumination and various kinds of underlying surfaces (for ground objects). Also, as an example, the scattering characteristics for one surface ship are given.

This book will be a valuable reference for scientists, engineers, and researchers of electromagnetic wave scattering, computational electrodynamics, and those working on radar detection and recognition algorithms for aerial and ground radar targets.

Scattering Characteristics of Aerial and Ground Radar Objects

Oleg I. Sukharevsky and Vitaly A. Vasilets

CRC Press
Taylor & Francis Group
Boca Raton London New York

CRC Press is an imprint of the
Taylor & Francis Group, an **informa** business

Cover image: © Alexey Sukharevskyi

First edition published 2024
by CRC Press
2385 NW Executive Center Drive, Suite 320, Boca Raton FL 33431

and by CRC Press
4 Park Square, Milton Park, Abingdon, Oxon, OX14 4RN

CRC Press is an imprint of Taylor & Francis Group, LLC

© 2024 Oleg I. Sukharevsky and Vitaly A. Vasilets

Reasonable efforts have been made to publish reliable data and information, but the author and publisher cannot assume responsibility for the validity of all materials or the consequences of their use. The authors and publishers have attempted to trace the copyright holders of all material reproduced in this publication and apologize to copyright holders if permission to publish in this form has not been obtained. If any copyright material has not been acknowledged please write and let us know so we may rectify in any future reprint.

Except as permitted under U.S. Copyright Law, no part of this book may be reprinted, reproduced, transmitted, or utilized in any form by any electronic, mechanical, or other means, now known or hereafter invented, including photocopying, microfilming, and recording, or in any information storage or retrieval system, without written permission from the publishers.

For permission to photocopy or use material electronically from this work, access www.copyright.com or contact the Copyright Clearance Center, Inc. (CCC), 222 Rosewood Drive, Danvers, MA 01923, 978-750-8400. For works that are not available on CCC please contact mpkbookspermissions@tandf.co.uk

Trademark notice: Product or corporate names may be trademarks or registered trademarks and are used only for identification and explanation without intent to infringe.

ISBN: 978-1-032-67639-5 (hbk)
ISBN: 978-1-032-67640-1 (pbk)
ISBN: 978-1-032-67642-5 (ebk)

DOI: 10.1201/9781032676425

Typeset in Times
by codeMantra

Access the Support Material: www.routledge.com/9781032676395

Contents

About the Authors viii
Introduction x
Abbreviations xii

Chapter 1 Brief Description of Calculation Methods for Scattering Characteristics of Aerial and Ground Objects 1

Chapter 2 Scattering Characteristics of Aerial Objects 10
    2.1   Multirole Fighter Su-27 12
    2.2   Front-Line Fighter MiG-29 18
    2.3   Multirole Fighter Su-57 24
    2.4   Tactical Fighter F-22 30
    2.5   Multirole Fighter F-35 36
    2.6   Multirole Fighter F-16 42
    2.7   Air Superiority Fighter F-15 48
    2.8   Multirole Attack and Fighter Aircraft F/A-18 54
    2.9   Multirole Combat Aircraft Tornado IDS 60
    2.10 Multirole Fighter EF-2000 Typhoon 66
    2.11 Attack Aircraft Su-25 72
    2.12 Attack Aircraft A-10 Thunderbolt II 78
    2.13 Tactical Bomber Su-24 84
    2.14 Long-Range Strategic Bomber Tu-22M3 90
    2.15 Strategic Bomber Tu-95 96
    2.16 Strategic Bomber Tu-160 102
    2.17 Tactical Bomber Su-34 108
    2.18 Strategic Bomber B-2 114
    2.19 Strategic Bomber B-52 120
    2.20 Strategic Bomber B-1B 126
    2.21 Airborne Early Warning and Control Aircraft A-50 132
    2.22 Airborne Early Warning and Control Aircraft E-3A 138
    2.23 Airborne Early Warning and Control Aircraft E-2C 144
    2.24 Jet Trainer Aircraft L-39 150
    2.25 Transport Aircraft An-26 156
    2.26 Strategic Airlifter Il-76 162
    2.27 Medium-Range Airliner Boeing 737-400 168
    2.28 Multi-Purpose Helicopter Mi-8 174
    2.29 Multi-Purpose Combat Helicopter Mi-24 180
    2.30 Unmanned Aerial Vehicle Tu-143 Reys 186
    2.31 Unmanned Aerial Vehicle Orlan-10 192
    2.32 Unmanned Aerial Vehicle RQ-1 Predator 198
    2.33 Unmanned Aerial Vehicle RQ-4 Global Hawk 204
    2.34 Unmanned Aerial Vehicle RQ-7 Shadow 210
    2.35 Unmanned Aerial Vehicle Bayraktar TB2 216
    2.36 Unmanned Aerial Vehicle Mohajer-6 222

2.37	Unmanned Aerial Vehicle IAI Searcher II (Forpost)	228
2.38	Unmanned Aerial Vehicle Dozor-600	234
2.39	Unmanned Aerial Vehicle Kronshtadt Orion	240
2.40	Loitering Munition Shahed 136	246
2.41	Cruise Missile Kh-555	252
2.42	Cruise Missile Kh-101	258
2.43	Cruise Missile P-700 Granit	264
2.44	Cruise Missile P-800 Oniks	270
2.45	Cruise Missile 3M-14 Kalibr	276
2.46	Cruise Missile 3M-54 Kalibr	282
2.47	Cruise Missile AGM-86C CALCM	288
2.48	Hypersonic Missile Kh-47M2 Kinzhal	294
2.49	Anti-Radiation Missile Kh-25MPU	300
2.50	Air-to-Surface Missile Kh-29T	306
2.51	Anti-Radiation Missile Kh-31PD	312
2.52	Cruise Missile Kh-32	318
2.53	Cruise Missile Kh-35	324
2.54	Air-to-Surface Missile Kh-38ML	330
2.55	Anti-Radiation Missile Kh-58UShKE	336
2.56	Air-to-Surface Missile Kh-59M	342
2.57	Air-to-Surface Missile AGM-65 Maverick	348
2.58	Air-to-Surface Missile AGM-114 Hellfire	354
2.59	Anti-Radiation Missile AGM-88 HARM	360
2.60	Cruise Missile Taurus KEPD 350	366
2.61	Cruise Missile Storm Shadow	372
2.62	Anti-Aircraft Missile 5V55R	378
2.63	Decoy Missile ADM-141C iTALD	384

Chapter 3 Scattering Characteristics of Ground Objects 390

3.1	Influence of the Underlying Surface on the Scattering Characteristics of a Ground Object	391
3.2	Main Battle Tank T-90	393
3.3	Main Battle Tank Leopard-2	399
3.4	Main Battle Tank M1 Abrams	405
3.5	Buk Target Acquisition Radar 9S18M1	411
3.6	Buk Command Post Vehicle 9S470	417
3.7	Buk Transporter Erector Launcher and Radar 9A310M1	423
3.8	S-300PS Transporter Erector Launcher 5P85D	429
3.9	S-300PS Transporter Erector Launcher 5P85S	435
3.10	S-300PS Fire Control System 30N6	441
3.11	S-300PS Low Altitude Acquisition Radar 76N6	447
3.12	S-300PS Command Post Vehicle 54K6	453
3.13	S-300V Transporter Erector Launcher and Radar 9A83	459
3.14	S-300V Command Post Vehicle 9S457	465
3.15	S-200 Launcher 5P72	471
3.16	Combat Vehicle 2S6 of Anti-Aircraft Gun-Missile System 2K22 Tunguska	477
3.17	Iskander Launcher 9P78-1	483

	3.18	Air Surveillance Radar ST-68U	489
	3.19	Radar P-18	495
	3.20	Ropucha-Class Landing Ship	501

Conclusion 507

References 508

Index 511

About the Authors

Oleg I. Sukharevsky has been working in the field of applied mathematics and computational electrodynamics since 1972 when he graduated from Kharkiv Gorky State University with a degree in computational mathematics and joined the Kharkiv Govorov Military Academy of Air Defence as an engineer. From 1977 he began his own study and development of numerical computational methods based on the solution of two-dimensional integral equations to obtain scattering characteristics of non-closed screen and antenna under frame radome. In 1983, he obtained the Candidate of Technical Science degree (Ph.D. analogue) from the All-Union Research Institute of Radiotechniques, and by that time he was promoted to the position of senior researcher. Since then, he has continued to study the scattering phenomenon and started to develop the scattering theory and technique for calculating the radar cross section of complex objects, with special emphasis on the objects that have surface fractures (sharp edges) and are (partially) covered with radar-absorbing materials. In 1993, he received a Doctor of Science degree in radar from the Kharkiv Govorov Military Academy of Air Defence.

Since the collapse of the Soviet Union in 1991, Oleg Sukharevsky has held not only academic positions but also teaching posts in applied mathematics and computational electrodynamics at the Govorov Academy, which was reorganized in 1993 as the Kharkiv Military University and later, in 2004, as the Ivan Kozhedub Kharkiv National Air Force University. From 1996 to 2001 and from 2004 to 2008, he was also employed as a senior researcher at the Usikov Institute of Radio Physics and Electronics of the National Academy of Sciences of Ukraine. During the turbulent period that followed the collapse of the Soviet Union, he founded a scientific school that has produced many prominent Ukrainian engineers and scientists. This was also the time when the explosive growth of computer power opened up entirely new possibilities for implementing theoretical electrodynamics methods in operational software. The appearance of personal computers facilitated the development of computer models of a number of air, ground, and underground objects, and Oleg Sukharevsky and the graduates of his academic school gained extensive experience in the simulation of scattering from real radar objects interrogated not only by narrow-band but also by wideband and ultra-wideband interrogation signals. He is the author and co-author of ten books and more than 250 papers in Russian, Ukrainian, and English. Oleg Sukharevsky is a senior member of IEEE and Honored Science Worker of Ukraine.

Currently, Dr. Sukharevsky is a leading researcher at the Scientific Center of Air Forces, Ivan Kozhedub Kharkiv National Air Force University, Kharkiv, Ukraine, professor. His main areas of interest include, but are not limited to, the mathematical theory of diffraction and scattering characteristics of radar targets.

About the Authors

Vitaly A. Vasilets has been working in the field of computational electrodynamics since 1992. In 1991, he graduated from Zhytomyr Higher Military School of Radio Electronics of Air Defence and entered the Kharkiv Govorov Military Academy of Air Defence as a postgraduate student. During that year, Vitaly sought his destiny under the scientific sun and found it in the person of Oleg Sukharevsky, who agreed to become his teacher. Since then, for 30 years, they have been developing applied electrodynamics together in many areas. In 1994, he received the degree of Candidate of Technical Sciences (Ph.D. analogue) from the Kharkiv Military University, at which time he was promoted to the position of senior researcher. Since then, he has continued to study scattering phenomena and started to develop a technique for calculating the radar cross section of complex objects located in free space and near the media interface. In 2006, he received a Doctor of Science degree in the field of radar from the Ivan Kozhedub Kharkiv National Air Force University.

After the collapse of the Soviet Union in 1991, Vitaly Vasilets worked in institutions that were the heirs of the Govorov Academy, which was reorganized in 1993 as the Kharkiv Military University, and later, in 2004, as the Ivan Kozhedub Kharkiv National Air Force University. This was also the time when the explosive growth of computer power opened up entirely new possibilities for implementing theoretical electrodynamics methods in operational software. The advent of personal computers facilitated the development of computer models of a number of air and ground objects, and Vitaly Vasilets gained extensive experience in the simulation of scattering from real radar objects interrogated not only by narrow-band but also by wideband and ultra-wideband interrogation signals. He has authored and co-authored five books and more than 150 papers in Russian, Ukrainian, and English.

Currently, Vitaly Vasilets is a leading scientific researcher at the Scientific Center of Air Forces, Ivan Kozhedub Kharkiv National Air Force University, Kharkiv, Ukraine, a senior scientist. His main areas of interests include the study of scattering characteristics of radar targets.

Introduction

Solving the problems of detection and recognition of objects in modern radar requires a priori information about their scattering characteristics, taking into account – in a complex – such complicating factors as the influence of an underlying surface, an irregularity of a boundary surface of an object, the presence of radar absorbing coatings (RAC) or composite structural elements.

Since it is very difficult and expensive to perform sufficiently accurate and statistically meaningful experimental studies of the scattering properties of radar targets, it is particularly important to develop theoretical foundations and computational methods for the mathematical modeling of air and ground objects, taking into account the complicating factors mentioned above.

We also note that mathematical modeling of the scattering properties of radar objects is relevant to analyze the effectiveness of target recognition in perspective radar systems. Such an analysis is carried out to determine the optimal composition, placement on the terrain, and parameters of the radars that will be part of such perspective systems.

Ten years have passed since the publication of the monograph "The Electromagnetic Wave Scattering by Aerial and Ground Radar Objects" [1], written by a group of authors from the Ivan Kozhedub Kharkiv National Air Force University. The monograph outlines a number of generalizations of the key theorems of classical electrodynamics, and the methods were developed by the authors on their basis for calculating the characteristics of the scattering of electromagnetic waves by radar objects. In Chapter 3 of the monograph [1], which has the character of a reference chapter, a large amount of actual calculation materials is provided: pie charts of backscattering, mean and median values of the radar cross section (RCS) of objects, distribution laws, amplitude scattering multipliers for different parameters of the probing signal, and types of the underlying surface (for ground objects).

After the publication of that monograph, the authors are being continued to work on improving calculation methods and increasing the volume of actual radar objects data.

In particular, the method for calculating the scattering characteristics of objects has been improved, taking into account re-reflections between smooth surface parts [2]. At the same time, areas of the object surface can be both perfectly conductive and covered with radar absorbing materials (RAM).

This consideration is very important if the object contains "corner reflectors", which are mutually orthogonal flat areas of the surface that collectively are wide-angle reflectors. At the same time, if there is a "trihedral corner reflector" (i.e., three flat, mutually orthogonal sections), after three times the reflection, a beam of rays is formed, which spreads in the direction opposite to the probing direction. Unlike the "dihedral corner reflector" (a wide scattering indicator in one plane), the "trihedral corner reflector" has a wide monostatic scattering indicator in both orthogonal planes.

Note that the presence of corner reflectors on the surface of the object can significantly increase its RCS. However, if the deviation from orthogonality between the flat sections of the surface forming the corner reflector reaches $1...2°$, then re-reflections between such sections can be neglected.

This book also uses the methods developed by the authors [3,4] to calculate the scattering characteristics of combined objects – objects having metal and composite elements under radar transparent (dielectric) shells. These methods make it possible to estimate the scattering characteristics of such complex (in terms of electrodynamics) objects as unmanned aerial vehicles (UAVs).

The computational methods, developed by the authors and used to obtain the scattering characteristics of radar objects, have been verified [1]. The verification of the methods was based on the comparison of the calculation results with the data of physical experiments and with the calculated values obtained for some specific cases using the FEKO software package [5]. In both cases, the comparison showed a quite satisfactory agreement.

Introduction

This book contains actual data for 63 aerial objects (fighters, attack aircraft, bombers, long-range radar detection aircraft, transport aircraft, helicopters, UAVs, and cruise missiles) and 18 ground objects, including anti-aircraft missile systems and tanks. The scattering characteristics of a surface ship are also given as an example.

The book contains an electronic appendix with a wide set of scattering characteristics data for the considered radar objects. The link to the electronic appendix is placed at the beginning of this book. The electronic appendix is also available on author's website at http://radar.dinos.net.

The authors would like to express their sincere gratitude to their colleague Serhiy Leshchenko. His advice and valuable comments helped to improve the quality of this book.

The authors would also like to thank Tanya Vereshchetina for her assistance in translating and preparing this book for publication.

Abbreviations

AWACS   Airborne Early Warning and Control System
RAC   radar absorbing coating
RAM   radar absorbing material
RCS   radar cross section
TEL   transporter erector launcher
TELAR   transporter erector launcher and radar
UAV   unmanned aerial vehicle

1 Brief Description of Calculation Methods for Scattering Characteristics of Aerial and Ground Objects

The calculations presented in this book are based on the methods developed by the authors for estimating the scattering characteristics of aerial and ground objects with complex shape (in general, with non-perfectly reflecting surfaces). Firstly, it is a method of calculating Kirchhoff-type integrals arising from the estimation of the contribution of smooth surfaces (possibly with RAC) to the full scattered field. In this case, the estimation of the contributions of smooth surface areas of an object is performed taking into account the re-reflections between separate surfaces. Secondly, it is a method of calculating the contribution of surface breaks (possibly also with RAC) to the full scattered field. It should be noted that for ground objects, the proposed method takes into account the electrodynamic interaction between the smooth parts of the object's surface and the ground, as well as between the edge-type parts of the object's surface and the ground. It should also be noted that the proposed method allows calculating the RCS of an object made entirely of dielectric or composite materials [1].

It should be knowledgeable that in the numerical calculation of the RCS using the proposed methods, most of the time is spent on calculating the field reflected from smooth surfaces. In this regard, the need to average the values in the frequency domain in order to obtain stable RCS estimates requires a significant increase in the computation time. One of the factors that makes the RCS a rapidly oscillating function of the frequency and the sounding angle is the dependence on these parameters of the phase difference with which the responses from different parts of the object surface are composed. To reduce the influence of this factor, it is proposed to use the sum of the RCS of separate sections of the object surface under study for a stable RCS estimation. Since this sum does not take into account the phase arrivals from different surface areas, such an estimate of the RCS will be called the "noncoherent" RCS. The value of the noncoherent RCS is a good and sufficiently stable estimate of the RCS over a given range of probing frequencies and aspect angles.

In addition, the authors' methods for calculating the scattering of electrodynamic waves on onboard antennas covered by dielectric nose radomes are used [1,4,6].

The estimation of the scattering characteristics of objects is based on a separate evaluation of the contributions of the smooth and edge parts of the object surface to the full scattered field. The field (projection on the direction $\vec{p}$) scattered by the smooth part S_1 (Figure 1.1) of the object surface in the direction $\vec{r}^0$ can be represented in the following form:

$$\vec{p} \cdot \vec{E}_{S_1} = -jk_0 \frac{\exp(jk_0 R)}{4\pi R} \int_{S_1} \left(\sqrt{\frac{\mu_0}{\varepsilon_0}} \left(\vec{p} \cdot \vec{H}_\perp \right) + \left(\vec{p} \times \vec{r}^0 \right) \cdot \vec{E}_\perp \right) \exp\left(-jk_0 \left(\vec{r}^0 \cdot \vec{x} \right)\right) ds. \quad (1.1)$$

Here R is the distance from the object to the observation point, $\vec{E}_\perp = \vec{n} \times \vec{E}$, $\vec{H}_\perp = \vec{n} \times \vec{H}$, $(\vec{E}, \vec{H})$ is the full field, $\vec{n}$ is the unit-vector of the external normal to the surface S_1 at point $\vec{x}$, k_0 is the wave number in free space. Note that everywhere in this book the time dependence $\exp(-j\omega t)$ was assumed. In the radar case, usually, smooth areas of the object surface have large electrical sizes

DOI: 10.1201/9781032676425-1

and small curvatures. Therefore, in the physical optics approximation, the tangential field components $\vec{E}_\perp(\vec{x})$, $\vec{H}_\perp(\vec{x})$ in (1.1) rotated by 90° in the tangential plane can be replaced by the corresponding values $\vec{\mathcal{E}}_\perp(\vec{x})$, $\vec{\mathcal{H}}_\perp(\vec{x})$ on a flat surface tangent to the surface S_1 at the point $\vec{x}$. In this case, for perfectly conducting parts of the object's surface covered with a layer of radio-absorbing material (RAM), the tangent surface is a plane-parallel layer of RAM located on a perfectly conducting plane (Figure 1.2a). At the points $\vec{x}$ of the surface bounding the parts of the object made entirely of the composite material, half-spaces with the electrodynamic properties of this material are constructed (Figure 1.2b). For example, the backscattering of a B-2 aircraft was calculated in this way.

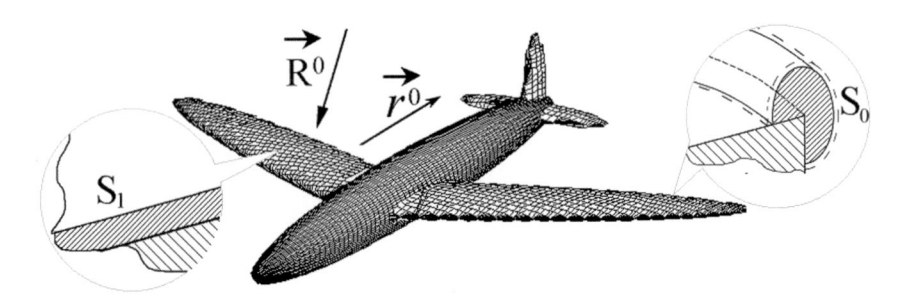

FIGURE 1.1 Example of the standalone object's surface description.

The solution to the first problem (Figure 1.2a) for an arbitrary polarization of the incident wave was obtained in [7]. Without going into details, we give only the final expression for the field reflected by the layer

$$\begin{pmatrix} \vec{\mathcal{E}}(x) \\ \vec{\mathcal{H}}(x) \end{pmatrix} = \begin{pmatrix} \vec{p}^0 \\ (\vec{R}^0 \times \vec{p}^0)\sqrt{\frac{\varepsilon_0}{\mu_0}} \end{pmatrix} \exp\left(jk_0\left(\vec{R}^0 \cdot \vec{x}\right)\right) + \begin{pmatrix} \vec{p}^1 \\ (\vec{R}^1 \times \vec{p}^0)\sqrt{\frac{\varepsilon_0}{\mu_0}} \end{pmatrix} \exp\left(jk_0\left(\vec{R}^1 \cdot \vec{x}\right)\right) \quad (1.2)$$

$$\vec{p}^1 = \vec{p}_T^1 - \vec{n}\frac{\left(\vec{p}_T^1 \cdot \vec{R}^0\right)}{\cos\theta}, \quad (1.3)$$

$$\vec{p}_T^1 = \frac{jc\cos\theta + 1}{jc\cos\theta - 1}\vec{p}_T^0 - \frac{2jc}{jc\cos\theta - 1}\left[\vec{R}_T^0\frac{\left(\vec{R}_T^0 \cdot \vec{p}^0\right)}{jc - \cos\theta} + \vec{R}_\perp^0\frac{\left(\vec{R}_\perp^0 \cdot \vec{p}^0\right)}{\varepsilon_1'\mu_1'\left(jc - \frac{\cos^2\theta_1}{\cos\theta}\right)}\right], \quad (1.4)$$

where $\vec{R}^1 = \vec{R}^0 - 2\vec{n}\cdot(\vec{R}^0\cdot\vec{n})$, $\vec{R}_T^0 = \vec{R}^0 - \vec{n}(\vec{R}^0\cdot\vec{n})$, $c = \sqrt{\frac{\mu_1'}{\varepsilon_1'}}\cos\theta_1\,tg(k_1\,\delta\cos\theta_1)$, $\vec{R}_\perp^0 = (\vec{n}\times\vec{R}^0)$, $\cos^2\theta_1 = 1 - \frac{\sin^2\theta}{\varepsilon_1'\mu_1'}$, (ε_1', μ_1') are relative permittivity and permeability of the RAM, θ is the wave incidence angle on the layer.

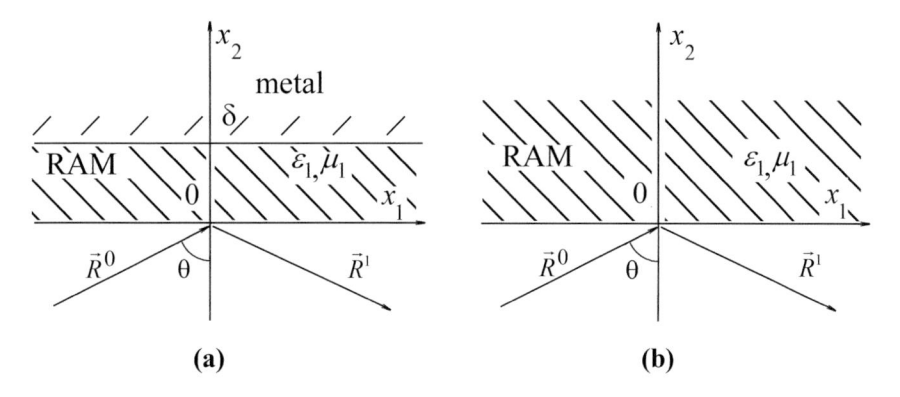

Description of Calculation Methods for Scattering Characteristics

FIGURE 1.2 Wave scattering at non-perfectly reflecting surface (a and b).

Note that the obtained expressions (1.2)-(1.4) are already suitable for calculating the scattered field using (1.1) for any polarization of the incident wave and any sounding directions (except those close to tangents). In particular, for angles close to zero, the obtained expression (1.3) of the complex vector reflection coefficient $\vec{p}^1$ contains no uncertainty, and at $\theta=0$, formulas (1.3) and (1.4) become the well-known formulas [8] for normal incidence.

If the thickness of the RAM layer is assumed to be $\delta \to \infty$, then a model problem can be considered as the reflection of an incident plane wave from a halfspace of the RAM (Figure 1.2b). In this case

$$\vec{p}_T^1 = \frac{c\cos\theta - 1}{c\cos\theta + 1}\vec{p}_T^0 + \frac{2c}{c\cos\theta + 1}\left[\vec{R}_T^0\frac{\left(\vec{R}_T^0 \cdot \vec{p}^0\right)}{c+\cos\theta} + \vec{R}_\perp^0\frac{\left(\vec{R}_\perp^0 \cdot \vec{p}^0\right)}{\varepsilon_1'\mu_1'\left(c+\frac{\cos^2\theta_1}{\cos\theta}\right)}\right], \quad (1.5)$$

where $c = \sqrt{\dfrac{\mu_1'}{\varepsilon_1'}}\cos\theta_1$.

Using expression (1.5), you can calculate the field scattered by an object or part of it made entirely of a dielectric or composite material.

In the physical-optical approximation, the field on the "unilluminated" surface of an object is exactly zero. Therefore, by replacing the surface S_1 with its illuminated part S_1' and substituting the obtained expressions for the field (1.2) into (1.1), we obtain an approximate expression for the scattered field:

$$\vec{p} \cdot \vec{E}^p\left(R\,\vec{r}^0\right) \approx -jk_0\frac{\exp\left(jk_0R\right)}{4\pi R}\int_{S_1'} f(\vec{x})\exp\left(jk_0\Omega(\vec{x})\right)dS, \quad (1.6)$$

where $f(\vec{x}) = \vec{h}(\vec{x})\cdot\vec{p} + \vec{e}(\vec{x})\cdot\left(\vec{p}\times\vec{r}^0\right)$, $\vec{e}(\vec{x}) = \vec{n}\times\left(\vec{p}^0 + \vec{p}^1\right)$, $\Omega(\vec{x}) = \left(\vec{R}^0 - \vec{r}^0\right)\cdot\vec{x}$, $\vec{h}(\vec{x}) = \vec{n}\times\left[\left(\vec{R}^0\times\vec{p}^0\right)+\left(\vec{R}^1\times\vec{p}^1\right)\right]$. The complex (in general) reflection coefficient $\vec{p}^1$ for the field scattered by a plane surface tangent to the surface S_1 at the point $\vec{x}$, can be obtained using the expressions given in [1]. Since the integrand function in (1.6) is rapidly oscillating, this integral was calculated using special cubature formulas. In this case, the surface of the object was triangulated – the area S_1' was covered by a system of triangles, and the integral (1.6) is represented as the sum of the integrals over these triangles. After linear interpolation of the functions $f(\vec{x}), \Omega(\vec{x})$, the integrals over the triangles can be obtained explicitly in barycentric coordinates [1].

The expression for the field scattered by edge local surface parts can be presented in the following form [1]

$$\vec{p} \cdot \vec{E}_{S_0} = -jk_0 \sqrt{\frac{\mu_0}{\varepsilon_0}} \frac{\exp(jk_0 R)}{4\pi R} \left(\vec{p} \cdot \vec{I}_{S_0}(\vec{r}^{\,0}) \right),$$

$$\vec{I}_{S_0}(\vec{r}^{\,0}) = \iint_{S_0} \left[\vec{H}_\perp - \sqrt{\frac{\varepsilon_0}{\mu_0}} \left(\vec{E}_\perp \times \vec{r}^{\,0} \right) \right] \exp\left(-jk_0 (\vec{r}^{\,0} \cdot \vec{x})\right) ds. \qquad (1.7)$$

A toroidal surface covering the edge is chosen as the integration surface S_0 (Figure 1.1). The field scattered by the edge parts of the surface is calculated using the solution of the model problem of the diffraction of an obliquely incident plane monochromatic wave on a perfectly conducting wedge with a radar-absorbing cylinder at the edge [9]. We also obtained an expression that allows us to calculate the field scattered by rectilinear edge local scattering parts with a RAC on an edge in the general case of bistatic reception [1].

The integral $\vec{I}_{S_0}(\vec{r}^{\,0})$ can be written in the following form [1]:

$$\vec{I}_{S_0}(\vec{r}_0) = \int_L \exp\left[jk_0 (\vec{R}^0 - \vec{r}^{\,0}) \vec{x}(l) \right] \vec{M}(l, \vec{r}^{\,0}) \, dl, \qquad (1.8)$$

where

$$\vec{M}(l, \vec{r}^{\,0}) = \int_{S_0'} \exp\left[-jk_0 (\vec{r}^{\,0} \vec{\xi}) \right] \vec{G}(\vec{\xi}) \, dq,$$

$\vec{x}(l)$ is the radius-vector of the point P with coordinate l on edge line L, S_0' is the curve (part of a circle) on the surface S_0 belonging to the plane orthogonal to L, $\vec{G}(\vec{\xi})$ is a vector-function containing the tangential components of the field $(\vec{E}, \vec{H})$ on the surface of the absorbing torus in the considered section, $dq = \rho_0 d\phi$ is an element of the curve arc S_0'.

By evaluating the integral (1.8) using the stationary phase method, we can show that for an edge line in the form of a convex smooth curve, there are at least two stationary phase points (except for some special cases, i.e., for a body of revolution with edges, this is the case of axial sensing and combined reception). In this "special" case, we can calculate $\vec{I}_{S_0}(\vec{r}^{\,0})$ by numerical integration. In other cases, the stationary phase method for (1.8) gives

$$\vec{I}_{S_0}(\vec{r}^{\,0}) \sim \sum_{(l_0)} \exp\left[jk_0 (\vec{R}^0 - \vec{r}^{\,0}) \cdot \vec{x}(l_0) + j\beta \frac{\pi}{4} \right] \cdot \vec{M}(l_0, \vec{r}^{\,0}) \sqrt{\frac{2\pi}{k_0 \mathfrak{x}(l_0) |(\vec{R}^0 - \vec{r}^{\,0}) \cdot \vec{v}(l_0)|}} \qquad (1.9)$$

where $\mathfrak{x}(l_0), \vec{v}(l_0)$ is curvature and orthogonal normal to the line L at the point P_0 with arc coordinate l_0, $\beta = \mathrm{sgn}\left[(\vec{R}^0 - \vec{r}^{\,0}) \cdot \vec{v}(l_0) \right]$, the symbol (l_0) under the sum indicates that the summation should be performed only for the "visible" points of the stationary phase (relative to the directions of probing and reception) [1].

Given the large electrical sizes of the scatterer and the small curvatures of the smooth parts of its surface, the tangential field components $(\vec{E}, \vec{H})$ on the line S_0' are approximated by the corresponding values on the surface of a circular absorbing cylinder covering the edge of the tangent at the point P_0 of the perfectly conducting wedge $(\rho = \rho_0)$. Thus, the main problem in formula (1.9) is the problem of oblique incidence of a plane electromagnetic wave on a perfectly conducting wedge with a radio-absorbing cylinder at the edge. The main difficulty is that this problem is significantly three-dimensional. Its solution cannot be presented as a superposition of two independent two-dimensional problems, such as the problem of the oblique incidence of a plane wave on a perfectly conducting wedge and the problem of the normal (to the edge) incidence of a plane wave on the structure under consideration. However, it can be shown that the problem reduces to a system of two two-dimensional problems whose solutions are connected by boundary conditions (using some matrix-differential operator) [1].

If, as in [10], we represent the third component of the full field in all regions of a piecewise homogeneous medium as $E_3 = u(x_1, x_2)\exp(jk_0 x_3 R_3^0)$, $H_3 = v(x_1, x_2)\exp(jk_0 x_3 R_3^0)$, ($R_3^0$ is the third component of the wedge probing direction of a plane wave) and introduce the vector $\vec{w} = \begin{pmatrix} u \\ v \end{pmatrix}$, then vector $\vec{w}$ can be represented as a Fourier-Bessel series with (2×2) matrix coefficients. For example, outside the absorbing cylinder

$$\vec{w} = \sum_{m=0}^{\infty}\left[B_m J_{v_m}(\chi_0 r) + C_m H_{v_m}^{(1)}(\chi_0 r)\right]\vec{f}_m(\phi), \tag{1.10}$$

where $J_{v_m}(z)$ is Bessel function, $H_{v_m}^{(1)}(z)$ is Hankel function,

$$\vec{f}_m(\phi) = \begin{pmatrix} \sin(v_m \phi) \\ \cos(v_m \phi) \end{pmatrix}, \chi_0 = k_0\sqrt{1-(R_3^0)^2}, v_m = m/\tau(l),$$

$\pi\tau(l)$ is the wedge opening angle $(0 \leq \phi \leq \pi\tau(l))$. The matrix coefficients B_m, C_m are determined from the boundary conditions for the functions u, v and their derivatives on the surface of the absorbing cylinder. The series of type (10) converge well for small values of $r(r \leq \rho_0)$ [1].

To calculate the scattering characteristics of an object placed near the ground surface, it is necessary to note the mutual influence of these objects among themselves, i.e., it is necessary to consider the system "object - half-space with ground parameters" (Figure 1.3), taking into account that the object surface S is probed, firstly, by a plane wave propagating in the direction $\vec{R}^0$ and, secondly, by a wave reflected by the surface D and propagating in the direction of $\vec{R}^1$ (Figure 1.4). Therefore, two mutually intersecting (in the general case) "illuminated" regions Q_0 and Q_1 (Figure 1.5) are localized on the surface of the object. The expression for the field scattered in the direction $-\vec{R}^0$ can be represented as

$$\vec{p} \cdot \vec{E}(\vec{R}^0) = -jk_0 \Omega(k_0 r)$$

$$\times \int_S \left[\sqrt{\frac{\mu_0}{\varepsilon_0}}\left[\vec{p}^0 \exp\left(jk_0(\vec{R}^0 \cdot \vec{x})\right) + \vec{p}^1 \exp\left(jk_0((\vec{R}^0 - \vec{R}^1)\cdot \vec{c} + \vec{R}^1 \cdot \vec{x})\right)\right]\vec{H}_\perp(\vec{x}) \tag{1.11}$$

$$+ \left[\vec{p}^{0\perp}\exp\left(jk_0(\vec{R}^0 \cdot \vec{x})\right) + \vec{p}^{1\perp}\exp\left(jk_0((\vec{R}^0 - \vec{R}^1)\cdot \vec{c} + \vec{R}^1 \cdot \vec{x})\right)\right]\vec{E}_\perp(\vec{x})\right]ds.$$

Here $\vec{c} = \vec{x} - \dfrac{(\vec{x}\cdot\vec{n})+h}{(\vec{R}^1 \cdot \vec{n})}\vec{R}^1$, $\vec{n}$ is the normal vector to surface D, h is the distance from the center O of the object's coordinate system to the surface D (Figure 1.4), $\vec{p}^{0\perp} = \vec{R}^0 \times \vec{p}^0$, $\vec{p}^{1\perp} = \vec{R}^1 \times \vec{p}^1$, and $\vec{p}^1$ is the complex (in general) reflection coefficient for the field scattered by surface D in direction $\vec{R}^1$, which can be obtained using the expressions given in [1]. M is the point on the object surface S with the radius-vector $\vec{x}$ (Figure 1.4). $\vec{E}_\perp(\vec{x}), \vec{H}_\perp(\vec{x})$ are the equivalent electric and magnetic currents on the object surface S. In the physical optics approximation, the calculation is reduced to the computation of four integrals of the form (1.6). These integrals can be interpreted as contributions from the "four-beam" sounding (Figure 1.3). By applying the cubic formula to the fast-oscillating integrals included in (1.11), it is possible to calculate the field scattered by the smooth part of the object surface.

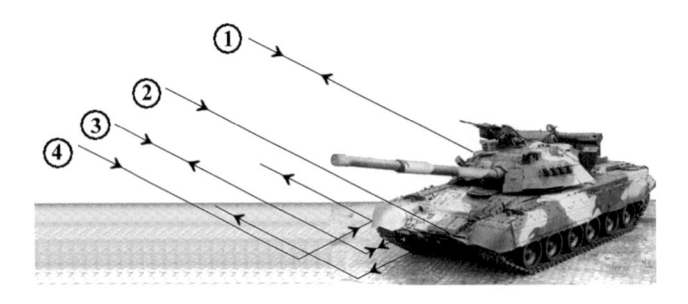

FIGURE 1.3 Basic paths of electromagnetic wave propagation for the case of ground object illumination.

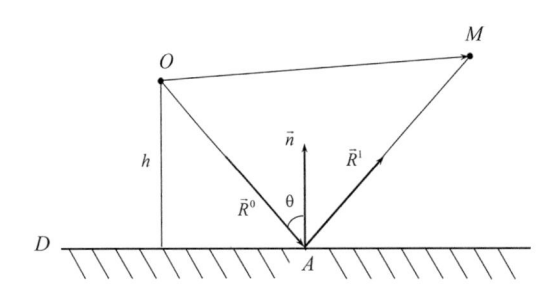

FIGURE 1.4 To the issue of incident wave reflection from underlying surface.

When calculating the field scattered by local edge surface parts of a ground object, the following integration surfaces are considered: W_0 is a set of toroidal surfaces covering the edges "illuminated" by sounding in the direction $\vec{R}^0$, W_1 is a set of toroidal surfaces covering the edges "illuminated" by a wave reflected from the surface D (Figure 1.6).

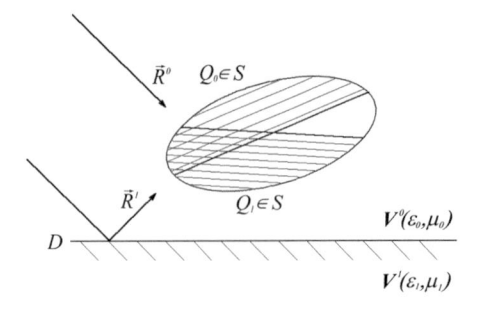

FIGURE 1.5 The "object–ground" system.

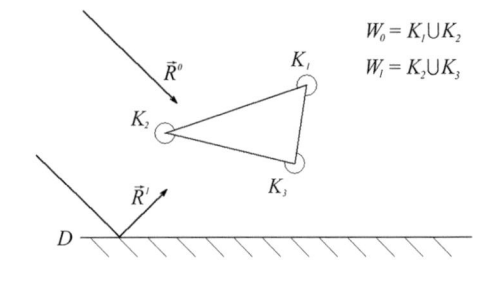

FIGURE 1.6 To the determination of surfaces W_0 and W_1.

The resulting final expressions allow us to take into account the contribution of edge local scattering areas (in general, covered by the RAC) to the field scattered by a ground object.

To calculate the scattering characteristics of an object, a geometric model of its surface must first be constructed. Smooth parts of the object's surface are approximated by parts of triaxial ellipsoids. When calculating the scattering characteristics for these parts, triangulation is performed.

The triangulation is covering the surface with a system of triangular facets, which are involved in the calculation process. The opening angle of the tangent wedge (at each break point) is taken into account. The method of constructing the surface of an object is described in [1].

The method of calculating the scattering characteristics of aerial and ground objects takes into consideration the re-reflection between smooth parts of the object surface. For this purpose, the smooth parts of the object surface are split into triangular facets. For each i-th facet of the k-th smooth part of the object surface, the normal vector $\vec{n}_i$ and the direction of the mirror reflection $\vec{R}^1$ are determined (Figure 1.7). Direction $\vec{R}^1$ corresponds to the direction $\vec{R}^0$ of the plane wave that is incident on the facet. From each vertex of the i-th facet, rays with direction $\vec{R}^1$ are sent to find the intersection of such reflection rays with other parts of the object surface. If all three rays intersect some single smooth part, a new facet is formed on this part. The vertices of this j-th facet are the intersection points of re-reflection rays with direction $\vec{R}^1$ and another m-th smooth part of object surface (Figure 1.7). There is an additional re-reflection plane wave from a direction $\vec{R}^1$ with a complex (generally) vector $\vec{p}^1$ obtained with physical optics approximation for i-th triangle facet. This additional field, which is incident on the j-th facet, also has an additional phase shift associated with the re-reflection from the i-th facet. The additional phase shift is determined similarly to the situation of phase shift in determining the field incident on a ground object when an electromagnetic wave is re-reflected from the underlying surface [1,2]. The field scattered by the j-th facet in the reception direction $\vec{r}^0$ (with accounting additional incident wave from direction $\vec{R}^1$) is calculated using the physical optics approximation. The described algorithm can be repeated to account for multiple re-reflections. The implementation of the proposed method allows to consider (especially double and triple) re-reflections between the elements of the object surface. This makes it possible to estimate the scattering characteristics of such objects as a triangular corner reflector.

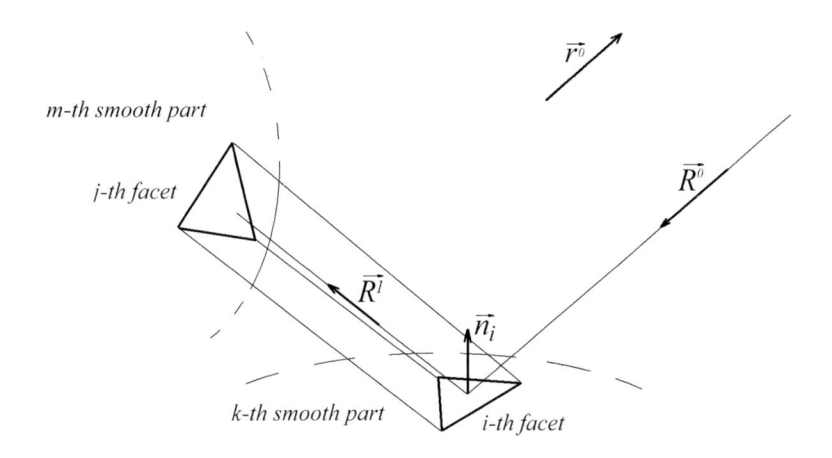

FIGURE 1.7 Re-reflection between smooth parts of the object surface.

Onboard forward-looking radars used on a number of fighters and bombers, as well as radars on Airborne Early Warning and Control System (AWACS) aircraft, significantly increase the total RCS of an aerial object. The development of a general method for calculating the RCS for this case includes a quantitative assessment of the contribution of such antenna systems to the RCS of an airborne object. The calculation has been performed as follows [11].

A model of a mirror antenna with a conical radome has been considered (Figure 1.8). Applying Lorentz's lemma to the desired full field $\left(\vec{E}, \vec{H}\right)$ and the auxiliary field $\left(\vec{\tilde{E}}, \vec{\tilde{H}}\left(\vec{x}|\vec{x}_0, \vec{p}\right)\right)$ generated by an electric dipole located at the point $\vec{x}_0$ with moment vector $\vec{p}$, in the presence of only single radome, allows us to obtain an integral representation for the desired field:

$$j\omega \vec{p} \cdot \vec{E}(\vec{x}_0) = j\omega \vec{p} \cdot \vec{E}_{rad}(\vec{x}_0) + \int_L \left(\vec{K}(\vec{x}) \cdot \vec{E}^T(\vec{x}|\vec{x}_0, \vec{p}) \right) dS, \qquad (1.12)$$

where $\vec{E}_{rad}(\vec{x}_0)$ is the field scattered only single radome, $\vec{K}(x)$ is the surface current density at the points of the antenna mirror L. The integral term of expression (1.12) represents the response of the antenna mirror to the probe wave, taking into account the electrodynamic interaction with the radome. Substituting $\vec{x}_0 = -r\vec{R}^0$ and directing $r \to \infty$, we obtain the expression for the full field scattered by the antenna-radome system in the far zone:

$$\vec{p} \cdot \vec{E}(\vec{R}^0) \sim \vec{p} \cdot \vec{E}_{rad}(\vec{R}^0) - jk_0 \frac{e^{jk_0 r}}{4\pi r} \sqrt{\frac{\mu_0}{\varepsilon_0}} \int_L \left(\vec{\tilde{E}}(\vec{x}) \cdot \vec{K}(\vec{x}) \right) dS. \qquad (1.13)$$

Here $\vec{\tilde{E}}(\vec{x})$ is the field generated by a plane monochromatic wave at the points of the mirror in the presence of only single radome. This field was calculated in the geometric-optical approximation. In the considered approximation, the field $\left(\vec{\tilde{E}}(\vec{x}), \vec{\tilde{H}}(\vec{x}) \right)$ is represented as the sum of the field on the mirror that passed directly through the illuminated surface of the radome (path 1 in Figure 1.9) and the field that hit the mirror after a single reflection from the inner surface of the radome (way 2 in Fig. 1.9).

It should be noted when a plane electromagnetic wave is reflected from the inner surface of the radome, a caustic surface (a zone of increased energy concentration) may be formed, which should be considered [4].

FIGURE 1.8 The "antenna-radome" system.

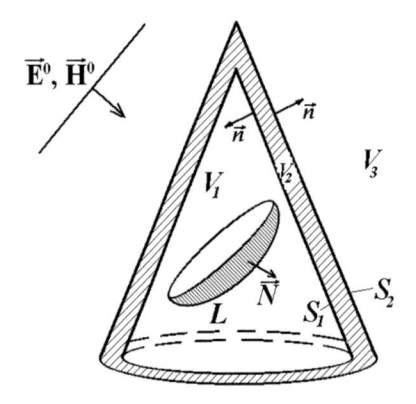

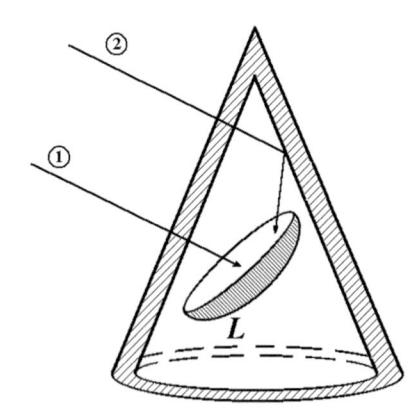

FIGURE 1.9 Paths of propagation of the incident wave.

This book uses the developed method for calculating the scattering characteristics of unmanned aerial vehicles (UAVs) [3,4]. The general scheme of a typical UAV design is shown in Fig. 1.10.

Description of Calculation Methods for Scattering Characteristics

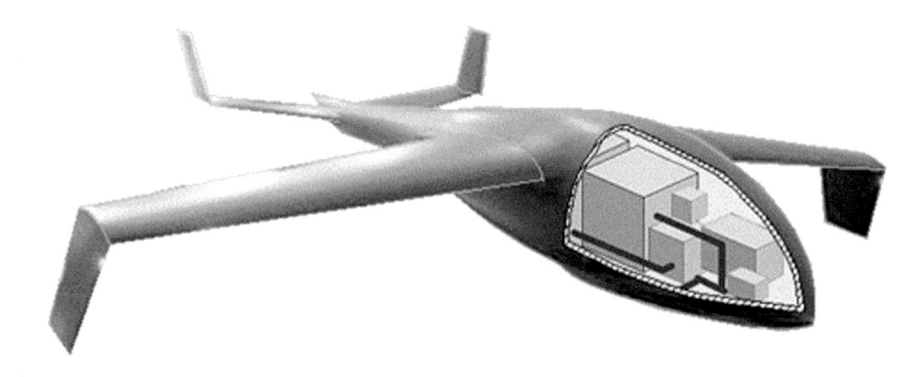

FIGURE 1.10 General scheme of a typical UAV design.

According to the above UAV design, it is necessary to calculate the scattering characteristics of perfectly conducting elements, equipment elements hidden under the fuselage dielectric shell, and hollow dielectric elements. The calculation algorithm is similar to the calculation of the scattering characteristics of an antenna under a dielectric radome: the scattering on the dielectric shell is calculated separately, and the scattering on the internal elements of the UAV is added to it.

2 Scattering Characteristics of Aerial Objects

Based on the developed methods for calculating the radar characteristics of aerial objects [1–4,6], mathematical modeling was performed, and data were obtained for 63 aerial objects (aircraft, helicopters, missiles, and UAVs).

The scattering characteristics of aerial objects have been obtained for the following sounding frequencies 10 GHz (wavelengths of 3 cm), 5 GHz (wavelengths of 6 cm), 3 GHz (wavelengths of 10 cm), 1.3 GHz (wavelengths of 23 cm), and 1 GHz (wavelengths of 30 cm). At the same time, this book contains data for frequencies of 10, 3, and 1 GHz and horizontal polarization of the sounding signal. The scattering characteristics for vertical polarization, as well as for the remaining frequencies in both polarizations, are given in the electronic appendix of this book. The sounding parameters were as follows: the step in the sounding azimuth was 0.02°, the azimuth (Figure 2.1) was counted from the nose angle (0° corresponds to sounding into the nose of the aerial object, 180° – sounding into the tail).

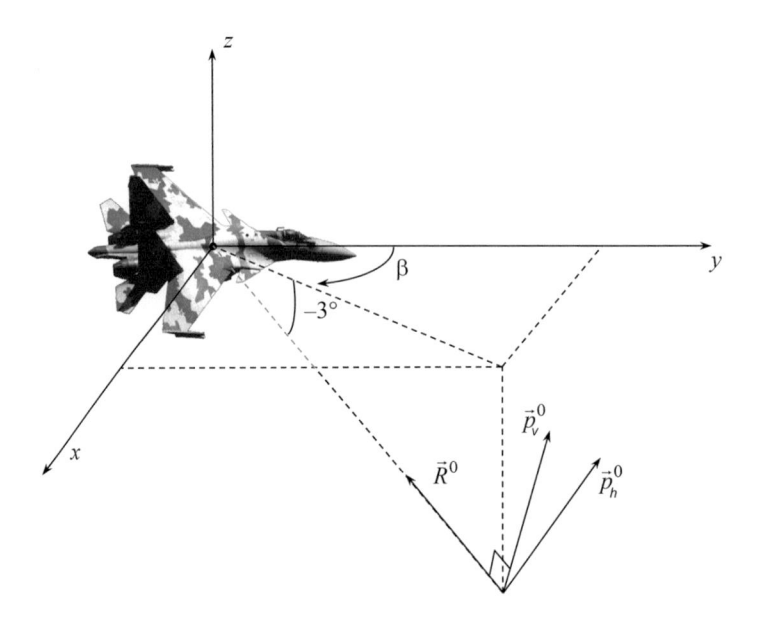

FIGURE 2.1 Scheme of sounding aerial object.

Noting that the aspect angle of the aircraft in the elevation plane has fluctuations during the flight, and the elevation angle of the sounding was taken randomly, uniformly distributed in the range of $-3° \pm 4°$ relative to the wing plane (the angle of $-3°$ corresponds to sounding from the lower hemisphere (Figure 2.1)).

The scattering characteristics are obtained for the case of monostatic reception in two orthogonal polarizations: horizontal – the electric field intensity vector of the incident wave $\vec{p}_h^0$ lies in the plane of the wing; and vertical – the electric field intensity vector of the incident wave $\vec{p}_v^0$ is orthogonal to $\vec{p}_h^0$ and lies in a plane that is perpendicular to the wing plane and passes through the direction $\vec{R}^0$ of the incident plane wave (Figure 2.1).

Since the radar target aspect can be regarded as random value, then its RCS at any given moment can be regarded as random value too. Probability distribution of such random value can be evaluated by diagrams of RCS obtained as a result of computation or experiment. Along with RCS, its square root is often used in radar theory, the later being proportional to the radar echo amplitude. Therefore, this chapter features the distribution histograms of radar echo amplitude multiplier given the object observation at the most common aspect angles. From a number of possible probability distributions (normal, Rayleigh, log-normal, Weibull, extreme value, β-distribution, and Γ-distribution), we chose the ones that fit best empirical probability distributions according to Kolmogorov–Smirnoff criterion (while doing this we also determined the parameters of theoretical probability distributions).

In all figures showing the histograms of the amplitude multiplier, the solid line shows the probability distributions (given in the margins of the figures) multiplied by the area of the corresponding histograms.

It should be noted that in some cases, despite a satisfactory agreement according to the Kolmogorov-Smirnov criterion, the curves of the theoretical probability distributions may differ significantly from the obtained histograms. In this case, the reader of this book can try to find other theoretical distributions (which give a more accurate result) using the scattering characteristics data provided in the electronic appendix of this book.

Taking into account the special importance of the attack ("nose") aspect angles of sounding aerial objects, this book contains distribution histograms of the amplitude multiplier of the scattered signal (square root of RCS) for the sounding azimuths $-20°$ to $+20°$ and the case of horizontal polarization of the incident electromagnetic plane wave.

The probability distributions of the RCS value of aerial objects, as well as the square root of the RCS (radar echo amplitude multiplier) have been obtained for $20°$ ranges of sounding azimuths, are given in the electronic appendix to this book for all five considered sounding frequencies and for horizontal and vertical polarizations.

2.1 MULTIROLE FIGHTER Su-27

Su-27 Flanker (Figure 2.2) is one-seat multirole fighter. The first flight took place in 1977.

General characteristics of Su-27 [12]: wingspan – 14.70 m, length – 21.93 m, height – 5.93 m, wing area – 62.04 m^2, weight – 16 300–30 000 kg, powerplant – 2 Saturn AL-31F afterburning turbofan engines, maximum speed – 2 500 km/h, range: at sea level – 1 370 km, at altitude – 3 680 km, and service ceiling – 18 500 m.

In accordance with the design of the Su-27, a model of its surface was created to obtaining scattering characteristics (in particular, RCS). The model is shown in Figure 2.3. In the modeling, the smooth parts of the aircraft surface were approximated by parts of 60 triaxial ellipsoids. The surface breaks were modeled using 42 straight edge scattering parts.

FIGURE 2.2 Multirole fighter Su-27.

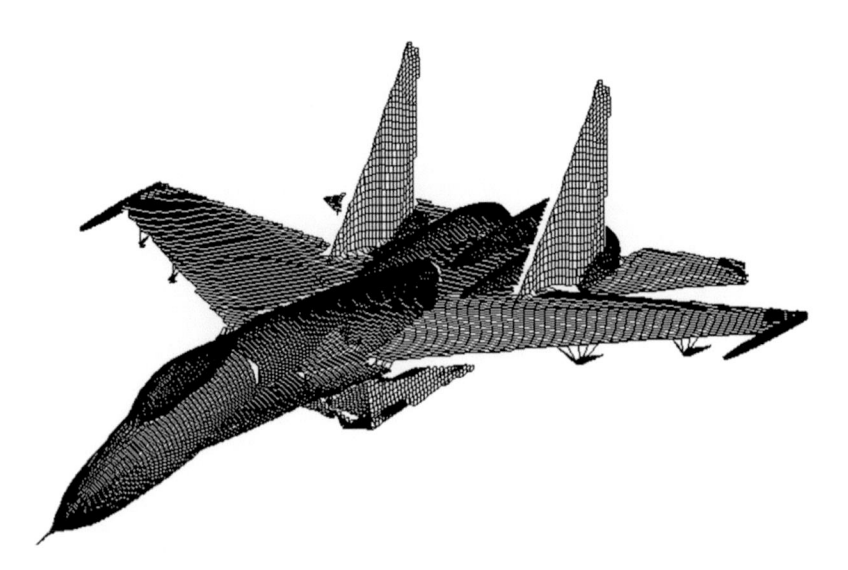

FIGURE 2.3 Surface model of multirole fighter Su-27.

Below are some scattering characteristics of the Su-27 model at sounding frequencies of 10, 3, and 1 GHz (wavelengths of 3, 10, and 30 cm, respectively) for horizontal polarization of the probing signal. The scattering characteristics for this model at two polarizations, as well as scattering characteristics at sounding frequencies of 5 and 1.3 GHz (wavelengths of 6 and 23 cm, respectively), are given in the electronic appendix of this book.

Sounding parameters: The elevation angle is to be a random value distributed uniformly in the range −3° ± 4° with respect to the wing plane (elevation angle of −3° corresponds to the radar observation from the lower hemisphere), azimuth aspect increment was 0.02°, the azimuth being counted off from the nose-on aspect (0° corresponds to the nose-on radar observation, 180° corresponds to the tail-on observation).

Scattering characteristics of Su-27 aircraft model for sounding frequency 10 GHz (wavelength 3 cm).

Figure 2.4 shows the RCS circular diagram of the Su-27. In Figure 2.5, the noncoherent RCS circular diagram of the Su-27 is represented.

The circular mean RCS of the Su-27 aircraft for horizontal polarization is 83.04 m². The circular median RCS (the RCS value used to calculate the detection range of an object with a probability of 0.5) for horizontal polarization is 2.06 m².

Figures 2.6 and 2.7 show the mean and median RCS for the main ranges of sounding azimuths (nose, side, and tail) and for ranges of 20°.

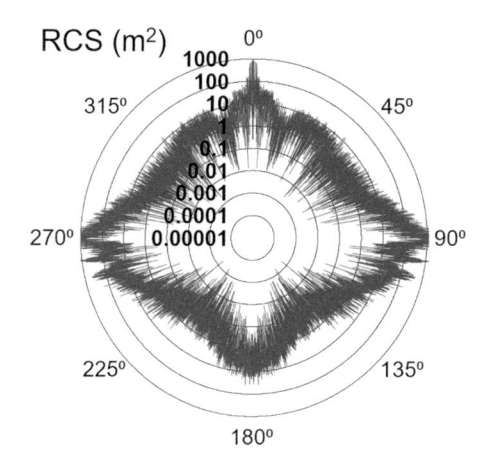

FIGURE 2.4 Circular diagram of RCS given radar observation of Su-27 aircraft model at a carrier frequency of 10 GHz (3 cm wavelength).

FIGURE 2.5 Circular diagram of noncoherent RCS given radar observation of Su-27 aircraft model at a carrier frequency of 10 GHz (3 cm wavelength).

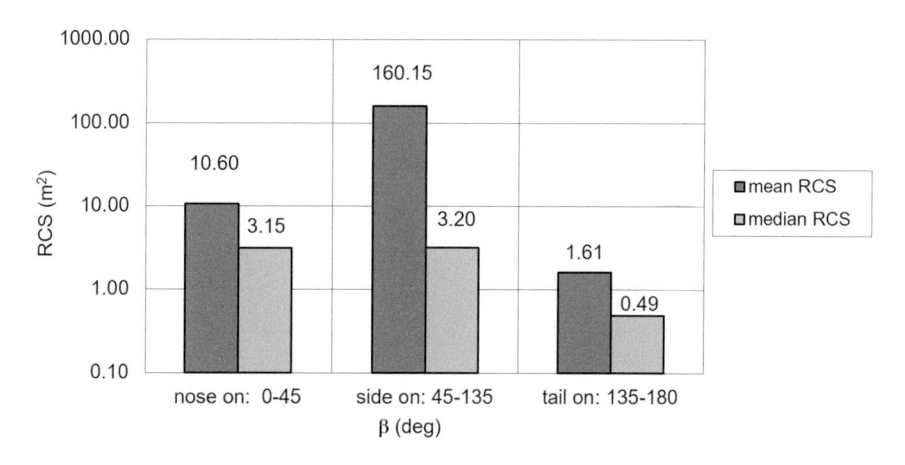

FIGURE 2.6 Diagrams of mean and median RCS of Su-27 aircraft model in three sectors of azimuth aspect given its radar observation at horizontal polarization and a carrier frequency of 10 GHz (3 cm wavelength).

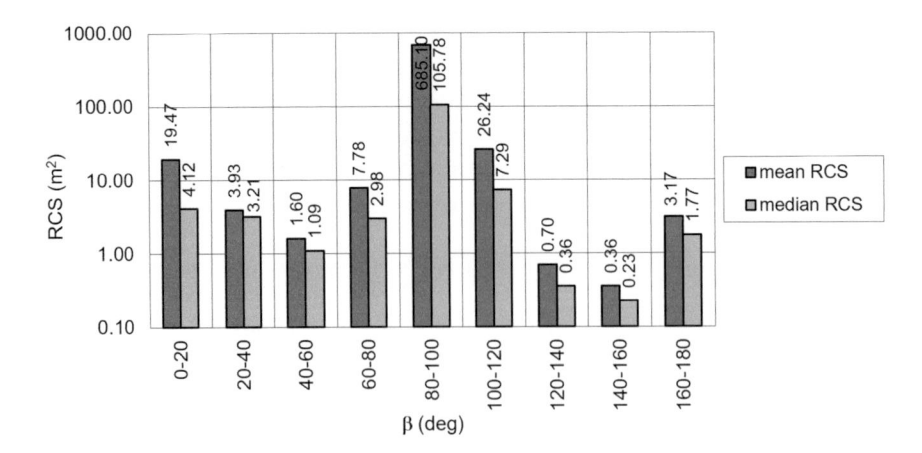

FIGURE 2.7 Diagrams of mean and median RCS of Su-27 aircraft model in 20-degree sectors of azimuth aspect given its radar observation at horizontal polarization and a carrier frequency of 10 GHz (3 cm wavelength).

Figures 2.8, 2.13, and 2.18 show histograms of the scattered signal amplitude (square root of the RCS) for the range of sounding azimuths −20° to +20° (sounding from the nose). The bold line shows the probability density functions of the distribution, which can be used to approximate the histogram of the signal amplitude.

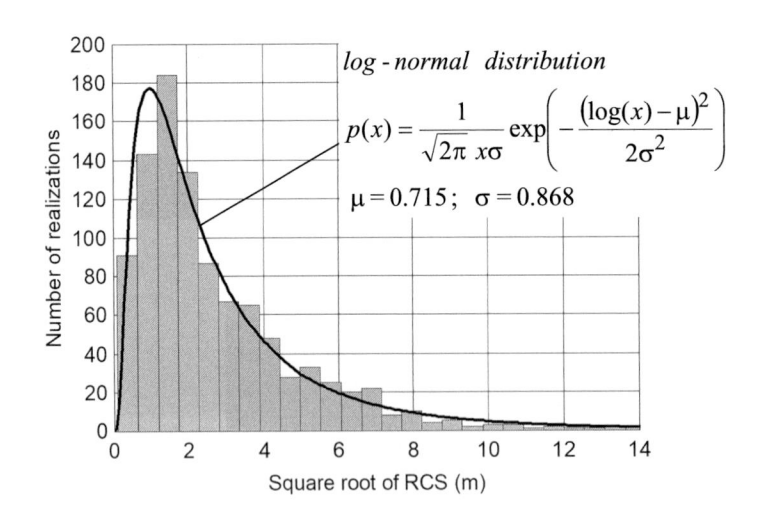

$$p(x) = \frac{1}{\sqrt{2\pi}\, x\sigma} \exp\left(-\frac{(\log(x)-\mu)^2}{2\sigma^2}\right)$$

$\mu = 0.715; \quad \sigma = 0.868$

FIGURE 2.8 Amplitude distribution of echo signal of Su-27 aircraft model at a carrier frequency of 10 GHz given its horizontal polarization.

Scattering characteristics of Su-27 aircraft model for sounding frequency 3 GHz (wavelength 10 cm).

Figure 2.9 shows the RCS circular diagram of the Su-27. In Figure 2.10, the noncoherent RCS circular diagram of the Su-27 is represented.

The circular mean RCS of the Su-27 aircraft for horizontal polarization is 69.65 m². The circular median RCS for horizontal polarization is 2.94 m².

Figures 2.11 and 2.12 show the mean and median RCS for the main ranges of sounding azimuths (nose, side, and tail) and for ranges of 20°.

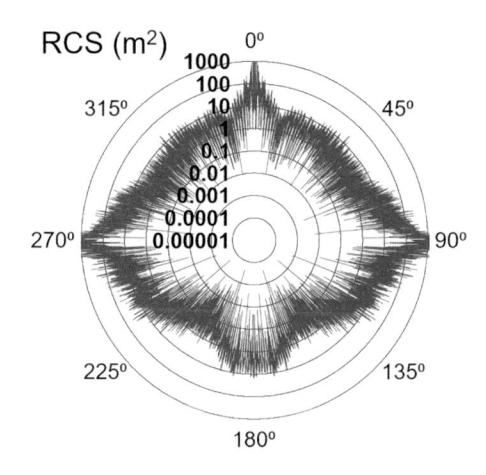

Scattering Characteristics of Aerial Objects

FIGURE 2.9 Circular diagram of RCS given radar observation of Su-27 aircraft model at a carrier frequency of 3 GHz (10 cm wavelength).

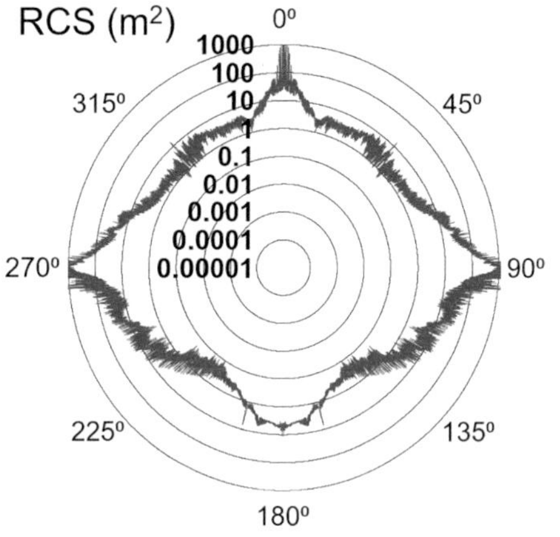

FIGURE 2.10 Circular diagram of noncoherent RCS given radar observation of Su-27 aircraft model at a carrier frequency of 3 GHz (10 cm wavelength).

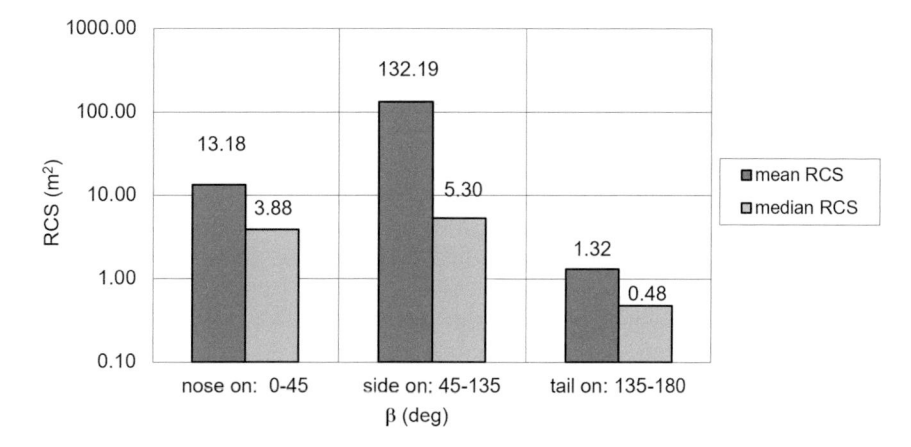

FIGURE 2.11 Diagrams of mean and median RCS of Su-27 aircraft model in three sectors of azimuth aspect given its radar observation at horizontal polarization and a carrier frequency of 3 GHz (10 cm wavelength).

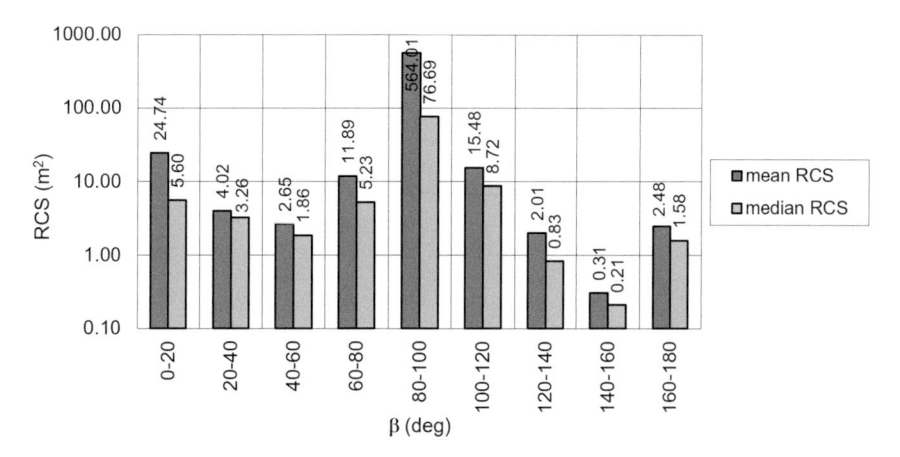

FIGURE 2.12 Diagrams of mean and median RCS of Su-27 aircraft model in 20-degree sectors of azimuth aspect given its radar observation at horizontal polarization and a carrier frequency of 3 GHz (10 cm wavelength).

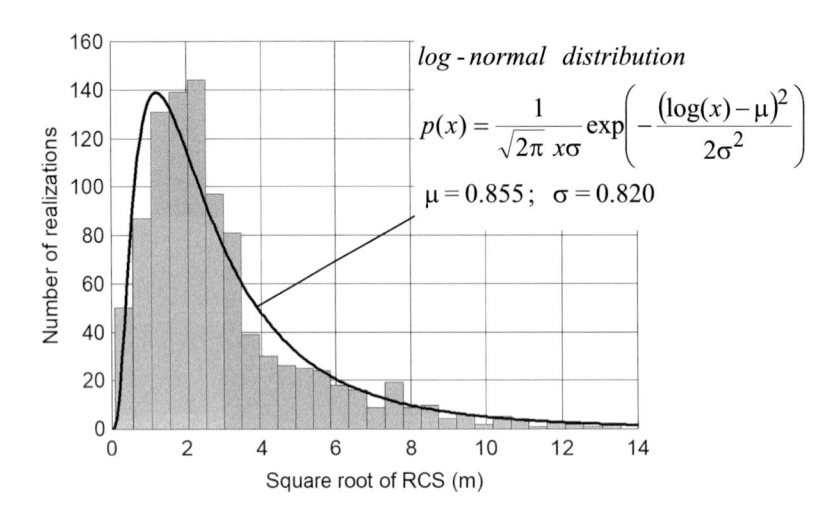

FIGURE 2.13 Amplitude distribution of echo signal of Su-27 aircraft model at a carrier frequency of 3 GHz given its horizontal polarization.

$$p(x) = \frac{1}{\sqrt{2\pi}\, x\sigma} \exp\left(-\frac{(\log(x)-\mu)^2}{2\sigma^2}\right)$$

log-normal distribution; $\mu = 0.855$; $\sigma = 0.820$

Scattering characteristics of Su-27 aircraft model for sounding frequency 1 GHz (wavelength 30 cm).

Figure 2.14 shows the RCS circular diagram of the Su-27. In Figure 2.15, the noncoherent RCS circular diagram of the Su-27 is represented.

The circular mean RCS of the Su-27 aircraft for horizontal polarization is 58.92 m². The circular median RCS for horizontal polarization is 2.51 m².

Figures 2.16 and 2.17 show the mean and median RCS for the main ranges of sounding azimuths (nose, side, and tail) and for ranges of 20°.

The expressions and parameters of probability distributions that are most consistent with the empirical distributions of the square root of the RCS for different ranges of sounding azimuths and polarizations are given in the electronic appendix to this book. The expressions and parameters of the probability distributions that are most consistent with the empirical distributions of the RCS (energy characteristic) for different ranges of sounding azimuths and polarizations are also given there.

FIGURE 2.14 Circular diagram of RCS given radar observation of Su-27 aircraft model at a carrier frequency of 1 GHz (30 cm wavelength).

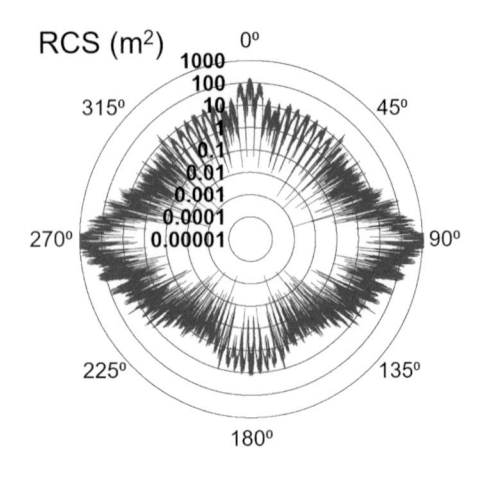

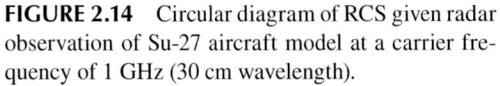

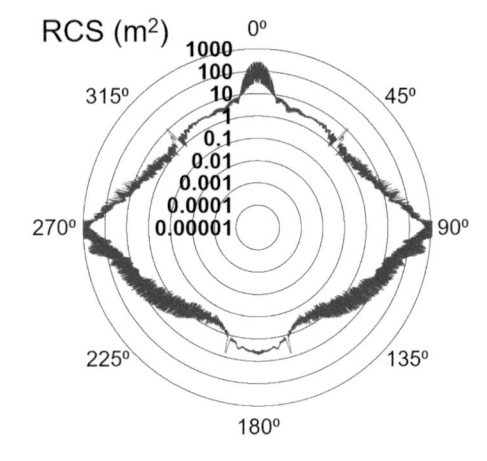

FIGURE 2.15 Circular diagram of noncoherent RCS given radar observation of Su-27 aircraft model at a carrier frequency of 1 GHz (30 cm wavelength).

Scattering Characteristics of Aerial Objects

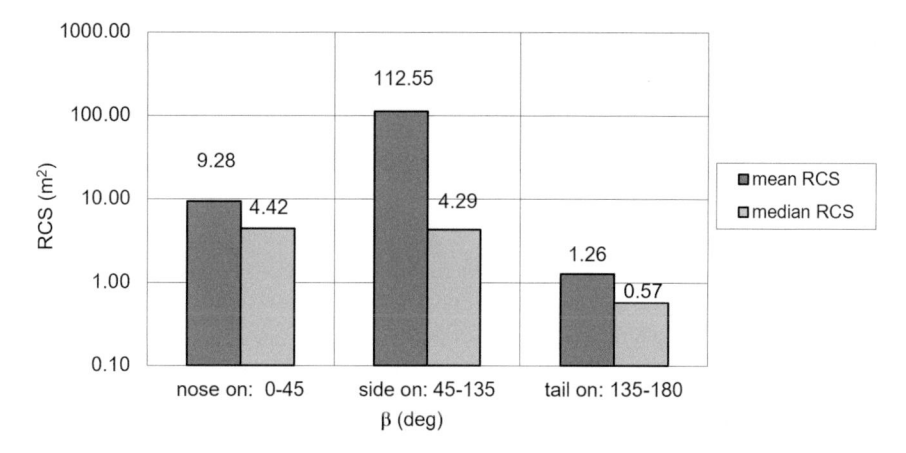

FIGURE 2.16 Diagrams of mean and median RCS of Su-27 aircraft model in three sectors of azimuth aspect given its radar observation at horizontal polarization and a carrier frequency of 1 GHz (30 cm wavelength).

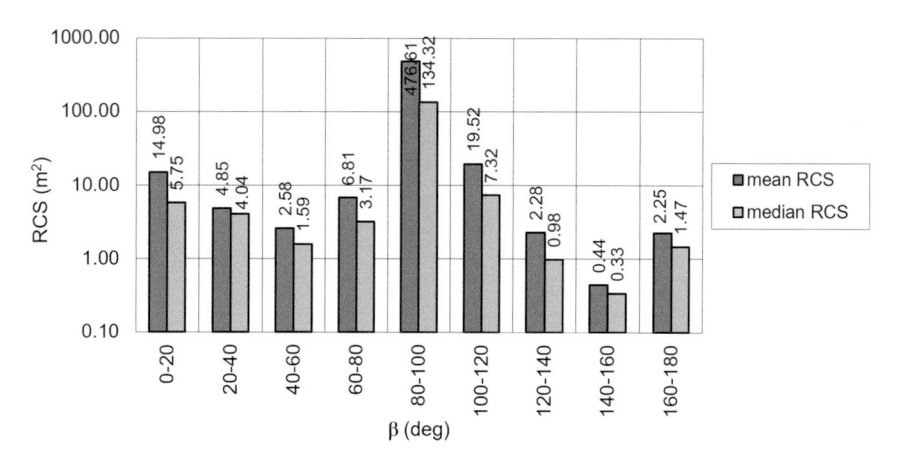

FIGURE 2.17 Diagrams of mean and median RCS of Su-27 aircraft model in 20-degree sectors of azimuth aspect given its radar observation at horizontal polarization and a carrier frequency of 1 GHz (30 cm wavelength).

log-normal distribution

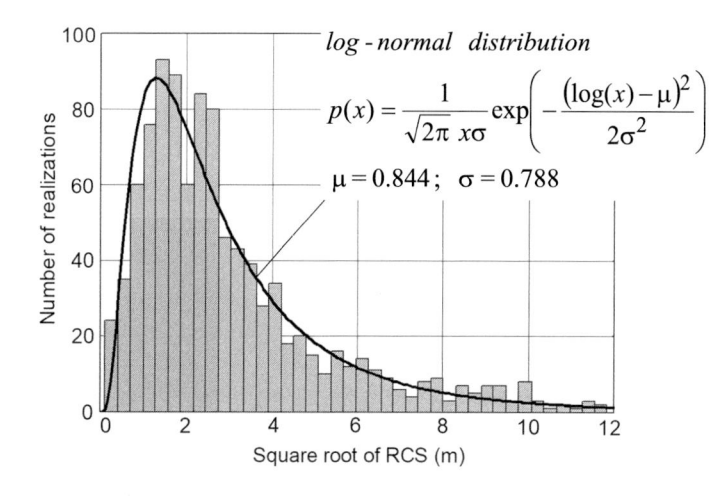

$$p(x) = \frac{1}{\sqrt{2\pi}\, x\sigma} \exp\left(-\frac{(\log(x)-\mu)^2}{2\sigma^2}\right)$$

$\mu = 0.844;\ \sigma = 0.788$

FIGURE 2.18 Amplitude distribution of echo signal of Su-27 aircraft model at a carrier frequency of 1 GHz given its horizontal polarization.

2.2 FRONT-LINE FIGHTER MiG-29

MiG-29 Fulcrum (Figure 2.19) is one-seat front-line fighter. The first flight took place in 1977.

General characteristics of MiG-29 [13]: wingspan – 11.36 m, length – 17.32 m, height – 4.73 m, wing area – 38.06 m², weight – 10 900–18 100 kg, powerplant – 2 Saturn AL-31F afterburning turbofan engines, maximum speed – 2 450 km/h, range – 1 430–2 100 km, and service ceiling – 18 000 m.

In accordance with the design of the MiG-29, a model of its surface was created to obtaining scattering characteristics (in particular, RCS). The model is shown in Figure 2.20. In the modeling, the smooth parts of the aircraft surface were approximated by parts of 40 triaxial ellipsoids. The surface breaks were modeled using 42 straight edge scattering parts.

FIGURE 2.19 Front-line fighter MiG-29.

FIGURE 2.20 Surface model of front-line fighter MiG-29.

Below are some scattering characteristics of the MiG-29 model at sounding frequencies of 10, 3, and 1 GHz (wavelengths of 3, 10, and 30 cm, respectively) for horizontal polarization of the probing signal. The scattering characteristics for this model at two polarizations, as well as scattering characteristics at sounding frequencies of 5 and 1.3 GHz (wavelengths of 6 and 23 cm, respectively), are given in the electronic appendix of this book.

Sounding parameters: The elevation angle is to be a random value distributed uniformly in the range −3° ± 4° with respect to the wing plane (elevation angle of −3° corresponds to the radar observation from the lower hemisphere), azimuth aspect increment was 0.02°, the azimuth being counted

Scattering Characteristics of Aerial Objects 19

off from the nose-on aspect (0° corresponds to the nose-on radar observation, 180° corresponds to the tail-on observation).

Scattering characteristics of MiG-29 aircraft model for sounding frequency 10 GHz (wavelength 3 cm).

Figure 2.21 shows the RCS circular diagram of the MiG-29. The noncoherent RCS circular diagram of the MiG-29 is shown in Figure 2.22.

The circular mean RCS of the MiG-29 aircraft for horizontal polarization is 55.68 m². The circular median RCS (the RCS value used to calculate the detection range of an object with a probability of 0.5) for horizontal polarization is 1.55 m².

Figures 2.23 and 2.24 show the mean and median RCS for the main ranges of sounding azimuths (nose, side, and tail) and for ranges of 20°.

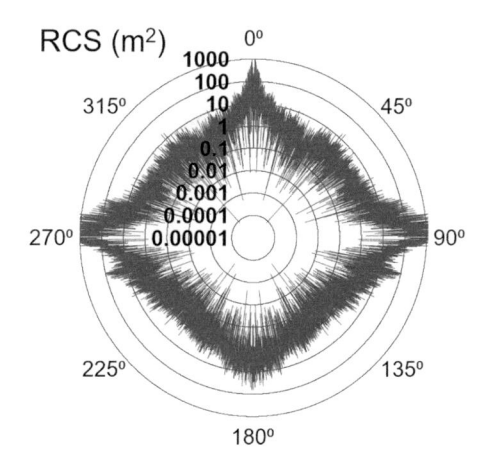

FIGURE 2.21 Circular diagram of RCS given radar observation of MiG-29 aircraft model at a carrier frequency of 10 GHz (3 cm wavelength).

FIGURE 2.22 Circular diagram of noncoherent RCS given radar observation of MiG-29 aircraft model at a carrier frequency of 10 GHz (3 cm wavelength).

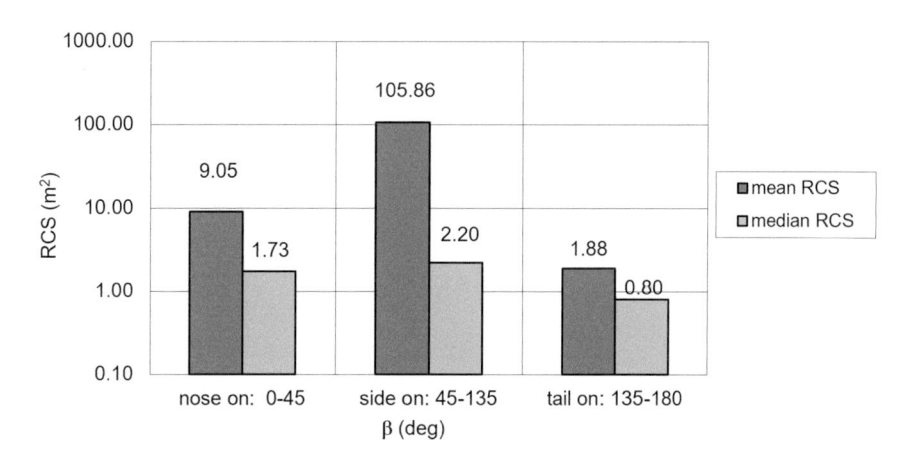

FIGURE 2.23 Diagrams of mean and median RCS of MiG-29 aircraft model in three sectors of azimuth aspect given its radar observation at horizontal polarization and a carrier frequency of 10 GHz (3 cm wavelength).

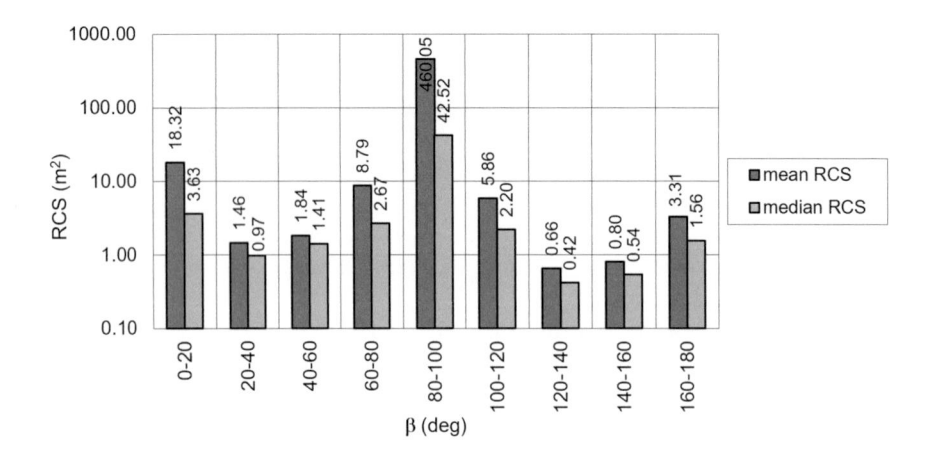

FIGURE 2.24 Diagrams of mean and median RCS of MiG-29 aircraft model in 20-degree sectors of azimuth aspect given its radar observation at horizontal polarization and a carrier frequency of 10 GHz (3 cm wavelength).

Figures 2.25, 2.30, and 2.35 show histograms of the scattered signal amplitude (square root of the RCS) for the range of sounding azimuths −20° to +20° (sounding from the nose). The bold line shows the probability density functions of the distribution, which can be used to approximate the histogram of the signal amplitude.

log - normal distribution

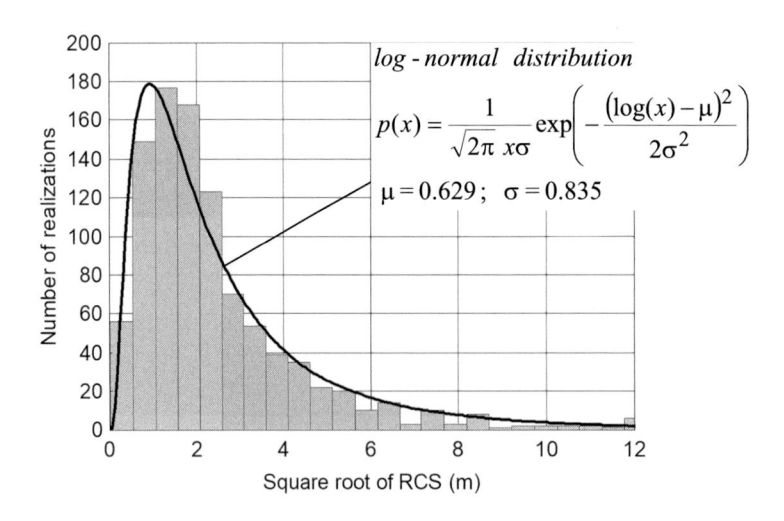

$$p(x) = \frac{1}{\sqrt{2\pi}\, x\sigma} \exp\left(-\frac{(\log(x)-\mu)^2}{2\sigma^2}\right)$$

$\mu = 0.629;\ \sigma = 0.835$

FIGURE 2.25 Amplitude distribution of echo signal of MiG-29 aircraft model at a carrier frequency of 10 GHz given its horizontal polarization.

Scattering characteristics of MiG-29 aircraft model for sounding frequency 3 GHz (wavelength 10 cm).

Figure 2.26 shows the RCS circular diagram of the MiG-29. The noncoherent RCS circular diagram of the MiG-29 is shown in Figure 2.27.

The circular mean RCS of the MiG-29 aircraft for horizontal polarization is 87.10 m². The circular median RCS for horizontal polarization is 1.55 m².

Figures 2.28 and 2.29 show the mean and median RCS for the main ranges of sounding azimuths (nose, side, and tail) and for ranges of 20°.

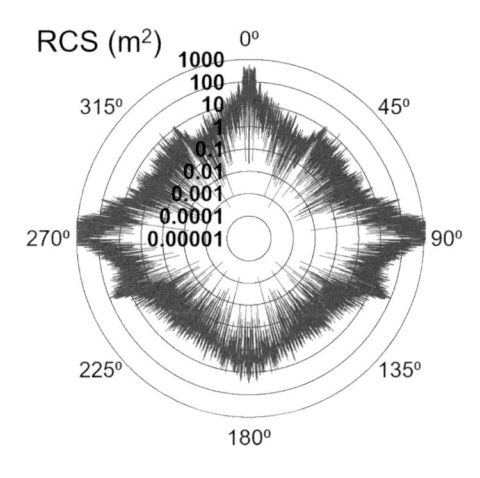

Scattering Characteristics of Aerial Objects

FIGURE 2.26 Circular diagram of RCS given radar observation of MiG-29 aircraft model at a carrier frequency of 3 GHz (10 cm wavelength).

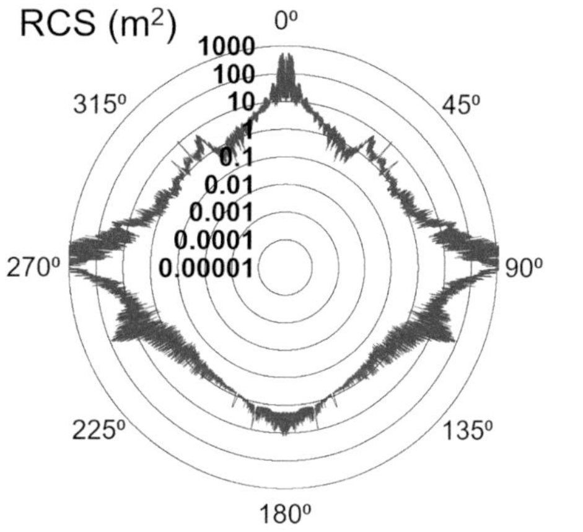

FIGURE 2.27 Circular diagram of noncoherent RCS given radar observation of MiG-29 aircraft model at a carrier frequency of 3 GHz (10 cm wavelength).

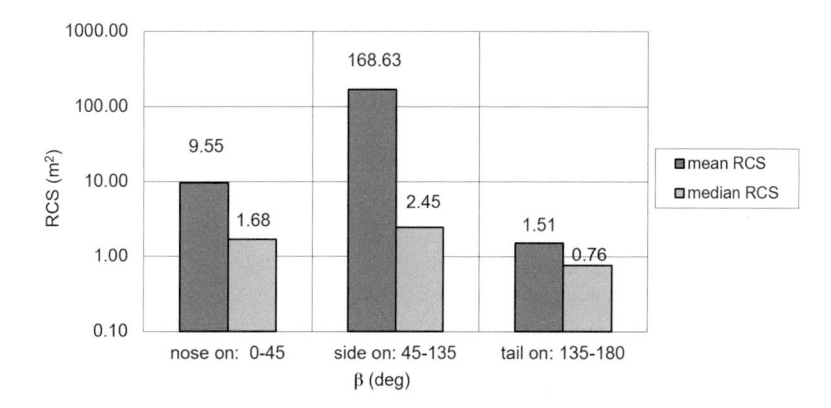

FIGURE 2.28 Diagrams of mean and median RCS of MiG-29 aircraft model in three sectors of azimuth aspect given its radar observation at horizontal polarization and a carrier frequency of 3 GHz (10 cm wavelength).

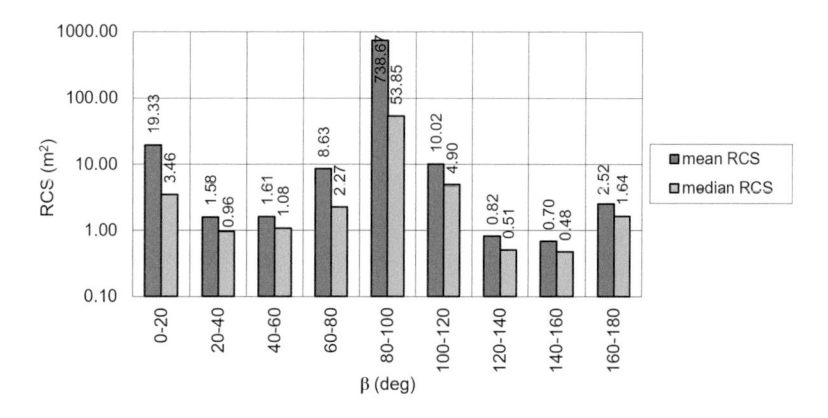

FIGURE 2.29 Diagrams of mean and median RCS of MiG-29 aircraft model in 20-degree sectors of azimuth aspect given its radar observation at horizontal polarization and a carrier frequency of 3 GHz (10 cm wavelength).

$$p(x) = \frac{1}{\sqrt{2\pi}\, x\sigma} \exp\left(-\frac{(\log(x)-\mu)^2}{2\sigma^2}\right)$$

log - normal distribution

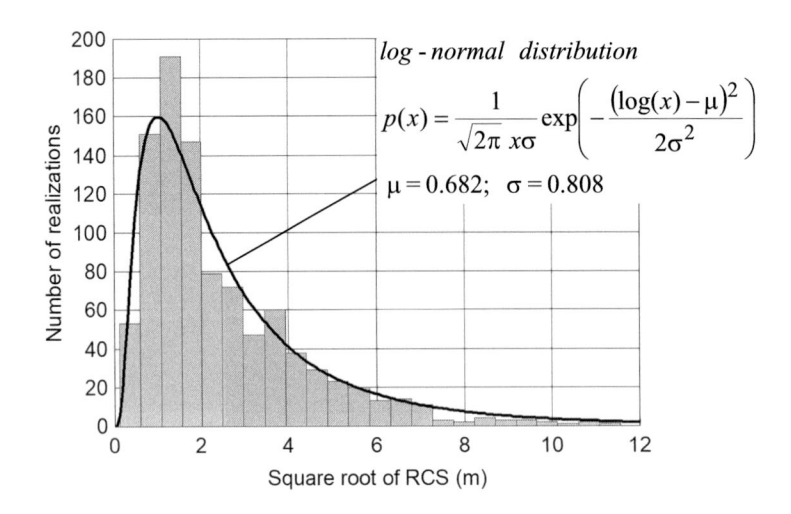

$\mu = 0.682; \quad \sigma = 0.808$

FIGURE 2.30 Amplitude distribution of echo signal of MiG-29 aircraft model at a carrier frequency of 3 GHz given its horizontal polarization.

Scattering characteristics of MiG-29 aircraft model for sounding frequency 1 GHz (wavelength 30 cm).

Figure 2.31 shows the RCS circular diagram of the MiG-29. The noncoherent RCS circular diagram of the MiG-29 is shown in Figure 2.32.

The circular mean RCS of the MiG-29 aircraft for horizontal polarization is 60.16 m². The circular median RCS for horizontal polarization is 1.98 m².

Figures 2.33 and 2.34 show the mean and median RCS for the main ranges of sounding azimuths (nose, side, and tail) and for ranges of 20°.

The expressions and parameters of probability distributions that are most consistent with the empirical distributions of the square root of the RCS for different ranges of sounding azimuths and polarizations are given in the electronic appendix to this book. The expressions and parameters of the probability distributions that are most consistent with the empirical distributions of the RCS (energy characteristic) for different ranges of sounding azimuths and polarizations are also given there.

FIGURE 2.31 Circular diagram of RCS given radar observation of MiG-29 aircraft model at a carrier frequency of 1 GHz (30 cm wavelength).

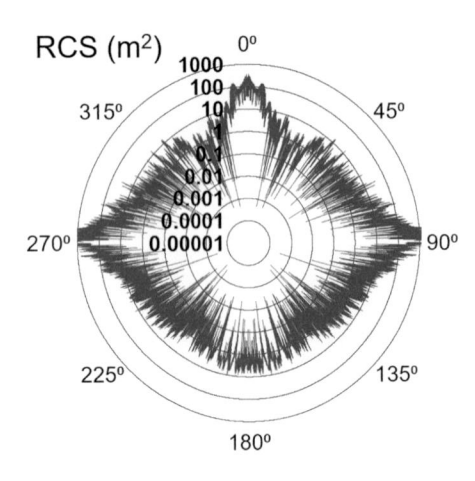

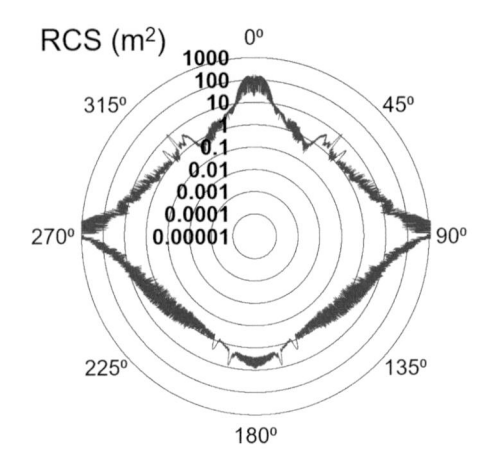

FIGURE 2.32 Circular diagram of noncoherent RCS given radar observation of MiG-29 aircraft model at a carrier frequency of 1 GHz (30 cm wavelength).

Scattering Characteristics of Aerial Objects

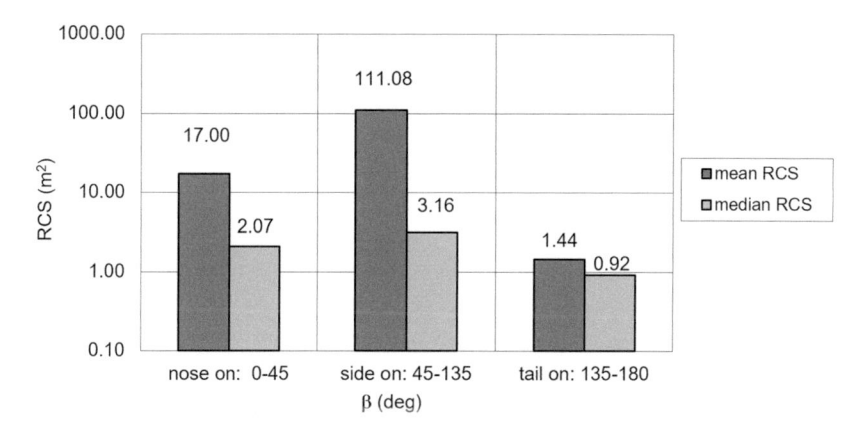

FIGURE 2.33 Diagrams of mean and median RCS of MiG-29 aircraft model in three sectors of azimuth aspect given its radar observation at horizontal polarization and a carrier frequency of 1 GHz (30 cm wavelength).

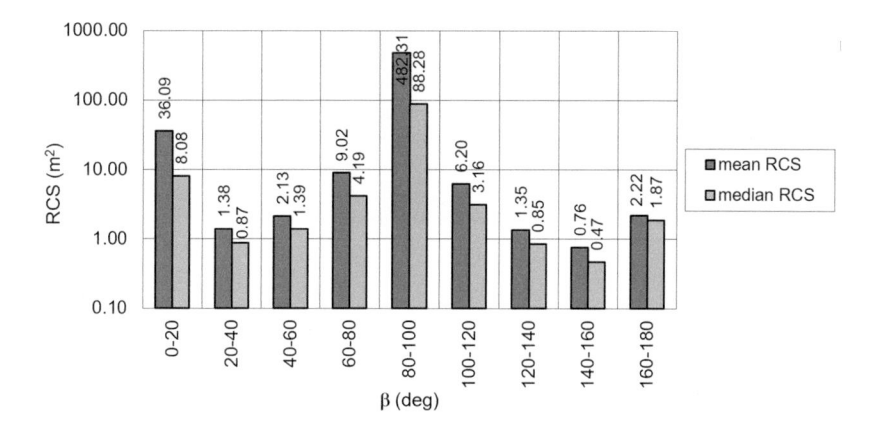

FIGURE 2.34 Diagrams of mean and median RCS of MiG-29 aircraft model in 20-degree sectors of azimuth aspect given its radar observation at horizontal polarization and a carrier frequency of 1 GHz (30 cm wavelength).

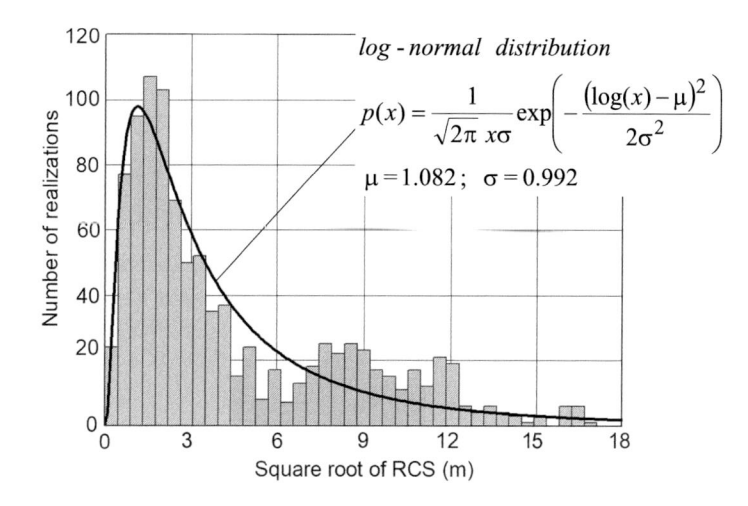

$$p(x) = \frac{1}{\sqrt{2\pi}\, x\sigma} \exp\left(-\frac{(\log(x)-\mu)^2}{2\sigma^2}\right)$$

$\mu = 1.082;\ \sigma = 0.992$

FIGURE 2.35 Amplitude distribution of echo signal of MiG-29 aircraft model at a carrier frequency of 1 GHz given its horizontal polarization.

2.3 MULTIROLE FIGHTER Su-57

Su-57 Felon (Figure 2.36) is one-seat multirole fighter. The first flight took place in 2010.

General characteristics of Su-57 [14]: wingspan – 14.00 m, length – 19.40 m, height – 4.80 m, wing area – 82.00 m^2, weight – 18 500–35 000 kg, powerplant – 2 Saturn AL41F1 turbofan engines, maximum speed – 2 550 km/h, range – 2 100–5 500 km, and service ceiling – 20 000 m.

In accordance with the design of the Su-57, a model of its surface was created to obtaining scattering characteristics (in particular, RCS). The model is shown in Figure 2.37. In the modeling, the smooth parts of the aircraft surface were approximated by parts of 55 triaxial ellipsoids. The surface breaks were modeled using 34 straight edge scattering parts. The model surface has a RAC that reduces the scattered signal by 18 dB at normal incidence for the 10 GHz sounding frequency and by 2.5 dB for the 3 GHz sounding frequency.

FIGURE 2.36 Multirole fighter Su-57.

FIGURE 2.37 Surface model of multirole fighter Su-57.

Below are some scattering characteristics of the Su-57 model at sounding frequencies of 10, 3, and 1 GHz (wavelengths of 3, 10, and 30 cm, respectively) for horizontal polarization of the probing signal. The scattering characteristics for this model at two polarizations, as well as scattering characteristics at sounding frequencies of 5 and 1.3 GHz (wavelengths of 6 and 23 cm, respectively), are given in the electronic appendix of this book.

Sounding parameters: The elevation angle is to be a random value distributed uniformly in the range −3° ± 4° with respect to the wing plane (elevation angle of −3° corresponds to the radar observation from the lower hemisphere), azimuth aspect increment was 0.02°, the azimuth being counted off from the nose-on aspect (0° corresponds to the nose-on radar observation, 180° corresponds to the tail-on observation).

Scattering Characteristics of Aerial Objects 25

Scattering characteristics of Su-57 aircraft model for sounding frequency 10 GHz (wavelength 3 cm).

Figure 2.38 shows the RCS circular diagram of the Su-57 model. The noncoherent RCS circular diagram of the Su-57 is shown in Figure 2.39.

The circular mean RCS of the Su-57 aircraft model for horizontal polarization is 4.61 m^2. The circular median RCS (the RCS value used to calculate the detection range of an object with a probability of 0.5) for horizontal polarization is 0.76 m^2.

Figures 2.40 and 2.41 show the mean and median RCS for the main ranges of sounding azimuths (nose, side, and tail) and for ranges of 20°.

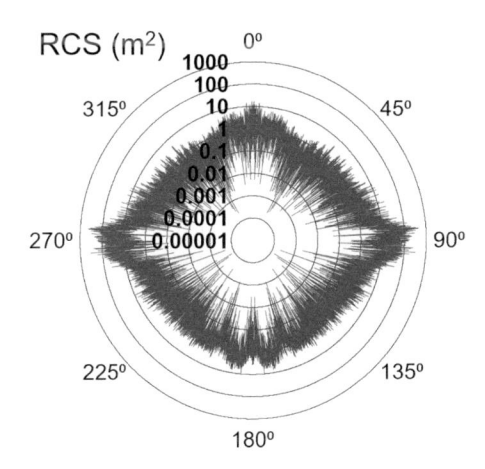

FIGURE 2.38 Circular diagram of RCS given radar observation of Su-57 aircraft model at a carrier frequency of 10 GHz (3 cm wavelength).

FIGURE 2.39 Circular diagram of noncoherent RCS given radar observation of Su-57 aircraft model at a carrier frequency of 10 GHz (3 cm wavelength).

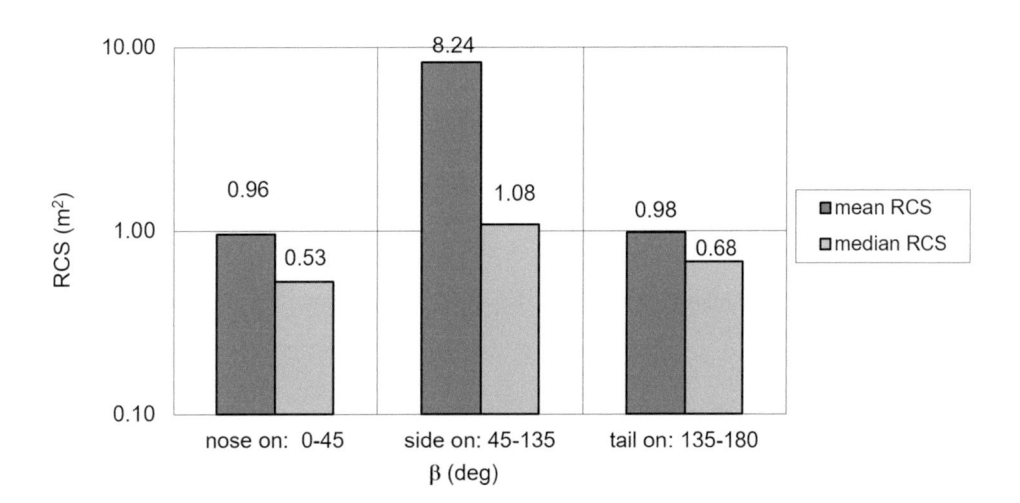

FIGURE 2.40 Diagrams of mean and median RCS of Su-57 aircraft model in three sectors of azimuth aspect given its radar observation at horizontal polarization and a carrier frequency of 10 GHz (3 cm wavelength).

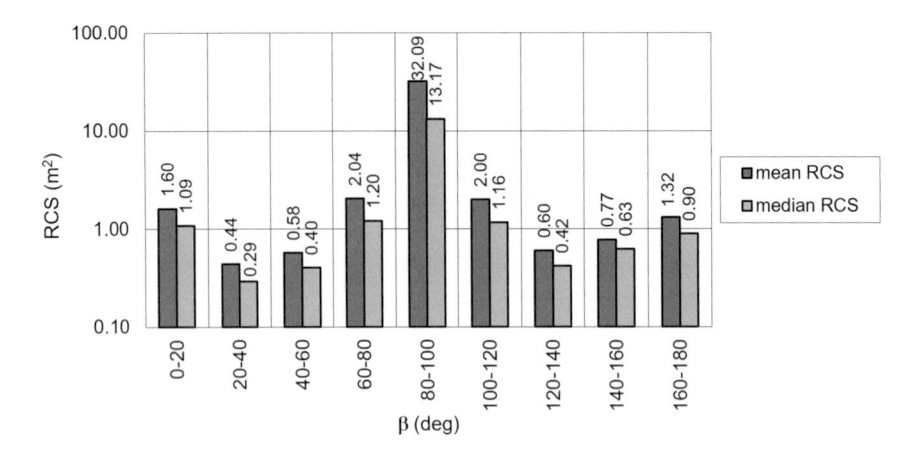

FIGURE 2.41 Diagrams of mean and median RCS of Su-57 aircraft model in 20-degree sectors of azimuth aspect given its radar observation at horizontal polarization and a carrier frequency of 10 GHz (3 cm wavelength).

Figures 2.42, 2.47, and 2.52 show histograms of the scattered signal amplitude (square root of the RCS) for the range of sounding azimuths −20° to +20° (sounding from the nose). The bold line shows the probability density functions of the distribution, which can be used to approximate the histogram of the signal amplitude.

extreme value distribution

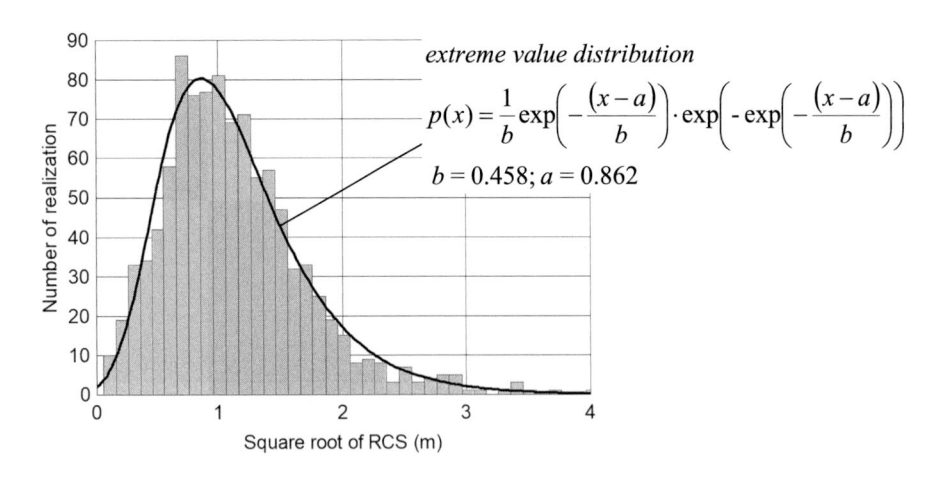

$$p(x) = \frac{1}{b} \exp\left(-\frac{(x-a)}{b}\right) \cdot \exp\left(-\exp\left(-\frac{(x-a)}{b}\right)\right)$$

$b = 0.458; a = 0.862$

FIGURE 2.42 Amplitude distribution of echo signal of Su-57 aircraft model at a carrier frequency of 10 GHz given its horizontal polarization.

Scattering characteristics of Su-57 aircraft model for sounding frequency 3 GHz (wavelength 10 cm).

Figure 2.43 shows the RCS circular diagram of the Su-57. The noncoherent RCS circular diagram of the Su-57 is shown in Figure 2.44.

The circular mean RCS of the Su-57 aircraft for horizontal polarization is 2.85 m². The circular median RCS for horizontal polarization is 0.58 m².

Figures 2.45 and 2.46 show the mean and median RCS for the main ranges of sounding azimuths (nose, side, and tail) and for ranges of 20°.

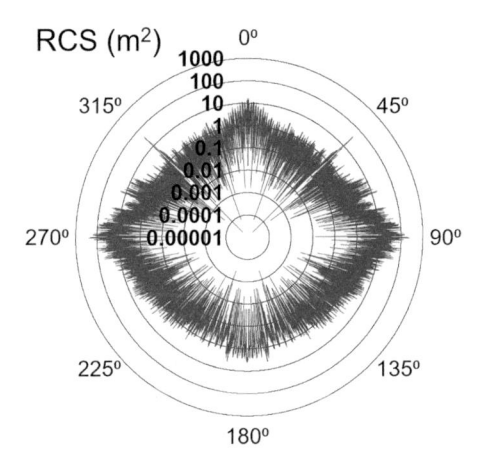

Scattering Characteristics of Aerial Objects

FIGURE 2.43 Circular diagram of RCS given radar observation of Su-57 aircraft model at a carrier frequency of 3 GHz (10 cm wavelength).

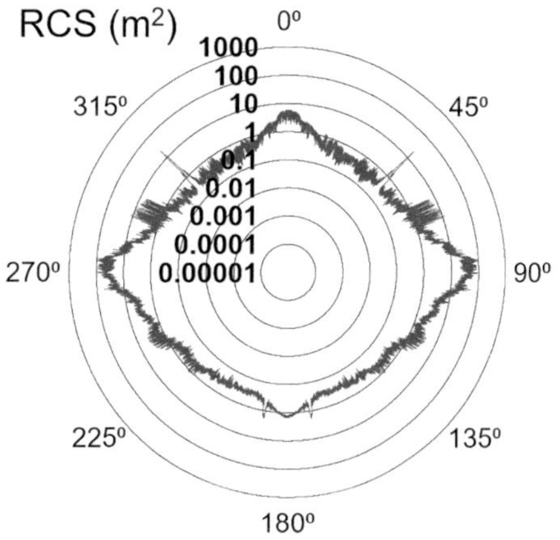

FIGURE 2.44 Circular diagram of noncoherent RCS given radar observation of Su-57 aircraft model at a carrier frequency of 3 GHz (10 cm wavelength).

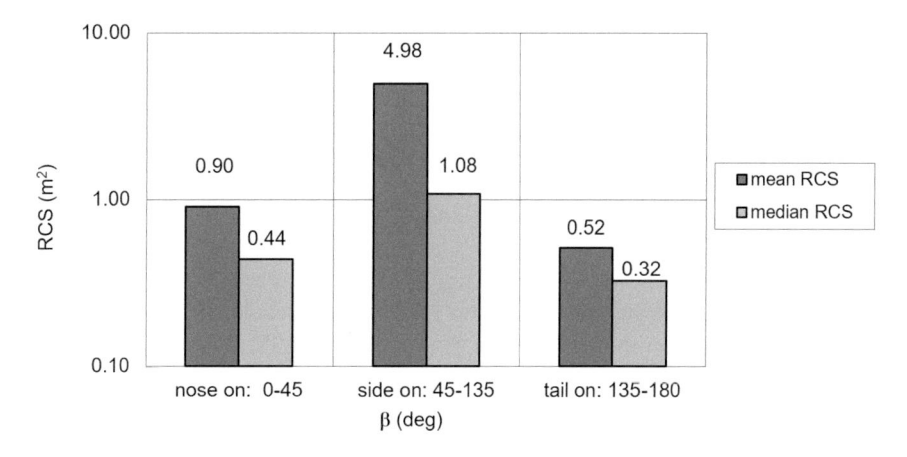

FIGURE 2.45 Diagrams of mean and median RCS of Su-57 aircraft model in three sectors of azimuth aspect given its radar observation at horizontal polarization and a carrier frequency of 3 GHz (10 cm wavelength).

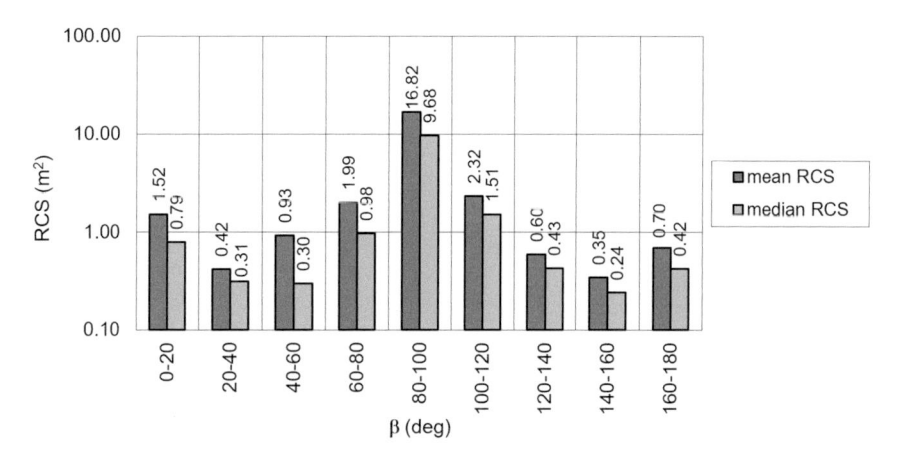

FIGURE 2.46 Diagrams of mean and median RCS of Su-57 aircraft model in 20-degree sectors of azimuth aspect given its radar observation at horizontal polarization and a carrier frequency of 3 GHz (10 cm wavelength).

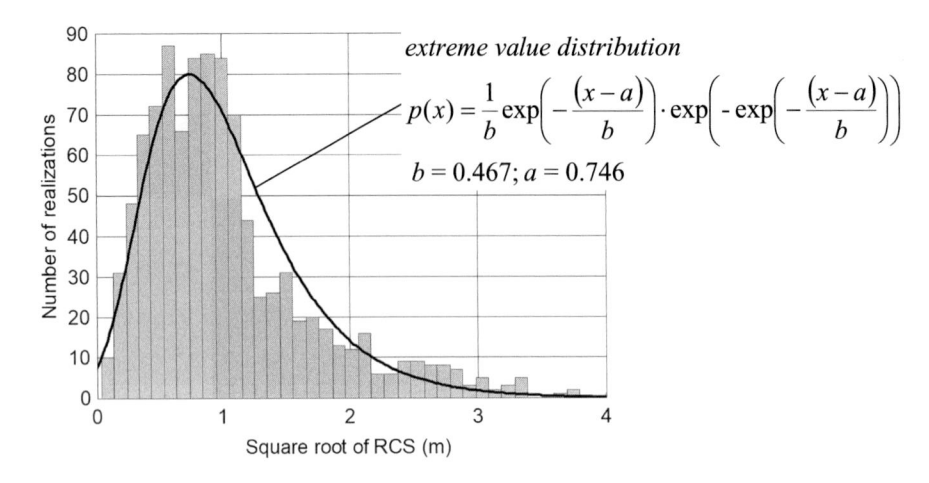

$$p(x) = \frac{1}{b}\exp\left(-\frac{(x-a)}{b}\right)\cdot\exp\left(-\exp\left(-\frac{(x-a)}{b}\right)\right)$$

extreme value distribution

$b = 0.467; a = 0.746$

FIGURE 2.47 Amplitude distribution of echo signal of Su-57 aircraft model at a carrier frequency of 3 GHz given its horizontal polarization.

Scattering characteristics of Su-57 aircraft model for sounding frequency 1 GHz (wavelength 30 cm).

Figure 2.48 shows the RCS circular diagram of the Su-57. The noncoherent RCS circular diagram of the Su-57 is shown in Figure 2.49.

The circular mean RCS of the Su-57 aircraft for horizontal polarization is 5.25 m². The circular median RCS for horizontal polarization is 0.74 m².

Figures 2.50 and 2.51 show the mean and median RCS for the main ranges of sounding azimuths (nose, side, and tail) and for ranges of 20°.

The expressions and parameters of probability distributions that are most consistent with the empirical distributions of the square root of the RCS for different ranges of sounding azimuths and polarizations are given in the electronic appendix to this book. The expressions and parameters of the probability distributions that are most consistent with the empirical distributions of the RCS (energy characteristic) for different ranges of sounding azimuths and polarizations are also given there.

FIGURE 2.48 Circular diagram of RCS given radar observation of Su-57 aircraft model at a carrier frequency of 1 GHz (30 cm wavelength).

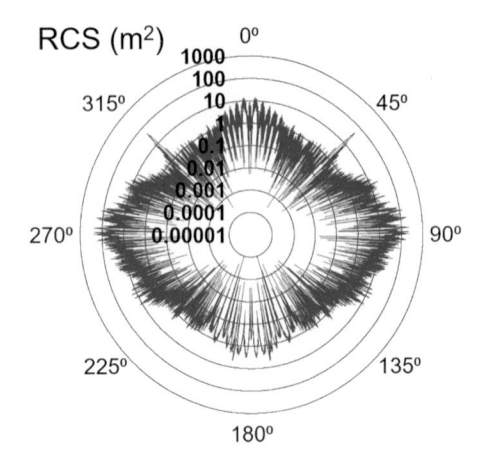

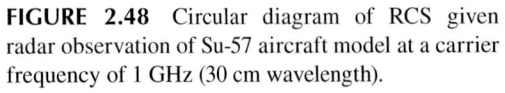

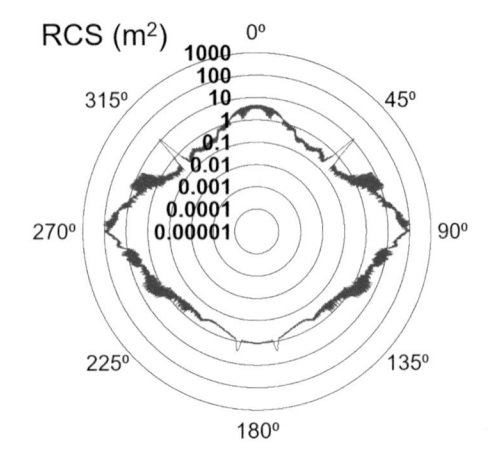

FIGURE 2.49 Circular diagram of noncoherent RCS given radar observation of Su-57 aircraft model at a carrier frequency of 1 GHz (30 cm wavelength).

Scattering Characteristics of Aerial Objects

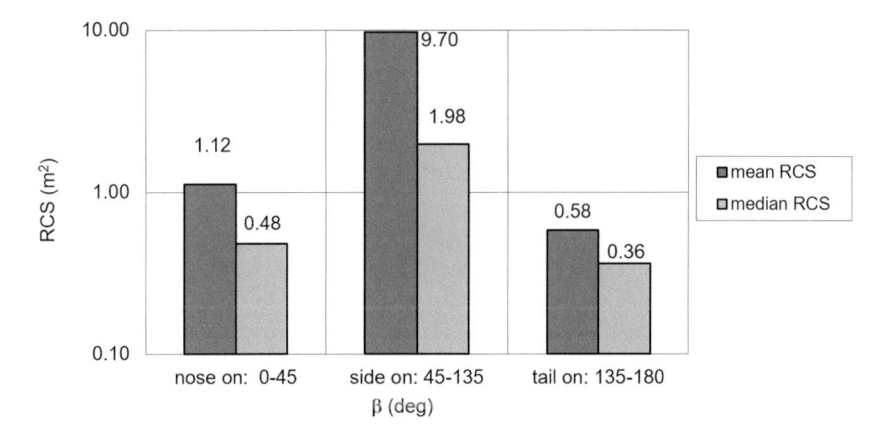

FIGURE 2.50 Diagrams of mean and median RCS of Su-57 aircraft model in three sectors of azimuth aspect given its radar observation at horizontal polarization and a carrier frequency of 1 GHz (30 cm wavelength).

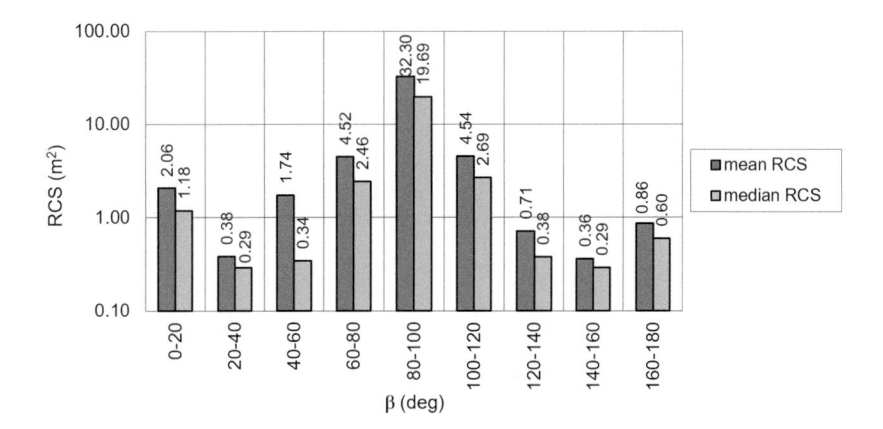

FIGURE 2.51 Diagrams of mean and median RCS of Su-57 aircraft model in 20-degree sectors of azimuth aspect given its radar observation at horizontal polarization and a carrier frequency of 1 GHz (30 cm wavelength).

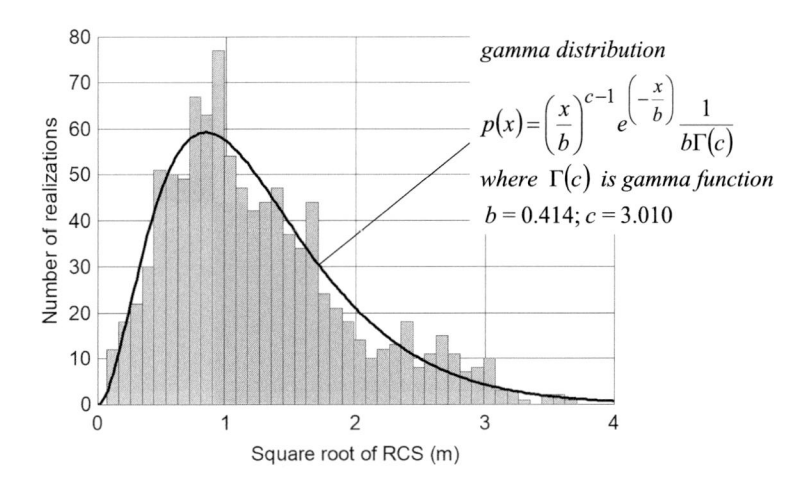

gamma distribution

$$p(x) = \left(\frac{x}{b}\right)^{c-1} e^{\left(-\frac{x}{b}\right)} \frac{1}{b\Gamma(c)}$$

where $\Gamma(c)$ is gamma function
$b = 0.414; c = 3.010$

FIGURE 2.52 Amplitude distribution of echo signal of Su-57 aircraft model at a carrier frequency of 1 GHz given its horizontal polarization.

2.4 TACTICAL FIGHTER F-22

F-22 Raptor (Figure 2.53) is one-seat tactical fighter. The first flight took place in 1997.

General characteristics of F-22 [15]: wingspan – 13.56 m, length – 18.92 m, height – 5.08 m, wing area – 78.04 m^2, weight – 19 700–38 000 kg, powerplant – 2 Pratt & Whitney F119-PW-100 augmented turbofans, maximum speed – 2 414 km/h, range – 1 083–3 220 km, and service ceiling – 20 000 m.

In accordance with the design of the F-22, a model of its surface was created to obtaining scattering characteristics (in particular, RCS). The model is shown in Figure 2.54. In the modeling, the smooth parts of the aircraft surface were approximated by parts of 65 triaxial ellipsoids. The surface breaks were modeled using 44 straight edge scattering parts. The model surface has a RAC that reduces the scattered signal by 18 dB at normal incidence for the 10 GHz sounding frequency and by 2.5 dB for the 3 GHz sounding frequency. The nose radome was modeled as perfectly conducting for all considered frequencies except 10 GHz.

FIGURE 2.53 Tactical fighter F-22.

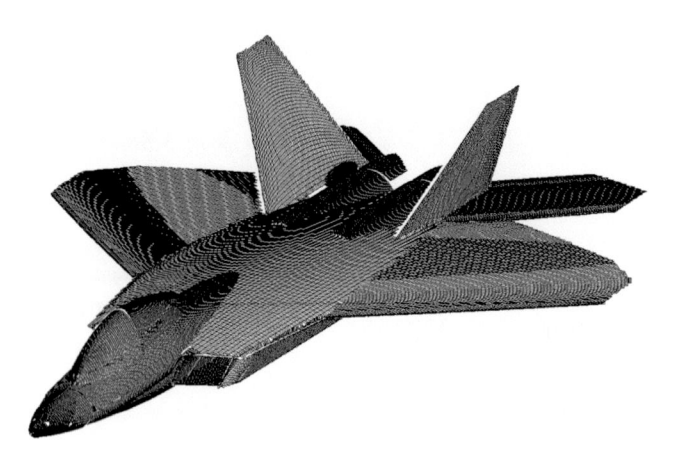

FIGURE 2.54 Surface model of tactical fighter F-22.

Below are some scattering characteristics of the F-22 model at sounding frequencies of 10, 3, and 1 GHz (wavelengths of 3, 10, and 30 cm, respectively) for horizontal polarization of the probing signal. The scattering characteristics for this model at two polarizations, as well as scattering

Scattering Characteristics of Aerial Objects

characteristics at sounding frequencies of 5 and 1.3 GHz (wavelengths of 6 and 23 cm, respectively), are given in the electronic appendix of this book.

Sounding parameters: The elevation angle is to be a random value distributed uniformly in the range −3° ± 4° with respect to the wing plane (elevation angle of −3° corresponds to the radar observation from the lower hemisphere), azimuth aspect increment was 0.02°, the azimuth being counted off from the nose-on aspect (0° corresponds to the nose-on radar observation, 180° corresponds to the tail-on observation).

Scattering characteristics of F-22 aircraft model for sounding frequency 10 GHz (wavelength 3 cm).

Figure 2.55 shows the RCS circular diagram of the F-22 model. The noncoherent RCS circular diagram of the F-22 is shown in Figure 2.56.

The circular mean RCS of the F-22 aircraft model for horizontal polarization is 1.63 m². The circular median RCS (the RCS value used to calculate the detection range of an object with a probability of 0.5) for horizontal polarization is 0.18 m².

Figures 2.57 and 2.58 show the mean and median RCS for the main ranges of sounding azimuths (nose, side, and tail) and for ranges of 20°.

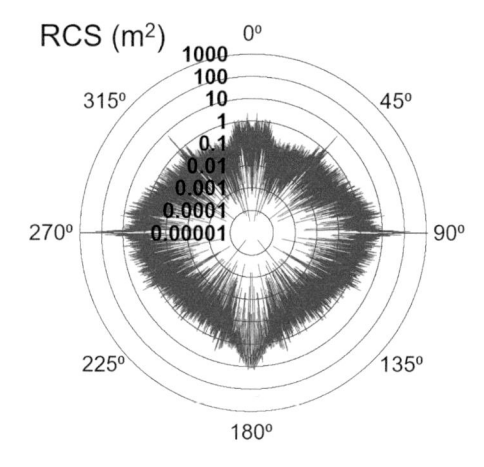

FIGURE 2.55 Circular diagram of RCS given radar observation of F-22 aircraft model at a carrier frequency of 10 GHz (3 cm wavelength).

FIGURE 2.56 Circular diagram of noncoherent RCS given radar observation of F-22 aircraft model at a carrier frequency of 10 GHz (3 cm wavelength).

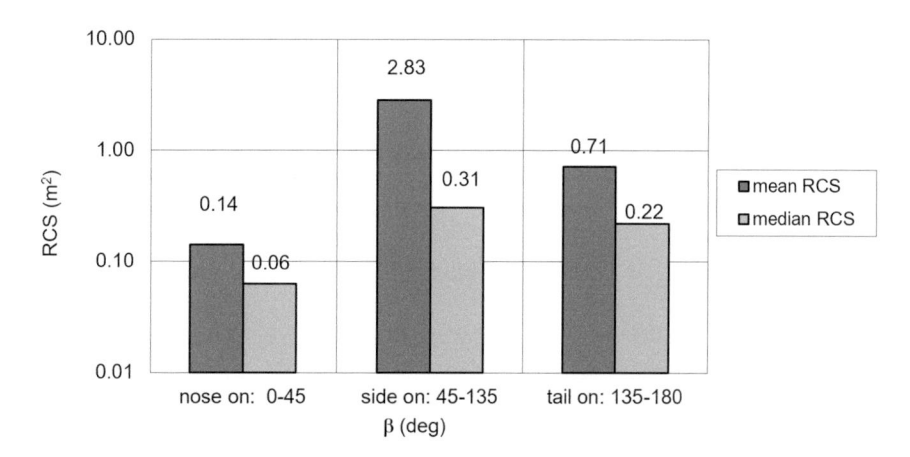

FIGURE 2.57 Diagrams of mean and median RCS of F-22 aircraft model in three sectors of azimuth aspect given its radar observation at horizontal polarization and a carrier frequency of 10 GHz (3 cm wavelength).

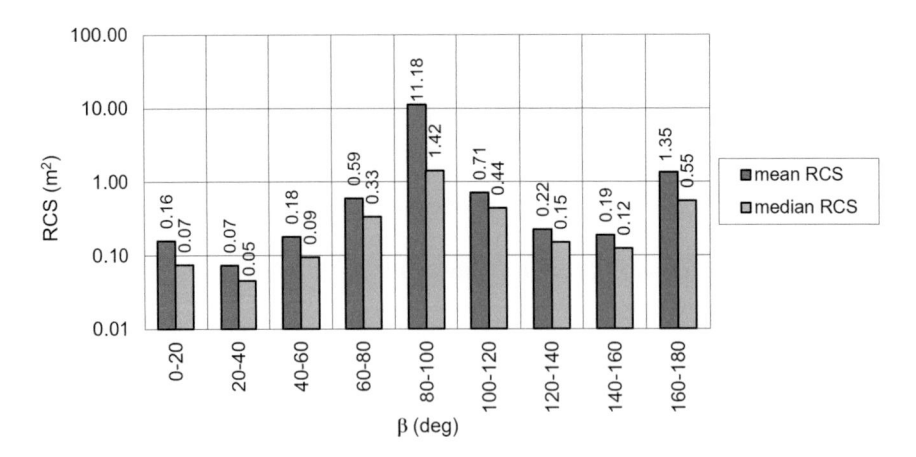

FIGURE 2.58 Diagrams of mean and median RCS of F-22 aircraft model in 20-degree sectors of azimuth aspect given its radar observation at horizontal polarization and a carrier frequency of 10 GHz (3 cm wavelength).

Figures 2.59, 2.64, and 2.69 show histograms of the scattered signal amplitude (square root of the RCS) for the range of sounding azimuths −20° to +20° (sounding from the nose). The bold line shows the probability density functions of the distribution, which can be used to approximate the histogram of the signal amplitude.

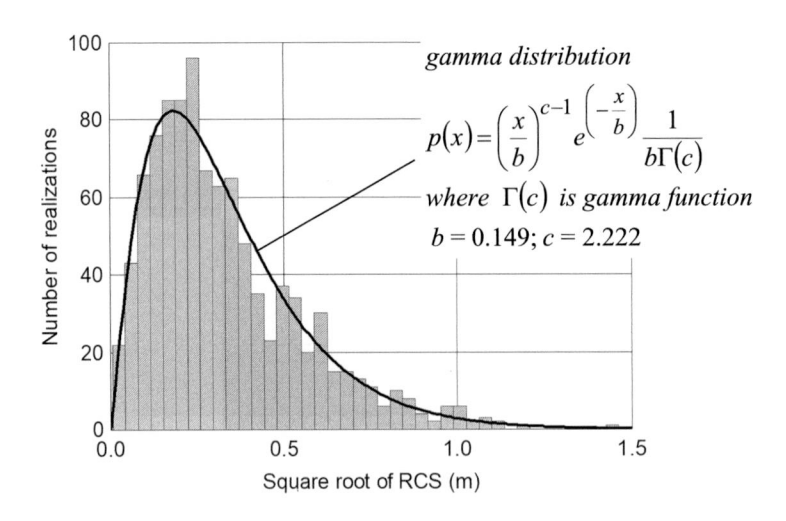

gamma distribution

$$p(x) = \left(\frac{x}{b}\right)^{c-1} e^{\left(-\frac{x}{b}\right)} \frac{1}{b\Gamma(c)}$$

where $\Gamma(c)$ is gamma function
$b = 0.149; c = 2.222$

FIGURE 2.59 Amplitude distribution of echo signal of F-22 aircraft model at a carrier frequency of 10 GHz given its horizontal polarization.

Scattering characteristics of F-22 aircraft model for sounding frequency 3 GHz (wavelength 10 cm).

Figure 2.60 shows the RCS circular diagram of the F-22. The noncoherent RCS circular diagram of the F-22 is shown in Figure 2.61.

The circular mean RCS of the F-22 aircraft for horizontal polarization is 2.26 m². The circular median RCS for horizontal polarization is 0.41 m².

Figures 2.62 and 2.63 show the mean and median RCS for the main ranges of sounding azimuths (nose, side, and tail) and for ranges of 20°.

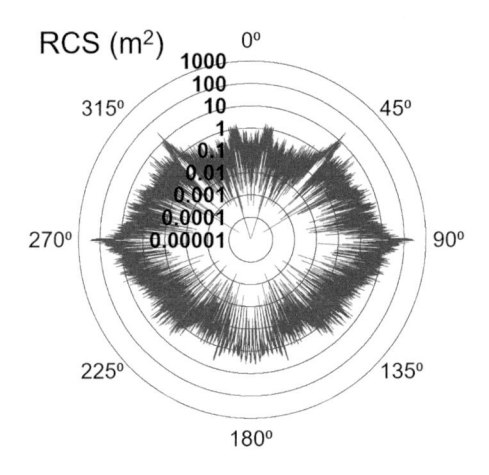

Scattering Characteristics of Aerial Objects

FIGURE 2.60 Circular diagram of RCS given radar observation of F-22 aircraft model at a carrier frequency of 3 GHz (10 cm wavelength).

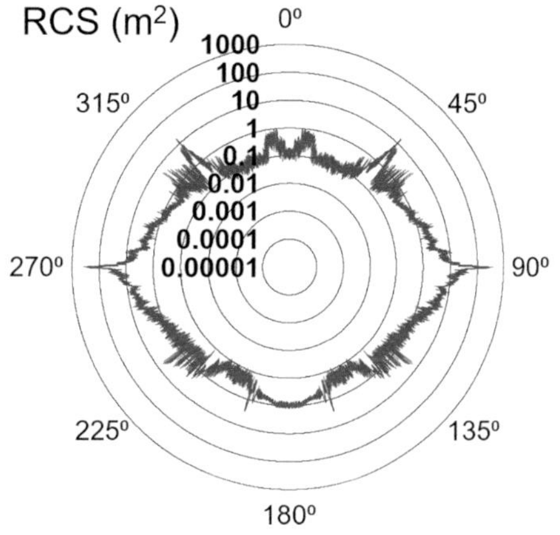

FIGURE 2.61 Circular diagram of noncoherent RCS given radar observation of F-22 aircraft model at a carrier frequency of 3 GHz (10 cm wavelength).

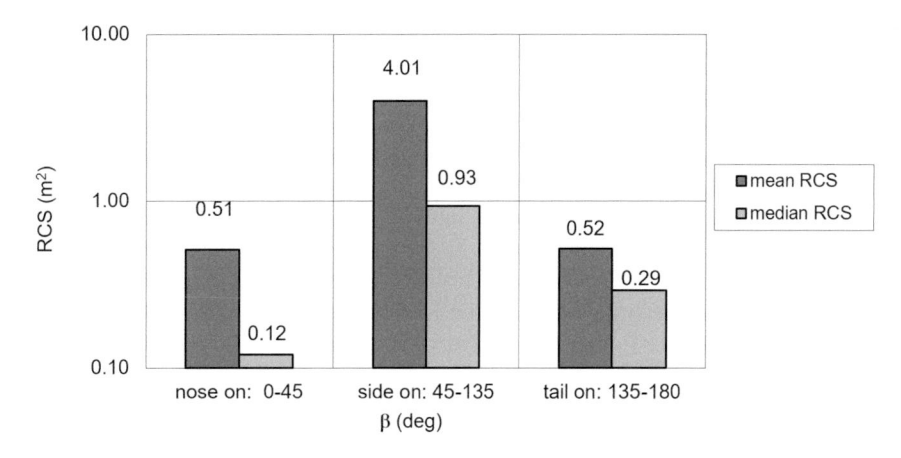

FIGURE 2.62 Diagrams of mean and median RCS of F-22 aircraft model in three sectors of azimuth aspect given its radar observation at horizontal polarization and a carrier frequency of 3 GHz (10 cm wavelength).

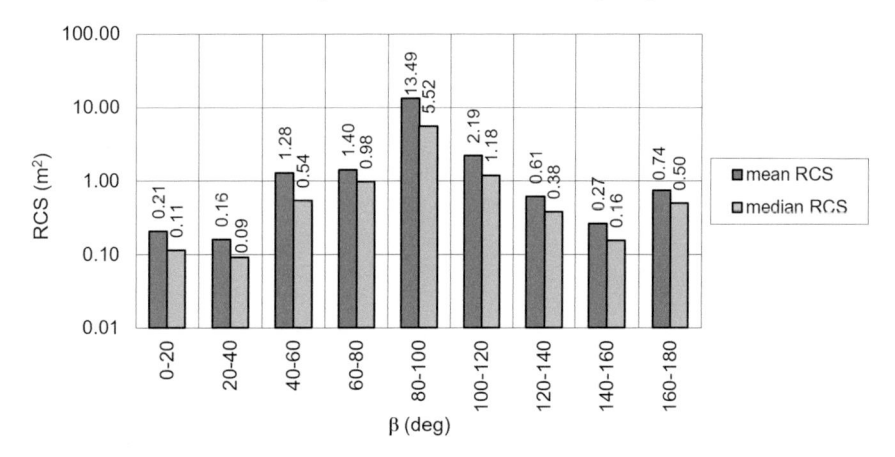

FIGURE 2.63 Diagrams of mean and median RCS of F-22 aircraft model in 20-degree sectors of azimuth aspect given its radar observation at horizontal polarization and a carrier frequency of 3 GHz (10 cm wavelength).

extreme value distribution

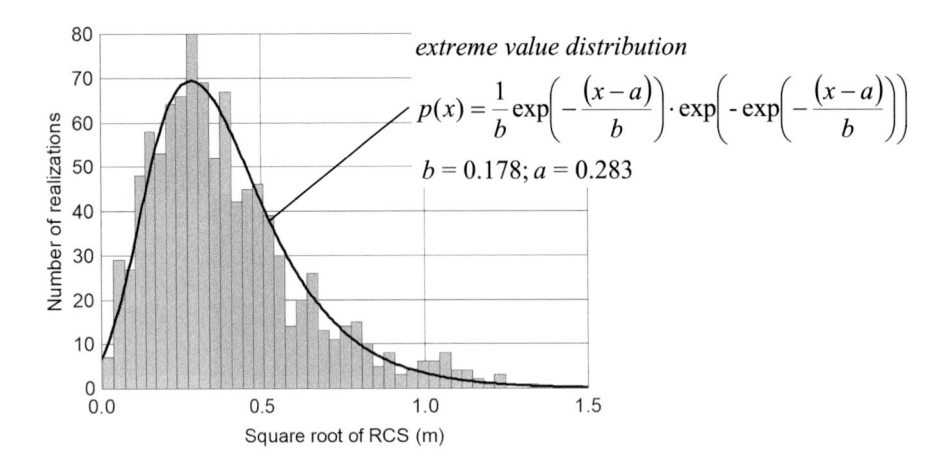

$$p(x) = \frac{1}{b}\exp\left(-\frac{(x-a)}{b}\right) \cdot \exp\left(-\exp\left(-\frac{(x-a)}{b}\right)\right)$$

$b = 0.178; a = 0.283$

FIGURE 2.64 Amplitude distribution of echo signal of F-22 aircraft model at a carrier frequency of 3 GHz given its horizontal polarization.

Scattering characteristics of F-22 aircraft model for sounding frequency 1 GHz (wavelength 30 cm).

Figure 2.65 shows the RCS circular diagram of the F-22. The noncoherent RCS circular diagram of the F-22 is shown in Figure 2.66.

The circular mean RCS of the F-22 aircraft for horizontal polarization is 2.90 m². The circular median RCS for horizontal polarization is 0.57 m².

Figures 2.67 and 2.68 show the mean and median RCS for the main ranges of sounding azimuths (nose, side, and tail) and for ranges of 20°.

The expressions and parameters of probability distributions that are most consistent with the empirical distributions of the square root of the RCS for different ranges of sounding azimuths and polarizations are given in the electronic appendix to this book. The expressions and parameters of the probability distributions that are most consistent with the empirical distributions of the RCS (energy characteristic) for different ranges of sounding azimuths and polarizations are also given there.

FIGURE 2.65 Circular diagram of RCS given radar observation of F-22 aircraft model at a carrier frequency of 1 GHz (30 cm wavelength).

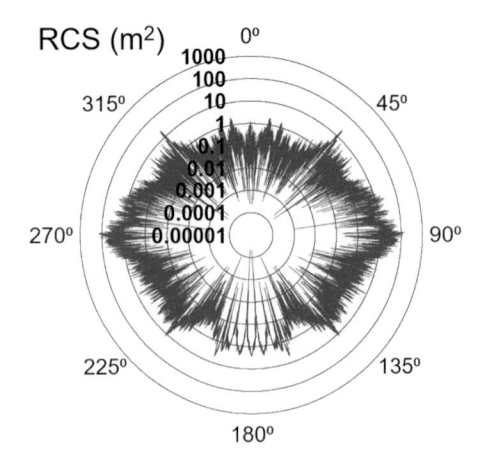

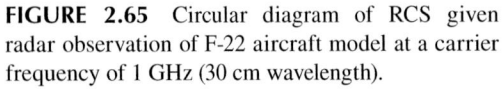

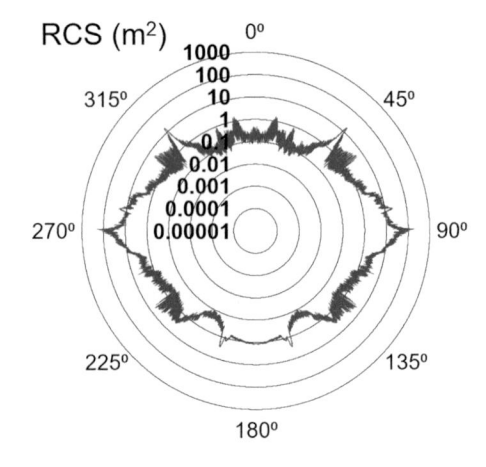

FIGURE 2.66 Circular diagram of noncoherent RCS given radar observation of F-22 aircraft model at a carrier frequency of 1 GHz (30 cm wavelength).

Scattering Characteristics of Aerial Objects

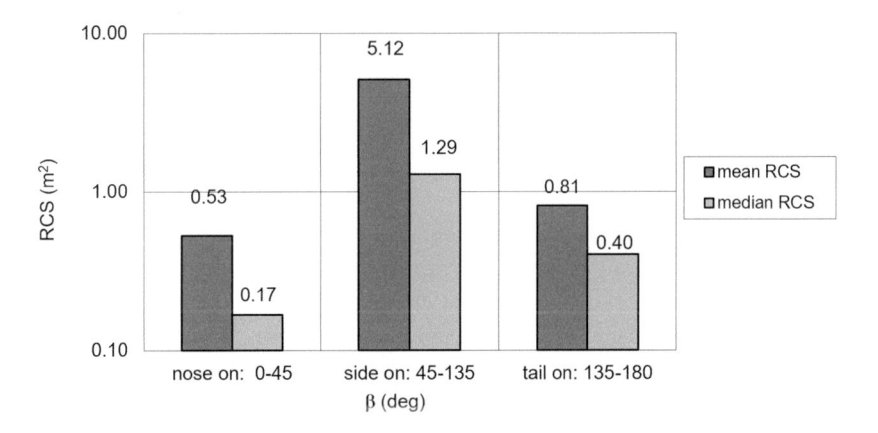

FIGURE 2.67 Diagrams of mean and median RCS of F-22 aircraft model in three sectors of azimuth aspect given its radar observation at horizontal polarization and a carrier frequency of 1 GHz (30 cm wavelength).

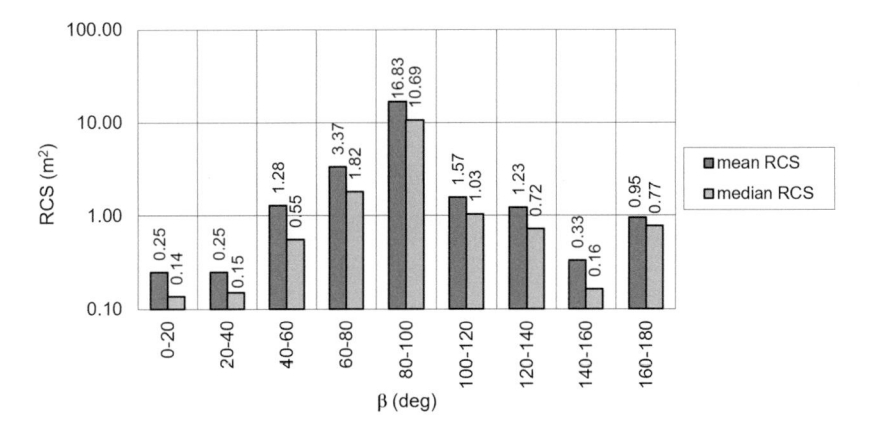

FIGURE 2.68 Diagrams of mean and median RCS of F-22 aircraft model in 20-degree sectors of azimuth aspect given its radar observation at horizontal polarization and a carrier frequency of 1 GHz (30 cm wavelength).

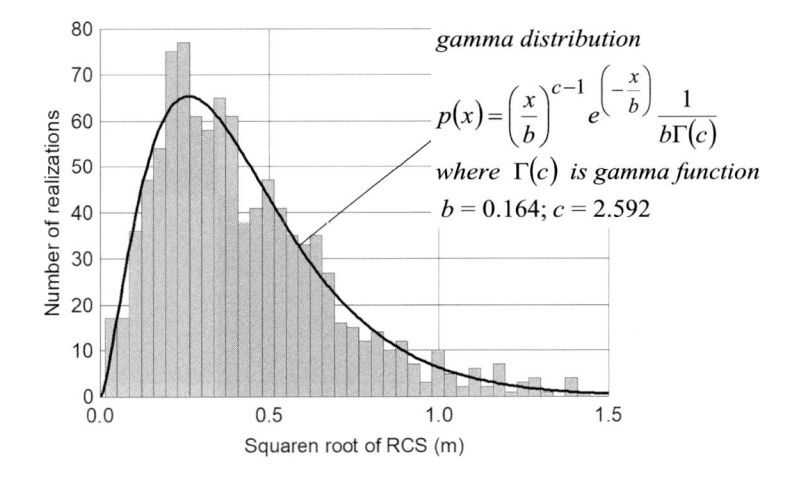

$$p(x) = \left(\frac{x}{b}\right)^{c-1} e^{\left(-\frac{x}{b}\right)} \frac{1}{b\Gamma(c)}$$

where $\Gamma(c)$ is gamma function
$b = 0.164; c = 2.592$

FIGURE 2.69 Amplitude distribution of echo signal of F-22 aircraft model at a carrier frequency of 1 GHz given its horizontal polarization.

2.5 MULTIROLE FIGHTER F-35

F-35 Lightning II (Figure 2.70) is one-seat multirole fighter. The first flight took place in 2006.

General characteristics of F-35 [16]: wingspan – 11 m, length – 15.7 m, height – 4.4 m, wing area – 43 m^2, weight – 13 300–29 900 kg, powerplant – 1 Pratt & Whitney F135-PW-100 afterburning turbofan, maximum speed – 1 950 km/h, range – 1 240–2 800 km, and service ceiling – 15 000 m.

In accordance with the design of the F-35, a model of its surface was created to obtaining scattering characteristics (in particular, RCS). The model is shown in Figure 2.71. In the modeling, the smooth parts of the aircraft surface were approximated by parts of 81 triaxial ellipsoids. The surface breaks were modeled using 30 straight edge scattering parts. The model surface has a RAC that reduces the scattered signal by 18 dB at normal incidence for the 10 GHz sounding frequency and by 2.5 dB for the 3 GHz sounding frequency.

FIGURE 2.70 Multirole fighter F-35.

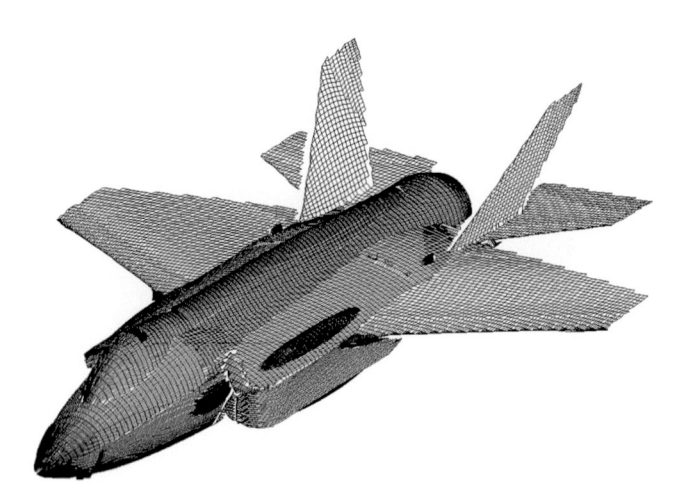

FIGURE 2.71 Surface model of multirole fighter F-35.

Below are some scattering characteristics of the F-35 model at sounding frequencies of 10, 3, and 1 GHz (wavelengths of 3, 10, and 30 cm, respectively) for horizontal polarization of the probing signal. The scattering characteristics for this model at two polarizations, as well as scattering characteristics at sounding frequencies of 5 and 1.3 GHz (wavelengths of 6 and 23 cm, respectively), are given in the electronic appendix of this book.

Sounding parameters: The elevation angle is to be a random value distributed uniformly in the range −3° ± 4° with respect to the wing plane (elevation angle of −3° corresponds to the radar observation from the lower hemisphere), azimuth aspect increment was 0.02°, the azimuth being counted off from the nose-on aspect (0° corresponds to the nose-on radar observation, 180° corresponds to the tail-on observation).

Scattering characteristics of F-35 aircraft model for sounding frequency 10 GHz (wavelength 3 cm).

Figure 2.72 shows the RCS circular diagram of the F-35 model. The noncoherent RCS circular diagram of the F-35 is shown in Figure 2.73.

The circular mean RCS of the F-35 aircraft model for horizontal polarization is 0.53 m². The circular median RCS (the RCS value used to calculate the detection range of an object with a probability of 0.5) for horizontal polarization is 0.15 m².

Figures 2.74 and 2.75 show the mean and median RCS for the main ranges of sounding azimuths (nose, side, and tail) and for ranges of 20°.

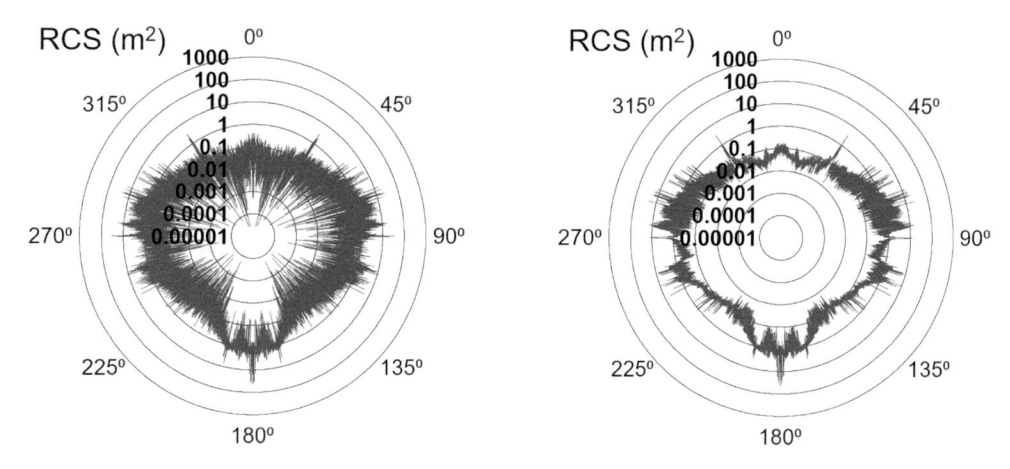

FIGURE 2.72 Circular diagram of RCS given radar observation of F-35 aircraft model at a carrier frequency of 10 GHz (3 cm wavelength).

FIGURE 2.73 Circular diagram of noncoherent RCS given radar observation of F-35 aircraft model at a carrier frequency of 10 GHz (3 cm wavelength).

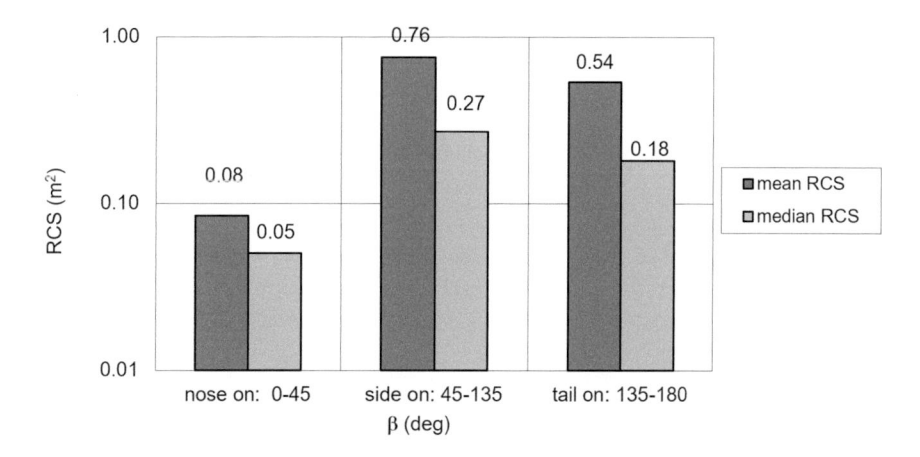

FIGURE 2.74 Diagrams of mean and median RCS of F-35 aircraft model in three sectors of azimuth aspect given its radar observation at horizontal polarization and a carrier frequency of 10 GHz (3 cm wavelength).

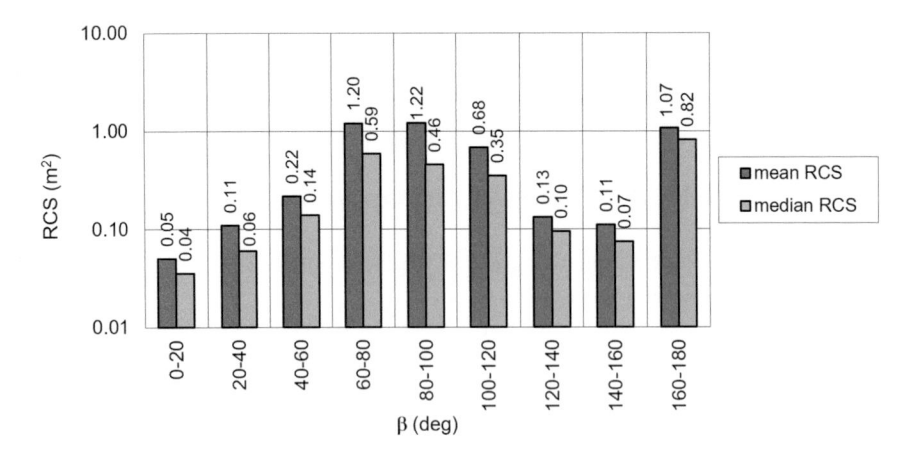

FIGURE 2.75 Diagrams of mean and median RCS of F-35 aircraft model in 20-degree sectors of azimuth aspect given its radar observation at horizontal polarization and a carrier frequency of 10 GHz (3 cm wavelength).

Figures 2.76, 2.81, and 2.86 show histograms of the scattered signal amplitude (square root of the RCS) for the range of sounding azimuths −20° to +20° (sounding from the nose). The bold line shows the probability density functions of the distribution, which can be used to approximate the histogram of the signal amplitude.

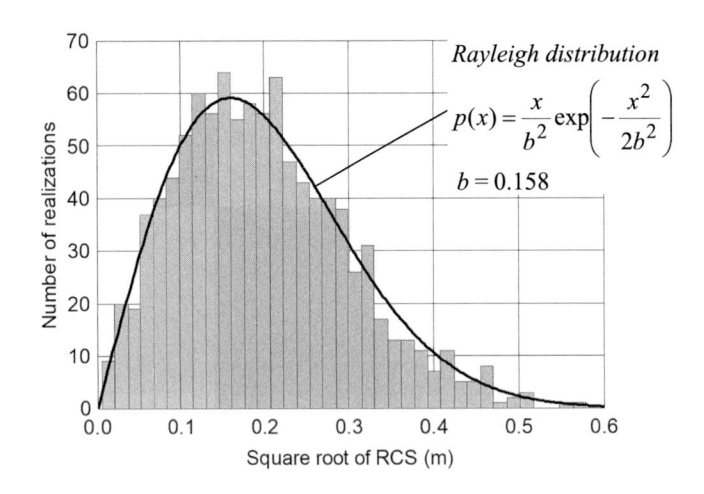

FIGURE 2.76 Amplitude distribution of echo signal of F-35 aircraft model at a carrier frequency of 10 GHz given its horizontal polarization.

Scattering characteristics of F-35 aircraft model for sounding frequency 3 GHz (wavelength 10 cm).

Figure 2.77 shows the RCS circular diagram of the F-35. The noncoherent RCS circular diagram of the F-35 is shown in Figure 2.78.

The circular mean RCS of the F-35 aircraft for horizontal polarization is 1.54 m². The circular median RCS for horizontal polarization is 0.31 m².

Figures 2.79 and 2.80 show the mean and median RCS for the main ranges of sounding azimuths (nose, side, and tail) and for ranges of 20°.

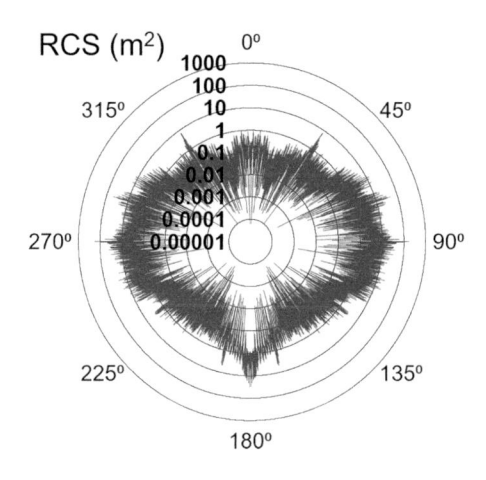

Scattering Characteristics of Aerial Objects

FIGURE 2.77 Circular diagram of RCS given radar observation of F-35 aircraft model at a carrier frequency of 3 GHz (10 cm wavelength).

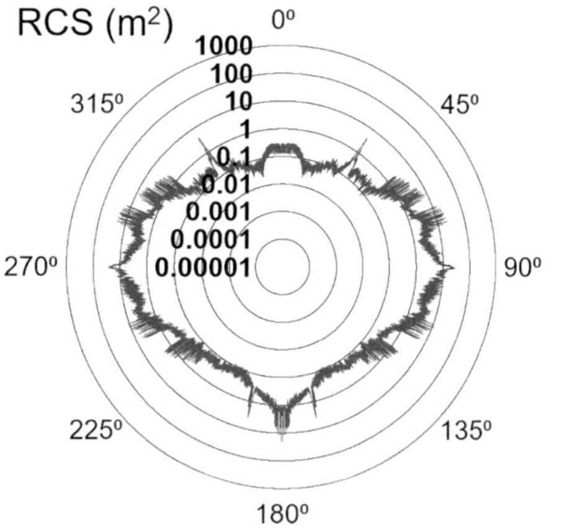

FIGURE 2.78 Circular diagram of noncoherent RCS given radar observation of F-35 aircraft model at a carrier frequency of 3 GHz (10 cm wavelength).

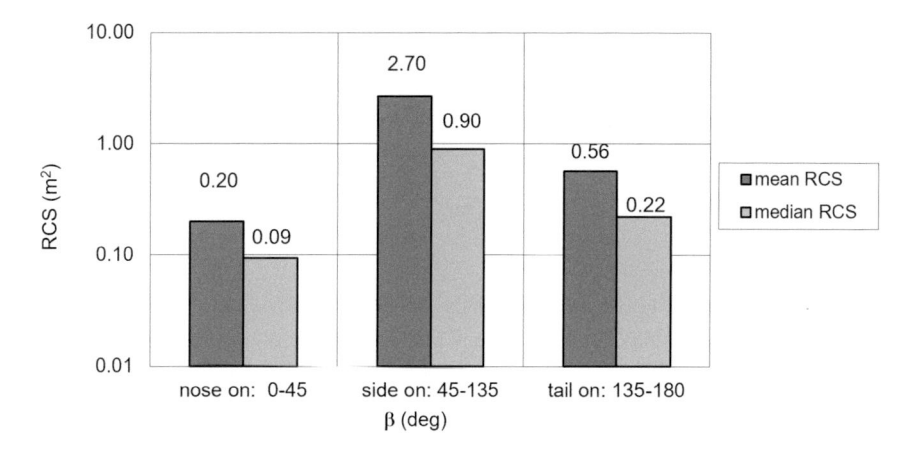

FIGURE 2.79 Diagrams of mean and median RCS of F-35 aircraft model in three sectors of azimuth aspect given its radar observation at horizontal polarization and a carrier frequency of 3 GHz (10 cm wavelength).

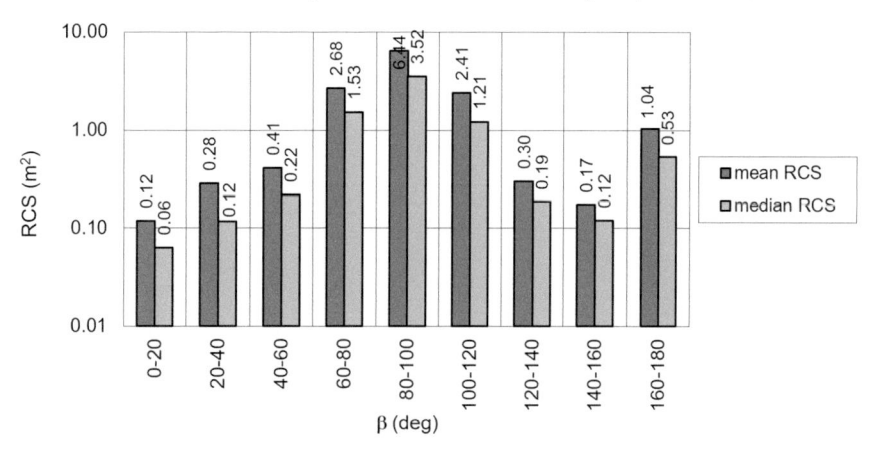

FIGURE 2.80 Diagrams of mean and median RCS of F-35 aircraft model in 20-degree sectors of azimuth aspect given its radar observation at horizontal polarization and a carrier frequency of 3 GHz (10 cm wavelength).

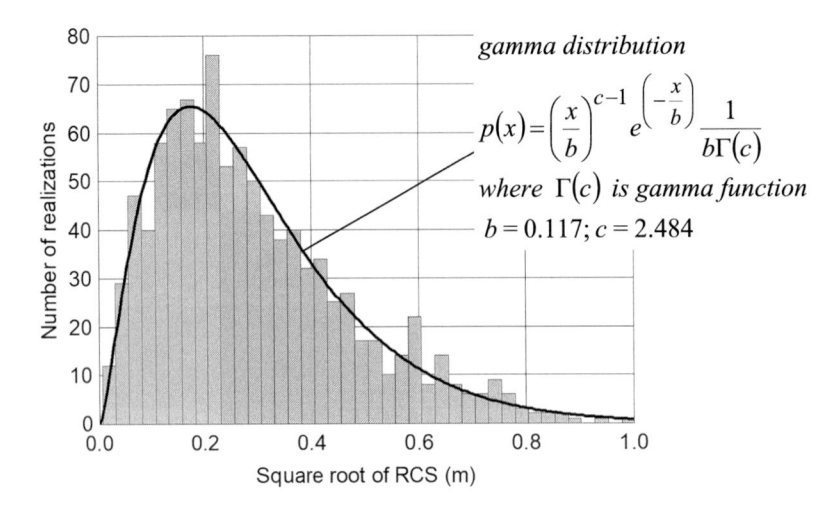

$$p(x) = \left(\frac{x}{b}\right)^{c-1} e^{\left(-\frac{x}{b}\right)} \frac{1}{b\Gamma(c)}$$

where $\Gamma(c)$ is gamma function

$b = 0.117; c = 2.484$

FIGURE 2.81 Amplitude distribution of echo signal of F-35 aircraft model at a carrier frequency of 3 GHz given its horizontal polarization.

Scattering characteristics of F-35 aircraft model for sounding frequency 1 GHz (wavelength 30 cm).

Figure 2.82 shows the RCS circular diagram of the F-35. The noncoherent RCS circular diagram of the F-35 is shown in Figure 2.83.

The circular mean RCS of the F-35 aircraft for horizontal polarization is 2.79 m². The circular median RCS for horizontal polarization is 0.46 m².

Figures 2.84 and 2.85 show the mean and median RCS for the main ranges of sounding azimuths (nose, side, and tail) and for ranges of 20°.

The expressions and parameters of probability distributions that are most consistent with the empirical distributions of the square root of the RCS for different ranges of sounding azimuths and polarizations are given in the electronic appendix to this book. The expressions and parameters of the probability distributions that are most consistent with the empirical distributions of the RCS (energy characteristic) for different ranges of sounding azimuths and polarizations are also given there.

FIGURE 2.82 Circular diagram of RCS given radar observation of F-35 aircraft model at a carrier frequency of 1 GHz (30 cm wavelength).

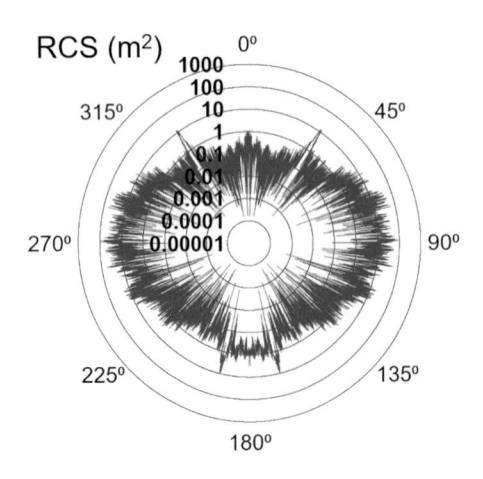

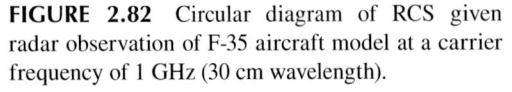

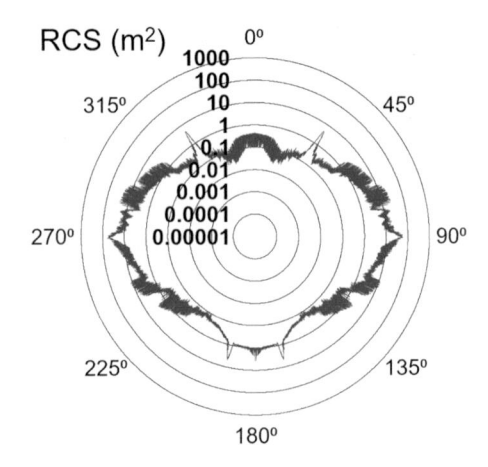

FIGURE 2.83 Circular diagram of noncoherent RCS given radar observation of F-35 aircraft model at a carrier frequency of 1 GHz (30 cm wavelength).

Scattering Characteristics of Aerial Objects 41

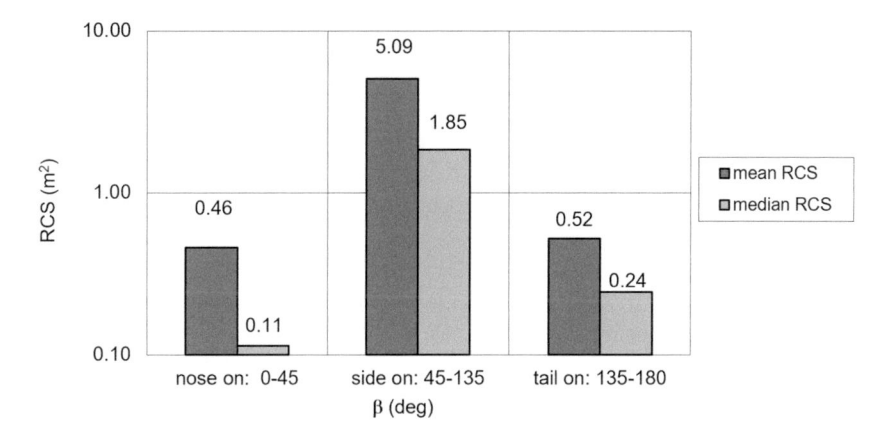

FIGURE 2.84 Diagrams of mean and median RCS of F-35 aircraft model in three sectors of azimuth aspect given its radar observation at horizontal polarization and a carrier frequency of 1 GHz (30 cm wavelength).

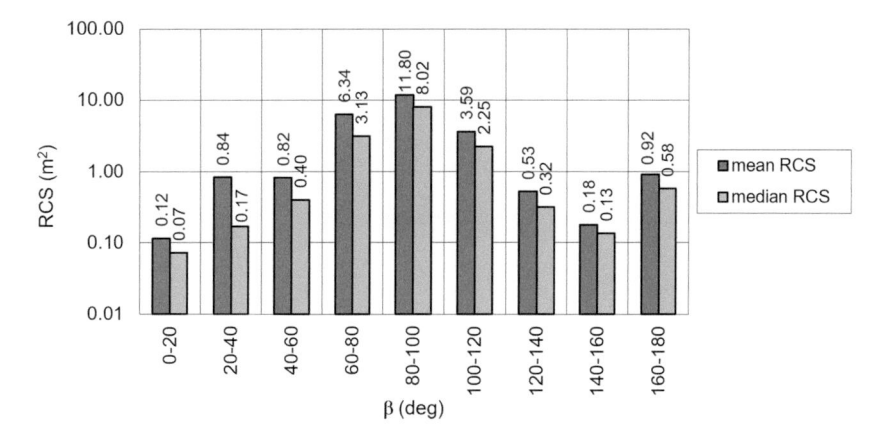

FIGURE 2.85 Diagrams of mean and median RCS of F-35 aircraft model in 20-degree sectors of azimuth aspect given its radar observation at horizontal polarization and a carrier frequency of 1 GHz (30 cm wavelength).

Weibull distribution

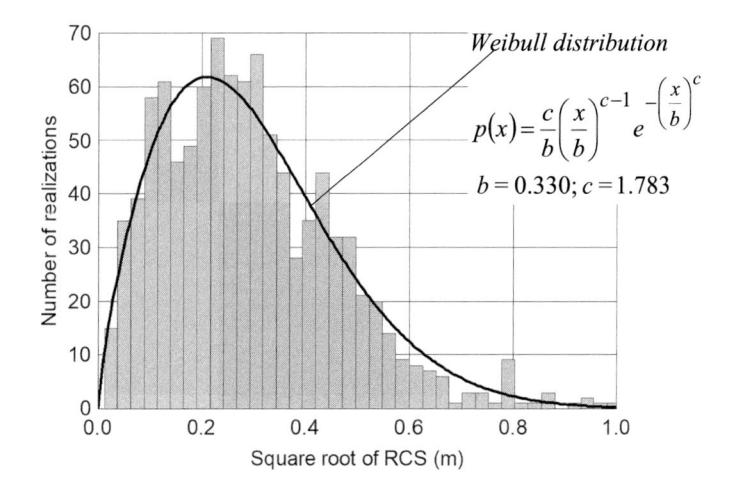

$$p(x) = \frac{c}{b}\left(\frac{x}{b}\right)^{c-1} e^{-\left(\frac{x}{b}\right)^c}$$

$b = 0.330; c = 1.783$

FIGURE 2.86 Amplitude distribution of echo signal of F-35 aircraft model at a carrier frequency of 1 GHz given its horizontal polarization.

2.6 MULTIROLE FIGHTER F-16

F-16 Fighting Falcon (Figure 2.87) is one-seat multirole fighter. The first flight took place in 1980.

General characteristics of F-16 [17]: wingspan – 9.45 m, length – 15.03 m, height – 5.09 m, wing area – 27.87 m^2, weight – 8 370–12 003 kg, powerplant – 1 Pratt & Whitney F100-PW-229 turbofan with afterburner, maximum speed – 2 145 km/h, combat range – 579 km, and service ceiling – 17 200 m.

In accordance with the design of the F-16, a model of its surface was created to obtaining scattering characteristics (in particular, RCS). The model is shown in Figure 2.88. In the modeling, the smooth parts of the aircraft surface were approximated by parts of 42 triaxial ellipsoids. The surface breaks were modeled using 30 straight edge scattering parts.

FIGURE 2.87 Multirole fighter F-16.

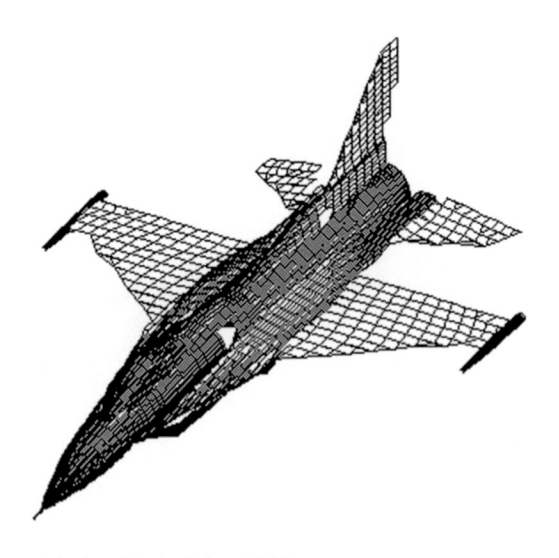

FIGURE 2.88 Surface model of multirole fighter F-16.

Below are some scattering characteristics of the F-16 model at sounding frequencies of 10, 3, and 1 GHz (wavelengths of 3, 10, and 30 cm, respectively) for horizontal polarization of the probing signal. The scattering characteristics for this model at two polarizations, as well as scattering characteristics at sounding frequencies of 5 and 1.3 GHz (wavelengths of 6 and 23 cm, respectively), are given in the electronic appendix of this book.

Scattering Characteristics of Aerial Objects

Sounding parameters: The elevation angle is to be a random value distributed uniformly in the range −3° ± 4° with respect to the wing plane (elevation angle of −3° corresponds to the radar observation from the lower hemisphere), azimuth aspect increment was 0.02°, the azimuth being counted off from the nose-on aspect (0° corresponds to the nose-on radar observation, 180° corresponds to the tail-on observation).

Scattering characteristics of F-16 aircraft model for sounding frequency 10 GHz (wavelength 3 cm).

Figure 2.89 shows the RCS circular diagram of the F-16 model. The noncoherent RCS circular diagram of the F-16 is shown in Figure 2.90.

The circular mean RCS of the F-16 aircraft model for horizontal polarization is 42.14 m². The circular median RCS (the RCS value used to calculate the detection range of an object with a probability of 0.5) for horizontal polarization is 0.63 m².

Figures 2.91 and 2.92 show the mean and median RCS for the main ranges of sounding azimuths (nose, side, and tail) and for ranges of 20°.

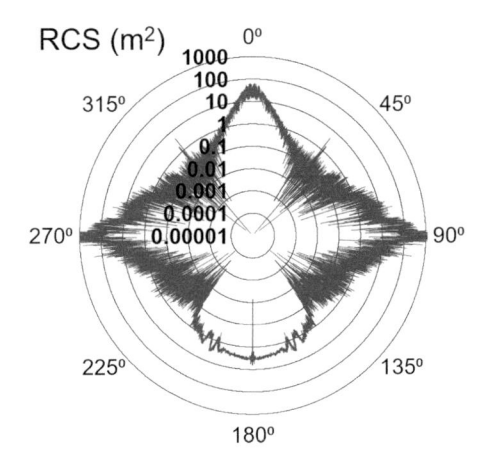

FIGURE 2.89 Circular diagram of RCS given radar observation of F-16 aircraft model at a carrier frequency of 10 GHz (3 cm wavelength).

FIGURE 2.90 Circular diagram of noncoherent RCS given radar observation of F-16 aircraft model at a carrier frequency of 10 GHz (3 cm wavelength).

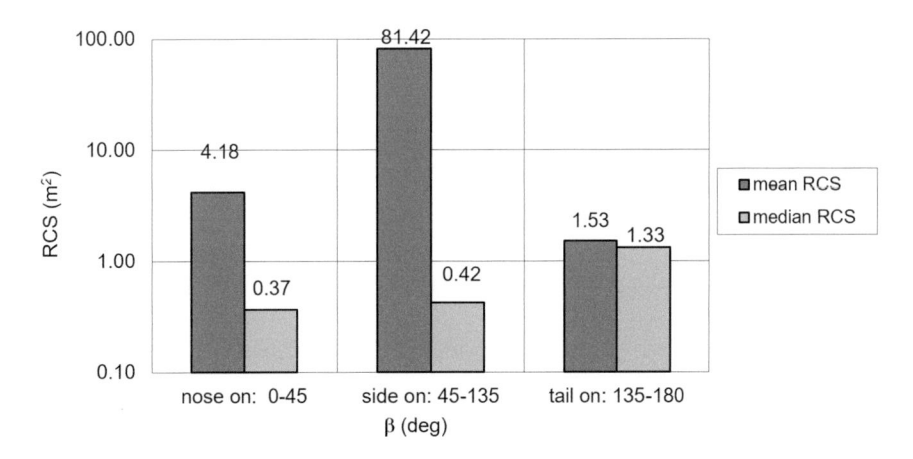

FIGURE 2.91 Diagrams of mean and median RCS of F-16 aircraft model in three sectors of azimuth aspect given its radar observation at horizontal polarization and a carrier frequency of 10 GHz (3 cm wavelength).

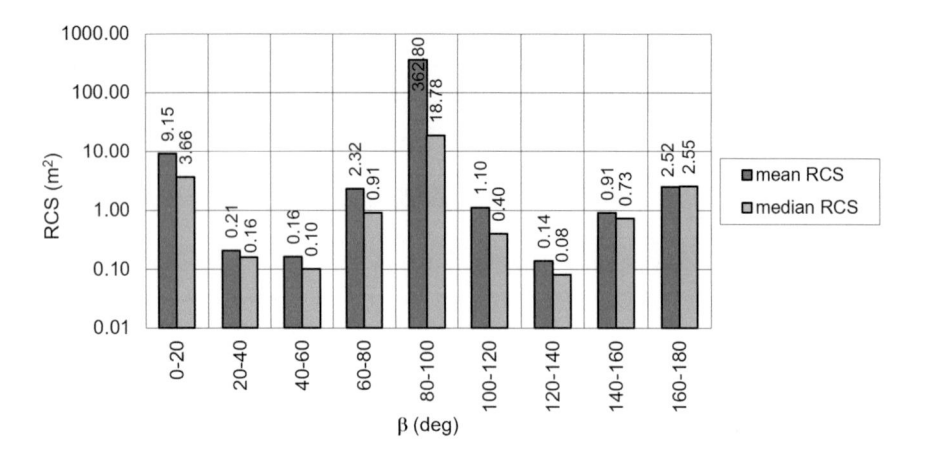

FIGURE 2.92 Diagrams of mean and median RCS of F-16 aircraft model in 20-degree sectors of azimuth aspect given its radar observation at horizontal polarization and a carrier frequency of 10 GHz (3 cm wavelength).

Figures 2.93, 2.98, and 2.103 show histograms of the scattered signal amplitude (square root of the RCS) for the range of sounding azimuths −20° to +20° (sounding from the nose). The bold line shows the probability density functions of the distribution, which can be used to approximate the histogram of the signal amplitude.

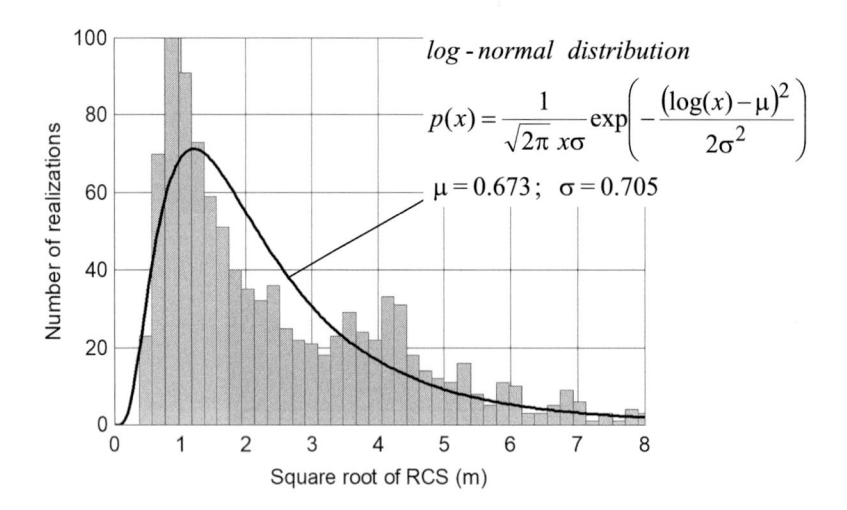

$$p(x) = \frac{1}{\sqrt{2\pi}\, x\sigma} \exp\left(-\frac{(\log(x)-\mu)^2}{2\sigma^2}\right)$$

$\mu = 0.673; \quad \sigma = 0.705$

FIGURE 2.93 Amplitude distribution of echo signal of F-16 aircraft model at a carrier frequency of 10 GHz given its horizontal polarization.

Scattering characteristics of F-16 aircraft model for sounding frequency 3 GHz (wavelength 10 cm).

Figure 2.94 shows the RCS circular diagram of the F-16. The noncoherent RCS circular diagram of the F-16 is shown in Figure 2.95.

The circular mean RCS of the F-16 aircraft for horizontal polarization is 35.36 m². The circular median RCS for horizontal polarization is 0.65 m².

Figures 2.96 and 2.97 show the mean and median RCS for the main ranges of sounding azimuths (nose, side, and tail) and for ranges of 20°.

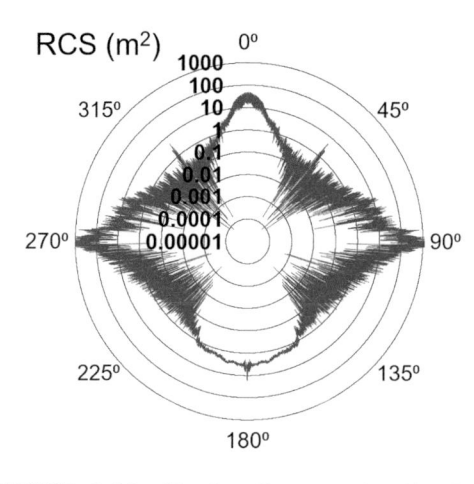

Scattering Characteristics of Aerial Objects

FIGURE 2.94 Circular diagram of RCS given radar observation of F-16 aircraft model at a carrier frequency of 3 GHz (10 cm wavelength).

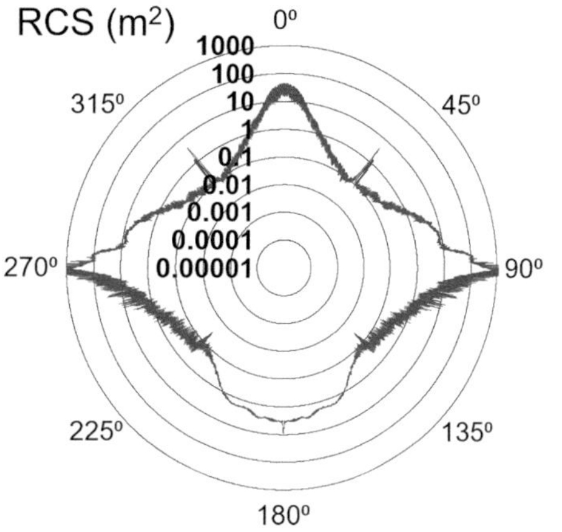

FIGURE 2.95 Circular diagram of noncoherent RCS given radar observation of F-16 aircraft model at a carrier frequency of 3 GHz (10 cm wavelength).

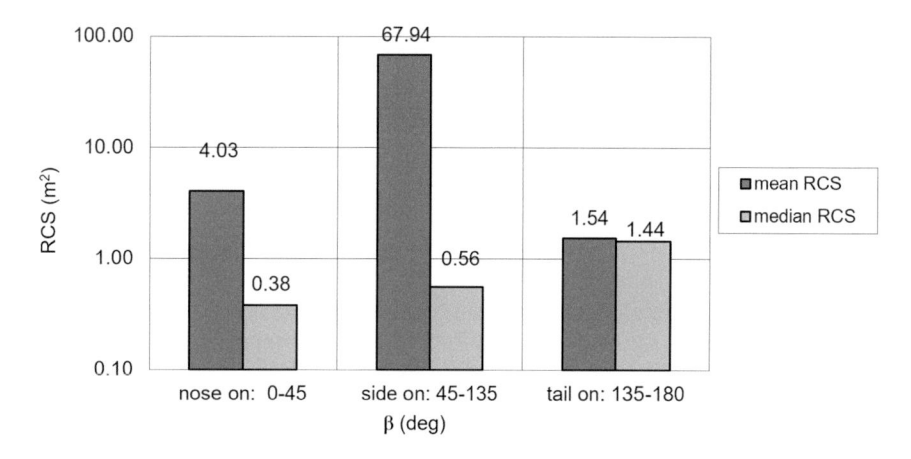

FIGURE 2.96 Diagrams of mean and median RCS of F-16 aircraft model in three sectors of azimuth aspect given its radar observation at horizontal polarization and a carrier frequency of 3 GHz (10 cm wavelength).

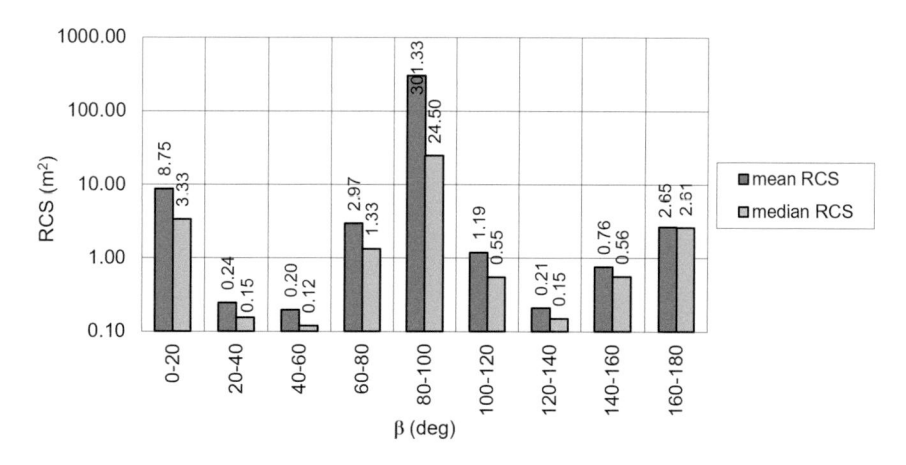

FIGURE 2.97 Diagrams of mean and median RCS of F-16 aircraft model in 20-degree sectors of azimuth aspect given its radar observation at horizontal polarization and a carrier frequency of 3 GHz (10 cm wavelength).

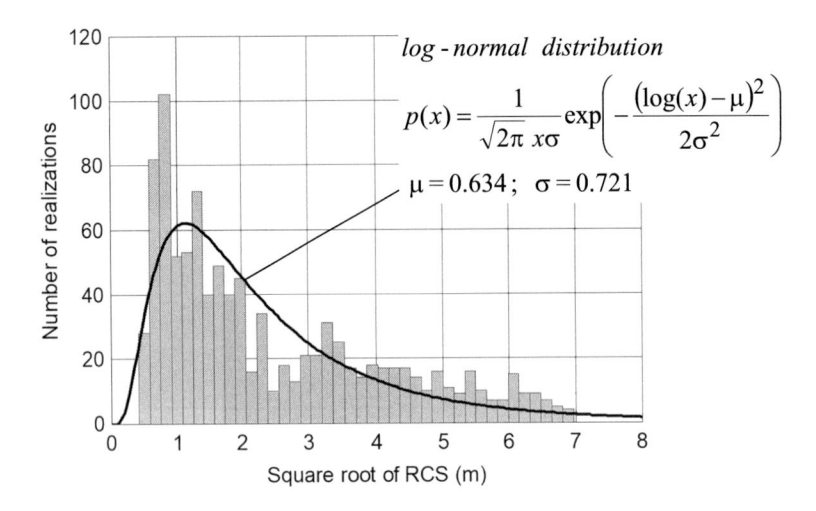

FIGURE 2.98 Amplitude distribution of echo signal of F-16 aircraft model at a carrier frequency of 3 GHz given its horizontal polarization.

$$p(x) = \frac{1}{\sqrt{2\pi}\, x\sigma} \exp\left(-\frac{(\log(x)-\mu)^2}{2\sigma^2}\right)$$

log-normal distribution, $\mu = 0.634$; $\sigma = 0.721$

Scattering characteristics of F-16 aircraft model for sounding frequency 1 GHz (wavelength 30 cm).

Figure 2.99 shows the RCS circular diagram of the F-16. The noncoherent RCS circular diagram of the F-16 is shown in Figure 2.100.

The circular mean RCS of the F-16 aircraft for horizontal polarization is 20.92 m². The circular median RCS for horizontal polarization is 0.67 m².

Figures 2.101 and 2.102 show the mean and median RCS for the main ranges of sounding azimuths (nose, side, and tail) and for ranges of 20°.

The expressions and parameters of probability distributions that are most consistent with the empirical distributions of the square root of the RCS for different ranges of sounding azimuths and polarizations are given in the electronic appendix to this book. The expressions and parameters of the probability distributions that are most consistent with the empirical distributions of the RCS (energy characteristic) for different ranges of sounding azimuths and polarizations are also given there.

FIGURE 2.99 Circular diagram of RCS given radar observation of F-16 aircraft model at a carrier frequency of 1 GHz (30 cm wavelength).

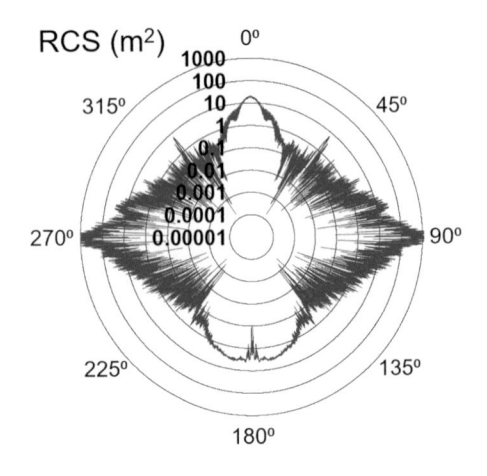

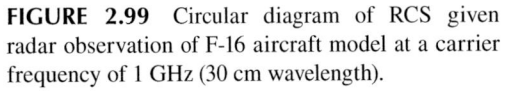

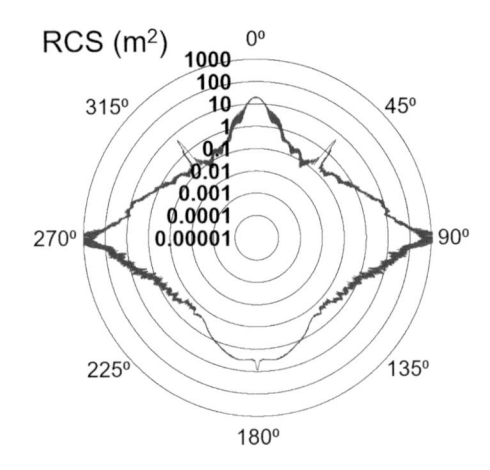

FIGURE 2.100 Circular diagram of noncoherent RCS given radar observation of F-16 aircraft model at a carrier frequency of 1 GHz (30 cm wavelength).

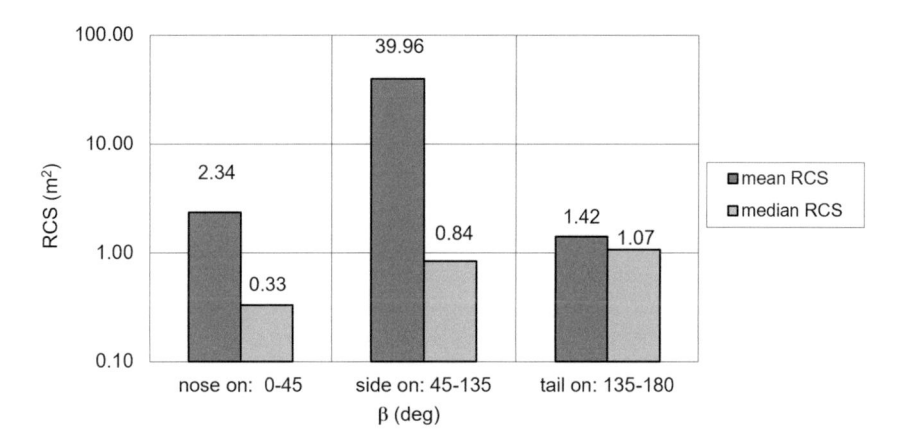

FIGURE 2.101 Diagrams of mean and median RCS of F-16 aircraft model in three sectors of azimuth aspect given its radar observation at horizontal polarization and a carrier frequency of 1 GHz (30 cm wavelength).

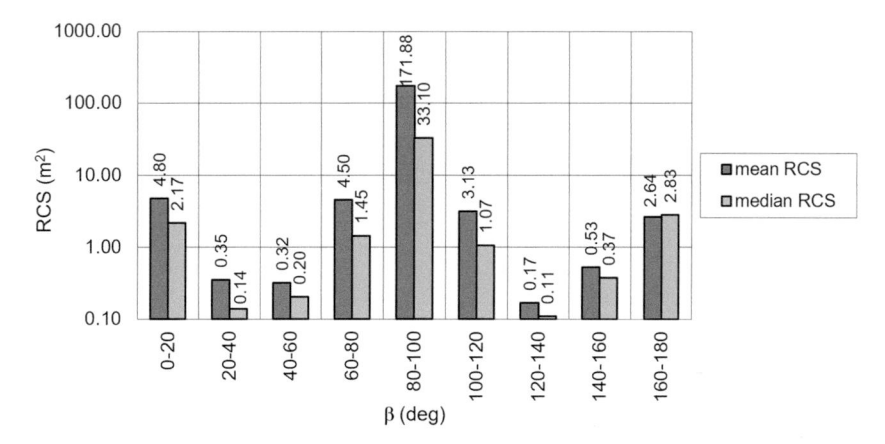

FIGURE 2.102 Diagrams of mean and median RCS of F-16 aircraft model in 20-degree sectors of azimuth aspect given its radar observation at horizontal polarization and a carrier frequency of 1 GHz (30 cm wavelength).

$$p(x) = \frac{1}{\sqrt{2\pi}\, x\sigma} \exp\left(-\frac{(\log(x)-\mu)^2}{2\sigma^2}\right)$$

$\mu = 0.377;\ \sigma = 0.685$

log-normal distribution

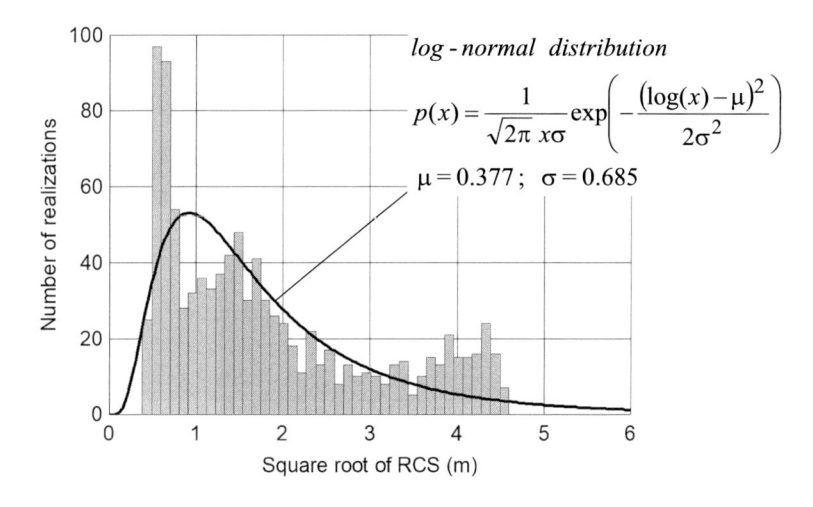

FIGURE 2.103 Amplitude distribution of echo signal of F-16 aircraft model at a carrier frequency of 1 GHz given its horizontal polarization.

2.7 AIR SUPERIORITY FIGHTER F-15

F-15 Eagle (Figure 2.104) is one-seat air superiority fighter. The first flight took place in 1972.

General characteristics of F-15 [18]: wingspan – 13.05 m, length – 19.05 m, height – 5.63 m, wing area – 56.6 m², weight – 12 970–30 850 kg, powerplant – 2 Pratt & Whitney F100-PW-220 afterburning turbofans, maximum speed – 2 650 km/h, combat range – 1 900 km, and service ceiling – 18 300 m.

In accordance with the design of the F-15, a model of its surface was created to obtaining scattering characteristics (in particular, RCS). The model is shown in Figure 2.105. In the modeling, the smooth parts of the aircraft surface were approximated by parts of 64 triaxial ellipsoids. The surface breaks were modeled using 22 straight edge scattering parts.

FIGURE 2.104 Air superiority fighter F-15.

FIGURE 2.105 Surface model of air superiority fighter F-15.

Below are some scattering characteristics of the F-15 model at sounding frequencies of 10, 3, and 1 GHz (wavelengths of 3, 10, and 30 cm, respectively) for horizontal polarization of the probing signal. The scattering characteristics for this model at two polarizations, as well as scattering characteristics at sounding frequencies of 5 and 1.3 GHz (wavelengths of 6 and 23 cm, respectively), are given in the electronic appendix of this book.

Sounding parameters: The elevation angle is to be a random value distributed uniformly in the range −3° ± 4° with respect to the wing plane (elevation angle of −3° corresponds to the radar observation from the lower hemisphere), azimuth aspect increment was 0.02°, the azimuth being counted

off from the nose-on aspect (0° corresponds to the nose-on radar observation, 180° corresponds to the tail-on observation).

Scattering characteristics of F-15 aircraft model for sounding frequency 10 GHz (wavelength 3 cm).

Figure 2.106 shows the RCS circular diagram of the F-15 model. The noncoherent RCS circular diagram of the F-15 is shown in Figure 2.107.

The circular mean RCS of the F-15 aircraft model for horizontal polarization is 120.56 m². The circular median RCS (the RCS value used to calculate the detection range of an object with a probability of 0.5) for horizontal polarization is 1.95 m².

Figures 2.108 and 2.109 show the mean and median RCS for the main ranges of sounding azimuths (nose, side, and tail) and for ranges of 20°.

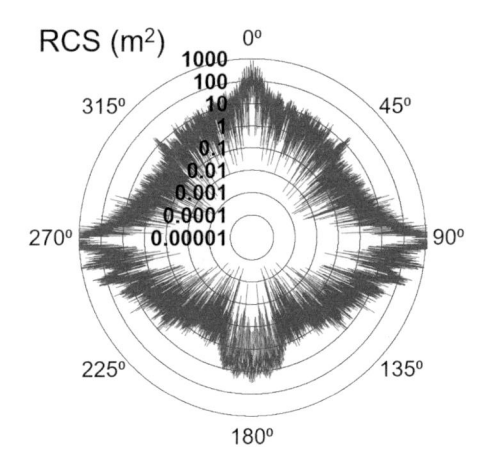

FIGURE 2.106 Circular diagram of RCS given radar observation of F-15 aircraft model at a carrier frequency of 10 GHz (3 cm wavelength).

FIGURE 2.107 Circular diagram of noncoherent RCS given radar observation of F-15 aircraft model at a carrier frequency of 10 GHz (3 cm wavelength).

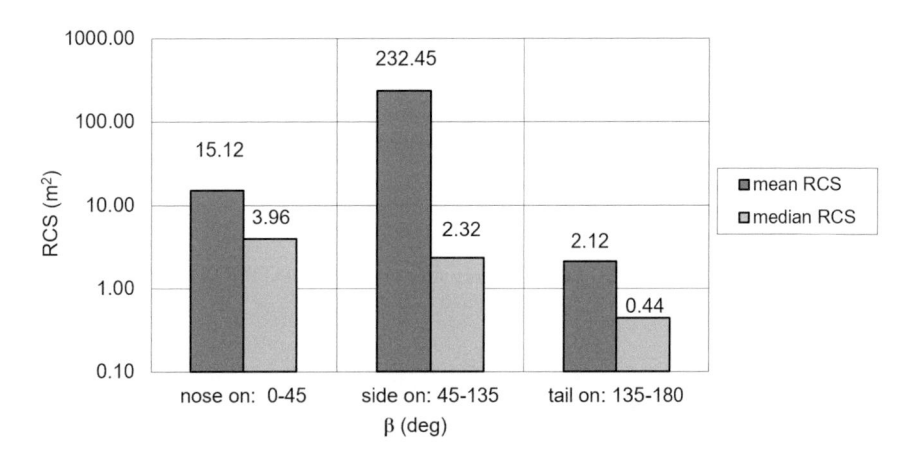

FIGURE 2.108 Diagrams of mean and median RCS of F-15 aircraft model in three sectors of azimuth aspect given its radar observation at horizontal polarization and a carrier frequency of 10 GHz (3 cm wavelength).

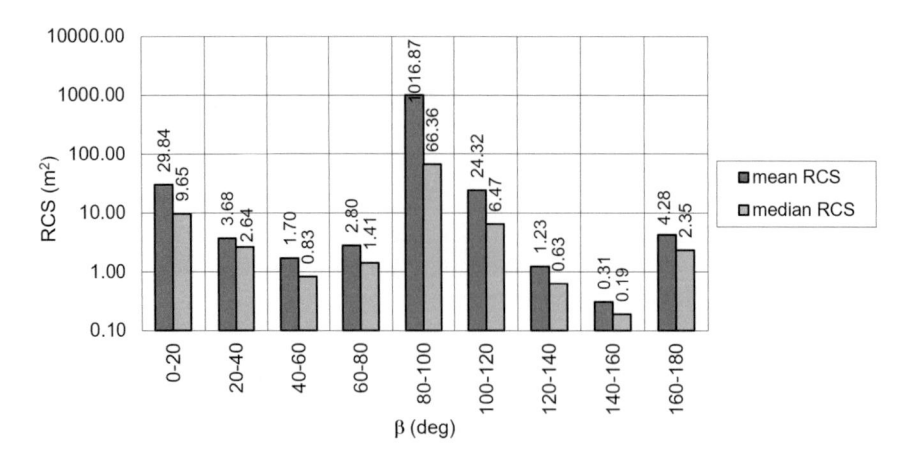

FIGURE 2.109 Diagrams of mean and median RCS of F-15 aircraft model in 20-degree sectors of azimuth aspect given its radar observation at horizontal polarization and a carrier frequency of 10 GHz (3 cm wavelength).

Figures 2.110, 2.115, and 2.120 show histograms of the scattered signal amplitude (square root of the RCS) for the range of sounding azimuths −20° to +20° (sounding from the nose). The bold line shows the probability density functions of the distribution, which can be used to approximate the histogram of the signal amplitude.

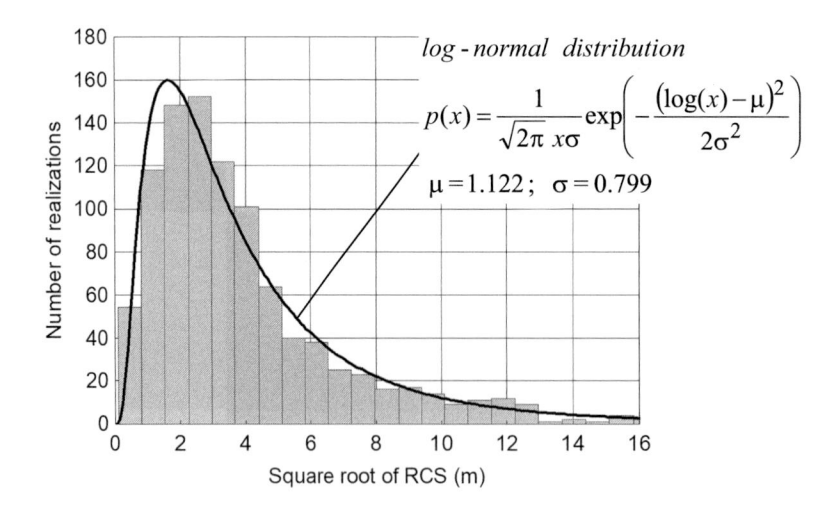

$$p(x) = \frac{1}{\sqrt{2\pi}\, x\sigma} \exp\left(-\frac{(\log(x)-\mu)^2}{2\sigma^2}\right)$$

$\mu = 1.122;\ \sigma = 0.799$

FIGURE 2.110 Amplitude distribution of echo signal of F-15 aircraft model at a carrier frequency of 10 GHz given its horizontal polarization.

Scattering characteristics of F-15 aircraft model for sounding frequency 3 GHz (wavelength 10 cm).

Figure 2.111 shows the RCS circular diagram of the F-15. The noncoherent RCS circular diagram of the F-15 is shown in Figure 2.112.

The circular mean RCS of the F-15 aircraft for horizontal polarization is 90.71 m². The circular median RCS for horizontal polarization is 2.07 m².

Figures 2.113 and 2.114 show the mean and median RCS for the main ranges of sounding azimuths (nose, side, and tail) and for ranges of 20°.

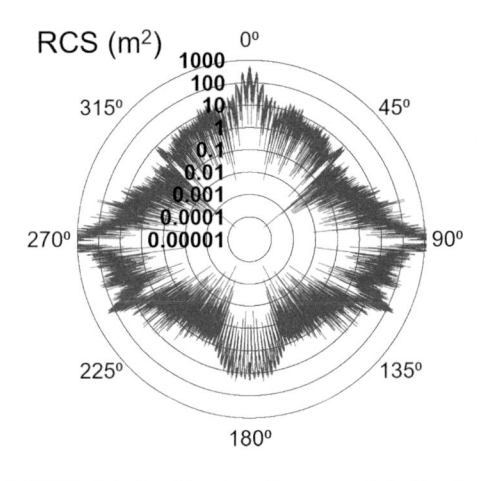

Scattering Characteristics of Aerial Objects

FIGURE 2.111 Circular diagram of RCS given radar observation of F-15 aircraft model at a carrier frequency of 3 GHz (10 cm wavelength).

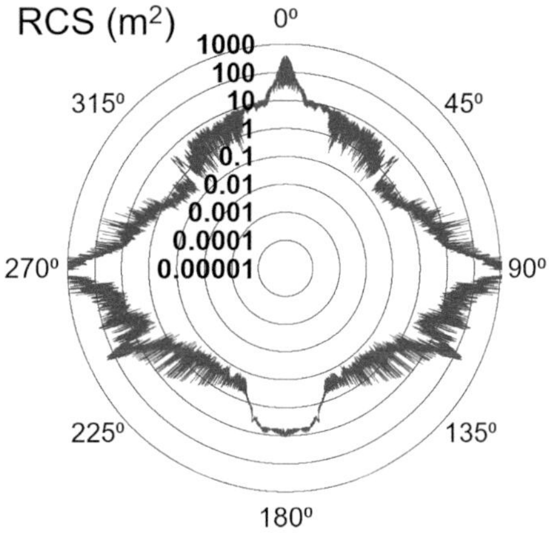

FIGURE 2.112 Circular diagram of noncoherent RCS given radar observation of F-15 aircraft model at a carrier frequency of 3 GHz (10 cm wavelength).

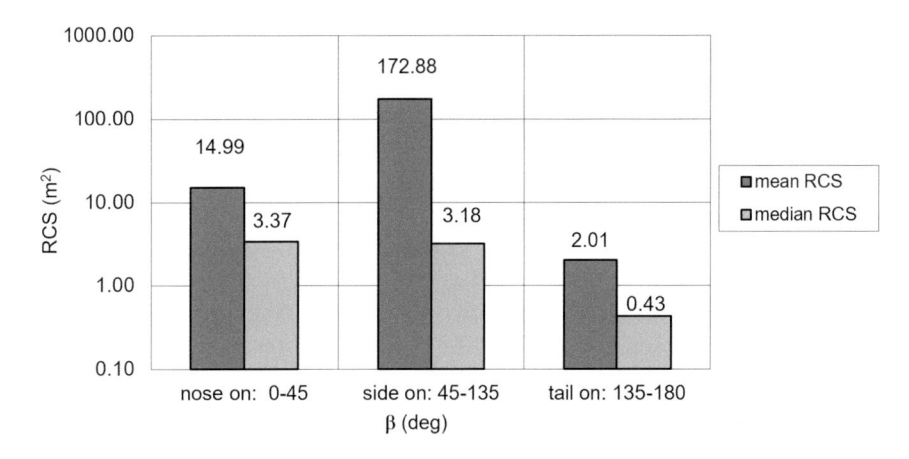

FIGURE 2.113 Diagrams of mean and median RCS of F-15 aircraft model in three sectors of azimuth aspect given its radar observation at horizontal polarization and a carrier frequency of 3 GHz (10 cm wavelength).

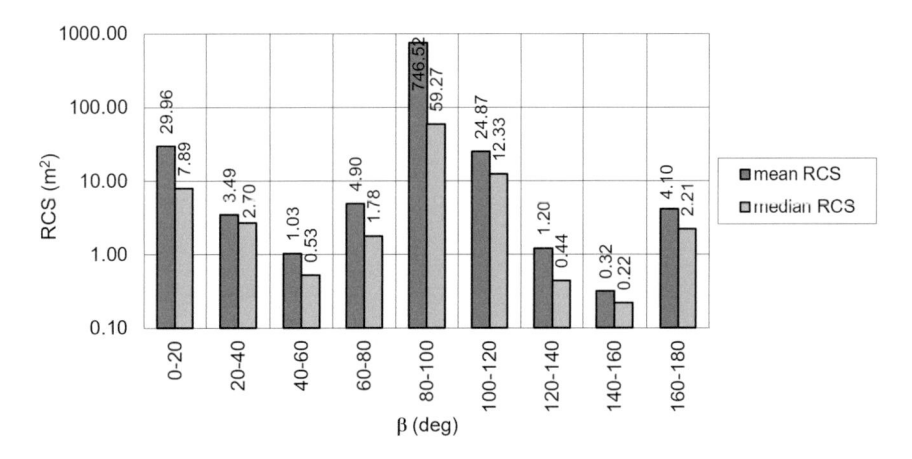

FIGURE 2.114 Diagrams of mean and median RCS of F-15 aircraft model in 20-degree sectors of azimuth aspect given its radar observation at horizontal polarization and a carrier frequency of 3 GHz (10 cm wavelength).

$$p(x) = \frac{1}{\sqrt{2\pi}\, x\sigma} \exp\left(-\frac{(\log(x)-\mu)^2}{2\sigma^2}\right)$$

log-normal distribution

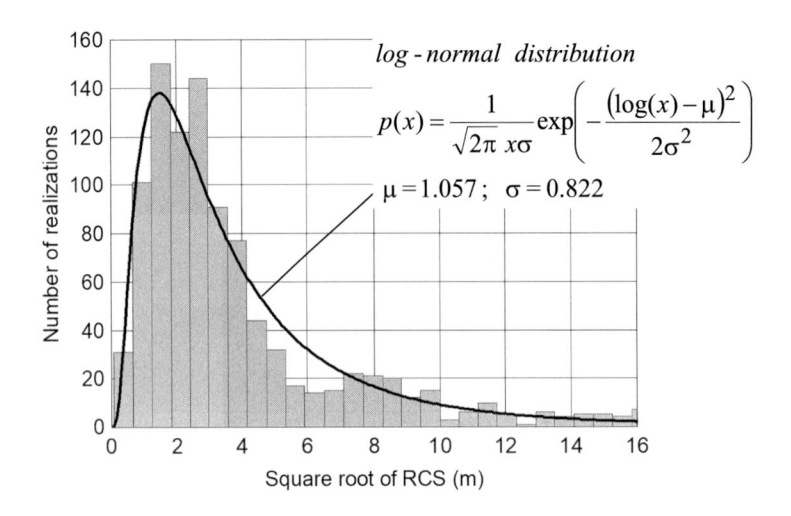

$\mu = 1.057;\ \sigma = 0.822$

FIGURE 2.115 Amplitude distribution of echo signal of F-15 aircraft model at a carrier frequency of 3 GHz given its horizontal polarization.

Scattering characteristics of F-15 aircraft model for sounding frequency 1 GHz (wavelength 30 cm).

Figure 2.116 shows the RCS circular diagram of the F-15. The noncoherent RCS circular diagram of the F-15 is shown in Figure 2.117.

The circular mean RCS of the F-15 aircraft for horizontal polarization is 72.84 m². The circular median RCS for horizontal polarization is 1.49 m².

Figures 2.118 and 2.119 show the mean and median RCS for the main ranges of sounding azimuths (nose, side, and tail) and for ranges of 20°.

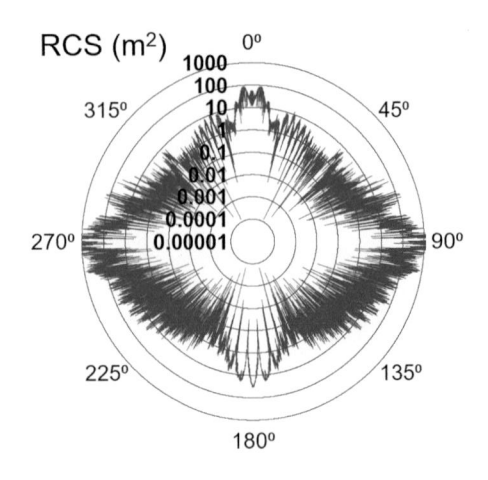

The expressions and parameters of probability distributions that are most consistent with the empirical distributions of the square root of the RCS for different ranges of sounding azimuths and polarizations are given in the electronic appendix to this book. The expressions and parameters of the probability distributions that are most consistent with the empirical distributions of the RCS (energy characteristic) for different ranges of sounding azimuths and polarizations are also given there.

FIGURE 2.116 Circular diagram of RCS given radar observation of F-15 aircraft model at a carrier frequency of 1 GHz (30 cm wavelength).

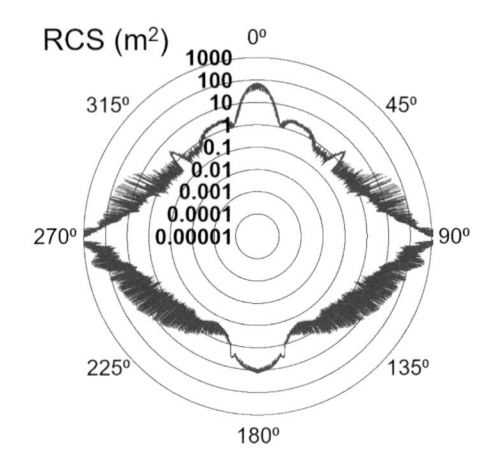

FIGURE 2.117 Circular diagram of noncoherent RCS given radar observation of F-15 aircraft model at a carrier frequency of 1 GHz (30 cm wavelength).

Scattering Characteristics of Aerial Objects

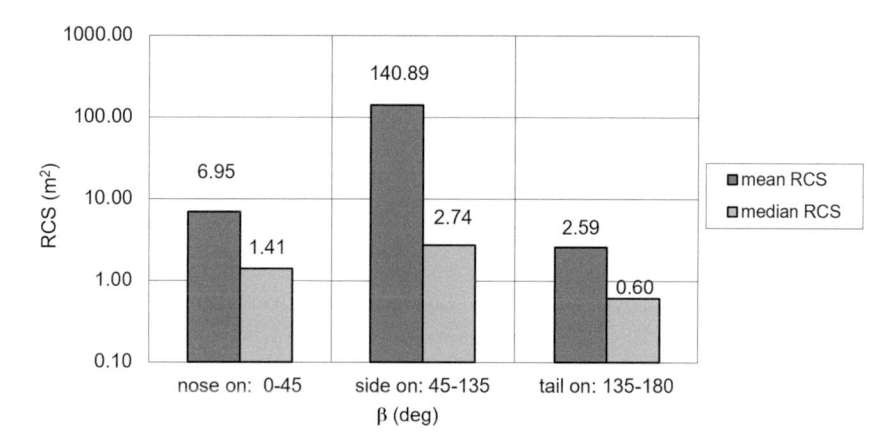

FIGURE 2.118 Diagrams of mean and median RCS of F-15 aircraft model in three sectors of azimuth aspect given its radar observation at horizontal polarization and a carrier frequency of 1 GHz (30 cm wavelength).

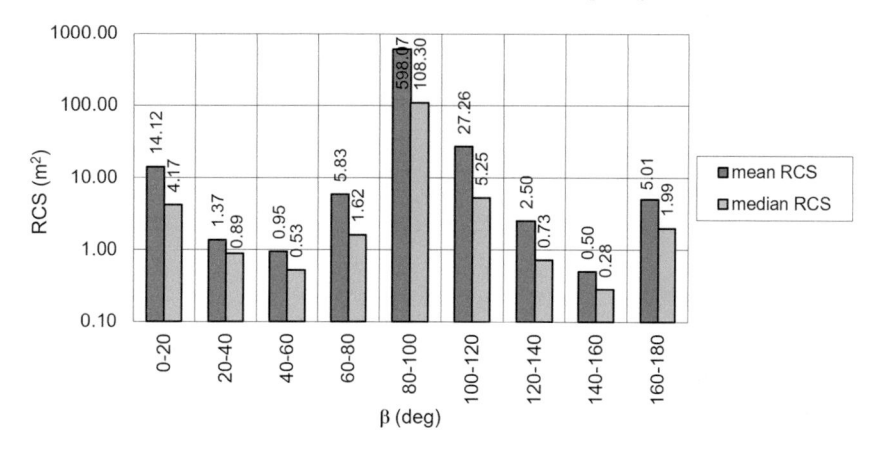

FIGURE 2.119 Diagrams of mean and median RCS of F-15 aircraft model in 20-degree sectors of azimuth aspect given its radar observation at horizontal polarization and a carrier frequency of 1 GHz (30 cm wavelength).

$$p(x) = \frac{1}{\sqrt{2\pi}\, x\sigma} \exp\left(-\frac{(\log(x)-\mu)^2}{2\sigma^2}\right)$$

$\mu = 0.779;\ \sigma = 0.821$

log-normal distribution

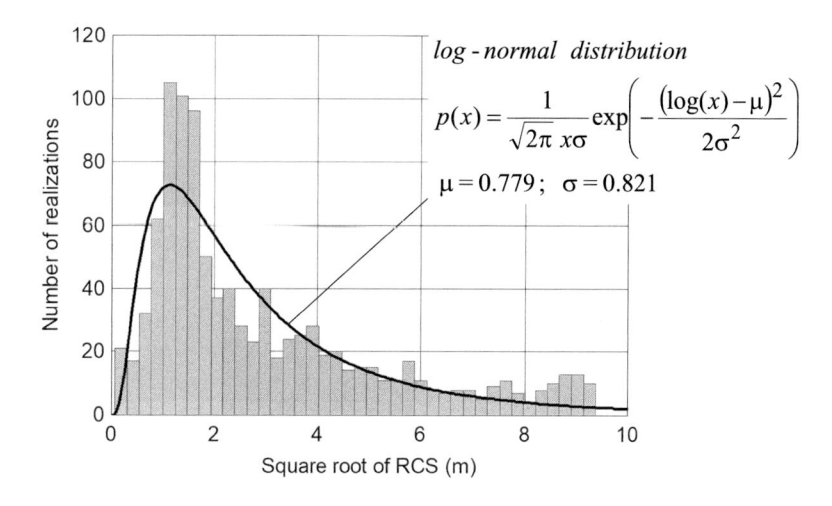

FIGURE 2.120 Amplitude distribution of echo signal of F-15 aircraft model at a carrier frequency of 1 GHz given its horizontal polarization.

2.8 MULTIROLE ATTACK AND FIGHTER AIRCRAFT F/A-18

F/A-18 Hornet (Figure 2.121) is carrier-capable multirole attack and fighter aircraft. The first flight took place in 1978.

General characteristics of F/A-18 [19]: wingspan – 12.3 m, length – 17.07 m, height – 4.6 m, wing area – 38.6 m², weight – 10 400–23 500 kg, powerplant – 2 General Electric F404-GE-402 afterburning turbofan engines, maximum speed – 2 000 km/h, combat range – 750–1 065 km, and service ceiling – 15 000 m.

In accordance with the design of the F/A-18, a model of its surface was created to obtaining scattering characteristics (in particular, RCS). The model is shown in Figure 2.122. In the modeling, the smooth parts of the aircraft surface were approximated by parts of 64 triaxial ellipsoids. The surface breaks were modeled using 22 straight edge scattering parts.

FIGURE 2.121 Multirole attack and fighter aircraft F/A-18

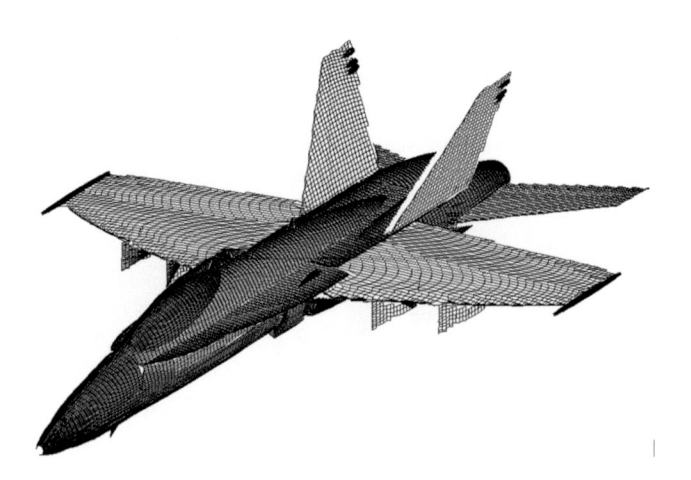

FIGURE 2.122 Surface model of multirole attack and fighter aircraft F/A-18.

Below are some scattering characteristics of the F/A-18 model at sounding frequencies of 10 GHz, 3 GHz, and 1 GHz (wavelengths of 3, 10, and 30 cm, respectively) for horizontal polarization of the probing signal. The scattering characteristics for this model at two polarizations, as well as scattering characteristics at sounding frequencies of 5 and 1.3 GHz (wavelengths of 6 and 23 cm, respectively), are given in the electronic appendix of this book.

Sounding parameters: The elevation angle is to be a random value distributed uniformly in the range −3° ± 4° with respect to the wing plane (elevation angle of −3° corresponds to the radar

Scattering Characteristics of Aerial Objects 55

observation from the lower hemisphere), azimuth aspect increment was 0.02°, the azimuth being counted off from the nose-on aspect (0° corresponds to the nose-on radar observation, 180° corresponds to the tail-on observation).

Scattering characteristics of F/A-18 aircraft model for sounding frequency 10 GHz (wavelength 3 cm).

Figure 2.123 shows the RCS circular diagram of the F/A-18 model. The noncoherent RCS circular diagram of the F/A-18 is shown in Figure 2.124.

The circular mean RCS of the F/A-18 aircraft model for horizontal polarization is 27.36 m². The circular median RCS (the RCS value used to calculate the detection range of an object with a probability of 0.5) for horizontal polarization is 2.14 m².

Figures 2.125 and 2.126 show the mean and median RCS for the main ranges of sounding azimuths (nose, side, and tail) and for ranges of 20°.

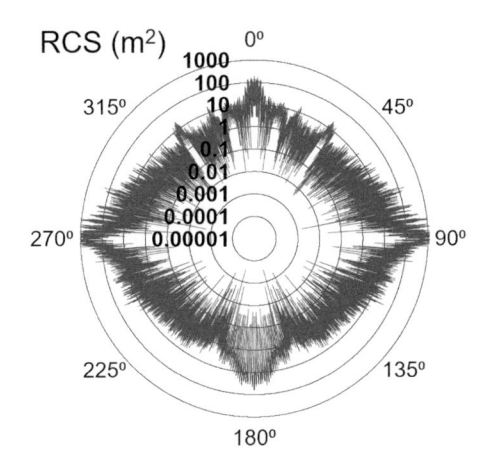

FIGURE 2.123 Circular diagram of RCS given radar observation of F/A-18 aircraft model at a carrier frequency of 10 GHz (3 cm wavelength).

FIGURE 2.124 Circular diagram of noncoherent RCS given radar observation of F/A-18 aircraft model at a carrier frequency of 10 GHz (3 cm wavelength).

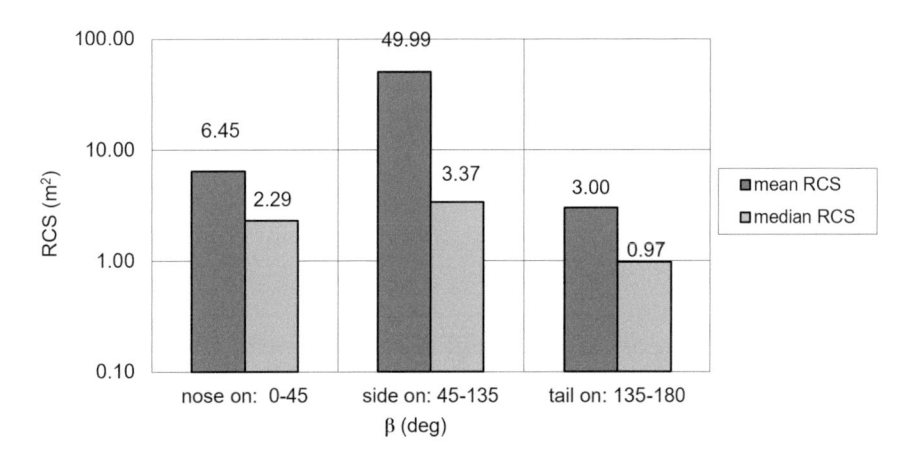

FIGURE 2.125 Diagrams of mean and median RCS of F/A-18 aircraft model in three sectors of azimuth aspect given its radar observation at horizontal polarization and a carrier frequency of 10 GHz (3 cm wavelength).

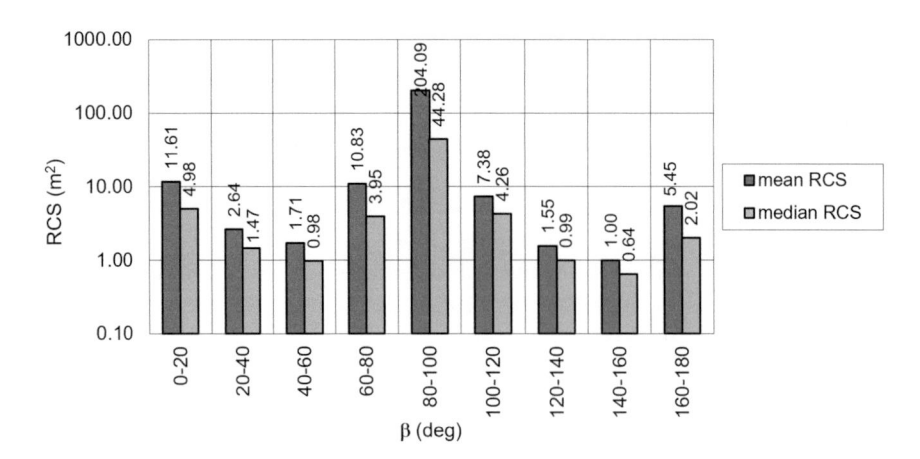

FIGURE 2.126 Diagrams of mean and median RCS of F/A-18 aircraft model in 20-degree sectors of azimuth aspect given its radar observation at horizontal polarization and a carrier frequency of 10 GHz (3 cm wavelength).

Figures 2.127, 2.132, and 2.137 show histograms of the scattered signal amplitude (square root of the RCS) for the range of sounding azimuths −20° to +20° (sounding from the nose). The bold line shows the probability density functions of the distribution, which can be used to approximate the histogram of the signal amplitude.

extreme value distribution

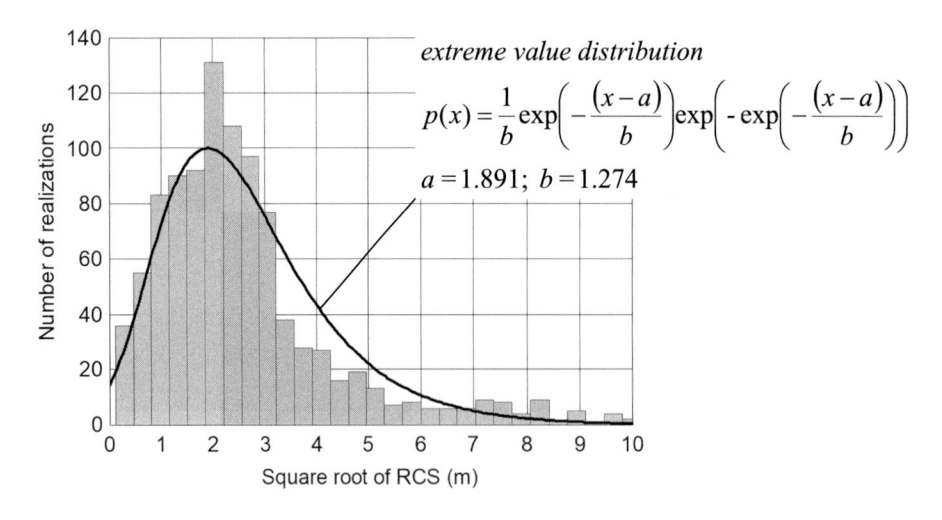

$$p(x) = \frac{1}{b}\exp\left(-\frac{(x-a)}{b}\right)\exp\left(-\exp\left(-\frac{(x-a)}{b}\right)\right)$$

$a = 1.891;\ b = 1.274$

FIGURE 2.127 Amplitude distribution of echo signal of F/A-18 aircraft model at a carrier frequency of 10 GHz given its horizontal polarization.

Scattering characteristics of F/A-18 aircraft model for sounding frequency 3 GHz (wavelength 10 cm).

Figure 2.128 shows the RCS circular diagram of the F/A-18. The noncoherent RCS circular diagram of the F/A-18 is shown in Figure 2.129.

The circular mean RCS of the F/A-18 aircraft for horizontal polarization is 19.01 m². The circular median RCS for horizontal polarization is 1.80 m².

Figures 2.130 and 2.131 show the mean and median RCS for the main ranges of sounding azimuths (nose, side, and tail) and for ranges of 20°.

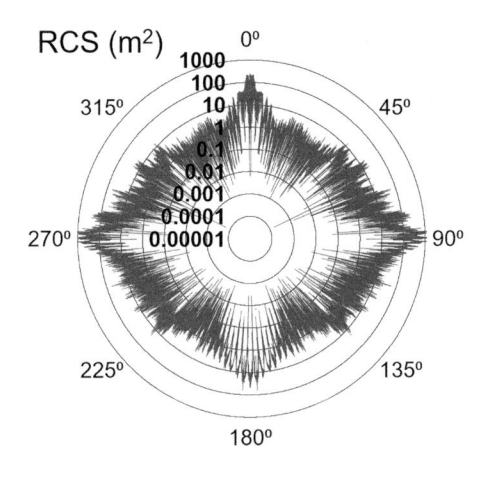

Scattering Characteristics of Aerial Objects

FIGURE 2.128 Circular diagram of RCS given radar observation of F/A-18 aircraft model at a carrier frequency of 3 GHz (10 cm wavelength).

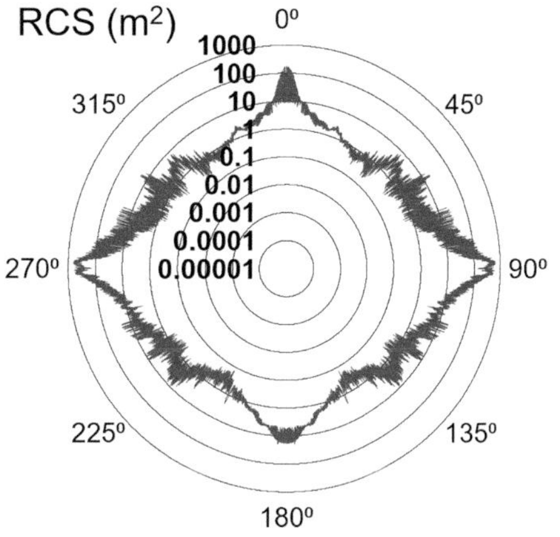

FIGURE 2.129 Circular diagram of noncoherent RCS given radar observation of F/A-18 aircraft model at a carrier frequency of 3 GHz (10 cm wavelength).

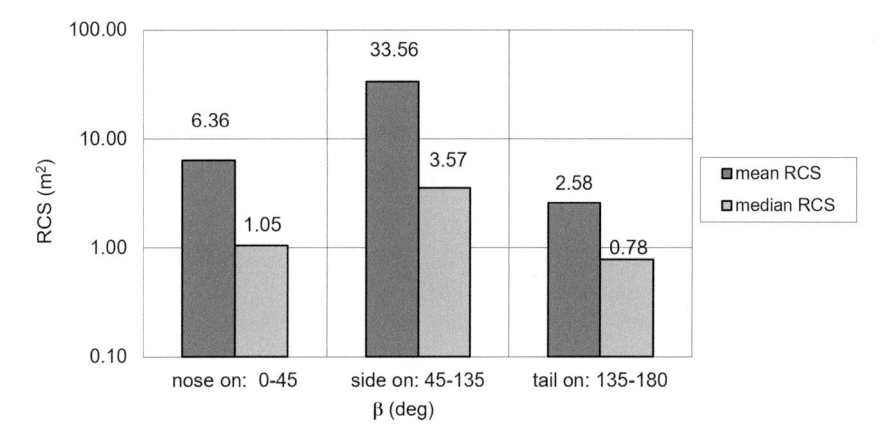

FIGURE 2.130 Diagrams of mean and median RCS of F/A-18 aircraft model in three sectors of azimuth aspect given its radar observation at horizontal polarization and a carrier frequency of 3 GHz (10 cm wavelength).

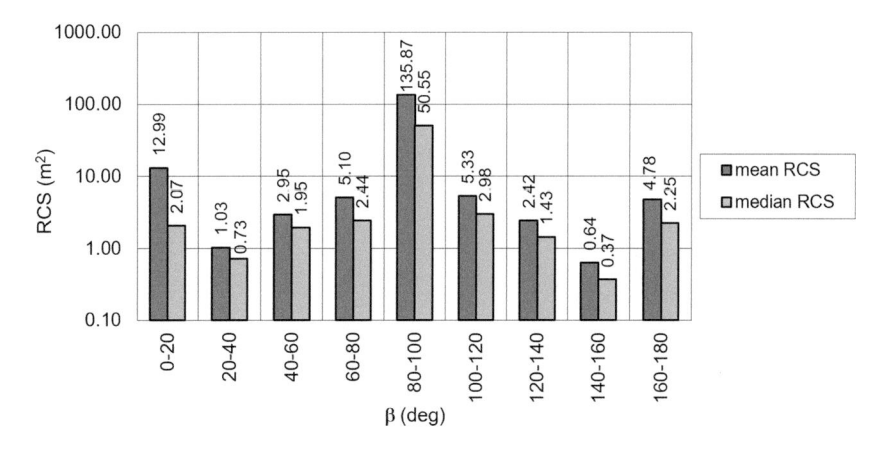

FIGURE 2.131 Diagrams of mean and median RCS of F/A-18 aircraft model in 20-degree sectors of azimuth aspect given its radar observation at horizontal polarization and a carrier frequency of 3 GHz (10 cm wavelength).

$$p(x) = \frac{1}{\sqrt{2\pi}\,x\sigma} \exp\left(-\frac{(\log(x)-\mu)^2}{2\sigma^2}\right)$$

log - normal distribution

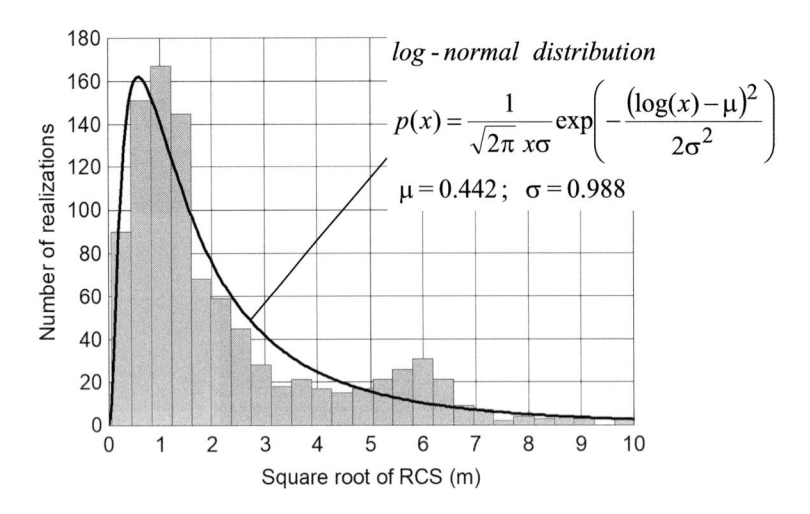

$\mu = 0.442$; $\sigma = 0.988$

FIGURE 2.132 Amplitude distribution of echo signal of F/A-18 aircraft model at a carrier frequency of 3 GHz given its horizontal polarization.

Scattering characteristics of F/A-18 aircraft model for sounding frequency 1 GHz (wavelength 30 cm).

Figure 2.133 shows the RCS circular diagram of the F/A-18. The noncoherent RCS circular diagram of the F/A-18 is shown in Figure 2.134.

The circular mean RCS of the F/A-18 aircraft for horizontal polarization is 15.94 m². The circular median RCS for horizontal polarization is 1.11 m².

Figures 2.135 and 2.136 show the mean and median RCS for the main ranges of sounding azimuths (nose, side, and tail) and for ranges of 20°.

The expressions and parameters of probability distributions that are most consistent with the empirical distributions of the square root of the RCS for different ranges of sounding azimuths and polarizations are given in the electronic appendix to this book. The expressions and parameters of the probability distributions that are most consistent with the empirical distributions of the RCS (energy characteristic) for different ranges of sounding azimuths and polarizations are also given there.

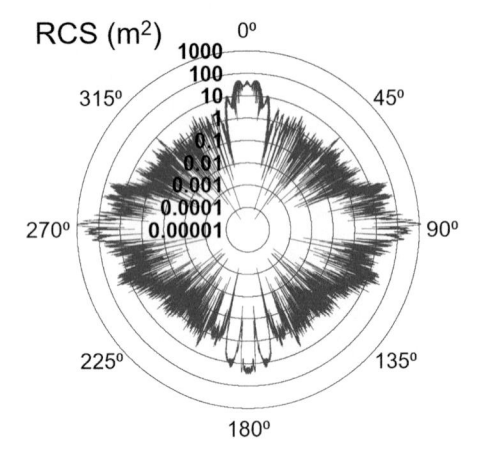

FIGURE 2.133 Circular diagram of RCS given radar observation of F/A-18 aircraft model at a carrier frequency of 1 GHz (30 cm wavelength).

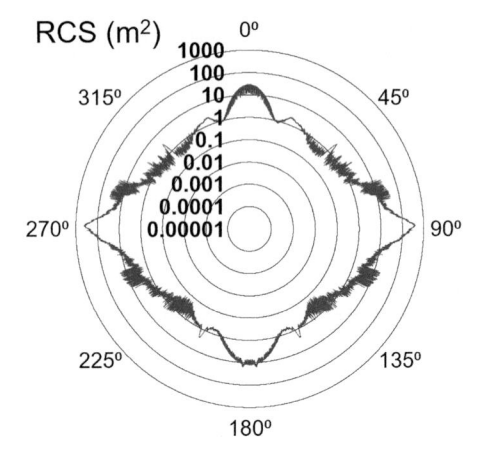

FIGURE 2.134 Circular diagram of noncoherent RCS given radar observation of F/A-18 aircraft model at a carrier frequency of 1 GHz (30 cm wavelength).

Scattering Characteristics of Aerial Objects

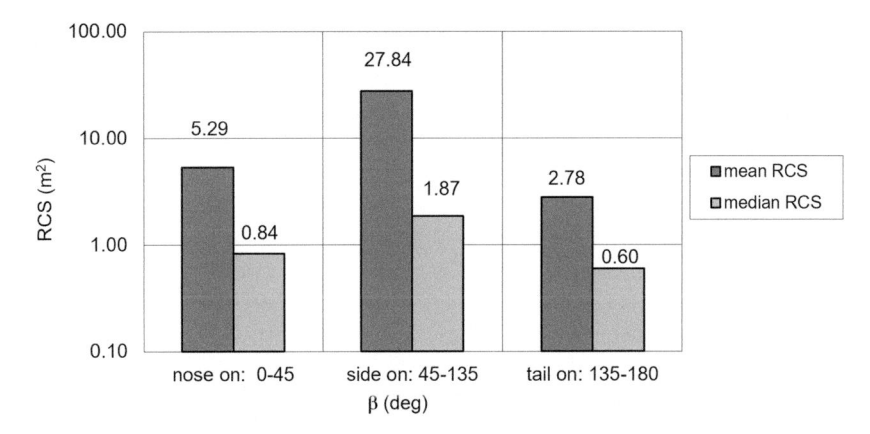

FIGURE 2.135 Diagrams of mean and median RCS of F/A-18 aircraft model in three sectors of azimuth aspect given its radar observation at horizontal polarization and a carrier frequency of 1 GHz (30 cm wavelength).

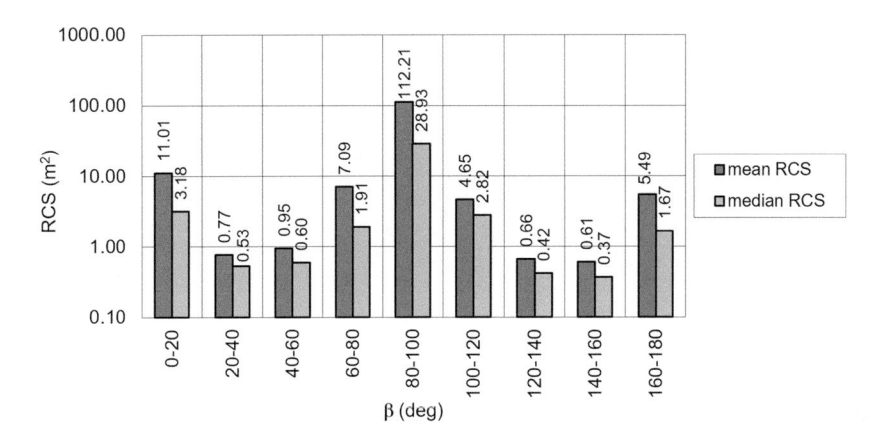

FIGURE 2.136 Diagrams of mean and median RCS of F/A-18 aircraft model in 20-degree sectors of azimuth aspect given its radar observation at horizontal polarization and a carrier frequency of 1 GHz (30 cm wavelength).

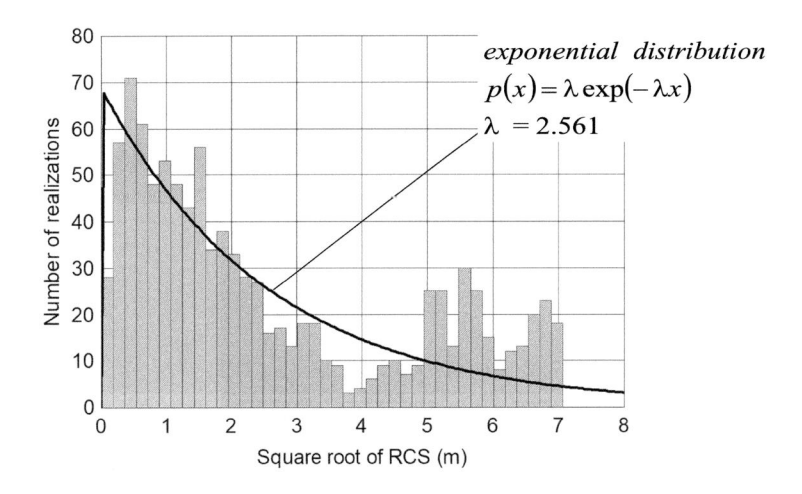

FIGURE 2.137 Amplitude distribution of echo signal of F/A-18 aircraft model at a carrier frequency of 1 GHz given its horizontal polarization.

2.9 MULTIROLE COMBAT AIRCRAFT TORNADO IDS

Tornado IDS (Figure 2.138) is мultirole combat aircraft. The first flight took place in 1974.

General characteristics of Tornado IDS [20]: wingspan – 8.60–13.92 m, length – 16.72 m, height – 5.95 m, wing area – 31 m^2, weight – 14 090–27 950 kg, powerplant – 2 Turbo-Union RB199-34R Mk 103 afterburning three-spool turbofan, maximum speed – 2 400 km/h, combat range – 1390 km, and service ceiling – 15 000 m.

In accordance with the design of the Tornado IDS, a model of its surface was created to obtaining scattering characteristics (in particular, RCS). The model is shown in Figure 2.139. In the modeling, the smooth parts of the aircraft surface were approximated by parts of 52 triaxial ellipsoids. The surface breaks were modeled using 17 straight edge scattering parts.

FIGURE 2.138 Multirole combat aircraft Tornado IDS.

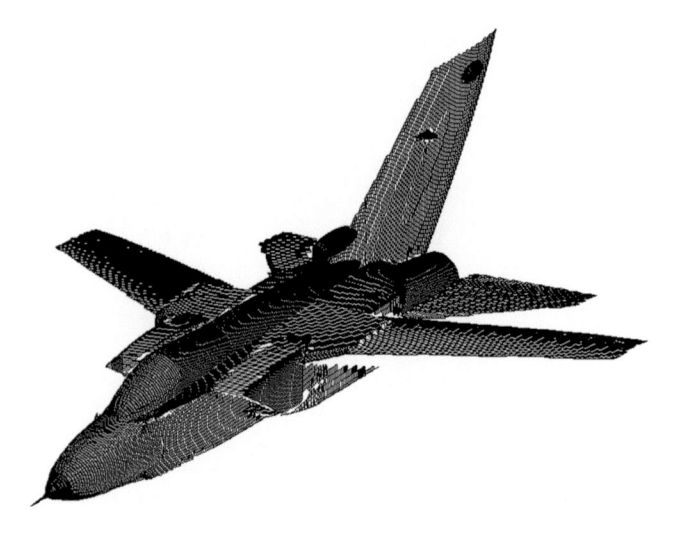

FIGURE 2.139 Surface model of multirole combat aircraft Tornado IDS.

Below are some scattering characteristics of the Tornado IDS model at sounding frequencies of 10, 3, and 1 GHz (wavelengths of 3, 10, and 30 cm, respectively) for horizontal polarization of the probing signal. The scattering characteristics for this model at two polarizations, as well as

Scattering Characteristics of Aerial Objects

scattering characteristics at sounding frequencies of 5 and 1.3 GHz (wavelengths of 6 and 23 cm, respectively), are given in the electronic appendix of this book.

Sounding parameters: The elevation angle is to be a random value distributed uniformly in the range $-3° \pm 4°$ with respect to the wing plane (elevation angle of $-3°$ corresponds to the radar observation from the lower hemisphere), azimuth aspect increment was $0.02°$, the azimuth being counted off from the nose-on aspect (0° corresponds to the nose-on radar observation, 180° corresponds to the tail-on observation).

Scattering characteristics of Tornado IDS aircraft model for sounding frequency 10 GHz (wavelength 3 cm).

Figure 2.140 shows the RCS circular diagram of the Tornado IDS model. The noncoherent RCS circular diagram of the Tornado IDS is shown in Figure 2.141.

The circular mean RCS of the Tornado IDS aircraft model for horizontal polarization is 102.49 m². The circular median RCS (the RCS value used to calculate the detection range of an object with a probability of 0.5) for horizontal polarization is 4.67 m².

Figures 2.142 and 2.143 show the mean and median RCS for the main ranges of sounding azimuths (nose, side, and tail) and for ranges of 20°.

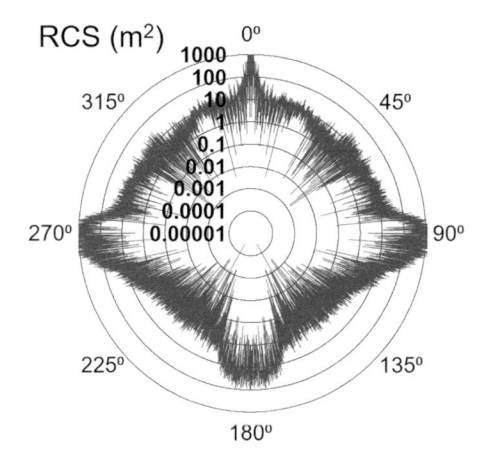

FIGURE 2.140 Circular diagram of RCS given radar observation of Tornado IDS aircraft model at a carrier frequency of 10 GHz (3 cm wavelength).

FIGURE 2.141 Circular diagram of noncoherent RCS given radar observation of Tornado IDS aircraft model at a carrier frequency of 10 GHz (3 cm wavelength).

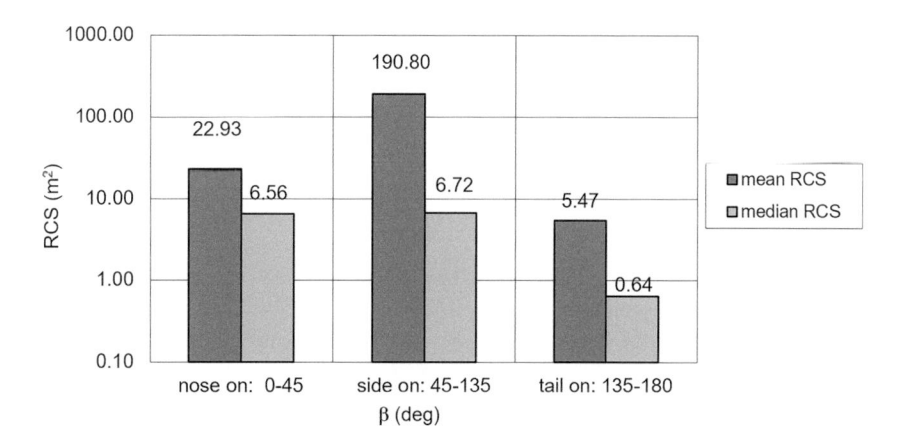

FIGURE 2.142 Diagrams of mean and median RCS of Tornado IDS aircraft model in three sectors of azimuth aspect given its radar observation at horizontal polarization and a carrier frequency of 10 GHz (3 cm wavelength).

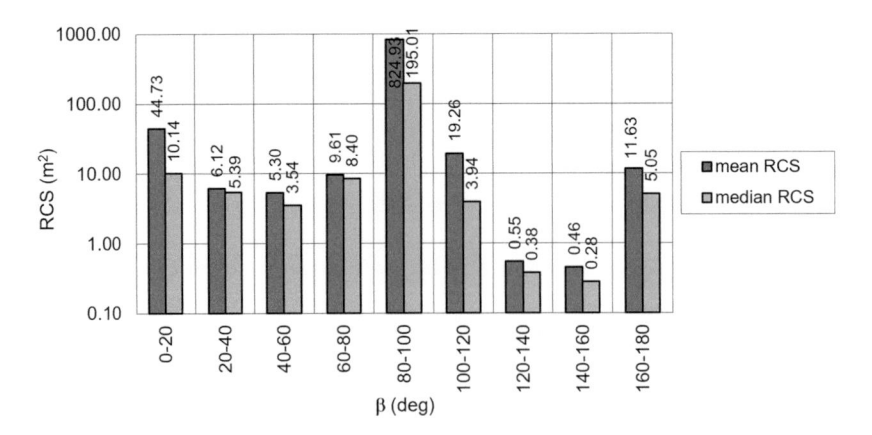

FIGURE 2.143 Diagrams of mean and median RCS of Tornado IDS aircraft model in 20-degree sectors of azimuth aspect given its radar observation at horizontal polarization and a carrier frequency of 10 GHz (3 cm wavelength).

Figures 2.144, 2.149, and 2.154 show histograms of the scattered signal amplitude (square root of the RCS) for the range of sounding azimuths −20° to +20° (sounding from the nose). The bold line shows the probability density functions of the distribution, which can be used to approximate the histogram of the signal amplitude.

log-normal distribution

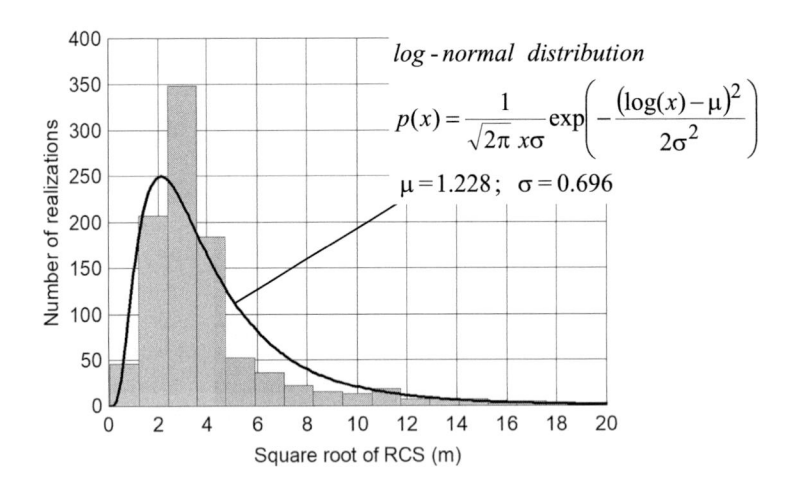

$$p(x) = \frac{1}{\sqrt{2\pi}\, x\sigma} \exp\left(-\frac{(\log(x)-\mu)^2}{2\sigma^2}\right)$$

$\mu = 1.228;\ \sigma = 0.696$

FIGURE 2.144 Amplitude distribution of echo signal of Tornado IDS aircraft model at a carrier frequency of 10 GHz given its horizontal polarization.

Scattering characteristics of Tornado IDS aircraft model for sounding frequency 3 GHz (wavelength 10 cm).

Figure 2.145 shows the RCS circular diagram of the Tornado IDS. The noncoherent RCS circular diagram of the Tornado IDS is shown in Figure 2.146.

The circular mean RCS of the Tornado IDS aircraft for horizontal polarization is 89.90 m². The circular median RCS for horizontal polarization is 5.05 m².

Figures 2.147 and 2.148 show the mean and median RCS for the main ranges of sounding azimuths (nose, side, and tail) and for ranges of 20°.

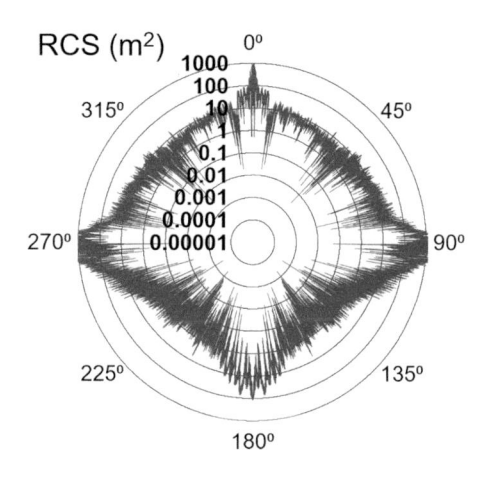

Scattering Characteristics of Aerial Objects

FIGURE 2.145 Circular diagram of RCS given radar observation of Tornado IDS aircraft model at a carrier frequency of 3 GHz (10 cm wavelength).

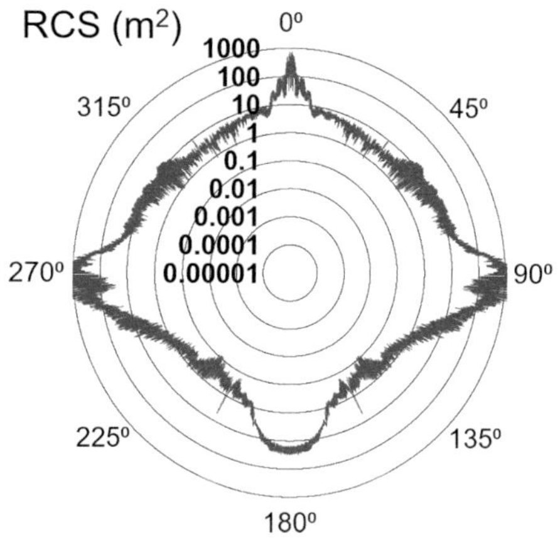

FIGURE 2.146 Circular diagram of noncoherent RCS given radar observation of Tornado IDS aircraft model at a carrier frequency of 3 GHz (10 cm wavelength).

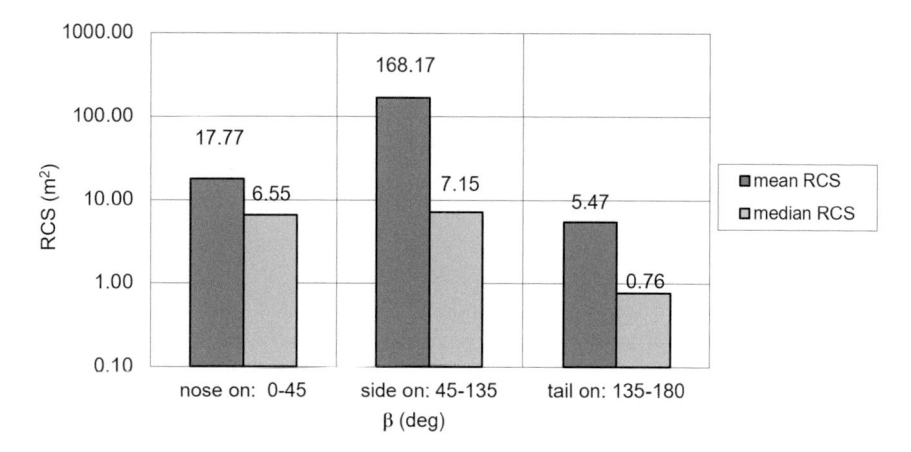

FIGURE 2.147 Diagrams of mean and median RCS of Tornado IDS aircraft model in three sectors of azimuth aspect given its radar observation at horizontal polarization and a carrier frequency of 3 GHz (10 cm wavelength).

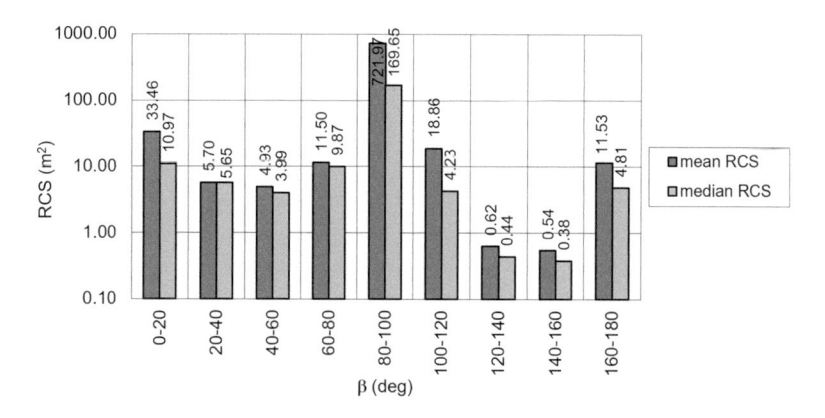

FIGURE 2.148 Diagrams of mean and median RCS of Tornado IDS aircraft model in 20-degree sectors of azimuth aspect given its radar observation at horizontal polarization and a carrier frequency of 3 GHz (10 cm wavelength).

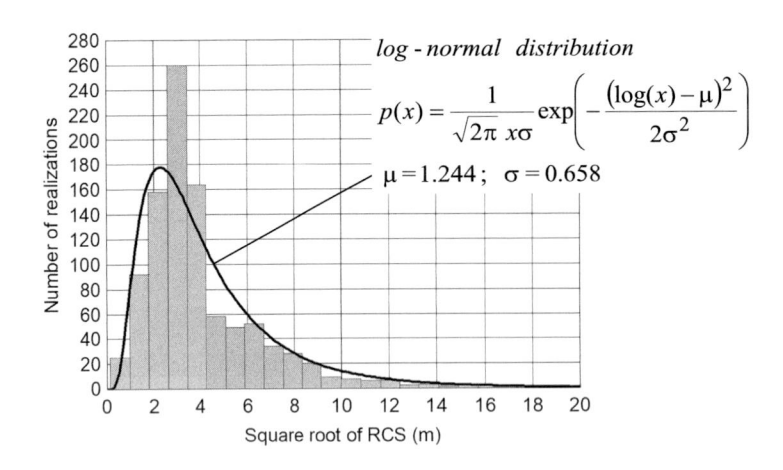

FIGURE 2.149 Amplitude distribution of echo signal of Tornado IDS aircraft model at a carrier frequency of 3 GHz given its horizontal polarization.

Scattering characteristics of Tornado IDS aircraft model for sounding frequency 1 GHz (wavelength 30 cm).

Figure 2.150 shows the RCS circular diagram of the Tornado IDS. The noncoherent RCS circular diagram of the Tornado IDS is shown in Figure 2.151.

The circular mean RCS of the Tornado IDS aircraft for horizontal polarization is 66.91 m². The circular median RCS for horizontal polarization is 3.87 m².

Figures 2.152 and 2.153 show the mean and median RCS for the main ranges of sounding azimuths (nose, side, and tail) and for ranges of 20°.

The expressions and parameters of probability distributions that are most consistent with the empirical distributions of the square root of the RCS for different ranges of sounding azimuths and polarizations are given in the electronic appendix to this book. The expressions and parameters of the probability distributions that are most consistent with the empirical distributions of the RCS (energy characteristic) for different ranges of sounding azimuths and polarizations are also given there.

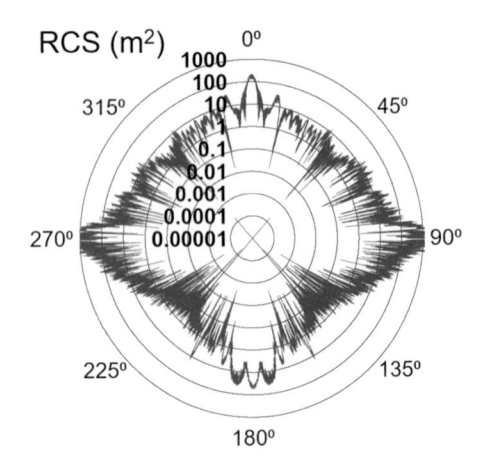

FIGURE 2.150 Circular diagram of RCS given radar observation of Tornado IDS aircraft model at a carrier frequency of 1 GHz (30 cm wavelength).

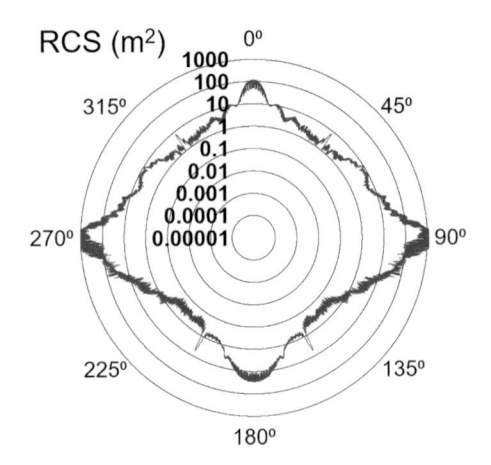

FIGURE 2.151 Circular diagram of noncoherent RCS given radar observation of Tornado IDS aircraft model at a carrier frequency of 1 GHz (30 cm wavelength).

Scattering Characteristics of Aerial Objects

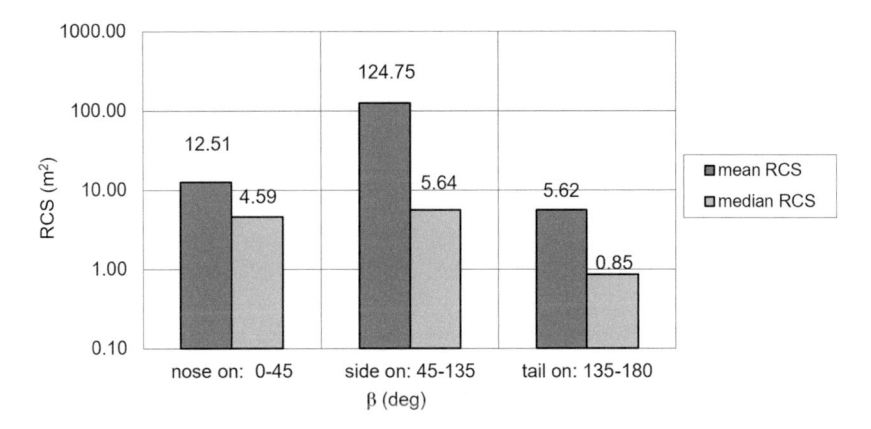

FIGURE 2.152 Diagrams of mean and median RCS of Tornado IDS aircraft model in three sectors of azimuth aspect given its radar observation at horizontal polarization and a carrier frequency of 1 GHz (30 cm wavelength).

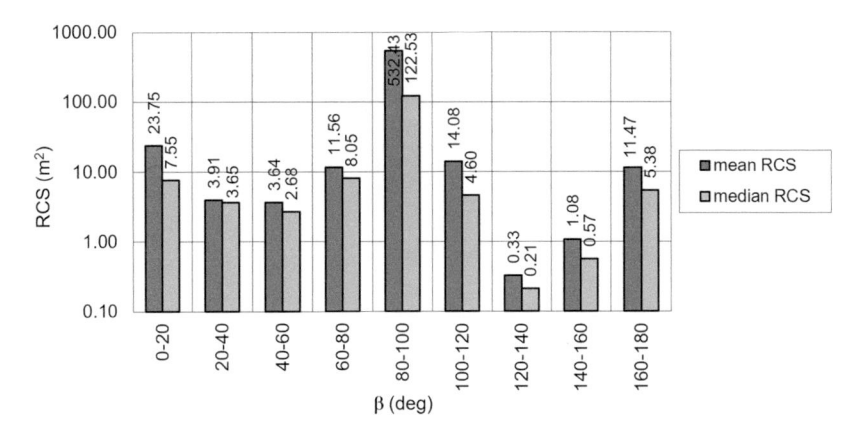

FIGURE 2.153 Diagrams of mean and median RCS of Tornado IDS aircraft model in 20-degree sectors of azimuth aspect given its radar observation at horizontal polarization and a carrier frequency of 1 GHz (30 cm wavelength).

$$p(x) = \frac{1}{\sqrt{2\pi}\, x\sigma} \exp\left(-\frac{(\log(x)-\mu)^2}{2\sigma^2}\right)$$

log-normal distribution

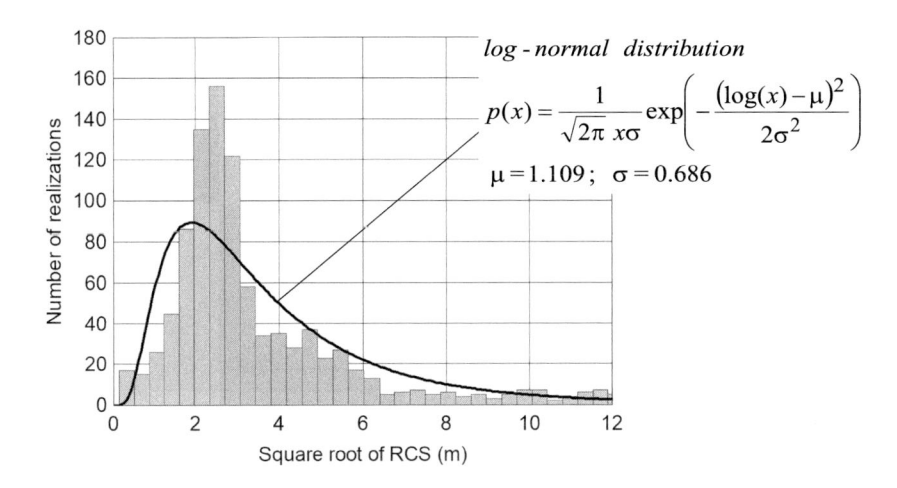

$\mu = 1.109;\ \sigma = 0.686$

FIGURE 2.154 Amplitude distribution of echo signal of Tornado IDS aircraft model at a carrier frequency of 1 GHz given its horizontal polarization.

2.10 MULTIROLE FIGHTER EF-2000 TYPHOON

EF-2000 Eurofighter Typhoon (Figure 2.155) is мultirole fighter. The first flight took place in 1979.

General characteristics of EF-2000 Typhoon [21]: wingspan –10.95 m, length – 15.98 m, height – 5.28 m, wing area – 51 m^2, weight – 11 000–23 500 kg, powerplant – 2 Eurojet EJ200 afterburning turbofan engines, maximum speed – 2 120 km/h, combat range – 601–1 390 km, and service ceiling – 19 800 m.

In accordance with the design of the EF-2000 Typhoon, a model of its surface was created to obtaining scattering characteristics (in particular, RCS). The model is shown in Figure 2.156. In the modeling, the smooth parts of the aircraft surface were approximated by parts of 104 triaxial ellipsoids. The surface breaks were modeled using 10 straight edge scattering parts.

FIGURE 2.155 Multirole combat aircraft EF-2000 Typhoon.

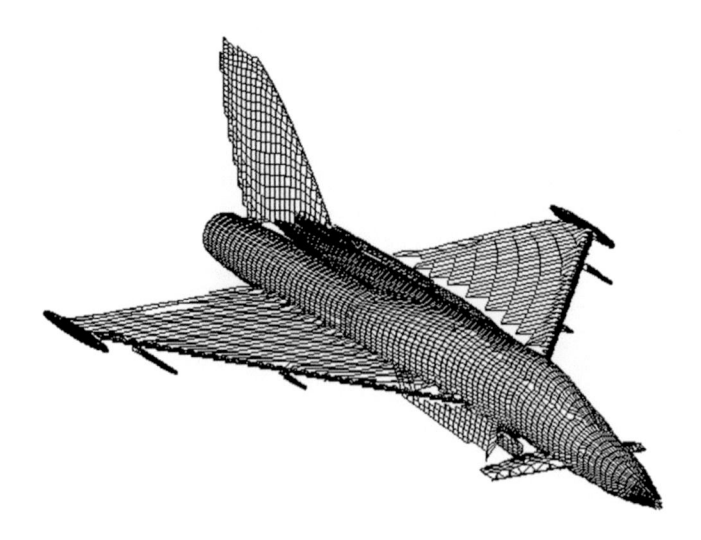

FIGURE 2.156 Surface model of multirole combat aircraft EF-2000 Typhoon.

Below are some scattering characteristics of the EF-2000 Typhoon model at sounding frequencies of 10, 3, and 1 GHz (wavelengths of 3, 10, and 30 cm, respectively) for horizontal polarization of the probing signal. The scattering characteristics for this model at two polarizations, as well as scattering characteristics at sounding frequencies of 5 and 1.3 GHz (wavelengths of 6 and 23 cm, respectively), are given in the electronic appendix of this book.

Scattering Characteristics of Aerial Objects 67

Sounding parameters: The elevation angle is to be a random value distributed uniformly in the range −3° ± 4° with respect to the wing plane (elevation angle of −3° corresponds to the radar observation from the lower hemisphere), azimuth aspect increment was 0.02°, the azimuth being counted off from the nose-on aspect (0° corresponds to the nose-on radar observation, 180° corresponds to the tail-on observation).

Scattering characteristics of EF-2000 Typhoon aircraft model for sounding frequency 10 GHz (wavelength 3 cm).

Figure 2.157 shows the RCS circular diagram of the EF-2000 Typhoon model. The noncoherent RCS circular diagram of the EF-2000 Typhoon is shown in Figure 2.158.

The circular mean RCS of the EF-2000 Typhoon aircraft model for horizontal polarization is 18.29 m^2. The circular median RCS (the RCS value used to calculate the detection range of an object with a probability of 0.5) for horizontal polarization is 2.00 m^2.

Figures 2.159 and 2.160 show the mean and median RCS for the main ranges of sounding azimuths (nose, side, and tail) and for ranges of 20°.

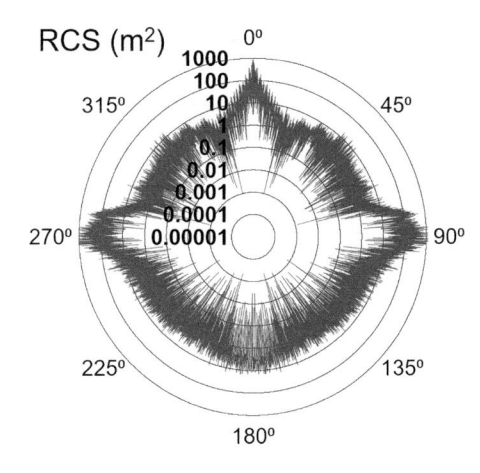

FIGURE 2.157 Circular diagram of RCS given radar observation of EF-2000 Typhoon aircraft model at a carrier frequency of 10 GHz (3 cm wavelength).

FIGURE 2.158 Circular diagram of noncoherent RCS given radar observation of EF-2000 Typhoon aircraft model at a carrier frequency of 10 GHz (3 cm wavelength).

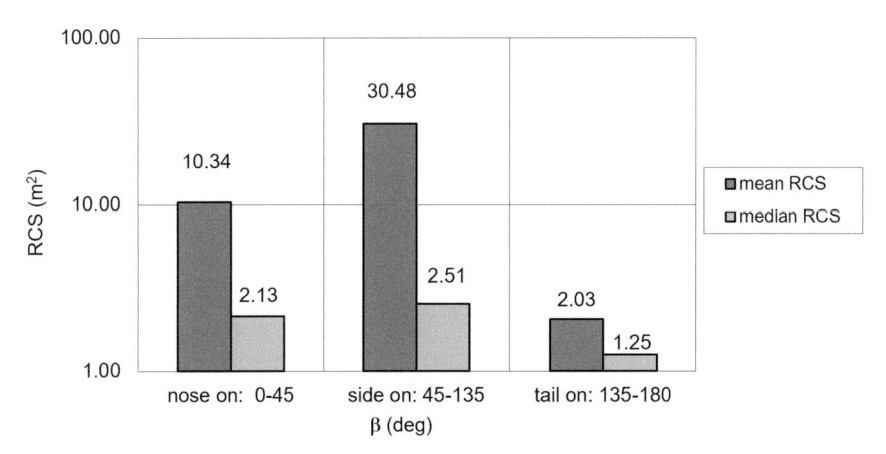

FIGURE 2.159 Diagrams of mean and median RCS of EF-2000 Typhoon aircraft model in three sectors of azimuth aspect given its radar observation at horizontal polarization and a carrier frequency of 10 GHz (3 cm wavelength).

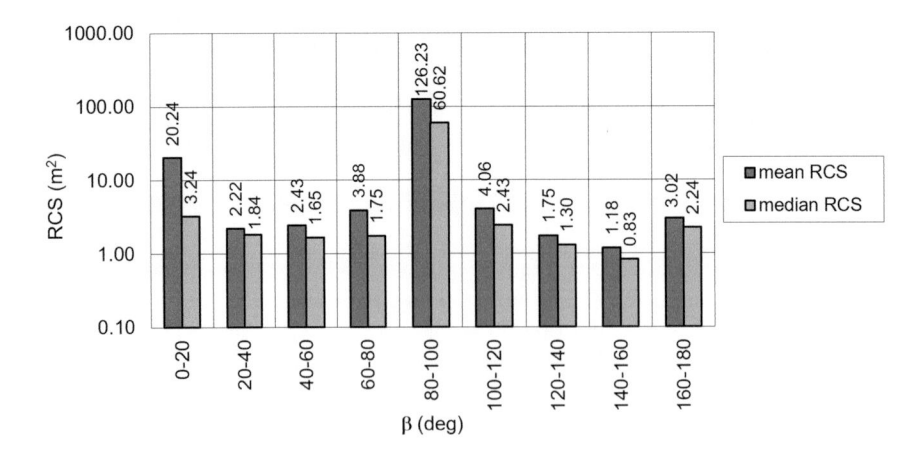

FIGURE 2.160 Diagrams of mean and median RCS of EF-2000 Typhoon aircraft model in 20-degree sectors of azimuth aspect given its radar observation at horizontal polarization and a carrier frequency of 10 GHz (3 cm wavelength).

Figures 2.161, 2.166, and 2.171 show histograms of the scattered signal amplitude (square root of the RCS) for the range of sounding azimuths −20° to +20° (sounding from the nose). The bold line shows the probability density functions of the distribution, which can be used to approximate the histogram of the signal amplitude.

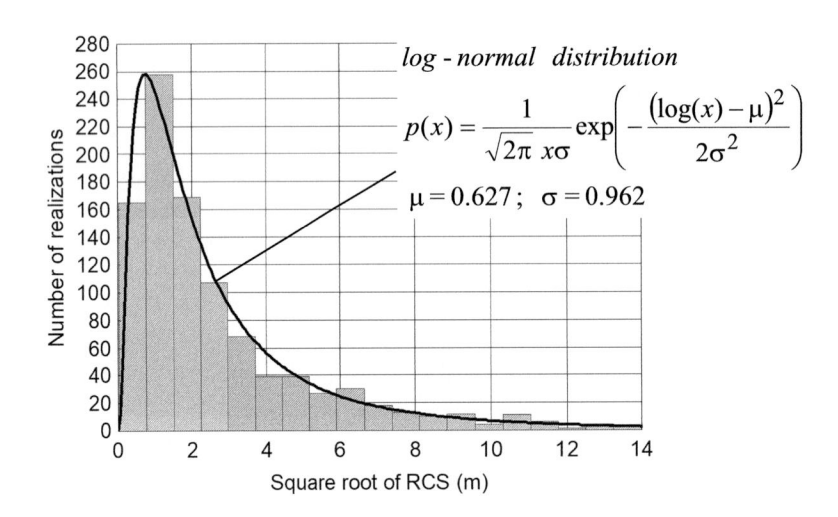

$$p(x) = \frac{1}{\sqrt{2\pi}\,x\sigma} \exp\left(-\frac{(\log(x)-\mu)^2}{2\sigma^2}\right)$$

$\mu = 0.627;\ \sigma = 0.962$

FIGURE 2.161 Amplitude distribution of echo signal of EF-2000 Typhoon aircraft model at a carrier frequency of 10 GHz given its horizontal polarization.

Scattering characteristics of EF-2000 Typhoon aircraft model for sounding frequency 3 GHz (wavelength 10 cm).

Figure 2.162 shows the RCS circular diagram of the EF-2000 Typhoon. The noncoherent RCS circular diagram of the EF-2000 Typhoon is shown in Figure 2.163.

The circular mean RCS of the EF-2000 Typhoon aircraft for horizontal polarization is 17.82 m². The circular median RCS for horizontal polarization is 1.76 m².

Figures 2.164 and 2.165 show the mean and median RCS for the main ranges of sounding azimuths (nose, side, and tail) and for ranges of 20°.

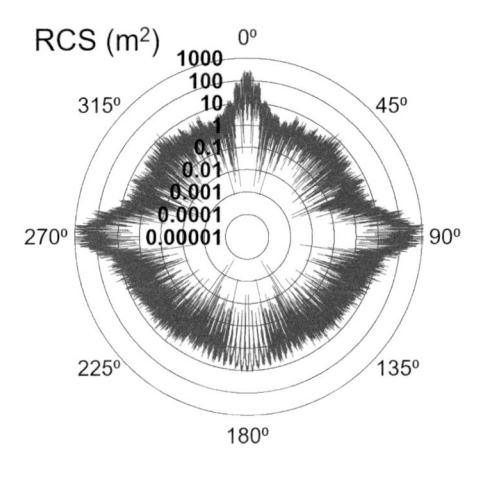

Scattering Characteristics of Aerial Objects 69

FIGURE 2.162 Circular diagram of RCS given radar observation of EF-2000 Typhoon aircraft model at a carrier frequency of 3 GHz (10 cm wavelength).

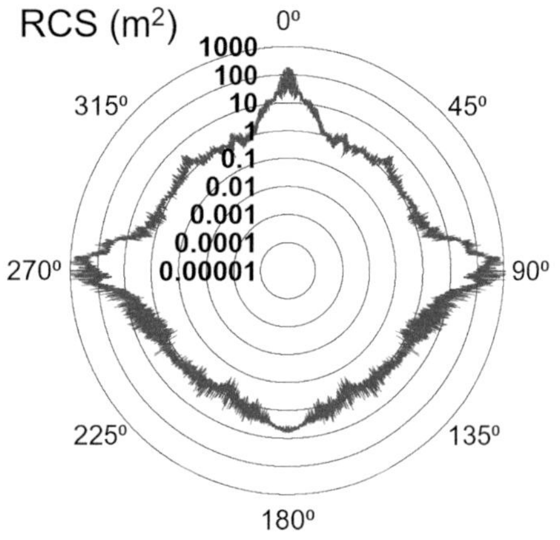

FIGURE 2.163 Circular diagram of noncoherent RCS given radar observation of EF-2000 Typhoon aircraft model at a carrier frequency of 3 GHz (10 cm wavelength).

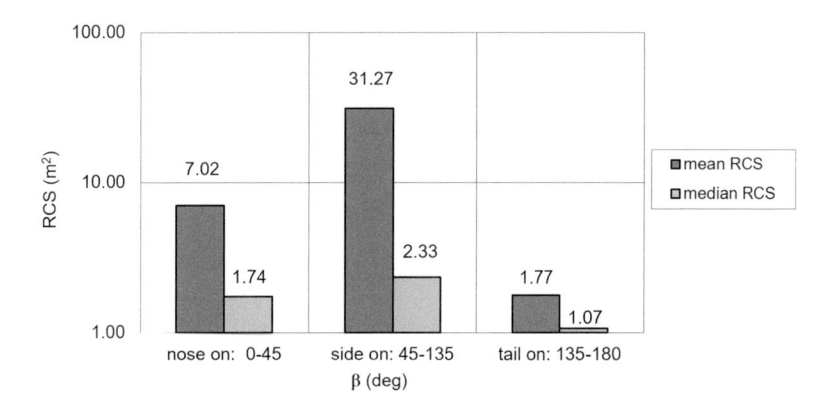

FIGURE 2.164 Diagrams of mean and median RCS of EF-2000 Typhoon aircraft model in three sectors of azimuth aspect given its radar observation at horizontal polarization and a carrier frequency of 3 GHz (10 cm wavelength).

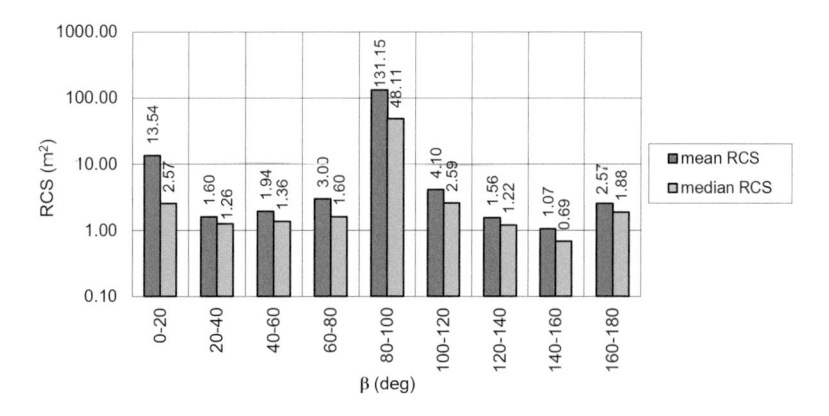

FIGURE 2.165 Diagrams of mean and median RCS of EF-2000 Typhoon aircraft model in 20-degree sectors of azimuth aspect given its radar observation at horizontal polarization and a carrier frequency of 3 GHz (10 cm wavelength).

$$p(x) = \frac{1}{\sqrt{2\pi}\, x\sigma} \exp\left(-\frac{(\log(x)-\mu)^2}{2\sigma^2}\right)$$

log-normal distribution

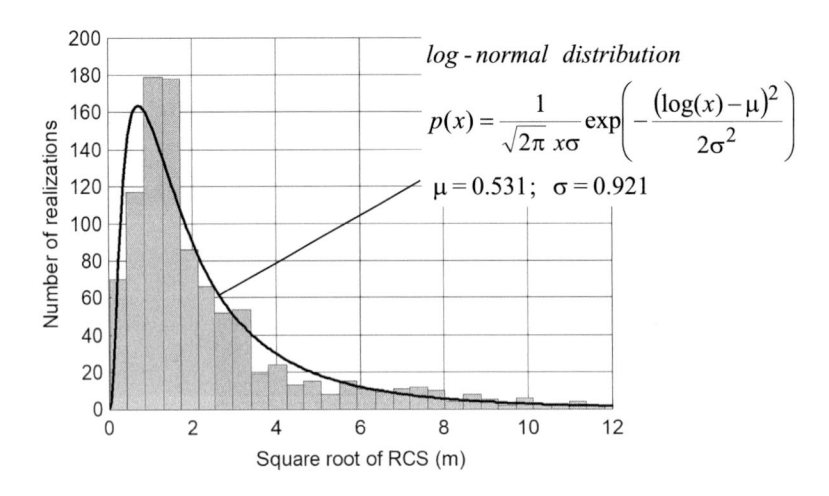

$\mu = 0.531;\ \sigma = 0.921$

FIGURE 2.166 Amplitude distribution of echo signal of EF-2000 Typhoon aircraft model at a carrier frequency of 3 GHz given its horizontal polarization.

Scattering characteristics of EF-2000 Typhoon aircraft model for sounding frequency 1 GHz (wavelength 30 cm).

Figure 2.167 shows the RCS circular diagram of the EF-2000 Typhoon. The noncoherent RCS circular diagram of the EF-2000 Typhoon is shown in Figure 2.168.

The circular mean RCS of the EF-2000 Typhoon aircraft for horizontal polarization is 12.95 m². The circular median RCS for horizontal polarization is 1.45 m².

Figures 2.169 and 2.170 show the mean and median RCS for the main ranges of sounding azimuths (nose, side, and tail) and for ranges of 20°.

The expressions and parameters of probability distributions that are most consistent with the empirical distributions of the square root of the RCS for different ranges of sounding azimuths and polarizations are given in the electronic appendix to this book. The expressions and parameters of the probability distributions that are most consistent with the empirical distributions of the RCS (energy characteristic) for different ranges of sounding azimuths and polarizations are also given there.

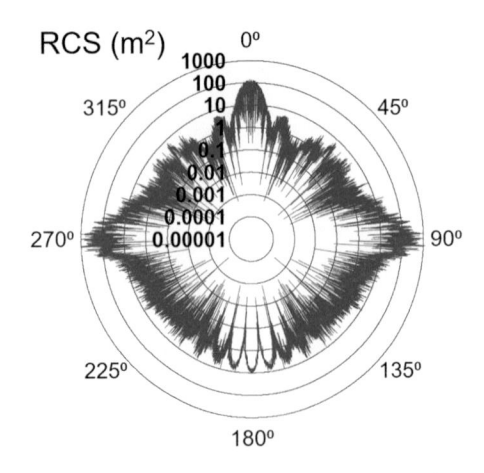

FIGURE 2.167 Circular diagram of RCS given radar observation of EF-2000 Typhoon aircraft model at a carrier frequency of 1 GHz (30 cm wavelength).

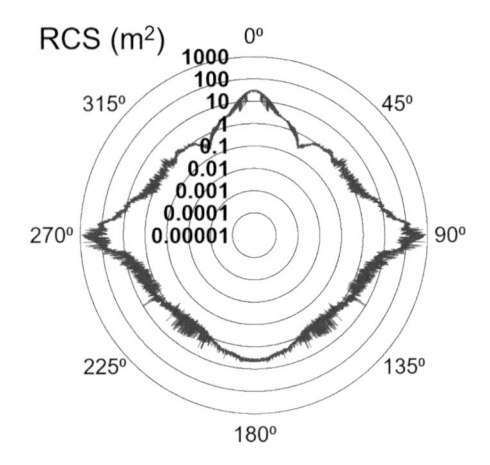

FIGURE 2.168 Circular diagram of noncoherent RCS given radar observation of EF-2000 Typhoon aircraft model at a carrier frequency of 1 GHz (30 cm wavelength).

Scattering Characteristics of Aerial Objects

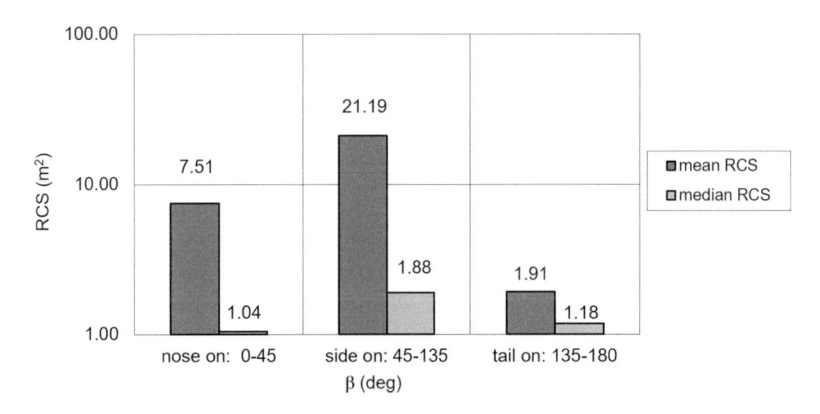

FIGURE 2.169 Diagrams of mean and median RCS of EF-2000 Typhoon aircraft model in three sectors of azimuth aspect given its radar observation at horizontal polarization and a carrier frequency of 1 GHz (30 cm wavelength).

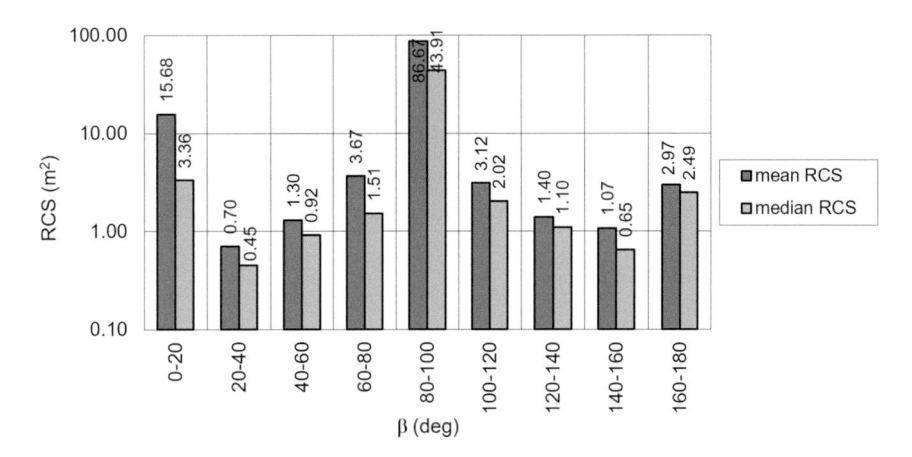

FIGURE 2.170 Diagrams of mean and median RCS of EF-2000 Typhoon aircraft model in 20-degree sectors of azimuth aspect given its radar observation at horizontal polarization and a carrier frequency of 1 GHz (30 cm wavelength).

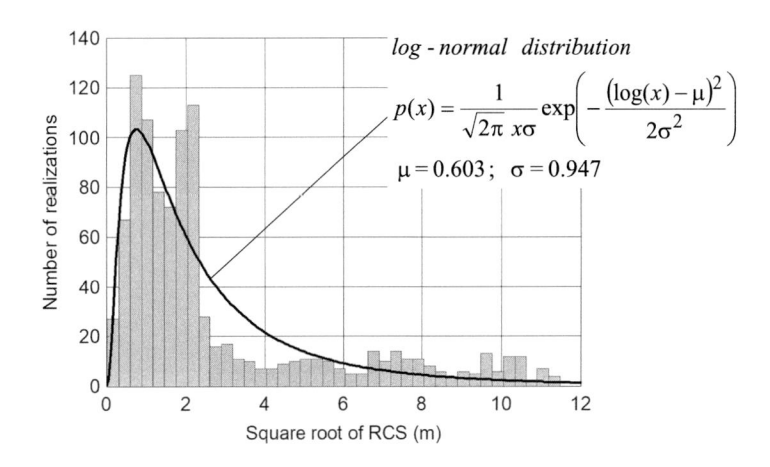

$$p(x) = \frac{1}{\sqrt{2\pi}\, x\sigma} \exp\left(-\frac{(\log(x) - \mu)^2}{2\sigma^2}\right)$$

$\mu = 0.603$; $\sigma = 0.947$

FIGURE 2.171 Amplitude distribution of echo signal of EF-2000 Typhoon aircraft model at a carrier frequency of 1 GHz given its horizontal polarization.

2.11 ATTACK AIRCRAFT Su-25

Su-25 (Figure 2.172) is an attack aircraft for close air support. The first flight took place in 1975.

General characteristics of Su-25 [22]: wingspan – 14.52 m, length – 15.33 m, height – 5.20 m, wing area – 30.1 m^2, weight – 9 500 – 19 500 kg, powerplant – 2 Soyuz/Tumansky R-195M turbojet engines, maximum speed – 950 km/h, combat range – 400–700 km, and service ceiling – 10 000 m.

In accordance with the design of the Su-25, a model of its surface was created to obtaining scattering characteristics (in particular, RCS). The model is shown in Figure 2.173. In the modeling, the smooth parts of the aircraft surface were approximated by parts of 63 triaxial ellipsoids. The surface breaks were modeled using 31 straight edge scattering parts.

FIGURE 2.172 Attack aircraft Su-25.

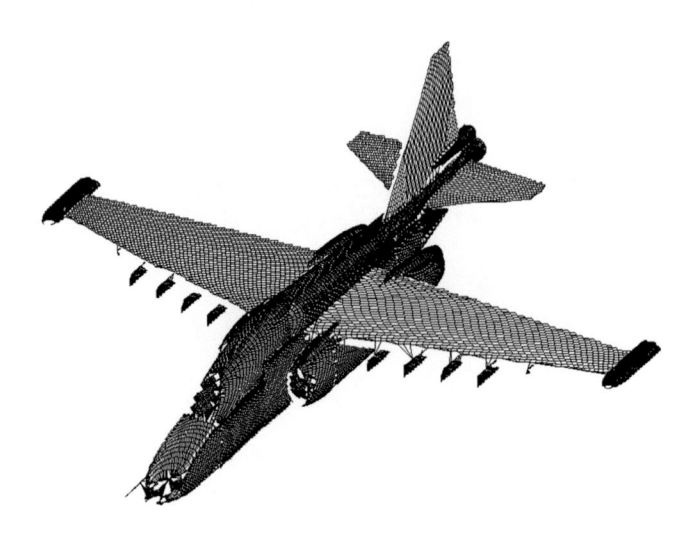

FIGURE 2.173 Surface model of attack aircraft Su-25.

Below are some scattering characteristics of the Su-25 model at sounding frequencies of 10, 3, and 1 GHz (wavelengths of 3, 10, and 30 cm, respectively) for horizontal polarization of the probing signal. The scattering characteristics for this model at two polarizations, as well as scattering characteristics at sounding frequencies of 5 and 1.3 GHz (wavelengths of 6 and 23 cm, respectively), are given in the electronic appendix of this book.

Sounding parameters: The elevation angle is to be a random value distributed uniformly in the range −3° ± 4° with respect to the wing plane (elevation angle of −3° corresponds to the radar observation from the lower hemisphere), azimuth aspect increment was 0.02°, the azimuth being counted off from the nose-on aspect (0° corresponds to the nose-on radar observation, 180° corresponds to the tail-on observation).

Scattering characteristics of Su-25 aircraft model for sounding frequency 10 GHz (wavelength 3 cm).

Figure 2.174 shows the RCS circular diagram of the Su-25 model. The noncoherent RCS circular diagram of the Su-25 is shown in Figure 2.175.

The circular mean RCS of the Su-25 aircraft model for horizontal polarization is 73.31 m². The circular median RCS (the RCS value used to calculate the detection range of an object with a probability of 0.5) for horizontal polarization is 2.86 m².

Figures 2.176 and 2.177 show the mean and median RCS for the main ranges of sounding azimuths (nose, side, and tail) and for ranges of 20°.

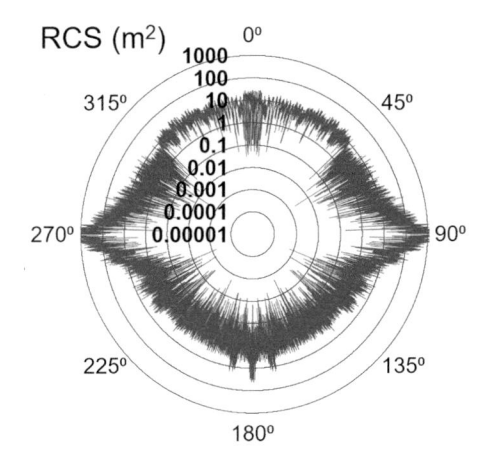

FIGURE 2.174 Circular diagram of RCS given radar observation of Su-25 aircraft model at a carrier frequency of 10 GHz (3 cm wavelength).

FIGURE 2.175 Circular diagram of noncoherent RCS given radar observation of Su-25 aircraft model at a carrier frequency of 10 GHz (3 cm wavelength).

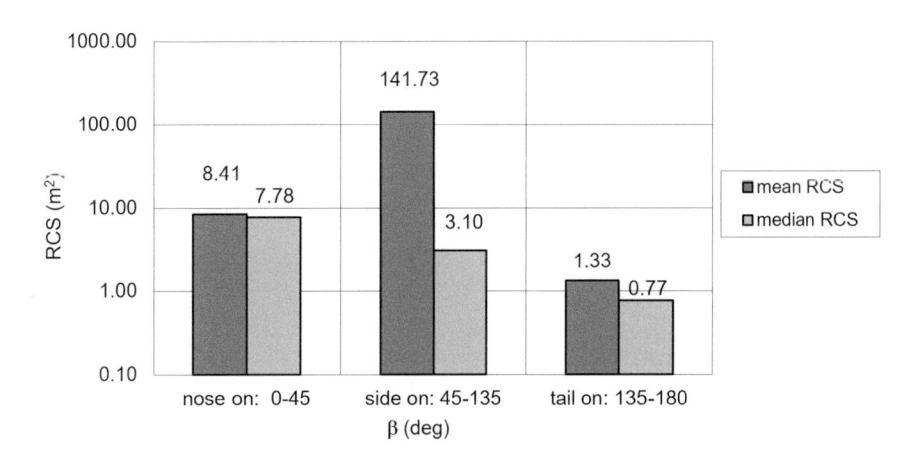

FIGURE 2.176 Diagrams of mean and median RCS of Su-25 aircraft model in three sectors of azimuth aspect given its radar observation at horizontal polarization and a carrier frequency of 10 GHz (3 cm wavelength).

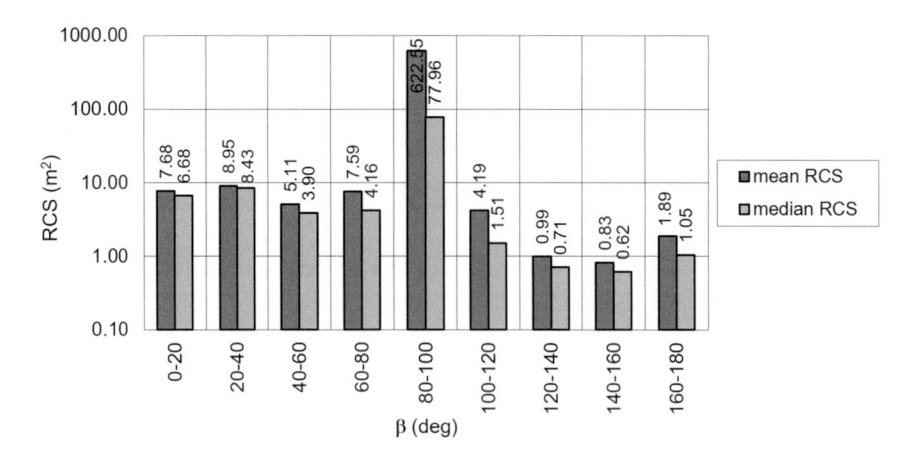

FIGURE 2.177 Diagrams of mean and median RCS of Su-25 aircraft model in 20-degree sectors of azimuth aspect given its radar observation at horizontal polarization and a carrier frequency of 10 GHz (3 cm wavelength).

Figures 2.178, 2.183, and 2.188 show histograms of the scattered signal amplitude (square root of the RCS) for the range of sounding azimuths −20° to +20° (sounding from the nose). The bold line shows the probability density functions of the distribution, which can be used to approximate the histogram of the signal amplitude.

normal distribution

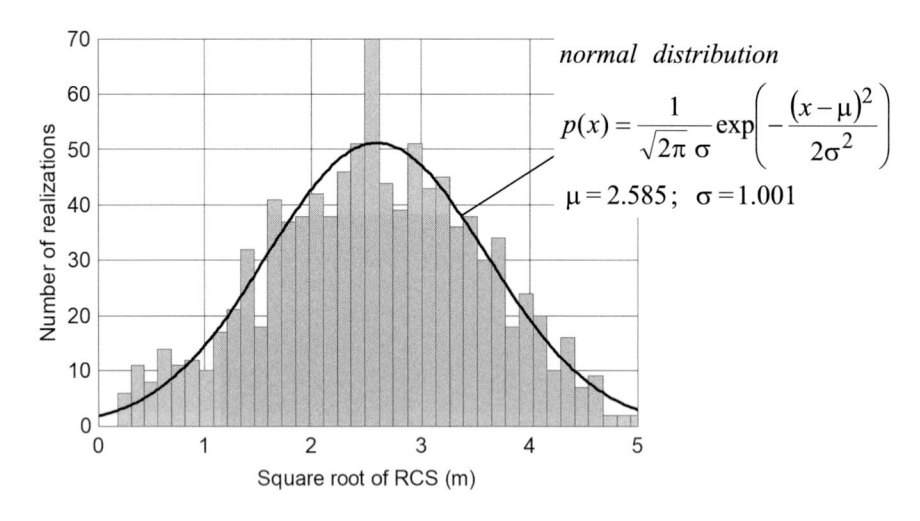

$$p(x) = \frac{1}{\sqrt{2\pi}\,\sigma} \exp\left(-\frac{(x-\mu)^2}{2\sigma^2}\right)$$

$\mu = 2.585; \quad \sigma = 1.001$

FIGURE 2.178 Amplitude distribution of echo signal of Su-25 aircraft model at a carrier frequency of 10 GHz given its horizontal polarization.

Scattering characteristics of Su-25 aircraft model for sounding frequency 3 GHz (wavelength 10 cm).

Figure 2.179 shows the RCS circular diagram of the Su-25. The noncoherent RCS circular diagram of the Su-25 is shown in Figure 2.180.

The circular mean RCS of the Su-25 aircraft for horizontal polarization is 79.06 m². The circular median RCS for horizontal polarization is 3.05 m².

Figures 2.181 and 2.182 show the mean and median RCS for the main ranges of sounding azimuths (nose, side, and tail) and for ranges of 20°.

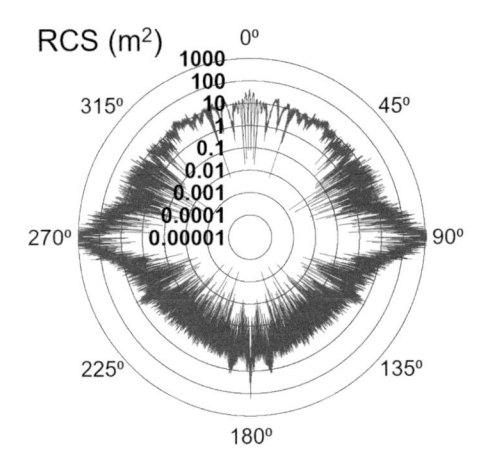

Scattering Characteristics of Aerial Objects

FIGURE 2.179 Circular diagram of RCS given radar observation of Su-25 aircraft model at a carrier frequency of 3 GHz (10 cm wavelength).

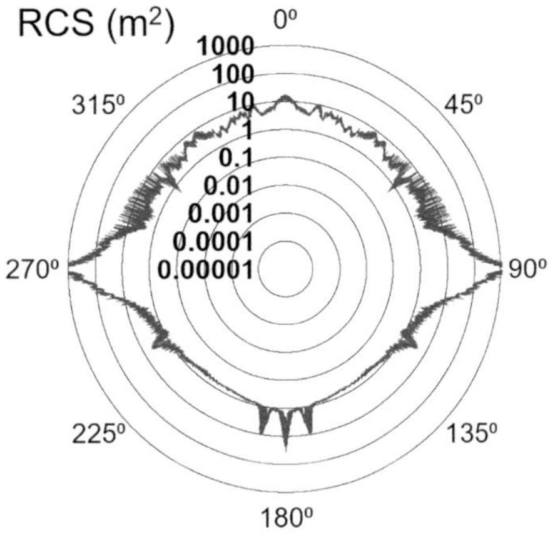

FIGURE 2.180 Circular diagram of noncoherent RCS given radar observation of Su-25 aircraft model at a carrier frequency of 3 GHz (10 cm wavelength).

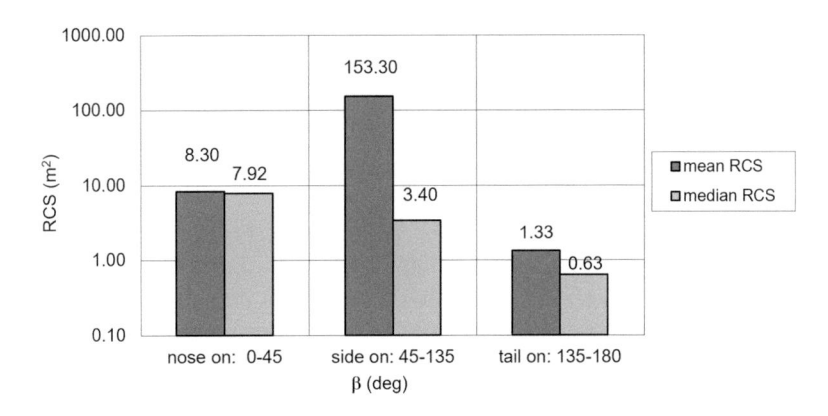

FIGURE 2.181 Diagrams of mean and median RCS of Su-25 aircraft model in three sectors of azimuth aspect given its radar observation at horizontal polarization and a carrier frequency of 3 GHz (10 cm wavelength).

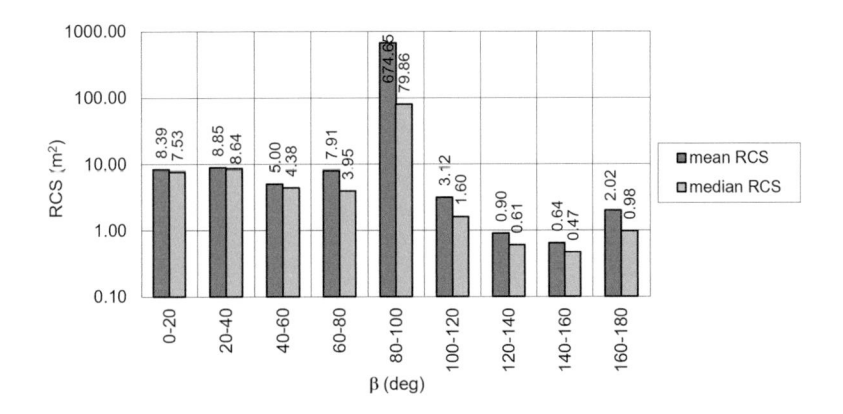

FIGURE 2.182 Diagrams of mean and median RCS of Su-25 aircraft model in 20-degree sectors of azimuth aspect given its radar observation at horizontal polarization and a carrier frequency of 3 GHz (10 cm wavelength).

$$p(x) = \frac{1}{\sqrt{2\pi}\,\sigma} \exp\left(-\frac{(x-\mu)^2}{2\sigma^2}\right)$$

normal distribution

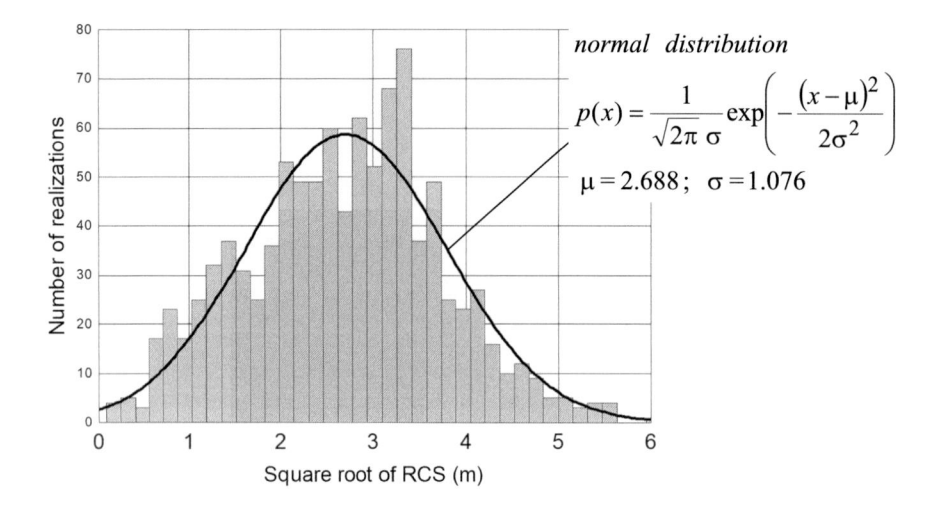

$\mu = 2.688;\quad \sigma = 1.076$

FIGURE 2.183 Amplitude distribution of echo signal of Su-25 aircraft model at a carrier frequency of 3 GHz given its horizontal polarization.

Scattering characteristics of Su-25 aircraft model for sounding frequency 1 GHz (wavelength 30 cm).

Figure 2.184 shows the RCS circular diagram of the Su-25. The noncoherent RCS circular diagram of the Su-25 is shown in Figure 2.185.

The circular mean RCS of the Su-25 aircraft for horizontal polarization is 55.13 m². The circular median RCS for horizontal polarization is 2.66 m².

Figures 2.186 and 2.187 show the mean and median RCS for the main ranges of sounding azimuths (nose, side, and tail) and for ranges of 20°.

The expressions and parameters of probability distributions that are most consistent with the empirical distributions of the square root of the RCS for different ranges of sounding azimuths and polarizations are given in the electronic appendix to this book. The expressions and parameters of the probability distributions that are most consistent with the empirical distributions of the RCS (energy characteristic) for different ranges of sounding azimuths and polarizations are also given there.

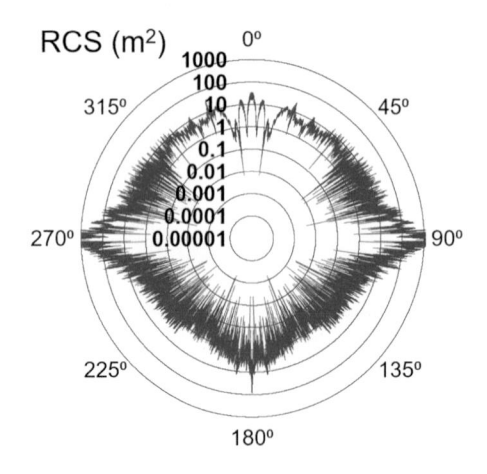

FIGURE 2.184 Circular diagram of RCS given radar observation of Su-25 aircraft model at a carrier frequency of 1 GHz (30 cm wavelength).

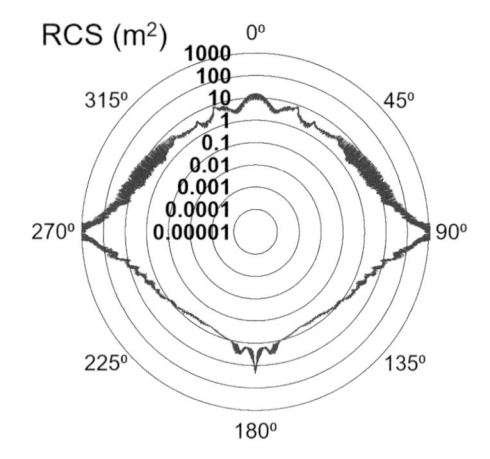

FIGURE 2.185 Circular diagram of noncoherent RCS given radar observation of Su-25 aircraft model at a carrier frequency of 1 GHz (30 cm wavelength).

Scattering Characteristics of Aerial Objects

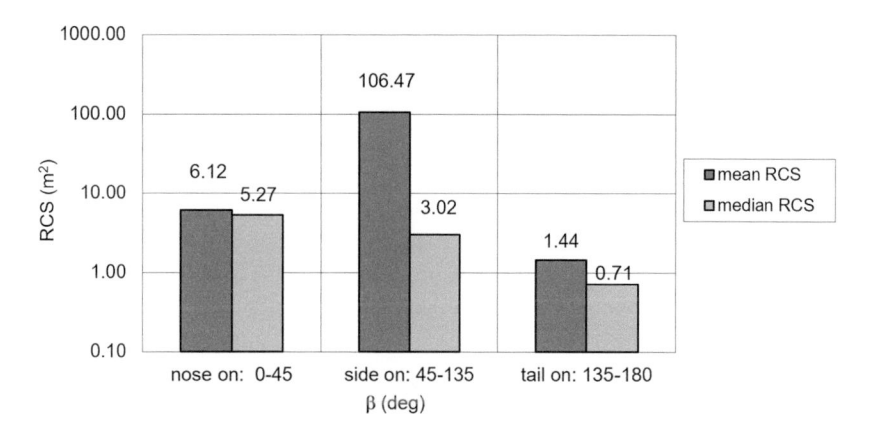

FIGURE 2.186 Diagrams of mean and median RCS of Su-25 aircraft model in three sectors of azimuth aspect given its radar observation at horizontal polarization and a carrier frequency of 1 GHz (30 cm wavelength).

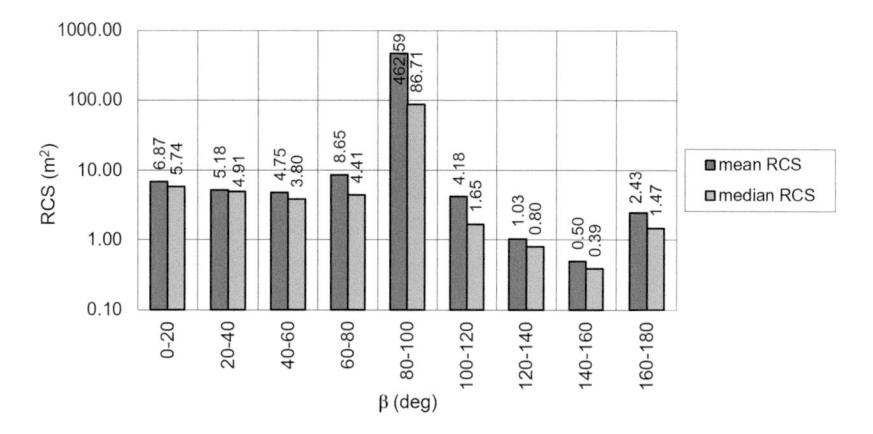

FIGURE 2.187 Diagrams of mean and median RCS of Su-25 aircraft model in 20-degree sectors of azimuth aspect given its radar observation at horizontal polarization and a carrier frequency of 1 GHz (30 cm wavelength).

Weibull distribution

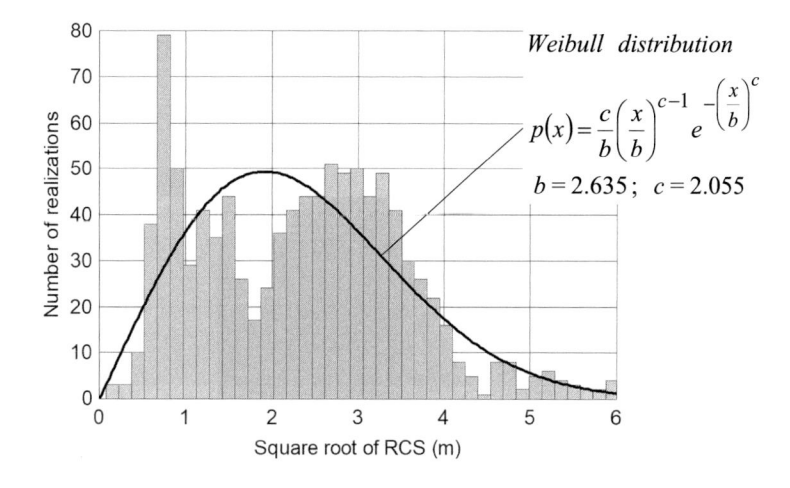

$$p(x) = \frac{c}{b}\left(\frac{x}{b}\right)^{c-1} e^{-\left(\frac{x}{b}\right)^c}$$

$b = 2.635;\ c = 2.055$

FIGURE 2.188 Amplitude distribution of echo signal of Su-25 aircraft model at a carrier frequency of 1 GHz given its horizontal polarization.

2.12 ATTACK AIRCRAFT A-10 THUNDERBOLT II

A-10 Thunderbolt II (Figure 2.189) is one-seat attack aircraft. The first flight took place in 1977.

General characteristics of A-10 [23]: wingspan –17.42 m, length – 16.16 m, height – 4.42 m, wing area – 47 m², weight – 11 300–22 950 kg, powerplant – 2 General Electric TF34-GE-100A turbofans, maximum speed – 833 km/h, range – 1280 km, and service ceiling – 13 600 m.

In accordance with the design of the A-10, a model of its surface was created to obtaining scattering characteristics (in particular, RCS). The model is shown in Figure 2.190. In the modeling, the smooth parts of the aircraft surface were approximated by parts of 41 triaxial ellipsoids. The surface breaks were modeled using 24 straight edge scattering parts.

FIGURE 2.189 Attack aircraft A-10.

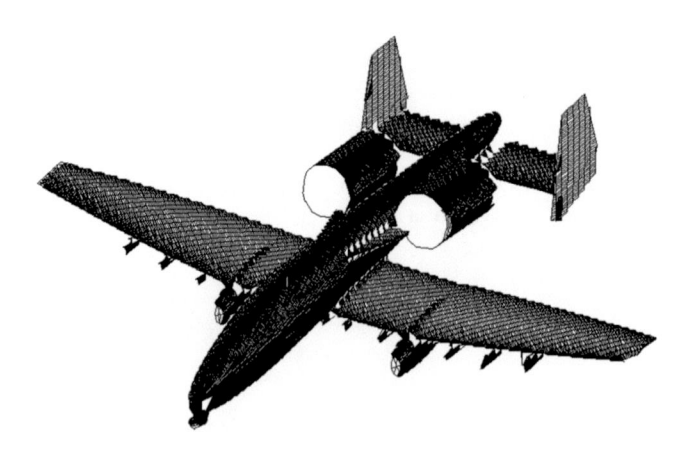

FIGURE 2.190 Surface model of attack aircraft A-10.

Below are some scattering characteristics of the A-10 model at sounding frequencies of 10, 3, and 1 GHz (wavelengths of 3, 10, and 30 cm, respectively) for horizontal polarization of the probing signal. The scattering characteristics for this model at two polarizations, as well as scattering characteristics at sounding frequencies of 5 and 1.3 GHz (wavelengths of 6 and 23 cm, respectively), are given in the electronic appendix of this book.

Sounding parameters: The elevation angle is to be a random value distributed uniformly in the range −3° ± 4° with respect to the wing plane (elevation angle of −3° corresponds to the radar observation from the lower hemisphere), azimuth aspect increment was 0.02°, the azimuth being counted off from the nose-on aspect (0° corresponds to the nose-on radar observation, 180° corresponds to the tail-on observation).

Scattering Characteristics of Aerial Objects

Scattering characteristics of A-10 aircraft model for sounding frequency 10 GHz (wavelength 3 cm).

Figure 2.191 shows the RCS circular diagram of the A-10 model. The noncoherent RCS circular diagram of the A-10 is shown in Figure 2.192.

The circular mean RCS of the A-10 aircraft model for horizontal polarization is 61.20 m^2. The circular median RCS (the RCS value used to calculate the detection range of an object with a probability of 0.5) for horizontal polarization is 8.04 m^2.

Figures 2.193 and 2.194 show the mean and median RCS for the main ranges of sounding azimuths (nose, side, and tail) and for ranges of 20°.

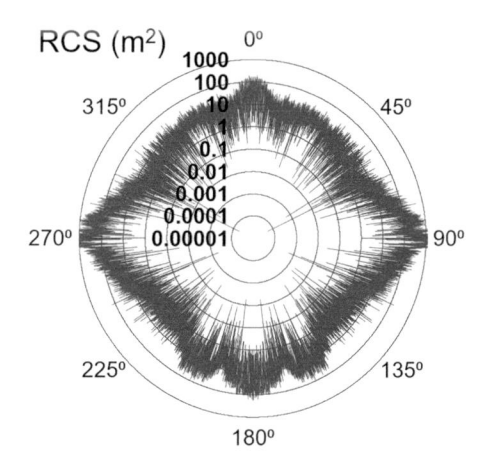

FIGURE 2.191 Circular diagram of RCS given radar observation of A-10 aircraft model at a carrier frequency of 10 GHz (3 cm wavelength).

FIGURE 2.192 Circular diagram of noncoherent RCS given radar observation of A-10 aircraft model at a carrier frequency of 10 GHz (3 cm wavelength).

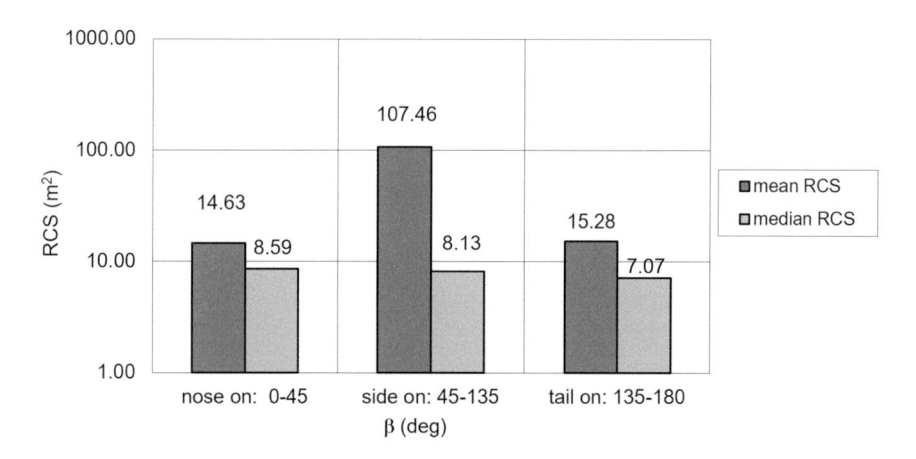

FIGURE 2.193 Diagrams of mean and median RCS of A-10 aircraft model in three sectors of azimuth aspect given its radar observation at horizontal polarization and a carrier frequency of 10 GHz (3 cm wavelength).

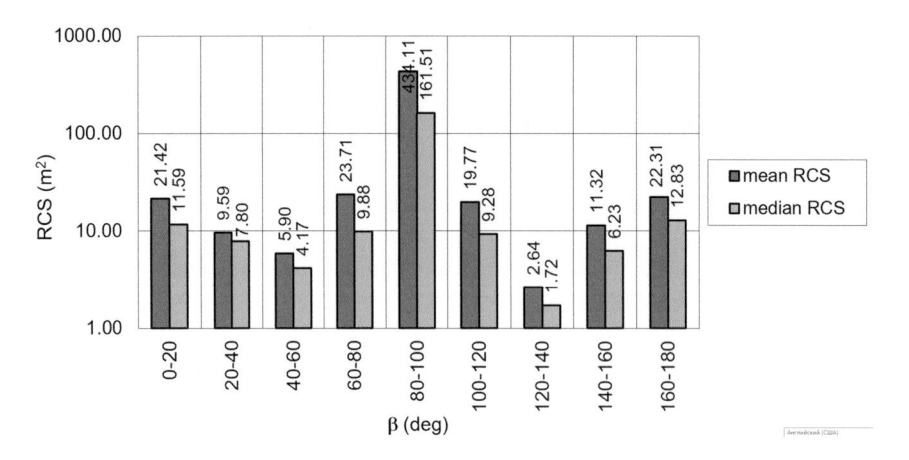

FIGURE 2.194 Diagrams of mean and median RCS of A-10 aircraft model in 20-degree sectors of azimuth aspect given its radar observation at horizontal polarization and a carrier frequency of 10 GHz (3 cm wavelength).

Figures 2.195, 2.200, and 2.205 show histograms of the scattered signal amplitude (square root of the RCS) for the range of sounding azimuths −20° to +20° (sounding from the nose). The bold line shows the probability density functions of the distribution, which can be used to approximate the histogram of the signal amplitude.

extreme value distribution

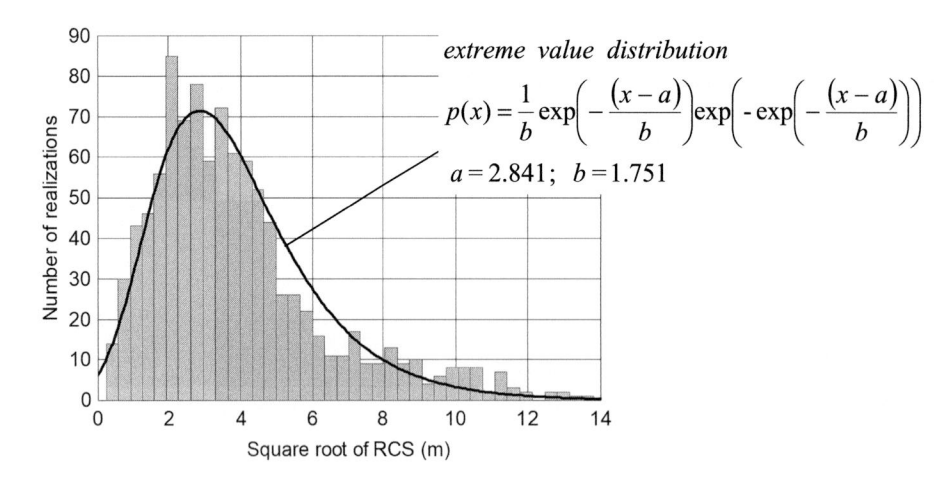

$$p(x) = \frac{1}{b} \exp\left(-\frac{(x-a)}{b}\right) \exp\left(-\exp\left(-\frac{(x-a)}{b}\right)\right)$$

$a = 2.841; \quad b = 1.751$

FIGURE 2.195 Amplitude distribution of echo signal of A-10 aircraft model at a carrier frequency of 10 GHz given its horizontal polarization.

Scattering characteristics of A-10 aircraft model for sounding frequency 3 GHz (wavelength 10 cm).

Figure 2.196 shows the RCS circular diagram of the A-10. The noncoherent RCS circular diagram of the A-10 is shown in Figure 2.197.

The circular mean RCS of the A-10 aircraft for horizontal polarization is 71.14 m². The circular median RCS for horizontal polarization is 7.40 m².

Figures 2.198 and 2.199 show the mean and median RCS for the main ranges of sounding azimuths (nose, side, and tail) and for ranges of 20°.

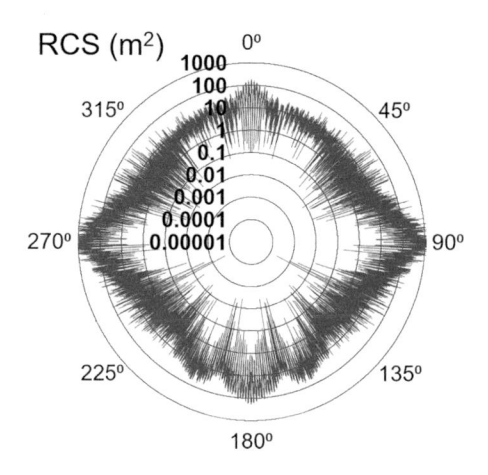

Scattering Characteristics of Aerial Objects

FIGURE 2.196 Circular diagram of RCS given radar observation of A-10 aircraft model at a carrier frequency of 3 GHz (10 cm wavelength).

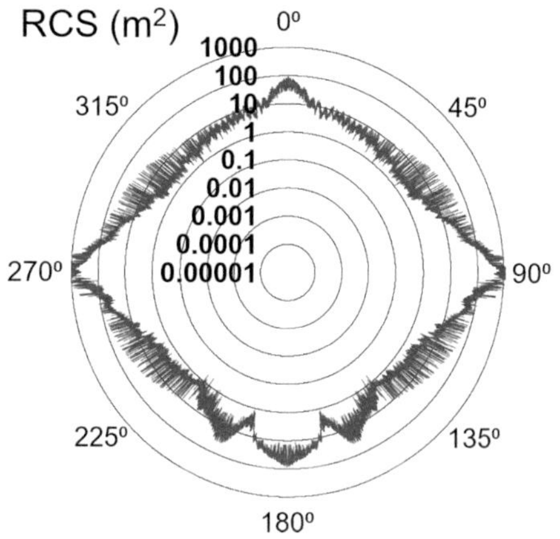

FIGURE 2.197 Circular diagram of noncoherent RCS given radar observation of A-10 aircraft model at a carrier frequency of 3 GHz (10 cm wavelength).

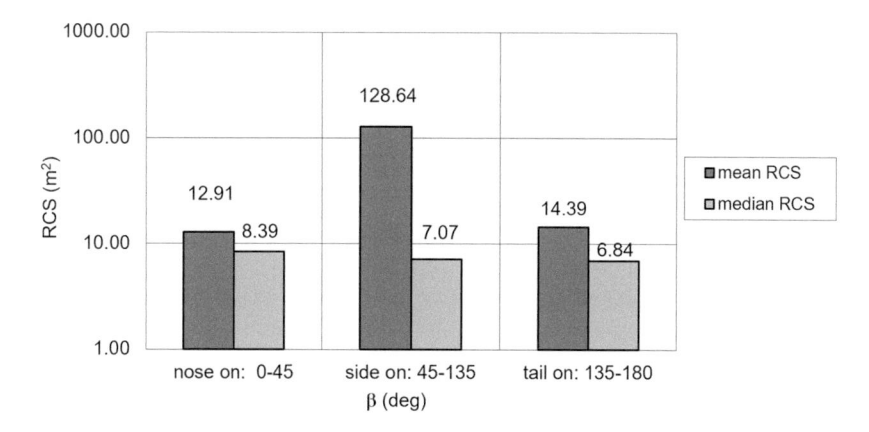

FIGURE 2.198 Diagrams of mean and median RCS of A-10 aircraft model in three sectors of azimuth aspect given its radar observation at horizontal polarization and a carrier frequency of 3 GHz (10 cm wavelength).

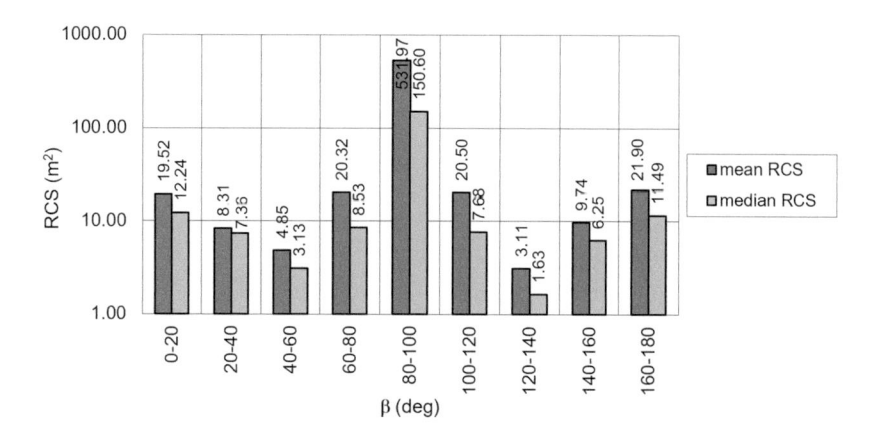

FIGURE 2.199 Diagrams of mean and median RCS of A-10 aircraft model in 20-degree sectors of azimuth aspect given its radar observation at horizontal polarization and a carrier frequency of 3 GHz (10 cm wavelength).

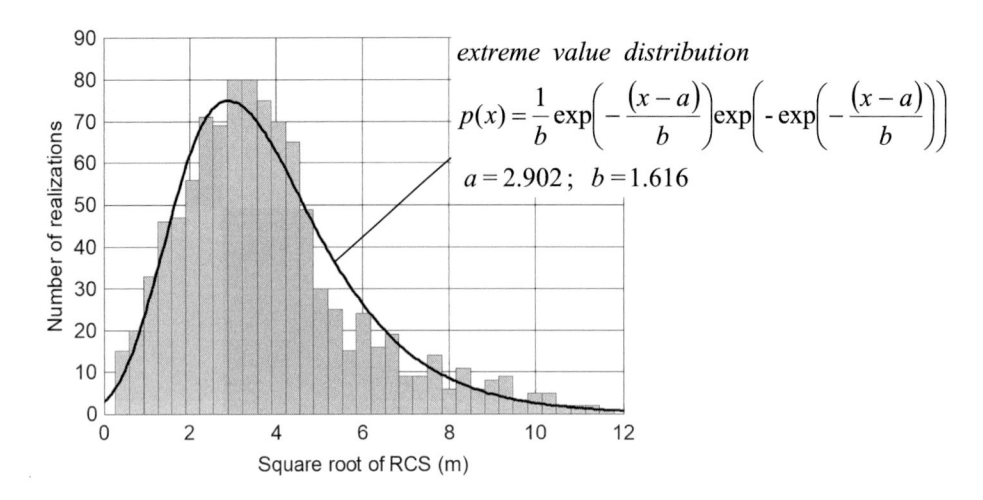

FIGURE 2.200 Amplitude distribution of echo signal of A-10 aircraft model at a carrier frequency of 3 GHz given its horizontal polarization.

extreme value distribution

$$p(x) = \frac{1}{b} \exp\left(-\frac{(x-a)}{b}\right) \exp\left(-\exp\left(-\frac{(x-a)}{b}\right)\right)$$

$a = 2.902; \quad b = 1.616$

Scattering characteristics of A-10 aircraft model for sounding frequency 1 GHz (wavelength 30 cm).

Figure 2.201 shows the RCS circular diagram of the A-10. The noncoherent RCS circular diagram of the A-10 is shown in Figure 2.202.

The circular mean RCS of the A-10 aircraft for horizontal polarization is 53.24 m². The circular median RCS for horizontal polarization is 5.88 m².

Figures 2.203 and 2.204 show the mean and median RCS for the main ranges of sounding azimuths (nose, side, and tail) and for ranges of 20°.

The expressions and parameters of probability distributions that are most consistent with the empirical distributions of the square root of the RCS for different ranges of sounding azimuths and polarizations are given in the electronic appendix to this book. The expressions and parameters of the probability distributions that are most consistent with the empirical distributions of the RCS (energy characteristic) for different ranges of sounding azimuths and polarizations are also given there.

FIGURE 2.201 Circular diagram of RCS given radar observation of A-10 aircraft model at a carrier frequency of 1 GHz (30 cm wavelength).

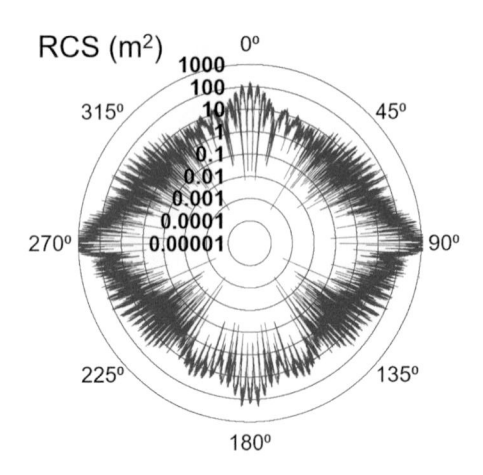

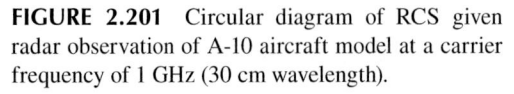

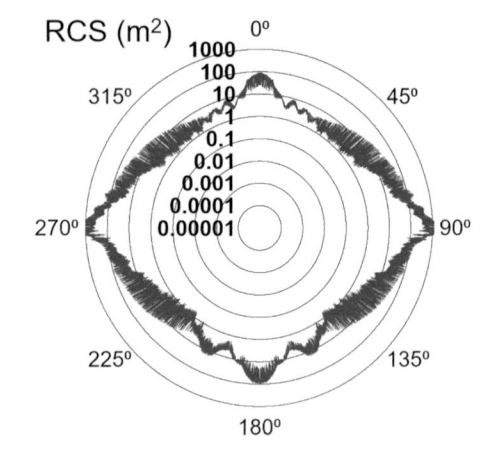

FIGURE 2.202 Circular diagram of noncoherent RCS given radar observation of A-10 aircraft model at a carrier frequency of 1 GHz (30 cm wavelength).

Scattering Characteristics of Aerial Objects

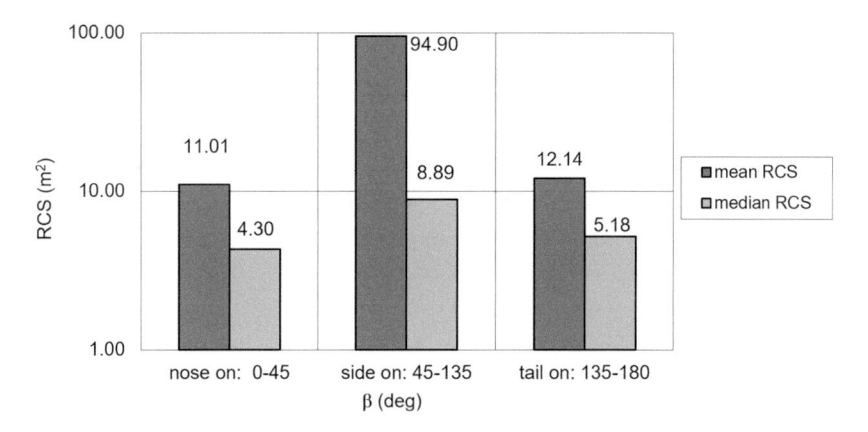

FIGURE 2.203 Diagrams of mean and median RCS of A-10 aircraft model in three sectors of azimuth aspect given its radar observation at horizontal polarization and a carrier frequency of 1 GHz (30 cm wavelength).

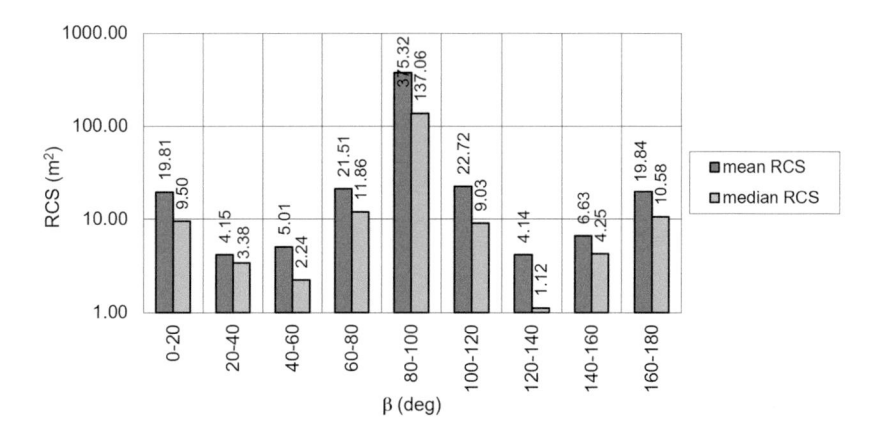

FIGURE 2.204 Diagrams of mean and median RCS of A-10 aircraft model in 20-degree sectors of azimuth aspect given its radar observation at horizontal polarization and a carrier frequency of 1 GHz (30 cm wavelength).

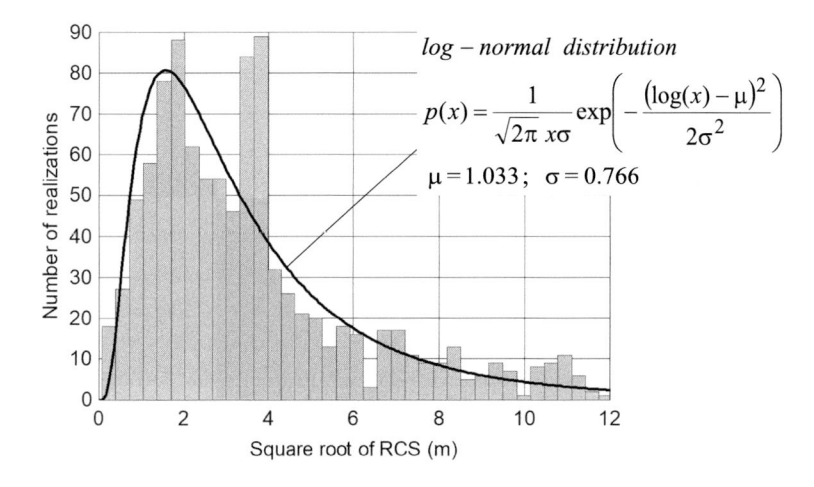

$$p(x) = \frac{1}{\sqrt{2\pi}\, x\sigma} \exp\left(-\frac{(\log(x) - \mu)^2}{2\sigma^2}\right)$$

$\mu = 1.033;\ \sigma = 0.766$

FIGURE 2.205 Amplitude distribution of echo signal of A-10 aircraft model at a carrier frequency of 1 GHz given its horizontal polarization.

2.13 TACTICAL BOMBER Su-24

Su-24 Fencer (Figure 2.206) is all-weather tactical bomber. The first flight took place in 1967.

General characteristics of Su-24 [24]: wingspan – 10.37–17.64 m, length – 22.67 m, height – 5.92 m, wing area – 51.02–55.16 m^2, weight – 21 200–39 750 kg, powerplant – 2 AL-21F-3A turbojet engines, maximum speed – 1 700 km/h, combat range – 600 km, and service ceiling – 11 000 m.

In accordance with the design of the Su-24, a model of its surface was created to obtaining scattering characteristics (in particular, RCS). The model is shown in Figure 2.207. In the modeling, the smooth parts of the aircraft surface were approximated by parts of 84 triaxial ellipsoids. The surface breaks were modeled using 33 straight edge scattering parts.

FIGURE 2.206 Tactical bomber Su-24.

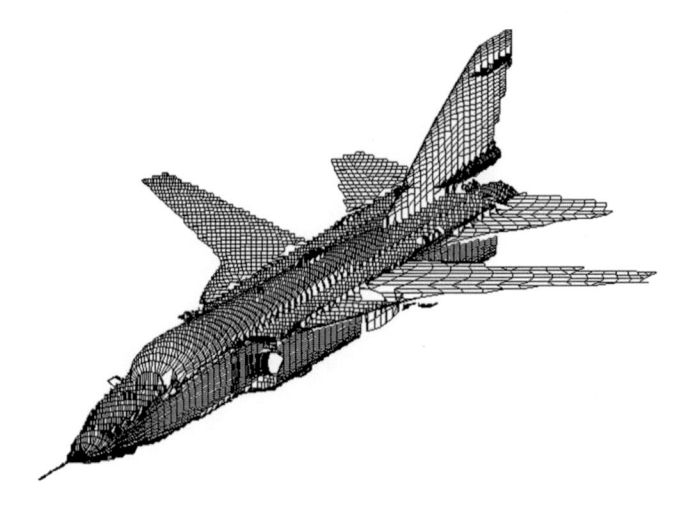

FIGURE 2.207 Surface model of tactical bomber Su-24.

Below are some scattering characteristics of the Su-24 model at sounding frequencies of 10, 3, and 1 GHz (wavelengths of 3, 10, and 30 cm, respectively) for horizontal polarization of the probing signal. The scattering characteristics for this model at two polarizations, as well as scattering characteristics at sounding frequencies of 5 and 1.3 GHz (wavelengths of 6 and 23 cm, respectively), are given in the electronic appendix of this book.

Sounding parameters: The elevation angle is to be a random value distributed uniformly in the range −3° ± 4° with respect to the wing plane (elevation angle of −3° corresponds to the radar observation from the lower hemisphere), azimuth aspect increment was 0.02°, the azimuth being counted off from the nose-on aspect (0° corresponds to the nose-on radar observation, 180° corresponds to the tail-on observation).

Scattering characteristics of Su-24 aircraft model for sounding frequency 10 GHz (wavelength 3 cm).

Figure 2.208 shows the RCS circular diagram of the Su-24 model. The noncoherent RCS circular diagram of the Su-24 is shown in Figure 2.209.

The circular mean RCS of the Su-24 aircraft model for horizontal polarization is 144.38 m². The circular median RCS (the RCS value used to calculate the detection range of an object with a probability of 0.5) for horizontal polarization is 3.71 m².

Figures 2.210 and 2.211 show the mean and median RCS for the main ranges of sounding azimuths (nose, side, and tail) and for ranges of 20°.

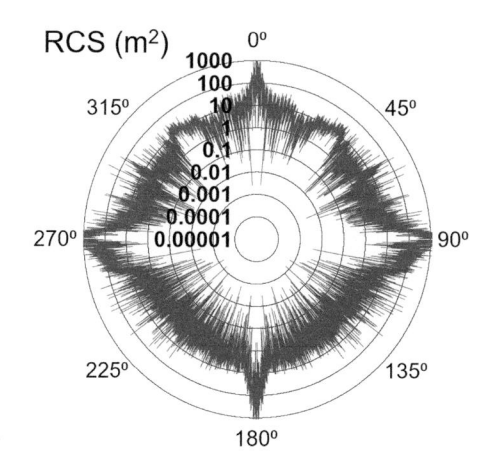

FIGURE 2.208 Circular diagram of RCS given radar observation of Su-24 aircraft model at a carrier frequency of 10 GHz (3 cm wavelength).

FIGURE 2.209 Circular diagram of noncoherent RCS given radar observation of Su-24 aircraft model at a carrier frequency of 10 GHz (3 cm wavelength).

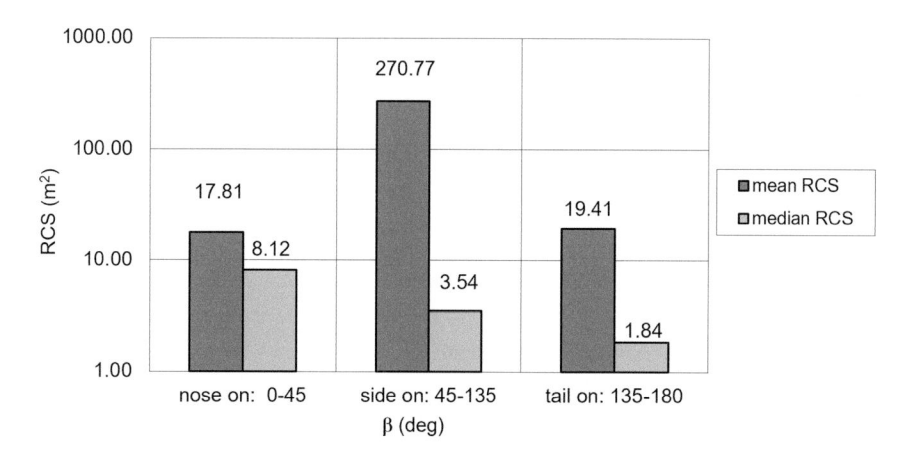

FIGURE 2.210 Diagrams of mean and median RCS of Su-24 aircraft model in three sectors of azimuth aspect given its radar observation at horizontal polarization and a carrier frequency of 10 GHz (3 cm wavelength).

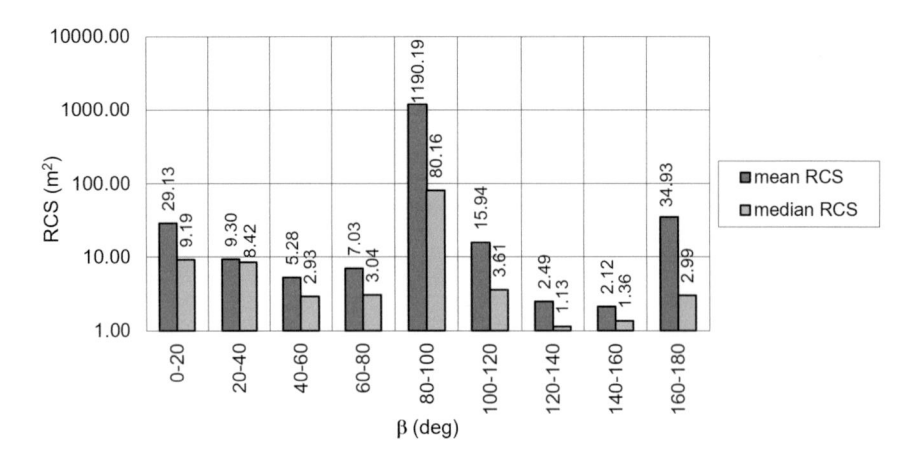

FIGURE 2.211 Diagrams of mean and median RCS of Su-24 aircraft model in 20-degree sectors of azimuth aspect given its radar observation at horizontal polarization and a carrier frequency of 10 GHz (3 cm wavelength).

Figures 2.212, 2.217, and 2.222 show histograms of the scattered signal amplitude (square root of the RCS) for the range of sounding azimuths −20° to +20° (sounding from the nose). The bold line shows the probability density functions of the distribution, which can be used to approximate the histogram of the signal amplitude.

$$p(x) = \frac{1}{\sqrt{2\pi}\, x\sigma} \exp\left(-\frac{(\log(x) - \mu)^2}{2\sigma^2}\right)$$

log − normal distribution

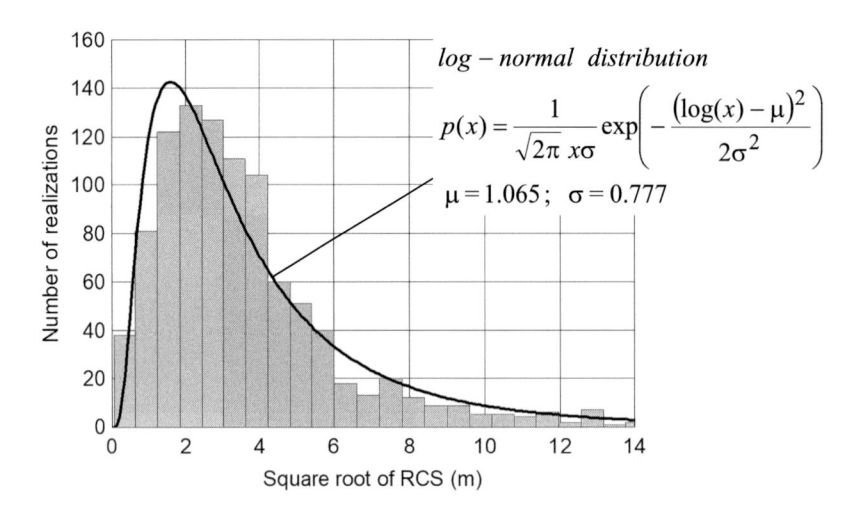

$\mu = 1.065;\ \sigma = 0.777$

FIGURE 2.212 Amplitude distribution of echo signal of Su-24 aircraft model at a carrier frequency of 10 GHz given its horizontal polarization.

Scattering characteristics of Su-24 aircraft model for sounding frequency 3 GHz (wavelength 10 cm).

Figure 2.213 shows the RCS circular diagram of the Su-24. The noncoherent RCS circular diagram of the Su-24 is shown in Figure 2.214.

The circular mean RCS of the Su-24 aircraft for horizontal polarization is 143.83 m². The circular median RCS for horizontal polarization is 5.39 m².

Figures 2.215 and 2.216 show the mean and median RCS for the main ranges of sounding azimuths (nose, side, and tail) and for ranges of 20°.

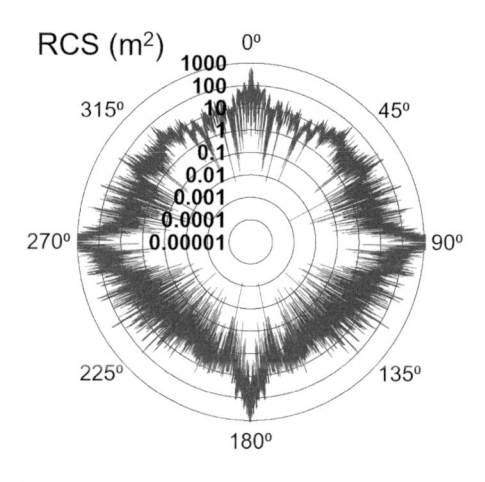

Scattering Characteristics of Aerial Objects

FIGURE 2.213 Circular diagram of RCS given radar observation of Su-24 aircraft model at a carrier frequency of 3 GHz (10 cm wavelength).

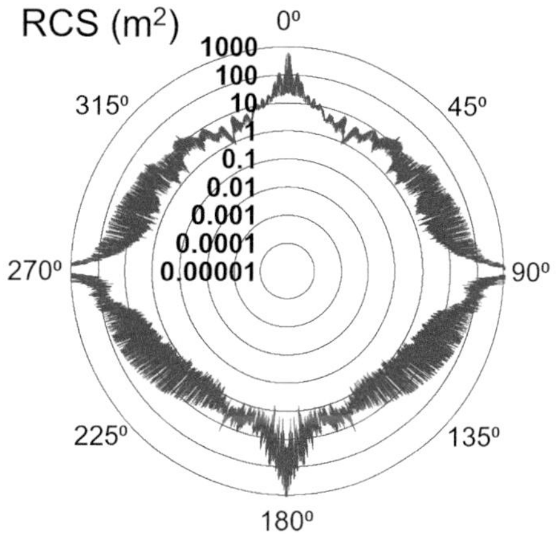

FIGURE 2.214 Circular diagram of noncoherent RCS given radar observation of Su-24 aircraft model at a carrier frequency of 3 GHz (10 cm wavelength).

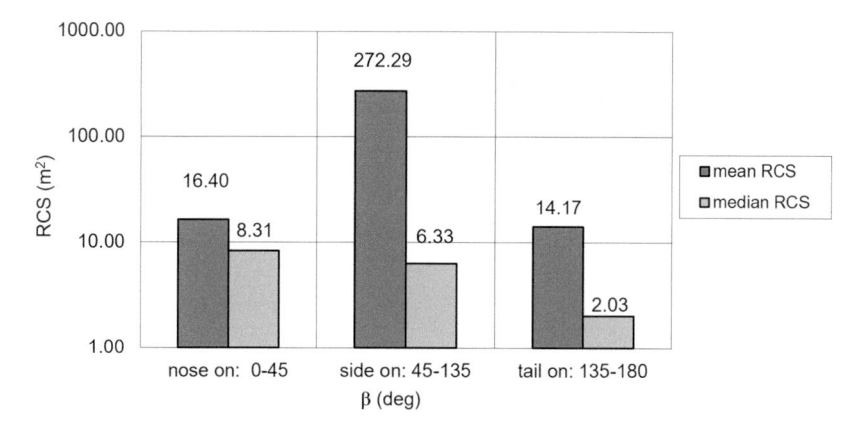

FIGURE 2.215 Diagrams of mean and median RCS of Su-24 aircraft model in three sectors of azimuth aspect given its radar observation at horizontal polarization and a carrier frequency of 3 GHz (10 cm wavelength).

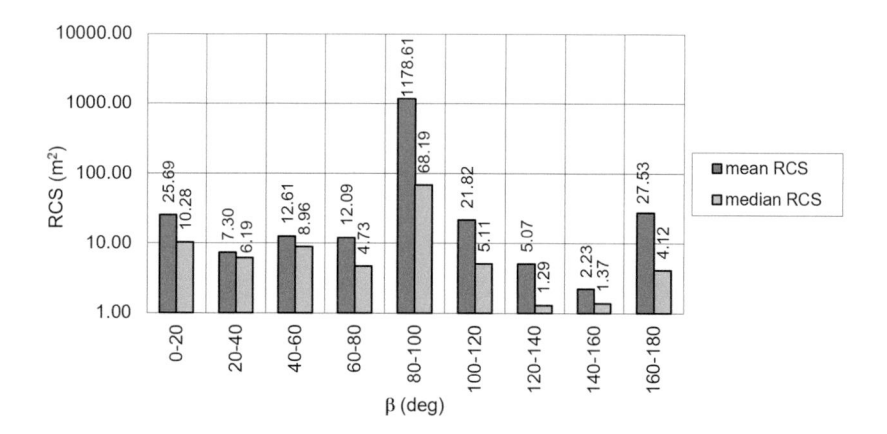

FIGURE 2.216 Diagrams of mean and median RCS of Su-24 aircraft model in 20-degree sectors of azimuth aspect given its radar observation at horizontal polarization and a carrier frequency of 3 GHz (10 cm wavelength).

$$p(x) = \frac{1}{\sqrt{2\pi}\, x\sigma} \exp\left(-\frac{(\log(x) - \mu)^2}{2\sigma^2}\right)$$

log – normal distribution

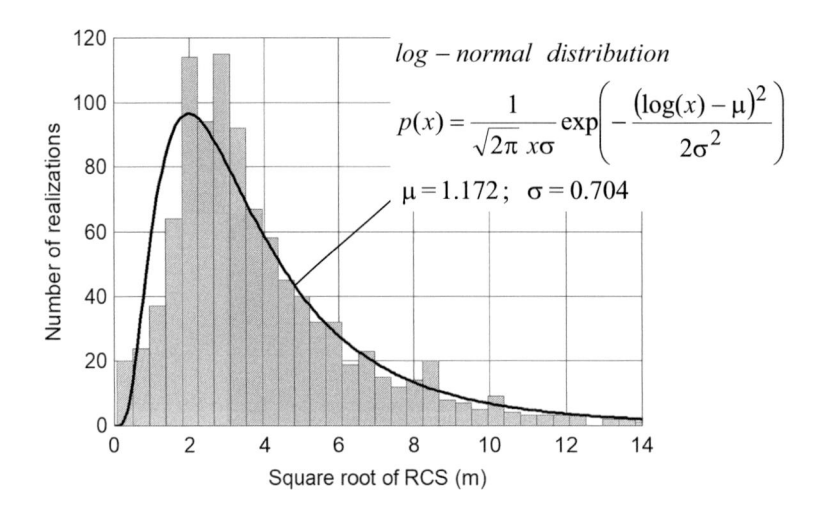

$\mu = 1.172;\ \sigma = 0.704$

FIGURE 2.217 Amplitude distribution of echo signal of Su-24 aircraft model at a carrier frequency of 3 GHz given its horizontal polarization.

Scattering characteristics of Su-24 aircraft model for sounding frequency 1 GHz (wavelength 30 cm).

Figure 2.218 shows the RCS circular diagram of the Su-24. The noncoherent RCS circular diagram of the Su-24 is shown in Figure 2.219.

The circular mean RCS of the Su-24 aircraft for horizontal polarization is 137.03 m². The circular median RCS for horizontal polarization is 5.96 m².

Figures 2.220 and 2.221 show the mean and median RCS for the main ranges of sounding azimuths (nose, side, and tail) and for ranges of 20°.

The expressions and parameters of probability distributions that are most consistent with the empirical distributions of the square root of the RCS for different ranges of sounding azimuths and polarizations are given in the electronic appendix to this book. The expressions and parameters of the probability distributions that are most consistent with the empirical distributions of the RCS (energy characteristic) for different ranges of sounding azimuths and polarizations are also given there.

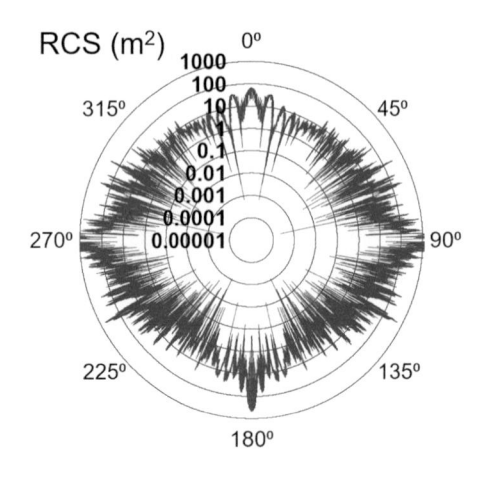

FIGURE 2.218 Circular diagram of RCS given radar observation of Su-24 aircraft model at a carrier frequency of 1 GHz (30 cm wavelength).

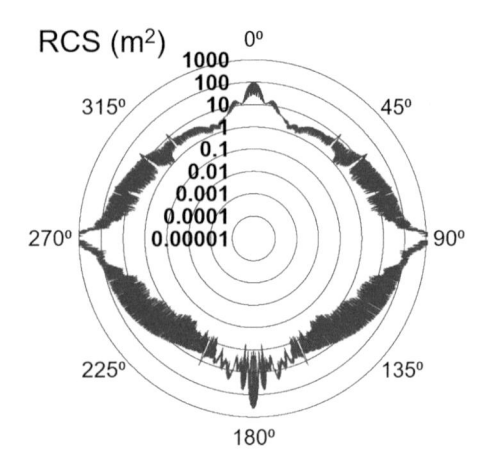

FIGURE 2.219 Circular diagram of noncoherent RCS given radar observation of Su-24 aircraft model at a carrier frequency of 1 GHz (30 cm wavelength).

Scattering Characteristics of Aerial Objects

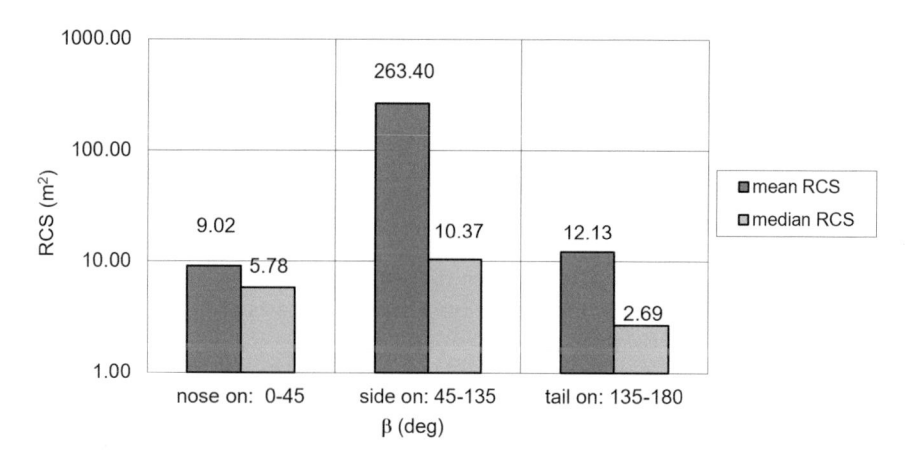

FIGURE 2.220 Diagrams of mean and median RCS of Su-24 aircraft model in three sectors of azimuth aspect given its radar observation at horizontal polarization and a carrier frequency of 1 GHz (30 cm wavelength).

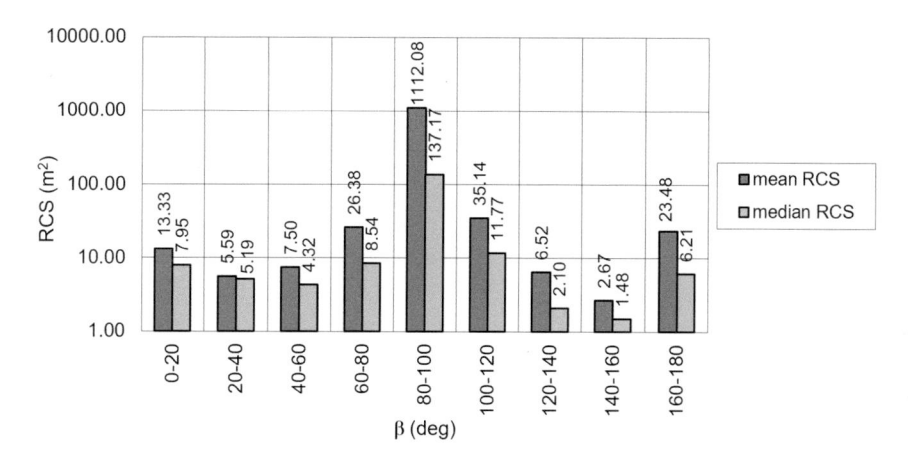

FIGURE 2.221 Diagrams of mean and median RCS of Su-24 aircraft model in 20-degree sectors of azimuth aspect given its radar observation at horizontal polarization and a carrier frequency of 1 GHz (30 cm wavelength).

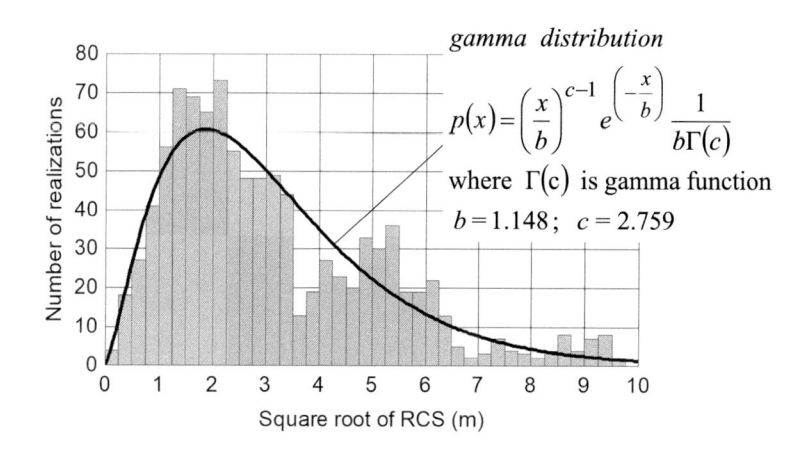

$$p(x) = \left(\frac{x}{b}\right)^{c-1} e^{\left(-\frac{x}{b}\right)} \frac{1}{b\Gamma(c)}$$

where $\Gamma(c)$ is gamma function
$b = 1.148$; $c = 2.759$

FIGURE 2.222 Amplitude distribution of echo signal of Su-24 aircraft model at a carrier frequency of 1 GHz given its horizontal polarization.

2.14 LONG-RANGE STRATEGIC BOMBER Tu-22M3

Tu-22M3 Backfire (Figure 2.223) is a long-range strategic bomber. The first flight took place in 1977.

General characteristics of Tu-22M3 [25]: wingspan –23.30–34.28 m, length – 41.46 m, height – 11.05 m, wing area – 175.80–183.57 m², weight – 78 000–126 000 kg, powerplant – 2 NK-25 afterburning turbofan engines, maximum speed – 2 000 km/h, combat range – 1 500–2 410 km, and service ceiling – 13 300 m.

In accordance with the design of the Tu-22M3, a model of its surface was created to obtaining scattering characteristics (in particular, RCS). The model is shown in Figure 2.224. In the modeling, the smooth parts of the aircraft surface were approximated by parts of 50 triaxial ellipsoids. The surface breaks were modeled using 25 straight edge scattering parts.

FIGURE 2.223 Long-range strategic bomber Tu-22M3.

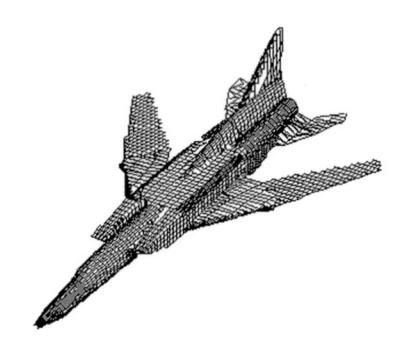

FIGURE 2.224 Surface model of long-range strategic bomber Tu-22M3.

Below are some scattering characteristics of the Tu-22M3 model at sounding frequencies of 10, 3, and 1 GHz (wavelengths of 3, 10, and 30 cm, respectively) for horizontal polarization of the probing signal. The scattering characteristics for this model at two polarizations, as well as scattering characteristics at sounding frequencies of 5 and 1.3 GHz (wavelengths of 6 and 23 cm, respectively), are given in the electronic appendix of this book.

Sounding parameters: The elevation angle is to be a random value distributed uniformly in the range $-3° \pm 4°$ with respect to the wing plane (elevation angle of $-3°$ corresponds to the radar observation from the lower hemisphere), azimuth aspect increment was 0.02°, the azimuth being counted off from the nose-on aspect (0° corresponds to the nose-on radar observation, 180° corresponds to the tail-on observation).

Scattering characteristics of Tu-22M3 aircraft model for sounding frequency 10 GHz (wavelength 3 cm).

Figure 2.225 shows the RCS circular diagram of the Tu-22M3 model. The noncoherent RCS circular diagram of the Tu-22M3 is shown in Figure 2.226.

The circular mean RCS of the Tu-22M3 aircraft model for horizontal polarization is 209.22 m². The circular median RCS (the RCS value used to calculate the detection range of an object with a probability of 0.5) for horizontal polarization is 1.66 m².

Figures 2.227 and 2.228 show the mean and median RCS for the main ranges of sounding azimuths (nose, side, and tail) and for ranges of 20°.

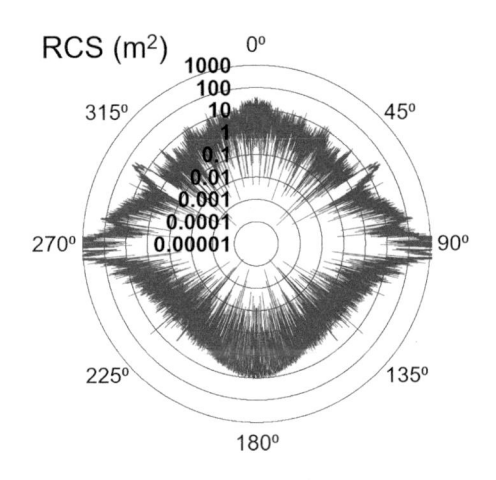

Scattering Characteristics of Aerial Objects

FIGURE 2.225 Circular diagram of RCS given radar observation of Tu-22M3 aircraft model at a carrier frequency of 10 GHz (3 cm wavelength).

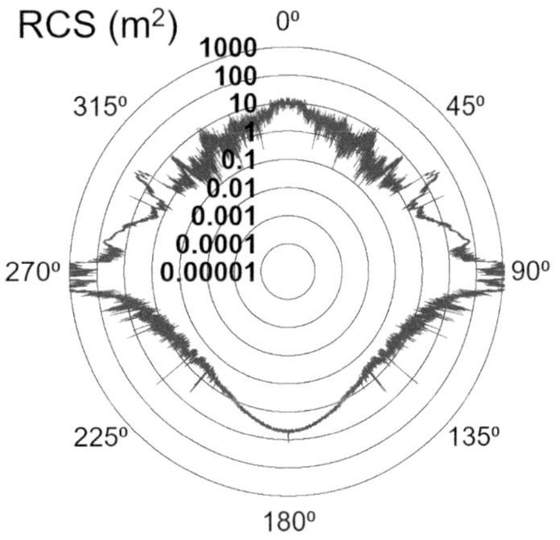

FIGURE 2.226 Circular diagram of noncoherent RCS given radar observation of Tu-22M3 aircraft model at a carrier frequency of 10 GHz (3 cm wavelength).

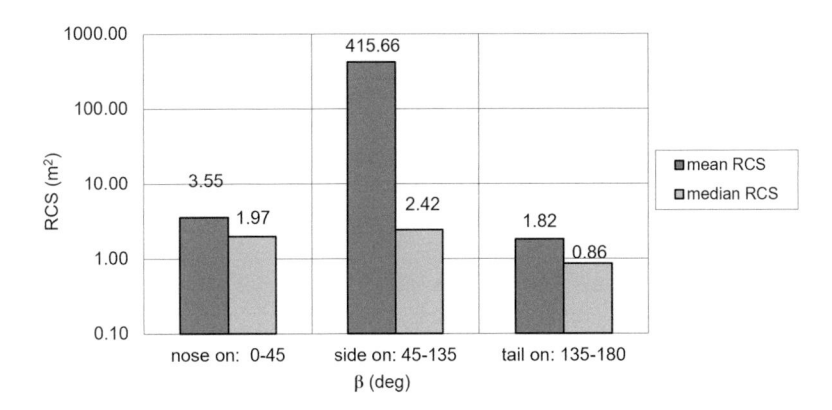

FIGURE 2.227 Diagrams of mean and median RCS of Tu-22M3 aircraft model in three sectors of azimuth aspect given its radar observation at horizontal polarization and a carrier frequency of 10 GHz (3 cm wavelength).

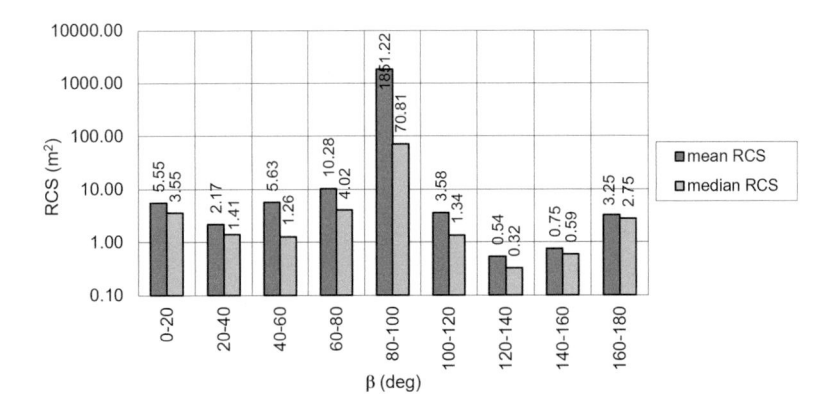

FIGURE 2.228 Diagrams of mean and median RCS of Tu-22M3 aircraft model in 20-degree sectors of azimuth aspect given its radar observation at horizontal polarization and a carrier frequency of 10 GHz (3 cm wavelength).

Figures 2.229, 2.234, and 2.239 show histograms of the scattered signal amplitude (square root of the RCS) for the range of sounding azimuths −20° to +20° (sounding from the nose). The bold line shows the probability density functions of the distribution, which can be used to approximate the histogram of the signal amplitude.

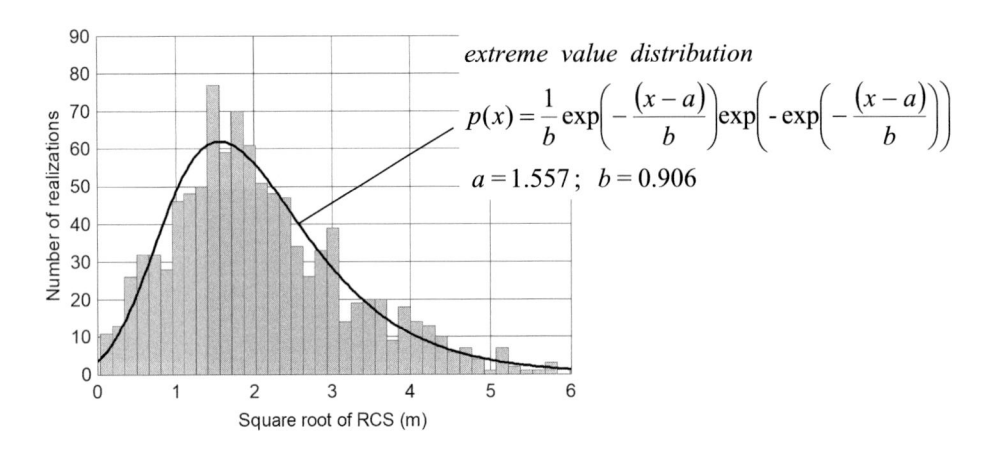

extreme value distribution

$$p(x) = \frac{1}{b} \exp\left(-\frac{(x-a)}{b}\right) \exp\left(-\exp\left(-\frac{(x-a)}{b}\right)\right)$$

$a = 1.557; \ b = 0.906$

FIGURE 2.229 Amplitude distribution of echo signal of Tu-22M3 aircraft model at a carrier frequency of 10 GHz given its horizontal polarization.

Scattering characteristics of Tu-22M3 aircraft model for sounding frequency 3 GHz (wavelength 10 cm).

Figure 2.230 shows the RCS circular diagram of the Tu-22M3. The noncoherent RCS circular diagram of the Tu-22M3 is shown in Figure 2.231.

The circular mean RCS of the Tu-22M3 aircraft for horizontal polarization is 298.22 m². The circular median RCS for horizontal polarization is 1.61 m².

Figures 2.232 and 2.233 show the mean and median RCS for the main ranges of sounding azimuths (nose, side, and tail) and for ranges of 20 degrees.

FIGURE 2.230 Circular diagram of RCS given radar observation of Tu-22M3 aircraft model at a carrier frequency of 3 GHz (10 cm wavelength).

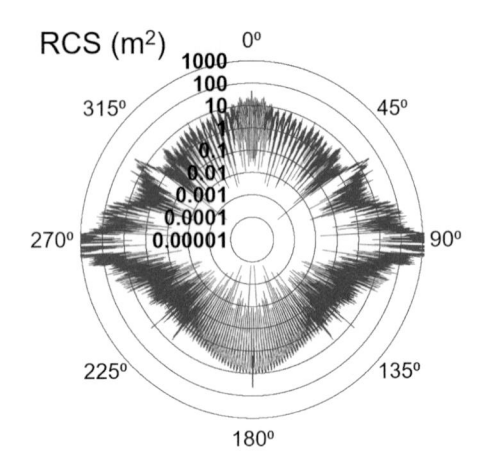

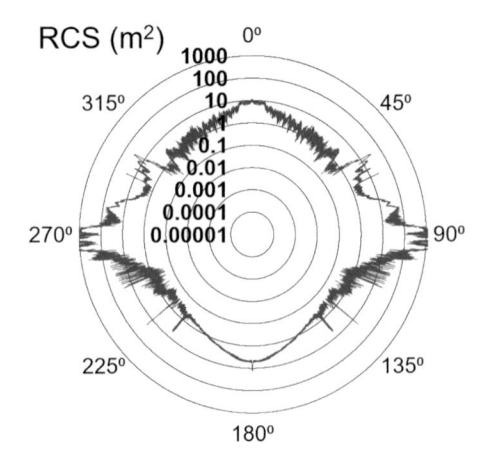

FIGURE 2.231 Circular diagram of noncoherent RCS given radar observation of Tu-22M3 aircraft model at a carrier frequency of 3 GHz (10 cm wavelength).

Scattering Characteristics of Aerial Objects 93

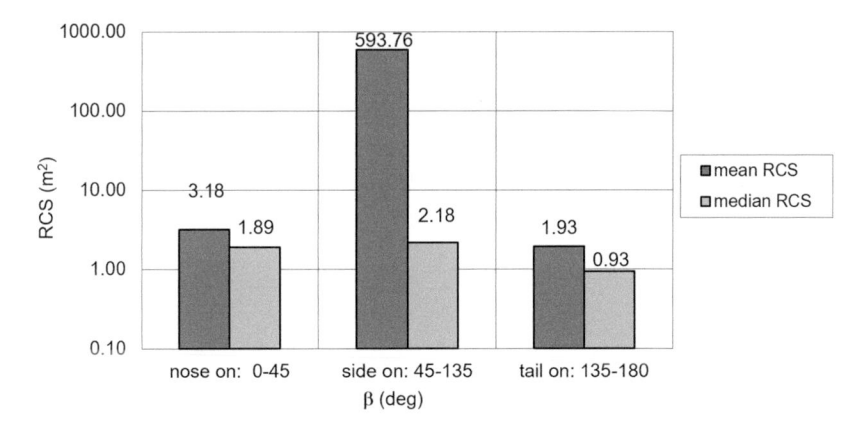

FIGURE 2.232 Diagrams of mean and median RCS of Tu-22M3 aircraft model in three sectors of azimuth aspect given its radar observation at horizontal polarization and a carrier frequency of 3 GHz (10 cm wavelength).

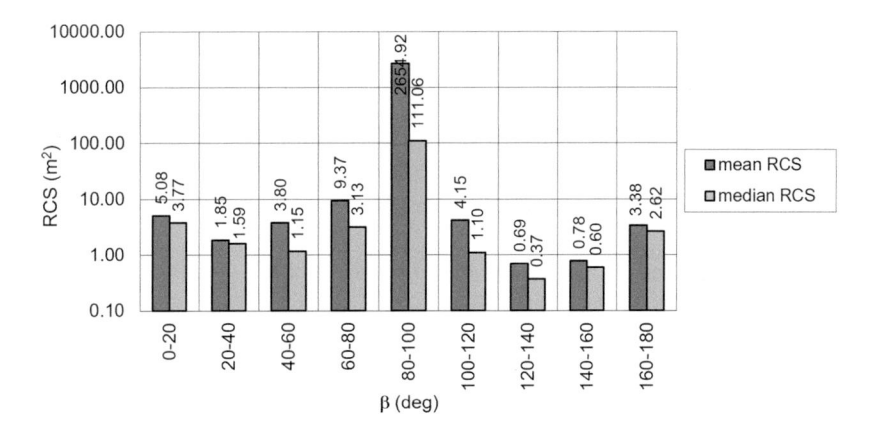

FIGURE 2.233 Diagrams of mean and median RCS of Tu-22M3 aircraft model in 20-degree sectors of azimuth aspect given its radar observation at horizontal polarization and a carrier frequency of 3 GHz (10 cm wavelength).

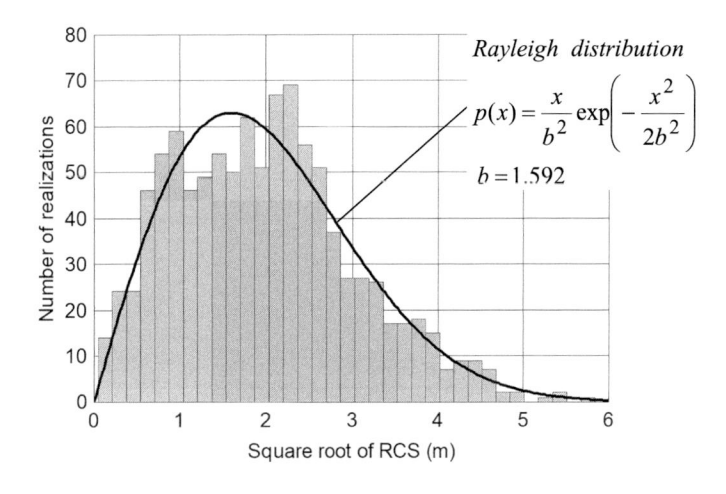

$$p(x) = \frac{x}{b^2} \exp\left(-\frac{x^2}{2b^2}\right)$$

$b = 1.592$

FIGURE 2.234 Amplitude distribution of echo signal of Tu-22M3 aircraft model at a carrier frequency of 3 GHz given its horizontal polarization.

Scattering characteristics of Tu-22M3 aircraft model for sounding frequency 1 GHz (wavelength 30 cm).

Figure 2.235 shows the RCS circular diagram of the Tu-22M3. The noncoherent RCS circular diagram of the Tu-22M3 is shown in Figure 2.236.

The circular mean RCS of the Tu-22M3 aircraft for horizontal polarization is 248.02 m². The circular median RCS for horizontal polarization is 1.88 m².

Figures 2.237 and 2.238 show the mean and median RCS for the main ranges of sounding azimuths (nose, side, and tail) and for ranges of 20°.

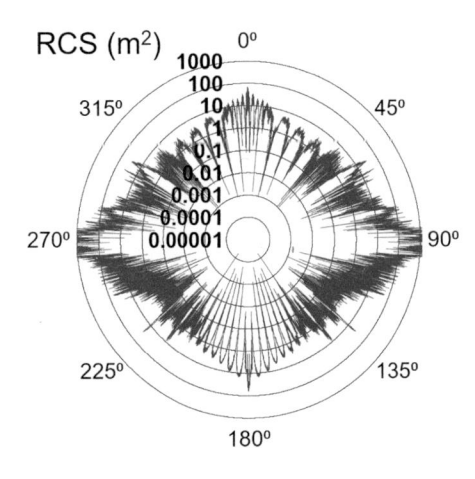

FIGURE 2.235 Circular diagram of RCS given radar observation of Tu-22M3 aircraft model at a carrier frequency of 1 GHz (30 cm wavelength).

FIGURE 2.236 Circular diagram of noncoherent RCS given radar observation of Tu-22M3 aircraft model at a carrier frequency of 1 GHz (30 cm wavelength).

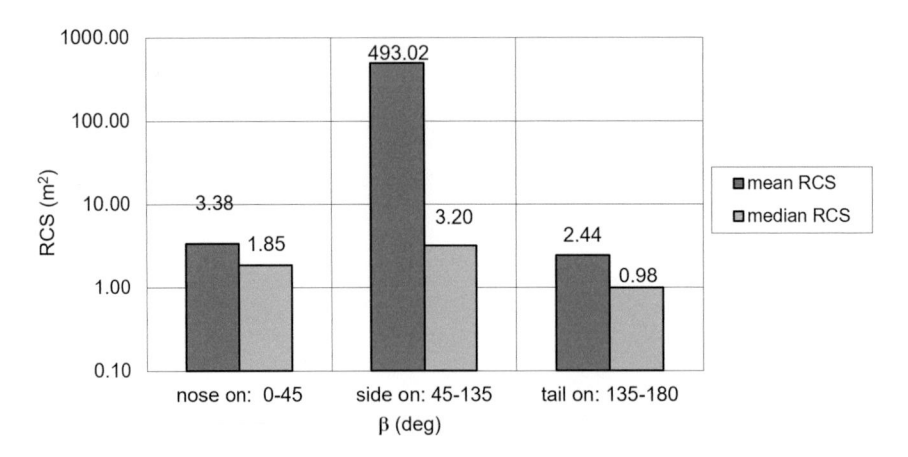

FIGURE 2.237 Diagrams of mean and median RCS of Tu-22M3 aircraft model in three sectors of azimuth aspect given its radar observation at horizontal polarization and a carrier frequency of 1 GHz (30 cm wavelength).

Scattering Characteristics of Aerial Objects

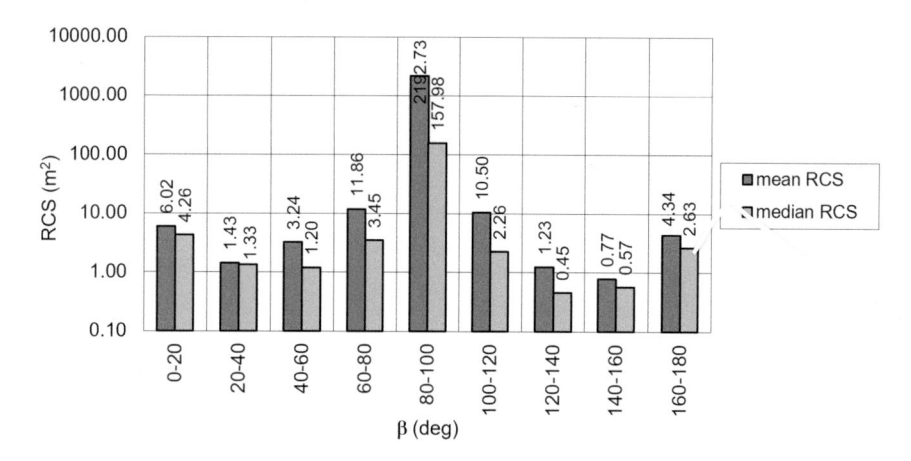

FIGURE 2.238 Diagrams of mean and median RCS of Tu-22M3 aircraft model in 20-degree sectors of azimuth aspect given its radar observation at horizontal polarization and a carrier frequency of 1 GHz (30 cm wavelength).

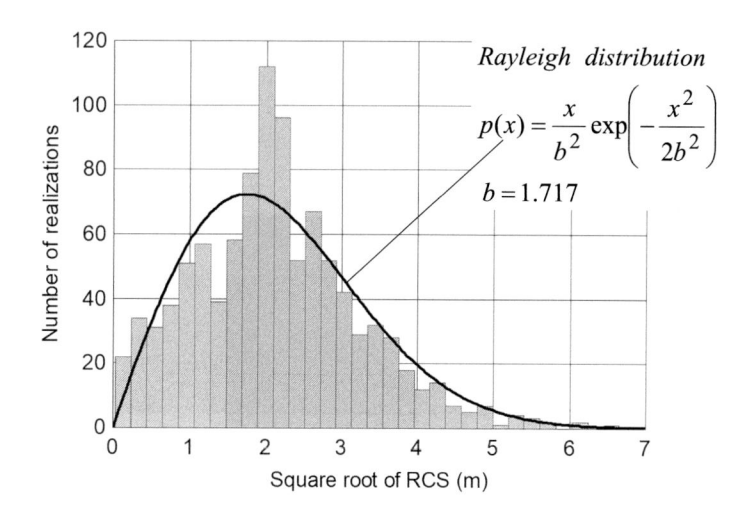

FIGURE 2.239 Amplitude distribution of echo signal of Tu-22M3 aircraft model at a carrier frequency of 1 GHz given its horizontal polarization.

$$p(x) = \frac{x}{b^2} \exp\left(-\frac{x^2}{2b^2}\right)$$

$b = 1.717$

The expressions and parameters of probability distributions that are most consistent with the empirical distributions of the square root of the RCS for different ranges of sounding azimuths and polarizations are given in the electronic appendix to this book. The expressions and parameters of the probability distributions that are most consistent with the empirical distributions of the RCS (energy characteristic) for different ranges of sounding azimuths and polarizations are also given there.

2.15 STRATEGIC BOMBER Tu-95

Tu-95 Bear (Figure 2.240) is a strategic bomber. The first flight took place in 1952.

General characteristics of Tu-95 [26]: wingspan –50.05 m, length – 47.09 m, height – 13.20 m, wing area – 295,00 m^2, weight – 94 400–187 700 kg, powerplant – 4 Kuznetsov NK-12 turboprop engines with 8-bladed contra-rotating fully feathering constant-speed propellers, maximum speed – 710 km/h, combat range – 6500 km, and service ceiling – 12 000 m.

In accordance with the design of the Tu-95, a model of its surface was created to obtaining scattering characteristics (in particular, RCS). The model is shown in Figure 2.241. In the modeling, the smooth parts of the aircraft surface were approximated by parts of 104 triaxial ellipsoids. The surface breaks were modeled using 10 straight edge scattering parts.

FIGURE 2.240 Long-range strategic bomber Tu-95.

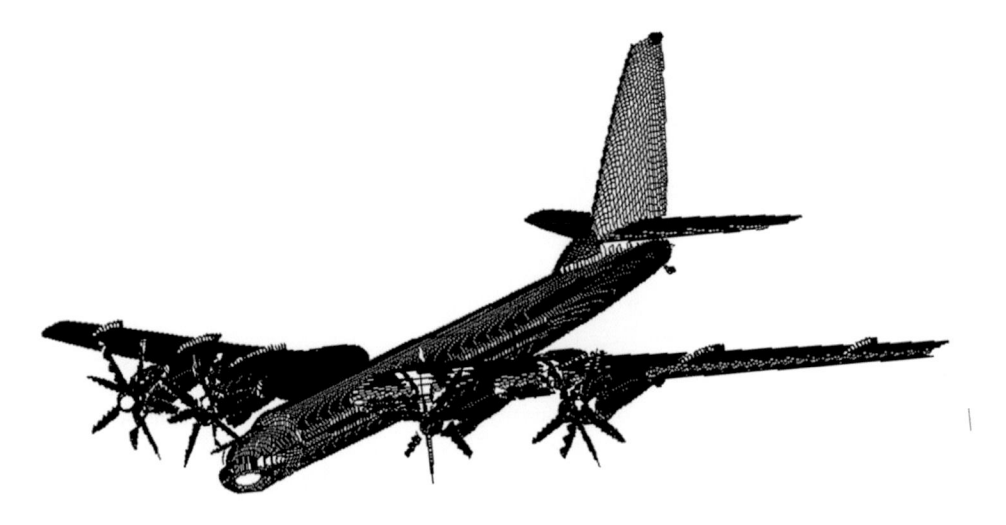

FIGURE 2.241 Surface model of long-range strategic bomber Tu-95.

Below are some scattering characteristics of the Tu-95 model at sounding frequencies of 10, 3, and 1 GHz (wavelengths of 3, 10, and 30 cm, respectively) for horizontal polarization of the probing signal. The scattering characteristics for this model at two polarizations, as well as scattering characteristics at sounding frequencies of 5 and 1.3 GHz (wavelengths of 6 and 23 cm, respectively), are given in the electronic appendix of this book.

Scattering Characteristics of Aerial Objects   97

Sounding parameters: The elevation angle is to be a random value distributed uniformly in the range −3° ± 4° with respect to the wing plane (elevation angle of −3° corresponds to the radar observation from the lower hemisphere), azimuth aspect increment was 0.02°, the azimuth being counted off from the nose-on aspect (0° corresponds to the nose-on radar observation, 180° corresponds to the tail-on observation).

Scattering characteristics of Tu-95 aircraft model for sounding frequency 10 GHz (wavelength 3 cm).

Figure 2.242 shows the RCS circular diagram of the Tu-95 model. The noncoherent RCS circular diagram of the Tu-95 is shown in Figure 2.243.

The circular mean RCS of the Tu-95 aircraft model for horizontal polarization is 238.90 m². The circular median RCS (the RCS value used to calculate the detection range of an object with a probability of 0.5) for horizontal polarization is 23.63 m².

Figures 2.244 and 2.245 show the mean and median RCS for the main ranges of sounding azimuths (nose, side, and tail) and for ranges of 20°.

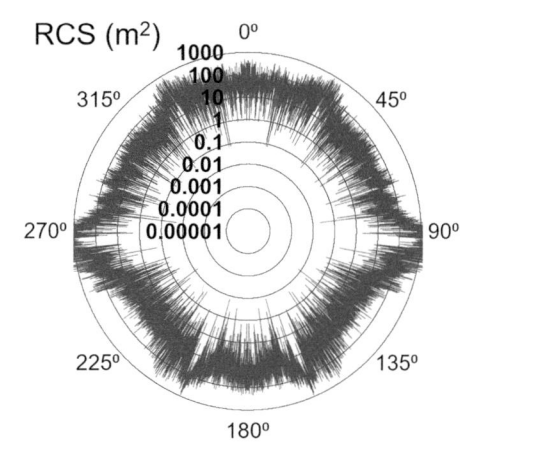

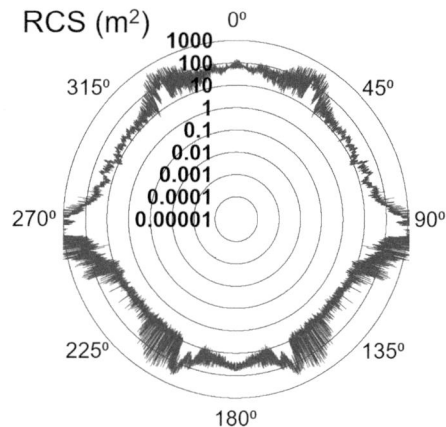

FIGURE 2.242 Circular diagram of RCS given radar observation of Tu-95 aircraft model at a carrier frequency of 10 GHz (3 cm wavelength).

FIGURE 2.243 Circular diagram of noncoherent RCS given radar observation of Tu-95 aircraft model at a carrier frequency of 10 GHz (3 cm wavelength).

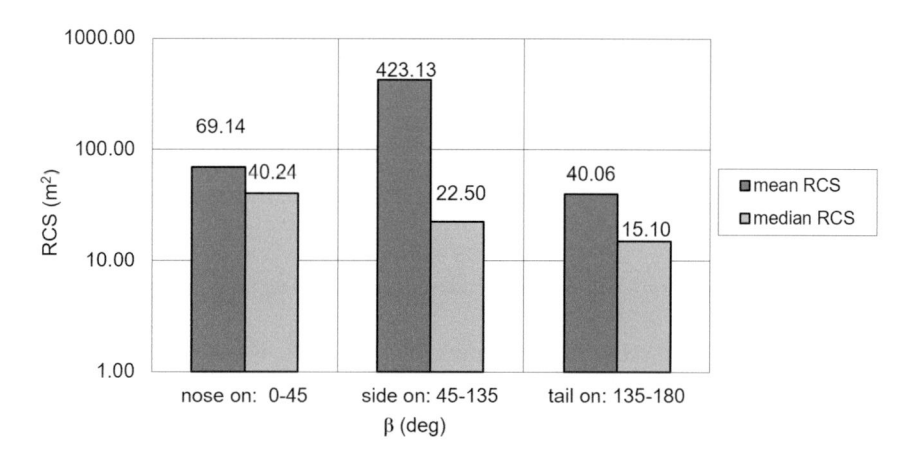

FIGURE 2.244 Diagrams of mean and median RCS of Tu-95 aircraft model in three sectors of azimuth aspect given its radar observation at horizontal polarization and a carrier frequency of 10 GHz (3 cm wavelength).

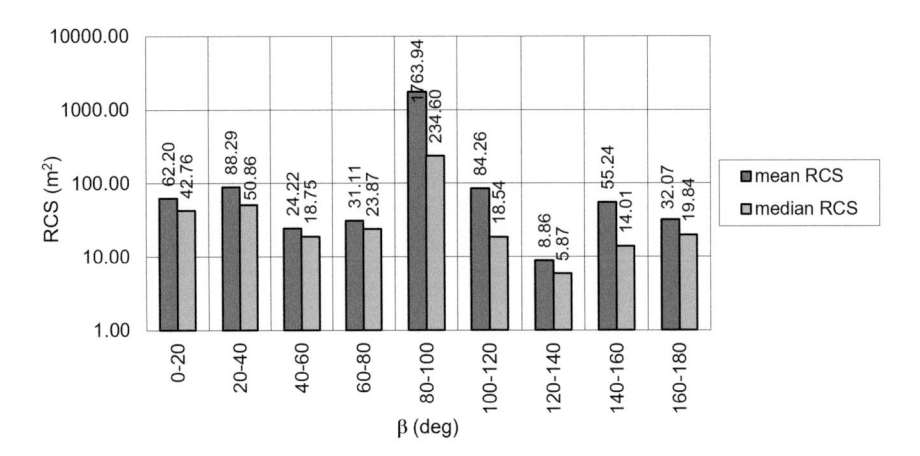

FIGURE 2.245 Diagrams of mean and median RCS of Tu-95 aircraft model in 20-degree sectors of azimuth aspect given its radar observation at horizontal polarization and a carrier frequency of 10 GHz (3 cm wavelength).

Figures 2.246, 2.251, and 2.256 show histograms of the scattered signal amplitude (square root of the RCS) for the range of sounding azimuths −20° to +20° (sounding from the nose). The bold line shows the probability density functions of the distribution, which can be used to approximate the histogram of the signal amplitude.

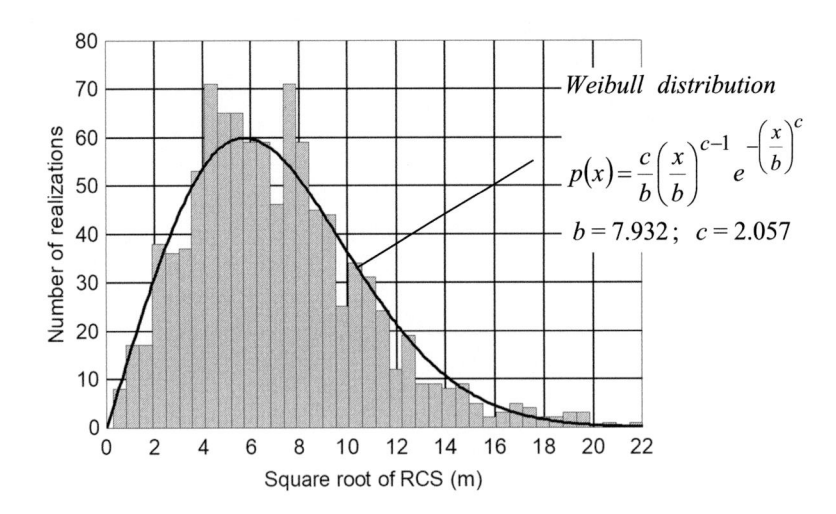

$$p(x) = \frac{c}{b}\left(\frac{x}{b}\right)^{c-1} e^{-\left(\frac{x}{b}\right)^c}$$

$b = 7.932; \quad c = 2.057$

FIGURE 2.246 Amplitude distribution of echo signal of Tu-95 aircraft model at a carrier frequency of 10 GHz given its horizontal polarization.

Scattering characteristics of Tu-95 aircraft model for sounding frequency 3 GHz (wavelength 10 cm).

Figure 2.247 shows the RCS circular diagram of the Tu-95. The noncoherent RCS circular diagram of the Tu-95 is shown in Figure 2.248.

The circular mean RCS of the Tu-95 aircraft for horizontal polarization is 199.83 m². The circular median RCS for horizontal polarization is 18.23 m².

Figures 2.249 and 2.250 show the mean and median RCS for the main ranges of sounding azimuths (nose, side, and tail) and for ranges of 20°.

Scattering Characteristics of Aerial Objects

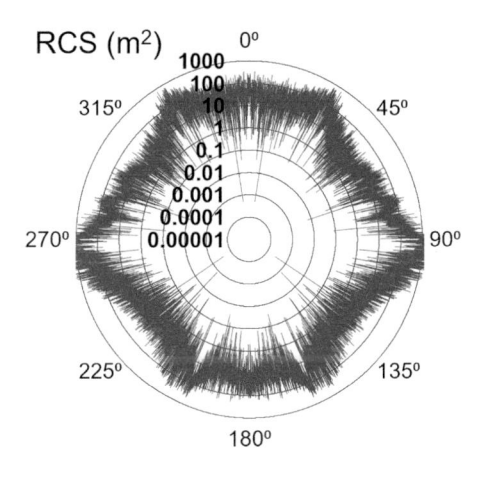

FIGURE 2.247 Circular diagram of RCS given radar observation of Tu-95 aircraft model at a carrier frequency of 3 GHz (10 cm wavelength).

FIGURE 2.248 Circular diagram of noncoherent RCS given radar observation of Tu-95 aircraft model at a carrier frequency of 3 GHz (10 cm wavelength).

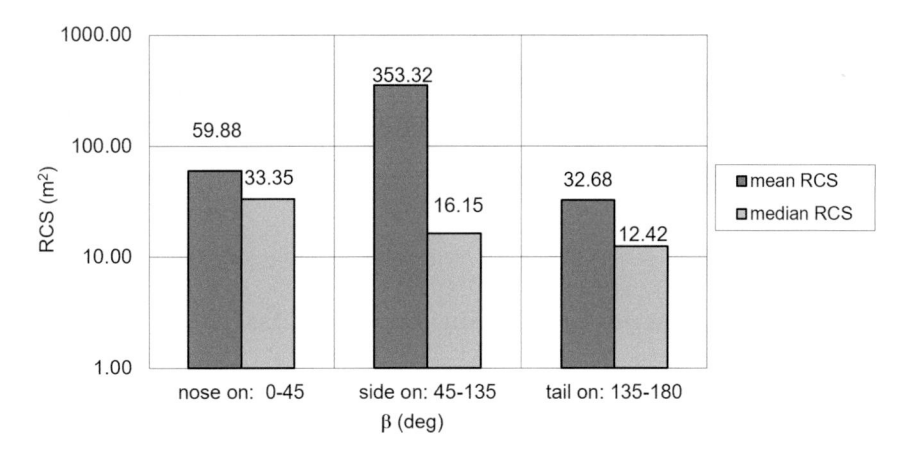

FIGURE 2.249 Diagrams of mean and median RCS of Tu-95 aircraft model in three sectors of azimuth aspect given its radar observation at horizontal polarization and a carrier frequency of 3 GHz (10 cm wavelength).

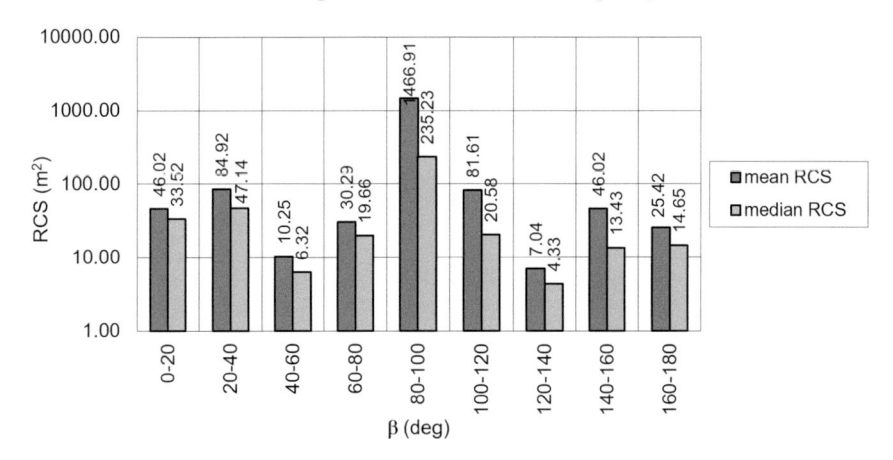

FIGURE 2.250 Diagrams of mean and median RCS of Tu-95 aircraft model in 20-degree sectors of azimuth aspect given its radar observation at horizontal polarization and a carrier frequency of 3 GHz (10 cm wavelength).

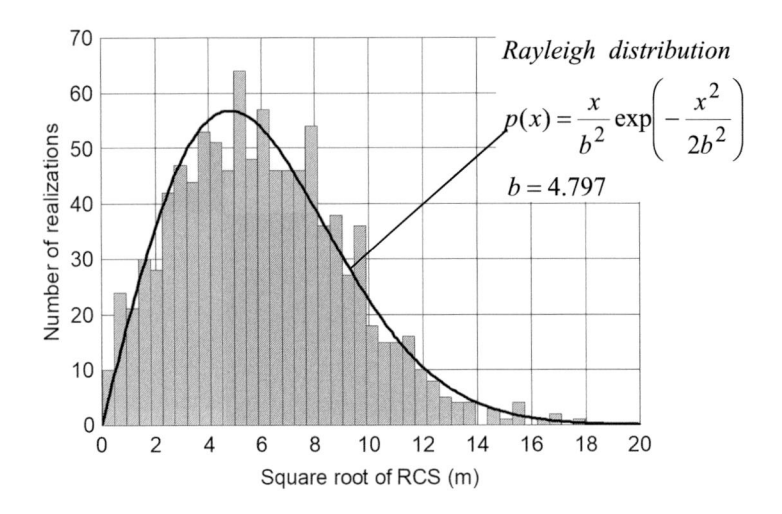

FIGURE 2.251 Amplitude distribution of echo signal of Tu-95 aircraft model at a carrier frequency of 3 GHz given its horizontal polarization.

Rayleigh distribution

$$p(x) = \frac{x}{b^2} \exp\left(-\frac{x^2}{2b^2}\right)$$

$b = 4.797$

Scattering characteristics of Tu-95 aircraft model for sounding frequency 1 GHz (wavelength 30 cm).

Figure 2.252 shows the RCS circular diagram of the Tu-95. The noncoherent RCS circular diagram of the Tu-95 is shown in Figure 2.253.

The circular mean RCS of the Tu-95 aircraft for horizontal polarization is 176.86 m². The circular median RCS for horizontal polarization is 20.96 m².

Figures 2.254 and 2.255 show the mean and median RCS for the main ranges of sounding azimuths (nose, side, and tail) and for ranges of 20°.

The expressions and parameters of probability distributions that are most consistent with the empirical distributions of the square root of the RCS for different ranges of sounding azimuths and polarizations are given in the electronic appendix to this book. The expressions and parameters of the probability distributions that are most consistent with the empirical distributions of the RCS (energy characteristic) for different ranges of sounding azimuths and polarizations are also given there.

FIGURE 2.252 Circular diagram of RCS given radar observation of Tu-95 aircraft model at a carrier frequency of 1 GHz (30 cm wavelength).

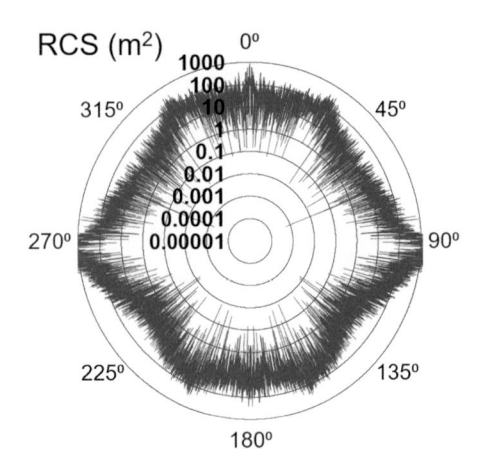

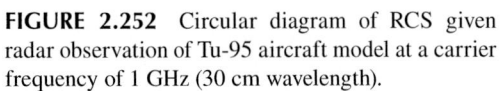

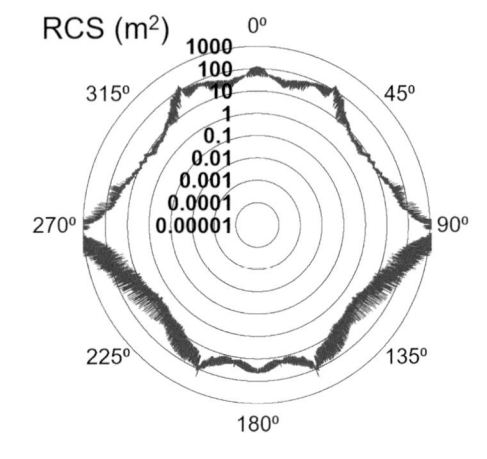

FIGURE 2.253 Circular diagram of noncoherent RCS given radar observation of Tu-95 aircraft model at a carrier frequency of 1 GHz (30 cm wavelength).

Scattering Characteristics of Aerial Objects 101

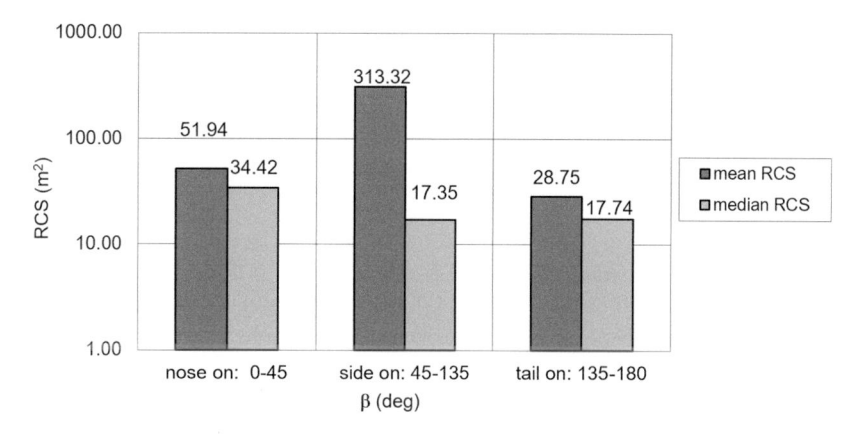

FIGURE 2.254 Diagrams of mean and median RCS of Tu-95 aircraft model in three sectors of azimuth aspect given its radar observation at horizontal polarization and a carrier frequency of 1 GHz (30 cm wavelength).

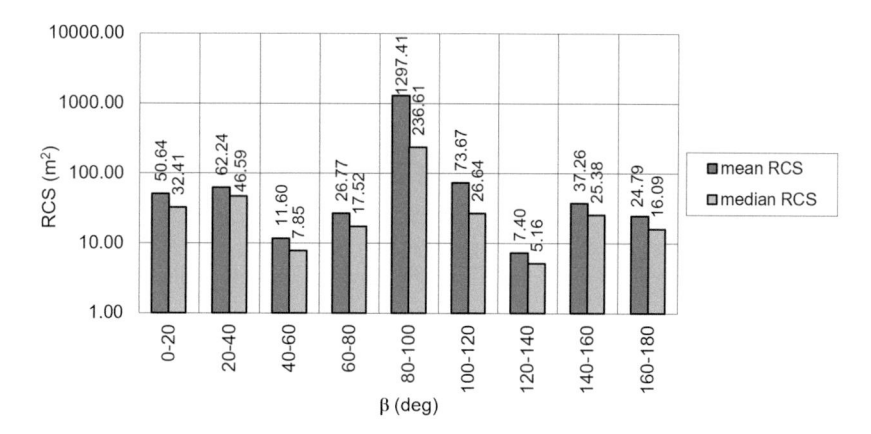

FIGURE 2.255 Diagrams of mean and median RCS of Tu-95 aircraft model in 20-degree sectors of azimuth aspect given its radar observation at horizontal polarization and a carrier frequency of 1 GHz (30 cm wavelength).

$$p(x) = \frac{c}{b}\left(\frac{x}{b}\right)^{c-1} e^{-\left(\frac{x}{b}\right)^c}$$

Weibull distribution; $b = 6.900$; $c = 1.779$

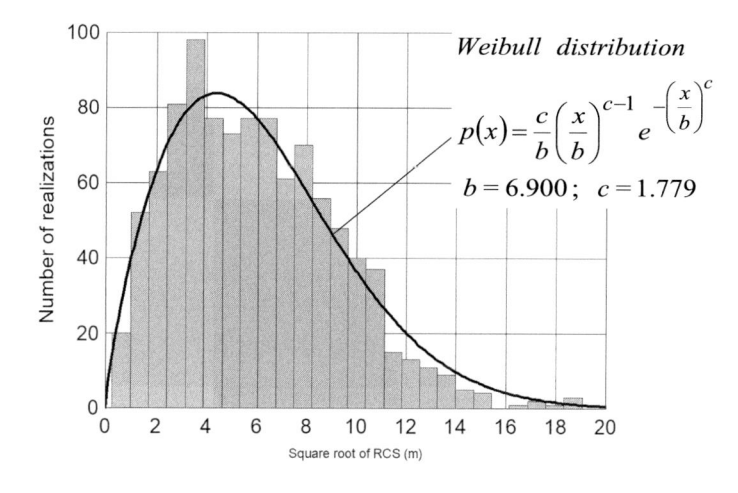

FIGURE 2.256 Amplitude distribution of echo signal of Tu-95 aircraft model at a carrier frequency of 1 GHz given its horizontal polarization.

2.16 STRATEGIC BOMBER Tu-160

Tu-160 Blackjack (Figure 2.257) is a strategic bomber. The first flight took place in 1984.

General characteristics of Tu-160 [22]: wingspan – 35.60–55.70 m, length – 54.10 m, height – 13.20 m, wing area – 360.00–400.00 m^2, weight – 110 000–275 700 kg, powerplant – 4 NK-321 afterburning turbofan engines, maximum speed – 2200 km/h, range – 10 500–14 000 km, and service ceiling – 15 600 m.

In accordance with the design of the Tu-160, a model of its surface was created to obtaining scattering characteristics (in particular, RCS). The model is shown in Figure 2.258. In the modeling, the smooth parts of the aircraft surface were approximated by parts of 103 triaxial ellipsoids. The surface breaks were modeled using 20 straight edge scattering parts.

FIGURE 2.257 Long-range strategic bomber Tu-160.

FIGURE 2.258 Surface model of long-range strategic bomber Tu-160.

Below are some scattering characteristics of the Tu-160 model at sounding frequencies of 10, 3, and 1 GHz (wavelengths of 3, 10, and 30 cm, respectively) for horizontal polarization of the probing signal. The scattering characteristics for this model at two polarizations, as well as scattering characteristics at sounding frequencies of 5 and 1.3 GHz (wavelengths of 6 and 23 cm, respectively), are given in the electronic appendix of this book.

Sounding parameters: The elevation angle is to be a random value distributed uniformly in the range −3° ± 4° with respect to the wing plane (elevation angle of −3° corresponds to the radar observation from the lower hemisphere), azimuth aspect increment was 0.02°, the azimuth being counted off from the nose-on aspect (0° corresponds to the nose-on radar observation, 180° corresponds to the tail-on observation).

Scattering characteristics of Tu-160 aircraft model for sounding frequency 10 GHz (wavelength 3 cm).

Figure 2.259 shows the RCS circular diagram of the Tu-160 model. The noncoherent RCS circular diagram of the Tu-160 is shown in Figure 2.260.

The circular mean RCS of the Tu-160 aircraft model for horizontal polarization is 460.03 m². The circular median RCS (the RCS value used to calculate the detection range of an object with a probability of 0.5) for horizontal polarization is 38.52 m².

Figures 2.261 and 2.262 show the mean and median RCS for the main ranges of sounding azimuths (nose, side, and tail) and for ranges of 20°.

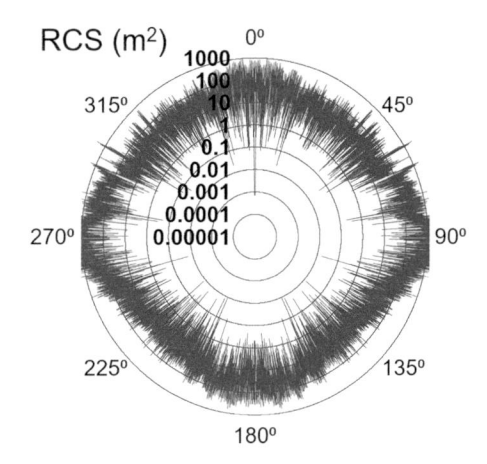

FIGURE 2.259 Circular diagram of RCS given radar observation of Tu-160 aircraft model at a carrier frequency of 10 GHz (3 cm wavelength).

FIGURE 2.260 Circular diagram of noncoherent RCS given radar observation of Tu-160 aircraft model at a carrier frequency of 10 GHz (3 cm wavelength).

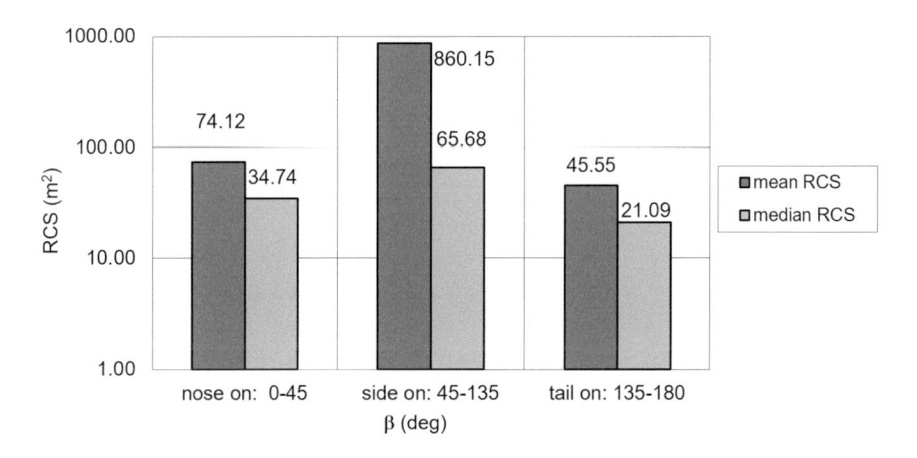

FIGURE 2.261 Diagrams of mean and median RCS of Tu-160 aircraft model in three sectors of azimuth aspect given its radar observation at horizontal polarization and a carrier frequency of 10 GHz (3 cm wavelength).

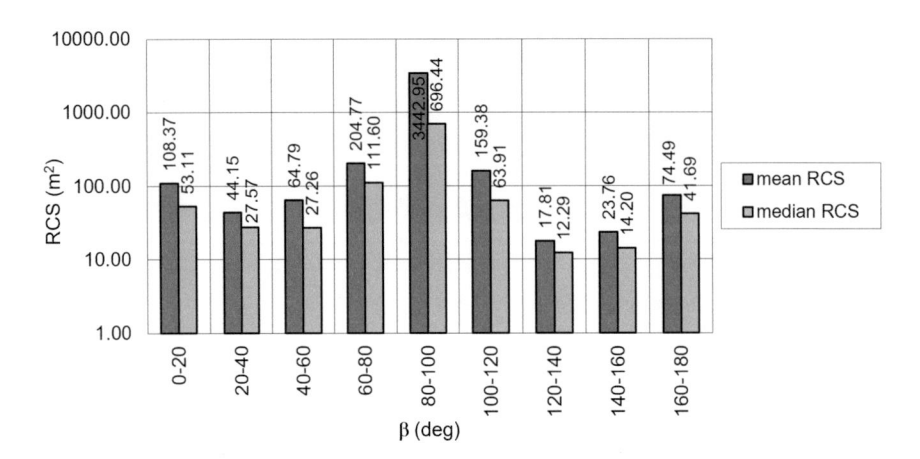

FIGURE 2.262 Diagrams of mean and median RCS of Tu-160 aircraft model in 20-degree sectors of azimuth aspect given its radar observation at horizontal polarization and a carrier frequency of 10 GHz (3 cm wavelength).

Figures 2.263, 2.268, and 2.273 show histograms of the scattered signal amplitude (square root of the RCS) for the range of sounding azimuths −20° to +20° (sounding from the nose). The bold line shows the probability density functions of the distribution, which can be used to approximate the histogram of the signal amplitude.

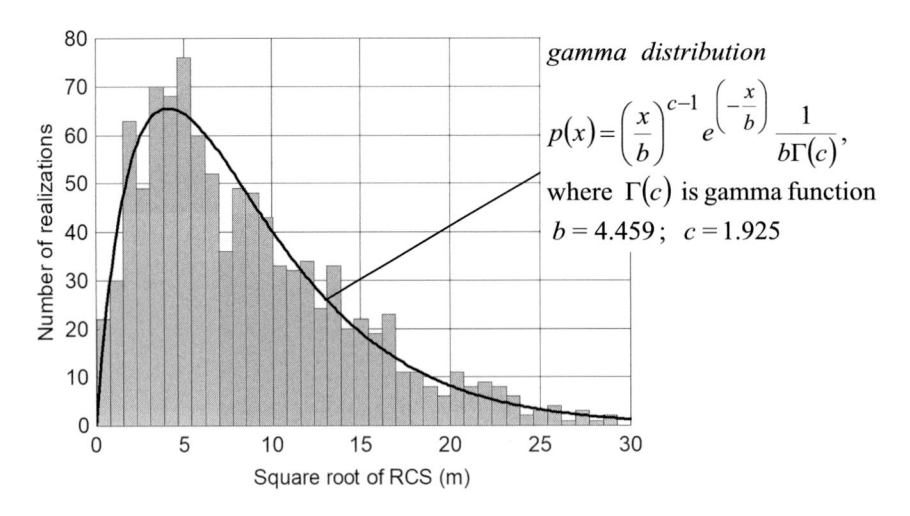

gamma distribution

$$p(x) = \left(\frac{x}{b}\right)^{c-1} e^{\left(-\frac{x}{b}\right)} \frac{1}{b\Gamma(c)},$$

where $\Gamma(c)$ is gamma function
$b = 4.459;\ c = 1.925$

FIGURE 2.263 Amplitude distribution of echo signal of Tu-160 aircraft model at a carrier frequency of 10 GHz given its horizontal polarization.

Scattering characteristics of Tu-160 aircraft model for sounding frequency 3 GHz (wavelength 10 cm).

Figure 2.264 shows the RCS circular diagram of the Tu-160. The noncoherent RCS circular diagram of the Tu-160 is shown in Figure 2.265.

The circular mean RCS of the Tu-160 aircraft for horizontal polarization is 377.92 m². The circular median RCS for horizontal polarization is 32.01 m².

Figures 2.266 and 2.267 show the mean and median RCS for the main ranges of sounding azimuths (nose, side, and tail) and for ranges of 20°.

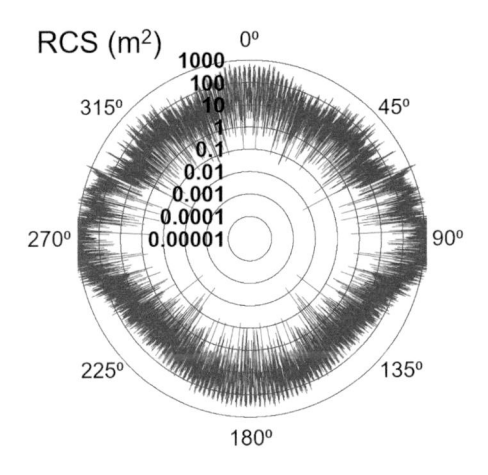

Scattering Characteristics of Aerial Objects

FIGURE 2.264 Circular diagram of RCS given radar observation of Tu-160 aircraft model at a carrier frequency of 3 GHz (10 cm wavelength).

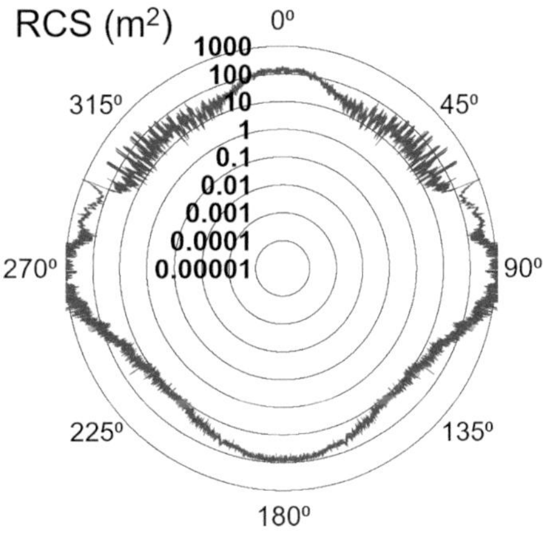

FIGURE 2.265 Circular diagram of noncoherent RCS given radar observation of Tu-160 aircraft model at a carrier frequency of 3 GHz (10 cm wavelength).

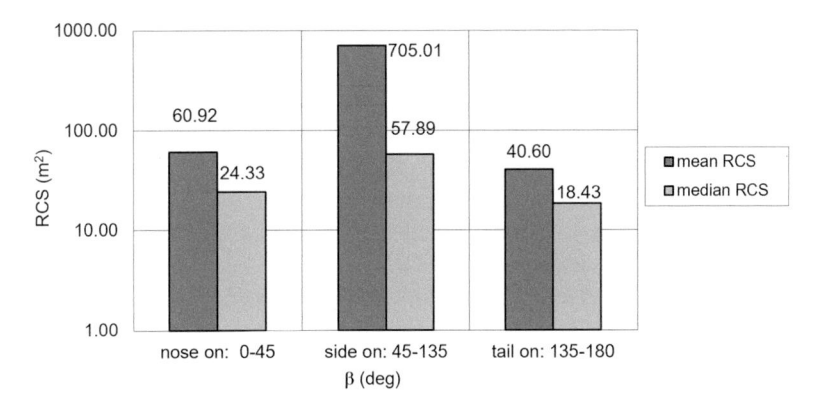

FIGURE 2.266 Diagrams of mean and median RCS of Tu-160 aircraft model in three sectors of azimuth aspect given its radar observation at horizontal polarization and a carrier frequency of 3 GHz (10 cm wavelength).

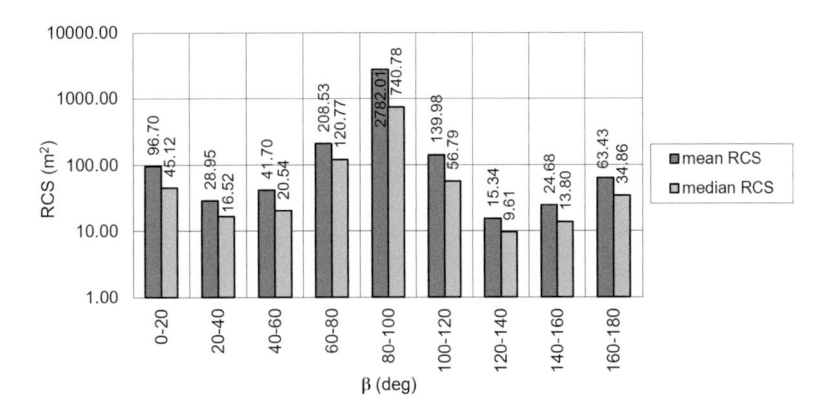

FIGURE 2.267 Diagrams of mean and median RCS of Tu-160 aircraft model in 20-degree sectors of azimuth aspect given its radar observation at horizontal polarization and a carrier frequency of 3 GHz (10 cm wavelength).

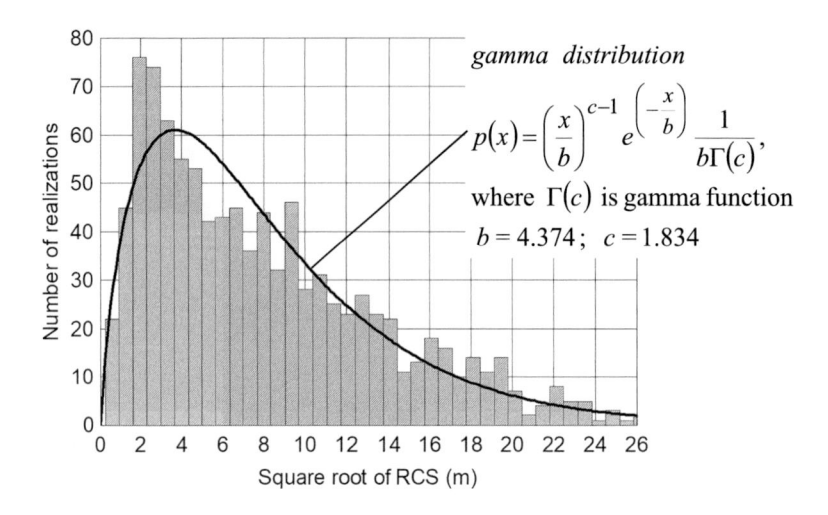

FIGURE 2.268 Amplitude distribution of echo signal of Tu-160 aircraft model at a carrier frequency of 3 GHz given its horizontal polarization.

Scattering characteristics of Tu-160 aircraft model for sounding frequency 1 GHz (wavelength 30 cm).

Figure 2.269 shows the RCS circular diagram of the Tu-160. The noncoherent RCS circular diagram of the Tu-160 is shown in Figure 2.270.

The circular mean RCS of the Tu-160 aircraft for horizontal polarization is 381.23 m². The circular median RCS for horizontal polarization is 36.42 m².

Figures 2.271 and 2.272 show the mean and median RCS for the main ranges of sounding azimuths (nose, side, and tail) and for ranges of 20°.

The expressions and parameters of probability distributions that are most consistent with the empirical distributions of the square root of the RCS for different ranges of sounding azimuths and polarizations are given in the electronic appendix to this book. The expressions and parameters of the probability distributions that are most consistent with the empirical distributions of the RCS (energy characteristic) for different ranges of sounding azimuths and polarizations are also given there.

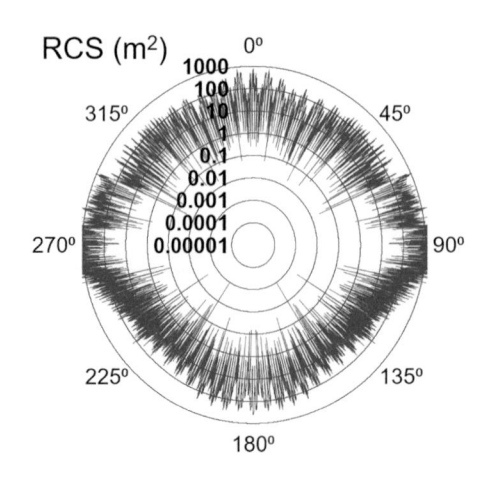

FIGURE 2.269 Circular diagram of RCS given radar observation of Tu-160 aircraft model at a carrier frequency of 1 GHz (30 cm wavelength).

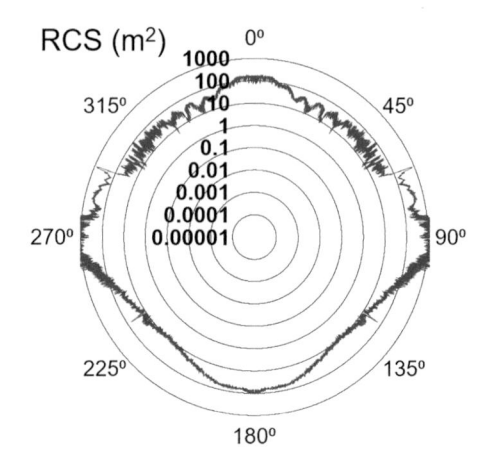

FIGURE 2.270 Circular diagram of noncoherent RCS given radar observation of Tu-160 aircraft model at a carrier frequency of 1 GHz (30 cm wavelength).

In the figure inset:

gamma distribution

$$p(x) = \left(\frac{x}{b}\right)^{c-1} e^{\left(-\frac{x}{b}\right)} \frac{1}{b\Gamma(c)},$$

where $\Gamma(c)$ is gamma function

$b = 4.374;\ c = 1.834$

Scattering Characteristics of Aerial Objects

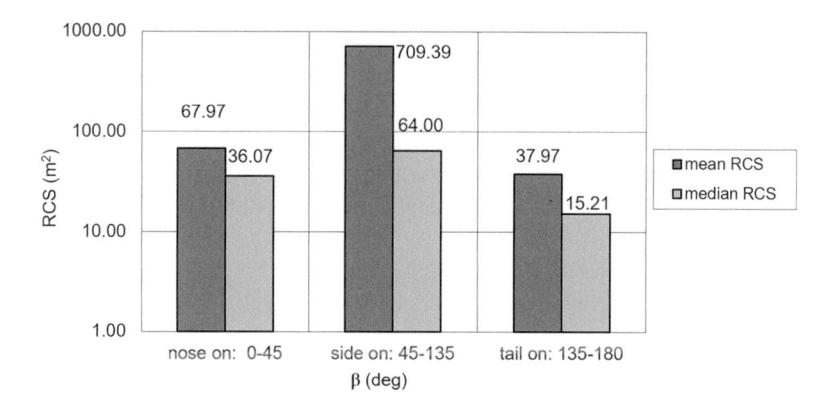

FIGURE 2.271 Diagrams of mean and median RCS of Tu-160 aircraft model in three sectors of azimuth aspect given its radar observation at horizontal polarization and a carrier frequency of 1 GHz (30 cm wavelength).

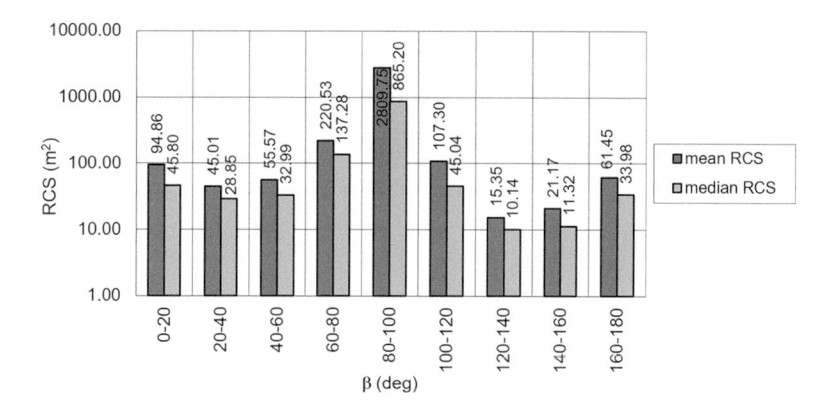

FIGURE 2.272 Diagrams of mean and median RCS of Tu-160 aircraft model in 20-degree sectors of azimuth aspect given its radar observation at horizontal polarization and a carrier frequency of 1 GHz (30 cm wavelength).

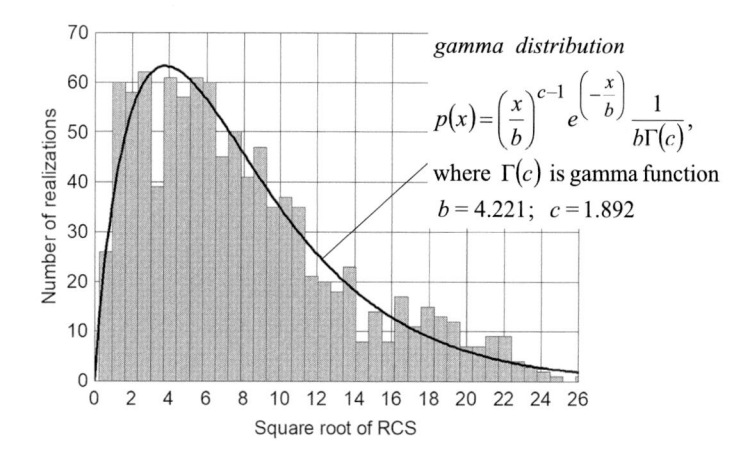

gamma distribution

$$p(x) = \left(\frac{x}{b}\right)^{c-1} e^{\left(-\frac{x}{b}\right)} \frac{1}{b\Gamma(c)},$$

where $\Gamma(c)$ is gamma function

$b = 4.221;\ c = 1.892$

FIGURE 2.273 Amplitude distribution of echo signal of Tu-160 aircraft model at a carrier frequency of 1 GHz given its horizontal polarization.

2.17 TACTICAL BOMBER Su-34

Su-34 Fullback (Figure 2.274) is an all-weather tactical bomber. The first flight took place in 1990.

General characteristics of Su-34 [27]: wingspan –14.70 m, length – 23.34 m, height – 6.09 m, wing area – 62.00 m^2, weight – 39 000–44 360 kg, powerplant – 2 Saturn AL31FM1 afterburning turbofan engines, maximum speed – 1 900 km/h, range – 4 500 km, and service ceiling – 17 000 m.

In accordance with the design of the Su-34, a model of its surface was created to obtaining scattering characteristics (in particular, RCS). The model is shown in Figure 2.275. In the modeling, the smooth parts of the aircraft surface were approximated by parts of 56 triaxial ellipsoids. The surface breaks were modeled using 48 straight edge scattering parts.

FIGURE 2.274 Tactical bomber Su-34.

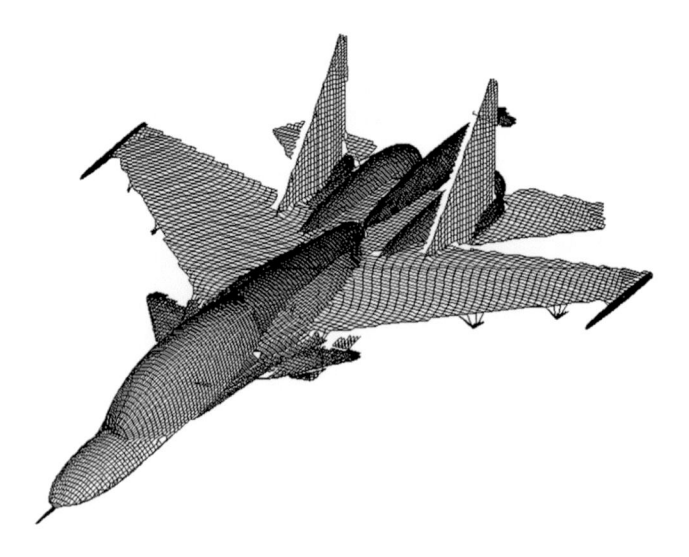

FIGURE 2.275 Surface model of tactical bomber Su-34.

Below are some scattering characteristics of the Su-34 model at sounding frequencies of 10, 3, and 1 GHz (wavelengths of 3, 10, and 30 cm, respectively) for horizontal polarization of the probing signal. The scattering characteristics for this model at two polarizations, as well as scattering characteristics at sounding frequencies of 5 and 1.3 GHz (wavelengths of 6 and 23 cm, respectively), are given in the electronic appendix of this book.

Scattering Characteristics of Aerial Objects

Sounding parameters: The elevation angle is to be a random value distributed uniformly in the range −3° ± 4° with respect to the wing plane (elevation angle of −3° corresponds to the radar observation from the lower hemisphere), azimuth aspect increment was 0.02°, the azimuth being counted off from the nose-on aspect (0° corresponds to the nose-on radar observation, 180° corresponds to the tail-on observation).

Scattering characteristics of Su-34 aircraft model for sounding frequency 10 GHz (wavelength 3 cm).

Figure 2.276 shows the RCS circular diagram of the Su-34 model. The noncoherent RCS circular diagram of the Su-34 is shown in Figure 2.277.

The circular mean RCS of the Su-34 aircraft model for horizontal polarization is 67.81 m². The circular median RCS (the RCS value used to calculate the detection range of an object with a probability of 0.5) for horizontal polarization is 2.00 m².

Figures 2.278 and 2.279 show the mean and median RCS for the main ranges of sounding azimuths (nose, side, and tail) and for ranges of 20°.

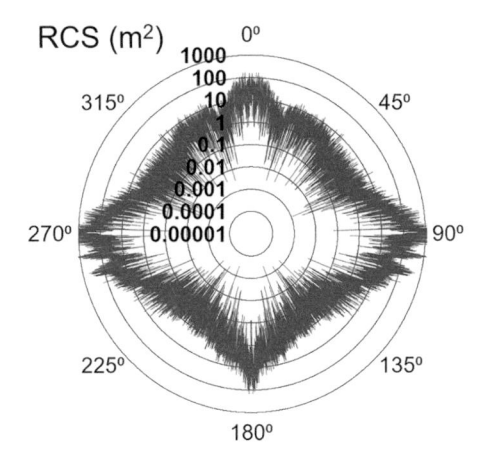

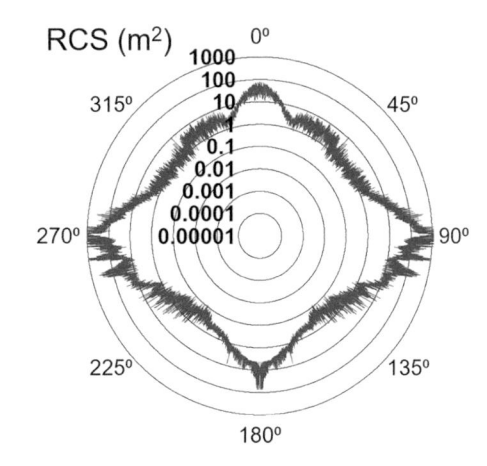

FIGURE 2.276 Circular diagram of RCS given radar observation of Su-34 aircraft model at a carrier frequency of 10 GHz (3 cm wavelength).

FIGURE 2.277 Circular diagram of noncoherent RCS given radar observation of Su-34 aircraft model at a carrier frequency of 10 GHz (3 cm wavelength).

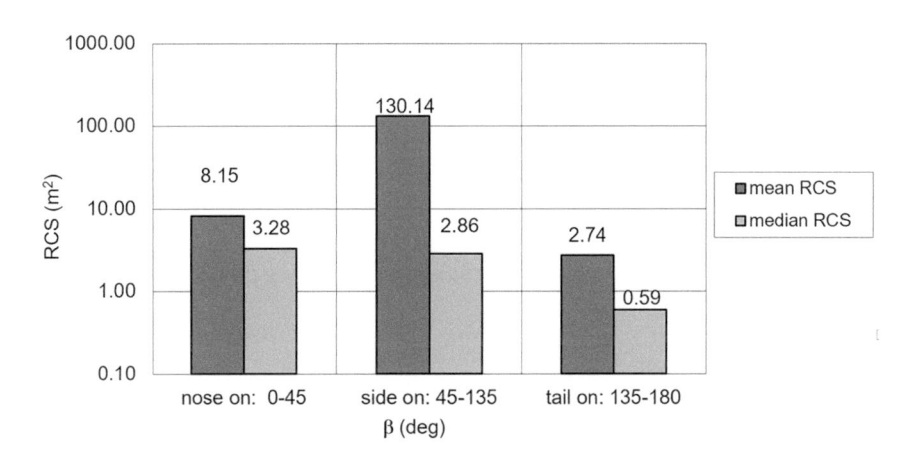

FIGURE 2.278 Diagrams of mean and median RCS of Su-34 aircraft model in three sectors of azimuth aspect given its radar observation at horizontal polarization and a carrier frequency of 10 GHz (3 cm wavelength).

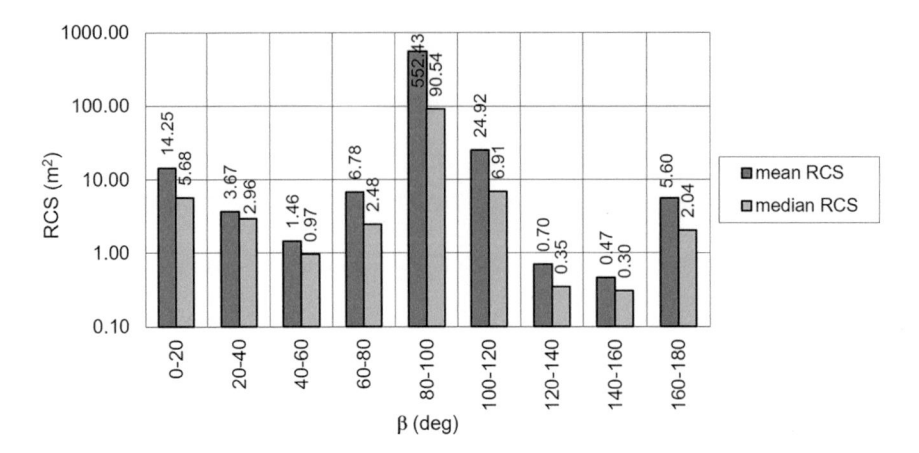

FIGURE 2.279 Diagrams of mean and median RCS of Su-34 aircraft model in 20-degree sectors of azimuth aspect given its radar observation at horizontal polarization and a carrier frequency of 10 GHz (3 cm wavelength).

Figures 2.280, 2.285, and 2.290 show histograms of the scattered signal amplitude (square root of the RCS) for the range of sounding azimuths −20° to +20° (sounding from the nose). The bold line shows the probability density functions of the distribution, which can be used to approximate the histogram of the signal amplitude.

$$p(x) = \frac{1}{\sqrt{2\pi}\, x\sigma} \exp\left(-\frac{(\log(x)-\mu)^2}{2\sigma^2}\right)$$

$\mu = 0.829;\ \sigma = 0.804$

log - normal distribution

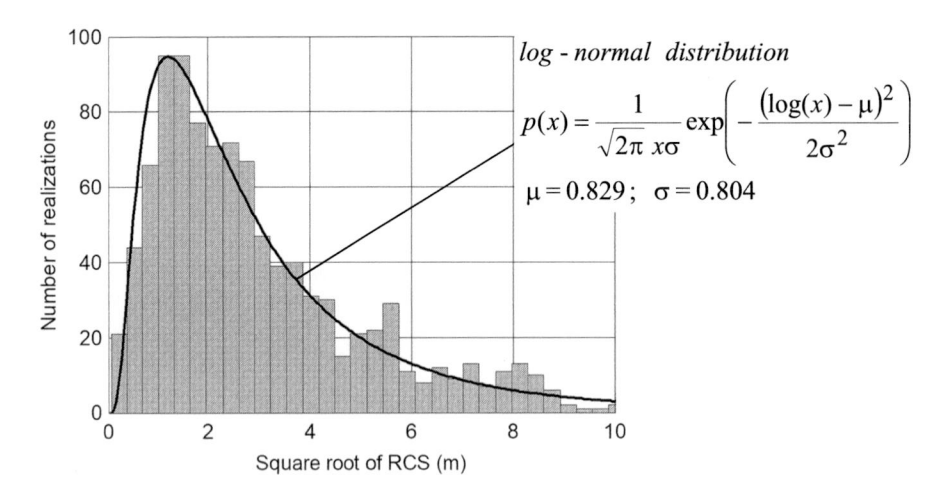

FIGURE 2.280 Amplitude distribution of echo signal of Su-34 aircraft model at a carrier frequency of 10 GHz given its horizontal polarization.

Scattering characteristics of Su-34 aircraft model for sounding frequency 3 GHz (wavelength 10 cm).

Figure 2.281 shows the RCS circular diagram of the Su-34. The noncoherent RCS circular diagram of the Su-34 is shown in Figure 2.282.

The circular mean RCS of the Su-34 aircraft for horizontal polarization is 57.43 m². The circular median RCS for horizontal polarization is 2.81 m².

Figures 2.283 and 2.284 show the mean and median RCS for the main ranges of sounding azimuths (nose, side, and tail) and for ranges of 20°.

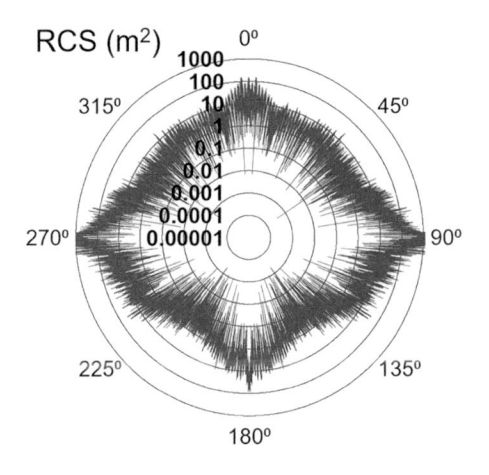

Scattering Characteristics of Aerial Objects 111

FIGURE 2.281 Circular diagram of RCS given radar observation of Su-34 aircraft model at a carrier frequency of 3 GHz (10 cm wavelength).

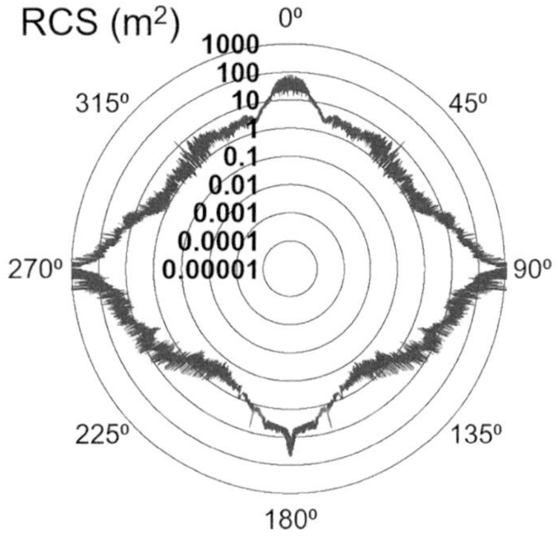

FIGURE 2.282 Circular diagram of noncoherent RCS given radar observation of Su-34 aircraft model at a carrier frequency of 3 GHz (10 cm wavelength).

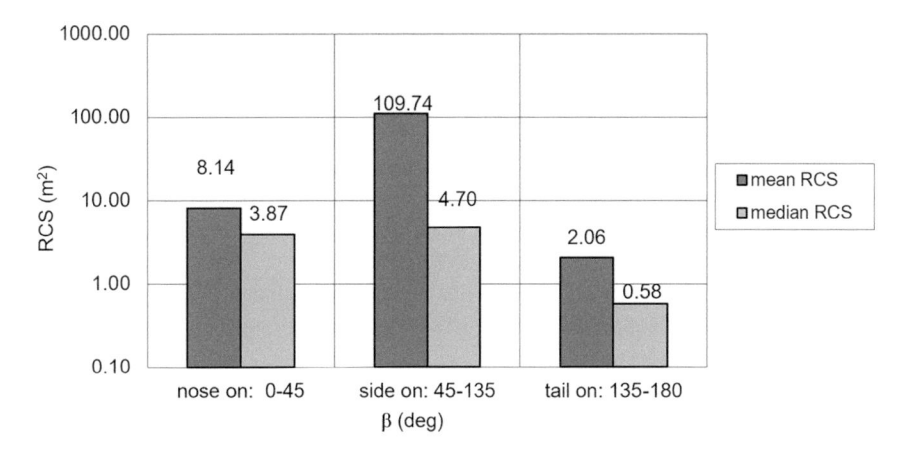

FIGURE 2.283 Diagrams of mean and median RCS of Su-34 aircraft model in three sectors of azimuth aspect given its radar observation at horizontal polarization and a carrier frequency of 3 GHz (10 cm wavelength).

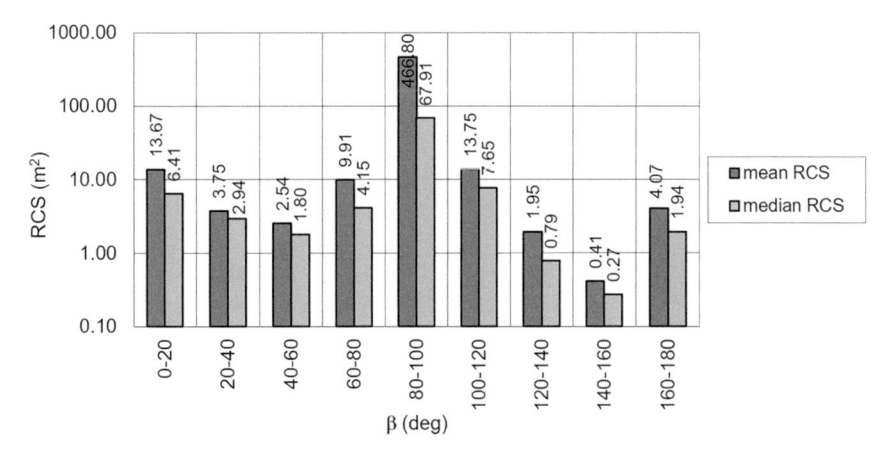

FIGURE 2.284 Diagrams of mean and median RCS of Su-34 aircraft model in 20-degree sectors of azimuth aspect given its radar observation at horizontal polarization and a carrier frequency of 3 GHz (10 cm wavelength).

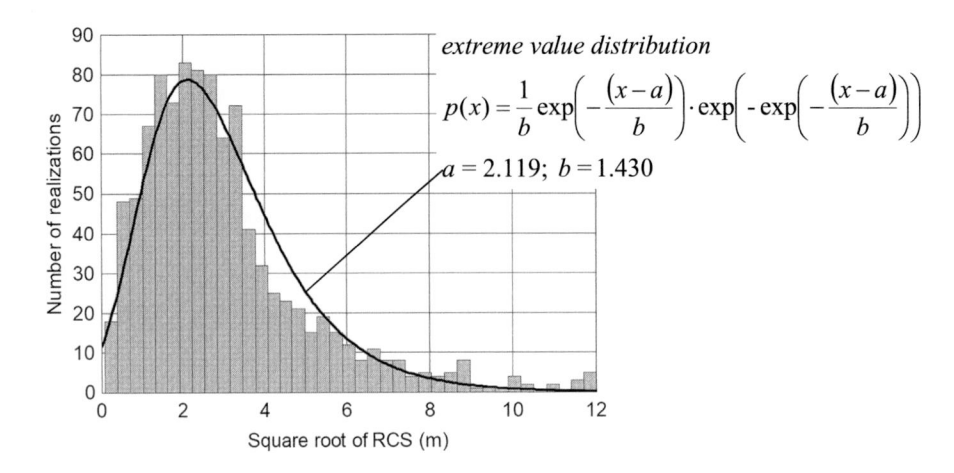

FIGURE 2.285 Amplitude distribution of echo signal of Su-34 aircraft model at a carrier frequency of 3 GHz given its horizontal polarization.

extreme value distribution

$$p(x) = \frac{1}{b}\exp\left(-\frac{(x-a)}{b}\right)\cdot\exp\left(-\exp\left(-\frac{(x-a)}{b}\right)\right)$$

$a = 2.119;\ b = 1.430$

Scattering characteristics of Su-34 aircraft model for sounding frequency 1 GHz (wavelength 30 cm).

Figure 2.286 shows the RCS circular diagram of the Su-34. The noncoherent RCS circular diagram of the Su-34 is shown in Figure 2.287.

The circular mean RCS of the Su-34 aircraft for horizontal polarization is 47.13 m². The circular median RCS for horizontal polarization is 2.63 m².

Figures 2.288 and 2.289 show the mean and median RCS for the main ranges of sounding azimuths (nose, side, and tail) and for ranges of 20°.

The expressions and parameters of probability distributions that are most consistent with the empirical distributions of the square root of the RCS for different ranges of sounding azimuths and polarizations are given in the electronic appendix to this book. The expressions and parameters of the probability distributions that are most consistent with the empirical distributions of the RCS (energy characteristic) for different ranges of sounding azimuths and polarizations are also given there.

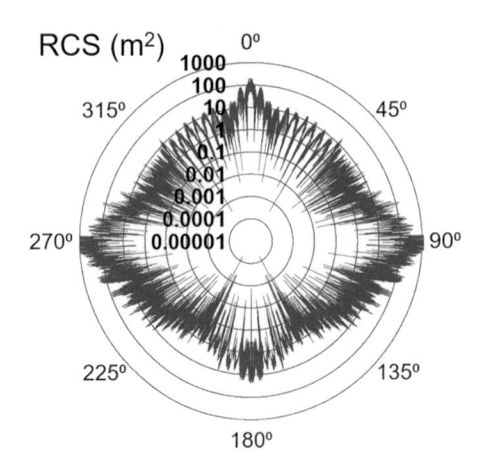

FIGURE 2.286 Circular diagram of RCS given radar observation of Su-34 aircraft model at a carrier frequency of 1 GHz (30 cm wavelength).

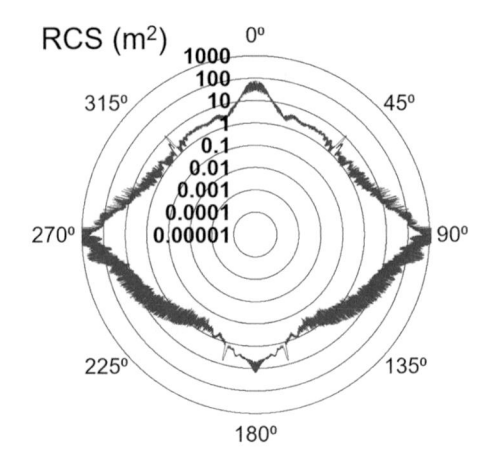

FIGURE 2.287 Circular diagram of noncoherent RCS given radar observation of Su-34 aircraft model at a carrier frequency of 1 GHz (30 cm wavelength).

Scattering Characteristics of Aerial Objects

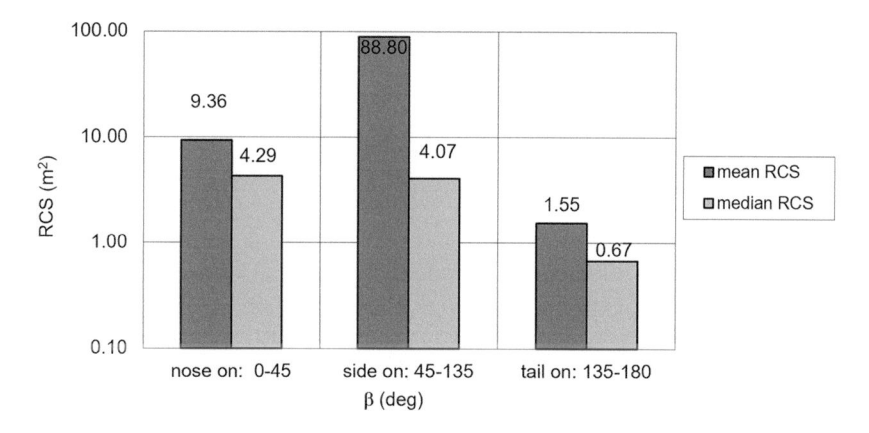

FIGURE 2.288 Diagrams of mean and median RCS of Su-34 aircraft model in three sectors of azimuth aspect given its radar observation at horizontal polarization and a carrier frequency of 1 GHz (30 cm wavelength).

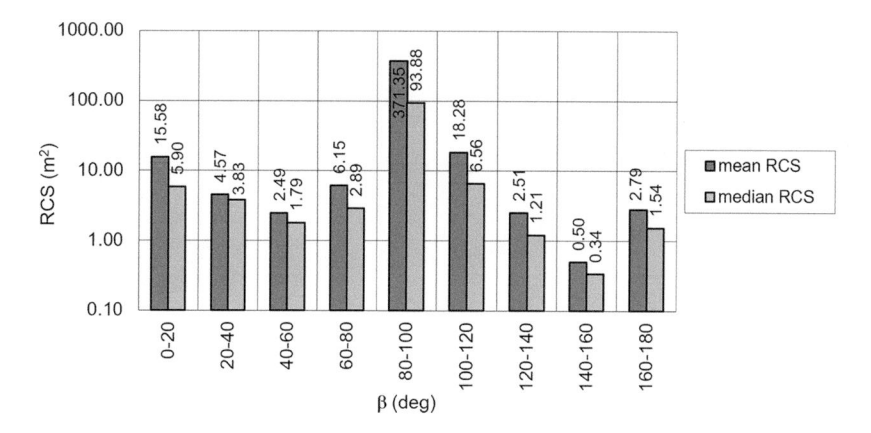

FIGURE 2.289 Diagrams of mean and median RCS of Su-34 aircraft model in 20-degree sectors of azimuth aspect given its radar observation at horizontal polarization and a carrier frequency of 1 GHz (30 cm wavelength).

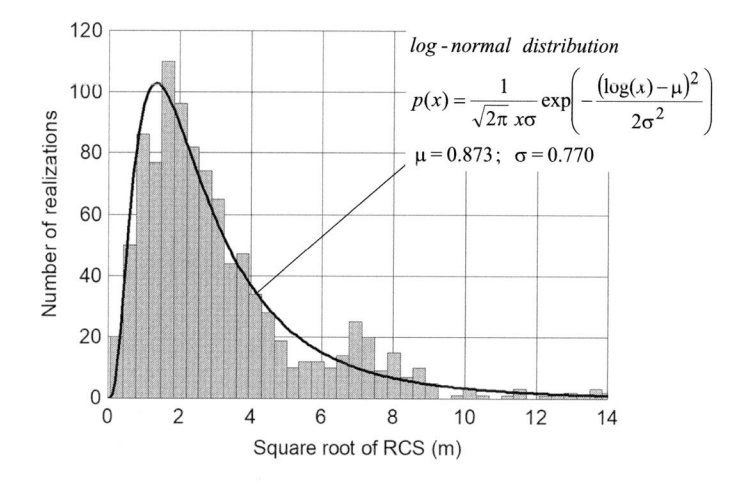

$$p(x) = \frac{1}{\sqrt{2\pi}\, x\sigma} \exp\left(-\frac{(\log(x)-\mu)^2}{2\sigma^2}\right)$$

$\mu = 0.873;\ \sigma = 0.770$

FIGURE 2.290 Amplitude distribution of echo signal of Su-34 aircraft model at a carrier frequency of 1 GHz given its horizontal polarization.

2.18 STRATEGIC BOMBER B-2

B-2 Spirit (Figure 2.291) is a stealth strategic bomber. The first flight took place in 1989.

General characteristics of B-2 [28]: wingspan –52.12 m, length – 20.90 m, height – 5.1 m, wing area – 477.50 m^2, weight – 72 575–152 630 kg, powerplant – 4 General Electric F118-GE-100 non-afterburning turbofans, maximum speed – 1 010 km/h, range – 11 000 km, and service ceiling – 15 250 m.

In accordance with the design of the B-2, a model of its surface was created to obtaining scattering characteristics (in particular, RCS). The model is shown in Figure 2.292. In the modeling, the smooth parts of the aircraft surface were approximated by parts of 26 triaxial ellipsoids. The surface breaks were modeled using 24 straight edge scattering parts. The model surface has a RAC of Sommefeld's type [29], which corresponds to some forms of existing RAM coatings. See [1] for more details on the material and coatings of this model.

FIGURE 2.291 Strategic bomber B-2.

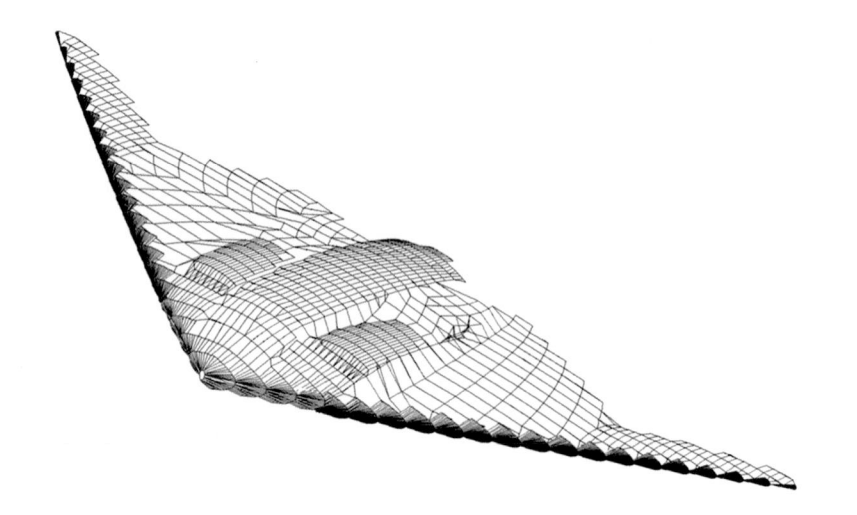

FIGURE 2.292 Surface model of strategic bomber B-2.

Below are some scattering characteristics of the B-2 model at sounding frequencies of 10, 3, and 1 GHz (wavelengths of 3, 10, and 30 cm, respectively) for horizontal polarization of the probing signal. The scattering characteristics for this model at two polarizations, as well as scattering characteristics at sounding frequencies of 5 and 1.3 GHz (wavelengths of 6 and 23 cm, respectively), are given in the electronic appendix of this book.

Scattering Characteristics of Aerial Objects 115

Sounding parameters: The elevation angle is to be a random value distributed uniformly in the range −3° ± 4° with respect to the wing plane (elevation angle of −3° corresponds to the radar observation from the lower hemisphere), azimuth aspect increment was 0.02°, the azimuth being counted off from the nose-on aspect (0° corresponds to the nose-on radar observation, 180° corresponds to the tail-on observation).

Scattering characteristics of B-2 aircraft model for sounding frequency 10 GHz (wavelength 3 cm).

Figure 2.293 shows the RCS circular diagram of the B-2 model. The noncoherent RCS circular diagram of the B-2 is shown in Figure 2.294.

The circular mean RCS of the B-2 aircraft model for horizontal polarization is 0.14 m². The circular median RCS (the RCS value used to calculate the detection range of an object with a probability of 0.5) for horizontal polarization is 0.07 m².

Figures 2.295 and 2.296 show the mean and median RCS for the main ranges of sounding azimuths (nose, side, and tail) and for ranges of 20°.

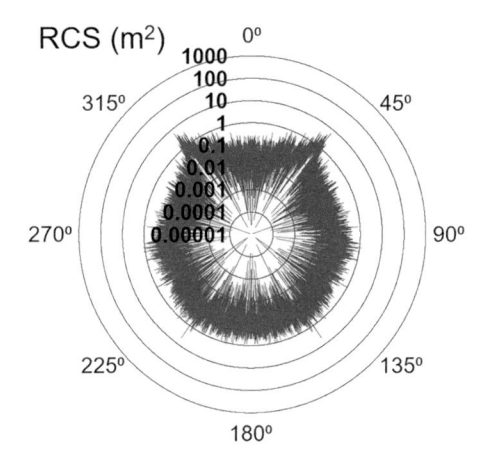

FIGURE 2.293 Circular diagram of RCS given radar observation of B-2 aircraft model at a carrier frequency of 10 GHz (3 cm wavelength).

FIGURE 2.294 Circular diagram of noncoherent RCS given radar observation of B-2 aircraft model at a carrier frequency of 10 GHz (3 cm wavelength).

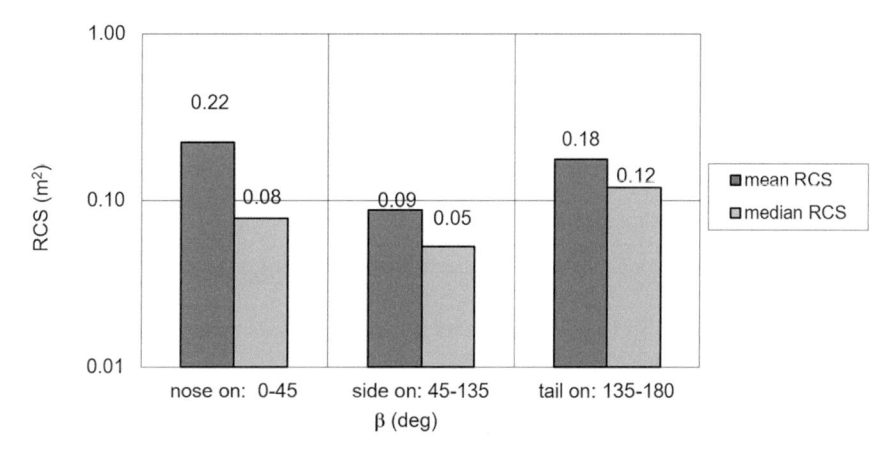

FIGURE 2.295 Diagrams of mean and median RCS of B-2 aircraft model in three sectors of azimuth aspect given its radar observation at horizontal polarization and a carrier frequency of 10 GHz (3 cm wavelength).

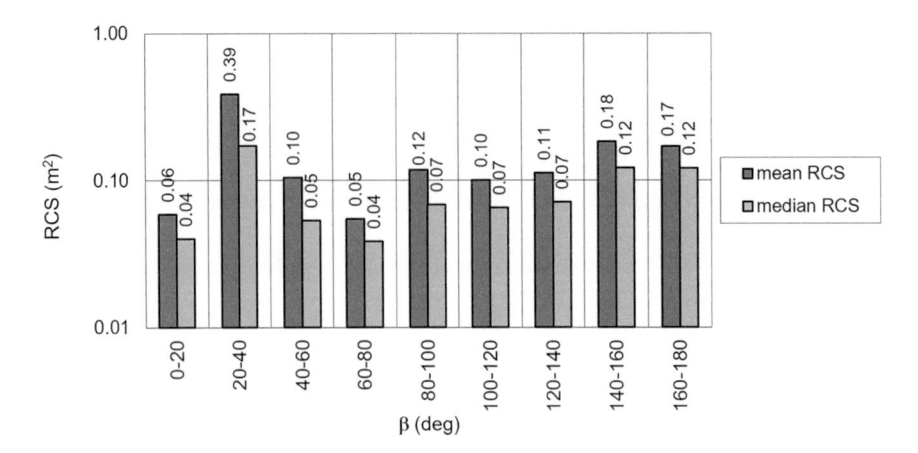

FIGURE 2.296 Diagrams of mean and median RCS of B-2 aircraft model in 20-degree sectors of azimuth aspect given its radar observation at horizontal polarization and a carrier frequency of 10 GHz (3 cm wavelength).

Figures 2.297, 2.302, and 2.307 show histograms of the scattered signal amplitude (square root of the RCS) for the range of sounding azimuths −20° to +20° (sounding from the nose). The bold line shows the probability density functions of the distribution, which can be used to approximate the histogram of the signal amplitude.

Weibull distribution

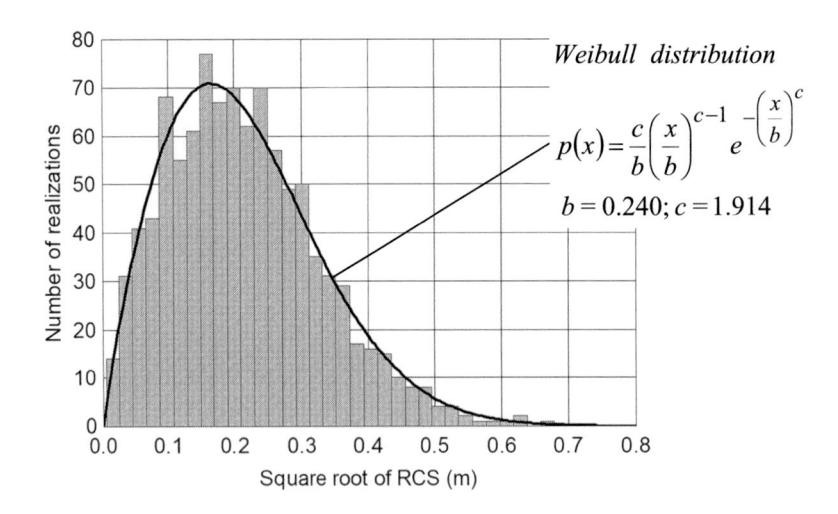

$$p(x) = \frac{c}{b}\left(\frac{x}{b}\right)^{c-1} e^{-\left(\frac{x}{b}\right)^c}$$

$b = 0.240; c = 1.914$

FIGURE 2.297 Amplitude distribution of echo signal of B-2 aircraft model at a carrier frequency of 10 GHz given its horizontal polarization.

Scattering characteristics of B-2 aircraft model for sounding frequency 3 GHz (wavelength 10 cm).

Figure 2.298 shows the RCS circular diagram of the B-2. The noncoherent RCS circular diagram of the B-2 is shown in Figure 2.299.

The circular mean RCS of the B-2 aircraft for horizontal polarization is 0.59 m². The circular median RCS for horizontal polarization is 0.10 m².

Figures 2.300 and 2.301 show the mean and median RCS for the main ranges of sounding azimuths (nose, side, and tail) and for ranges of 20°.

Scattering Characteristics of Aerial Objects

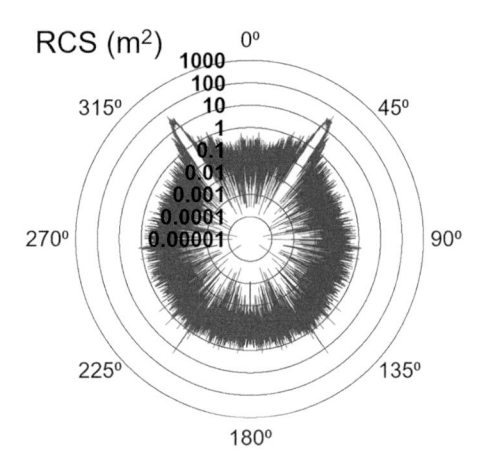

FIGURE 2.298 Circular diagram of RCS given radar observation of B-2 aircraft model at a carrier frequency of 3 GHz (10 cm wavelength).

FIGURE 2.299 Circular diagram of noncoherent RCS given radar observation of B-2 aircraft model at a carrier frequency of 3 GHz (10 cm wavelength).

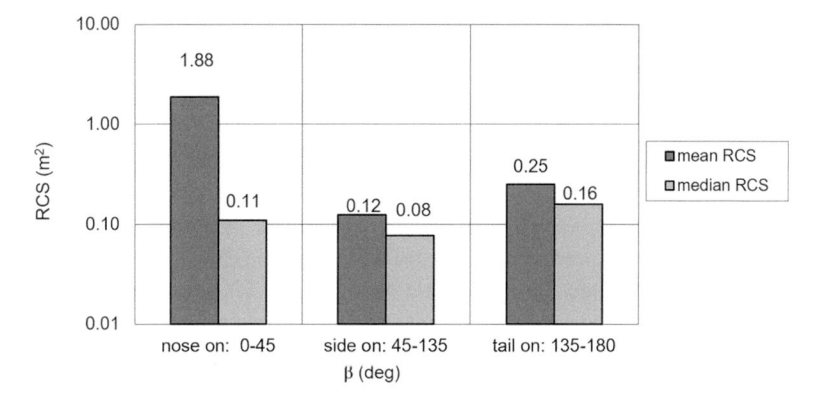

FIGURE 2.300 Diagrams of mean and median RCS of B-2 aircraft model in three sectors of azimuth aspect given its radar observation at horizontal polarization and a carrier frequency of 3 GHz (10 cm wavelength).

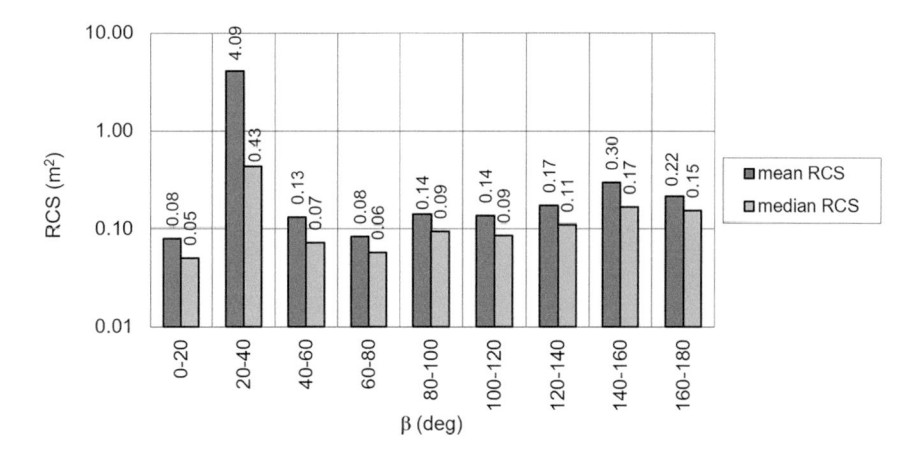

FIGURE 2.301 Diagrams of mean and median RCS of B-2 aircraft model in 20-degree sectors of azimuth aspect given its radar observation at horizontal polarization and a carrier frequency of 3 GHz (10 cm wavelength).

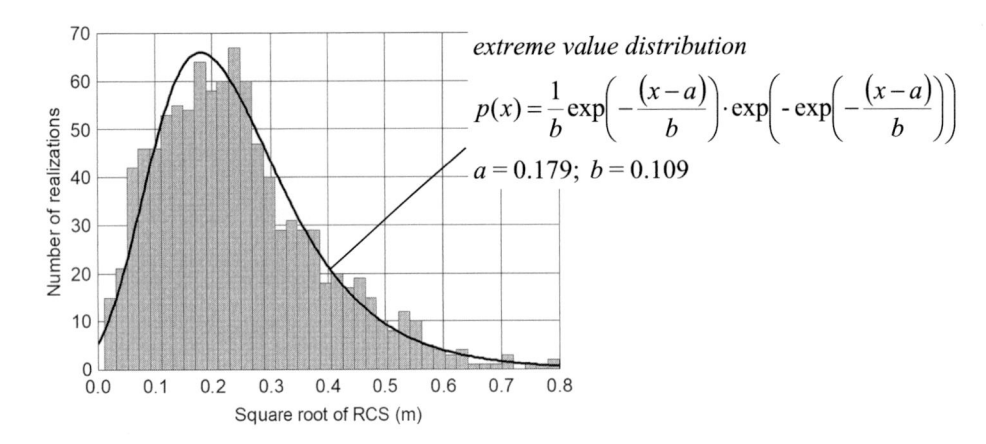

FIGURE 2.302 Amplitude distribution of echo signal of B-2 aircraft model at a carrier frequency of 3 GHz given its horizontal polarization.

extreme value distribution

$$p(x) = \frac{1}{b}\exp\left(-\frac{(x-a)}{b}\right)\cdot\exp\left(-\exp\left(-\frac{(x-a)}{b}\right)\right)$$

$a = 0.179;\ b = 0.109$

Scattering characteristics of B-2 aircraft model for sounding frequency 1 GHz (wavelength 30 cm).

Figure 2.303 shows the RCS circular diagram of the B-2. The noncoherent RCS circular diagram of the B-2 is shown in Figure 2.304.

The circular mean RCS of the B-2 aircraft for horizontal polarization is 1.81 m². The circular median RCS for horizontal polarization is 0.19 m².

Figures 2.305 and 2.306 show the mean and median RCS for the main ranges of sounding azimuths (nose, side, and tail) and for ranges of 20°.

The expressions and parameters of probability distributions that are most consistent with the empirical distributions of the square root of the RCS for different ranges of sounding azimuths and polarizations are given in the electronic appendix to this book. The expressions and parameters of the probability distributions that are most consistent with the empirical distributions of the RCS (energy characteristic) for different ranges of sounding azimuths and polarizations are also given there.

FIGURE 2.303 Circular diagram of RCS given radar observation of B-2 aircraft model at a carrier frequency of 1 GHz (30 cm wavelength).

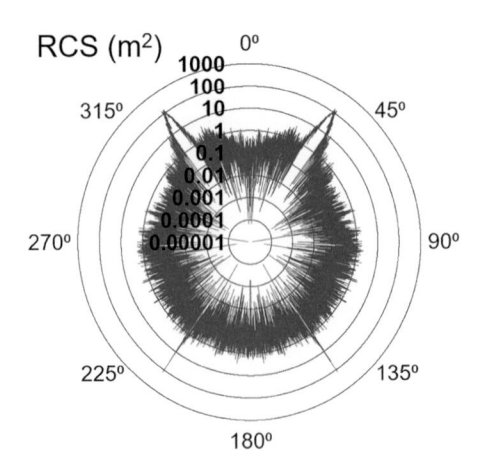

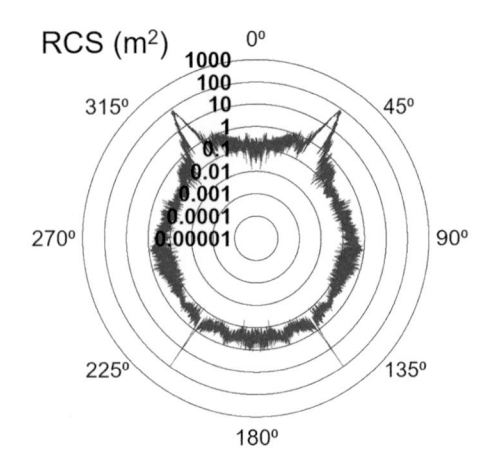

FIGURE 2.304 Circular diagram of noncoherent RCS given radar observation of B-2 aircraft model at a carrier frequency of 1 GHz (30 cm wavelength).

Scattering Characteristics of Aerial Objects

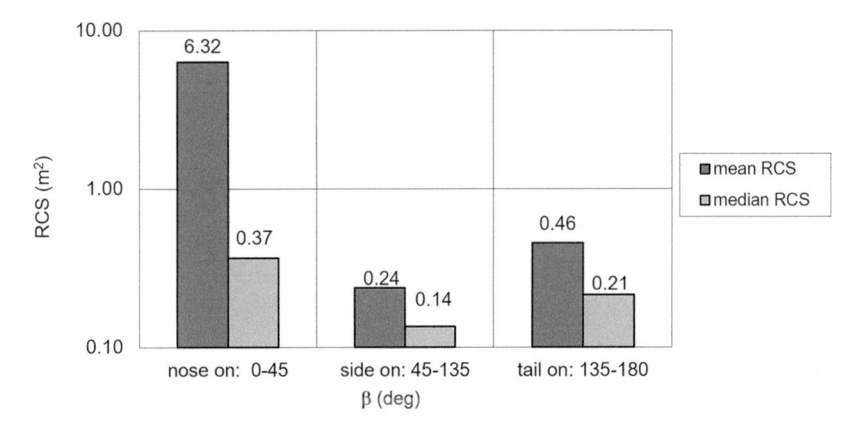

FIGURE 2.305 Diagrams of mean and median RCS of B-2 aircraft model in three sectors of azimuth aspect given its radar observation at horizontal polarization and a carrier frequency of 1 GHz (30 cm wavelength).

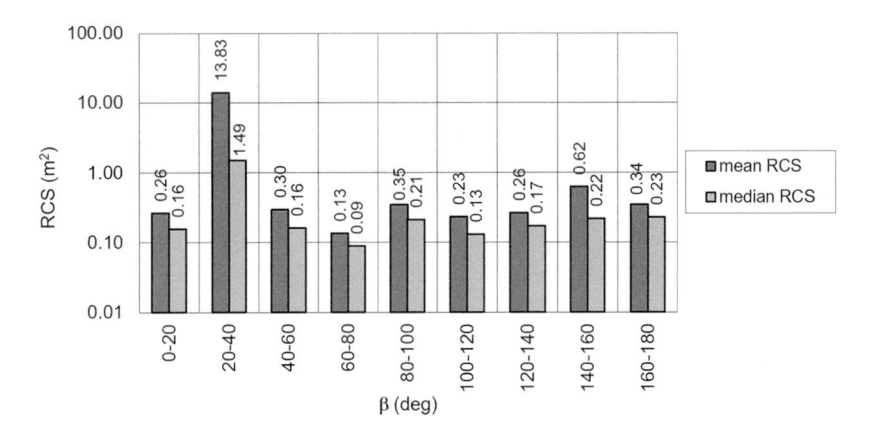

FIGURE 2.306 Diagrams of mean and median RCS of B-2 aircraft model in 20-degree sectors of azimuth aspect given its radar observation at horizontal polarization and a carrier frequency of 1 GHz (30 cm wavelength).

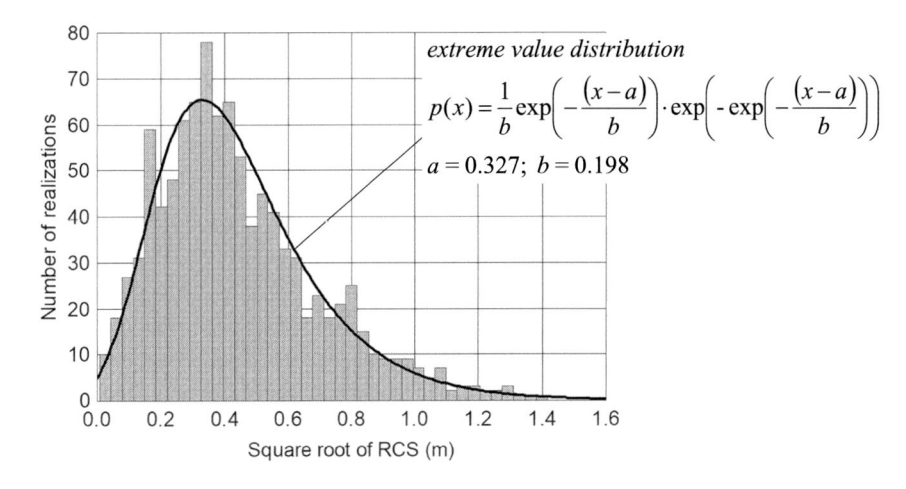

extreme value distribution

$$p(x) = \frac{1}{b}\exp\left(-\frac{(x-a)}{b}\right) \cdot \exp\left(-\exp\left(-\frac{(x-a)}{b}\right)\right)$$

$a = 0.327;\ b = 0.198$

FIGURE 2.307 Amplitude distribution of echo signal of B-2 aircraft model at a carrier frequency of 1 GHz given its horizontal polarization.

2.19 STRATEGIC BOMBER B-52

B-52 Stratofortress (Figure 2.308) is a strategic bomber. The first flight took place in 1952.

General characteristics of B-52 [30]: wingspan –56.39 m, length – 48.50 m, height – 12.40 m, wing area – 371.60 m^2, weight – 83 250–219 600 kg, powerplant – 8 Pratt & Whitney TF33-P-3/103 turbofans, maximum speed – 1 070 km/h, combat range – 5 875 km, and service ceiling – 15 150 m.

In accordance with the design of the B-52, a model of its surface was created to obtaining scattering characteristics (in particular, RCS). The model is shown in Figure 2.309. In the modeling, the smooth parts of the aircraft surface were approximated by parts of 95 triaxial ellipsoids. The surface breaks were modeled using 7 straight edge scattering parts.

FIGURE 2.308 Strategic bomber B-52.

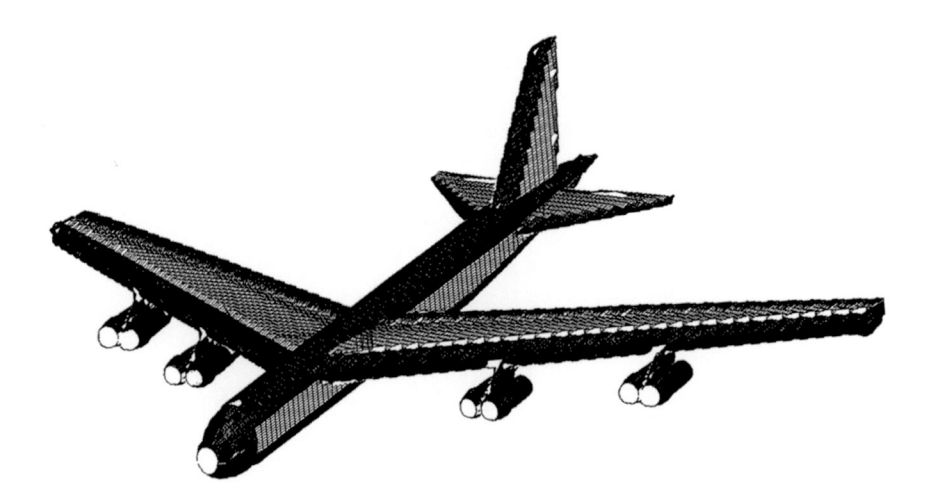

FIGURE 2.309 Surface model of strategic bomber B-52.

Below are some scattering characteristics of the B-52 model at sounding frequencies of 10, 3, and 1 GHz (wavelengths of 3, 10, and 30 cm, respectively) for horizontal polarization of the probing signal. The scattering characteristics for this model at two polarizations, as well as scattering characteristics at sounding frequencies of 5 and 1.3 GHz (wavelengths of 6 and 23 cm, respectively), are given in the electronic appendix of this book.

Sounding parameters: The elevation angle is to be a random value distributed uniformly in the range −3° ± 4° with respect to the wing plane (elevation angle of −3° corresponds to the radar observation from the lower hemisphere), azimuth aspect increment was 0.02°, the azimuth being counted

off from the nose-on aspect (0° corresponds to the nose-on radar observation, 180° corresponds to the tail-on observation).

Scattering characteristics of B-52 aircraft model for sounding frequency 10 GHz (wavelength 3 cm).

Figure 2.310 shows the RCS circular diagram of the B-52 model. The noncoherent RCS circular diagram of the B-52 is shown in Figure 2.311.

The circular mean RCS of the B-52 aircraft model for horizontal polarization is 372.46 m². The circular median RCS (the RCS value used to calculate the detection range of an object with a probability of 0.5) for horizontal polarization is 19.18 m².

Figures 2.312 and 2.313 show the mean and median RCS for the main ranges of sounding azimuths (nose, side, and tail) and for ranges of 20°.

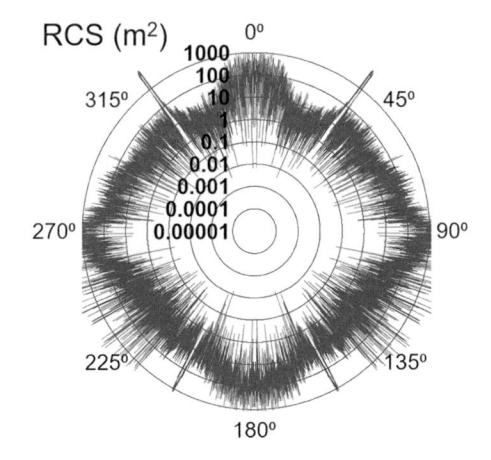

FIGURE 2.310 Circular diagram of RCS given radar observation of B-52 aircraft model at a carrier frequency of 10 GHz (3 cm wavelength).

FIGURE 2.311 Circular diagram of noncoherent RCS given radar observation of B-52 aircraft model at a carrier frequency of 10 GHz (3 cm wavelength).

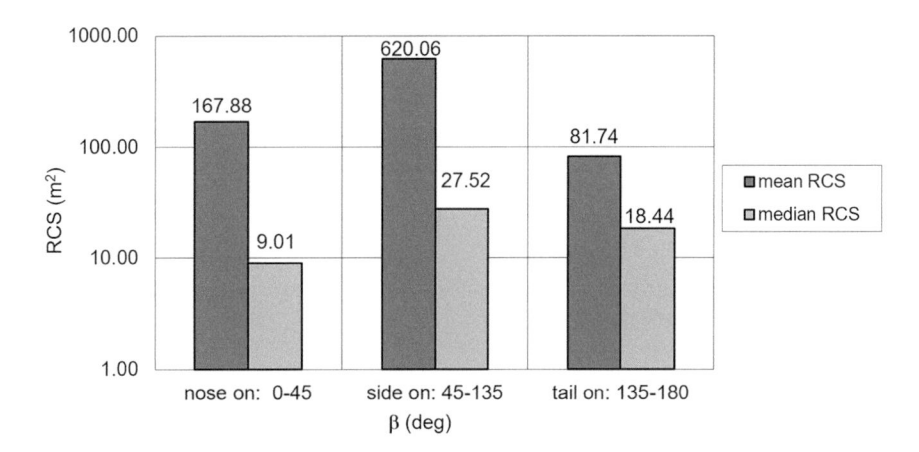

FIGURE 2.312 Diagrams of mean and median RCS of B-52 aircraft model in three sectors of azimuth aspect given its radar observation at horizontal polarization and a carrier frequency of 10 GHz (3 cm wavelength).

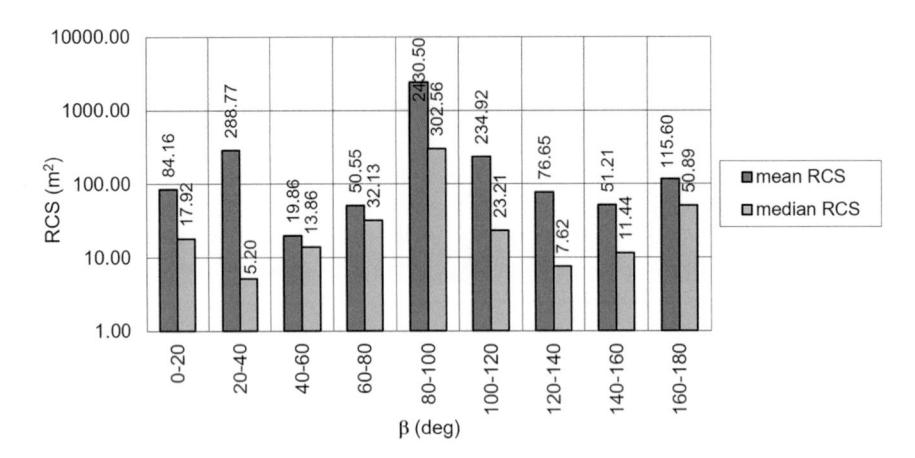

FIGURE 2.313 Diagrams of mean and median RCS of B-52 aircraft model in 20-degree sectors of azimuth aspect given its radar observation at horizontal polarization and a carrier frequency of 10 GHz (3 cm wavelength).

Figures 2.314, 2.319, and 2.324 show histograms of the scattered signal amplitude (square root of the RCS) for the range of sounding azimuths −20° to +20° (sounding from the nose). The bold line shows the probability density functions of the distribution, which can be used to approximate the histogram of the signal amplitude.

$$p(x) = \frac{1}{\sqrt{2\pi}\, x\sigma} \exp\left(-\frac{(\log(x)-\mu)^2}{2\sigma^2}\right)$$

log - normal distribution

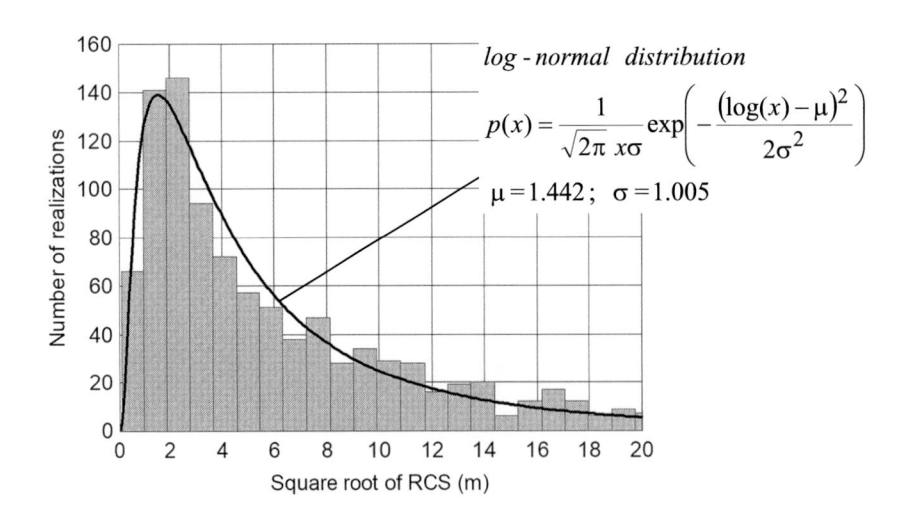

$\mu = 1.442;\ \sigma = 1.005$

FIGURE 2.314 Amplitude distribution of echo signal of B-52 aircraft model at a carrier frequency of 10 GHz given its horizontal polarization.

Scattering characteristics of B-52 aircraft model for sounding frequency 3 GHz (wavelength 10 cm).

Figure 2.315 shows the RCS circular diagram of the B-52. The noncoherent RCS circular diagram of the B-52 is shown in Figure 2.316.

The circular mean RCS of the B-52 aircraft for horizontal polarization is 421.30 m². The circular median RCS for horizontal polarization is 14.44 m².

Figures 2.317 and 2.318 show the mean and median RCS for the main ranges of sounding azimuths (nose, side, and tail) and for ranges of 20°.

Scattering Characteristics of Aerial Objects 123

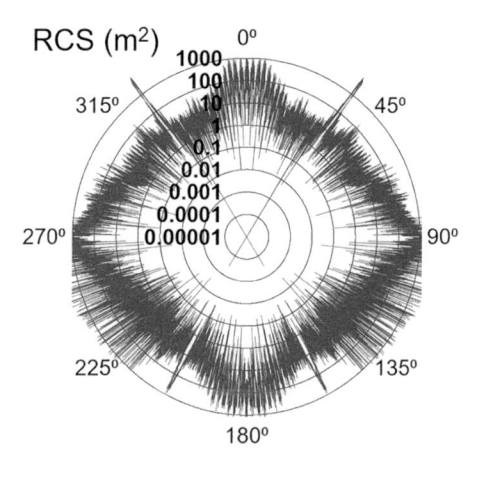

FIGURE 2.315 Circular diagram of RCS given radar observation of B-52 aircraft model at a carrier frequency of 3 GHz (10 cm wavelength).

FIGURE 2.316 Circular diagram of noncoherent RCS given radar observation of B-52 aircraft model at a carrier frequency of 3 GHz (10 cm wavelength).

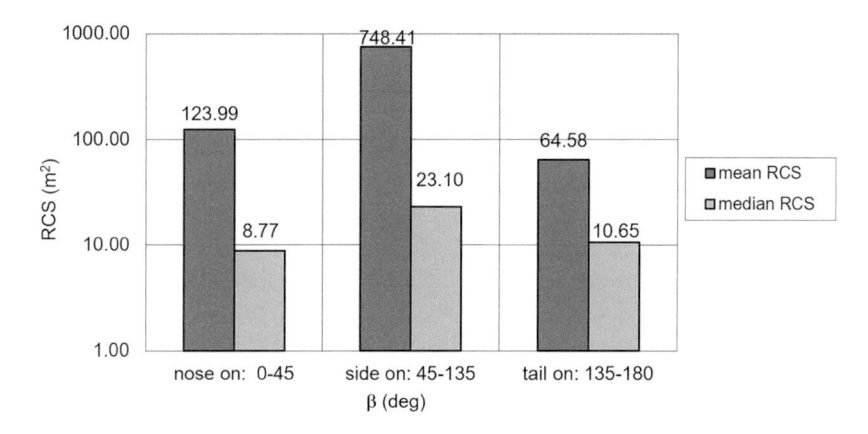

FIGURE 2.317 Diagrams of mean and median RCS of B-52 aircraft model in three sectors of azimuth aspect given its radar observation at horizontal polarization and a carrier frequency of 3 GHz (10 cm wavelength).

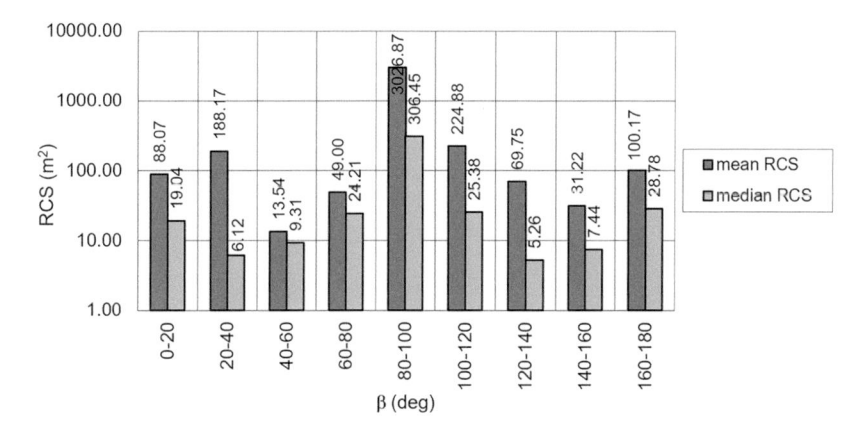

FIGURE 2.318 Diagrams of mean and median RCS of B-52 aircraft model in 20-degree sectors of azimuth aspect given its radar observation at horizontal polarization and a carrier frequency of 3 GHz (10 cm wavelength).

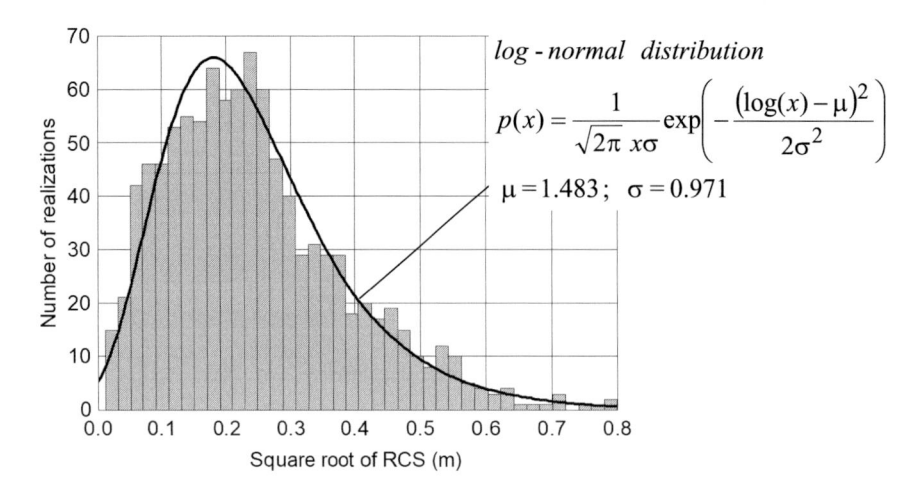

FIGURE 2.319 Amplitude distribution of echo signal of B-52 aircraft model at a carrier frequency of 3 GHz given its horizontal polarization.

$$p(x) = \frac{1}{\sqrt{2\pi}\, x\sigma} \exp\left(-\frac{(\log(x)-\mu)^2}{2\sigma^2}\right)$$

$\mu = 1.483; \quad \sigma = 0.971$

Scattering characteristics of B-52 aircraft model for sounding frequency 1 GHz (wavelength 30 cm).

Figure 2.320 shows the RCS circular diagram of the B-52. The noncoherent RCS circular diagram of the B-52 is shown in Figure 2.321.

The circular mean RCS of the B-52 aircraft for horizontal polarization is 297.14 m². The circular median RCS for horizontal polarization is 14.95 m².

Figures 2.322 and 2.323 show the mean and median RCS for the main ranges of sounding azimuths (nose, side, and tail) and for ranges of 20°.

The expressions and parameters of probability distributions that are most consistent with the empirical distributions of the square root of the RCS for different ranges of sounding azimuths and polarizations are given in the electronic appendix to this book. The expressions and parameters of the probability distributions that are most consistent with the empirical distributions of the RCS (energy characteristic) for different ranges of sounding azimuths and polarizations are also given there.

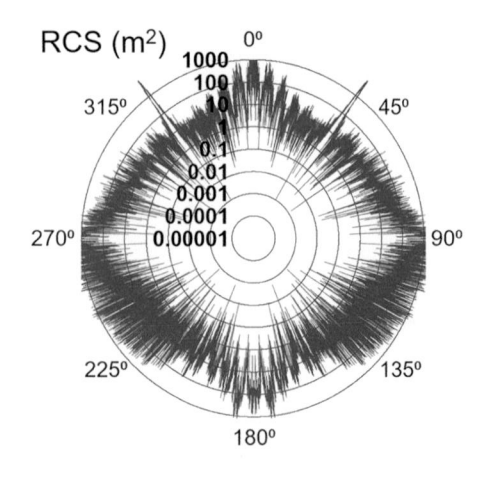

FIGURE 2.320 Circular diagram of RCS given radar observation of B-52 aircraft model at a carrier frequency of 1 GHz (30 cm wavelength).

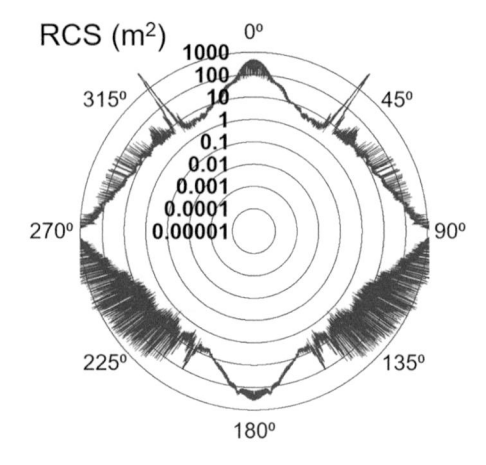

FIGURE 2.321 Circular diagram of noncoherent RCS given radar observation of B-52 aircraft model at a carrier frequency of 1 GHz (30 cm wavelength).

Scattering Characteristics of Aerial Objects

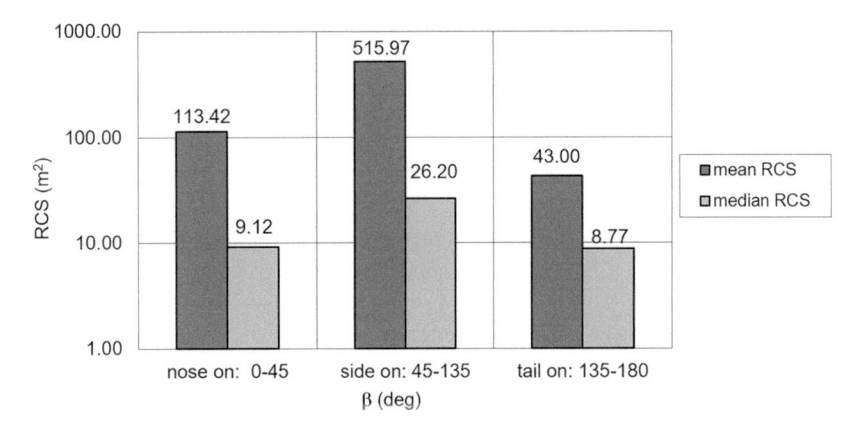

FIGURE 2.322 Diagrams of mean and median RCS of B-52 aircraft model in three sectors of azimuth aspect given its radar observation at horizontal polarization and a carrier frequency of 1 GHz (30 cm wavelength).

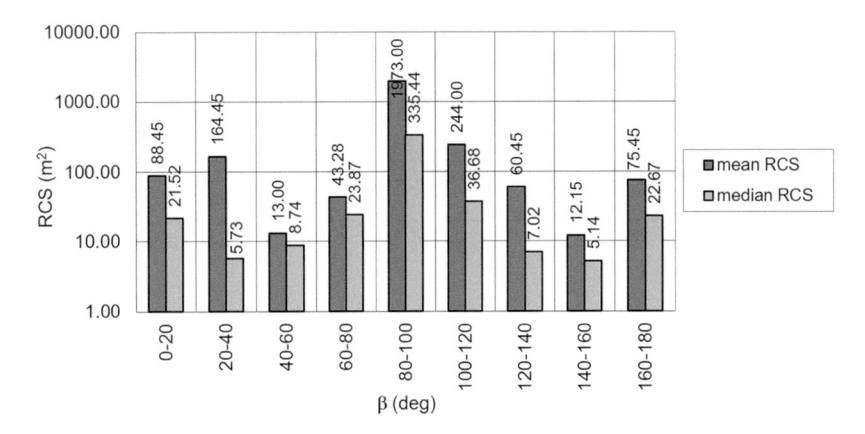

FIGURE 2.323 Diagrams of mean and median RCS of B-52 aircraft model in 20-degree sectors of azimuth aspect given its radar observation at horizontal polarization and a carrier frequency of 1 GHz (30 cm wavelength).

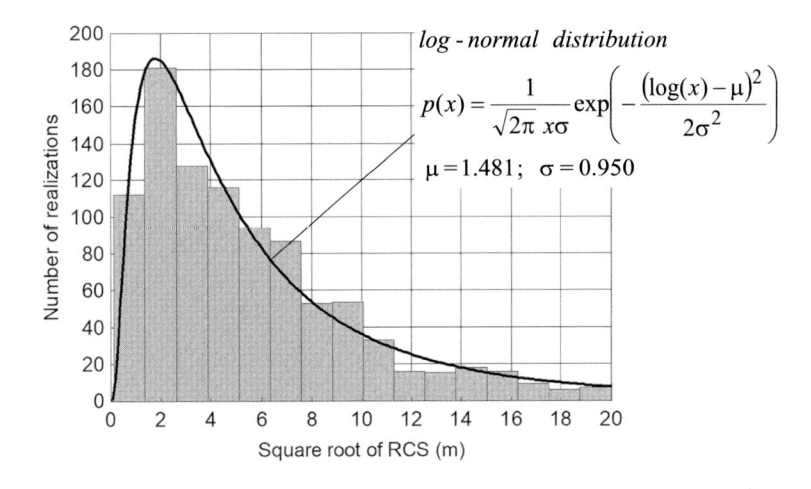

$$p(x) = \frac{1}{\sqrt{2\pi}\, x\sigma} \exp\left(-\frac{(\log(x) - \mu)^2}{2\sigma^2}\right)$$

$\mu = 1.481; \quad \sigma = 0.950$

FIGURE 2.324 Amplitude distribution of echo signal of B-52 aircraft model at a carrier frequency of 1 GHz given its horizontal polarization.

2.20 STRATEGIC BOMBER B-1B

B-1B Lancer (Figure 2.325) is a strategic bomber. The first flight took place in 1984.

General characteristics of B-1B [31]: wingspan – 24.10–41.80 m, length – 44.50 m, height – 10.36 m, wing area – 181.16 m^2, weight – 86 200–214 600 kg, powerplant – 4 General Electric F101-GE-102 afterburning turbofan engines, maximum speed – 1 328 km/h, range – 8 200–12500 km, and service ceiling – 18 300 m.

In accordance with the design of the B-1B, a model of its surface was created to obtaining scattering characteristics (in particular, RCS). The model is shown in Figure 2.326. In the modeling, the smooth parts of the aircraft surface were approximated by parts of 105 triaxial ellipsoids. The surface breaks were modeled using 22 straight edge scattering parts.

FIGURE 2.325 Strategic bomber B-1B.

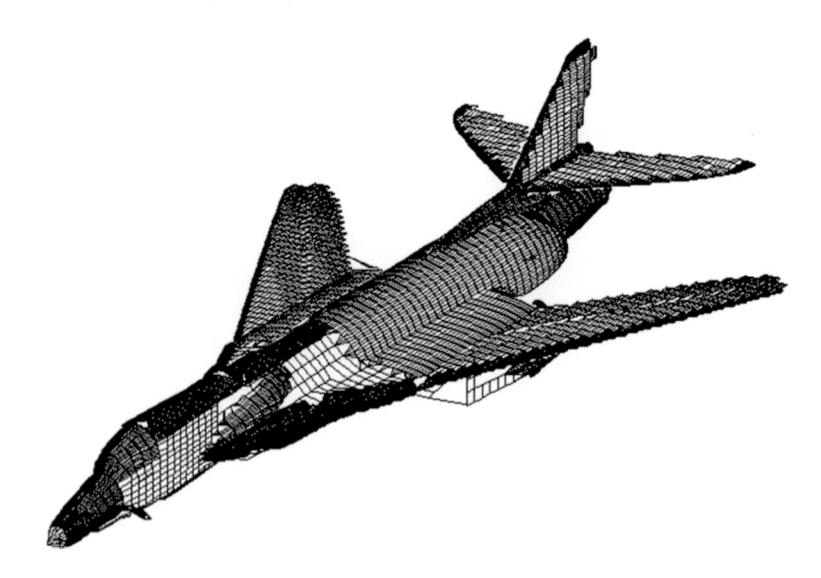

FIGURE 2.326 Surface model of strategic bomber B-1B.

Below are some scattering characteristics of the B-1B model at sounding frequencies of 10, 3, and 1 GHz (wavelengths of 3, 10, and 30 cm, respectively) for horizontal polarization of the probing signal. The scattering characteristics for this model at two polarizations, as well as scattering characteristics at sounding frequencies of 5 and 1.3 GHz (wavelengths of 6 and 23 cm, respectively), are given in the electronic appendix of this book.

Sounding parameters: The elevation angle is to be a random value distributed uniformly in the range −3° ± 4° with respect to the wing plane (elevation angle of −3° corresponds to the radar observation from the lower hemisphere), azimuth aspect increment was 0.02°, the azimuth being counted off from the nose-on aspect (0° corresponds to the nose-on radar observation, 180° corresponds to the tail-on observation).

Scattering characteristics of B-1B aircraft model for sounding frequency 10 GHz (wavelength 3 cm).

Figure 2.327 shows the RCS circular diagram of the B-1B model. The noncoherent RCS circular diagram of the B-1B is shown in Figure 2.328.

The circular mean RCS of the B-1B aircraft model for horizontal polarization is 293.89 m². The circular median RCS (the RCS value used to calculate the detection range of an object with a probability of 0.5) for horizontal polarization is 19.03 m².

Figures 2.329 and 2.330 show the mean and median RCS for the main ranges of sounding azimuths (nose, side, and tail) and for ranges of 20°.

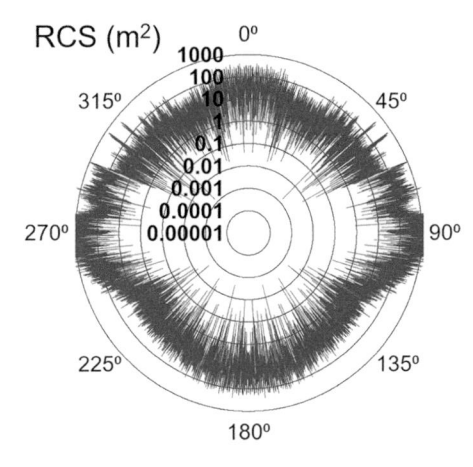

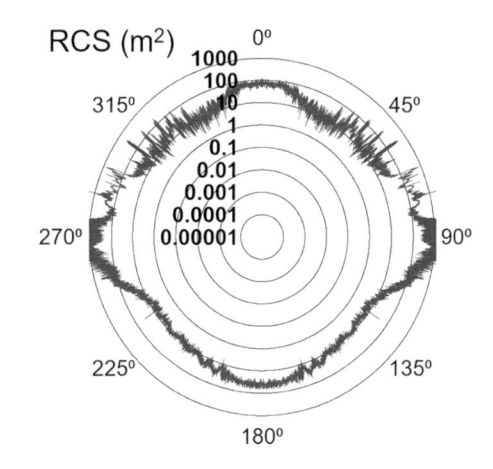

FIGURE 2.327 Circular diagram of RCS given radar observation of B-1B aircraft model at a carrier frequency of 10 GHz (3 cm wavelength).

FIGURE 2.328 Circular diagram of noncoherent RCS given radar observation of B-1B aircraft model at a carrier frequency of 10 GHz (3 cm wavelength).

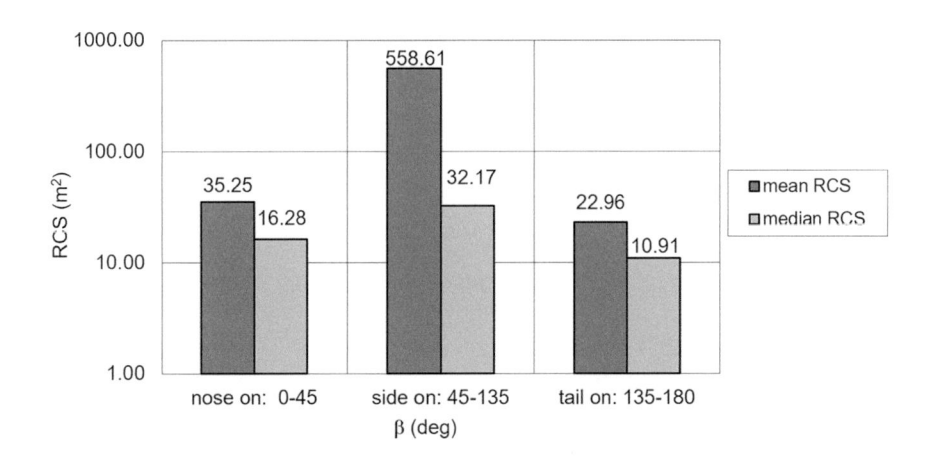

FIGURE 2.329 Diagrams of mean and median RCS of B-1B aircraft model in three sectors of azimuth aspect given its radar observation at horizontal polarization and a carrier frequency of 10 GHz (3 cm wavelength).

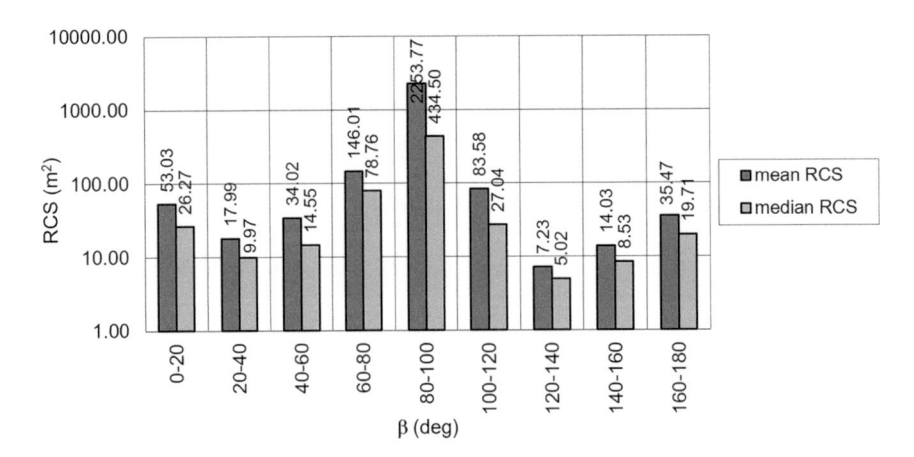

FIGURE 2.330 Diagrams of mean and median RCS of B-1B aircraft model in 20-degree sectors of azimuth aspect given its radar observation at horizontal polarization and a carrier frequency of 10 GHz (3 cm wavelength).

Figures 2.331, 2.336, and 2.341 show histograms of the scattered signal amplitude (square root of the RCS) for the range of sounding azimuths −20° to +20° (sounding from the nose). The bold line shows the probability density functions of the distribution, which can be used to approximate the histogram of the signal amplitude.

Weibull distribution

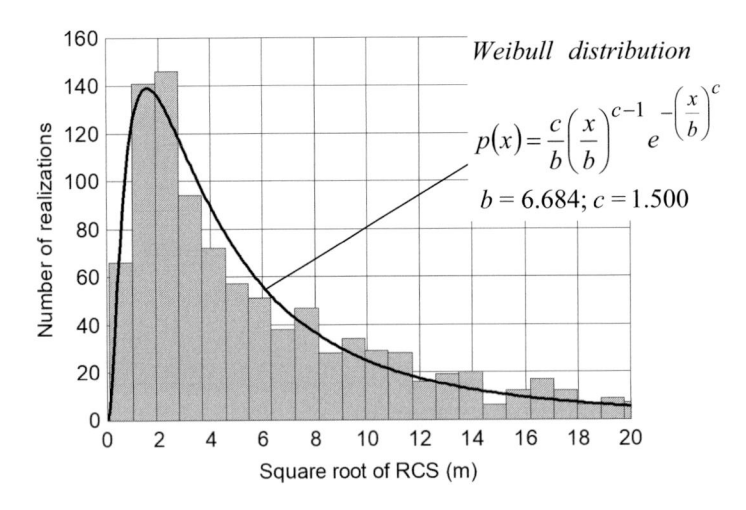

$$p(x) = \frac{c}{b}\left(\frac{x}{b}\right)^{c-1} e^{-\left(\frac{x}{b}\right)^c}$$

$b = 6.684;\ c = 1.500$

FIGURE 2.331 Amplitude distribution of echo signal of B-1B aircraft model at a carrier frequency of 10 GHz given its horizontal polarization.

Scattering characteristics of B-1B aircraft model for sounding frequency 3 GHz (wavelength 10 cm).

Figure 2.332 shows the RCS circular diagram of the B-1B. The noncoherent RCS circular diagram of the B-1B is shown in Figure 2.333.

The circular mean RCS of the B-1B aircraft for horizontal polarization is 215.14 m². The circular median RCS for horizontal polarization is 17.21 m².

Figures 2.334 and 2.335 show the mean and median RCS for the main ranges of sounding azimuths (nose, side, and tail) and for ranges of 20°.

Scattering Characteristics of Aerial Objects   129

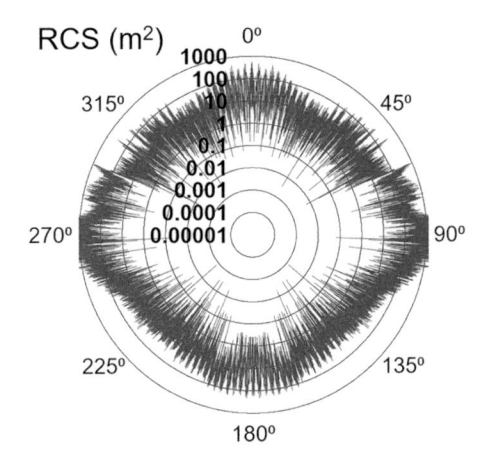

FIGURE 2.332 Circular diagram of RCS given radar observation of B-1B aircraft model at a carrier frequency of 3 GHz (10 cm wavelength).

FIGURE 2.333 Circular diagram of noncoherent RCS given radar observation of B-1B aircraft model at a carrier frequency of 3 GHz (10 cm wavelength).

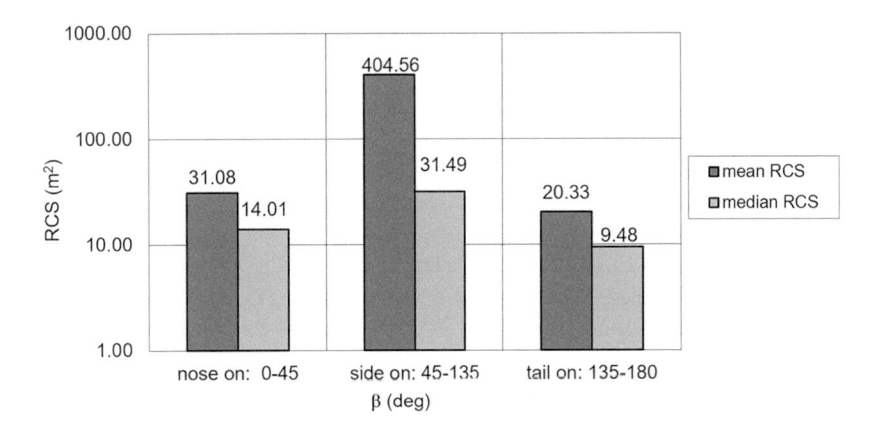

FIGURE 2.334 Diagrams of mean and median RCS of B-1B aircraft model in three sectors of azimuth aspect given its radar observation at horizontal polarization and a carrier frequency of 3 GHz (10 cm wavelength).

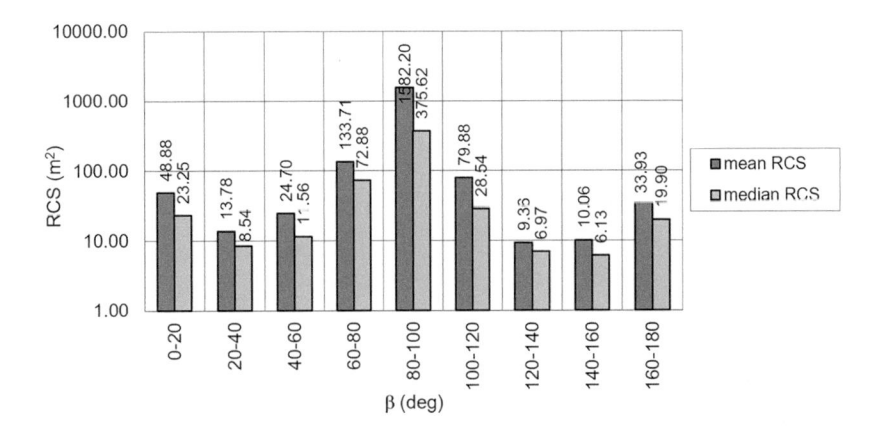

FIGURE 2.335 Diagrams of mean and median RCS of B-1B aircraft model in 20-degree sectors of azimuth aspect given its radar observation at horizontal polarization and a carrier frequency of 3 GHz (10 cm wavelength).

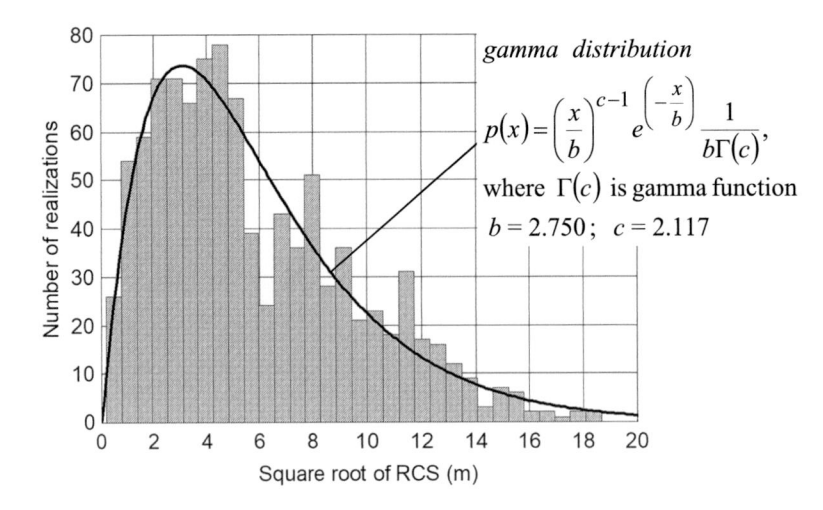

FIGURE 2.336 Amplitude distribution of echo signal of B-1B aircraft model at a carrier frequency of 3 GHz given its horizontal polarization.

$$p(x) = \left(\frac{x}{b}\right)^{c-1} e^{\left(-\frac{x}{b}\right)} \frac{1}{b\Gamma(c)},$$

where $\Gamma(c)$ is gamma function
$b = 2.750; \quad c = 2.117$

Scattering characteristics of B-1B aircraft model for sounding frequency 1 GHz (wavelength 30 cm).

Figure 2.337 shows the RCS circular diagram of the B-1B. The noncoherent RCS circular diagram of the B-1B is shown in Figure 2.338.

The circular mean RCS of the B-1B aircraft for horizontal polarization is 200.80 m². The circular median RCS for horizontal polarization is 15.79 m².

Figures 2.339 and 2.340 show the mean and median RCS for the main ranges of sounding azimuths (nose, side, and tail) and for ranges of 20°.

The expressions and parameters of probability distributions that are most consistent with the empirical distributions of the square root of the RCS for different ranges of sounding azimuths and polarizations are given in the electronic appendix to this book. The expressions and parameters of the probability distributions that are most consistent with the empirical distributions of the RCS (energy characteristic) for different ranges of sounding azimuths and polarizations are also given there.

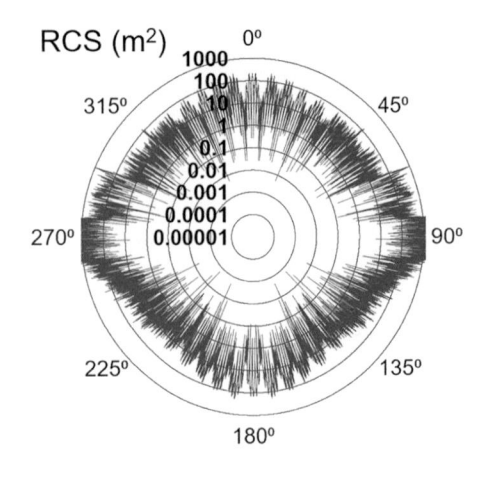

FIGURE 2.337 Circular diagram of RCS given radar observation of B-1B aircraft model at a carrier frequency of 1 GHz (30 cm wavelength).

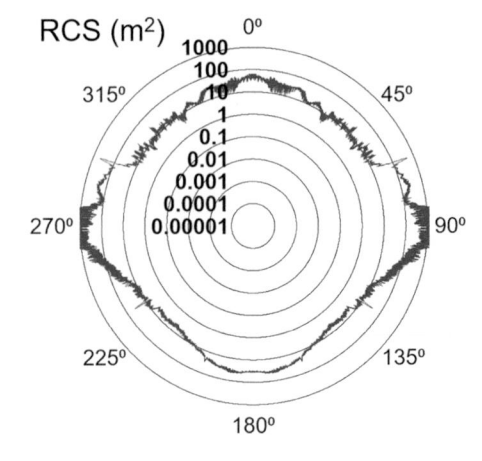

FIGURE 2.338 Circular diagram of noncoherent RCS given radar observation of B-1B aircraft model at a carrier frequency of 1 GHz (30 cm wavelength).

Scattering Characteristics of Aerial Objects 131

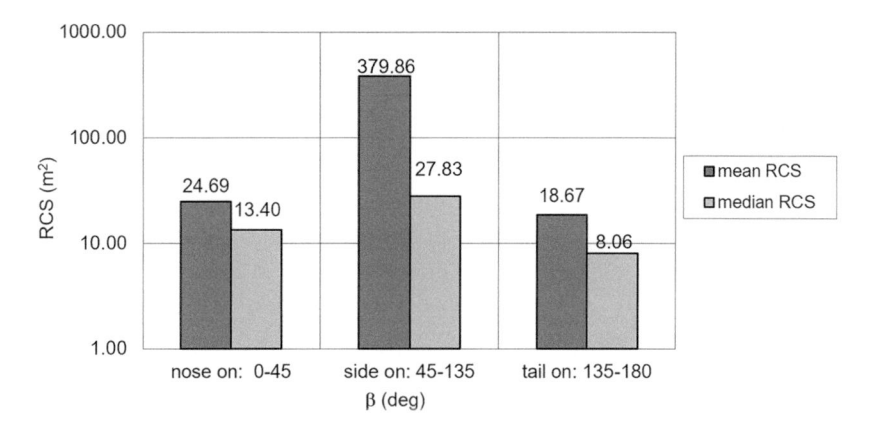

FIGURE 2.339 Diagrams of mean and median RCS of B-1B aircraft model in three sectors of azimuth aspect given its radar observation at horizontal polarization and a carrier frequency of 1 GHz (30 cm wavelength).

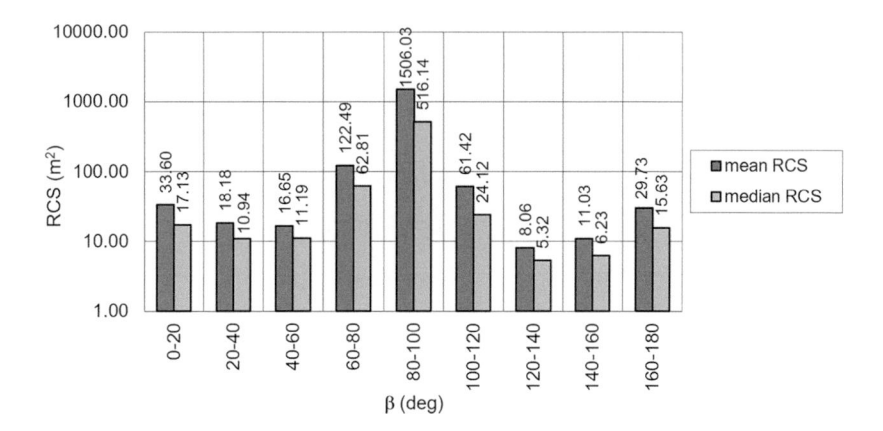

FIGURE 2.340 Diagrams of mean and median RCS of B-1B aircraft model in 20-degree sectors of azimuth aspect given its radar observation at horizontal polarization and a carrier frequency of 1 GHz (30 cm wavelength).

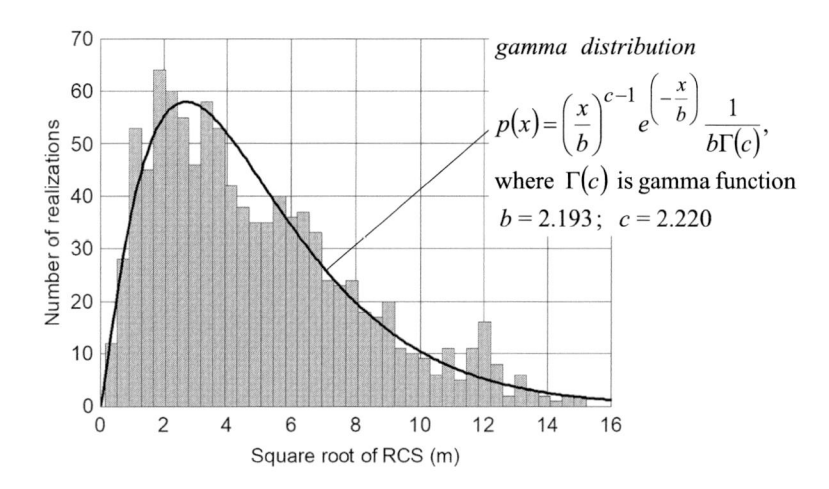

gamma distribution

$$p(x) = \left(\frac{x}{b}\right)^{c-1} e^{\left(-\frac{x}{b}\right)} \frac{1}{b\Gamma(c)},$$

where $\Gamma(c)$ is gamma function
$b = 2.193; \quad c = 2.220$

FIGURE 2.341 Amplitude distribution of echo signal of B-1B aircraft model at a carrier frequency of 1 GHz given its horizontal polarization.

2.21 AIRBORNE EARLY WARNING AND CONTROL AIRCRAFT A-50

A-50 Mainstay (Figure 2.342) is an airborne early warning and control aircraft. The first flight took place in 1978.

General characteristics of A-50 [32]: wingspan –50.50 m, length – 46.50 m, height – 14.80 m, wing area – 300.00 m², weight – 190 000 kg, powerplant – 4 D-30KP turbofan engines, speed – 800 km/h, range – 7 500 km, and service ceiling – 12 000 m.

In accordance with the design of the A-50, a model of its surface was created to obtaining scattering characteristics (in particular, RCS). The model is shown in Figure 2.343. In the modeling, the smooth parts of the aircraft surface were approximated by parts of 81 triaxial ellipsoids. The surface breaks were modeled using 18 straight edge scattering parts. The main antenna is pointed to the front.

FIGURE 2.342 Airborne early warning and control aircraft A-50.

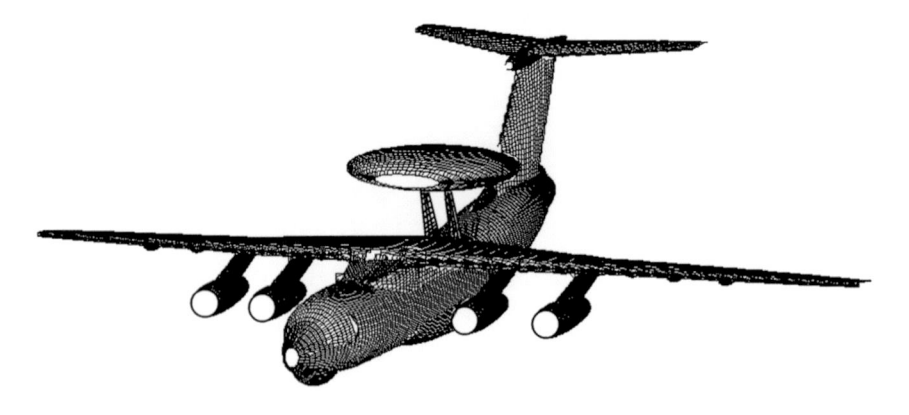

FIGURE 2.343 Surface model of airborne early warning and control aircraft A-50.

Below are some scattering characteristics of the A-50 model at sounding frequencies of 10, 3, and 1 GHz (wavelengths of 3, 10, and 30 cm, respectively) for horizontal polarization of the probing signal. The scattering characteristics for this model at two polarizations, as well as scattering characteristics at sounding frequencies of 5 and 1.3 GHz (wavelengths of 6 and 23 cm, respectively), are given in the electronic appendix of this book.

Sounding parameters: The elevation angle is to be a random value distributed uniformly in the range −3° ± 4° with respect to the wing plane (elevation angle of −3° corresponds to the radar observation from the lower hemisphere), azimuth aspect increment was 0.02°, the azimuth being counted off from the nose-on aspect (0° corresponds to the nose-on radar observation, 180° corresponds to the tail-on observation).

Scattering characteristics of A-50 aircraft model for sounding frequency 10 GHz (wavelength 3 cm).

Figure 2.344 shows the RCS circular diagram of the A-50 model. The noncoherent RCS circular diagram of the A-50 is shown in Figure 2.345.

The circular mean RCS of the A-50 aircraft model for horizontal polarization is 441.81 m². The circular median RCS (the RCS value used to calculate the detection range of an object with a probability of 0.5) for horizontal polarization is 24.52 m².

Figures 2.346 and 2.347 show the mean and median RCS for the main ranges of sounding azimuths (nose, side, and tail) and for ranges of 20°.

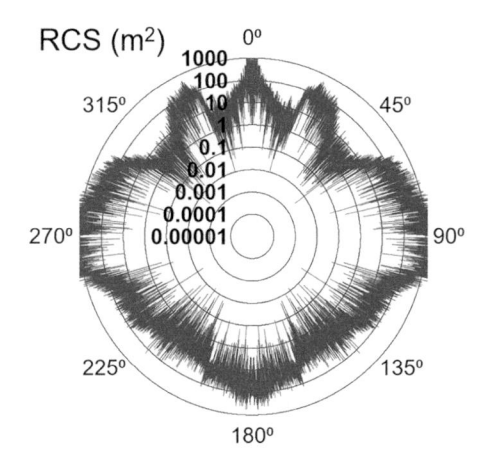

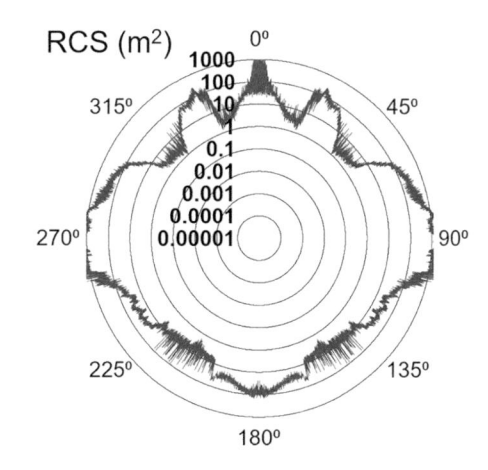

FIGURE 2.344 Circular diagram of RCS given radar observation of A-50 aircraft model at a carrier frequency of 10 GHz (3 cm wavelength).

FIGURE 2.345 Circular diagram of noncoherent RCS given radar observation of A-50 aircraft model at a carrier frequency of 10 GHz (3 cm wavelength).

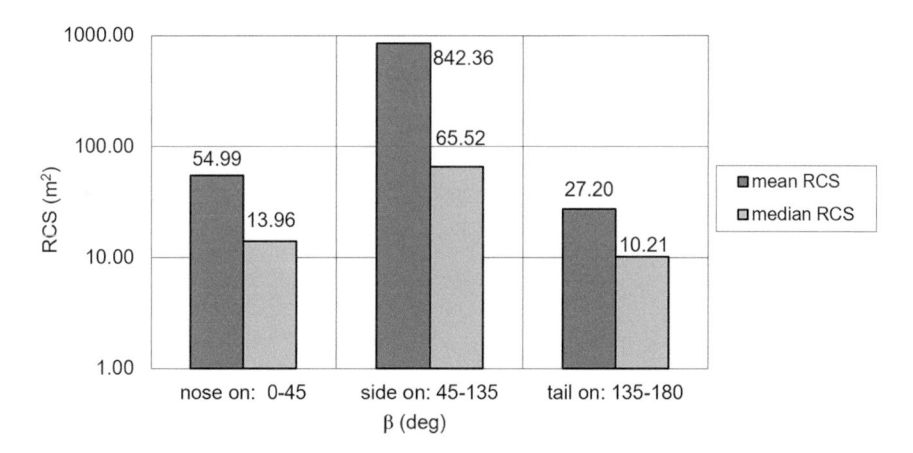

FIGURE 2.346 Diagrams of mean and median RCS of A-50 aircraft model in three sectors of azimuth aspect given its radar observation at horizontal polarization and a carrier frequency of 10 GHz (3 cm wavelength).

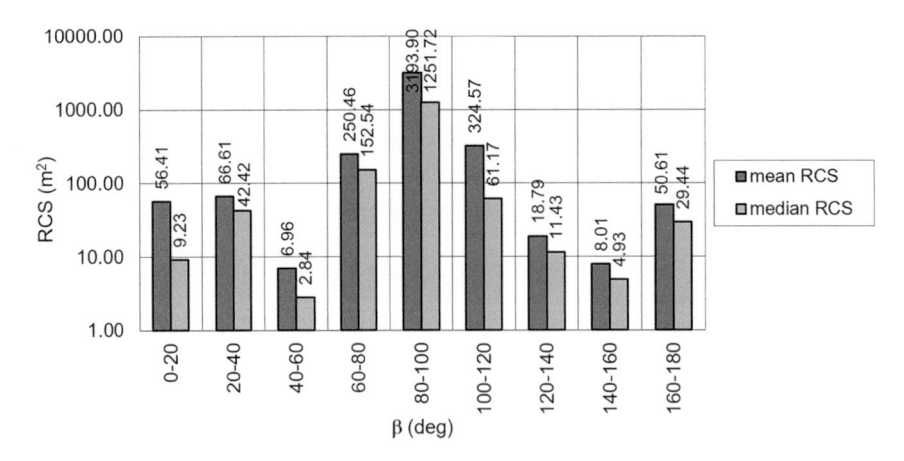

FIGURE 2.347 Diagrams of mean and median RCS of A-50 aircraft model in 20-degree sectors of azimuth aspect given its radar observation at horizontal polarization and a carrier frequency of 10 GHz (3 cm wavelength).

Figures 2.348, 2.353, and 2.358 show histograms of the scattered signal amplitude (square root of the RCS) for the range of sounding azimuths −20° to +20° (sounding from the nose). The bold line shows the probability density functions of the distribution, which can be used to approximate the histogram of the signal amplitude.

$$p(x) = \frac{1}{\sqrt{2\pi}\, x\sigma} \exp\left(-\frac{(\log(x) - \mu)^2}{2\sigma^2}\right)$$

log - normal distribution

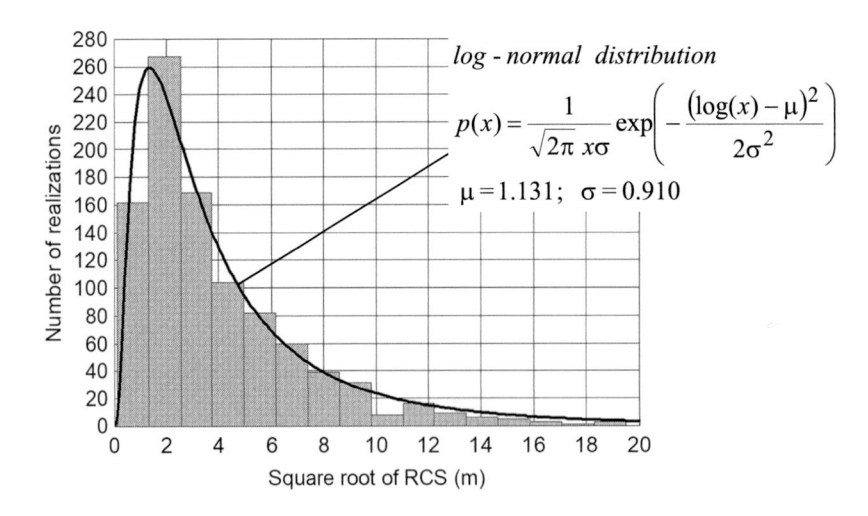

$\mu = 1.131;\quad \sigma = 0.910$

FIGURE 2.348 Amplitude distribution of echo signal of A-50 aircraft model at a carrier frequency of 10 GHz given its horizontal polarization.

Scattering characteristics of A-50 aircraft model for sounding frequency 3 GHz (wavelength 10 cm).

Figure 2.349 shows the RCS circular diagram of the A-50. The noncoherent RCS circular diagram of the A-50 is shown in Figure 2.350.

The circular mean RCS of the A-50 aircraft for horizontal polarization is 421.77 m². The circular median RCS for horizontal polarization is 10.93 m².

Figures 2.351 and 2.352 show the mean and median RCS for the main ranges of sounding azimuths (nose, side, and tail) and for ranges of 20°.

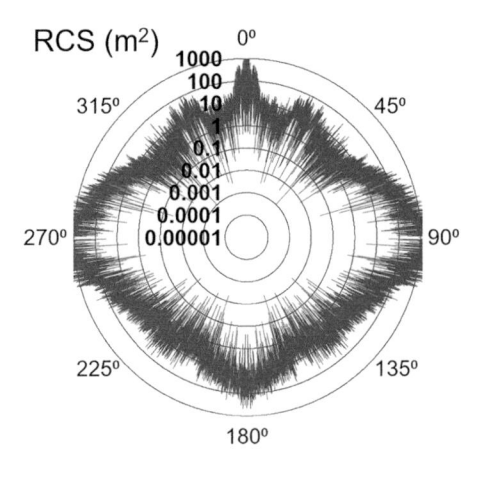

Scattering Characteristics of Aerial Objects

FIGURE 2.349 Circular diagram of RCS given radar observation of A-50 aircraft model at a carrier frequency of 3 GHz (10 cm wavelength).

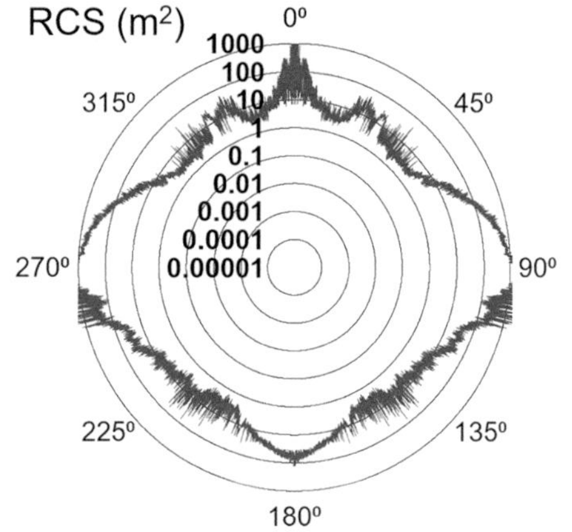

FIGURE 2.350 Circular diagram of noncoherent RCS given radar observation of A-50 aircraft model at a carrier frequency of 3 GHz (10 cm wavelength).

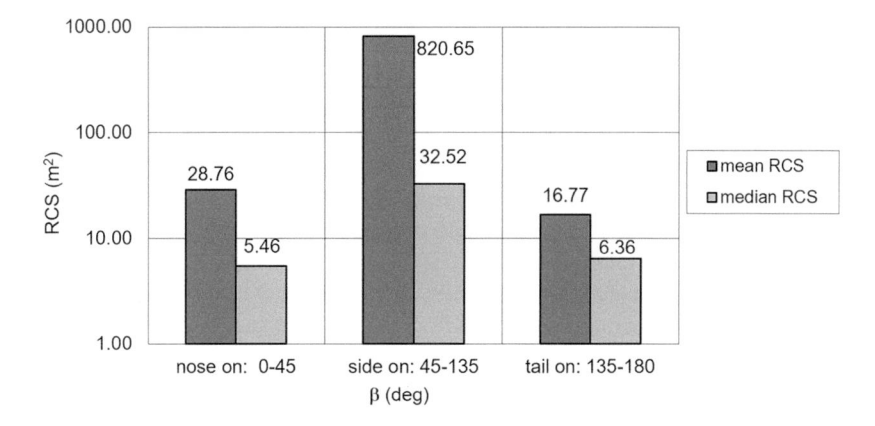

FIGURE 2.351 Diagrams of mean and median RCS of A-50 aircraft model in three sectors of azimuth aspect given its radar observation at horizontal polarization and a carrier frequency of 3 GHz (10 cm wavelength).

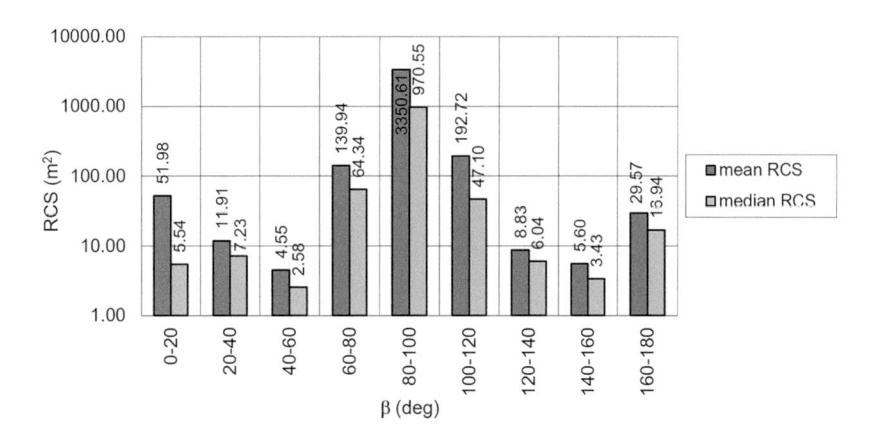

FIGURE 2.352 Diagrams of mean and median RCS of A-50 aircraft model in 20-degree sectors of azimuth aspect given its radar observation at horizontal polarization and a carrier frequency of 3 GHz (10 cm wavelength).

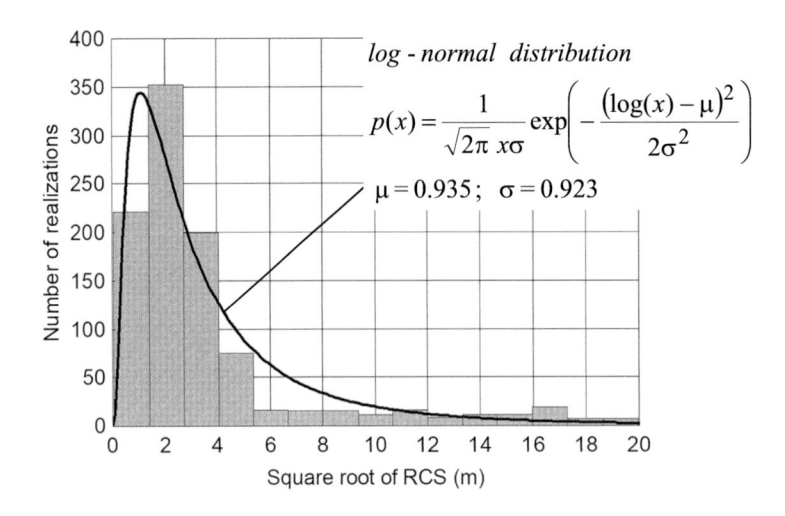

FIGURE 2.353 Amplitude distribution of echo signal of A-50 aircraft model at a carrier frequency of 3 GHz given its horizontal polarization.

$$p(x) = \frac{1}{\sqrt{2\pi}\, x\sigma} \exp\left(-\frac{(\log(x) - \mu)^2}{2\sigma^2}\right)$$

log-normal distribution; $\mu = 0.935$; $\sigma = 0.923$

Scattering characteristics of A-50 aircraft model for sounding frequency 1 GHz (wavelength 30 cm).

Figure 2.354 shows the RCS circular diagram of the A-50. The noncoherent RCS circular diagram of the A-50 is shown in Figure 2.355.

The circular mean RCS of the A-50 aircraft for horizontal polarization is 380.58 m². The circular median RCS for horizontal polarization is 13.18 m².

Figures 2.356 and 2.357 show the mean and median RCS for the main ranges of sounding azimuths (nose, side, and tail) and for ranges of 20°.

The expressions and parameters of probability distributions that are most consistent with the empirical distributions of the square root of the RCS for different ranges of sounding azimuths and polarizations are given in the electronic appendix to this book. The expressions and parameters of the probability distributions that are most consistent with the empirical distributions of the RCS (energy characteristic) for different ranges of sounding azimuths and polarizations are also given there.

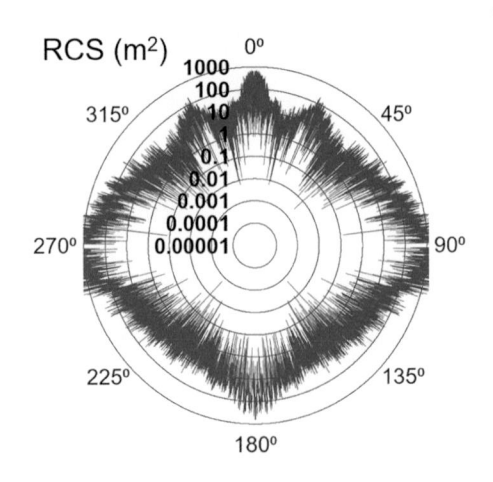

FIGURE 2.354 Circular diagram of RCS given radar observation of A-50 aircraft model at a carrier frequency of 1 GHz (30 cm wavelength).

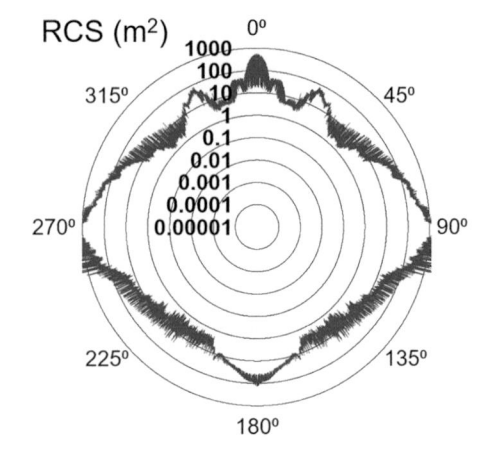

FIGURE 2.355 Circular diagram of noncoherent RCS given radar observation of A-50 aircraft model at a carrier frequency of 1 GHz (30 cm wavelength).

Scattering Characteristics of Aerial Objects

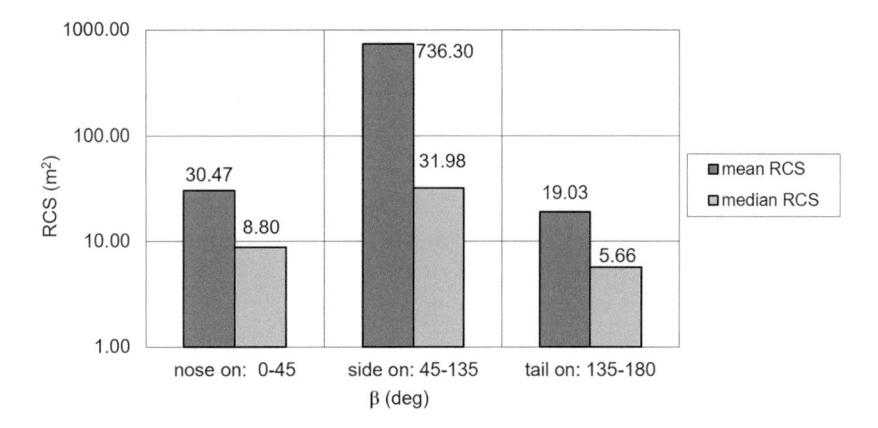

FIGURE 2.356 Diagrams of mean and median RCS of A-50 aircraft model in three sectors of azimuth aspect given its radar observation at horizontal polarization and a carrier frequency of 1 GHz (30 cm wavelength).

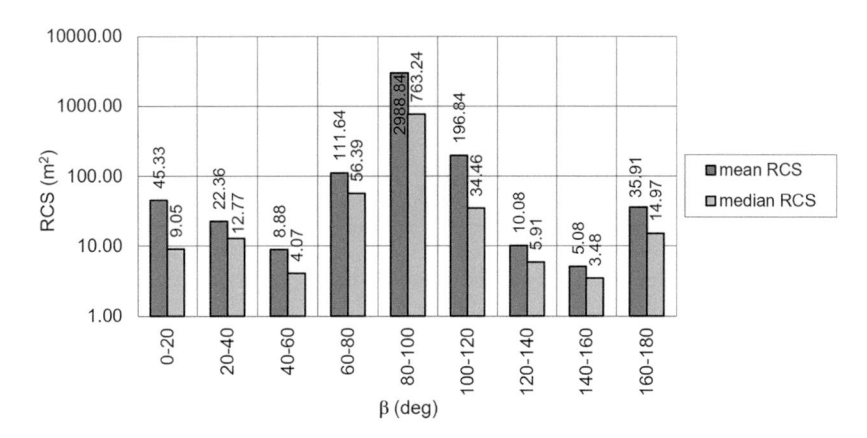

FIGURE 2.357 Diagrams of mean and median RCS of A-50 aircraft model in 20-degree sectors of azimuth aspect given its radar observation at horizontal polarization and a carrier frequency of 1 GHz (30 cm wavelength).

$$p(x) = \frac{1}{\sqrt{2\pi}\, x\sigma} \exp\left(-\frac{(\log(x)-\mu)^2}{2\sigma^2}\right)$$

$\mu = 1.140;\ \sigma = 0.889$

log - normal distribution

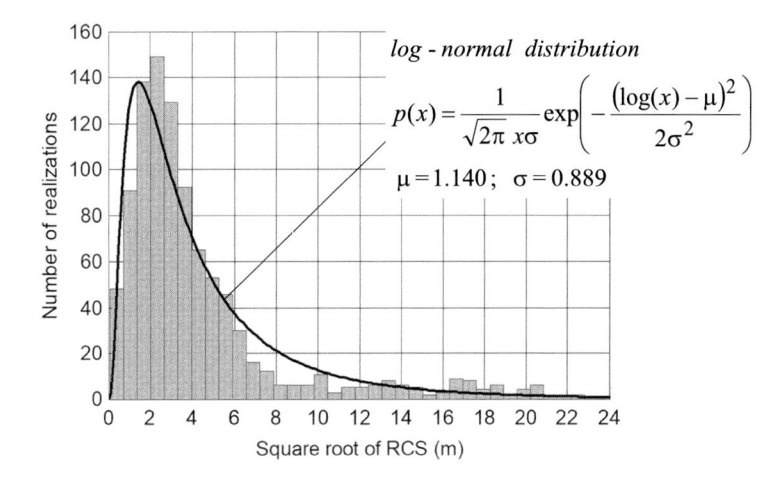

FIGURE 2.358 Amplitude distribution of echo signal of A-50 aircraft model at a carrier frequency of 1 GHz given its horizontal polarization.

2.22 AIRBORNE EARLY WARNING AND CONTROL AIRCRAFT E-3A

E-3A Sentry (Figure 2.359) is an airborne early warning and control aircraft. The first flight took place in 1972.

General characteristics of E-3A [33]: wingspan −44.42 m, length − 46.61 m, height − 12.6 m, wing area − 283.75 m^2, weight − 156 000 kg, powerplant − 4 Pratt and Whitney TF33-PW-100A turbofan, maximum speed − 854 km/h, range − 7 400 km, and service ceiling − 8 850 m.

In accordance with the design of the E-3A, a model of its surface was created to obtaining scattering characteristics (in particular, RCS). The model is shown in Figure 2.360. In the modeling, the smooth parts of the aircraft surface were approximated by parts of 52 triaxial ellipsoids. The surface breaks were modeled using 19 straight edge scattering parts. The main antenna is pointed to the front.

FIGURE 2.359 Airborne early warning and control aircraft E-3A.

FIGURE 2.360 Surface model of airborne early warning and control aircraft E-3A.

Below are some scattering characteristics of the E-3A model at sounding frequencies of 10, 3, and 1 GHz (wavelengths of 3, 10, and 30 cm, respectively) for horizontal polarization of the probing signal. The scattering characteristics for this model at two polarizations, as well as scattering characteristics at sounding frequencies of 5 and 1.3 GHz (wavelengths of 6 and 23 cm, respectively), are given in the electronic appendix of this book.

Scattering Characteristics of Aerial Objects 139

Sounding parameters: The elevation angle is to be a random value distributed uniformly in the range −3° ± 4° with respect to the wing plane (elevation angle of −3° corresponds to the radar observation from the lower hemisphere), azimuth aspect increment was 0.02°, the azimuth being counted off from the nose-on aspect (0° corresponds to the nose-on radar observation, 180° corresponds to the tail-on observation).

Scattering characteristics of E-3A aircraft model for sounding frequency 10 GHz (wavelength 3 cm).

Figure 2.361 shows the RCS circular diagram of the E-3A model. The noncoherent RCS circular diagram of the E-3A is shown in Figure 2.362.

The circular mean RCS of the E-3A aircraft model for horizontal polarization is 519.70 m². The circular median RCS (the RCS value used to calculate the detection range of an object with a probability of 0.5) for horizontal polarization is 17.74 m².

Figures 2.363 and 2.364 show the mean and median RCS for the main ranges of sounding azimuths (nose, side, and tail) and for ranges of 20°.

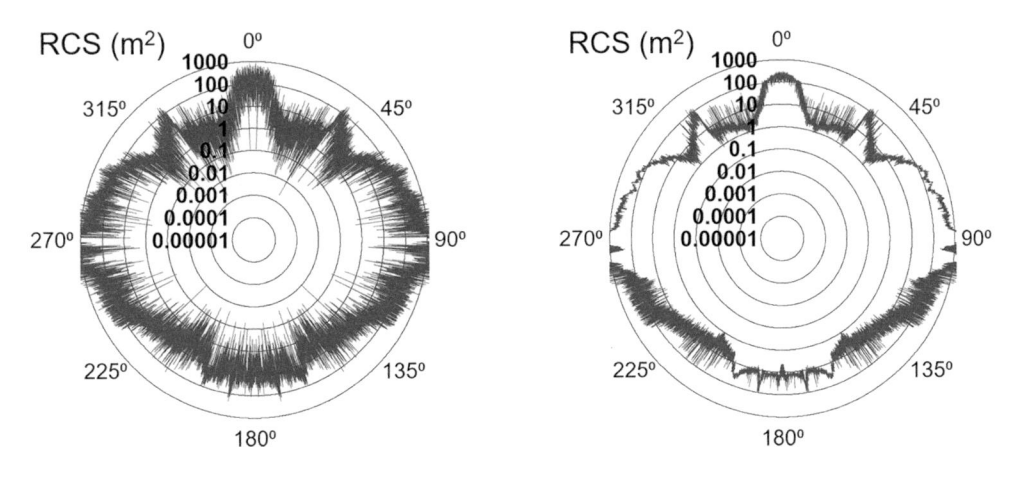

FIGURE 2.361 Circular diagram of RCS given radar observation of E-3A aircraft model at a carrier frequency of 10 GHz (3 cm wavelength).

FIGURE 2.362 Circular diagram of noncoherent RCS given radar observation of E-3A aircraft model at a carrier frequency of 10 GHz (3 cm wavelength).

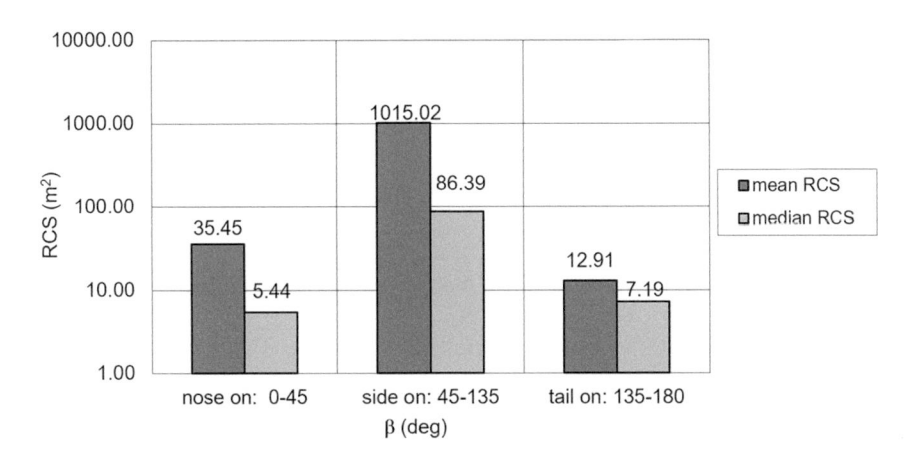

FIGURE 2.363 Diagrams of mean and median RCS of E-3A aircraft model in three sectors of azimuth aspect given its radar observation at horizontal polarization and a carrier frequency of 10 GHz (3 cm wavelength).

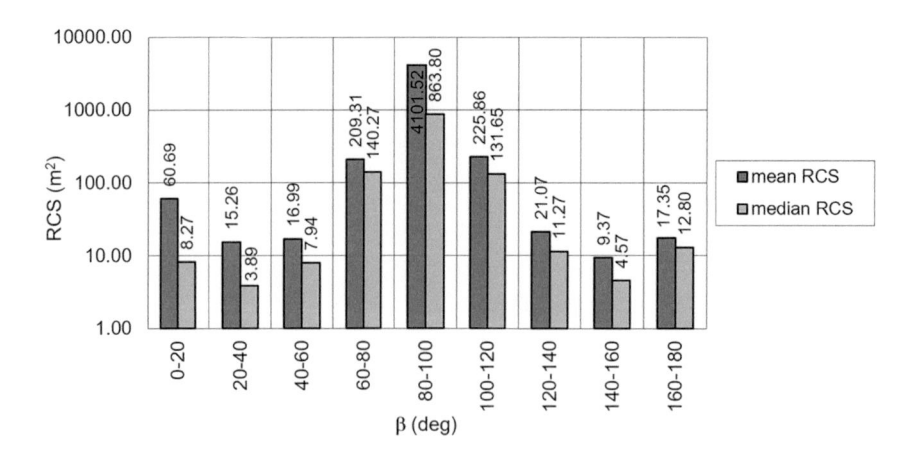

FIGURE 2.364 Diagrams of mean and median RCS of E-3A aircraft model in 20-degree sectors of azimuth aspect given its radar observation at horizontal polarization and a carrier frequency of 10 GHz (3 cm wavelength).

Figures 2.365, 2.370, and 2.375 show histograms of the scattered signal amplitude (square root of the RCS) for the range of sounding azimuths −20° to +20° (sounding from the nose). The bold line shows the probability density functions of the distribution, which can be used to approximate the histogram of the signal amplitude.

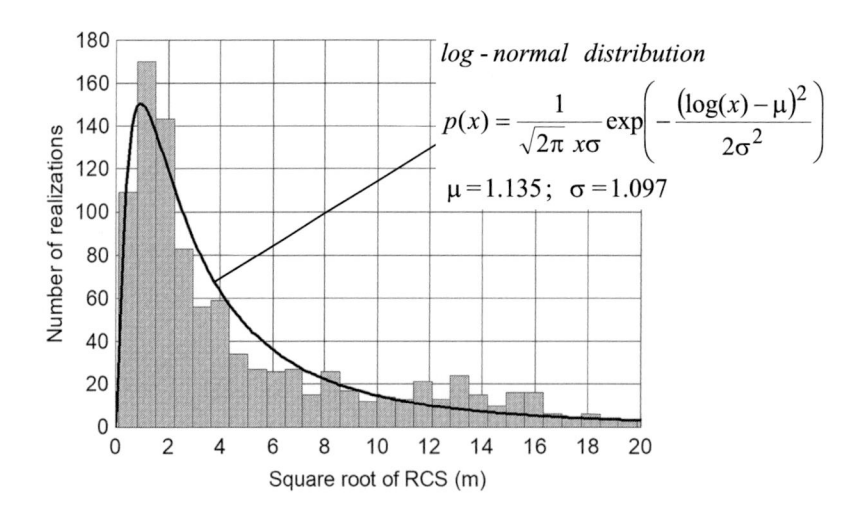

$$\text{log - normal distribution}$$
$$p(x) = \frac{1}{\sqrt{2\pi}\, x\sigma} \exp\left(-\frac{(\log(x) - \mu)^2}{2\sigma^2}\right)$$
$$\mu = 1.135;\quad \sigma = 1.097$$

FIGURE 2.365 Amplitude distribution of echo signal of E-3A aircraft model at a carrier frequency of 10 GHz given its horizontal polarization.

Scattering characteristics of E-3A aircraft model for sounding frequency 3 GHz (wavelength 10 cm).

Figure 2.366 shows the RCS circular diagram of the E-3A. The noncoherent RCS circular diagram of the E-3A is shown in Figure 2.367.

The circular mean RCS of the E-3A aircraft for horizontal polarization is 369.99 m². The circular median RCS for horizontal polarization is 9.75 m².

Figures 2.368 and 2.369 show the mean and median RCS for the main ranges of sounding azimuths (nose, side, and tail) and for ranges of 20°.

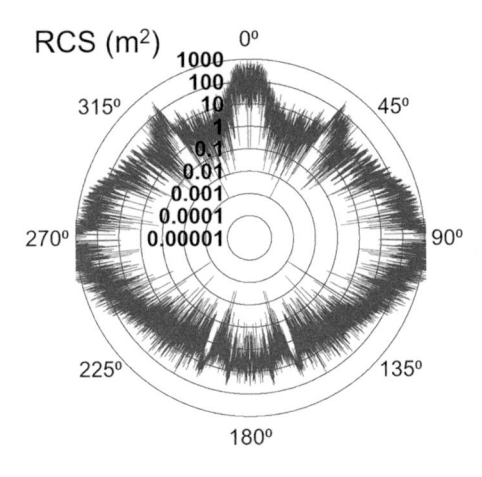

Scattering Characteristics of Aerial Objects   141

FIGURE 2.366 Circular diagram of RCS given radar observation of E-3A aircraft model at a carrier frequency of 3 GHz (10 cm wavelength).

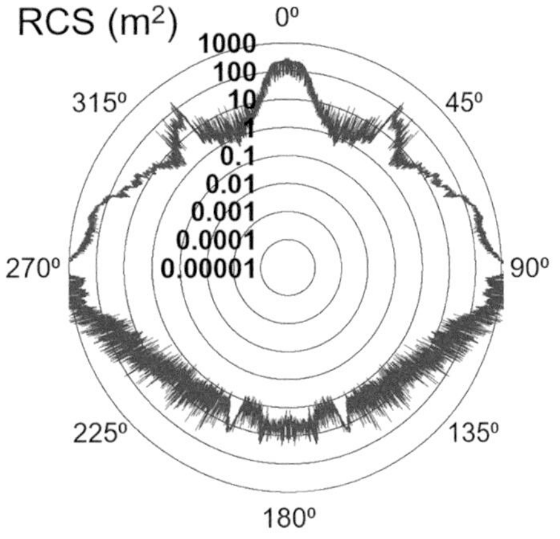

FIGURE 2.367 Circular diagram of noncoherent RCS given radar observation of E-3A aircraft model at a carrier frequency of 3 GHz (10 cm wavelength).

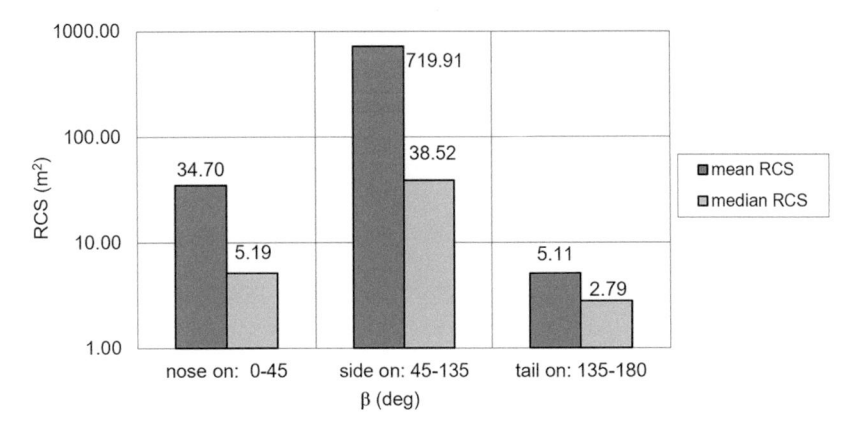

FIGURE 2.368 Diagrams of mean and median RCS of E-3A aircraft model in three sectors of azimuth aspect given its radar observation at horizontal polarization and a carrier frequency of 3 GHz (10 cm wavelength).

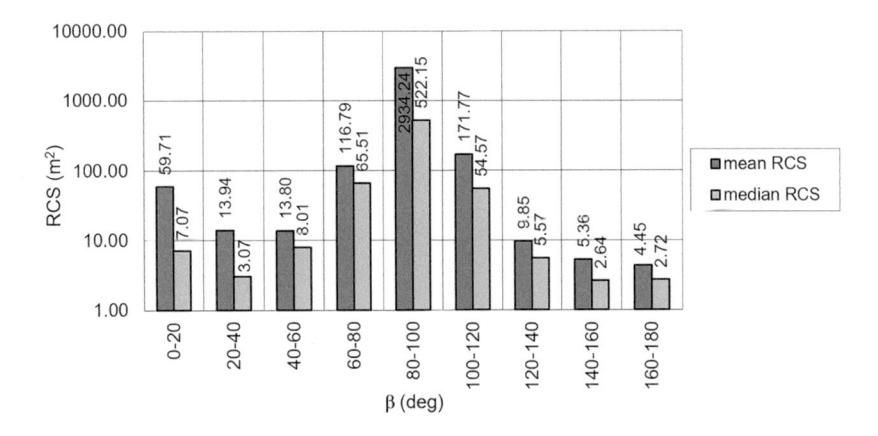

FIGURE 2.369 Diagrams of mean and median RCS of E-3A aircraft model in 20-degree sectors of azimuth aspect given its radar observation at horizontal polarization and a carrier frequency of 3 GHz (10 cm wavelength).

$$p(x) = \frac{1}{\sqrt{2\pi}\, x\sigma} \exp\left(-\frac{(\log(x)-\mu)^2}{2\sigma^2}\right)$$

log - normal distribution

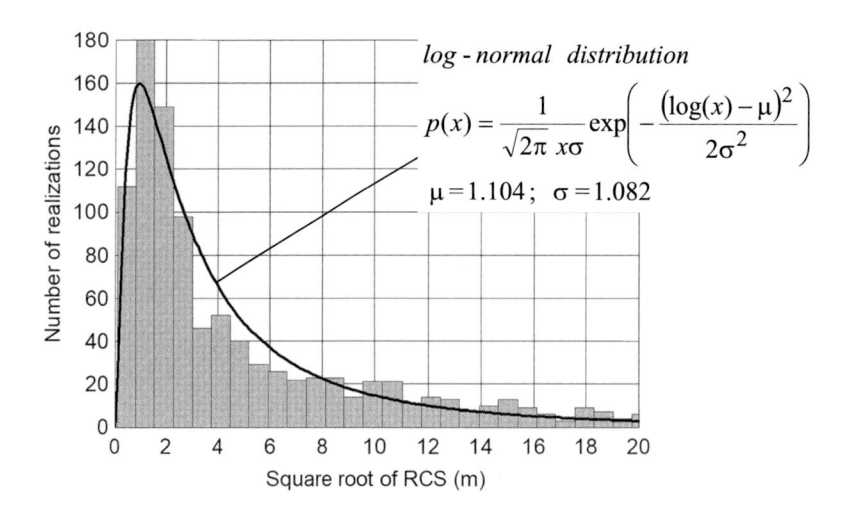

$\mu = 1.104; \quad \sigma = 1.082$

FIGURE 2.370 Amplitude distribution of echo signal of E-3A aircraft model at a carrier frequency of 3 GHz given its horizontal polarization.

Scattering characteristics of E-3A aircraft model for sounding frequency 1 GHz (wavelength 30 cm).

Figure 2.371 shows the RCS circular diagram of the E-3A. The noncoherent RCS circular diagram of the E-3A is shown in Figure 2.372.

The circular mean RCS of the E-3A aircraft for horizontal polarization is 246.05 m². The circular median RCS for horizontal polarization is 6.34 m².

Figures 2.373 and 2.374 show the mean and median RCS for the main ranges of sounding azimuths (nose, side, and tail) and for ranges of 20°.

The expressions and parameters of probability distributions that are most consistent with the empirical distributions of the square root of the RCS for different ranges of sounding azimuths and polarizations are given in the electronic appendix to this book. The expressions and parameters of the probability distributions that are most consistent with the empirical distributions of the RCS (energy characteristic) for different ranges of sounding azimuths and polarizations are also given there.

FIGURE 2.371 Circular diagram of RCS given radar observation of E-3A aircraft model at a carrier frequency of 1 GHz (30 cm wavelength).

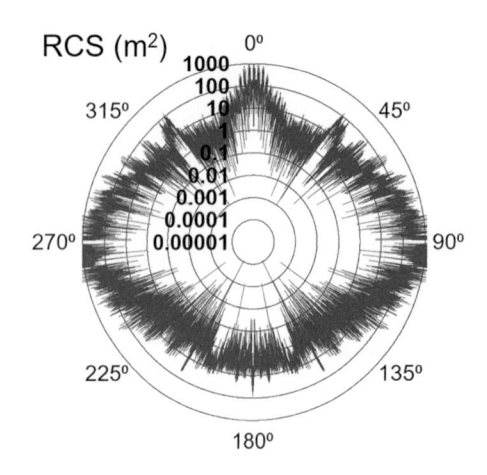

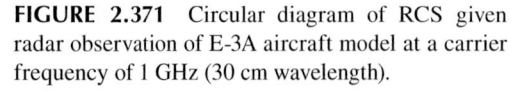

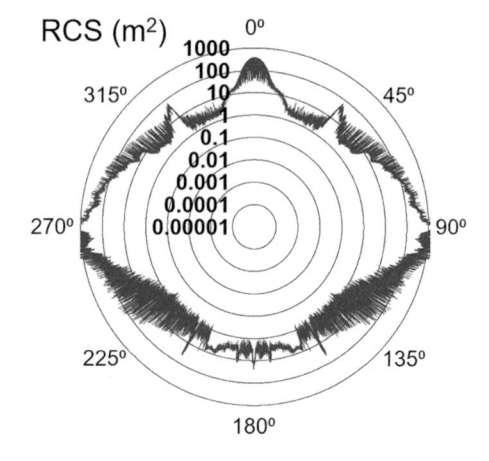

FIGURE 2.372 Circular diagram of noncoherent RCS given radar observation of E-3A aircraft model at a carrier frequency of 1 GHz (30 cm wavelength).

Scattering Characteristics of Aerial Objects

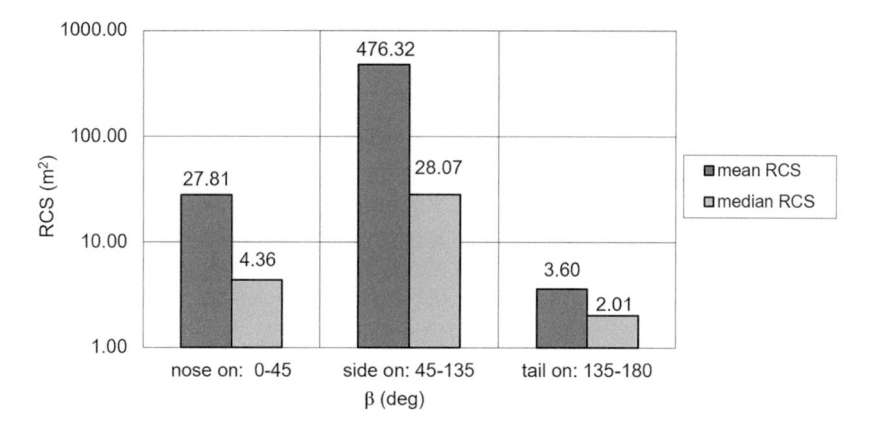

FIGURE 2.373 Diagrams of mean and median RCS of E-3A aircraft model in three sectors of azimuth aspect given its radar observation at horizontal polarization and a carrier frequency of 1 GHz (30 cm wavelength).

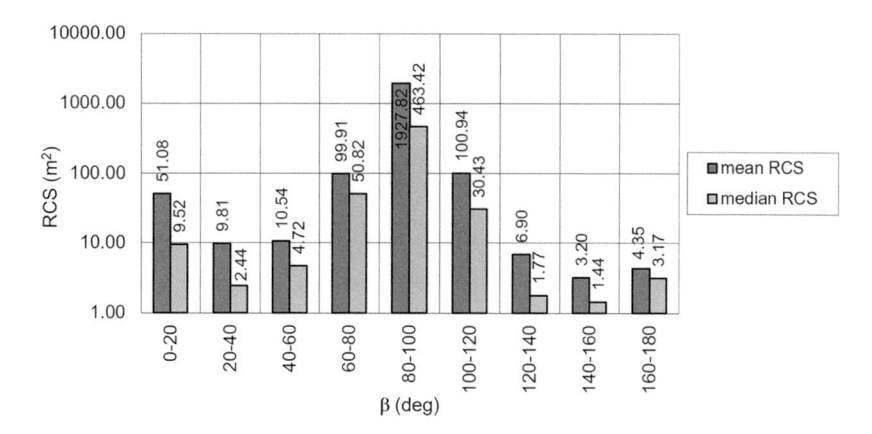

FIGURE 2.374 Diagrams of mean and median RCS of E-3A aircraft model in 20-degree sectors of azimuth aspect given its radar observation at horizontal polarization and a carrier frequency of 1 GHz (30 cm wavelength).

$$p(x) = \frac{1}{\sqrt{2\pi}\, x\sigma} \exp\left(-\frac{(\log(x) - \mu)^2}{2\sigma^2}\right)$$

log - normal distribution

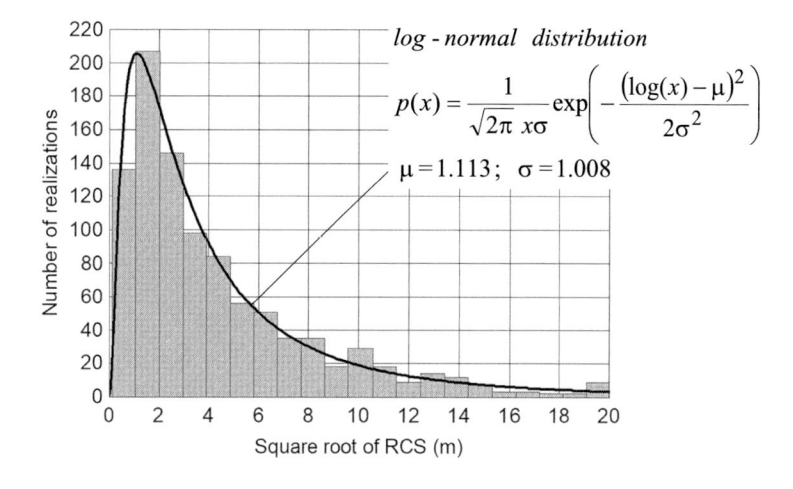

$\mu = 1.113; \quad \sigma = 1.008$

FIGURE 2.375 Amplitude distribution of echo signal of E-3A aircraft model at a carrier frequency of 1 GHz given its horizontal polarization.

2.23 AIRBORNE EARLY WARNING AND CONTROL AIRCRAFT E-2C

E-2C Hawkeye (Figure 2.376) is a carrier-capable tactical airborne early warning aircraft. The first flight took place in 1971.

General characteristics of E-2C [34]: wingspan –24.56 m, length – 17.54 m, height – 5.58 m, wing area – 65.03 m^2, weight – 18 265–23 856 kg, powerplant – 2 Allison/Rolls-Royce T-56-A427 turboprop engines, speed – 552 km/h, range – 2 700 km, and service ceiling – 9 400 m.

In accordance with the design of the E-2C, a model of its surface was created to obtaining scattering characteristics (in particular, RCS). The model is shown in Figure 2.377. In the modeling, the smooth parts of the aircraft surface were approximated by parts of 58 triaxial ellipsoids. The surface breaks were modeled using 8 straight edge scattering parts. The main antenna is pointed to the front.

FIGURE 2.376 Airborne early warning aircraft E-2C.

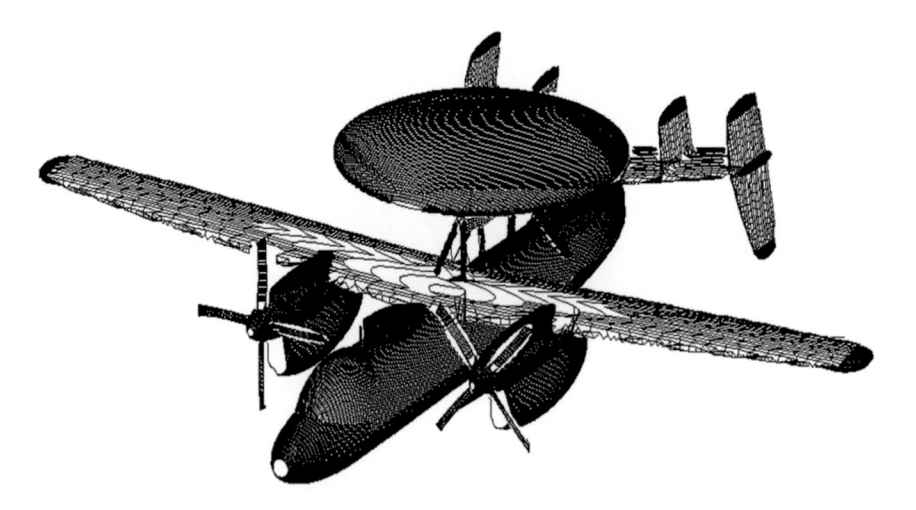

FIGURE 2.377 Surface model of airborne early warning aircraft E-2C.

Below are some scattering characteristics of the E-2C model at sounding frequencies of 10, 3, and 1 GHz (wavelengths of 3, 10, and 30 cm, respectively) for horizontal polarization of the probing signal. The scattering characteristics for this model at two polarizations, as well as scattering characteristics at sounding frequencies of 5 and 1.3 GHz (wavelengths of 6 and 23 cm, respectively), are given in the electronic appendix of this book.

Scattering Characteristics of Aerial Objects 145

Sounding parameters: The elevation angle is to be a random value distributed uniformly in the range –3° ± 4° with respect to the wing plane (elevation angle of –3° corresponds to the radar observation from the lower hemisphere), azimuth aspect increment was 0.02°, the azimuth being counted off from the nose-on aspect (0° corresponds to the nose-on radar observation, 180° corresponds to the tail-on observation).

Scattering characteristics of E-2C aircraft model for sounding frequency 10 GHz (wavelength 3 cm).

Figure 2.378 shows the RCS circular diagram of the E-2C model. The noncoherent RCS circular diagram of the E-2C is shown in Figure 2.379.

The circular mean RCS of the E-2C aircraft model for horizontal polarization is 51.21 m². The circular median RCS (the RCS value used to calculate the detection range of an object with a probability of 0.5) for horizontal polarization is 12.90 m².

Figures 2.380 and 2.381 show the mean and median RCS for the main ranges of sounding azimuths (nose, side, and tail) and for ranges of 20°.

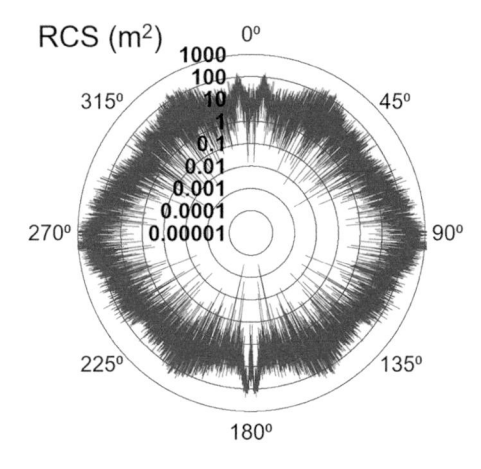

FIGURE 2.378 Circular diagram of RCS given radar observation of E-2C aircraft model at a carrier frequency of 10 GHz (3 cm wavelength).

FIGURE 2.379 Circular diagram of noncoherent RCS given radar observation of E-2C aircraft model at a carrier frequency of 10 GHz (3 cm wavelength).

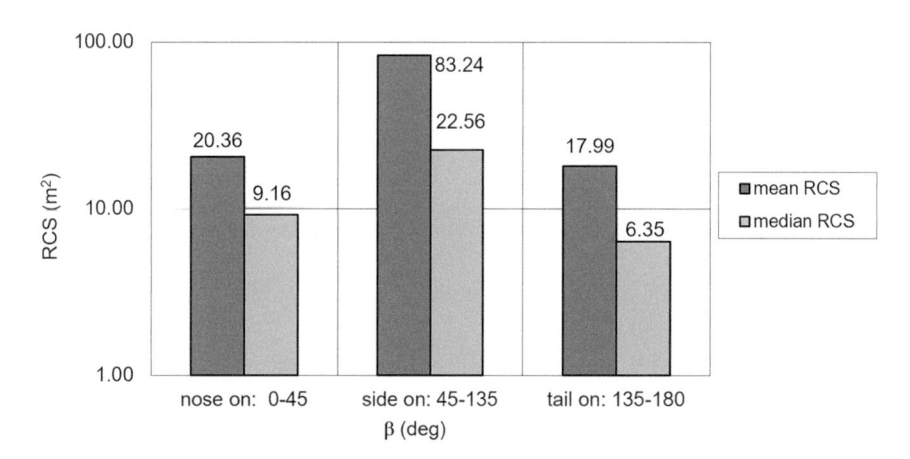

FIGURE 2.380 Diagrams of mean and median RCS of E-2C aircraft model in three sectors of azimuth aspect given its radar observation at horizontal polarization and a carrier frequency of 10 GHz (3 cm wavelength).

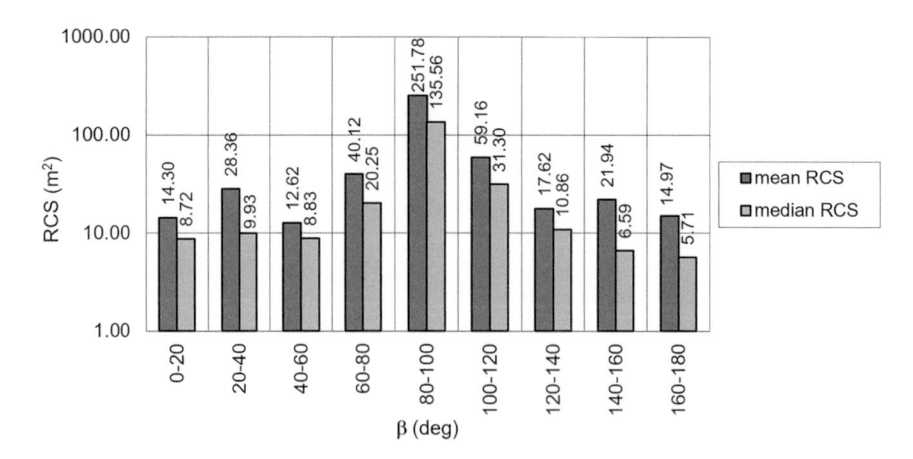

FIGURE 2.381 Diagrams of mean and median RCS of E-2C aircraft model in 20-degree sectors of azimuth aspect given its radar observation at horizontal polarization and a carrier frequency of 10 GHz (3 cm wavelength).

Figures 2.382, 2.387, and 2.392 show histograms of the scattered signal amplitude (square root of the RCS) for the range of sounding azimuths −20° to +20° (sounding from the nose). The bold line shows the probability density functions of the distribution, which can be used to approximate the histogram of the signal amplitude.

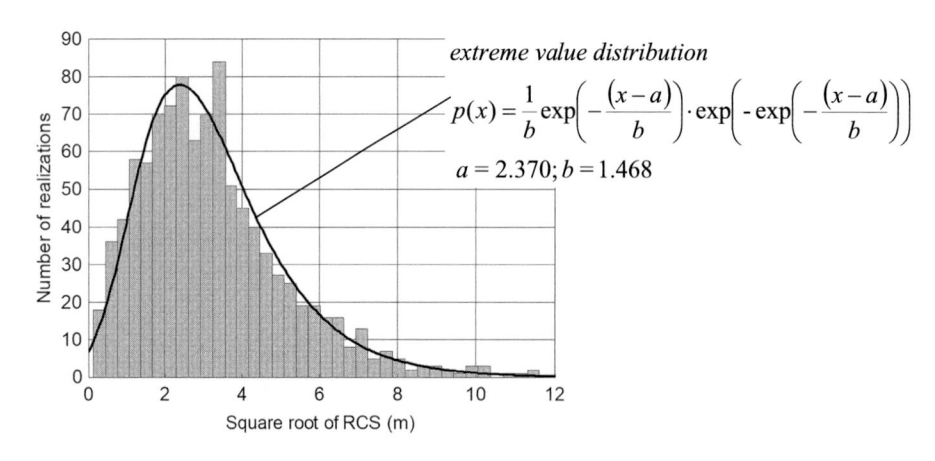

$$p(x) = \frac{1}{b} \exp\left(-\frac{(x-a)}{b}\right) \cdot \exp\left(-\exp\left(-\frac{(x-a)}{b}\right)\right)$$

$a = 2.370; b = 1.468$

FIGURE 2.382 Amplitude distribution of echo signal of E-2C aircraft model at a carrier frequency of 10 GHz given its horizontal polarization.

Scattering characteristics of E-2C aircraft model for sounding frequency 3 GHz (wavelength 10 cm).

Figure 2.383 shows the RCS circular diagram of the E-2C. The noncoherent RCS circular diagram of the E-2C is shown in Figure 2.384.

The circular mean RCS of the E-2C aircraft for horizontal polarization is 36.42 m². The circular median RCS for horizontal polarization is 10.80 m².

Figures 2.385 and 2.386 show the mean and median RCS for the main ranges of sounding azimuths (nose, side, and tail) and for ranges of 20°.

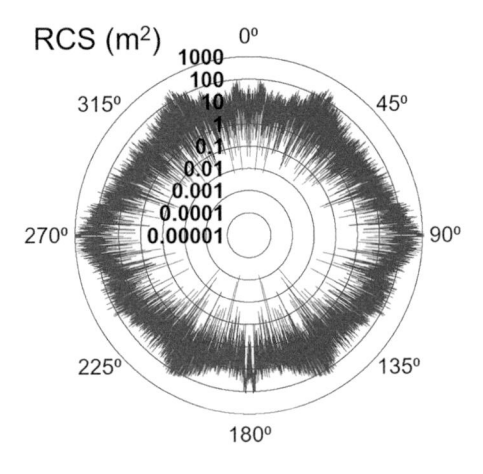

Scattering Characteristics of Aerial Objects 147

FIGURE 2.383 Circular diagram of RCS given radar observation of E-2C aircraft model at a carrier frequency of 3 GHz (10 cm wavelength).

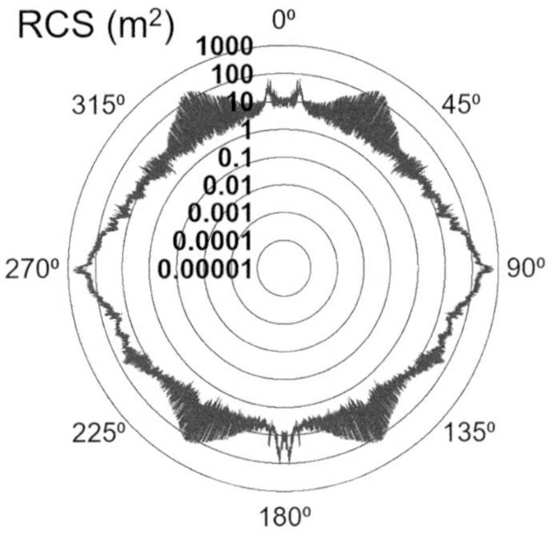

FIGURE 2.384 Circular diagram of noncoherent RCS given radar observation of E-2C aircraft model at a carrier frequency of 3 GHz (10 cm wavelength).

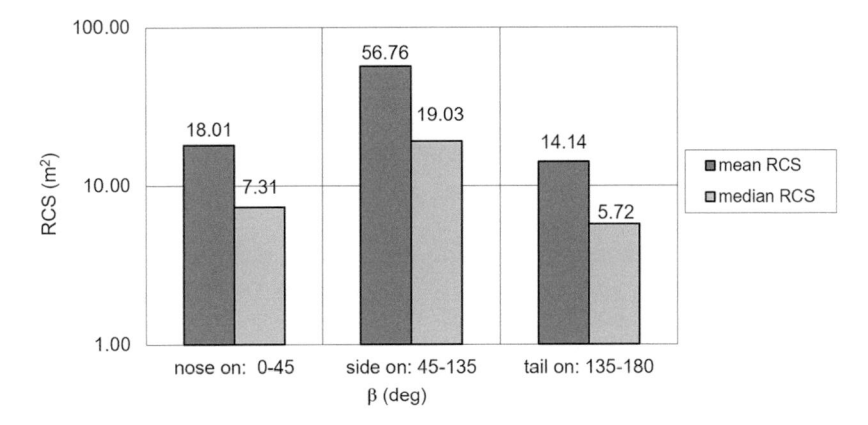

FIGURE 2.385 Diagrams of mean and median RCS of E-2C aircraft model in three sectors of azimuth aspect given its radar observation at horizontal polarization and a carrier frequency of 3 GHz (10 cm wavelength).

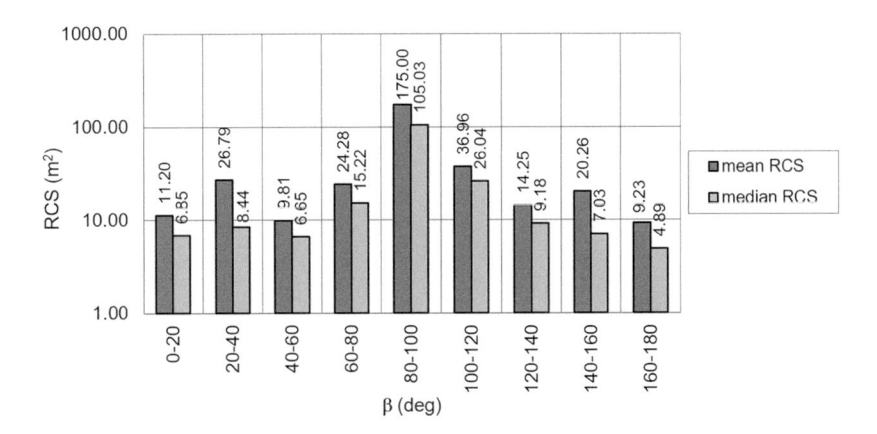

FIGURE 2.386 Diagrams of mean and median RCS of E-2C aircraft model in 20-degree sectors of azimuth aspect given its radar observation at horizontal polarization and a carrier frequency of 3 GHz (10 cm wavelength).

extreme value distribution

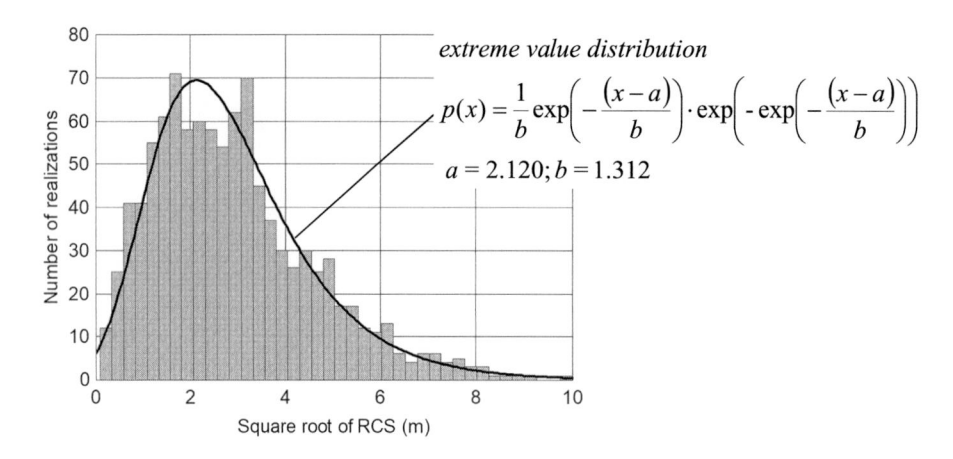

$$p(x) = \frac{1}{b}\exp\left(-\frac{(x-a)}{b}\right) \cdot \exp\left(-\exp\left(-\frac{(x-a)}{b}\right)\right)$$

$a = 2.120; b = 1.312$

FIGURE 2.387 Amplitude distribution of echo signal of E-2C aircraft model at a carrier frequency of 3 GHz given its horizontal polarization.

Scattering characteristics of E-2C aircraft model for sounding frequency 1 GHz (wavelength 30 cm).

Figure 2.388 shows the RCS circular diagram of the E-2C. The noncoherent RCS circular diagram of the E-2C is shown in Figure 2.389.

The circular mean RCS of the E-2C aircraft for horizontal polarization is 38.90 m². The circular median RCS for horizontal polarization is 10.93 m².

Figures 2.390 and 2.391 show the mean and median RCS for the main ranges of sounding azimuths (nose, side, and tail) and for ranges of 20°.

The expressions and parameters of probability distributions that are most consistent with the empirical distributions of the square root of the RCS for different ranges of sounding azimuths and polarizations are given in the electronic appendix to this book. The expressions and parameters of the probability distributions that are most consistent with the empirical distributions of the RCS (energy characteristic) for different ranges of sounding azimuths and polarizations are also given there.

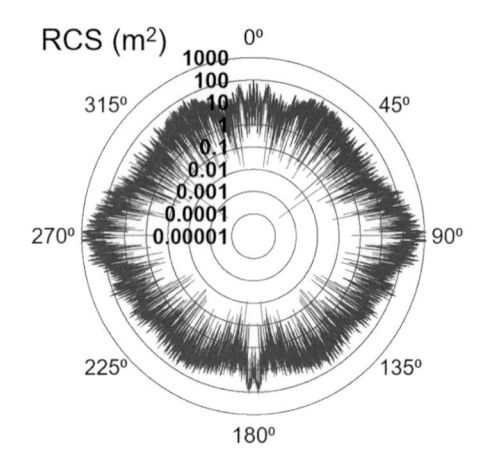

FIGURE 2.388 Circular diagram of RCS given radar observation of E-2C aircraft model at a carrier frequency of 1 GHz (30 cm wavelength).

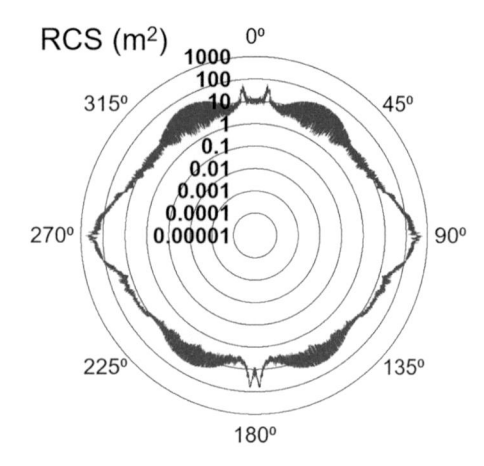

FIGURE 2.389 Circular diagram of noncoherent RCS given radar observation of E-2C aircraft model at a carrier frequency of 1 GHz (30 cm wavelength).

Scattering Characteristics of Aerial Objects

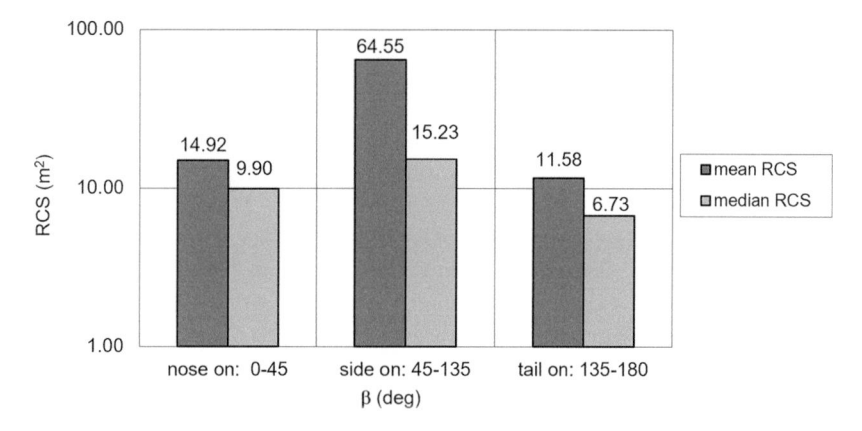

FIGURE 2.390 Diagrams of mean and median RCS of E-2C aircraft model in three sectors of azimuth aspect given its radar observation at horizontal polarization and a carrier frequency of 1 GHz (30 cm wavelength).

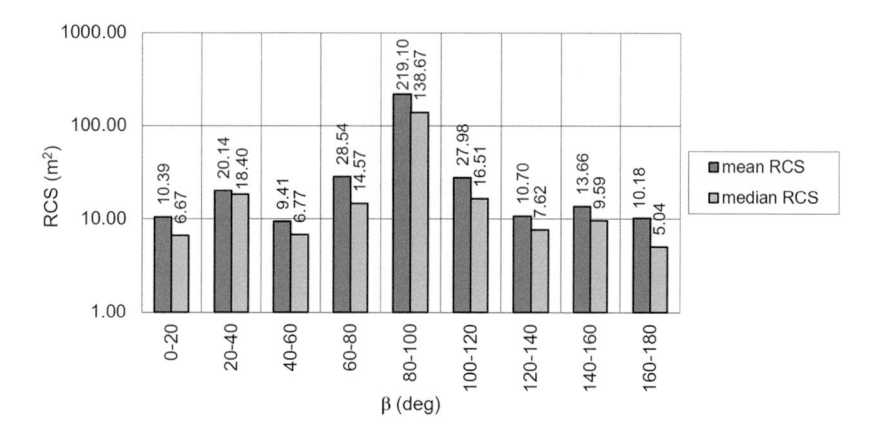

FIGURE 2.391 Diagrams of mean and median RCS of E-2C aircraft model in 20-degree sectors of azimuth aspect given its radar observation at horizontal polarization and a carrier frequency of 1 GHz (30 cm wavelength).

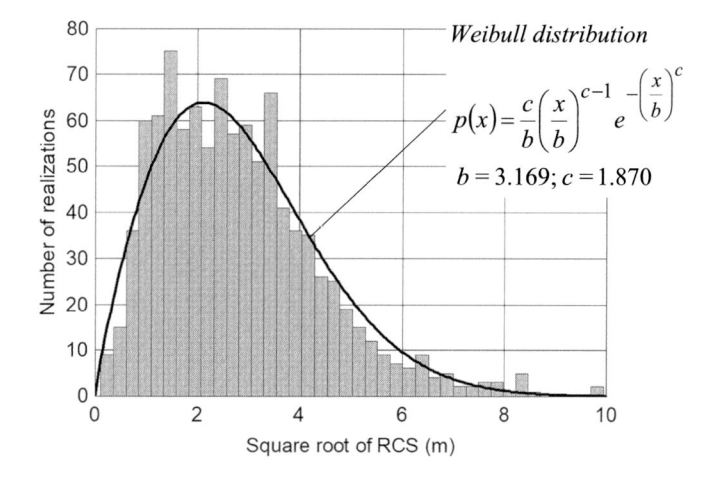

$$p(x) = \frac{c}{b}\left(\frac{x}{b}\right)^{c-1} e^{-\left(\frac{x}{b}\right)^c}$$

$b = 3.169; c = 1.870$

FIGURE 2.392 Amplitude distribution of echo signal of E-2C aircraft model at a carrier frequency of 1 GHz given its horizontal polarization.

2.24 JET TRAINER AIRCRAFT L-39

L-39 Albatros (Figure 2.393) is a high-performance jet trainer aircraft. The first flight took place in 1968.

General characteristics of L-39 [35]: wingspan –9.44 m, length – 12.15 m, height – 4.47 m, wing area – 18.80 m^2, weight – 3 395–4 600 kg, powerplant – 1 Ivchenko AI-25TL turbofan engine, speed – 757 km/h, range – 1 000–1 750 km, and service ceiling – 11 500 m.

In accordance with the design of the L-39, a model of its surface was created to obtaining scattering characteristics (in particular, RCS). The model is shown in Figure 2.394. In the modeling, the smooth parts of the aircraft surface were approximated by parts of 41 triaxial ellipsoids. The surface breaks were modeled using 9 straight edge scattering parts.

FIGURE 2.393 Jet trainer aircraft L-39.

FIGURE 2.394 Surface model of jet trainer aircraft L-39.

Below are some scattering characteristics of the L-39 model at sounding frequencies of 10, 3, and 1 GHz (wavelengths of 3, 10, and 30 cm, respectively) for horizontal polarization of the probing signal. The scattering characteristics for this model at two polarizations, as well as scattering characteristics at sounding frequencies of 5 and 1.3 GHz (wavelengths of 6 and 23 cm, respectively), are given in the electronic appendix of this book.

Scattering Characteristics of Aerial Objects 151

Sounding parameters: The elevation angle is to be a random value distributed uniformly in the range −3° ± 4° with respect to the wing plane (elevation angle of −3° corresponds to the radar observation from the lower hemisphere), azimuth aspect increment was 0.02°, the azimuth being counted off from the nose-on aspect (0° corresponds to the nose-on radar observation, 180° corresponds to the tail-on observation).

Scattering characteristics of L-39 aircraft model for sounding frequency 10 GHz (wavelength 3 cm).

Figure 2.395 shows the RCS circular diagram of the L-39 model. The noncoherent RCS circular diagram of the L-39 is shown in Figure 2.396.

The circular mean RCS of the L-39 aircraft model for horizontal polarization is 20.09 m². The circular median RCS (the RCS value used to calculate the detection range of an object with a probability of 0.5) for horizontal polarization is 1.38 m².

Figures 2.397 and 2.398 show the mean and median RCS for the main ranges of sounding azimuths (nose, side, and tail) and for ranges of 20°.

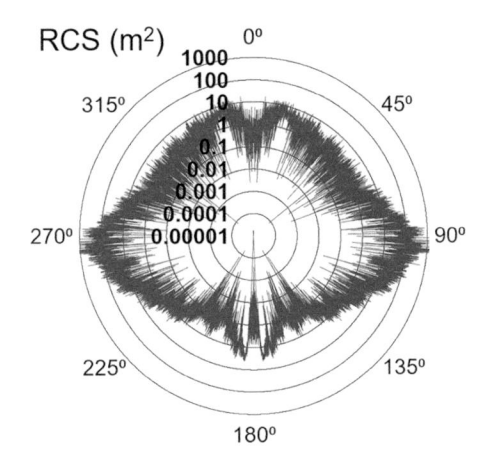

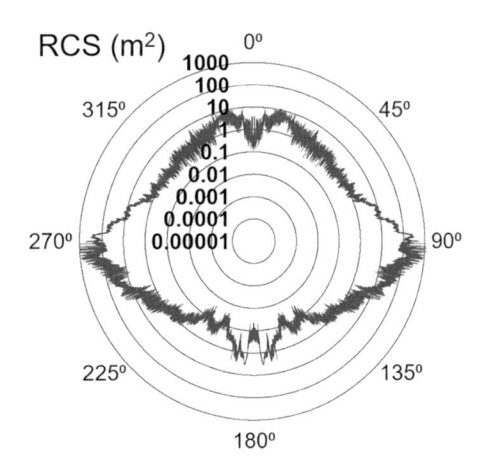

FIGURE 2.395 Circular diagram of RCS given radar observation of L-39 aircraft model at a carrier frequency of 10 GHz (3 cm wavelength).

FIGURE 2.396 Circular diagram of noncoherent RCS given radar observation of L-39 aircraft model at a carrier frequency of 10 GHz (3 cm wavelength).

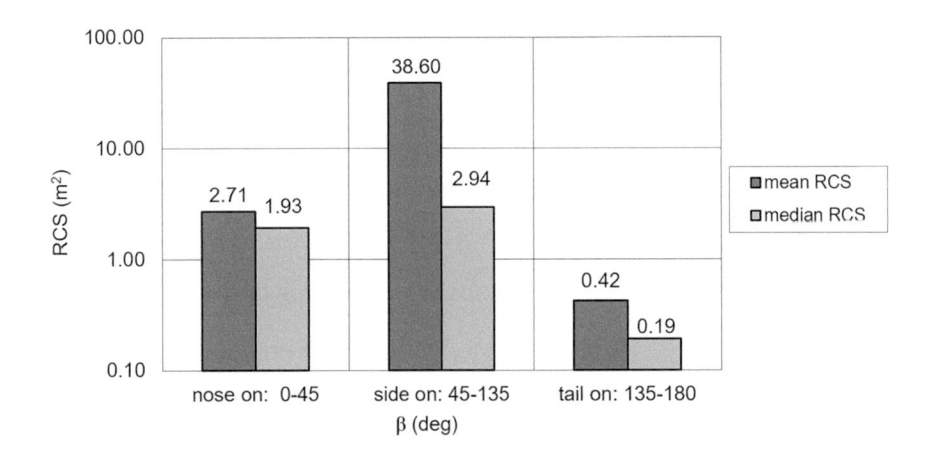

FIGURE 2.397 Diagrams of mean and median RCS of L-39 aircraft model in three sectors of azimuth aspect given its radar observation at horizontal polarization and a carrier frequency of 10 GHz (3 cm wavelength).

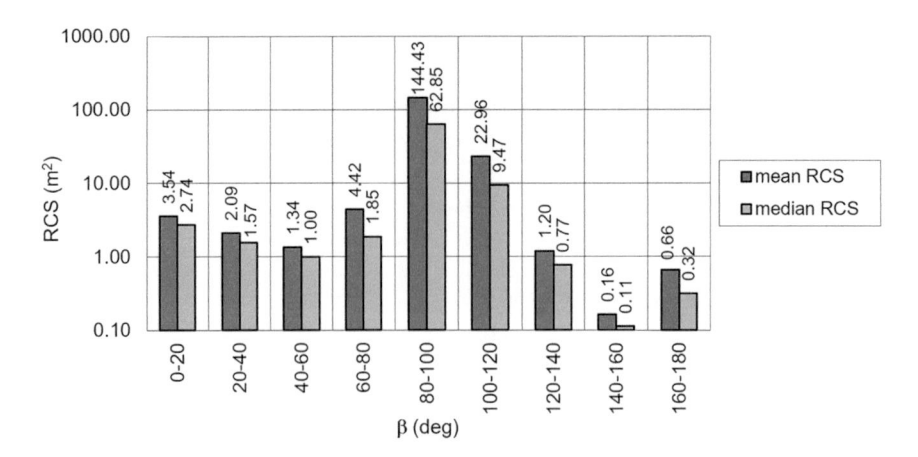

FIGURE 2.398 Diagrams of mean and median RCS of L-39 aircraft model in 20-degree sectors of azimuth aspect given its radar observation at horizontal polarization and a carrier frequency of 10 GHz (3 cm wavelength).

Figures 2.399, 2.404, and 2.409 show histograms of the scattered signal amplitude (square root of the RCS) for the range of sounding azimuths −20° to +20° (sounding from the nose). The bold line shows the probability density functions of the distribution, which can be used to approximate the histogram of the signal amplitude.

normal distribution

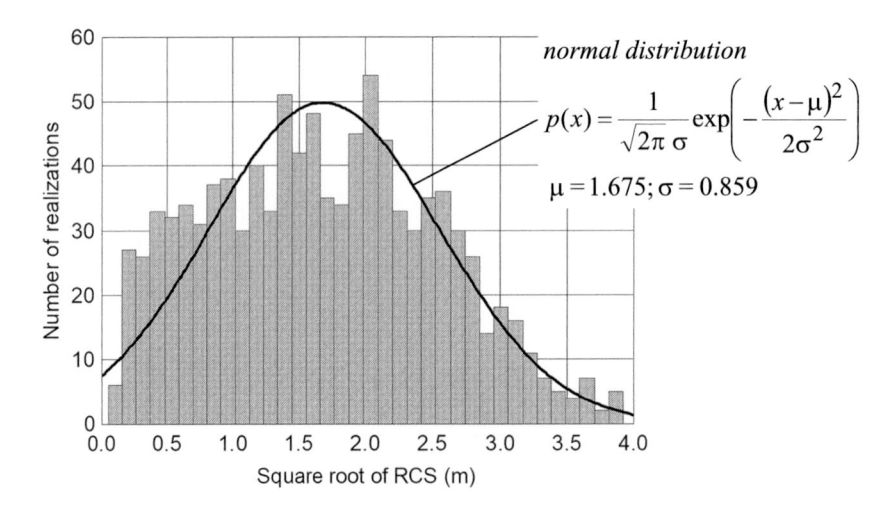

$$p(x) = \frac{1}{\sqrt{2\pi}\,\sigma} \exp\left(-\frac{(x-\mu)^2}{2\sigma^2}\right)$$

$\mu = 1.675;\ \sigma = 0.859$

FIGURE 2.399 Amplitude distribution of echo signal of L-39 aircraft model at a carrier frequency of 10 GHz given its horizontal polarization.

Scattering characteristics of L-39 aircraft model for sounding frequency 3 GHz (wavelength 10 cm).

Figure 2.400 shows the RCS circular diagram of the L-39. The noncoherent RCS circular diagram of the L-39 is shown in Figure 2.401.

The circular mean RCS of the L-39 aircraft for horizontal polarization is 16.95 m². The circular median RCS for horizontal polarization is 1.20 m².

Figures 2.402 and 2.403 show the mean and median RCS for the main ranges of sounding azimuths (nose, side, and tail) and for ranges of 20°.

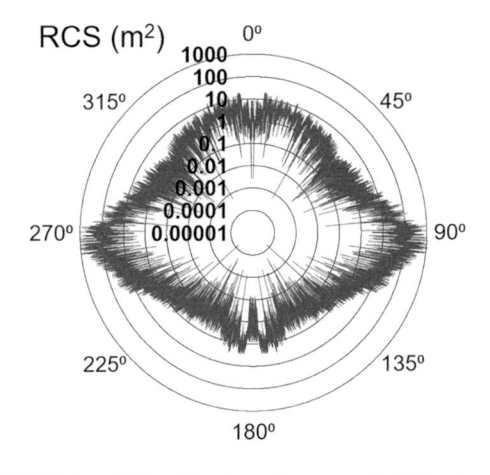

Scattering Characteristics of Aerial Objects

FIGURE 2.400 Circular diagram of RCS given radar observation of L-39 aircraft model at a carrier frequency of 3 GHz (10 cm wavelength).

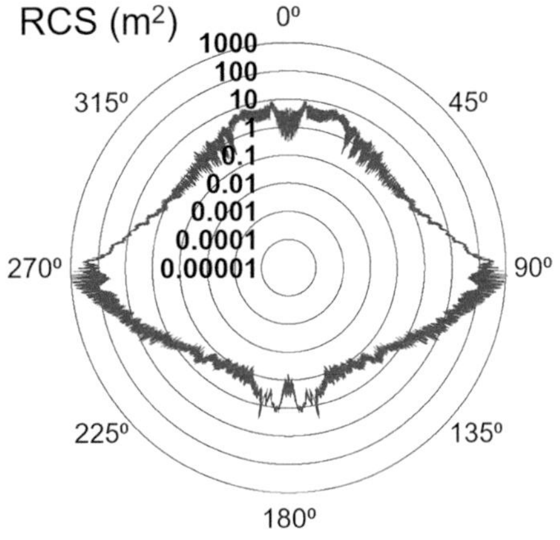

FIGURE 2.401 Circular diagram of noncoherent RCS given radar observation of L-39 aircraft model at a carrier frequency of 3 GHz (10 cm wavelength).

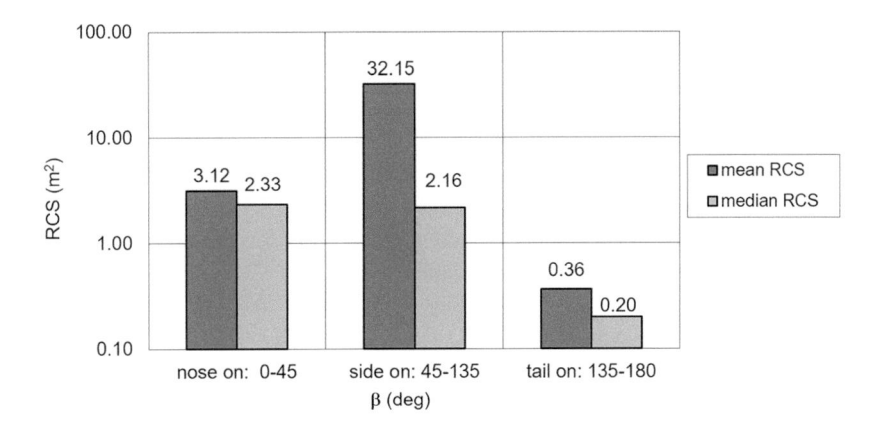

FIGURE 2.402 Diagrams of mean and median RCS of L-39 aircraft model in three sectors of azimuth aspect given its radar observation at horizontal polarization and a carrier frequency of 3 GHz (10 cm wavelength).

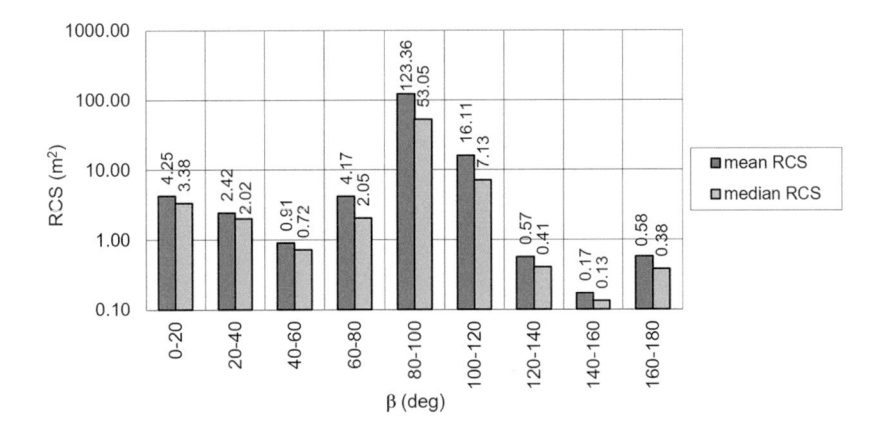

FIGURE 2.403 Diagrams of mean and median RCS of L-39 aircraft model in 20-degree sectors of azimuth aspect given its radar observation at horizontal polarization and a carrier frequency of 3 GHz (10 cm wavelength).

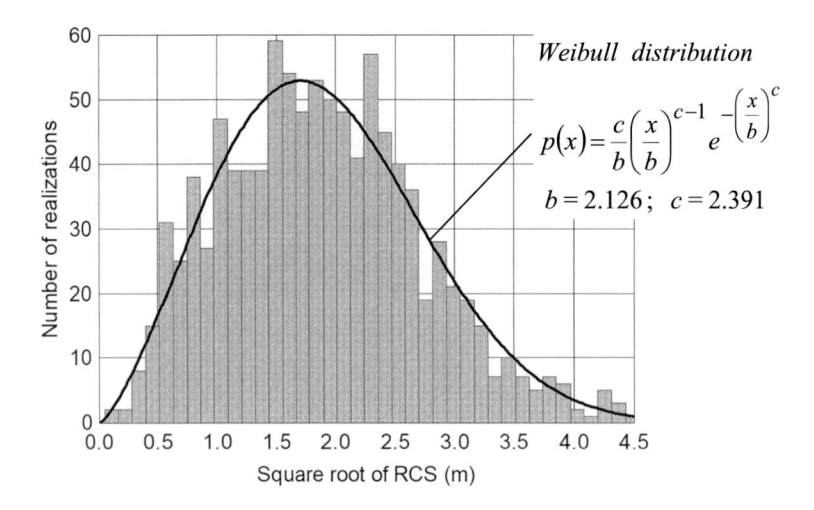

FIGURE 2.404 Amplitude distribution of echo signal of L-39 aircraft model at a carrier frequency of 3 GHz given its horizontal polarization.

Weibull distribution
$$p(x) = \frac{c}{b}\left(\frac{x}{b}\right)^{c-1} e^{-\left(\frac{x}{b}\right)^c}$$
$b = 2.126;\ c = 2.391$

Scattering characteristics of L-39 aircraft model for sounding frequency 1 GHz (wavelength 30 cm).

Figure 2.405 shows the RCS circular diagram of the L-39. The noncoherent RCS circular diagram of the L-39 is shown in Figure 2.406.

The circular mean RCS of the L-39 aircraft for horizontal polarization is 17.38 m². The circular median RCS for horizontal polarization is 1.18 m².

Figures 2.407 and 2.408 show the mean and median RCS for the main ranges of sounding azimuths (nose, side, and tail) and for ranges of 20°.

The expressions and parameters of probability distributions that are most consistent with the empirical distributions of the square root of the RCS for different ranges of sounding azimuths and polarizations are given in the electronic appendix to this book. The expressions and parameters of the probability distributions that are most consistent with the empirical distributions of the RCS (energy characteristic) for different ranges of sounding azimuths and polarizations are also given there.

FIGURE 2.405 Circular diagram of RCS given radar observation of L-39 aircraft model at a carrier frequency of 1 GHz (30 cm wavelength).

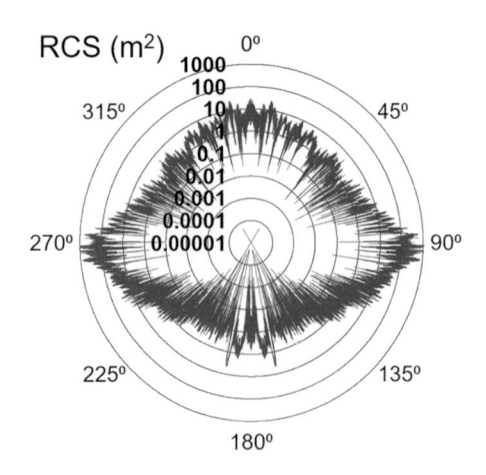

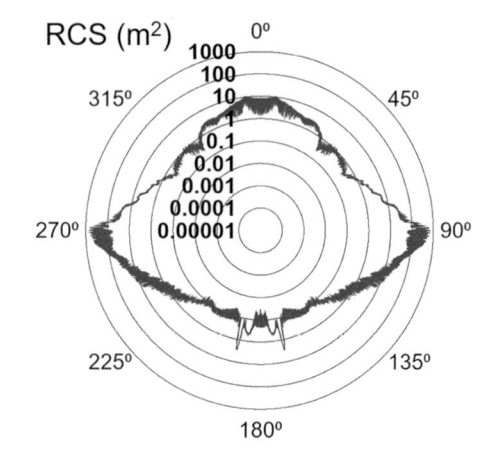

FIGURE 2.406 Circular diagram of noncoherent RCS given radar observation of L-39 aircraft model at a carrier frequency of 1 GHz (30 cm wavelength).

Scattering Characteristics of Aerial Objects 155

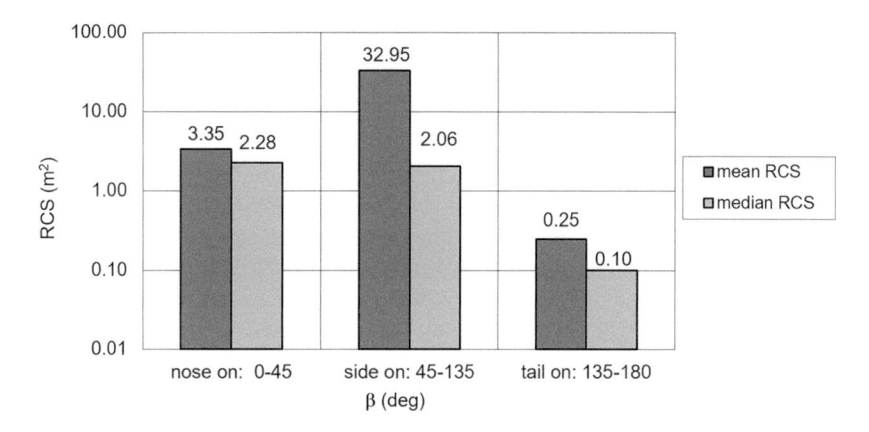

FIGURE 2.407 Diagrams of mean and median RCS of L-39 aircraft model in three sectors of azimuth aspect given its radar observation at horizontal polarization and a carrier frequency of 1 GHz (30 cm wavelength).

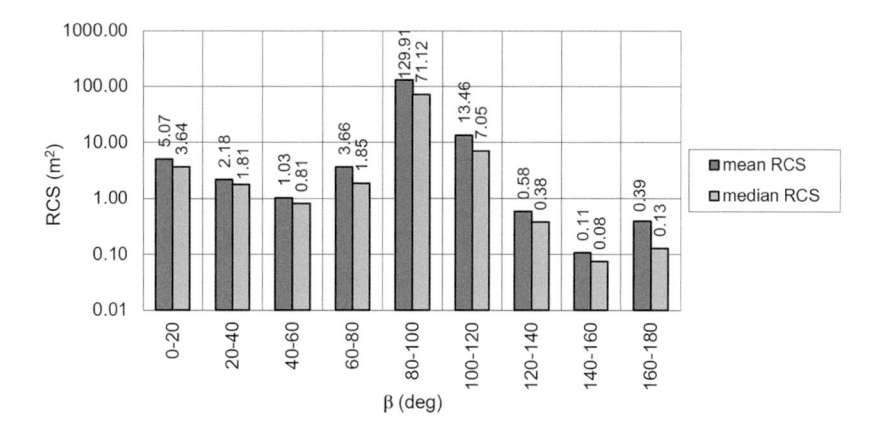

FIGURE 2.408 Diagrams of mean and median RCS of L-39 aircraft model in 20-degree sectors of azimuth aspect given its radar observation at horizontal polarization and a carrier frequency of 1 GHz (30 cm wavelength).

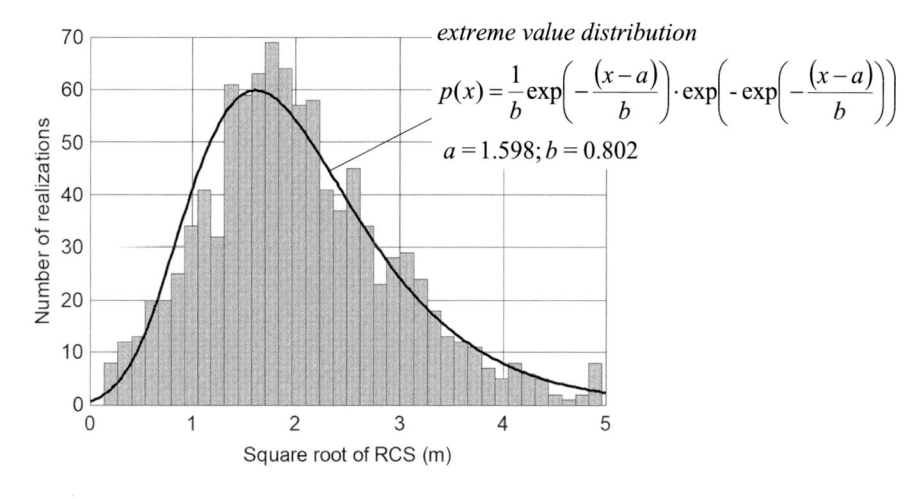

$$p(x) = \frac{1}{b}\exp\left(-\frac{(x-a)}{b}\right)\cdot\exp\left(-\exp\left(-\frac{(x-a)}{b}\right)\right)$$

$a = 1.598; b = 0.802$

FIGURE 2.409 Amplitude distribution of echo signal of L-39 aircraft model at a carrier frequency of 1 GHz given its horizontal polarization.

2.25 TRANSPORT AIRCRAFT An-26

An-26 Curl (Figure 2.410) is a multi-purpose transport aircraft. The first flight took place in 1969.

General characteristics of An-26 [36]: wingspan –29.20 m, length – 23.6 m, height – 8.68m, wing area – 76.98 m², weight – 15 000–23 800 kg, powerplant – 2 Progress AI-24VT Turboprop engines and 1 Tumansky Ru-19-A300 Turbojet booster, speed – 440 km/h, range – 3 600 km, and service ceiling – 7 500 m.

In accordance with the design of the An-26, a model of its surface was created to obtaining scattering characteristics (in particular, RCS). The model is shown in Figure 2.411. In the modeling, the smooth parts of the aircraft surface were approximated by parts of 40 triaxial ellipsoids. The surface breaks were modeled using 18 straight edge scattering parts.

FIGURE 2.410 Transport aircraft An-26.

FIGURE 2.411 Surface model of transport aircraft An-26.

Below are some scattering characteristics of the An-26 model at sounding frequencies of 10, 3, and 1 GHz (wavelengths of 3, 10, and 30 cm, respectively) for horizontal polarization of the probing signal. The scattering characteristics for this model at two polarizations, as well as scattering characteristics at sounding frequencies of 5 and 1.3 GHz (wavelengths of 6 and 23 cm, respectively), are given in the electronic appendix of this book.

Sounding parameters: The elevation angle is to be a random value distributed uniformly in the range −3° ± 4° with respect to the wing plane (elevation angle of −3° corresponds to the radar

Scattering Characteristics of Aerial Objects

observation from the lower hemisphere), azimuth aspect increment was 0.02°, the azimuth being counted off from the nose-on aspect (0° corresponds to the nose-on radar observation, 180° corresponds to the tail-on observation).

Scattering characteristics of An-26 aircraft model for sounding frequency 10 GHz (wavelength 3 cm).

Figure 2.412 shows the RCS circular diagram of the An-26 model. The noncoherent RCS circular diagram of the An-26 is shown in Figure 2.413.

The circular mean RCS of the An-26 aircraft model for horizontal polarization is 64.18 m^2. The circular median RCS (the RCS value used to calculate the detection range of an object with a probability of 0.5) for horizontal polarization is 7.56 m^2.

Figures 2.414 and 2.415 show the mean and median RCS for the main ranges of sounding azimuths (nose, side, and tail) and for ranges of 20°.

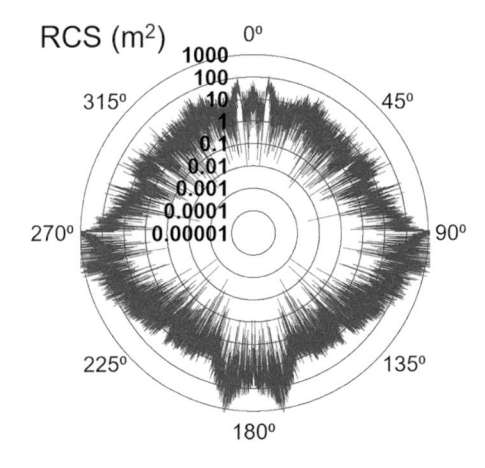

FIGURE 2.412 Circular diagram of RCS given radar observation of An-26 aircraft model at a carrier frequency of 10 GHz (3 cm wavelength).

FIGURE 2.413 Circular diagram of noncoherent RCS given radar observation of An-26 aircraft model at a carrier frequency of 10 GHz (3 cm wavelength).

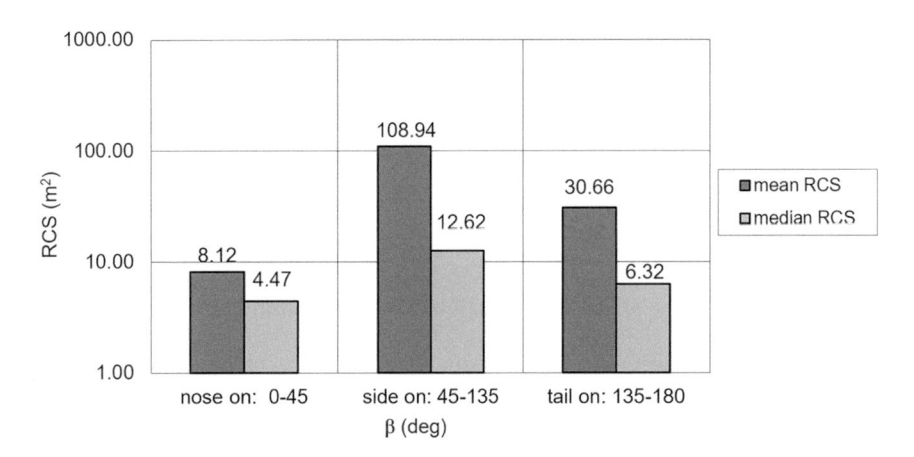

FIGURE 2.414 Diagrams of mean and median RCS of An-26 aircraft model in three sectors of azimuth aspect given its radar observation at horizontal polarization and a carrier frequency of 10 GHz (3 cm wavelength).

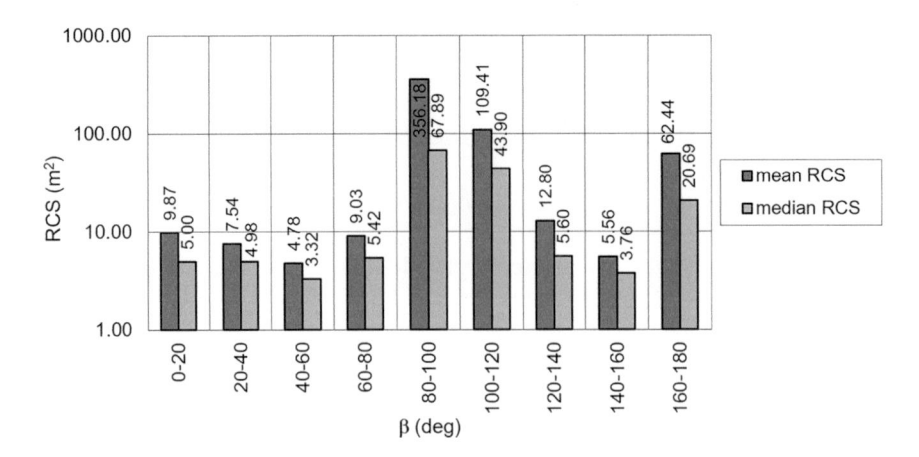

FIGURE 2.415 Diagrams of mean and median RCS of An-26 aircraft model in 20-degree sectors of azimuth aspect given its radar observation at horizontal polarization and a carrier frequency of 10 GHz (3 cm wavelength).

Figures 2.416, 2.421, and 2.426 show histograms of the scattered signal amplitude (square root of the RCS) for the range of sounding azimuths −20° to +20° (sounding from the nose). The bold line shows the probability density functions of the distribution, which can be used to approximate the histogram of the signal amplitude.

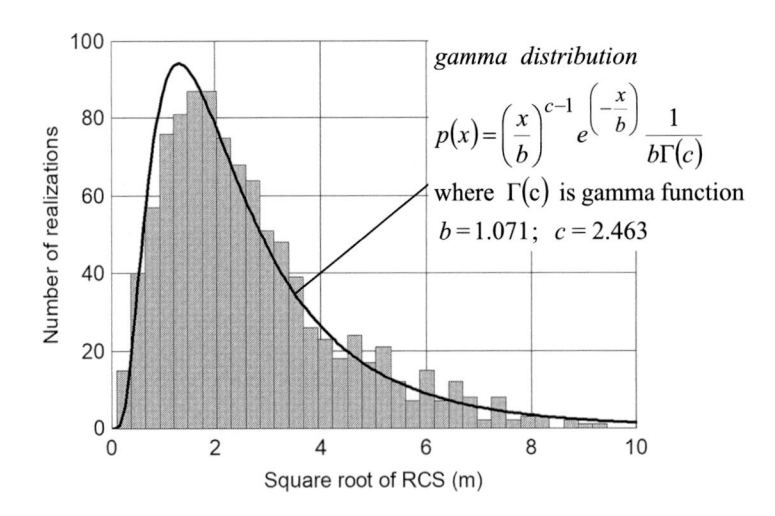

gamma distribution

$$p(x) = \left(\frac{x}{b}\right)^{c-1} e^{\left(-\frac{x}{b}\right)} \frac{1}{b\Gamma(c)}$$

where $\Gamma(c)$ is gamma function
$b = 1.071; \quad c = 2.463$

FIGURE 2.416 Amplitude distribution of echo signal of An-26 aircraft model at a carrier frequency of 10 GHz given its horizontal polarization.

Scattering characteristics of An-26 aircraft model for sounding frequency 3 GHz (wavelength 10 cm).

Figure 2.417 shows the RCS circular diagram of the An-26. The noncoherent RCS circular diagram of the An-26 is shown in Figure 2.418.

The circular mean RCS of the An-26 aircraft for horizontal polarization is 60.33 m². The circular median RCS for horizontal polarization is 6.42 m².

Figures 2.419 and 2.420 show the mean and median RCS for the main ranges of sounding azimuths (nose, side, and tail) and for ranges of 20°.

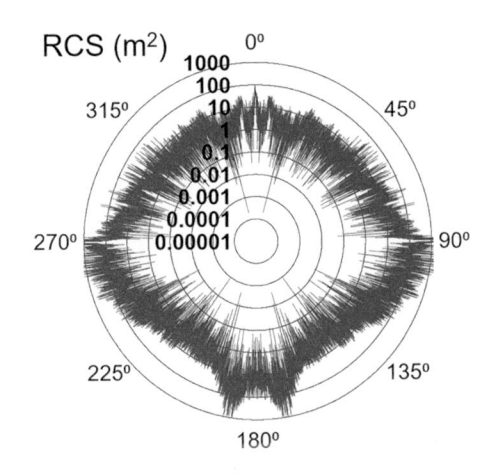

Scattering Characteristics of Aerial Objects

FIGURE 2.417 Circular diagram of RCS given radar observation of An-26 aircraft model at a carrier frequency of 3 GHz (10 cm wavelength).

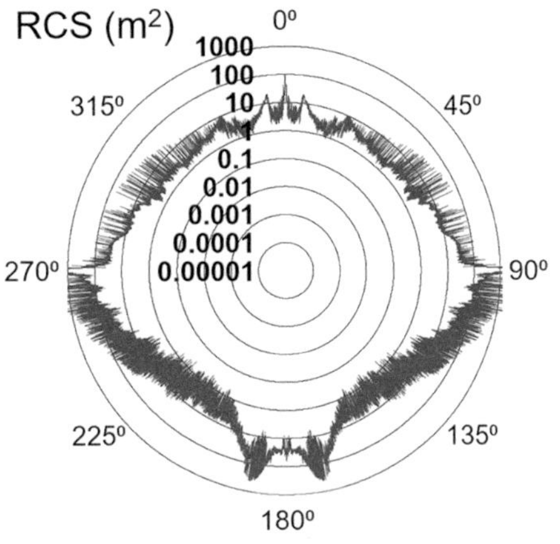

FIGURE 2.418 Circular diagram of noncoherent RCS given radar observation of An-26 aircraft model at a carrier frequency of 3 GHz (10 cm wavelength).

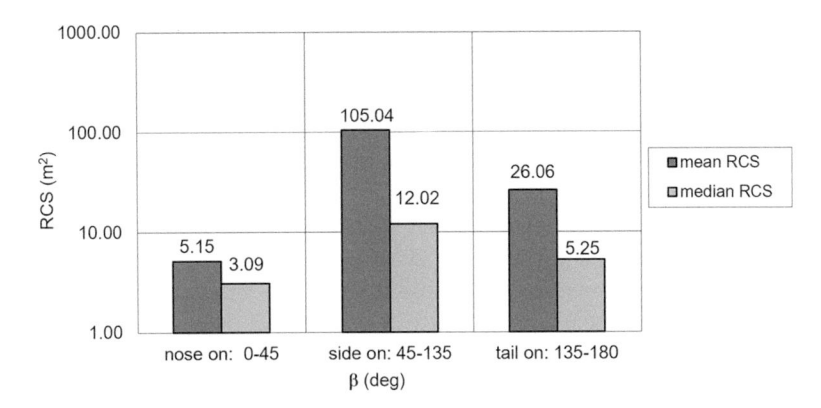

FIGURE 2.419 Diagrams of mean and median RCS of An-26 aircraft model in three sectors of azimuth aspect given its radar observation at horizontal polarization and a carrier frequency of 3 GHz (10 cm wavelength).

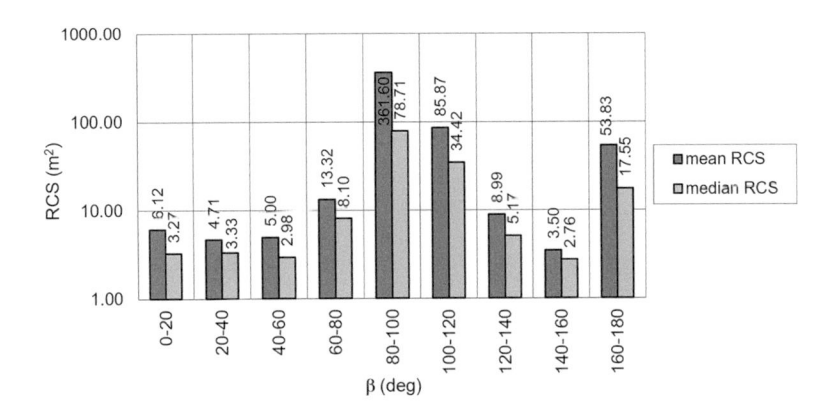

FIGURE 2.420 Diagrams of mean and median RCS of An-26 aircraft model in 20-degree sectors of azimuth aspect given its radar observation at horizontal polarization and a carrier frequency of 3 GHz (10 cm wavelength).

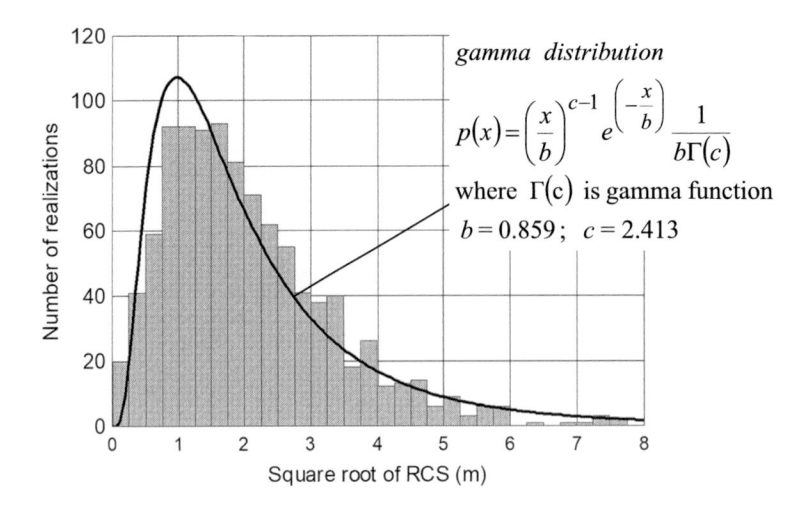

$$p(x) = \left(\frac{x}{b}\right)^{c-1} e^{\left(-\frac{x}{b}\right)} \frac{1}{b\Gamma(c)}$$

where $\Gamma(c)$ is gamma function
$b = 0.859$; $c = 2.413$

FIGURE 2.421 Amplitude distribution of echo signal of An-26 aircraft model at a carrier frequency of 3 GHz given its horizontal polarization.

Scattering characteristics of An-26 aircraft model for sounding frequency 1 GHz (wavelength 30 cm).

Figure 2.422 shows the RCS circular diagram of the An-26. The noncoherent RCS circular diagram of the An-26 is shown in Figure 2.423.

The circular mean RCS of the An-26 aircraft for horizontal polarization is 64.46 m². The circular median RCS for horizontal polarization is 6.66 m².

Figures 2.424 and 2.425 show the mean and median RCS for the main ranges of sounding azimuths (nose, side, and tail) and for ranges of 20°.

The expressions and parameters of probability distributions that are most consistent with the empirical distributions of the square root of the RCS for different ranges of sounding azimuths and polarizations are given in the electronic appendix to this book. The expressions and parameters of the probability distributions that are most consistent with the empirical distributions of the RCS (energy characteristic) for different ranges of sounding azimuths and polarizations are also given there.

FIGURE 2.422 Circular diagram of RCS given radar observation of An-26 aircraft model at a carrier frequency of 1 GHz (30 cm wavelength).

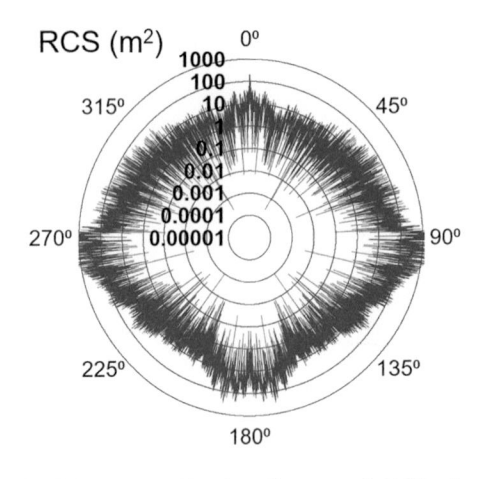

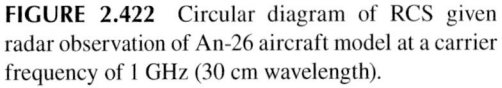

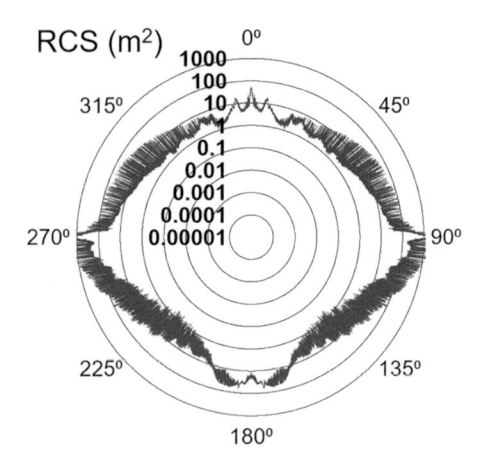

FIGURE 2.423 Circular diagram of noncoherent RCS given radar observation of An-26 aircraft model at a carrier frequency of 1 GHz (30 cm wavelength).

Scattering Characteristics of Aerial Objects 161

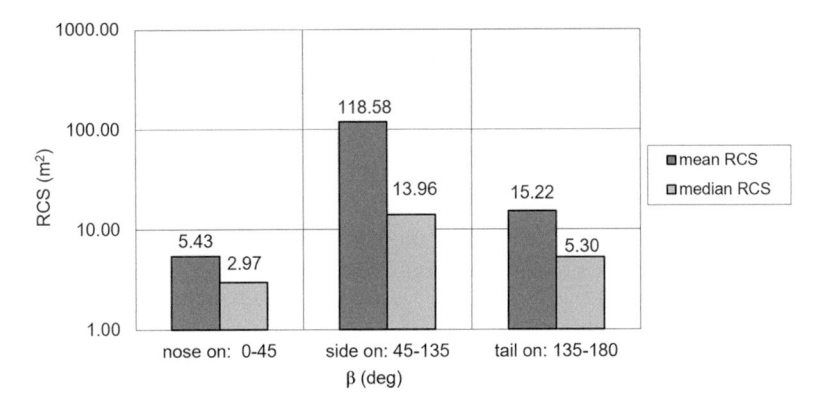

FIGURE 2.424 Diagrams of mean and median RCS of An-26 aircraft model in three sectors of azimuth aspect given its radar observation at horizontal polarization and a carrier frequency of 1 GHz (30 cm wavelength).

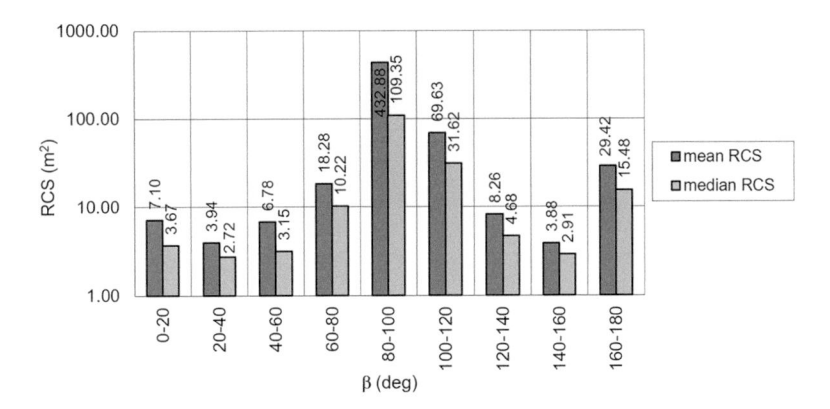

FIGURE 2.425 Diagrams of mean and median RCS of An-26 aircraft model in 20-degree sectors of azimuth aspect given its radar observation at horizontal polarization and a carrier frequency of 1 GHz (30 cm wavelength).

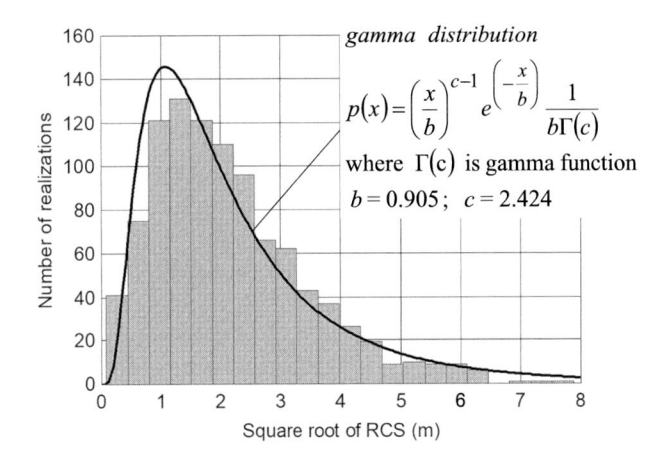

$$p(x) = \left(\frac{x}{b}\right)^{c-1} e^{\left(-\frac{x}{b}\right)} \frac{1}{b\Gamma(c)}$$

where $\Gamma(c)$ is gamma function

$b = 0.905;\ c = 2.424$

FIGURE 2.426 Amplitude distribution of echo signal of An-26 aircraft model at a carrier frequency of 1 GHz given its horizontal polarization.

2.26 STRATEGIC AIRLIFTER Il-76

Il-76 Candid (Figure 2.427) is a strategic airlifter. The first flight took place in 1971.

General characteristics of Il-76 [37]: wingspan –50.50 m, length – 46.59 m, height – 14.76m, wing area – 300.00 m^2, weight – 92 000–190 000 kg, powerplant – 4 D-30KP turbofans, speed – 850 km/h, range – 4 200–7 200 km, and service ceiling – 12 000 m.

In accordance with the design of the Il-76, a model of its surface was created to obtaining scattering characteristics (in particular, RCS). The model is shown in Figure 2.428. In the modeling, the smooth parts of the aircraft surface were approximated by parts of 76 triaxial ellipsoids. The surface breaks were modeled using 12 straight edge scattering parts.

FIGURE 2.427 Strategic airlifter Il-76.

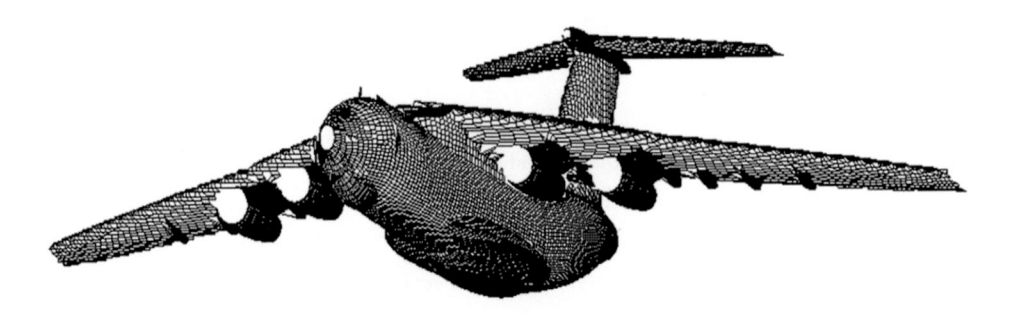

FIGURE 2.428 Surface model of t strategic airlifter Il-76.

Below are some scattering characteristics of the Il-76 model at sounding frequencies of 10, 3, and 1 GHz (wavelengths of 3, 10, and 30 cm, respectively) for horizontal polarization of the probing signal. The scattering characteristics for this model at two polarizations, as well as scattering characteristics at sounding frequencies of 5 and 1.3 GHz (wavelengths of 6 and 23 cm, respectively), are given in the electronic appendix of this book.

Sounding parameters: The elevation angle is to be a random value distributed uniformly in the range −3° ± 4° with respect to the wing plane (elevation angle of −3° corresponds to the radar observation from the lower hemisphere), azimuth aspect increment was 0.02°, the azimuth being counted off from the nose-on aspect (0° corresponds to the nose-on radar observation, 180° corresponds to the tail-on observation).

Scattering characteristics of Il-76 aircraft model for sounding frequency 10 GHz (wavelength 3 cm).

Figure 2.429 shows the RCS circular diagram of the Il-76 model. The noncoherent RCS circular diagram of the Il-76 is shown in Figure 2.430.

The circular mean RCS of the Il-76 aircraft model for horizontal polarization is 420.74 m². The circular median RCS (the RCS value used to calculate the detection range of an object with a probability of 0.5) for horizontal polarization is 20.27 m².

Figures 2.431 and 2.432 show the mean and median RCS for the main ranges of sounding azimuths (nose, side, and tail) and for ranges of 20°.

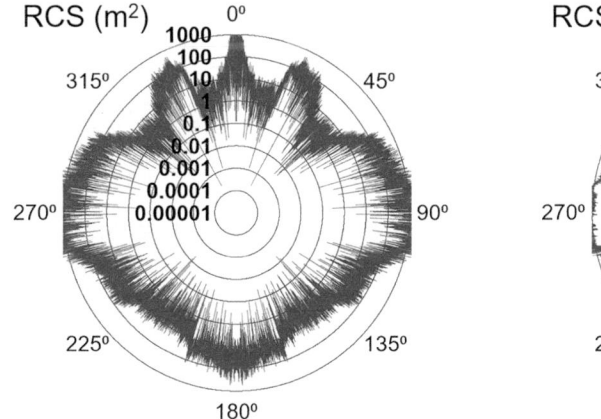

FIGURE 2.429 Circular diagram of RCS given radar observation of Il-76 aircraft model at a carrier frequency of 10 GHz (3 cm wavelength).

FIGURE 2.430 Circular diagram of noncoherent RCS given radar observation of Il-76 aircraft model at a carrier frequency of 10 GHz (3 cm wavelength).

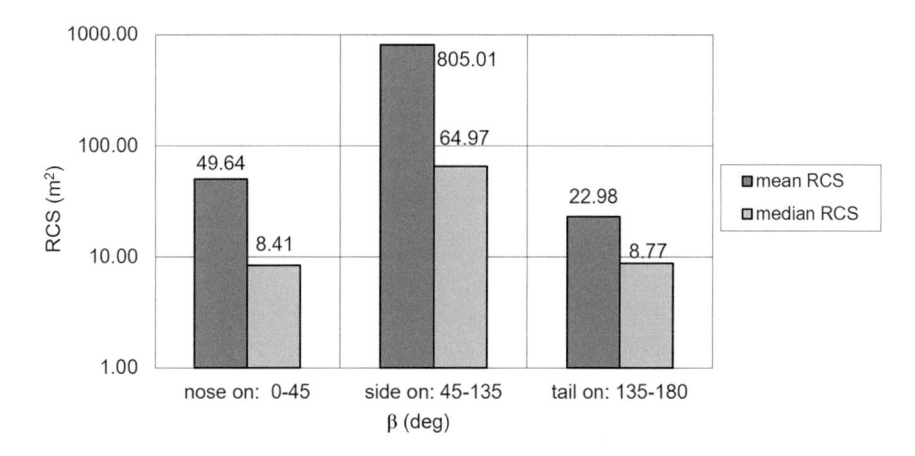

FIGURE 2.431 Diagrams of mean and median RCS of Il-76 aircraft model in three sectors of azimuth aspect given its radar observation at horizontal polarization and a carrier frequency of 10 GHz (3 cm wavelength).

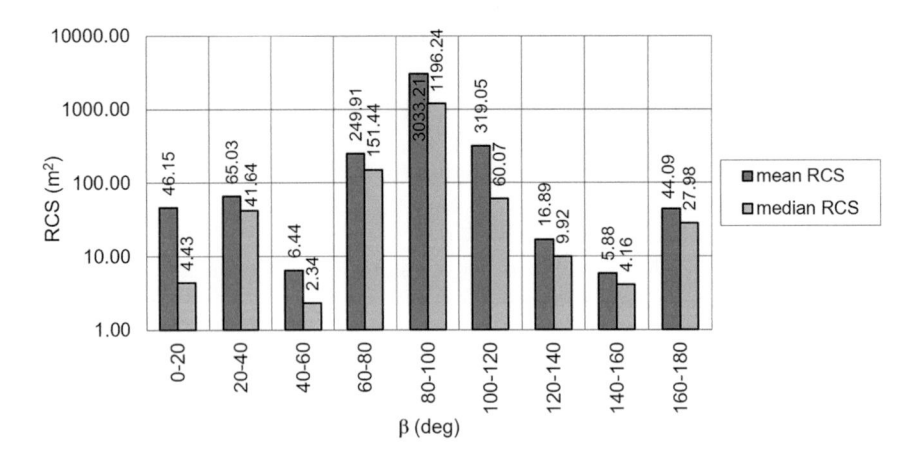

FIGURE 2.432 Diagrams of mean and median RCS of Il-76 aircraft model in 20-degree sectors of azimuth aspect given its radar observation at horizontal polarization and a carrier frequency of 10 GHz (3 cm wavelength).

Figures 2.433, 2.438, and 2.439 show histograms of the scattered signal amplitude (square root of the RCS) for the range of sounding azimuths −20° to +20° (sounding from the nose). The bold line shows the probability density functions of the distribution, which can be used to approximate the histogram of the signal amplitude.

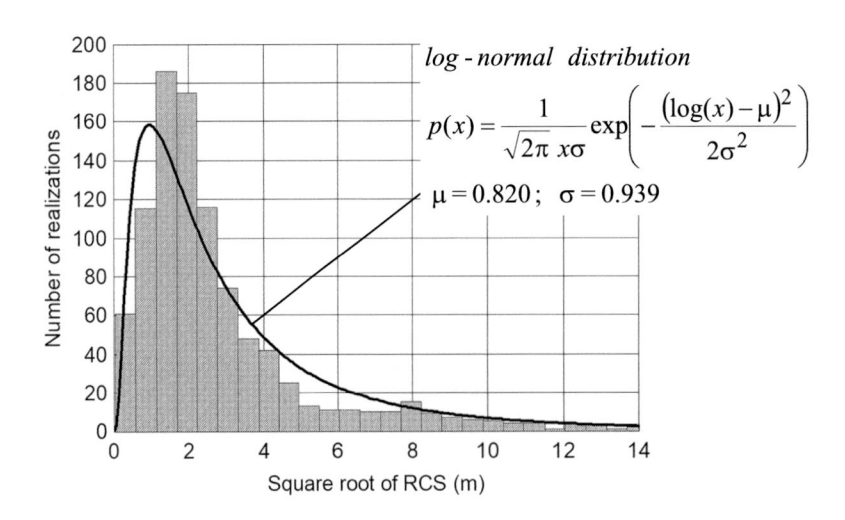

$$\text{log - normal distribution}$$
$$p(x) = \frac{1}{\sqrt{2\pi}\, x\sigma} \exp\left(-\frac{(\log(x)-\mu)^2}{2\sigma^2}\right)$$
$$\mu = 0.820;\quad \sigma = 0.939$$

FIGURE 2.433 Amplitude distribution of echo signal of Il-76 aircraft model at a carrier frequency of 10 GHz given its horizontal polarization.

Scattering characteristics of Il-76 aircraft model for sounding frequency 3 GHz (wavelength 10 cm).

Figure 2.434 shows the RCS circular diagram of the Il-76. The noncoherent RCS circular diagram of the Il-76 is shown in Figure 2.435.

The circular mean RCS of the Il-76 aircraft for horizontal polarization is 383.96 m². The circular median RCS for horizontal polarization is 8.94 m².

Figures 2.436 and 2.437 show the mean and median RCS for the main ranges of sounding azimuths (nose, side, and tail) and for ranges of 20°.

Scattering Characteristics of Aerial Objects

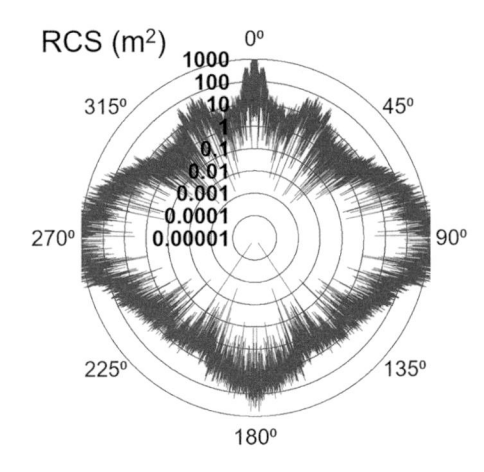

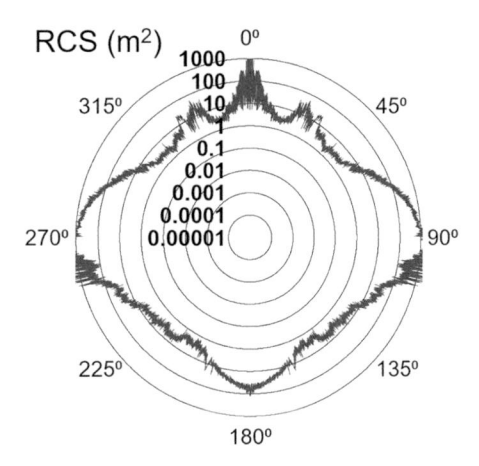

FIGURE 2.434 Circular diagram of RCS given radar observation of Il-76 aircraft model at a carrier frequency of 3 GHz (10 cm wavelength).

FIGURE 2.435 Circular diagram of noncoherent RCS given radar observation of Il-76 aircraft model at a carrier frequency of 3 GHz (10 cm wavelength).

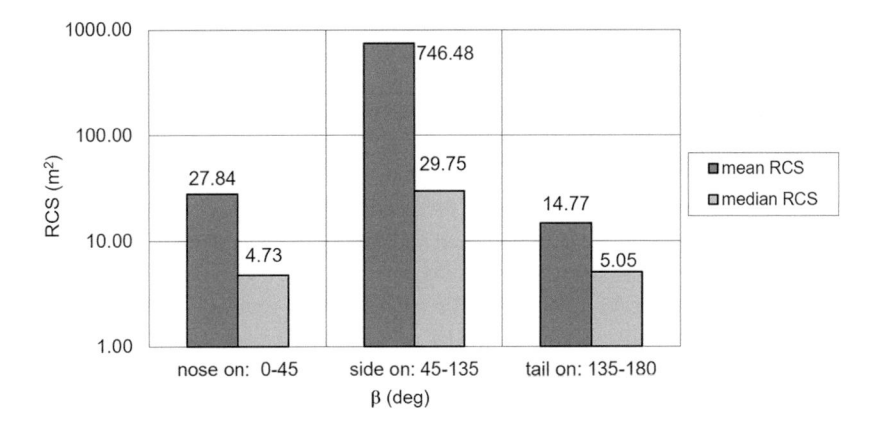

FIGURE 2.436 Diagrams of mean and median RCS of Il-76 aircraft model in three sectors of azimuth aspect given its radar observation at horizontal polarization and a carrier frequency of 3 GHz (10 cm wavelength).

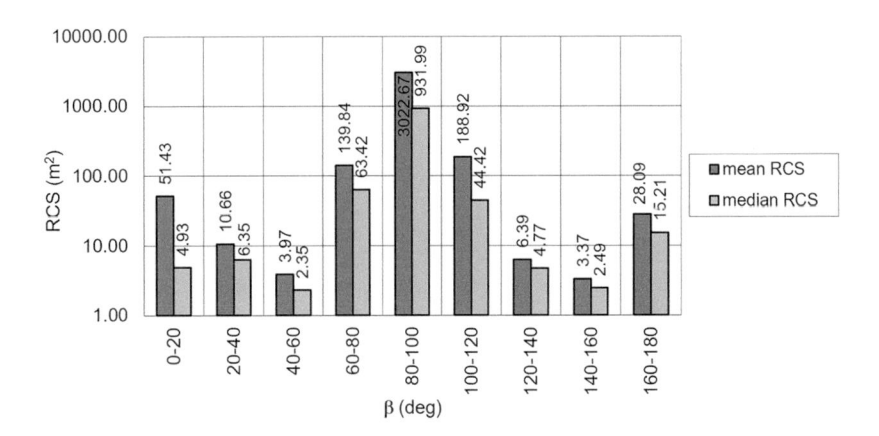

FIGURE 2.437 Diagrams of mean and median RCS of Il-76 aircraft model in 20-degree sectors of azimuth aspect given its radar observation at horizontal polarization and a carrier frequency of 3 GHz (10 cm wavelength).

$$p(x) = \frac{1}{\sqrt{2\pi}\,x\sigma} \exp\left(-\frac{(\log(x)-\mu)^2}{2\sigma^2}\right)$$

log - normal distribution

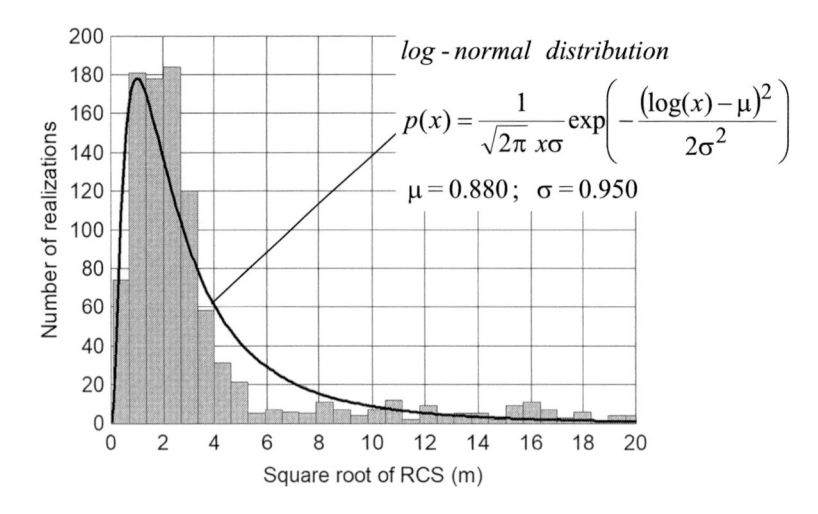

$\mu = 0.880; \quad \sigma = 0.950$

FIGURE 2.438 Amplitude distribution of echo signal of Il-76 aircraft model at a carrier frequency of 3 GHz given its horizontal polarization.

Scattering characteristics of Il-76 aircraft model for sounding frequency 1 GHz (wavelength 30 cm).

Figure 2.439 shows the RCS circular diagram of the Il-76. The noncoherent RCS circular diagram of the Il-76 is shown in Figure 2.440.

The circular mean RCS of the Il-76 aircraft for horizontal polarization is 311.52 m². The circular median RCS for horizontal polarization is 9.80 m².

Figures 2.441 and 2.442 show the mean and median RCS for the main ranges of sounding azimuths (nose, side, and tail) and for ranges of 20°.

The expressions and parameters of probability distributions that are most consistent with the empirical distributions of the square root of the RCS for different ranges of sounding azimuths and polarizations are given in the electronic appendix to this book. The expressions and parameters of the probability distributions that are most consistent with the empirical distributions of the RCS (energy characteristic) for different ranges of sounding azimuths and polarizations are also given there.

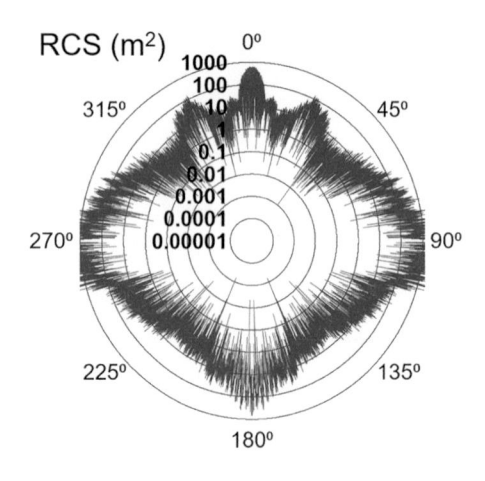

FIGURE 2.439 Circular diagram of RCS given radar observation of Il-76 aircraft model at a carrier frequency of 1 GHz (30 cm wavelength).

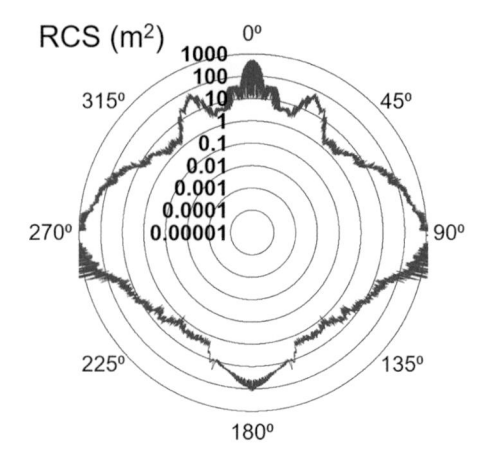

FIGURE 2.440 Circular diagram of noncoherent RCS given radar observation of Il-76 aircraft model at a carrier frequency of 1 GHz (30 cm wavelength).

Scattering Characteristics of Aerial Objects 167

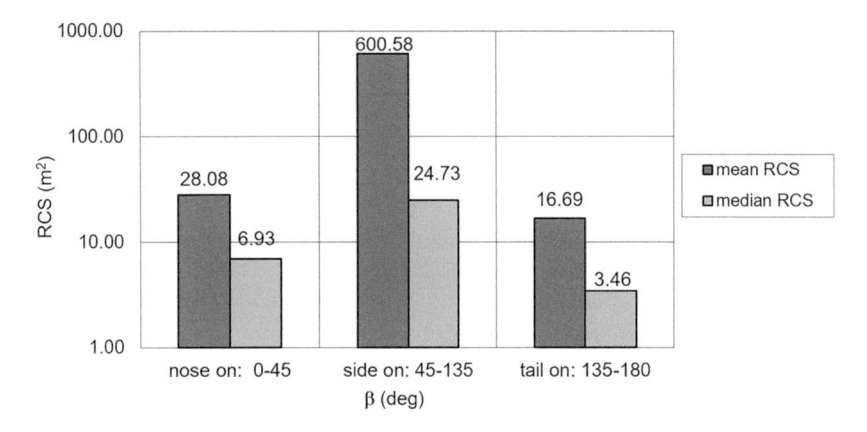

FIGURE 2.441 Diagrams of mean and median RCS of Il-76 aircraft model in three sectors of azimuth aspect given its radar observation at horizontal polarization and a carrier frequency of 1 GHz (30 cm wavelength).

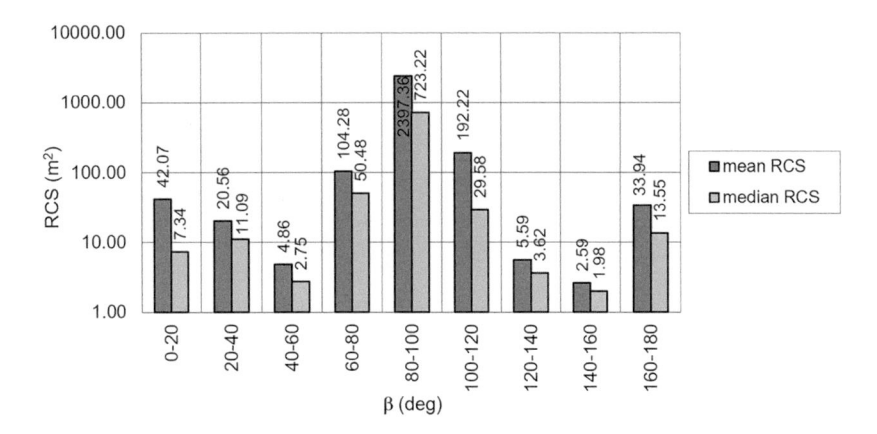

FIGURE 2.442 Diagrams of mean and median RCS of Il-76 aircraft model in 20-degree sectors of azimuth aspect given its radar observation at horizontal polarization and a carrier frequency of 1 GHz (30 cm wavelength).

$$p(x) = \frac{1}{\sqrt{2\pi}\, x\sigma} \exp\left(-\frac{(\log(x)-\mu)^2}{2\sigma^2}\right)$$

$\mu = 1.059;\ \sigma = 0.865$

log - normal distribution

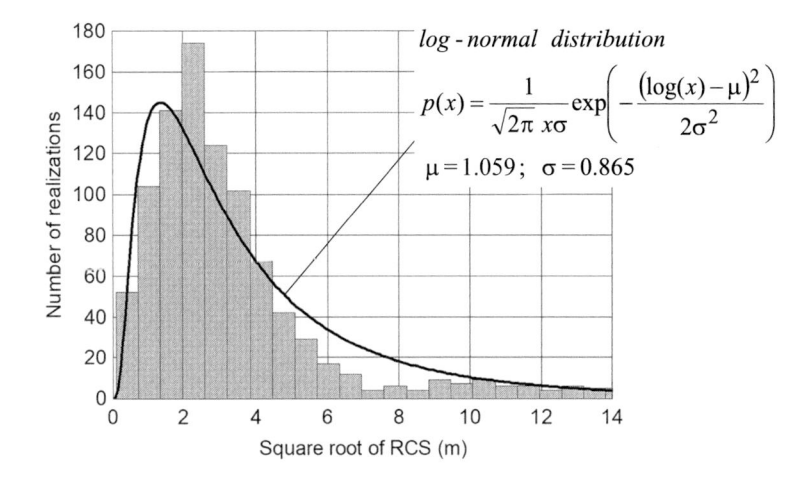

FIGURE 2.443 Amplitude distribution of echo signal of Il-76 aircraft model at a carrier frequency of 1 GHz given its horizontal polarization.

2.27 MEDIUM-RANGE AIRLINER BOEING 737-400

Boeing 737-400 Classic (Figure 2.444) is a medium-range airliner. The first flight took place in 1984. Scattering characteristics of this airplane have been obtained because Boeing 737 is the most mass produced civil aircraft.

General characteristics of Boeing 737-400 [38]: wingspan – 28.88 m, length – 36.40 m, height – 11.13 m, wing area – 91.04 m^2, weight – 34 560– 68 040 kg, powerplant – 2 CFM56-3C-1 engines, speed – 876 km/h, range – 3 800 km, and service ceiling – 11 300 m.

In accordance with the design of the Boeing 737, a model of its surface was created to obtaining scattering characteristics (in particular, RCS). The model is shown in Figure 2.445. In the modeling, the smooth parts of the aircraft surface were approximated by parts of 40 triaxial ellipsoids. The surface breaks were modeled using 18 straight edge scattering parts.

FIGURE 2.444 Medium-range airliner Boeing 737-400.

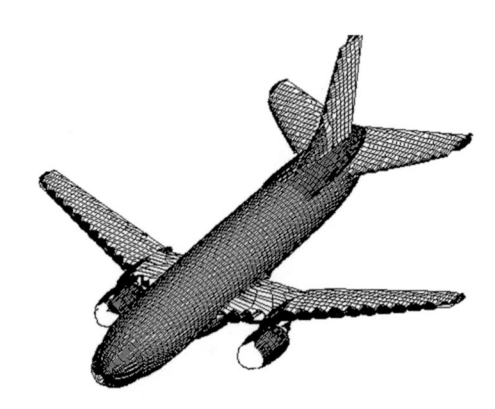

FIGURE 2.445 Surface model of medium-range airliner Boeing 737-400.

Below are some scattering characteristics of the Boeing 737 model at sounding frequencies of 10, 3, and 1 GHz (wavelengths of 3, 10, and 30 cm, respectively) for horizontal polarization of the probing signal. The scattering characteristics for this model at two polarizations, as well as scattering characteristics at sounding frequencies of 5 and 1.3 GHz (wavelengths of 6 and 23 cm, respectively), are given in the electronic appendix of this book.

Sounding parameters: The elevation angle is to be a random value distributed uniformly in the range −3° ± 4° with respect to the wing plane (elevation angle of −3° corresponds to the radar observation from the lower hemisphere), azimuth aspect increment was 0.02°, the azimuth being counted off from the nose-on aspect (0° corresponds to the nose-on radar observation, 180° corresponds to the tail-on observation).

Scattering characteristics of Boeing 737 aircraft model for sounding frequency 10 GHz (wavelength 3 cm).

Figure 2.446 shows the RCS circular diagram of the Boeing 737 model. The noncoherent RCS circular diagram of the Boeing 737 is shown in Figure 2.447.

The circular mean RCS of the Boeing 737 aircraft model for horizontal polarization is 147.94 m^2. The circular median RCS (the RCS value used to calculate the detection range of an object with a probability of 0.5) for horizontal polarization is 15.81 m^2.

Figures 2.448 and 2.449 show the mean and median RCS for the main ranges of sounding azimuths (nose, side, and tail) and for ranges of 20°.

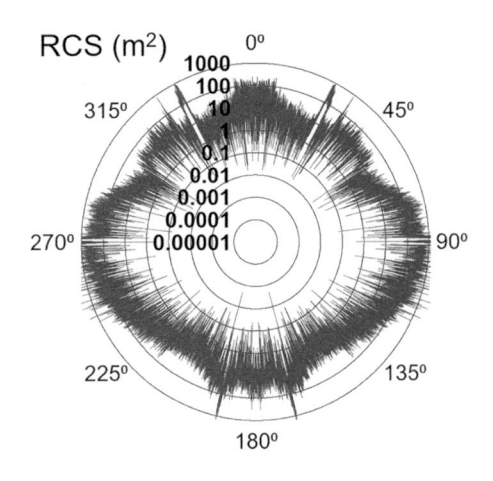

Scattering Characteristics of Aerial Objects

FIGURE 2.446 Circular diagram of RCS given radar observation of Boeing 737 aircraft model at a carrier frequency of 10 GHz (3 cm wavelength).

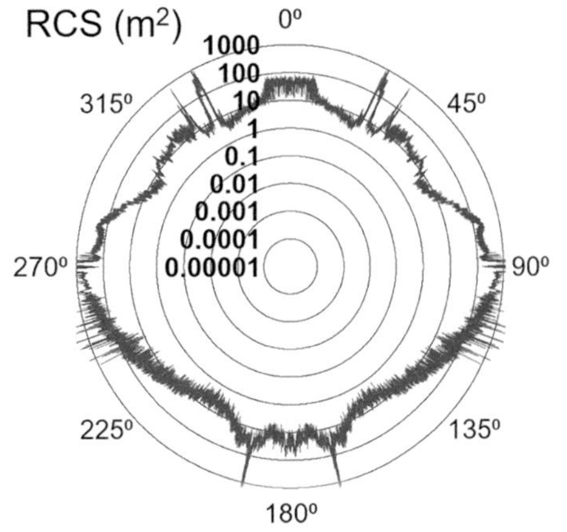

FIGURE 2.447 Circular diagram of noncoherent RCS given radar observation of Boeing 737 aircraft model at a carrier frequency of 10 GHz (3 cm wavelength).

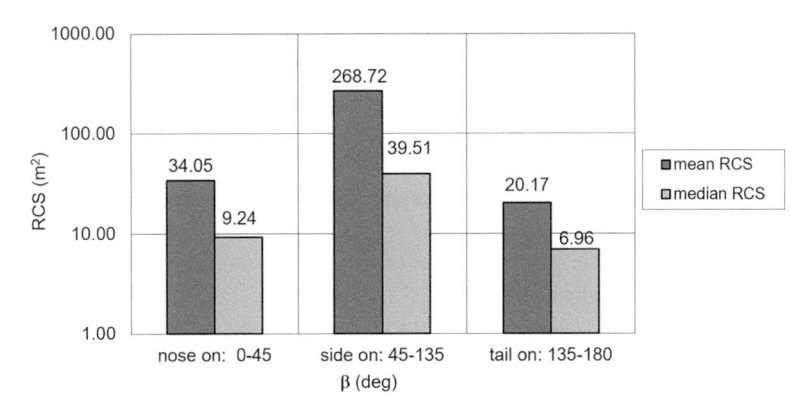

FIGURE 2.448 Diagrams of mean and median RCS of Boeing 737 aircraft model in three sectors of azimuth aspect given its radar observation at horizontal polarization and a carrier frequency of 10 GHz (3 cm wavelength).

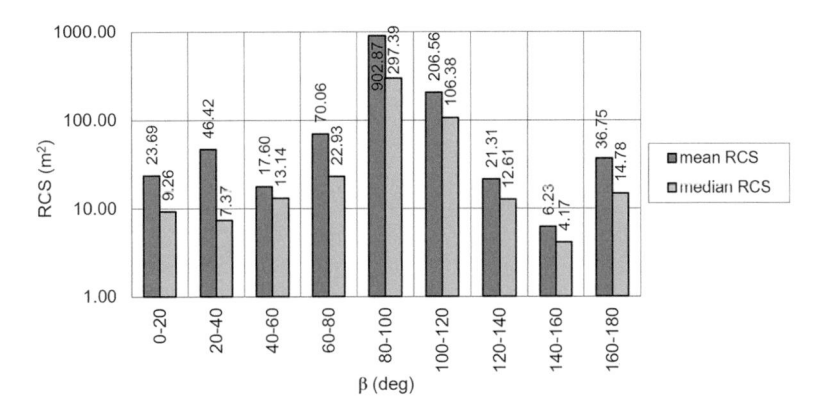

FIGURE 2.449 Diagrams of mean and median RCS of Boeing 737 aircraft model in 20-degree sectors of azimuth aspect given its radar observation at horizontal polarization and a carrier frequency of 10 GHz (3 cm wavelength).

Figures 2.450, 2.455, and 2.460 show histograms of the scattered signal amplitude (square root of the RCS) for the range of sounding azimuths −20° to +20° (sounding from the nose). The bold line shows the probability density functions of the distribution, which can be used to approximate the histogram of the signal amplitude.

$$p(x) = \frac{1}{\sqrt{2\pi}\, x\sigma} \exp\left(-\frac{(\log(x)-\mu)^2}{2\sigma^2}\right)$$

log-normal distribution

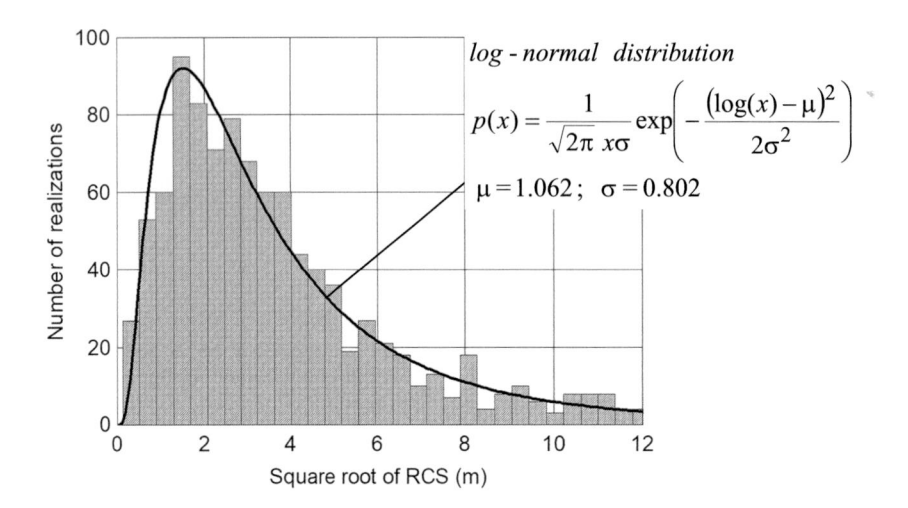

$\mu = 1.062;\ \sigma = 0.802$

FIGURE 2.450 Amplitude distribution of echo signal of Boeing 737 aircraft model at a carrier frequency of 10 GHz given its horizontal polarization.

Scattering characteristics of Boeing 737 aircraft model for sounding frequency 3 GHz (wavelength 10 cm).

Figure 2.451 shows the RCS circular diagram of the Boeing 737. The noncoherent RCS circular diagram of the Boeing 737 is shown in Figure 2.452.

The circular mean RCS of the Boeing 737 aircraft for horizontal polarization is 101.40 m². The circular median RCS for horizontal polarization is 12.37 m².

Figures 2.453 and 2.454 show the mean and median RCS for the main ranges of sounding azimuths (nose, side, and tail) and for ranges of 20°.

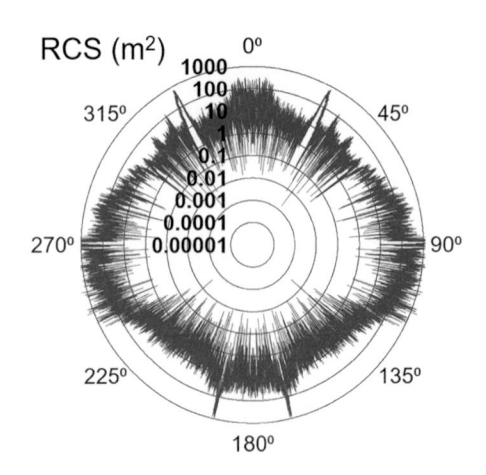

FIGURE 2.451 Circular diagram of RCS given radar observation of Boeing 737 aircraft model at a carrier frequency of 3 GHz (10 cm wavelength).

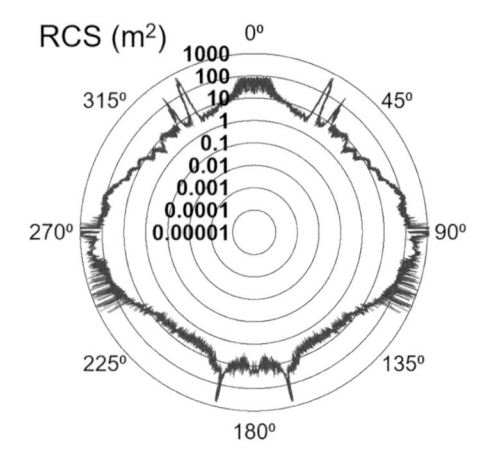

FIGURE 2.452 Circular diagram of noncoherent RCS given radar observation of Boeing 737 aircraft model at a carrier frequency of 3 GHz (10 cm wavelength).

Scattering Characteristics of Aerial Objects

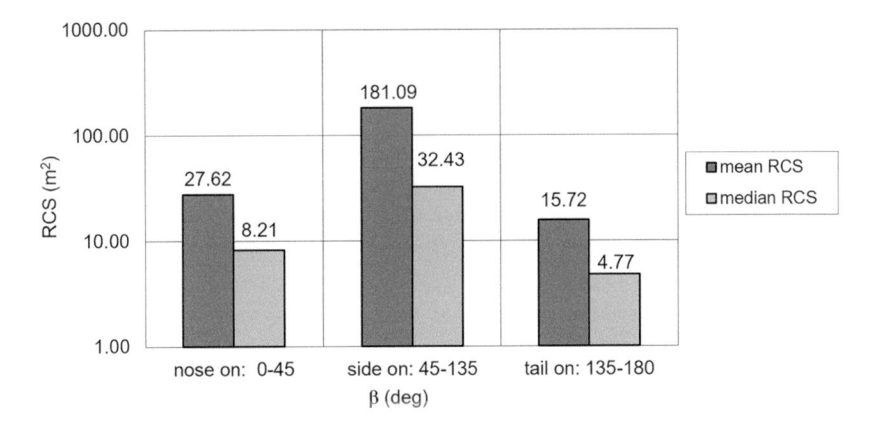

FIGURE 2.453 Diagrams of mean and median RCS of Boeing 737 aircraft model in three sectors of azimuth aspect given its radar observation at horizontal polarization and a carrier frequency of 3 GHz (10 cm wavelength).

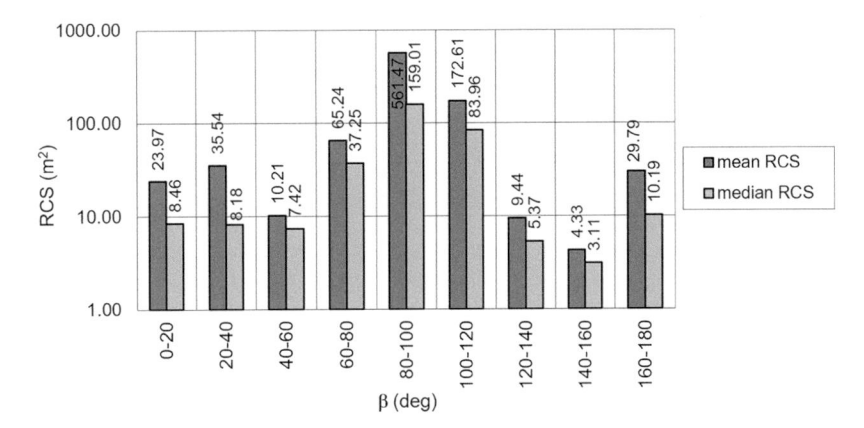

FIGURE 2.454 Diagrams of mean and median RCS of Boeing 737 aircraft model in 20-degree sectors of azimuth aspect given its radar observation at horizontal polarization and a carrier frequency of 3 GHz (10 cm wavelength).

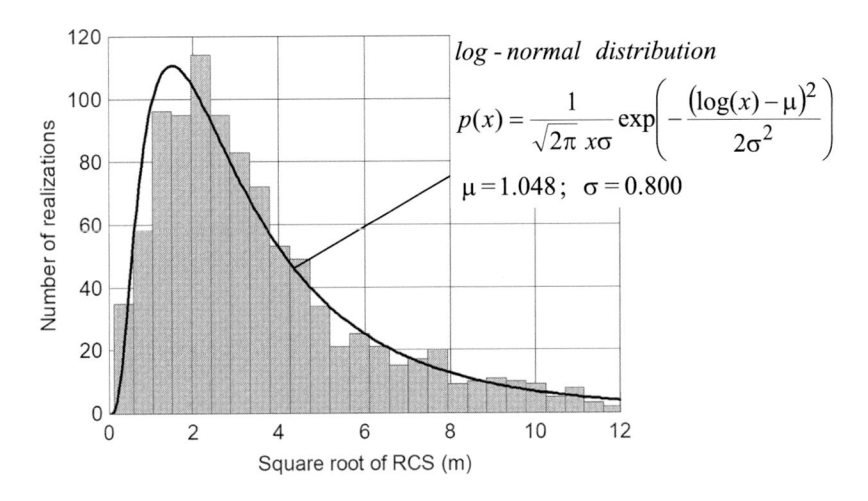

$$\text{log - normal distribution}$$
$$p(x) = \frac{1}{\sqrt{2\pi}\, x\sigma} \exp\left(-\frac{(\log(x)-\mu)^2}{2\sigma^2}\right)$$
$$\mu = 1.048;\quad \sigma = 0.800$$

FIGURE 2.455 Amplitude distribution of echo signal of Boeing 737 aircraft model at a carrier frequency of 3 GHz given its horizontal polarization.

Scattering characteristics of Boeing 737 aircraft model for sounding frequency 1 GHz (wavelength 30 cm).

Figure 2.456 shows the RCS circular diagram of the Boeing 737. The noncoherent RCS circular diagram of the Boeing 737 is shown in Figure 2.457.

The circular mean RCS of the Boeing 737 aircraft for horizontal polarization is 98.51 m². The circular median RCS for horizontal polarization is 14.25 m².

Figures 2.458 and 2.459 show the mean and median RCS for the main ranges of sounding azimuths (nose, side, and tail) and for ranges of 20°.

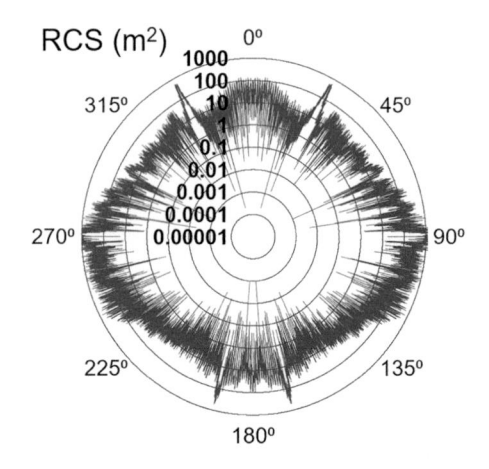

FIGURE 2.456 Circular diagram of RCS given radar observation of Boeing 737 aircraft model at a carrier frequency of 1 GHz (30 cm wavelength).

FIGURE 2.457 Circular diagram of noncoherent RCS given radar observation of Boeing 737 aircraft model at a carrier frequency of 1 GHz (30 cm wavelength).

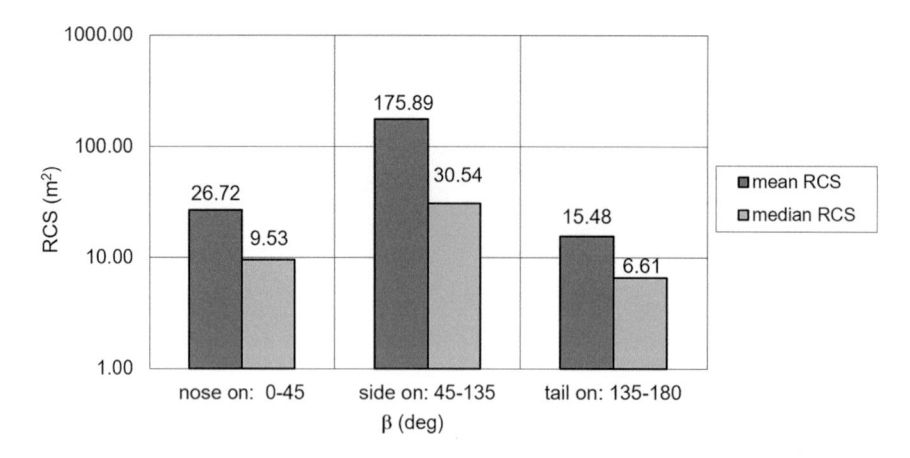

FIGURE 2.458 Diagrams of mean and median RCS of Boeing 737 aircraft model in three sectors of azimuth aspect given its radar observation at horizontal polarization and a carrier frequency of 1 GHz (30 cm wavelength).

Scattering Characteristics of Aerial Objects

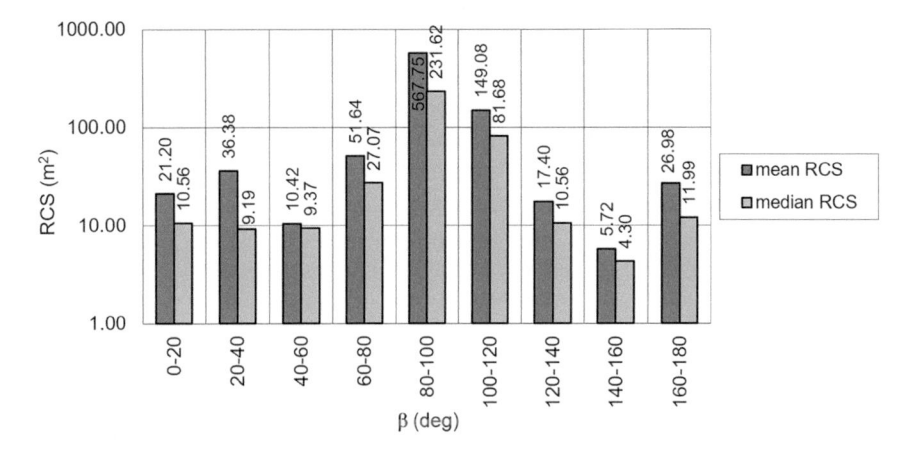

FIGURE 2.459 Diagrams of mean and median RCS of Boeing 737 aircraft model in 20-degree sectors of azimuth aspect given its radar observation at horizontal polarization and a carrier frequency of 1 GHz (30 cm wavelength).

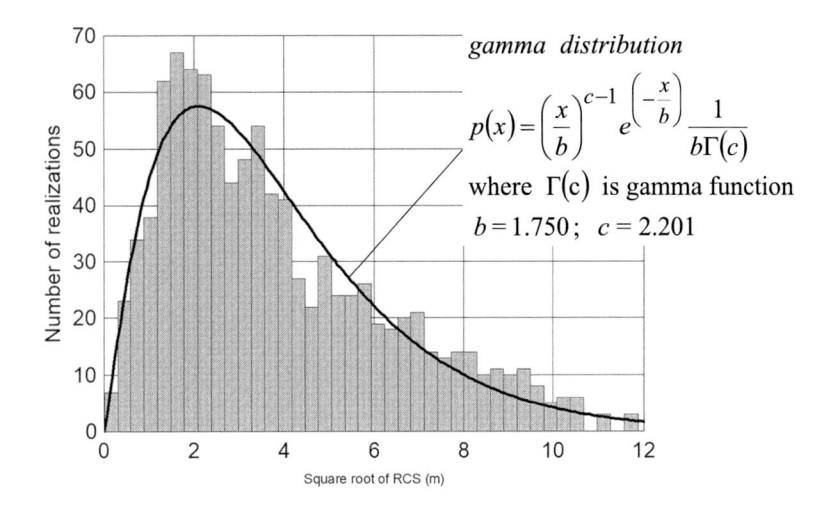

gamma distribution

$$p(x) = \left(\frac{x}{b}\right)^{c-1} e^{\left(-\frac{x}{b}\right)} \frac{1}{b\Gamma(c)}$$

where $\Gamma(c)$ is gamma function
$b = 1.750; \quad c = 2.201$

FIGURE 2.460 Amplitude distribution of echo signal of Boeing 737 aircraft model at a carrier frequency of 1 GHz given its horizontal polarization.

The expressions and parameters of probability distributions that are most consistent with the empirical distributions of the square root of the RCS for different ranges of sounding azimuths and polarizations are given in the electronic appendix to this book. The expressions and parameters of the probability distributions that are most consistent with the empirical distributions of the RCS (energy characteristic) for different ranges of sounding azimuths and polarizations are also given there.

2.28 MULTI-PURPOSE HELICOPTER Mi-8

Mi-8 Hip (Figure 2.461) is a multi-purpose helicopter. The first flight took place in 1961.

General characteristics of Mi-8 [39]: main rotor diameter – 21.3 m, length – 18.42 m, height – 5.34 m, weight – 7 100– 13 000 kg, powerplant – 2 TV3–117MT turboshaft engines, speed – 250 km/h, range – 500 km, and service ceiling – 5 000 m.

In accordance with the design of the Mi-8, a model of its surface was created to obtaining scattering characteristics (in particular, RCS). The model is shown in Figure 2.462. In the modeling, the smooth parts of the helicopter surface were approximated by parts of 96 triaxial ellipsoids. The surface breaks were modeled using 11 straight edge scattering parts.

FIGURE 2.461 Multi-purpose helicopter Mi8.

FIGURE 2.462 Surface model of multi-purpose helicopter Mi-8.

Below are some scattering characteristics of the Mi-8 model at sounding frequencies of 10, 3, and 1 GHz (wavelengths of 3, 10, and 30 cm, respectively) for horizontal polarization of the probing signal. The scattering characteristics for this model at two polarizations, as well as scattering characteristics at sounding frequencies of 5 and 1.3 GHz (wavelengths of 6 and 23 cm, respectively), are given in the electronic appendix of this book.

Scattering Characteristics of Aerial Objects   175

Sounding parameters: The elevation angle is to be a random value distributed uniformly in the range −3° ± 4° with respect to the wing plane (elevation angle of −3° corresponds to the radar observation from the lower hemisphere), azimuth aspect increment was 0.02°, the azimuth being counted off from the nose-on aspect (0° corresponds to the nose-on radar observation, 180° corresponds to the tail-on observation).

Scattering characteristics of Mi-8 helicopter model for sounding frequency 10 GHz (wavelength 3 cm).

Figure 2.463 shows the RCS circular diagram of the Mi-8 model. The noncoherent RCS circular diagram of the Mi-8 is shown in Figure 2.464.

The circular mean RCS of the Mi-8 helicopter model for horizontal polarization is 41.18 m². The circular median RCS (the RCS value used to calculate the detection range of an object with a probability of 0.5) for horizontal polarization is 12.54 m².

Figures 2.465 and 2.466 show the mean and median RCS for the main ranges of sounding azimuths (nose, side, and tail) and for ranges of 20°.

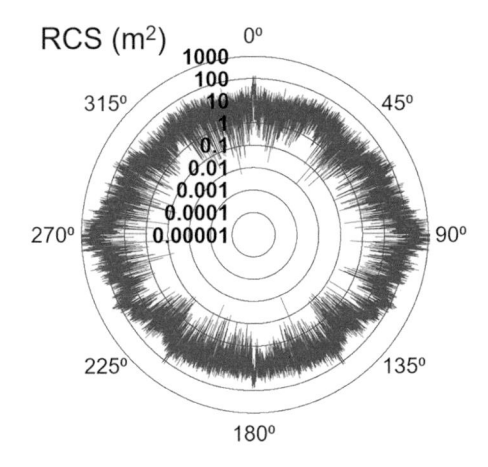

FIGURE 2.463 Circular diagram of RCS given radar observation of Mi-8 helicopter model at a carrier frequency of 10 GHz (3 cm wavelength).

FIGURE 2.464 Circular diagram of noncoherent RCS given radar observation of Mi-8 helicopter model at a carrier frequency of 10 GHz (3 cm wavelength).

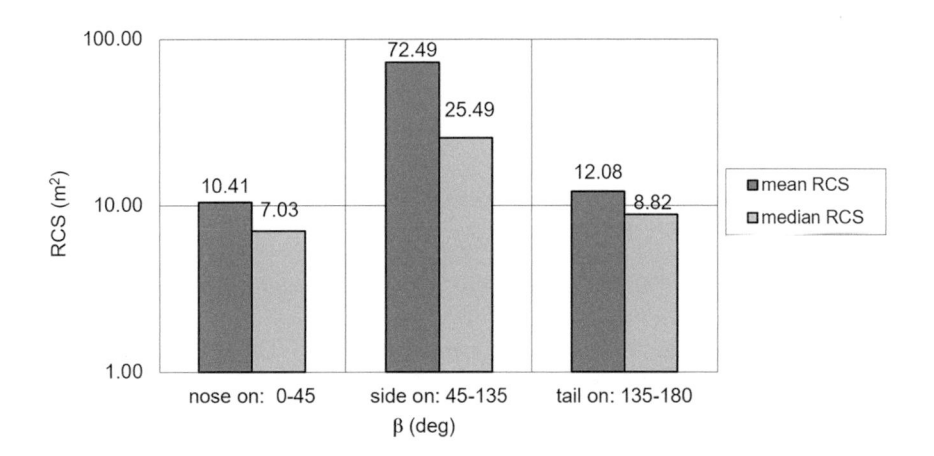

FIGURE 2.465 Diagrams of mean and median RCS of Mi-8 helicopter model in three sectors of azimuth aspect given its radar observation at horizontal polarization and a carrier frequency of 10 GHz (3 cm wavelength).

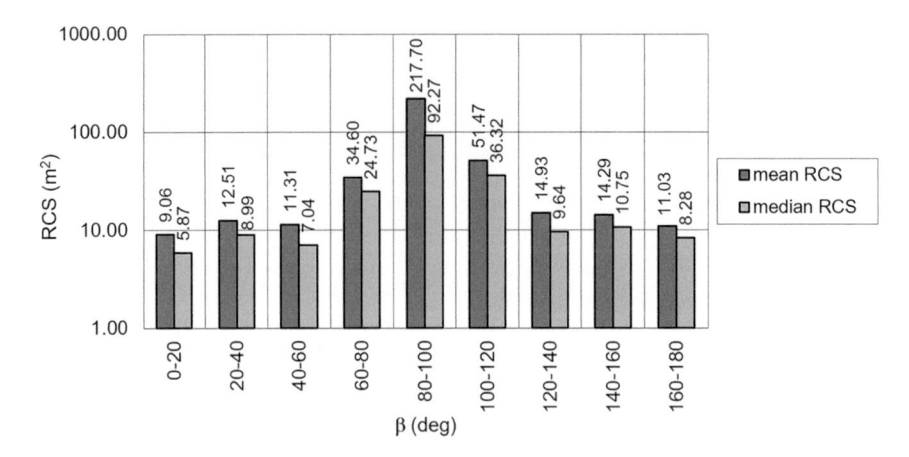

FIGURE 2.466 Diagrams of mean and median RCS of Mi-8 helicopter model in 20-degree sectors of azimuth aspect given its radar observation at horizontal polarization and a carrier frequency of 10 GHz (3 cm wavelength).

Figures 2.467, 2.472, and 2.477 show histograms of the scattered signal amplitude (square root of the RCS) for the range of sounding azimuths −20° to +20° (sounding from the nose). The bold line shows the probability density functions of the distribution, which can be used to approximate the histogram of the signal amplitude.

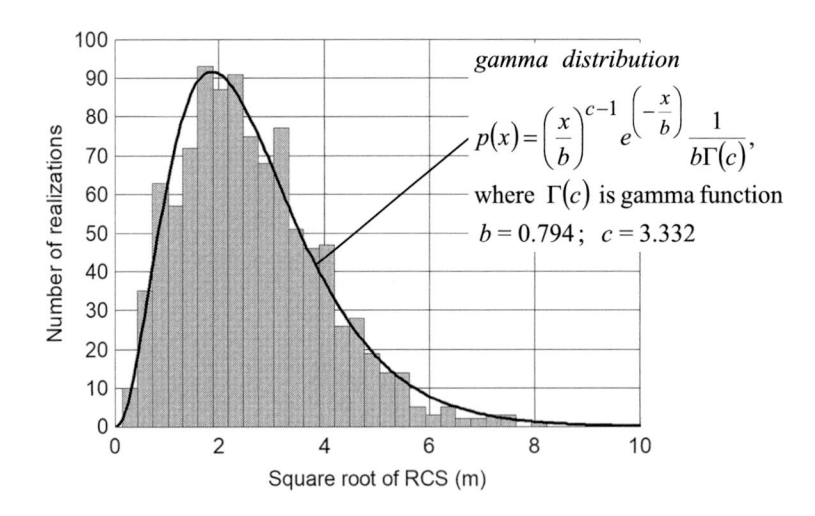

gamma distribution

$$p(x) = \left(\frac{x}{b}\right)^{c-1} e^{\left(-\frac{x}{b}\right)} \frac{1}{b\Gamma(c)},$$

where $\Gamma(c)$ is gamma function
$b = 0.794; \quad c = 3.332$

FIGURE 2.467 Amplitude distribution of echo signal of Mi-8 helicopter model at a carrier frequency of 10 GHz given its horizontal polarization.

Scattering characteristics of Mi-8 helicopter model for sounding frequency 3 GHz (wavelength 10 cm).

Figure 2.468 shows the RCS circular diagram of the Mi-8. The noncoherent RCS circular diagram of the Mi-8 is shown in Figure 2.469.

The circular mean RCS of the Mi-8 helicopter for horizontal polarization is 36.80 m². The circular median RCS for horizontal polarization is 11.25 m².

Figures 2.470 and 2.471 show the mean and median RCS for the main ranges of sounding azimuths (nose, side, and tail) and for ranges of 20°.

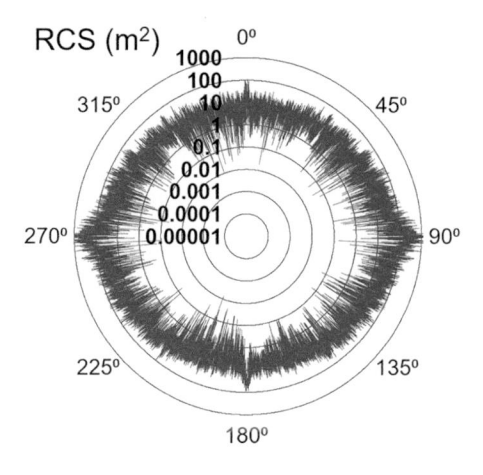

Scattering Characteristics of Aerial Objects 177

FIGURE 2.468 Circular diagram of RCS given radar observation of Mi-8 helicopter model at a carrier frequency of 3 GHz (10 cm wavelength).

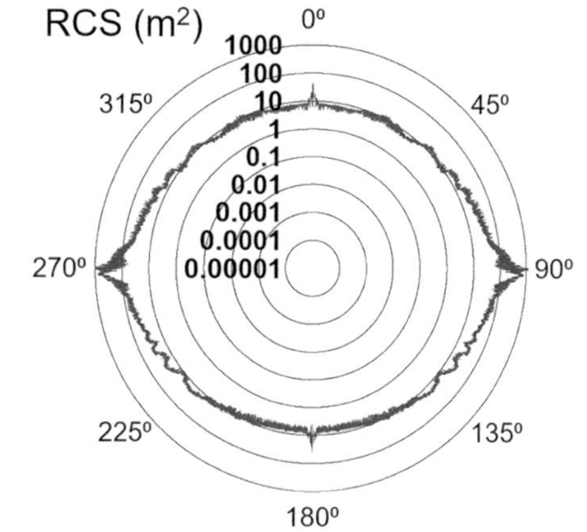

FIGURE 2.469 Circular diagram of noncoherent RCS given radar observation of Mi-8 helicopter model at a carrier frequency of 3 GHz (10 cm wavelength).

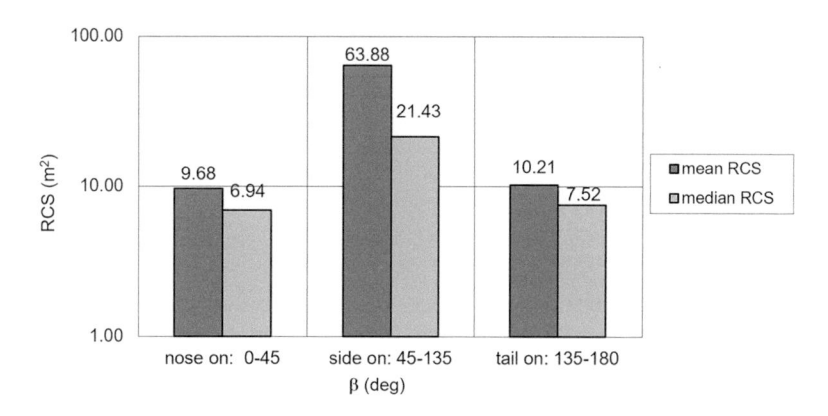

FIGURE 2.470 Diagrams of mean and median RCS of Mi-8 helicopter model in three sectors of azimuth aspect given its radar observation at horizontal polarization and a carrier frequency of 3 GHz (10 cm wavelength).

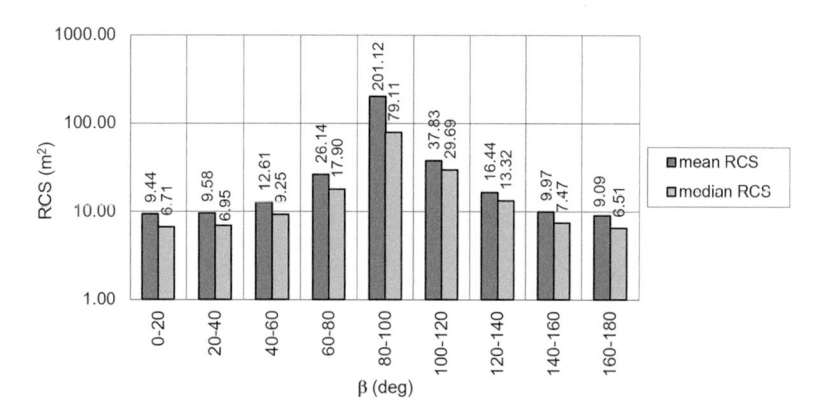

FIGURE 2.471 Diagrams of mean and median RCS of Mi-8 helicopter model in 20-degree sectors of azimuth aspect given its radar observation at horizontal polarization and a carrier frequency of 3 GHz (10 cm wavelength).

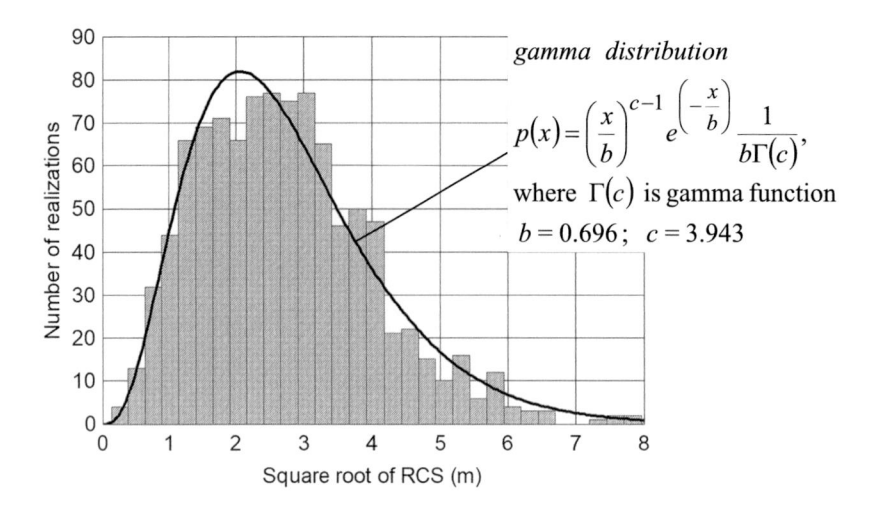

FIGURE 2.472 Amplitude distribution of echo signal of Mi-8 helicopter model at a carrier frequency of 3 GHz given its horizontal polarization.

$$p(x) = \left(\frac{x}{b}\right)^{c-1} e^{\left(-\frac{x}{b}\right)} \frac{1}{b\Gamma(c)},$$

where $\Gamma(c)$ is gamma function
$b = 0.696; \quad c = 3.943$

Scattering characteristics of Mi-8 helicopter model for sounding frequency 1 GHz (wavelength 30 cm).

Figure 2.473 shows the RCS circular diagram of the Mi-8. The noncoherent RCS circular diagram of the Mi-8 is shown in Figure 2.474.

The circular mean RCS of the Mi-8 helicopter for horizontal polarization is 38.48 m². The circular median RCS for horizontal polarization is 12.43 m².

Figures 2.475 and 2.476 show the mean and median RCS for the main ranges of sounding azimuths (nose, side, and tail) and for ranges of 20°.

The expressions and parameters of probability distributions that are most consistent with the empirical distributions of the square root of the RCS for different ranges of sounding azimuths and polarizations are given in the electronic appendix to this book. The expressions and parameters of the probability distributions that are most consistent with the empirical distributions of the RCS (energy characteristic) for different ranges of sounding azimuths and polarizations are also given there.

FIGURE 2.473 Circular diagram of RCS given radar observation of Mi-8 helicopter model at a carrier frequency of 1 GHz (30 cm wavelength).

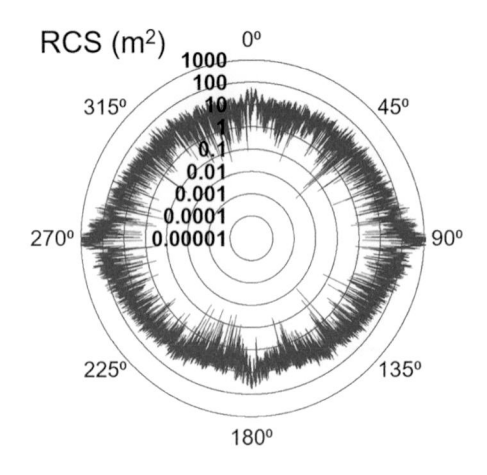

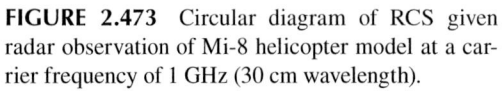

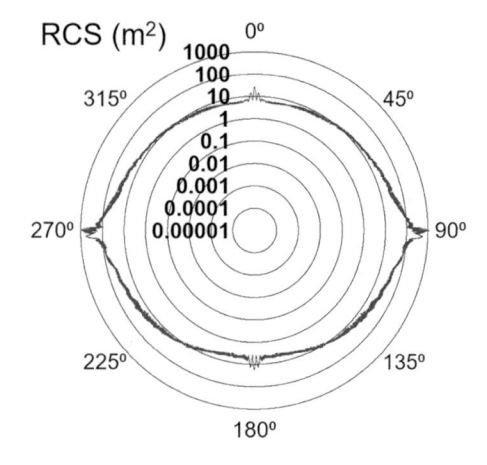

FIGURE 2.474 Circular diagram of noncoherent RCS given radar observation of Mi-8 helicopter model at a carrier frequency of 1 GHz (30 cm wavelength).

Scattering Characteristics of Aerial Objects

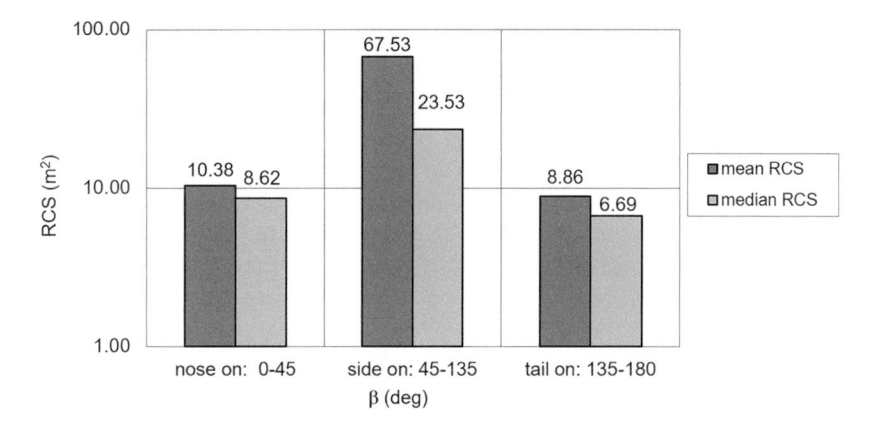

FIGURE 2.475 Diagrams of mean and median RCS of Mi-8 helicopter model in three sectors of azimuth aspect given its radar observation at horizontal polarization and a carrier frequency of 1 GHz (30 cm wavelength).

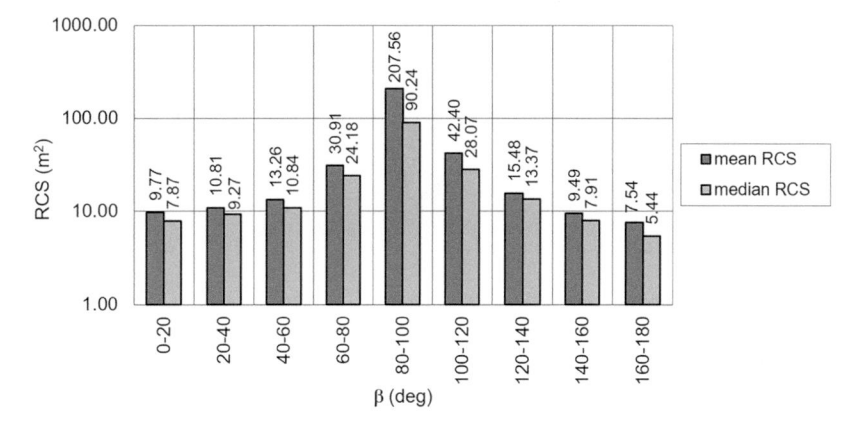

FIGURE 2.476 Diagrams of mean and median RCS of Mi-8 helicopter model in 20-degree sectors of azimuth aspect given its radar observation at horizontal polarization and a carrier frequency of 1 GHz (30 cm wavelength).

Weibull distribution

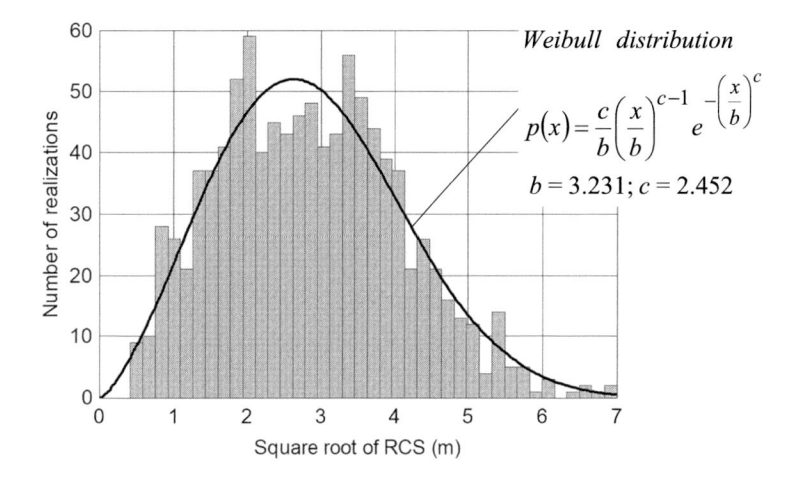

$$p(x) = \frac{c}{b}\left(\frac{x}{b}\right)^{c-1} e^{-\left(\frac{x}{b}\right)^c}$$

$b = 3.231; c = 2.452$

FIGURE 2.477 Amplitude distribution of echo signal of Mi-8 helicopter model at a carrier frequency of 1 GHz given its horizontal polarization.

2.29 MULTI-PURPOSE COMBAT HELICOPTER Mi-24

Mi-24 Hind (Figure 2.478) is a multi-purpose combat helicopter. The first flight took place in 1970.

General characteristics of Mi-24 [40]: main rotor diameter – 17.3 m, length – 17.51 m, height – 3.90 m, weight – 8 570– 12 500 kg, powerplant – 2 TV3–117V turboshaft engines, speed – 320 km/h, range – 500 km, and service ceiling – 4 500 m.

In accordance with the design of the Mi-24, a model of its surface was created to obtaining scattering characteristics (in particular, RCS). The model is shown in Figure 2.479. In the modeling, the smooth parts of the helicopter surface were approximated by parts of 86 triaxial ellipsoids. The surface breaks were modeled using 12 straight edge scattering parts.

FIGURE 2.478 Multi-purpose combat helicopter Mi-24.

FIGURE 2.479 Surface model of multi-purpose combat helicopter Mi-24.

Below are some scattering characteristics of the Mi-24 model at sounding frequencies of 10, 3, and 1 GHz (wavelengths of 3, 10, and 30 cm, respectively) for horizontal polarization of the probing signal. The scattering characteristics for this model at two polarizations, as well as scattering characteristics at sounding frequencies of 5 and 1.3 GHz (wavelengths of 6 and 23 cm, respectively), are given in the electronic appendix of this book.

Sounding parameters: The elevation angle is to be a random value distributed uniformly in the range −3° ± 4° with respect to the wing plane (elevation angle of −3° corresponds to the radar observation from the lower hemisphere), azimuth aspect increment was 0.02°, the azimuth being counted off from the nose-on aspect (0° corresponds to the nose-on radar observation, 180° corresponds to the tail-on observation).

Scattering characteristics of Mi-24 helicopter model for sounding frequency 10 GHz (wavelength 3 cm).

Figure 2.480 shows the RCS circular diagram of the Mi-24 model. The noncoherent RCS circular diagram of the Mi-24 is shown in Figure 2.481.

The circular mean RCS of the Mi-24 helicopter model for horizontal polarization is 55.83 m². The circular median RCS (the RCS value used to calculate the detection range of an object with a probability of 0.5) for horizontal polarization is 8.59 m².

Figures 2.482 and 2.483 show the mean and median RCS for the main ranges of sounding azimuths (nose, side, and tail) and for ranges of 20°.

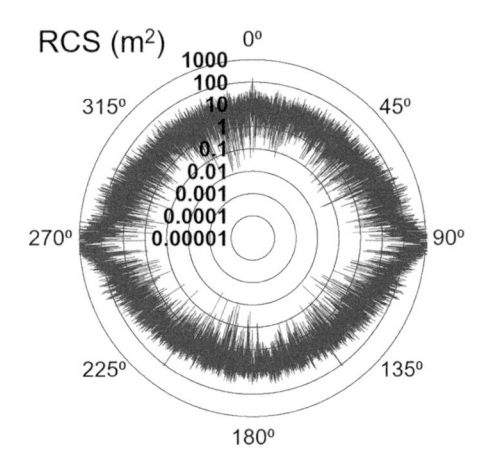

FIGURE 2.480 Circular diagram of RCS given radar observation of Mi-24 helicopter model at a carrier frequency of 10 GHz (3 cm wavelength).

FIGURE 2.481 Circular diagram of noncoherent RCS given radar observation of Mi-24 helicopter model at a carrier frequency of 10 GHz (3 cm wavelength).

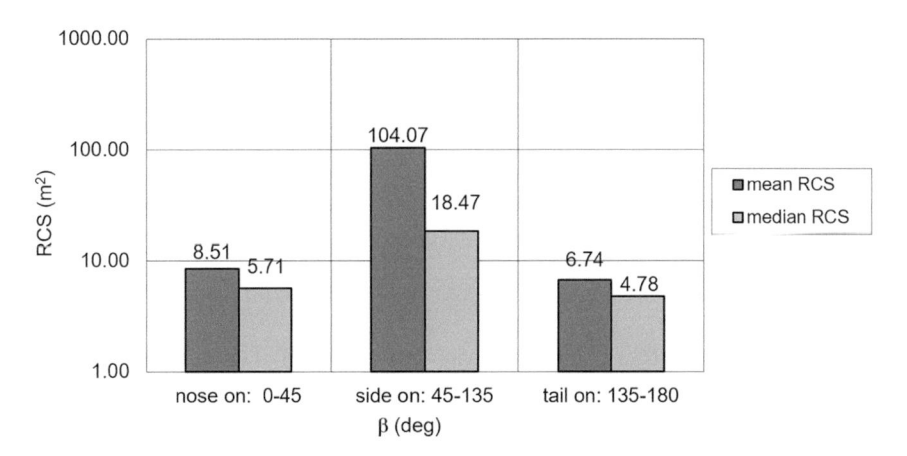

FIGURE 2.482 Diagrams of mean and median RCS of Mi-24 helicopter model in three sectors of azimuth aspect given its radar observation at horizontal polarization and a carrier frequency of 10 GHz (3 cm wavelength).

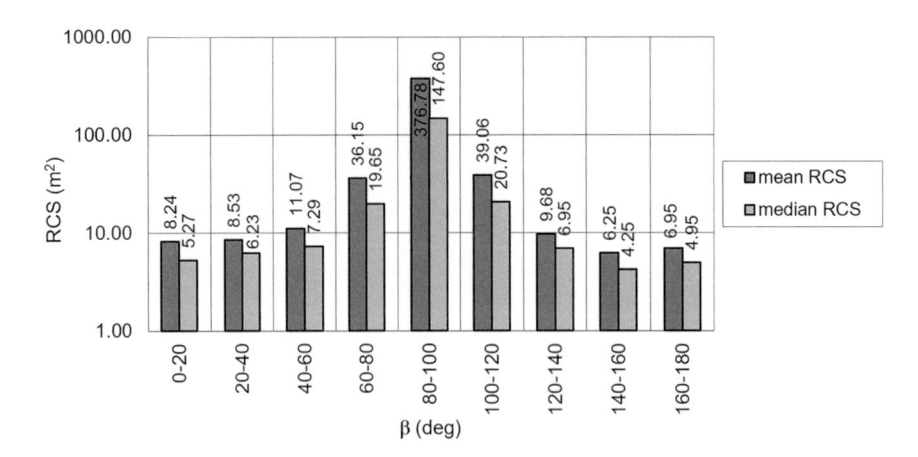

FIGURE 2.483 Diagrams of mean and median RCS of Mi-24 helicopter model in 20-degree sectors of azimuth aspect given its radar observation at horizontal polarization and a carrier frequency of 10 GHz (3 cm wavelength).

Figures 2.484, 2.489, and 2.494 show histograms of the scattered signal amplitude (square root of the RCS) for the range of sounding azimuths −20° to +20° (sounding from the nose). The bold line shows the probability density functions of the distribution, which can be used to approximate the histogram of the signal amplitude.

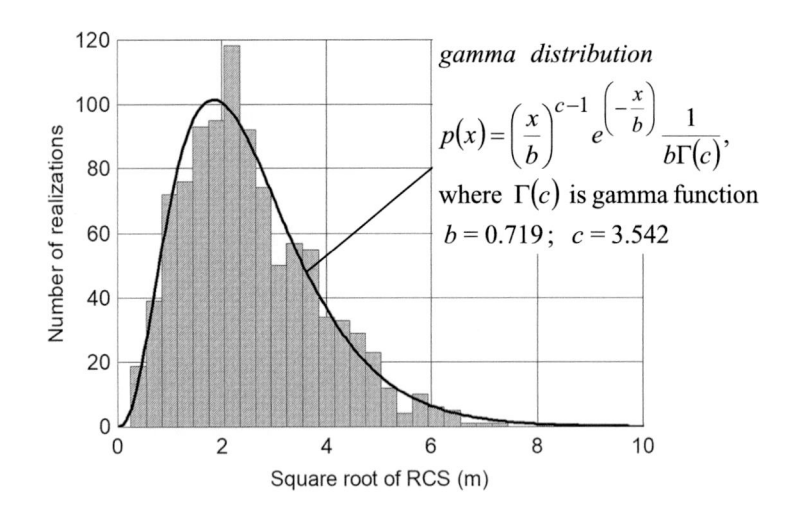

gamma distribution

$$p(x) = \left(\frac{x}{b}\right)^{c-1} e^{\left(-\frac{x}{b}\right)} \frac{1}{b\Gamma(c)},$$

where $\Gamma(c)$ is gamma function
$b = 0.719;\ c = 3.542$

FIGURE 2.484 Amplitude distribution of echo signal of Mi-24 helicopter model at a carrier frequency of 10 GHz given its horizontal polarization.

Scattering characteristics of Mi-24 helicopter model for sounding frequency 3 GHz (wavelength 10 cm).

Figure 2.485 shows the RCS circular diagram of the Mi-24. The noncoherent RCS circular diagram of the Mi-24 is shown in Figure 2.486.

The circular mean RCS of the Mi-24 helicopter for horizontal polarization is 48.37 m². The circular median RCS for horizontal polarization is 7.87 m².

Figures 2.487 and 2.488 show the mean and median RCS for the main ranges of sounding azimuths (nose, side, and tail) and for ranges of 20°.

Scattering Characteristics of Aerial Objects

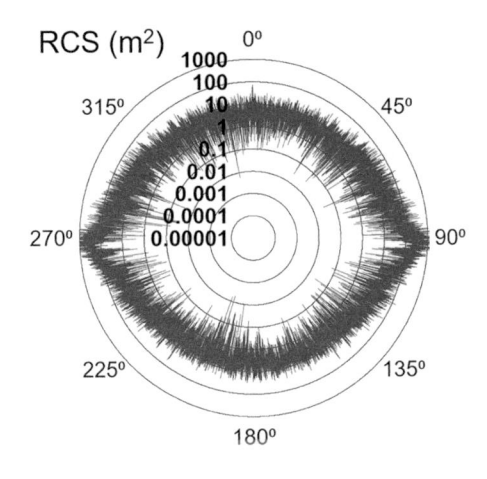

FIGURE 2.485 Circular diagram of RCS given radar observation of Mi-24 helicopter model at a carrier frequency of 3 GHz (10 cm wavelength).

FIGURE 2.486 Circular diagram of noncoherent RCS given radar observation of Mi-24 helicopter model at a carrier frequency of 3 GHz (10 cm wavelength).

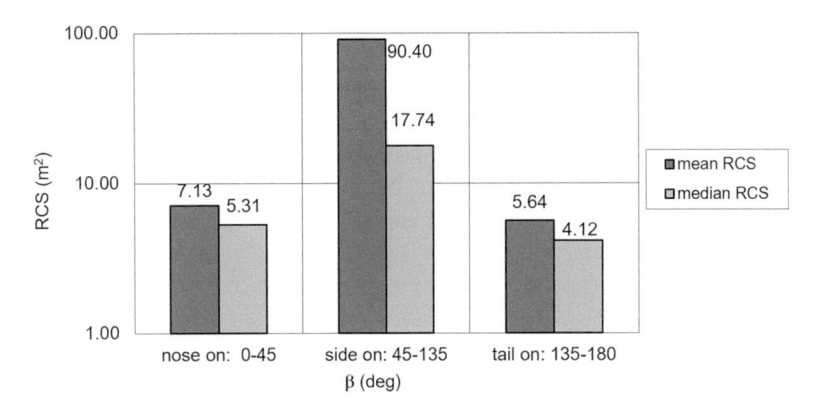

FIGURE 2.487 Diagrams of mean and median RCS of Mi-24 helicopter model in three sectors of azimuth aspect given its radar observation at horizontal polarization and a carrier frequency of 3 GHz (10 cm wavelength).

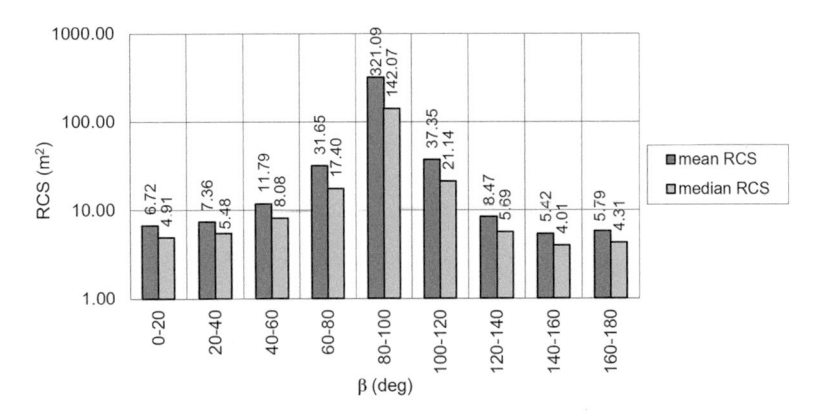

FIGURE 2.488 Diagrams of mean and median RCS of Mi-24 helicopter model in 20-degree sectors of azimuth aspect given its radar observation at horizontal polarization and a carrier frequency of 3 GHz (10 cm wavelength).

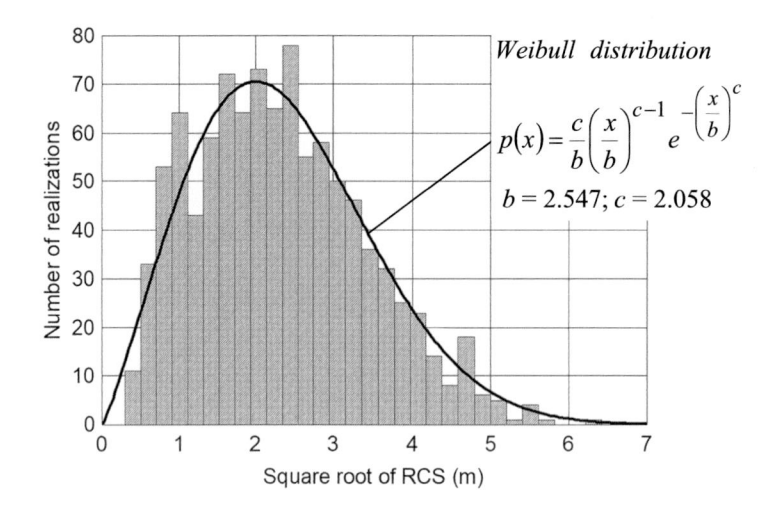

FIGURE 2.489 Amplitude distribution of echo signal of Mi-24 helicopter model at a carrier frequency of 3 GHz given its horizontal polarization.

Weibull distribution

$$p(x) = \frac{c}{b}\left(\frac{x}{b}\right)^{c-1} e^{-\left(\frac{x}{b}\right)^c}$$

$b = 2.547; c = 2.058$

Scattering characteristics of Mi-24 helicopter model for sounding frequency 1 GHz (wavelength 30 cm).

Figure 2.490 shows the RCS circular diagram of the Mi-24. The noncoherent RCS circular diagram of the Mi-24 is shown in Figure 2.491.

The circular mean RCS of the Mi-24 helicopter for horizontal polarization is 48.69 m². The circular median RCS for horizontal polarization is 7.08 m².

Figures 2.492 and 2.493 show the mean and median RCS for the main ranges of sounding azimuths (nose, side, and tail) and for ranges of 20°.

The expressions and parameters of probability distributions that are most consistent with the empirical distributions of the square root of the RCS for different ranges of sounding azimuths and polarizations are given in the electronic appendix to this book. The expressions and parameters of the probability distributions that are most consistent with the empirical distributions of the RCS (energy characteristic) for different ranges of sounding azimuths and polarizations are also given there.

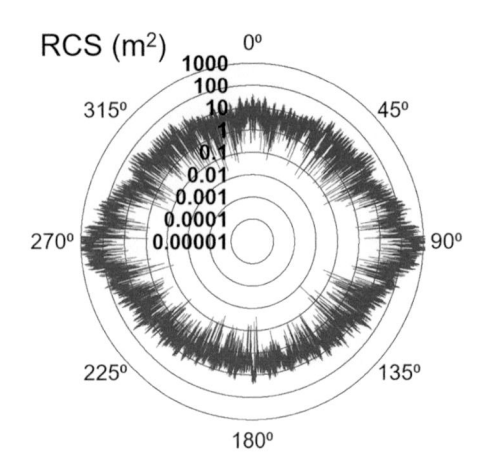

FIGURE 2.490 Circular diagram of RCS given radar observation of Mi-24 helicopter model at a carrier frequency of 1 GHz (30 cm wavelength).

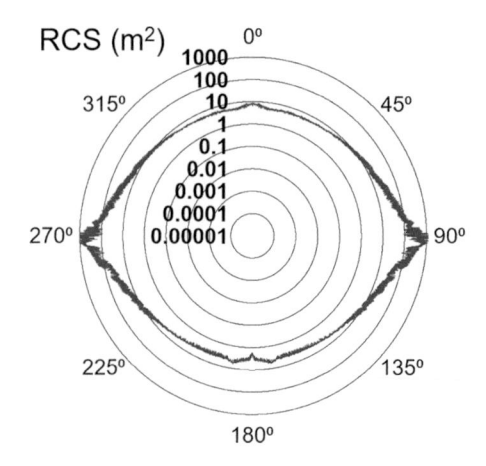

FIGURE 2.491 Circular diagram of noncoherent RCS given radar observation of Mi-24 helicopter model at a carrier frequency of 1 GHz (30 cm wavelength).

Scattering Characteristics of Aerial Objects

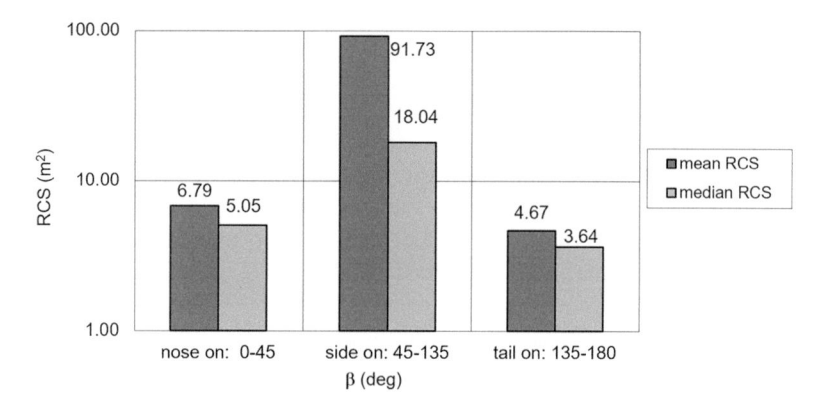

FIGURE 2.492 Diagrams of mean and median RCS of Mi-24 helicopter model in three sectors of azimuth aspect given its radar observation at horizontal polarization and a carrier frequency of 1 GHz (30 cm wavelength).

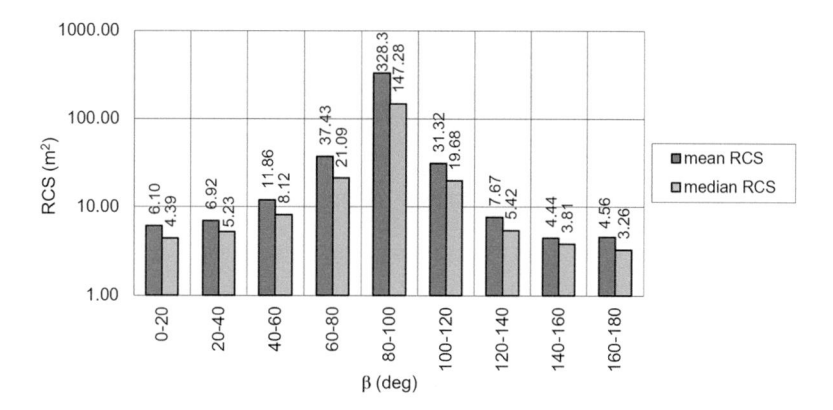

FIGURE 2.493 Diagrams of mean and median RCS of Mi-24 helicopter model in 20-degree sectors of azimuth aspect given its radar observation at horizontal polarization and a carrier frequency of 1 GHz (30 cm wavelength).

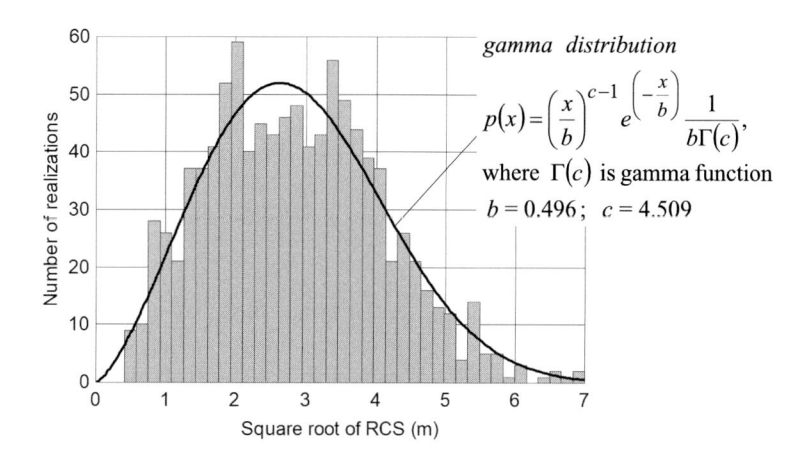

gamma distribution

$$p(x) = \left(\frac{x}{b}\right)^{c-1} e^{\left(-\frac{x}{b}\right)} \frac{1}{b\Gamma(c)},$$

where $\Gamma(c)$ is gamma function

$b = 0.496;\ c = 4.509$

FIGURE 2.494 Amplitude distribution of echo signal of Mi-24 helicopter model at a carrier frequency of 1 GHz given its horizontal polarization.

2.30 UNMANNED AERIAL VEHICLE Tu-143 REYS

Tu-143 Reys (Figure 2.495) is a tactical unmanned aerial reconnaissance vehicle. The first flight took place in 1970.

General characteristics of Tu-143 [41]: wingspan –2.24 m, length – 8.06 m, height – 1.545 m, wing area – 2.9 m^2, weight – 1 230 kg, powerplant – 1 turbojet TR3–117 engine, speed – 950 km/h, range – 200 km, and service ceiling – 5 000 m.

In accordance with the design of the Tu-143, a model of its surface was created to obtaining scattering characteristics (in particular, RCS). The model is shown in Figure 2.496. In the modeling, the smooth parts of the UAV surface were approximated by parts of 17 triaxial ellipsoids. The surface breaks were modeled using 15 straight edge scattering parts.

FIGURE 2.495 Unmanned aerial vehicle Tu143 Reys.

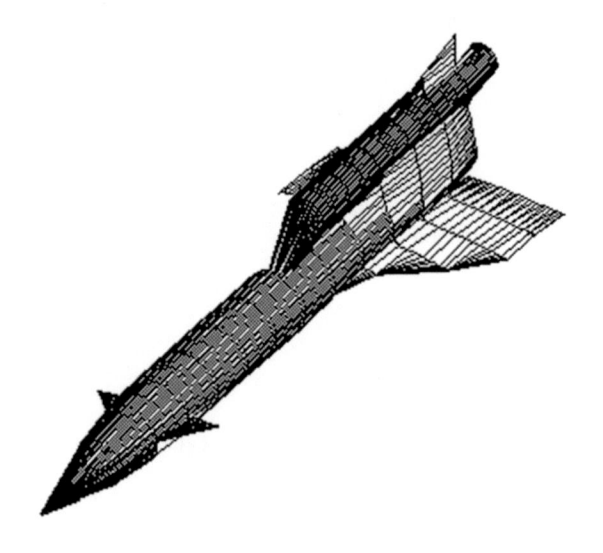

FIGURE 2.496 Surface model of unmanned aerial vehicle Tu-143 Reys.

Below are some scattering characteristics of the Tu-143 model at sounding frequencies of 10, 3, and 1 GHz (wavelengths of 3, 10, and 30 cm, respectively) for horizontal polarization of the probing signal. The scattering characteristics for this model at two polarizations, as well as scattering characteristics at sounding frequencies of 5 and 1.3 GHz (wavelengths of 6 and 23 cm, respectively), are given in the electronic appendix of this book.

Sounding parameters: The elevation angle is to be a random value distributed uniformly in the range −3° ± 4° with respect to the wing plane (elevation angle of −3° corresponds to the radar

Scattering Characteristics of Aerial Objects   187

observation from the lower hemisphere), azimuth aspect increment was 0.02°, the azimuth being counted off from the nose-on aspect (0° corresponds to the nose-on radar observation, 180° corresponds to the tail-on observation).

Scattering characteristics of UAV Tu-143 Reys model for sounding frequency 10 GHz (wavelength 3 cm).

Figure 2.497 shows the RCS circular diagram of the Tu-143 model. The noncoherent RCS circular diagram of the Tu-143 is shown in Figure 2.408.

The circular mean RCS of the UAV Tu-143 Reys model for horizontal polarization is 4.07 m². The circular median RCS (the RCS value used to calculate the detection range of an object with a probability of 0.5) for horizontal polarization is 0.15 m².

Figures 2.482 and 2.483 show the mean and median RCS for the main ranges of sounding azimuths (nose, side, and tail) and for ranges of 20°.

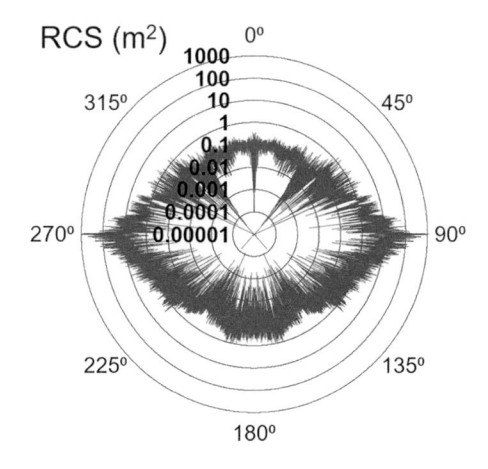

FIGURE 2.497 Circular diagram of RCS given radar observation of UAV Tu-143 Reys model at a carrier frequency of 10 GHz (3 cm wavelength).

FIGURE 2.498 Circular diagram of noncoherent RCS given radar observation of UAV Tu-143 Reys model at a carrier frequency of 10 GHz (3 cm wavelength).

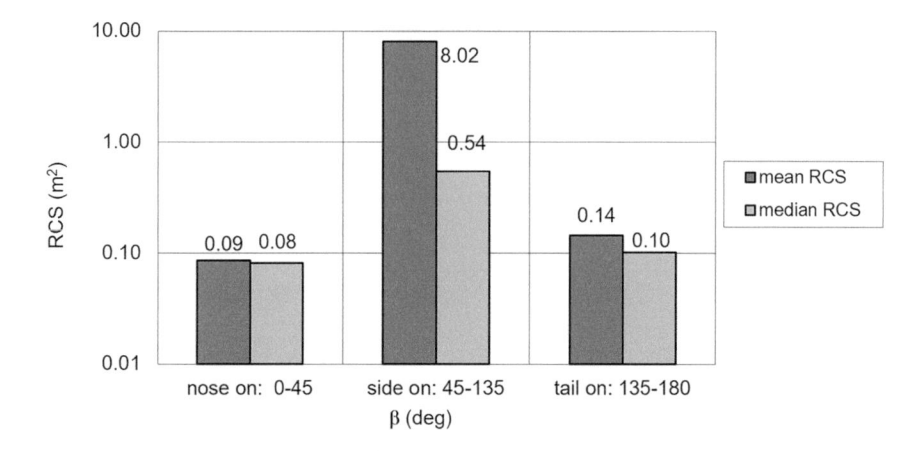

FIGURE 2.499 Diagrams of mean and median RCS of UAV Tu-143 Reys model in three sectors of azimuth aspect given its radar observation at horizontal polarization and a carrier frequency of 10 GHz (3 cm wavelength).

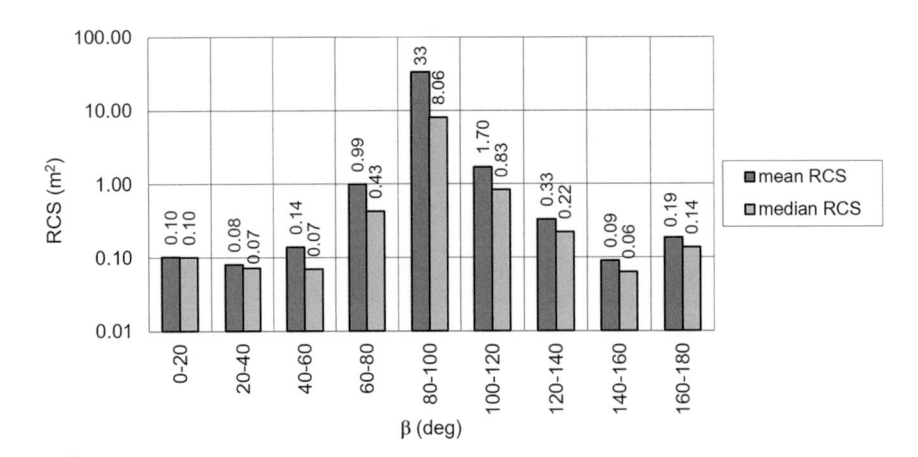

FIGURE 2.500 Diagrams of mean and median RCS of UAV Tu-143 Reys model in 20-degree sectors of azimuth aspect given its radar observation at horizontal polarization and a carrier frequency of 10 GHz (3 cm wavelength).

Figures 2.501, 2.506, and 2.511 show histograms of the scattered signal amplitude (square root of the RCS) for the range of sounding azimuths −20° to +20° (sounding from the nose). The bold line shows the probability density functions of the distribution, which can be used to approximate the histogram of the signal amplitude.

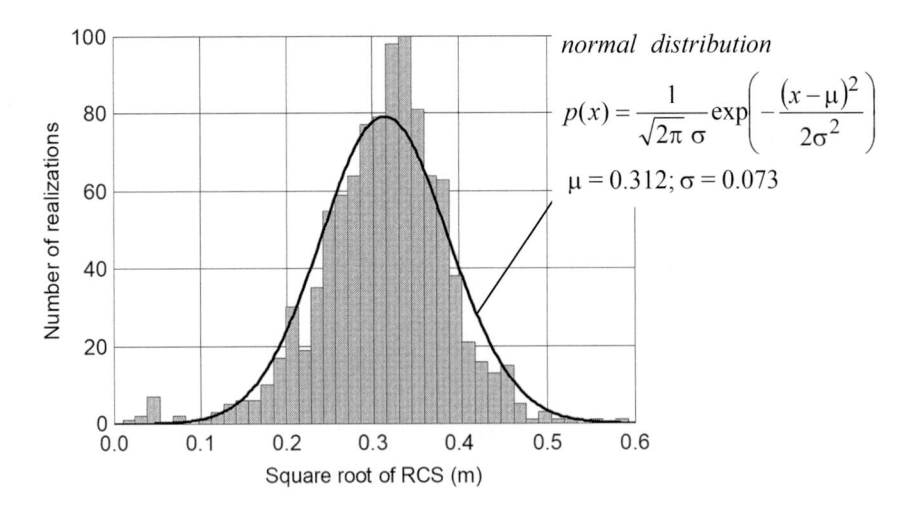

FIGURE 2.501 Amplitude distribution of echo signal of UAV Tu-143 Reys model at a carrier frequency of 10 GHz given its horizontal polarization.

Scattering characteristics of UAV Tu-143 Reys model for sounding frequency 3 GHz (wavelength 10 cm).

Figure 2.502 shows the RCS circular diagram of the Tu-143. The noncoherent RCS circular diagram of the Tu-143 is shown in Figure 2.503.

The circular mean RCS of the UAV Tu-143 Reys for horizontal polarization is 4.09 m². The circular median RCS for horizontal polarization is 0.15 m².

Figures 2.504 and 2.505 show the mean and median RCS for the main ranges of sounding azimuths (nose, side, and tail) and for ranges of 20°.

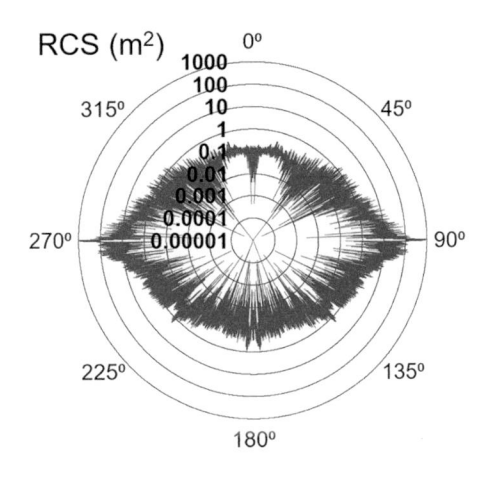

Scattering Characteristics of Aerial Objects

FIGURE 2.502 Circular diagram of RCS given radar observation of UAV Tu-143 Reys model at a carrier frequency of 3 GHz (10 cm wavelength).

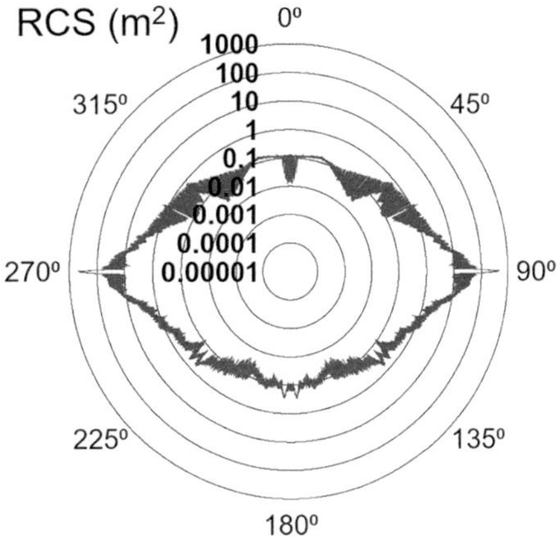

FIGURE 2.503 Circular diagram of noncoherent RCS given radar observation of UAV Tu-143 Reys model at a carrier frequency of 3 GHz (10 cm wavelength).

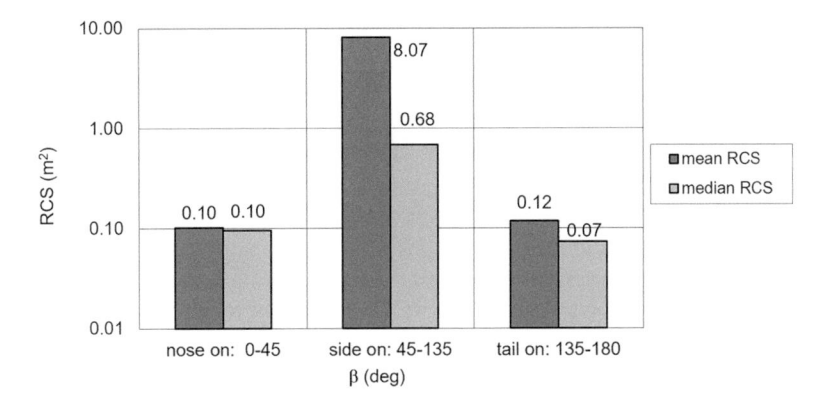

FIGURE 2.504 Diagrams of mean and median RCS of UAV Tu-143 Reys model in three sectors of azimuth aspect given its radar observation at horizontal polarization and a carrier frequency of 3 GHz (10 cm wavelength).

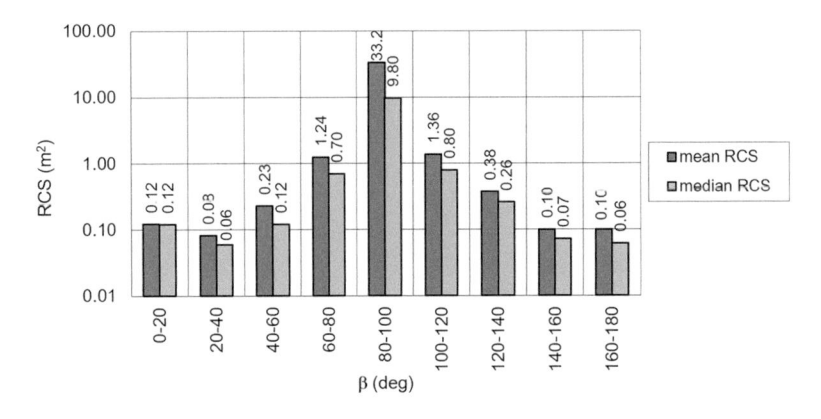

FIGURE 2.505 Diagrams of mean and median RCS of UAV Tu-143 Reys model in 20-degree sectors of azimuth aspect given its radar observation at horizontal polarization and a carrier frequency of 3 GHz (10 cm wavelength).

$$p(x) = \frac{1}{\sqrt{2\pi}\,\sigma} \exp\left(-\frac{(x-\mu)^2}{2\sigma^2}\right)$$

normal distribution

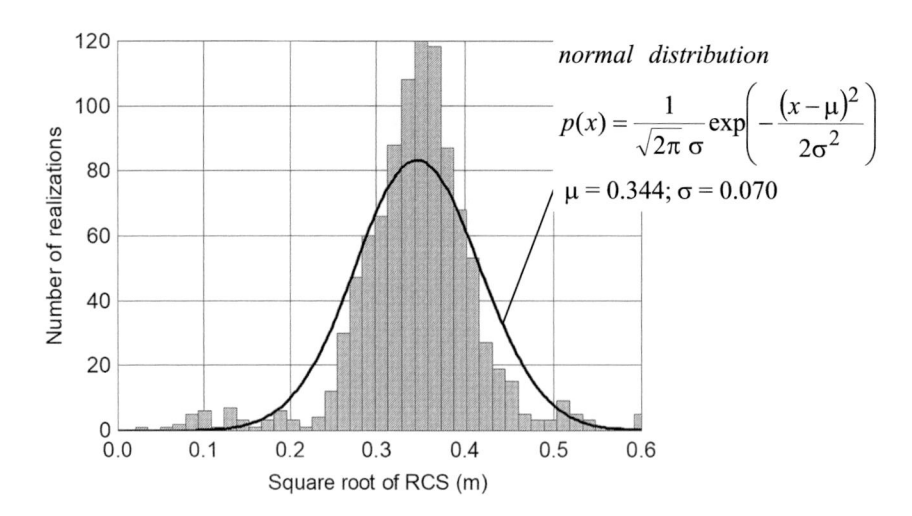

$\mu = 0.344;\ \sigma = 0.070$

FIGURE 2.506 Amplitude distribution of echo signal of UAV Tu-143 Reys model at a carrier frequency of 3 GHz given its horizontal polarization.

Scattering characteristics of UAV Tu-143 Reys model for sounding frequency 1 GHz (wavelength 30 cm).

Figure 2.507 shows the RCS circular diagram of the Tu-143. The noncoherent RCS circular diagram of the Tu-143 is shown in Figure 2.508.

The circular mean RCS of the UAV Tu-143 Reys for horizontal polarization is 3.37 m². The circular median RCS for horizontal polarization is 0.20 m².

Figures 2.509 and 2.510 show the mean and median RCS for the main ranges of sounding azimuths (nose, side, and tail) and for ranges of 20°.

The expressions and parameters of probability distributions that are most consistent with the empirical distributions of the square root of the RCS for different ranges of sounding azimuths and polarizations are given in the electronic appendix to this book. The expressions and parameters of the probability distributions that are most consistent with the empirical distributions of the RCS (energy characteristic) for different ranges of sounding azimuths and polarizations are also given there.

FIGURE 2.507 Circular diagram of RCS given radar observation of UAV Tu-143 Reys model at a carrier frequency of 1 GHz (30 cm wavelength).

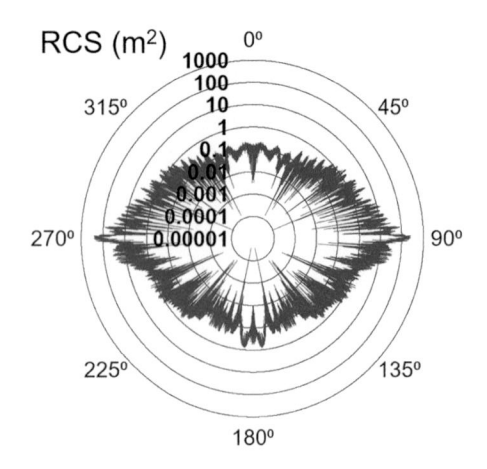

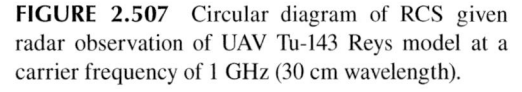

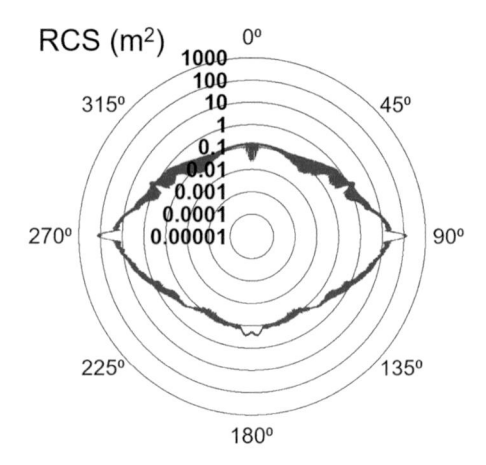

FIGURE 2.508 Circular diagram of noncoherent RCS given radar observation of UAV Tu-143 Reys model at a carrier frequency of 1 GHz (30 cm wavelength).

Scattering Characteristics of Aerial Objects

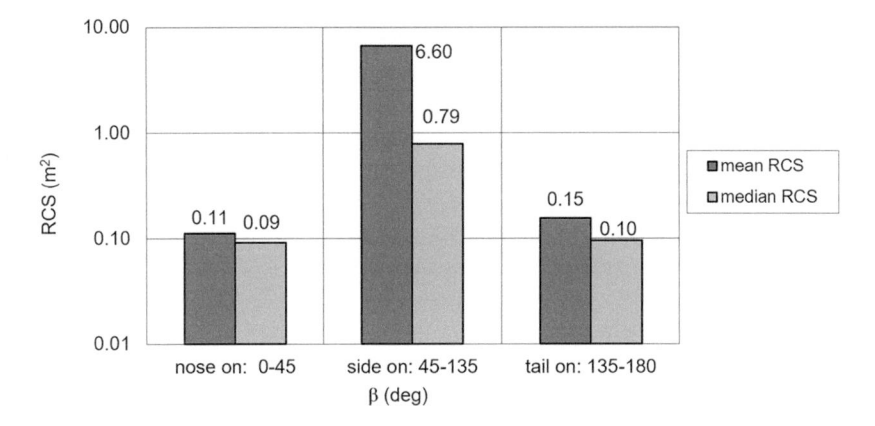

FIGURE 2.509 Diagrams of mean and median RCS of UAV Tu-143 Reys model in three sectors of azimuth aspect given its radar observation at horizontal polarization and a carrier frequency of 1 GHz (30 cm wavelength).

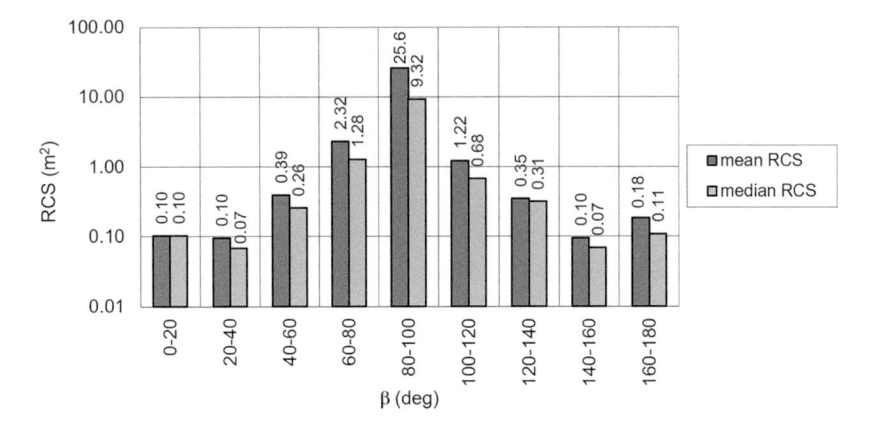

FIGURE 2.510 Diagrams of mean and median RCS of UAV Tu-143 Reys model in 20-degree sectors of azimuth aspect given its radar observation at horizontal polarization and a carrier frequency of 1 GHz (30 cm wavelength).

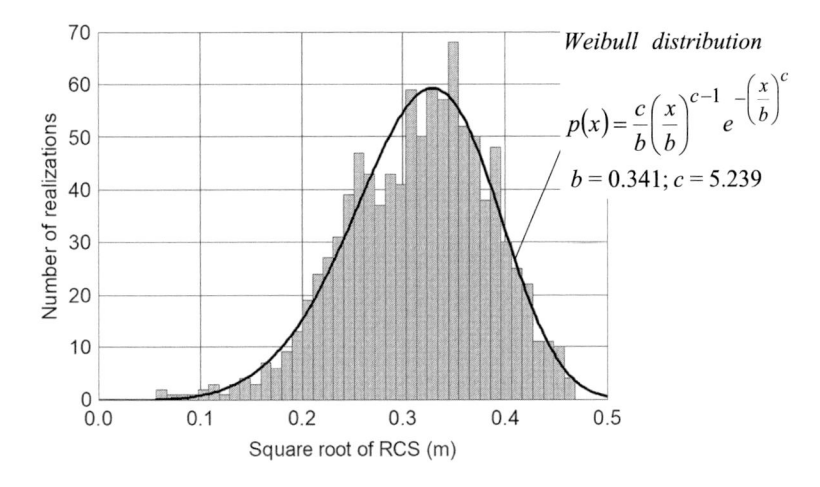

$$p(x) = \frac{c}{b}\left(\frac{x}{b}\right)^{c-1} e^{-\left(\frac{x}{b}\right)^c}$$

$b = 0.341; c = 5.239$

FIGURE 2.511 Amplitude distribution of echo signal of UAV Tu-143 Reys model at a carrier frequency of 1 GHz given its horizontal polarization.

2.31 UNMANNED AERIAL VEHICLE ORLAN-10

Orlan-10 (Figure 2.512) is a medium-range, multi-purpose unmanned aerial vehicle. The first flight took place in 2010.

General characteristics of Orlan-10 [42]: wingspan –3.10 m, length – 1.1 m, weight – 18 kg, powerplant – 1 Saito Manufacturing FA-62B single-cylinder four-stroke glow fuel piston engine, speed – 100–150 km/h, range – 50–120 km, and service ceiling – 6 000 m.

In accordance with the design of the Orlan-10, a model of its surface was created to obtaining scattering characteristics (in particular, RCS). The model is shown in Figure 2.513. In the modeling, the plastic envelope and propeller were approximated by parts of 23 triaxial ellipsoids. The smooth parts of the inner equipment surfaces were approximated by parts of 22 triaxial ellipsoids. The surface breaks of the inner equipment were modeled using 16 straight edge scattering parts.

FIGURE 2.512 Unmanned aerial vehicle Orlan-10.

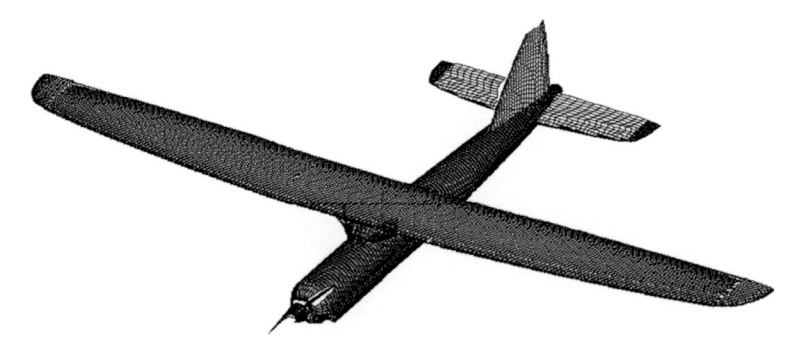

FIGURE 2.513 Surface model of unmanned aerial vehicle Orlan-10.

Below are some scattering characteristics of the Orlan-10 model at sounding frequencies of 10, 3, and 1 GHz (wavelengths of 3, 10, and 30 cm, respectively) for horizontal polarization of the probing signal. The scattering characteristics for this model at two polarizations, as well as scattering characteristics at sounding frequencies of 5 and 1.3 GHz (wavelengths of 6 and 23 cm, respectively), are given in the electronic appendix of this book.

Sounding parameters: The elevation angle is to be a random value distributed uniformly in the range $-3° \pm 4°$ with respect to the wing plane (elevation angle of $-3°$ corresponds to the radar observation from the lower hemisphere), azimuth aspect increment was $0.02°$, the azimuth being counted off from the nose-on aspect ($0°$ corresponds to the nose-on radar observation, $180°$ corresponds to the tail-on observation).

Scattering characteristics of UAV Orlan-10 model for sounding frequency 10 GHz (wavelength 3 cm).

Figure 2.514 shows the RCS circular diagram of the Orlan-10 model. The noncoherent RCS circular diagram of the Orlan-10 is shown in Figure 2.515.

The circular mean RCS of the UAV Orlan-10 model for horizontal polarization is 0.218 m². The circular median RCS (the RCS value used to calculate the detection range of an object with a probability of 0.5) for horizontal polarization is 0.035 m².

Figures 2.516 and 2.517 show the mean and median RCS for the main ranges of sounding azimuths (nose, side, and tail) and for ranges of 20°.

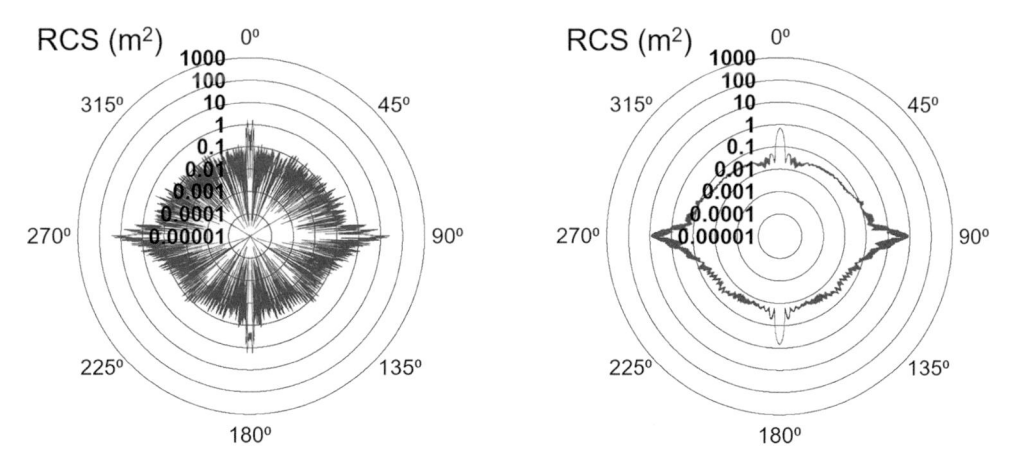

FIGURE 2.514 Circular diagram of RCS given radar observation of UAV Orlan-10 model at a carrier frequency of 10 GHz (3 cm wavelength).

FIGURE 2.515 Circular diagram of noncoherent RCS given radar observation of UAV Orlan-10 model at a carrier frequency of 10 GHz (3 cm wavelength).

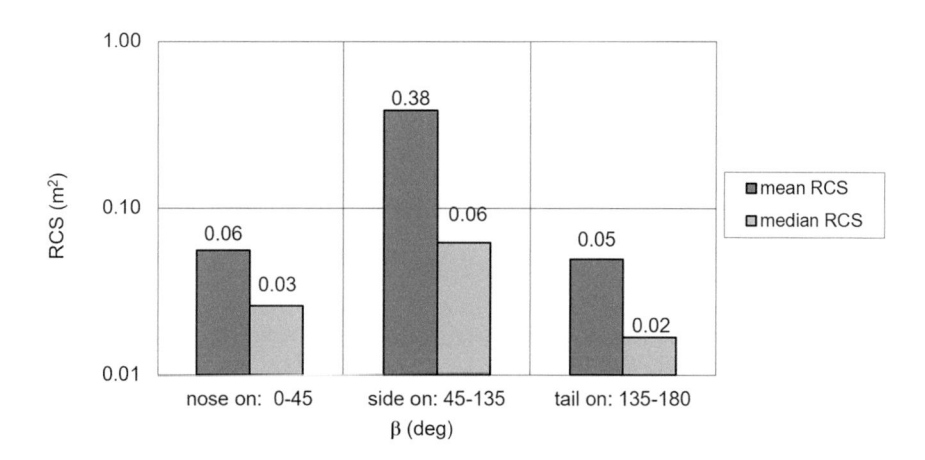

FIGURE 2.516 Diagrams of mean and median RCS of UAV Orlan-10 model in three sectors of azimuth aspect given its radar observation at horizontal polarization and a carrier frequency of 10 GHz (3 cm wavelength).

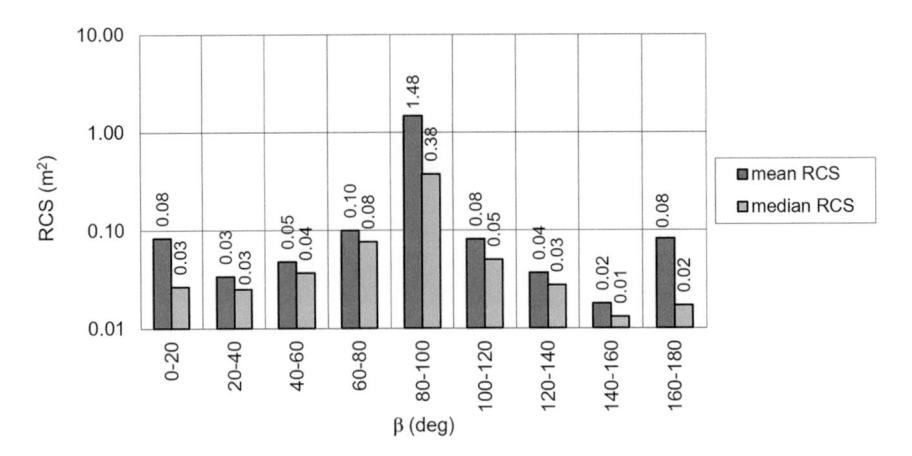

FIGURE 2.517 Diagrams of mean and median RCS of UAV Orlan-10 model in 20-degree sectors of azimuth aspect given its radar observation at horizontal polarization and a carrier frequency of 10 GHz (3 cm wavelength).

Figures 2.518, 2.523, and 2.528 show histograms of the scattered signal amplitude (square root of the RCS) for the range of sounding azimuths −20° to +20° (sounding from the nose). The bold line shows the probability density functions of the distribution, which can be used to approximate the histogram of the signal amplitude.

log - normal distribution

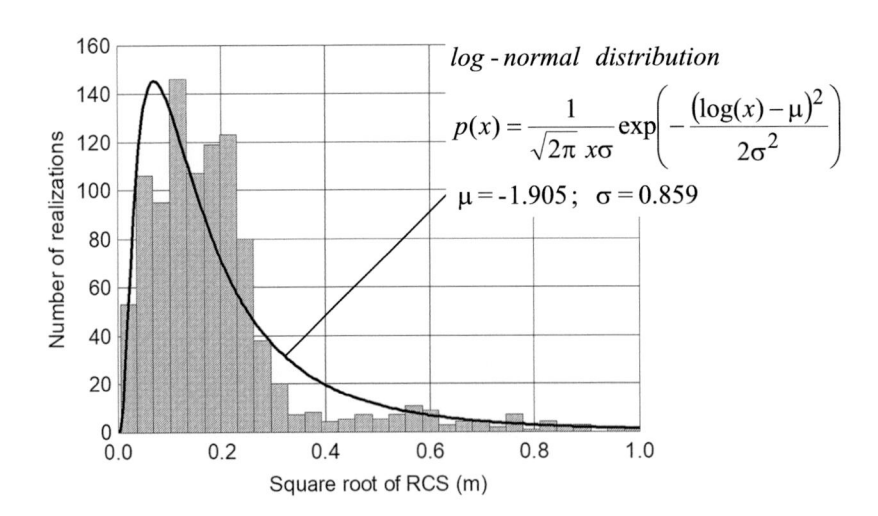

$$p(x) = \frac{1}{\sqrt{2\pi}\, x\sigma} \exp\left(-\frac{(\log(x)-\mu)^2}{2\sigma^2}\right)$$

$\mu = -1.905;\ \sigma = 0.859$

FIGURE 2.518 Amplitude distribution of echo signal of UAV Orlan-10 model at a carrier frequency of 10 GHz given its horizontal polarization.

Scattering characteristics of UAV Orlan-10 model for sounding frequency 3 GHz (wavelength 10 cm).

Figure 2.519 shows the RCS circular diagram of the Orlan-10. The noncoherent RCS circular diagram of the Orlan-10 is shown in Figure 2.520.

The circular mean RCS of the UAV Orlan-10 for horizontal polarization is 0.118 m². The circular median RCS for horizontal polarization is 0.030 m².

Figures 2.521 and 2.522 show the mean and median RCS for the main ranges of sounding azimuths (nose, side, and tail) and for ranges of 20°.

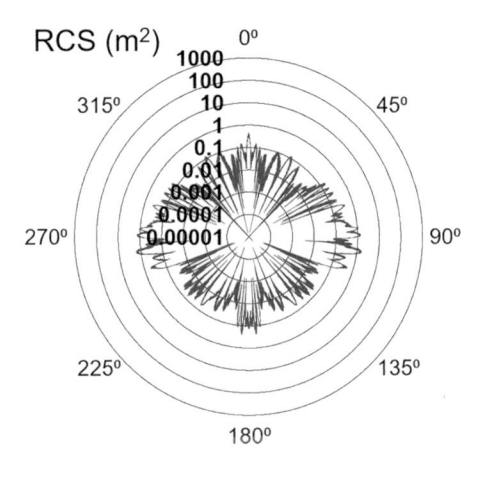

Scattering Characteristics of Aerial Objects

FIGURE 2.519 Circular diagram of RCS given radar observation of UAV Orlan-10 model at a carrier frequency of 3 GHz (10 cm wavelength).

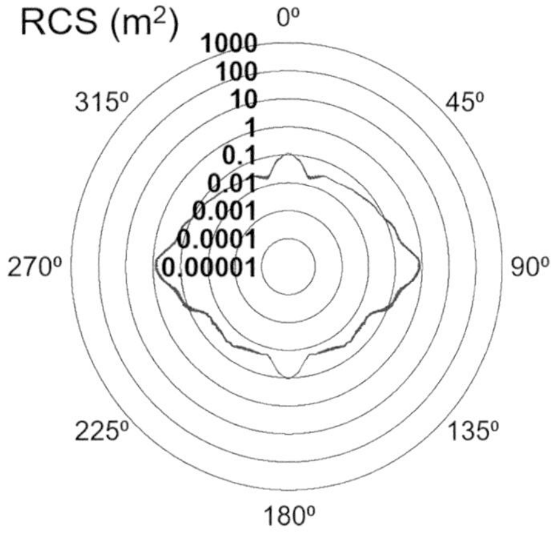

FIGURE 2.520 Circular diagram of noncoherent RCS given radar observation of UAV Orlan-10 model at a carrier frequency of 3 GHz (10 cm wavelength).

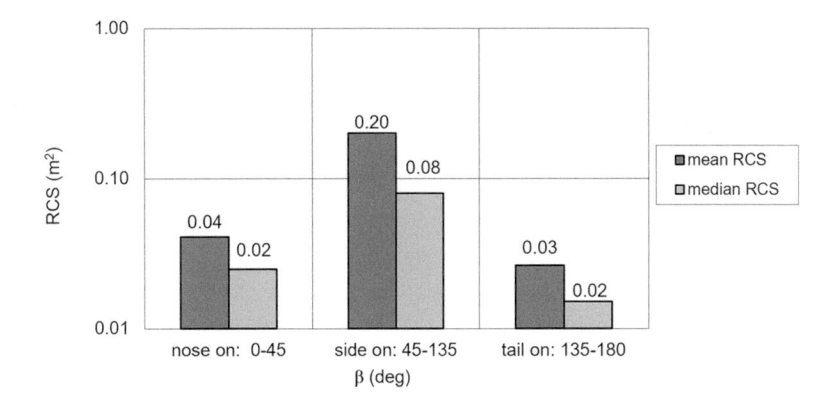

FIGURE 2.521 Diagrams of mean and median RCS of UAV Orlan-10 model in three sectors of azimuth aspect given its radar observation at horizontal polarization and a carrier frequency of 3 GHz (10 cm wavelength).

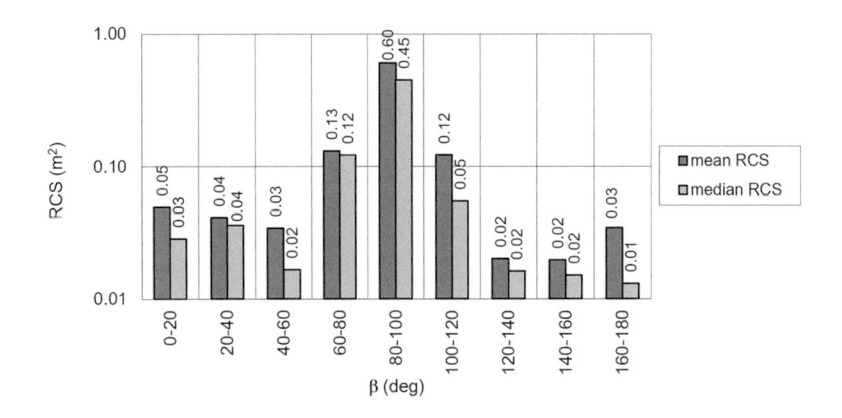

FIGURE 2.522 Diagrams of mean and median RCS of UAV Orlan-10 model in 20-degree sectors of azimuth aspect given its radar observation at horizontal polarization and a carrier frequency of 3 GHz (10 cm wavelength).

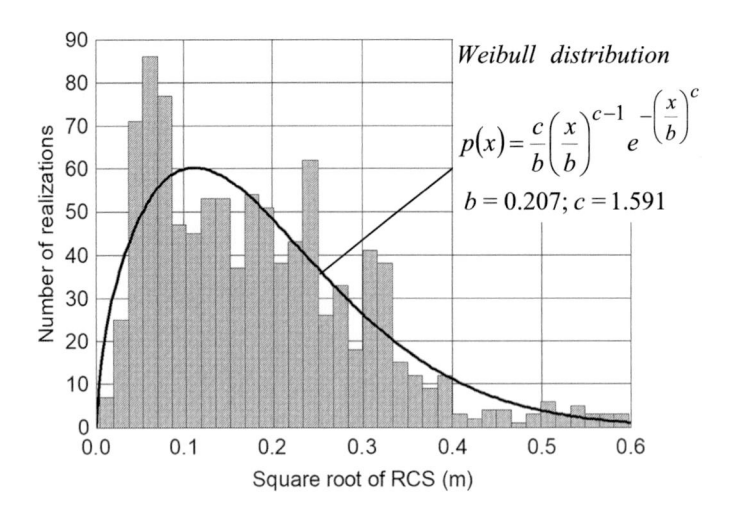

FIGURE 2.523 Amplitude distribution of echo signal of UAV Orlan-10 model at a carrier frequency of 3 GHz given its horizontal polarization.

Weibull distribution

$$p(x) = \frac{c}{b}\left(\frac{x}{b}\right)^{c-1} e^{-\left(\frac{x}{b}\right)^c}$$

$b = 0.207; c = 1.591$

Scattering characteristics of UAV Orlan-10 model for sounding frequency 1 GHz (wavelength 30 cm).

Figure 2.524 shows the RCS circular diagram of the Orlan-10. The noncoherent RCS circular diagram of the Orlan-10 is shown in Figure 2.525.

The circular mean RCS of the UAV Orlan-10 for horizontal polarization is 0.099 m². The circular median RCS for horizontal polarization is 0.048 m².

Figures 2.526 and 2.527 show the mean and median RCS for the main ranges of sounding azimuths (nose, side, and tail) and for ranges of 20°.

The expressions and parameters of probability distributions that are most consistent with the empirical distributions of the square root of the RCS for different ranges of sounding azimuths and polarizations are given in the electronic appendix to this book. The expressions and parameters of the probability distributions that are most consistent with the empirical distributions of the RCS (energy characteristic) for different ranges of sounding azimuths and polarizations are also given there.

FIGURE 2.524 Circular diagram of RCS given radar observation of UAV Orlan-10 model at a carrier frequency of 1 GHz (30 cm wavelength).

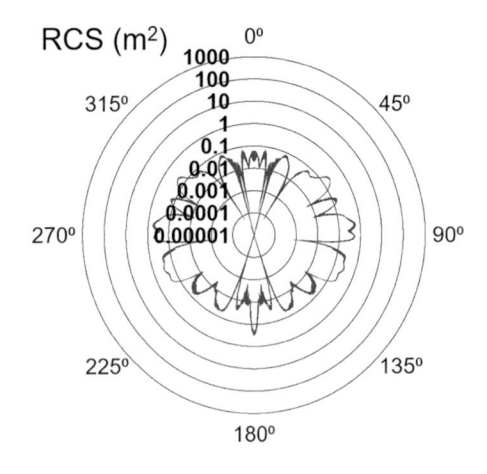

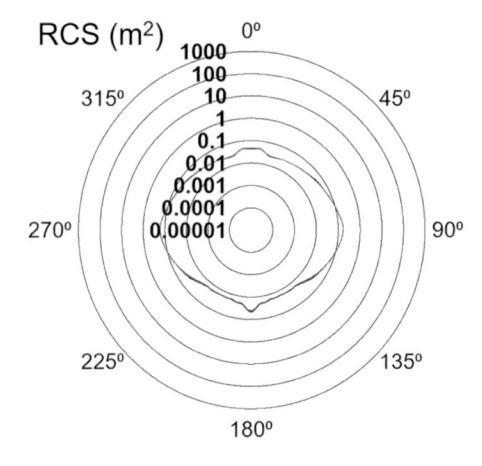

FIGURE 2.525 Circular diagram of noncoherent RCS given radar observation of UAV Orlan-10 model at a carrier frequency of 1 GHz (30 cm wavelength).

Scattering Characteristics of Aerial Objects 197

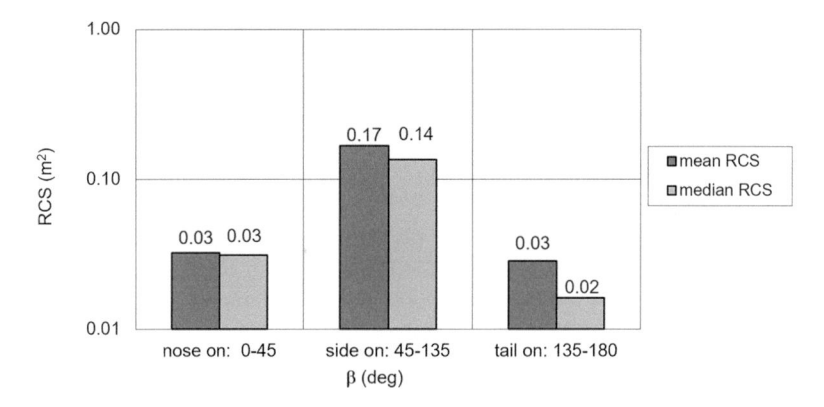

FIGURE 2.526 Diagrams of mean and median RCS of UAV Orlan-10 model in three sectors of azimuth aspect given its radar observation at horizontal polarization and a carrier frequency of 1 GHz (30 cm wavelength).

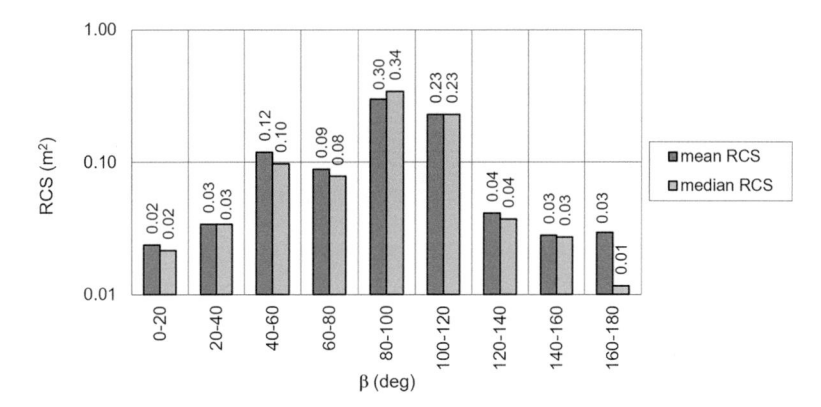

FIGURE 2.527 Diagrams of mean and median RCS of UAV Orlan-10 model in 20-degree sectors of azimuth aspect given its radar observation at horizontal polarization and a carrier frequency of 1 GHz (30 cm wavelength).

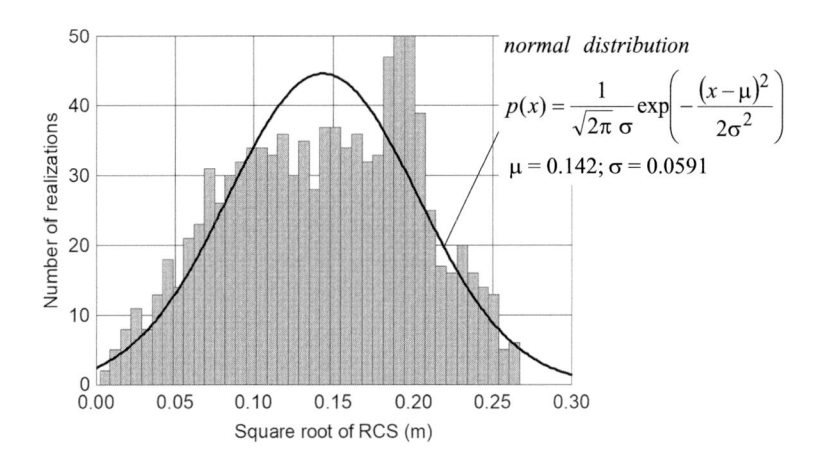

normal distribution
$$p(x) = \frac{1}{\sqrt{2\pi}\,\sigma} \exp\left(-\frac{(x-\mu)^2}{2\sigma^2}\right)$$
$\mu = 0.142; \sigma = 0.0591$

FIGURE 2.528 Amplitude distribution of echo signal of UAV Orlan-10 model at a carrier frequency of 1 GHz given its horizontal polarization.

2.32 UNMANNED AERIAL VEHICLE RQ-1 PREDATOR

RQ-1 Predator (Figure 2.529) is a long-endurance, medium-altitude unmanned aerial vehicle for surveillance and reconnaissance missions. The first flight took place in 1994.

General characteristics of RQ-1 [43]: wingspan –14.84 m, length – 8.23 m, height – 2.21 m, weight – 430–1 020 kg, powerplant – 1 Rotax 914F piston engine, speed – 100–130 km/h, range – 740 km, and service ceiling – 6 000 m.

In accordance with the design of the RQ-1, a model of its surface was created to obtaining scattering characteristics (in particular, RCS). The model is shown in Figure 2.530. The model surface was assumed to be perfectly conducting. The smooth parts of the model surface were approximated by parts of 26 triaxial ellipsoids. The surface breaks were modeled using 15 straight edge scattering parts.

FIGURE 2.529 Unmanned aerial vehicle RQ-1 Predator.

FIGURE 2.530 Surface model of unmanned aerial vehicle RQ-1 Predator.

Below are some scattering characteristics of the RQ-1 model at sounding frequencies of 10, 3, and 1 GHz (wavelengths of 3, 10, and 30 cm, respectively) for horizontal polarization of the probing signal. The scattering characteristics for this model at two polarizations, as well as scattering characteristics at sounding frequencies of 5 and 1.3 GHz (wavelengths of 6 and 23 cm, respectively), are given in the electronic appendix of this book.

Sounding parameters: The elevation angle is to be a random value distributed uniformly in the range $-3° \pm 4°$ with respect to the wing plane (elevation angle of $-3°$ corresponds to the radar observation from the lower hemisphere), azimuth aspect increment was $0.02°$, the azimuth being counted off from the nose-on aspect ($0°$ corresponds to the nose-on radar observation, $180°$ corresponds to the tail-on observation).

Scattering characteristics of UAV RQ-1 model for sounding frequency 10 GHz (wavelength 3 cm).

Figure 2.531 shows the RCS circular diagram of the RQ-1 model. The noncoherent RCS circular diagram of the RQ-1 is shown in Figure 2.532.

The circular mean RCS of the UAV RQ-1 model for horizontal polarization is 3.334 m². The circular median RCS (the RCS value used to calculate the detection range of an object with a probability of 0.5) for horizontal polarization is 1.054 m².

Figures 2.533 and 2.534 show the mean and median RCS for the main ranges of sounding azimuths (nose, side, and tail) and for ranges of 20°.

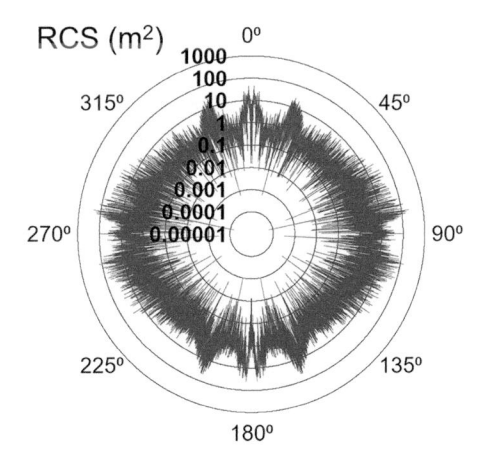

FIGURE 2.531 Circular diagram of RCS given radar observation of UAV RQ-1 model at a carrier frequency of 10 GHz (3 cm wavelength).

FIGURE 2.532 Circular diagram of noncoherent RCS given radar observation of UAV RQ-1 model at a carrier frequency of 10 GHz (3 cm wavelength).

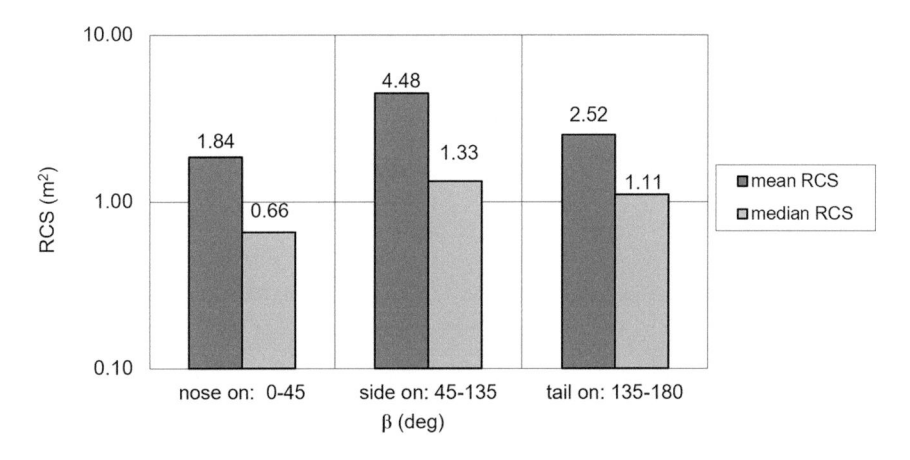

FIGURE 2.533 Diagrams of mean and median RCS of UAV RQ-1 model in three sectors of azimuth aspect given its radar observation at horizontal polarization and a carrier frequency of 10 GHz (3 cm wavelength).

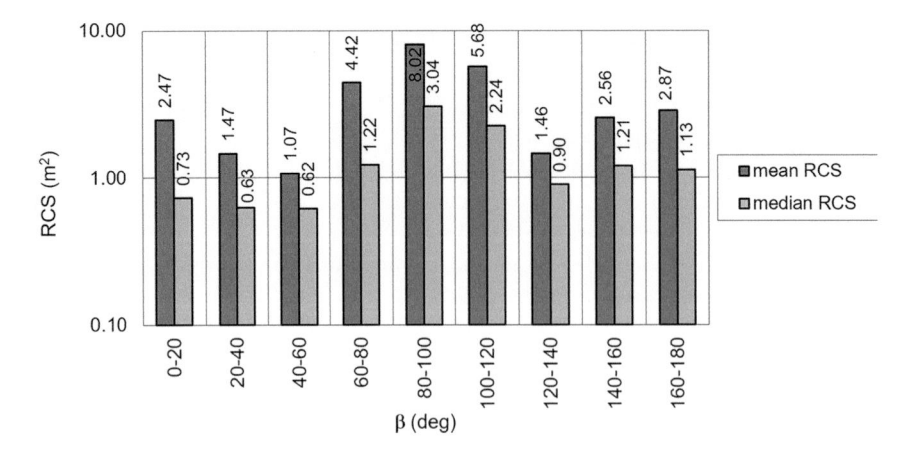

FIGURE 2.534 Diagrams of mean and median RCS of UAV RQ-1 model in 20-degree sectors of azimuth aspect given its radar observation at horizontal polarization and a carrier frequency of 10 GHz (3 cm wavelength).

Figures 2.535, 2.540, and 2.545 show histograms of the scattered signal amplitude (square root of the RCS) for the range of sounding azimuths −20° to +20° (sounding from the nose). The bold line shows the probability density functions of the distribution, which can be used to approximate the histogram of the signal amplitude.

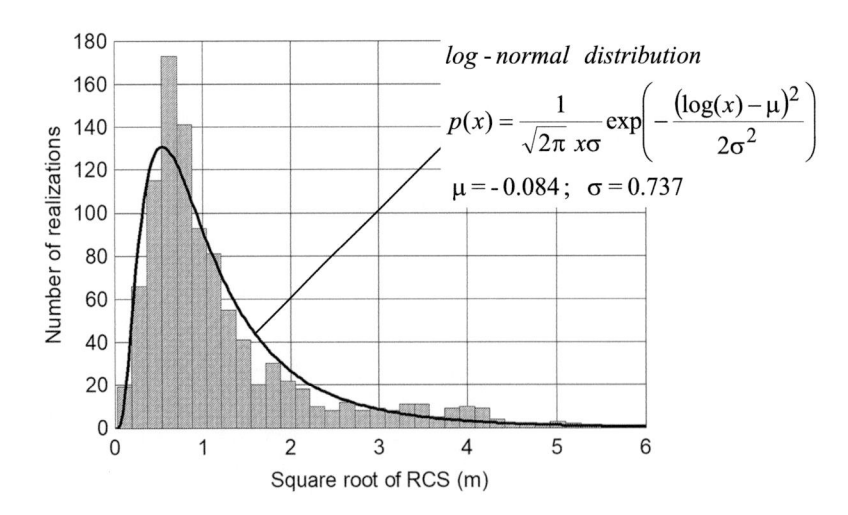

$$\text{log-normal distribution}$$
$$p(x) = \frac{1}{\sqrt{2\pi}\, x\sigma} \exp\left(-\frac{(\log(x)-\mu)^2}{2\sigma^2}\right)$$
$$\mu = -0.084;\quad \sigma = 0.737$$

FIGURE 2.535 Amplitude distribution of echo signal of UAV RQ-1 model at a carrier frequency of 10 GHz given its horizontal polarization.

Scattering characteristics of UAV RQ-1 model for sounding frequency 3 GHz (wavelength 10 cm).

Figure 2.536 shows the RCS circular diagram of the RQ-1. The noncoherent RCS circular diagram of the RQ-1 is shown in Figure 2.537.

The circular mean RCS of the UAV RQ-1 for horizontal polarization is 2.668 m². The circular median RCS for horizontal polarization is 0.971 m².

Figures 2.538 and 2.539 show the mean and median RCS for the main ranges of sounding azimuths (nose, side, and tail) and for ranges of 20°.

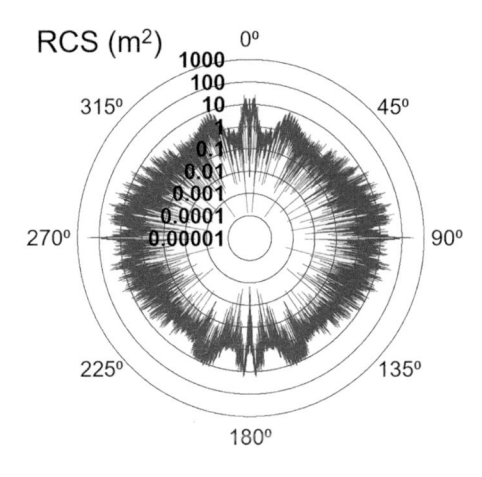

Scattering Characteristics of Aerial Objects

FIGURE 2.536 Circular diagram of RCS given radar observation of UAV RQ-1 model at a carrier frequency of 3 GHz (10 cm wavelength).

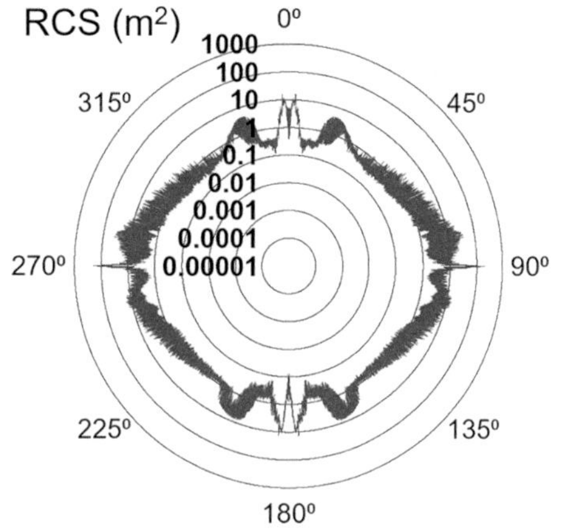

FIGURE 2.537 Circular diagram of noncoherent RCS given radar observation of UAV RQ-1 model at a carrier frequency of 3 GHz (10 cm wavelength).

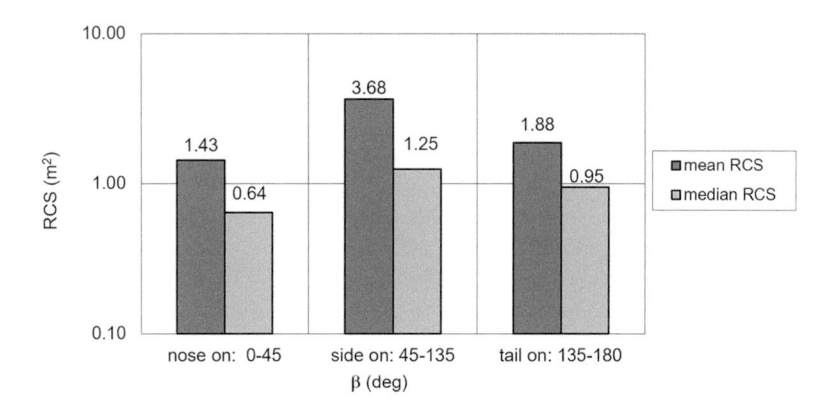

FIGURE 2.538 Diagrams of mean and median RCS of UAV RQ-1 model in three sectors of azimuth aspect given its radar observation at horizontal polarization and a carrier frequency of 3 GHz (10 cm wavelength).

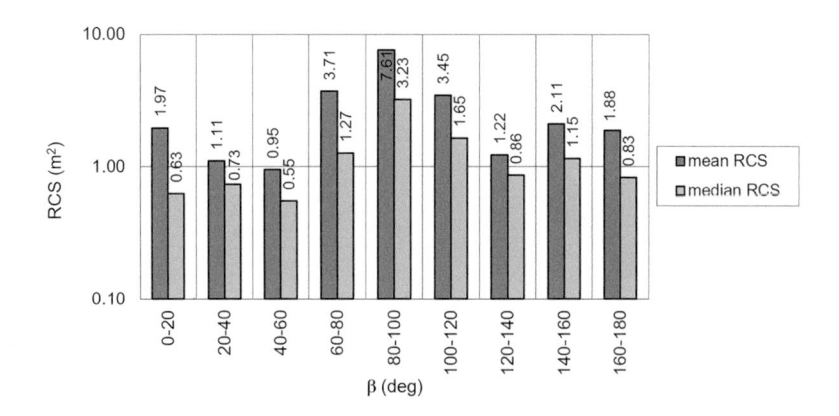

FIGURE 2.539 Diagrams of mean and median RCS of UAV RQ-1 model in 20-degree sectors of azimuth aspect given its radar observation at horizontal polarization and a carrier frequency of 3 GHz (10 cm wavelength).

$$p(x) = \frac{1}{\sqrt{2\pi}\, x\sigma} \exp\left(-\frac{(\log(x)-\mu)^2}{2\sigma^2}\right)$$

log-normal distribution

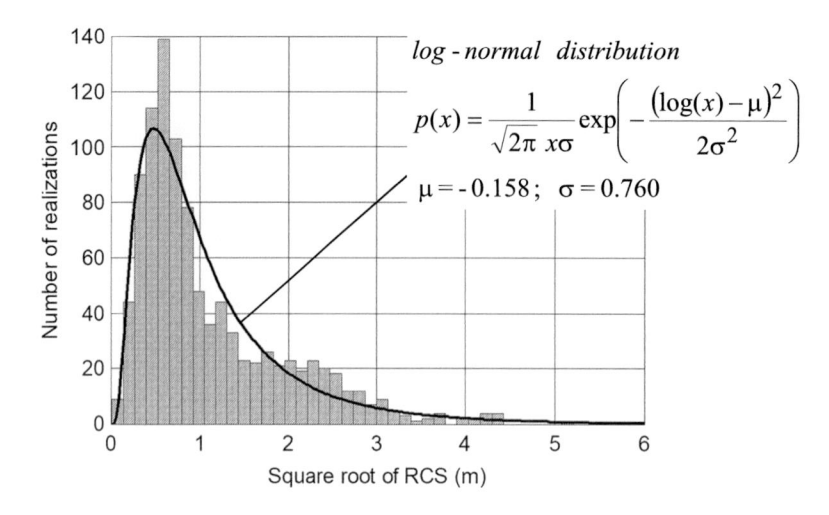

$\mu = -0.158;\quad \sigma = 0.760$

FIGURE 2.540 Amplitude distribution of echo signal of UAV RQ-1 model at a carrier frequency of 3 GHz given its horizontal polarization.

Scattering characteristics of UAV RQ-1 model for sounding frequency 1 GHz (wavelength 30 cm).

Figure 2.541 shows the RCS circular diagram of the RQ-1. The noncoherent RCS circular diagram of the RQ-1 is shown in Figure 2.542.

The circular mean RCS of the UAV RQ-1 for horizontal polarization is 2.507 m². The circular median RCS for horizontal polarization is 0.673 m².

Figures 2.543 and 2.544 show the mean and median RCS for the main ranges of sounding azimuths (nose, side, and tail) and for ranges of 20°.

The expressions and parameters of probability distributions that are most consistent with the empirical distributions of the square root of the RCS for different ranges of sounding azimuths and polarizations are given in the electronic appendix to this book. The expressions and parameters of the probability distributions that are most consistent with the empirical distributions of the RCS (energy characteristic) for different ranges of sounding azimuths and polarizations are also given there.

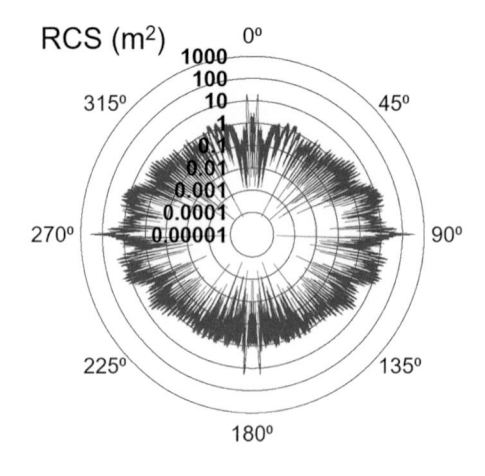

FIGURE 2.541 Circular diagram of RCS given radar observation of UAV RQ-1 model at a carrier frequency of 1 GHz (30 cm wavelength).

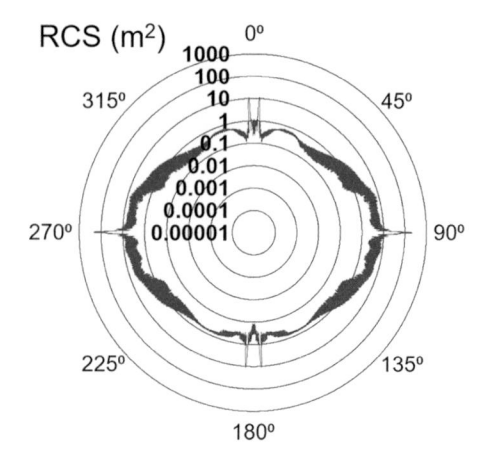

FIGURE 2.542 Circular diagram of noncoherent RCS given radar observation of UAV RQ-1 model at a carrier frequency of 1 GHz (30 cm wavelength).

Scattering Characteristics of Aerial Objects

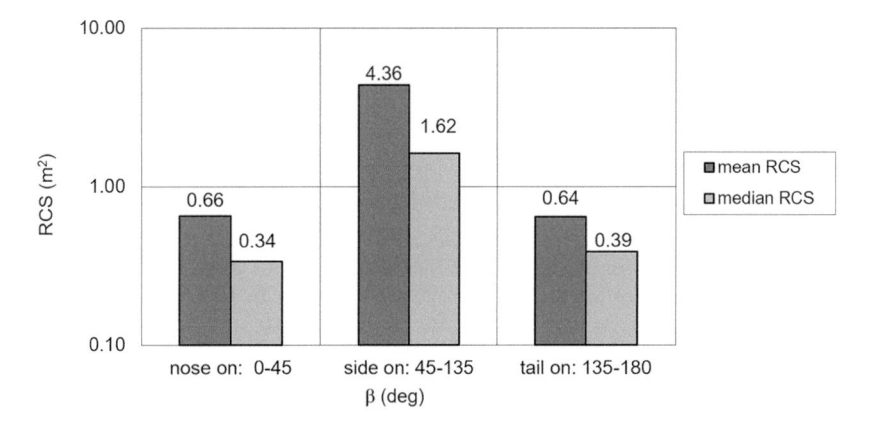

FIGURE 2.543 Diagrams of mean and median RCS of UAV RQ-1 model in three sectors of azimuth aspect given its radar observation at horizontal polarization and a carrier frequency of 1 GHz (30 cm wavelength).

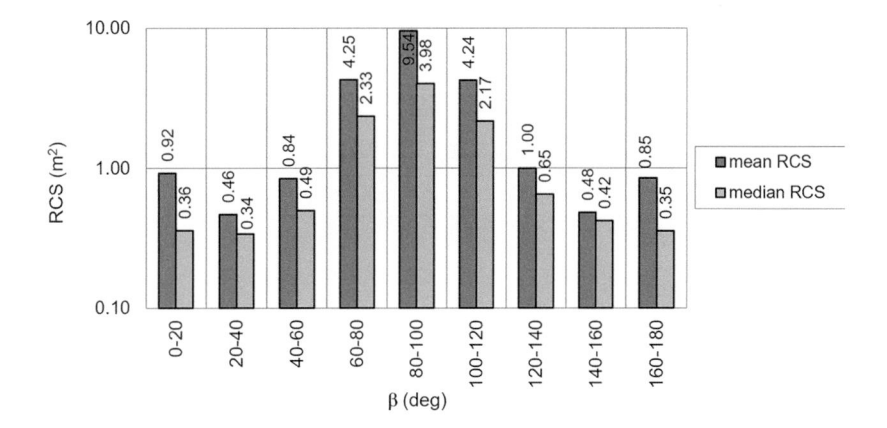

FIGURE 2.544 Diagrams of mean and median RCS of UAV RQ-1 model in 20-degree sectors of azimuth aspect given its radar observation at horizontal polarization and a carrier frequency of 1 GHz (30 cm wavelength).

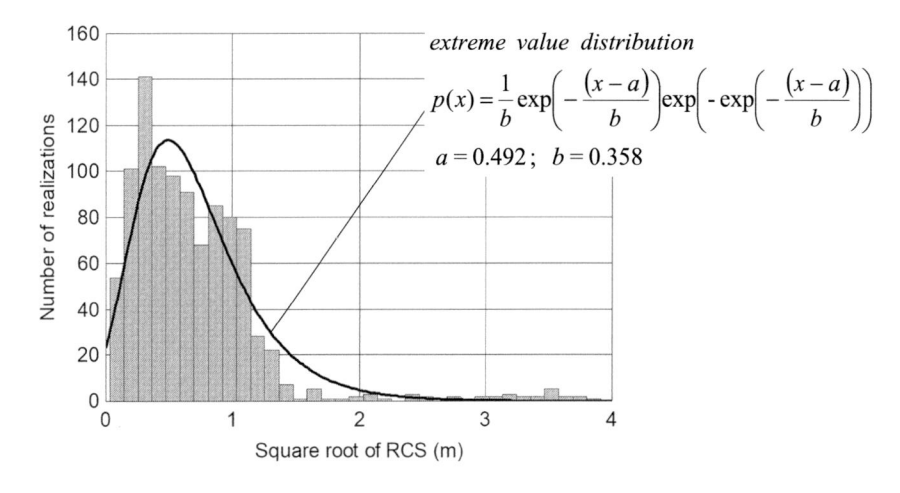

$$p(x) = \frac{1}{b}\exp\left(-\frac{(x-a)}{b}\right)\exp\left(-\exp\left(-\frac{(x-a)}{b}\right)\right)$$

$a = 0.492; \ b = 0.358$

FIGURE 2.545 Amplitude distribution of echo signal of UAV RQ-1 model at a carrier frequency of 1 GHz given its horizontal polarization.

2.33 UNMANNED AERIAL VEHICLE RQ-4 GLOBAL HAWK

RQ-4 Global Hawk (Figure 2.546) is a high-altitude, long-endurance unmanned aerial vehicle for intelligence, surveillance, and reconnaissance missions. The first flight took place in 1998.

General characteristics of RQ-4 [44]: wingspan –39.90 m, length – 14.50 m, height – 4.70 m, wing area – 50.00 m², weight – 6 780–14 620 kg, powerplant – 1 Rolls-Royce-North American F137-RR-100 turbofan engine, maximum speed – 630 km/h, range – 22 800 km, and service ceiling – 18 300 m.

In accordance with the design of the RQ-4, a model of its surface was created to obtaining scattering characteristics (in particular, RCS). The model is shown in Figure 2.547. The figure inset shows the model fuselage without dielectric shells. In the modeling, the UAV metal surfaces were approximated by parts of 50 triaxial ellipsoids. The UAV dielectric surfaces were approximated by parts of 29 triaxial ellipsoids.

FIGURE 2.546 Unmanned aerial vehicle RQ-4 Global Hawk.

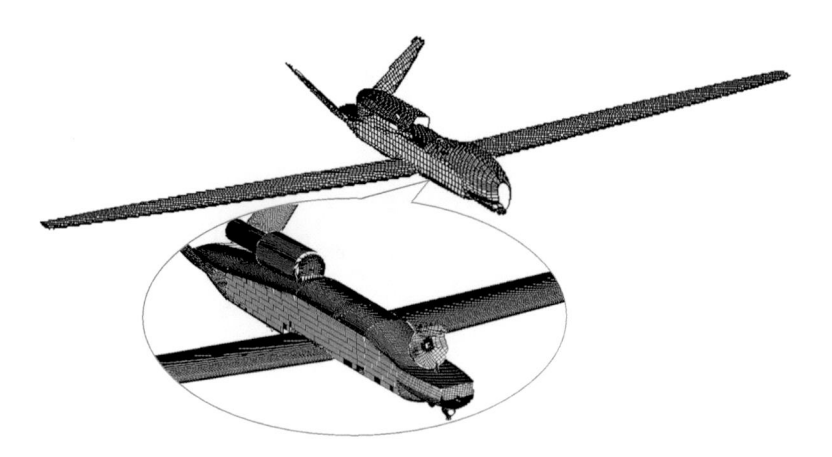

FIGURE 2.547 Surface model of unmanned aerial vehicle RQ-4 Global Hawk.

Below are some scattering characteristics of the RQ-4 model at sounding frequencies of 10, 3, and 1 GHz (wavelengths of 3, 10, and 30 cm, respectively) for horizontal polarization of the probing signal. The scattering characteristics for this model at two polarizations, as well as scattering characteristics at sounding frequencies of 5 and 1.3 GHz (wavelengths of 6 and 23 cm, respectively), are given in the electronic appendix of this book.

Scattering Characteristics of Aerial Objects   205

Sounding parameters: The elevation angle is to be a random value distributed uniformly in the range −3° ± 4° with respect to the wing plane (elevation angle of −3° corresponds to the radar observation from the lower hemisphere), azimuth aspect increment was 0.02°, the azimuth being counted off from the nose-on aspect (0° corresponds to the nose-on radar observation, 180° corresponds to the tail-on observation).

Scattering characteristics of UAV RQ-4 model for sounding frequency 10 GHz (wavelength 3 cm).

Figure 2.548 shows the RCS circular diagram of the RQ-4 model. The noncoherent RCS circular diagram of the RQ-4 is shown in Figure 2.549.

The circular mean RCS of the UAV RQ-4 model for horizontal polarization is 44.36 m². The circular median RCS (the RCS value used to calculate the detection range of an object with a probability of 0.5) for horizontal polarization is 2.16 m².

Figures 2.550 and 2.551 show the mean and median RCS for the main ranges of sounding azimuths (nose, side, and tail) and for ranges of 20°.

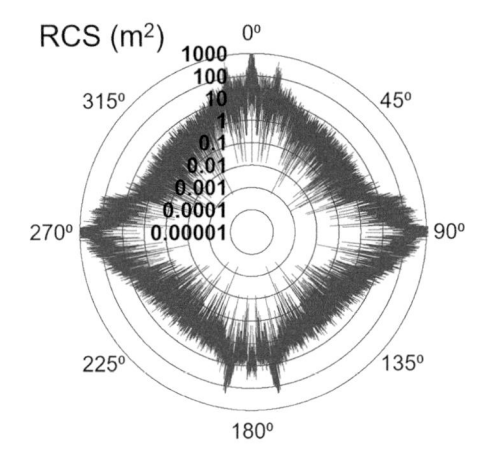

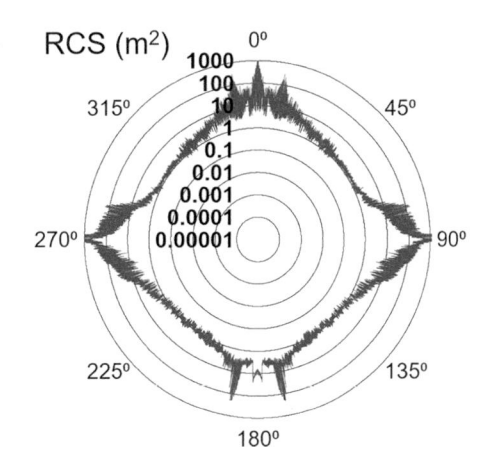

FIGURE 2.548 Circular diagram of RCS given radar observation of UAV RQ-4 model at a carrier frequency of 10 GHz (3 cm wavelength).

FIGURE 2.549 Circular diagram of noncoherent RCS given radar observation of UAV RQ-4 model at a carrier frequency of 10 GHz (3 cm wavelength).

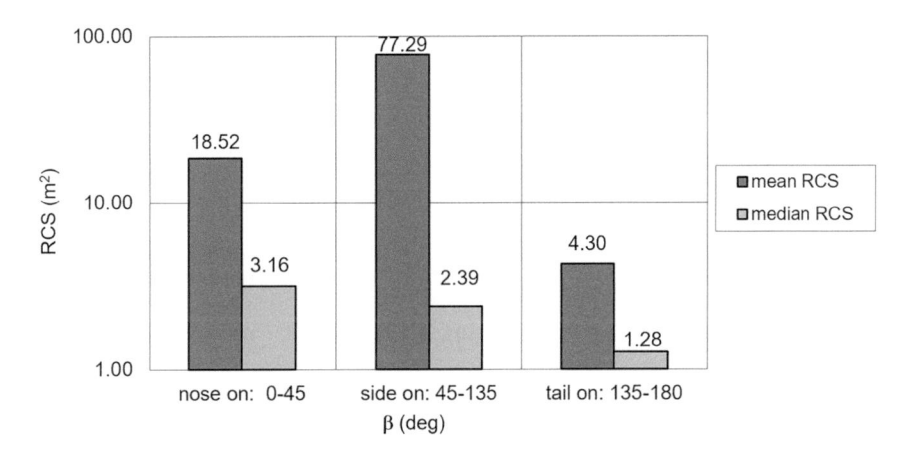

FIGURE 2.550 Diagrams of mean and median RCS of UAV RQ-4 model in three sectors of azimuth aspect given its radar observation at horizontal polarization and a carrier frequency of 10 GHz (3 cm wavelength).

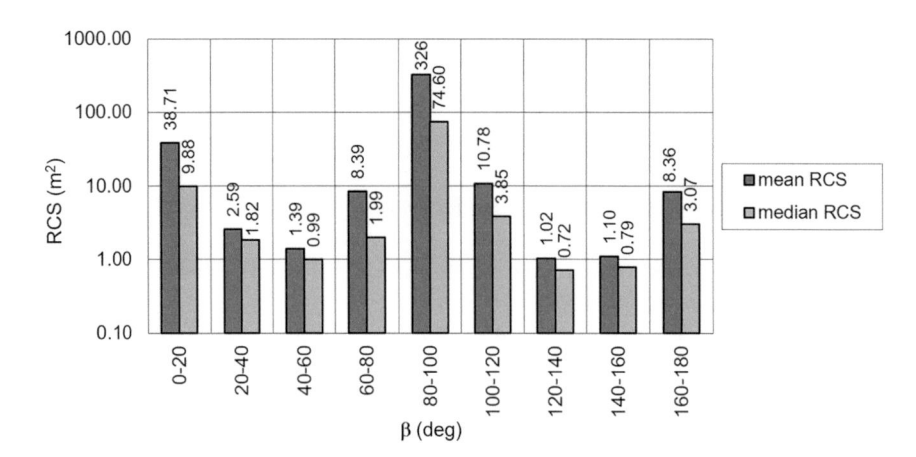

FIGURE 2.551 Diagrams of mean and median RCS of UAV RQ-4 model in 20-degree sectors of azimuth aspect given its radar observation at horizontal polarization and a carrier frequency of 10 GHz (3 cm wavelength).

Figures 2.552, 2.557, and 2.562 show histograms of the scattered signal amplitude (square root of the RCS) for the range of sounding azimuths −20° to +20° (sounding from the nose). The bold line shows the probability density functions of the distribution, which can be used to approximate the histogram of the signal amplitude.

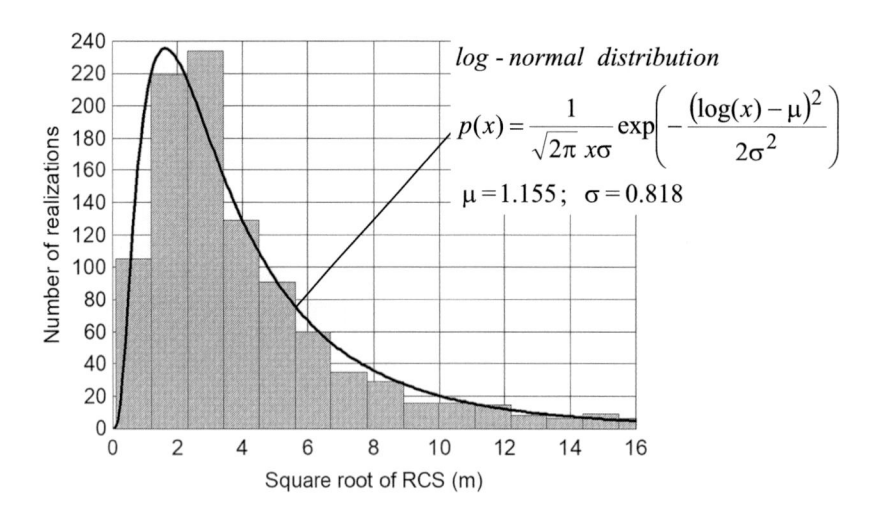

$$p(x) = \frac{1}{\sqrt{2\pi}\, x\sigma} \exp\left(-\frac{(\log(x)-\mu)^2}{2\sigma^2}\right)$$

$\mu = 1.155; \quad \sigma = 0.818$

FIGURE 2.552 Amplitude distribution of echo signal of UAV RQ-4 model at a carrier frequency of 10 GHz given its horizontal polarization.

Scattering characteristics of UAV RQ-4 model for sounding frequency 3 GHz (wavelength 10 cm).

Figure 2.553 shows the RCS circular diagram of the RQ-4. The noncoherent RCS circular diagram of the RQ-4 is shown in Figure 2.554.

The circular mean RCS of the UAV RQ-4 for horizontal polarization is 40.23 m². The circular median RCS for horizontal polarization is 1.32 m².

Figures 2.555 and 2.556 show the mean and median RCS for the main ranges of sounding azimuths (nose, side, and tail) and for ranges of 20°.

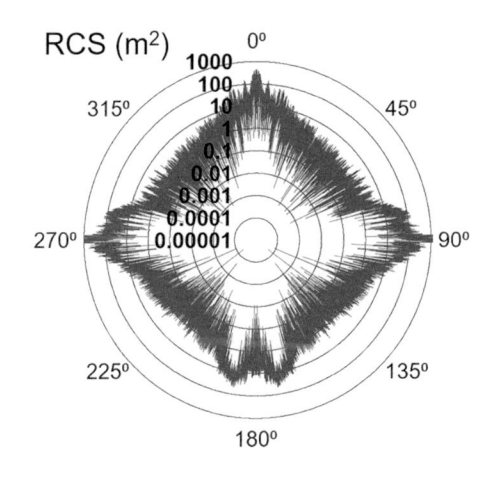

Scattering Characteristics of Aerial Objects

FIGURE 2.553 Circular diagram of RCS given radar observation of UAV RQ-4 model at a carrier frequency of 3 GHz (10 cm wavelength).

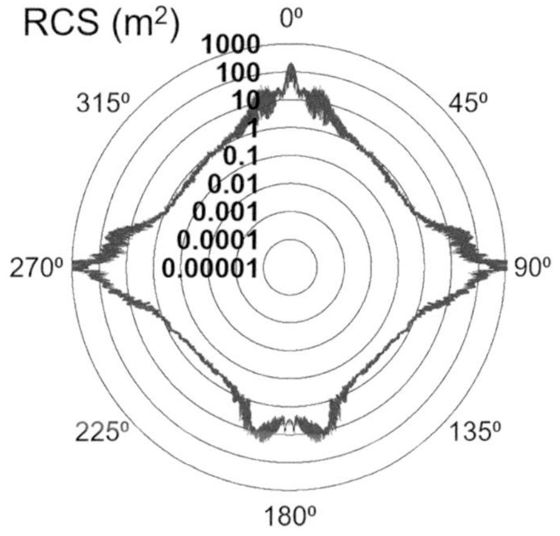

FIGURE 2.554 Circular diagram of noncoherent RCS given radar observation of UAV RQ-4 model at a carrier frequency of 3 GHz (10 cm wavelength).

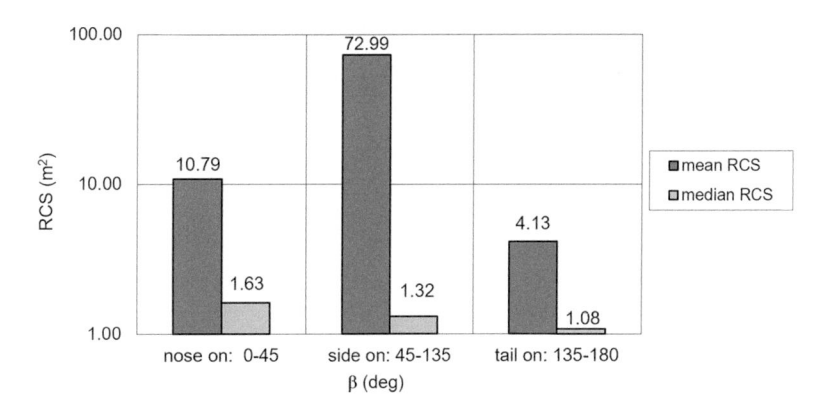

FIGURE 2.555 Diagrams of mean and median RCS of UAV RQ-4 model in three sectors of azimuth aspect given its radar observation at horizontal polarization and a carrier frequency of 3 GHz (10 cm wavelength).

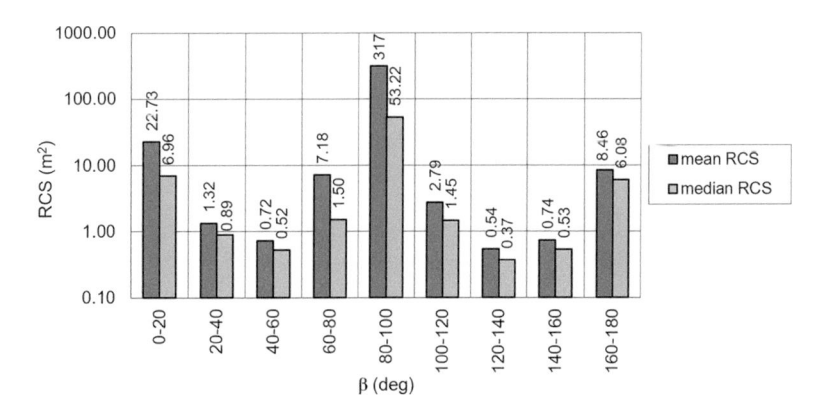

FIGURE 2.556 Diagrams of mean and median RCS of UAV RQ-4 model in 20-degree sectors of azimuth aspect given its radar observation at horizontal polarization and a carrier frequency of 3 GHz (10 cm wavelength).

$$p(x) = \frac{1}{\sqrt{2\pi}\, x\sigma} \exp\left(-\frac{(\log(x)-\mu)^2}{2\sigma^2}\right)$$

log - normal distribution

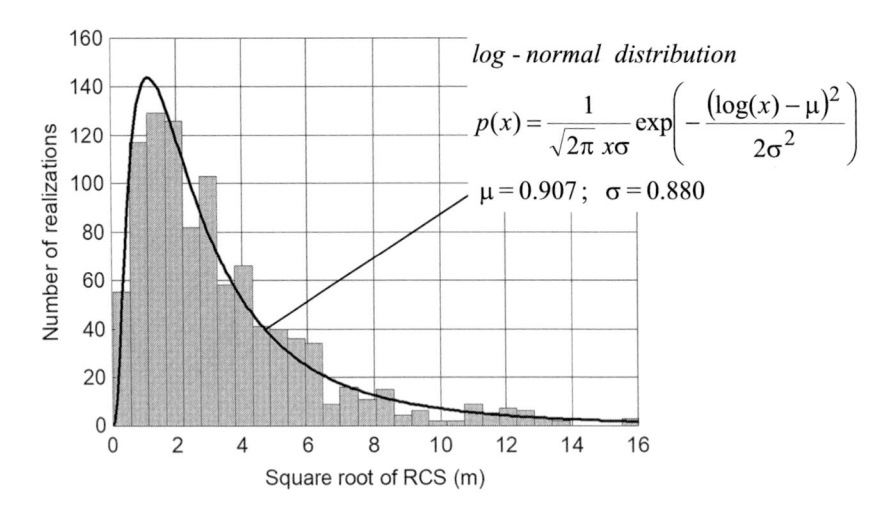

$\mu = 0.907;\ \sigma = 0.880$

FIGURE 2.557 Amplitude distribution of echo signal of UAV RQ-4 model at a carrier frequency of 3 GHz given its horizontal polarization.

Scattering characteristics of UAV RQ-4 model for sounding frequency 1 GHz (wavelength 30 cm).

Figure 2.558 shows the RCS circular diagram of the RQ-4. The noncoherent RCS circular diagram of the RQ-4 is shown in Figure 2.559.

The circular mean RCS of the UAV RQ-4 for horizontal polarization is 41.35 m². The circular median RCS for horizontal polarization is 1.35 m².

Figures 2.560 and 2.561 show the mean and median RCS for the main ranges of sounding azimuths (nose, side, and tail) and for ranges of 20°.

The expressions and parameters of probability distributions that are most consistent with the empirical distributions of the square root of the RCS for different ranges of sounding azimuths and polarizations are given in the electronic appendix to this book. The expressions and parameters of the probability distributions that are most consistent with the empirical distributions of the RCS (energy characteristic) for different ranges of sounding azimuths and polarizations are also given there.

FIGURE 2.558 Circular diagram of RCS given radar observation of UAV RQ-4 model at a carrier frequency of 1 GHz (30 cm wavelength).

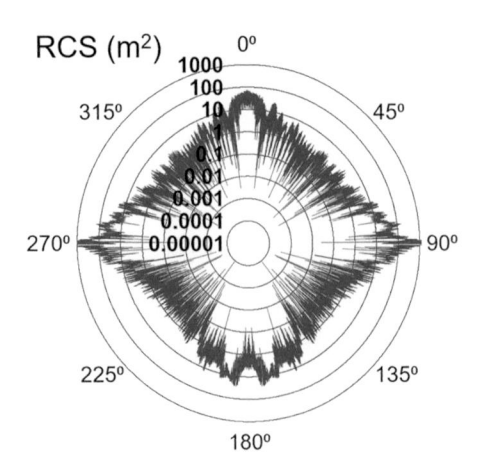

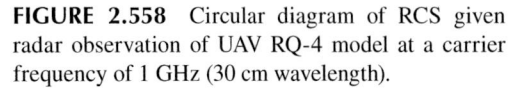

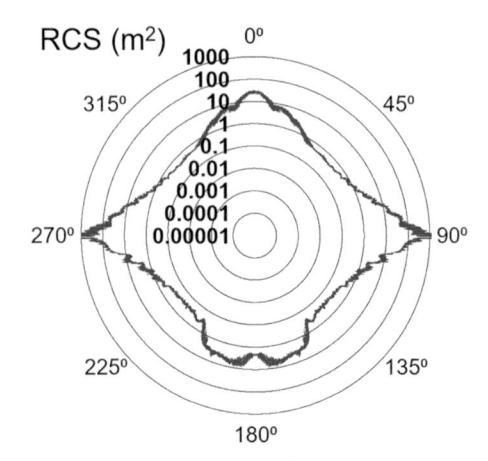

FIGURE 2.559 Circular diagram of noncoherent RCS given radar observation of UAV RQ-4 model at a carrier frequency of 1 GHz (30 cm wavelength).

Scattering Characteristics of Aerial Objects

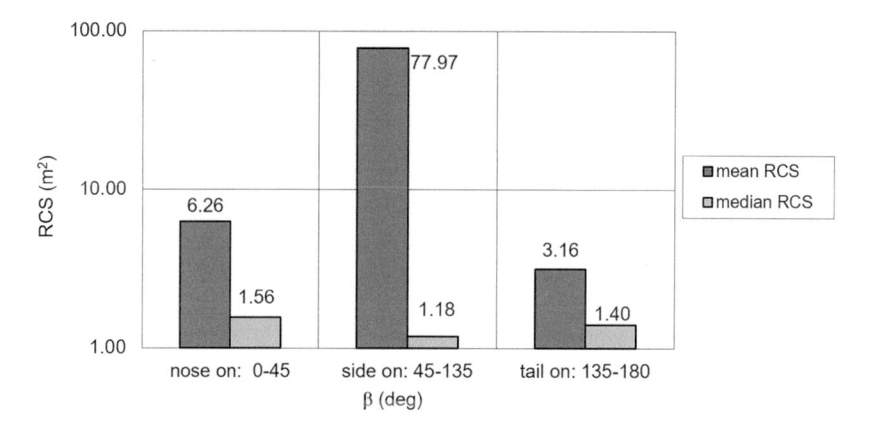

FIGURE 2.560 Diagrams of mean and median RCS of UAV RQ-4 model in three sectors of azimuth aspect given its radar observation at horizontal polarization and a carrier frequency of 1 GHz (30 cm wavelength).

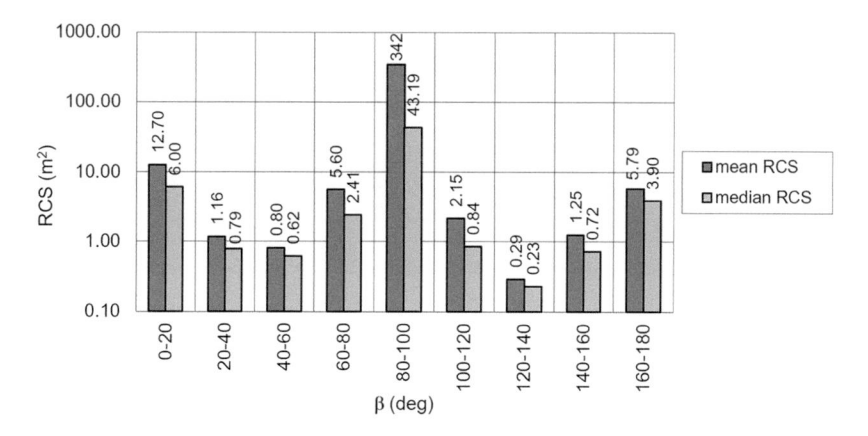

FIGURE 2.561 Diagrams of mean and median RCS of UAV RQ-4 model in 20-degree sectors of azimuth aspect given its radar observation at horizontal polarization and a carrier frequency of 1 GHz (30 cm wavelength).

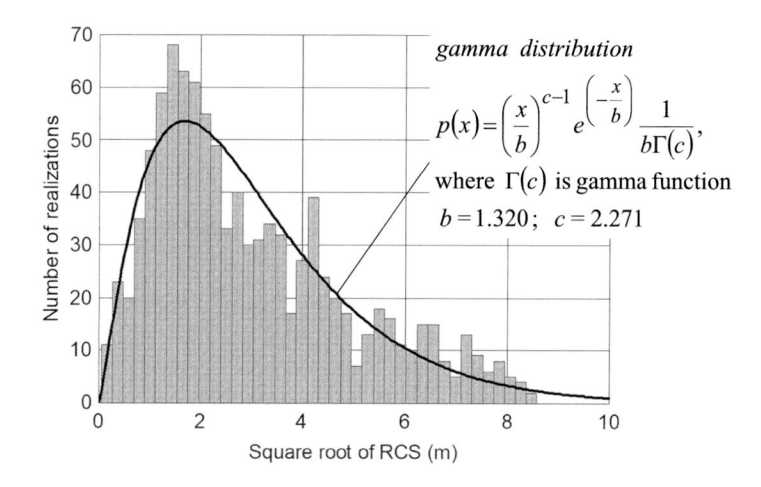

$$p(x) = \left(\frac{x}{b}\right)^{c-1} e^{\left(-\frac{x}{b}\right)} \frac{1}{b\Gamma(c)},$$

where $\Gamma(c)$ is gamma function
$b = 1.320; \quad c = 2.271$

FIGURE 2.562 Amplitude distribution of echo signal of UAV RQ-4 model at a carrier frequency of 1 GHz given its horizontal polarization.

2.34 UNMANNED AERIAL VEHICLE RQ-7 SHADOW

RQ-7 Shadow (Figure 2.563) is a tactical reconnaissance unmanned aerial vehicle. The first flight took place in 1991.

General characteristics of RQ-7 [45]: wingspan –3.89 m, length – 3.40 m, height – 0.91 m, weight – 75–170 kg, powerplant – 1 Wankel UAV Engine 741 used only with Silkolene Synthetic Oil, maximum speed – 227 km/h, cruise speed – 130 km/h, range – 125 km, and service ceiling – 4 570 m.

In accordance with the design of the RQ-7, a model of its surface was created to obtaining scattering characteristics (in particular, RCS). The model is shown in Figure 2.564. In the modeling, the UAV metal surfaces were approximated by parts of 57 triaxial ellipsoids. The UAV dielectric surfaces were approximated by parts of 37 triaxial ellipsoids.

FIGURE 2.563 Unmanned aerial vehicle RQ-7 Shadow.

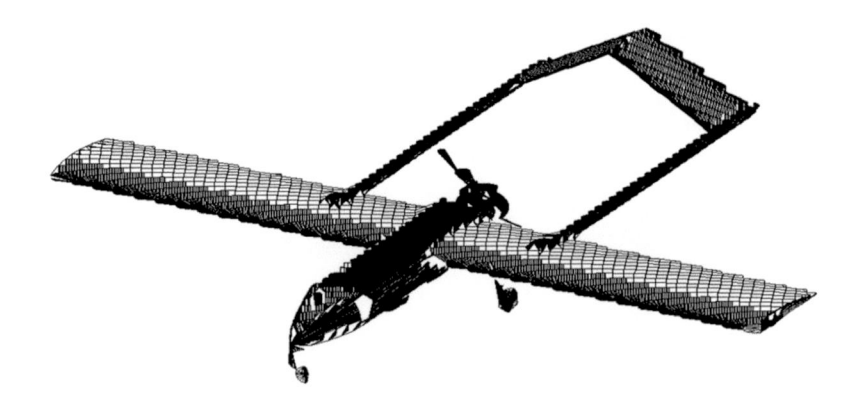

FIGURE 2.564 Surface model of unmanned aerial vehicle RQ-7 Shadow.

Below are some scattering characteristics of the RQ-7 model at sounding frequencies of 10, 3, and 1 GHz (wavelengths of 3, 10, and 30 cm, respectively) for horizontal polarization of the probing signal. The scattering characteristics for this model at two polarizations, as well as scattering characteristics at sounding frequencies of 5 and 1.3 GHz (wavelengths of 6 and 23 cm, respectively), are given in the electronic appendix of this book.

Sounding parameters: The elevation angle is to be a random value distributed uniformly in the range −3° ± 4° with respect to the wing plane (elevation angle of −3° corresponds to the radar observation from the lower hemisphere), azimuth aspect increment was 0.02°, the azimuth being counted

Scattering Characteristics of Aerial Objects

off from the nose-on aspect (0° corresponds to the nose-on radar observation, 180° corresponds to the tail-on observation).

Scattering characteristics of UAV RQ-7 model for sounding frequency 10 GHz (wavelength 3 cm).

Figure 2.565 shows the RCS circular diagram of the RQ-7 model. The noncoherent RCS circular diagram of the RQ-7 is shown in Figure 2.566.

The circular mean RCS of the UAV RQ-7 model for horizontal polarization is 0.329 m^2. The circular median RCS (the RCS value used to calculate the detection range of an object with a probability of 0.5) for horizontal polarization is 0.099 m^2.

Figures 2.567 and 2.568 show the mean and median RCS for the main ranges of sounding azimuths (nose, side, and tail) and for ranges of 20°.

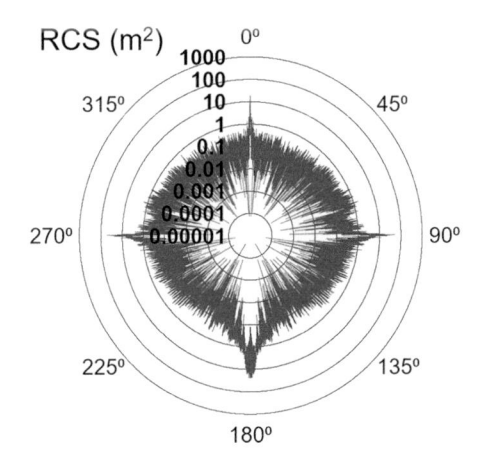

FIGURE 2.565 Circular diagram of RCS given radar observation of UAV RQ-7 model at a carrier frequency of 10 GHz (3 cm wavelength).

FIGURE 2.566 Circular diagram of noncoherent RCS given radar observation of UAV RQ-7 model at a carrier frequency of 10 GHz (3 cm wavelength).

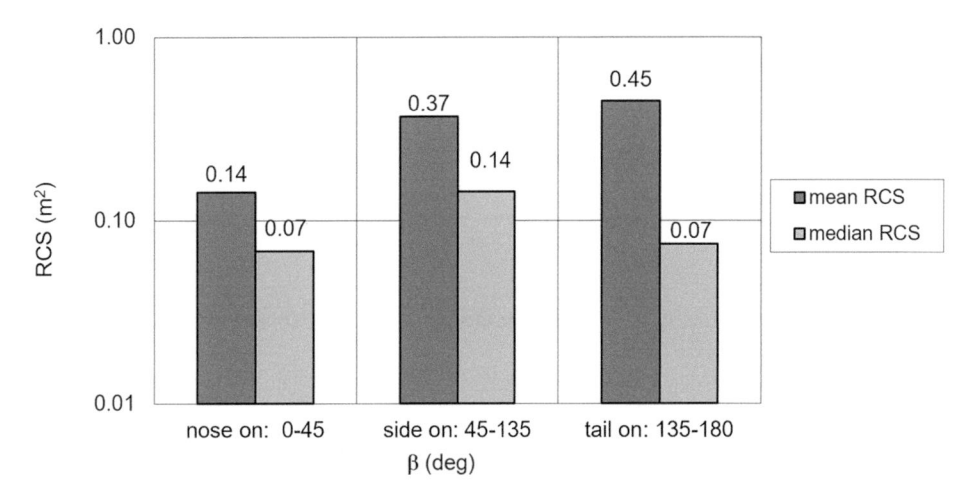

FIGURE 2.567 Diagrams of mean and median RCS of UAV RQ-7 model in three sectors of azimuth aspect given its radar observation at horizontal polarization and a carrier frequency of 10 GHz (3 cm wavelength).

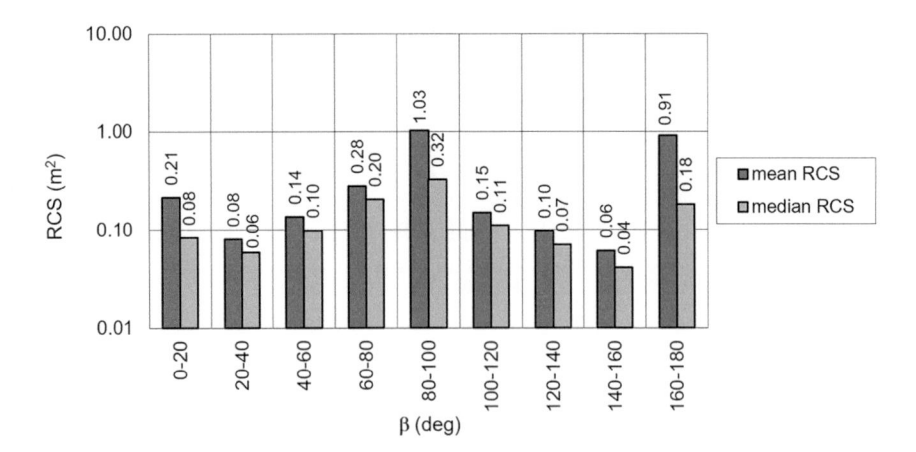

FIGURE 2.568 Diagrams of mean and median RCS of UAV RQ-7 model in 20-degree sectors of azimuth aspect given its radar observation at horizontal polarization and a carrier frequency of 10 GHz (3 cm wavelength).

Figures 2.569, 2.574, and 2.579 show histograms of the scattered signal amplitude (square root of the RCS) for the range of sounding azimuths −20° to +20° (sounding from the nose). The bold line shows the probability density functions of the distribution, which can be used to approximate the histogram of the signal amplitude.

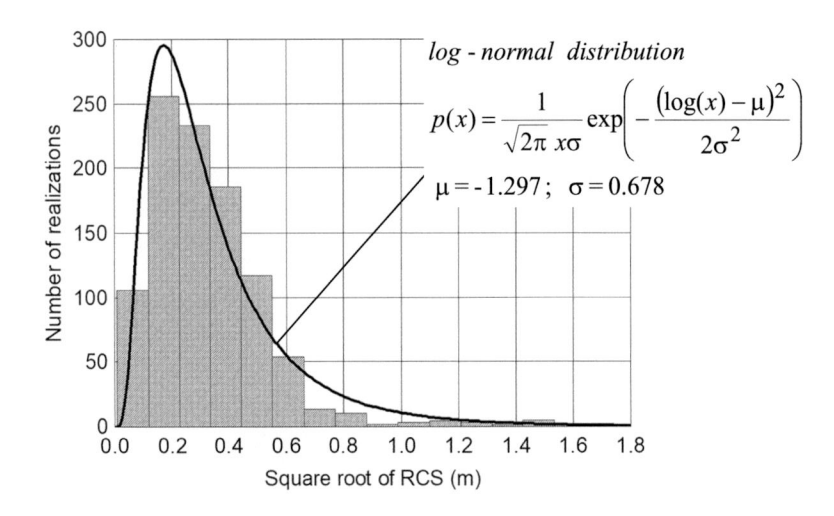

$$\text{log - normal distribution}$$
$$p(x) = \frac{1}{\sqrt{2\pi}\, x\sigma} \exp\left(-\frac{(\log(x) - \mu)^2}{2\sigma^2}\right)$$
$$\mu = -1.297;\ \sigma = 0.678$$

FIGURE 2.569 Amplitude distribution of echo signal of UAV RQ-7 model at a carrier frequency of 10 GHz given its horizontal polarization.

Scattering characteristics of UAV RQ-7 model for sounding frequency 3 GHz (wavelength 10 cm).

Figure 2.570 shows the RCS circular diagram of the RQ-7. The noncoherent RCS circular diagram of the RQ-7 is shown in Figure 2.571.

The circular mean RCS of the UAV RQ-7 for horizontal polarization is 0.201 m². The circular median RCS for horizontal polarization is 0.111 m².

Figures 2.572 and 2.573 show the mean and median RCS for the main ranges of sounding azimuths (nose, side, and tail) and for ranges of 20°.

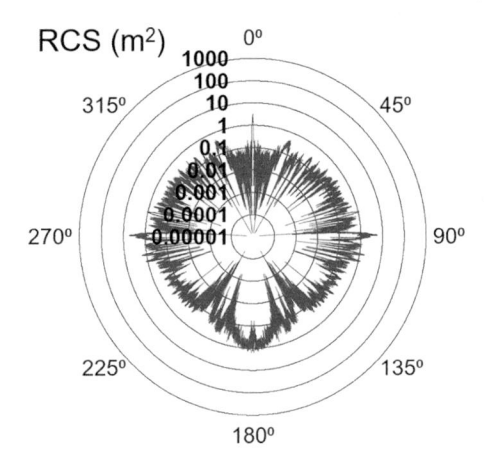

Scattering Characteristics of Aerial Objects

FIGURE 2.570 Circular diagram of RCS given radar observation of UAV RQ-7 model at a carrier frequency of 3 GHz (10 cm wavelength).

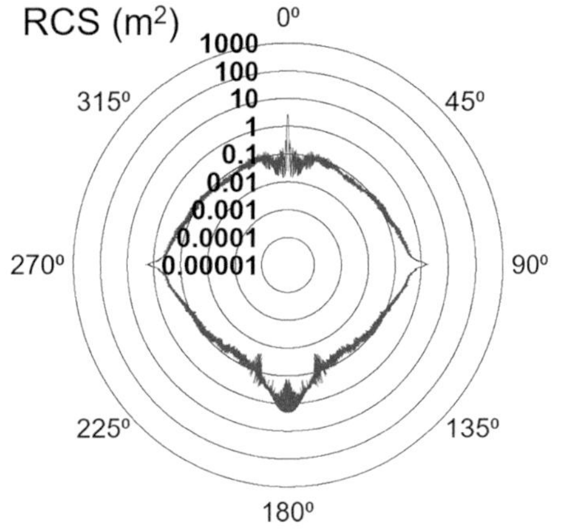

FIGURE 2.571 Circular diagram of noncoherent RCS given radar observation of UAV RQ-7 model at a carrier frequency of 3 GHz (10 cm wavelength).

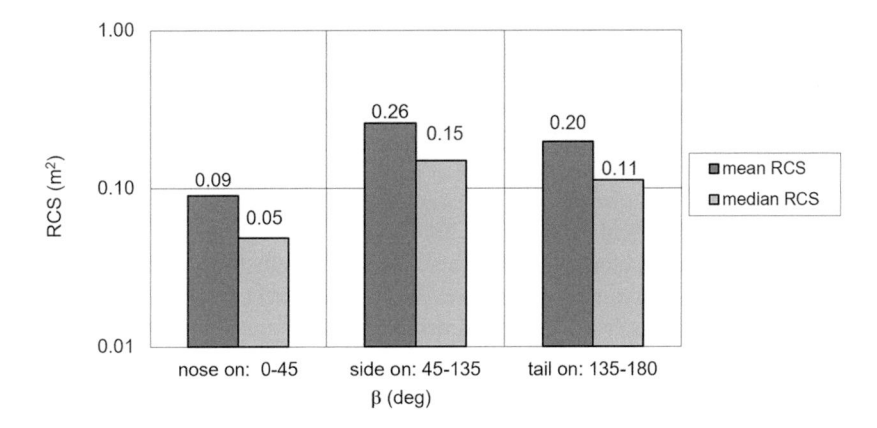

FIGURE 2.572 Diagrams of mean and median RCS of UAV RQ-7 model in three sectors of azimuth aspect given its radar observation at horizontal polarization and a carrier frequency of 3 GHz (10 cm wavelength).

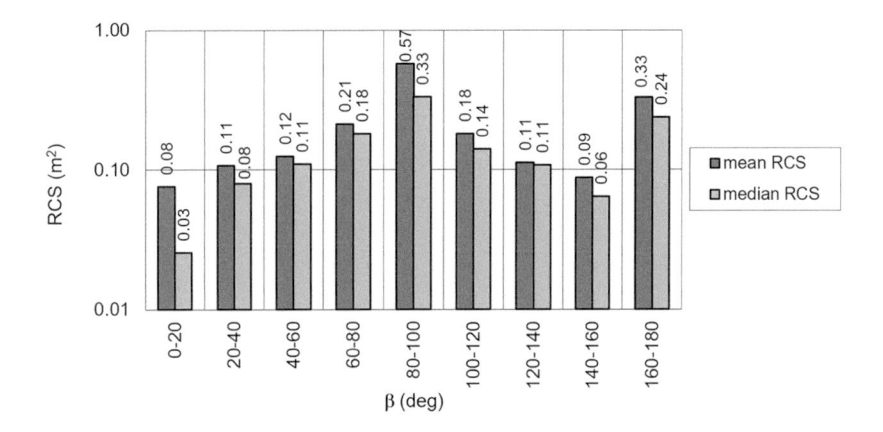

FIGURE 2.573 Diagrams of mean and median RCS of UAV RQ-7 model in 20-degree sectors of azimuth aspect given its radar observation at horizontal polarization and a carrier frequency of 3 GHz (10 cm wavelength).

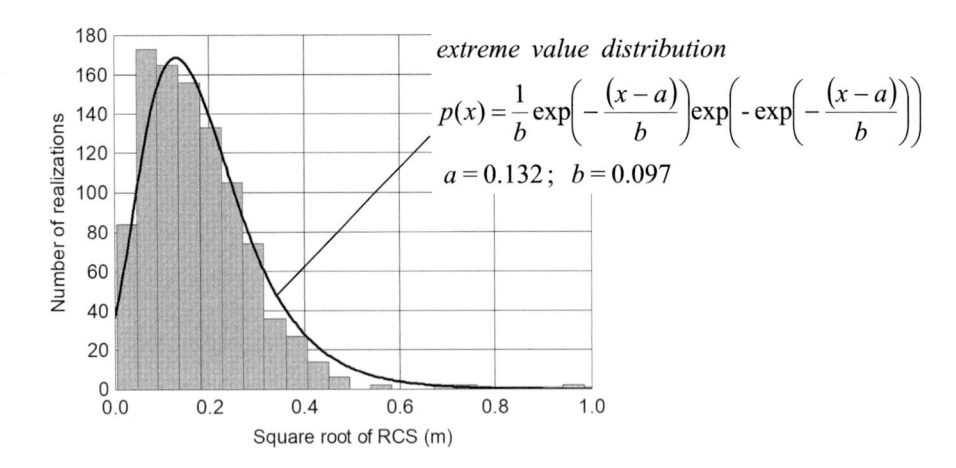

$$p(x) = \frac{1}{b}\exp\left(-\frac{(x-a)}{b}\right)\exp\left(-\exp\left(-\frac{(x-a)}{b}\right)\right)$$

extreme value distribution

$a = 0.132;\quad b = 0.097$

FIGURE 2.574 Amplitude distribution of echo signal of UAV RQ-7 model at a carrier frequency of 3 GHz given its horizontal polarization.

Scattering characteristics of UAV RQ-7 model for sounding frequency 1 GHz (wavelength 30 cm).

Figure 2.575 shows the RCS circular diagram of the RQ-7. The noncoherent RCS circular diagram of the RQ-7 is shown in Figure 2.576.

The circular mean RCS of the UAV RQ-7 for horizontal polarization is 0.146 m². The circular median RCS for horizontal polarization is 0.110 m².

Figures 2.577 and 2.578 show the mean and median RCS for the main ranges of sounding azimuths (nose, side, and tail) and for ranges of 20°.

The expressions and parameters of probability distributions that are most consistent with the empirical distributions of the square root of the RCS for different ranges of sounding azimuths and polarizations are given in the electronic appendix to this book. The expressions and parameters of the probability distributions that are most consistent with the empirical distributions of the RCS (energy characteristic) for different ranges of sounding azimuths and polarizations are also given there.

FIGURE 2.575 Circular diagram of RCS given radar observation of UAV RQ-7 model at a carrier frequency of 1 GHz (30 cm wavelength).

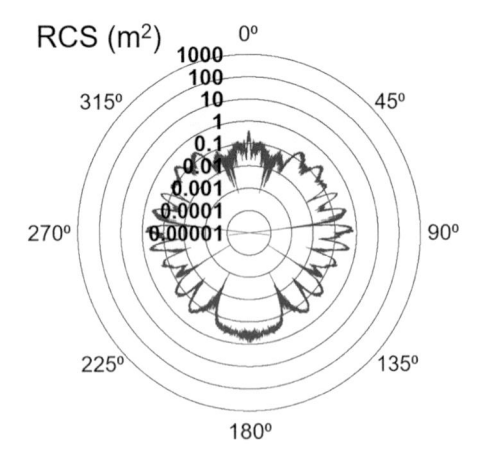

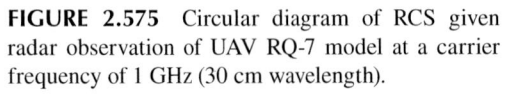

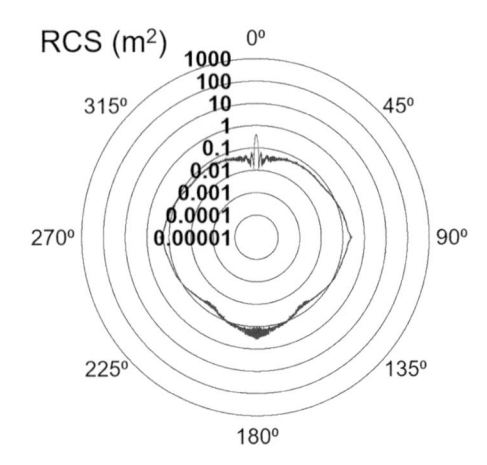

FIGURE 2.576 Circular diagram of noncoherent RCS given radar observation of UAV RQ-7 model at a carrier frequency of 1 GHz (30 cm wavelength).

Scattering Characteristics of Aerial Objects

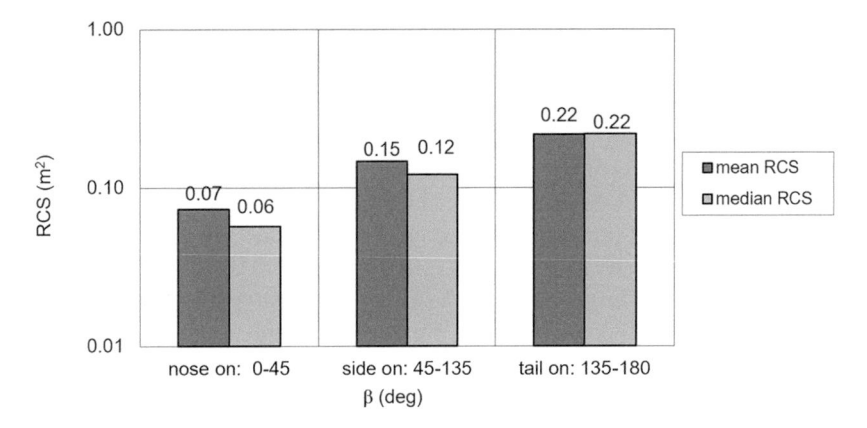

FIGURE 2.577 Diagrams of mean and median RCS of UAV RQ-7 model in three sectors of azimuth aspect given its radar observation at horizontal polarization and a carrier frequency of 1 GHz (30 cm wavelength).

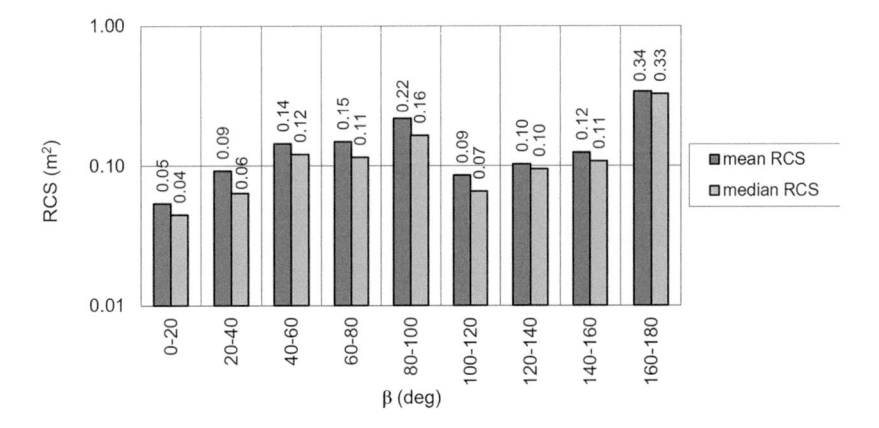

FIGURE 2.578 Diagrams of mean and median RCS of UAV RQ-7 model in 20-degree sectors of azimuth aspect given its radar observation at horizontal polarization and a carrier frequency of 1 GHz (30 cm wavelength).

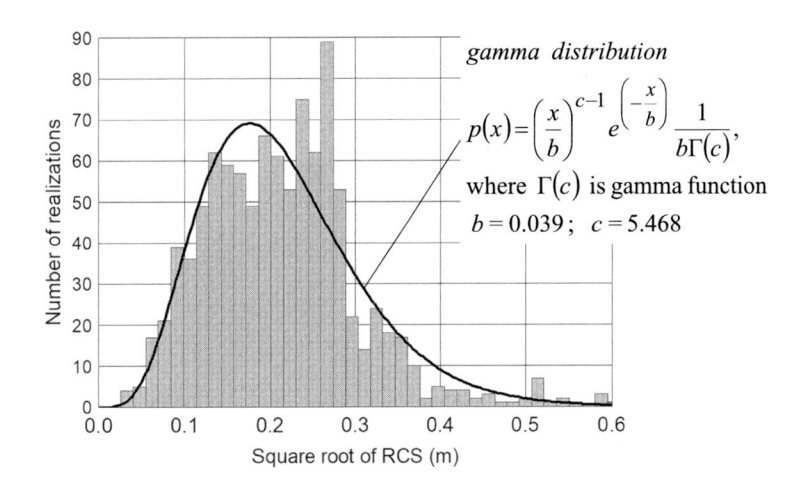

gamma distribution

$$p(x) = \left(\frac{x}{b}\right)^{c-1} e^{\left(-\frac{x}{b}\right)} \frac{1}{b\Gamma(c)},$$

where $\Gamma(c)$ is gamma function
$b = 0.039; \quad c = 5.468$

FIGURE 2.579 Amplitude distribution of echo signal of UAV RQ-7 model at a carrier frequency of 1 GHz given its horizontal polarization.

2.35 UNMANNED AERIAL VEHICLE BAYRAKTAR TB2

Bayraktar TB2 (Figure 2.580) is a medium-altitude long-endurance unmanned combat aerial vehicle. The first flight took place in 2009.

General characteristics of Bayraktar TB2 [46]: wingspan –12.00 m, length – 6.5 m, weight – 310–700 kg, powerplant – 1 Rotax 912-iS internal combustion engine with injection, speed – 130–230 km/h, range – 300 km, and service ceiling – 7 600 m.

In accordance with the design of the Bayraktar TB2, a model of its surface was created to obtaining scattering characteristics (in particular, RCS). The model is shown in Figure 2.581. The model surface was assumed to be perfectly conducting. The smooth parts of the model surface were approximated by parts of 64 triaxial ellipsoids. The surface breaks were modeled using 12 straight edge scattering parts.

FIGURE 2.580 Unmanned aerial vehicle Bayraktar TB2.

FIGURE 2.581 Surface model of unmanned aerial vehicle Bayraktar TB2.

Below are some scattering characteristics of the Bayraktar TB2 model at sounding frequencies of 10, 3, and 1 GHz (wavelengths of 3, 10, and 30 cm, respectively) for horizontal polarization of the probing signal. The scattering characteristics for this model at two polarizations, as well as scattering characteristics at sounding frequencies of 5 and 1.3 GHz (wavelengths of 6 and 23 cm, respectively), are given in the electronic appendix of this book.

Sounding parameters: The elevation angle is to be a random value distributed uniformly in the range −3° ± 4° with respect to the wing plane (elevation angle of −3° corresponds to the radar observation from the lower hemisphere), azimuth aspect increment was 0.02°, the azimuth being counted off from the nose-on aspect (0° corresponds to the nose-on radar observation, 180° corresponds to the tail-on observation).

Scattering Characteristics of Aerial Objects 217

Scattering characteristics of UAV Bayraktar TB2 model for sounding frequency 10 GHz (wavelength 3 cm).

Figure 2.582 shows the RCS circular diagram of the Bayraktar TB2 model. The noncoherent RCS circular diagram of the Bayraktar TB2 is shown in Figure 2.583.

The circular mean RCS of the UAV Bayraktar TB2 model for horizontal polarization is 3.236 m². The circular median RCS (the RCS value used to calculate the detection range of an object with a probability of 0.5) for horizontal polarization is 0.940 m².

Figures 2.584 and 2.585 show the mean and median RCS for the main ranges of sounding azimuths (nose, side, and tail) and for ranges of 20°.

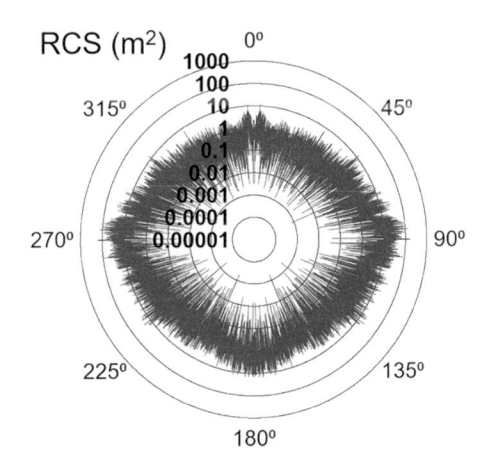

FIGURE 2.582 Circular diagram of RCS given radar observation of UAV Bayraktar TB2 model at a carrier frequency of 10 GHz (3 cm wavelength).

FIGURE 2.583 Circular diagram of noncoherent RCS given radar observation of UAV Bayraktar TB2 model at a carrier frequency of 10 GHz (3 cm wavelength).

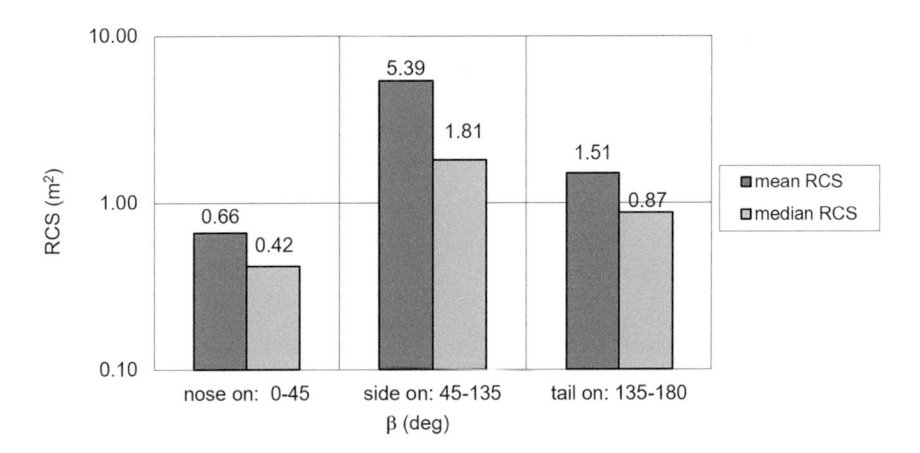

FIGURE 2.584 Diagrams of mean and median RCS of UAV Bayraktar TB2 model in three sectors of azimuth aspect given its radar observation at horizontal polarization and a carrier frequency of 10 GHz (3 cm wavelength).

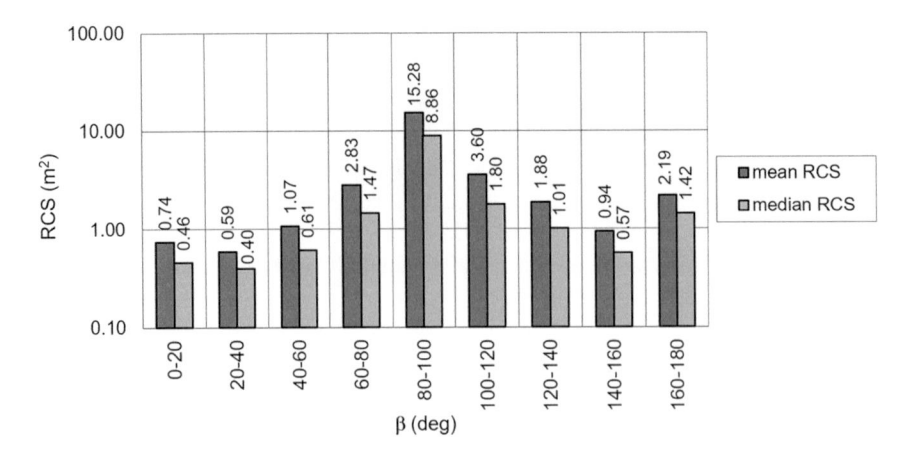

FIGURE 2.585 Diagrams of mean and median RCS of UAV Bayraktar TB2 model in 20-degree sectors of azimuth aspect given its radar observation at horizontal polarization and a carrier frequency of 10 GHz (3 cm wavelength).

Figures 2.586, 2.591, and 2.596 show histograms of the scattered signal amplitude (square root of the RCS) for the range of sounding azimuths −20° to +20° (sounding from the nose). The bold line shows the probability density functions of the distribution, which can be used to approximate the histogram of the signal amplitude.

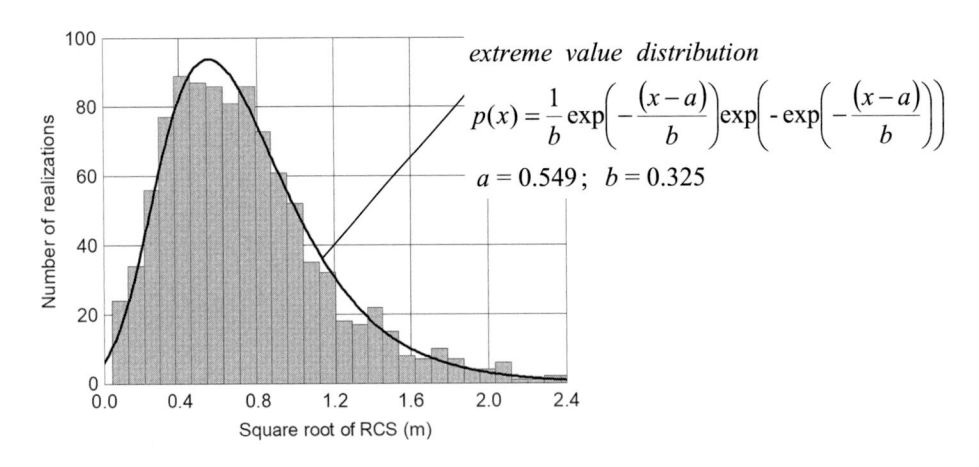

extreme value distribution
$$p(x) = \frac{1}{b}\exp\left(-\frac{(x-a)}{b}\right)\exp\left(-\exp\left(-\frac{(x-a)}{b}\right)\right)$$
$a = 0.549;\ b = 0.325$

FIGURE 2.586 Amplitude distribution of echo signal of UAV Bayraktar TB2 model at a carrier frequency of 10 GHz given its horizontal polarization.

Scattering characteristics of UAV Bayraktar TB2 model for sounding frequency 3 GHz (wavelength 10 cm).

Figure 2.587 shows the RCS circular diagram of the Bayraktar TB2. The noncoherent RCS circular diagram of the Bayraktar TB2 is shown in Figure 2.588.

The circular mean RCS of the UAV Bayraktar TB2 for horizontal polarization is 2.516 m². The circular median RCS for horizontal polarization is 0.709 m².

Figures 2.589 and 2.590 show the mean and median RCS for the main ranges of sounding azimuths (nose, side, and tail) and for ranges of 20°.

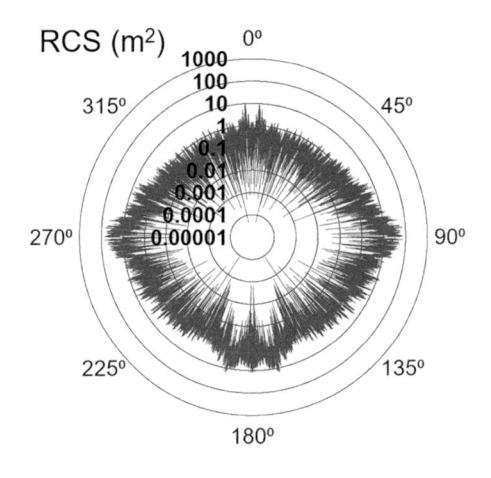

Scattering Characteristics of Aerial Objects 219

FIGURE 2.587 Circular diagram of RCS given radar observation of UAV Bayraktar TB2 model at a carrier frequency of 3 GHz (10 cm wavelength).

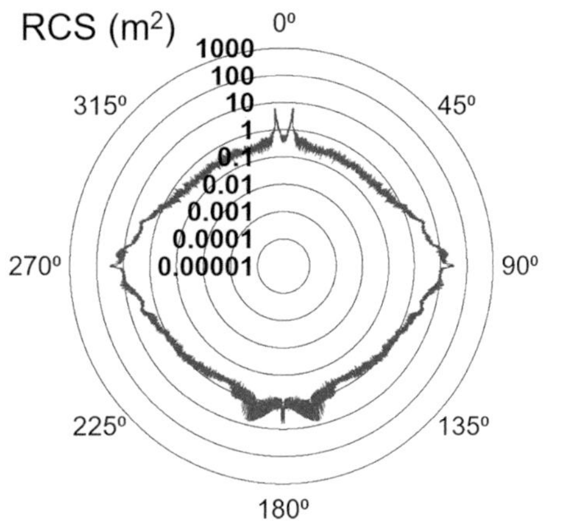

FIGURE 2.588 Circular diagram of noncoherent RCS given radar observation of UAV Bayraktar TB2 model at a carrier frequency of 3 GHz (10 cm wavelength).

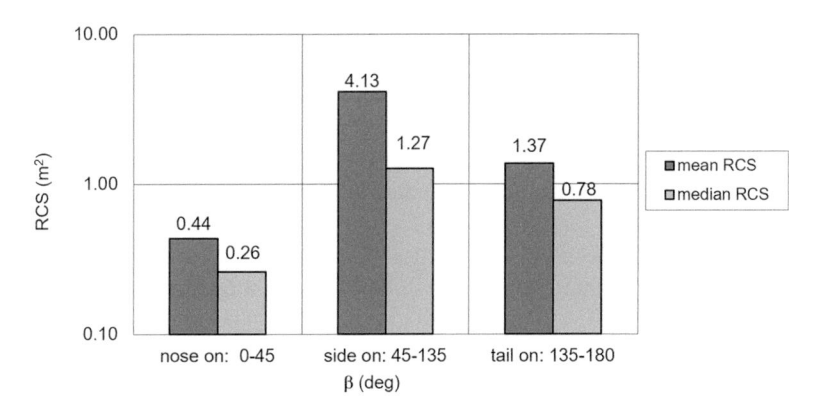

FIGURE 2.589 Diagrams of mean and median RCS of UAV Bayraktar TB2 model in three sectors of azimuth aspect given its radar observation at horizontal polarization and a carrier frequency of 3 GHz (10 cm wavelength).

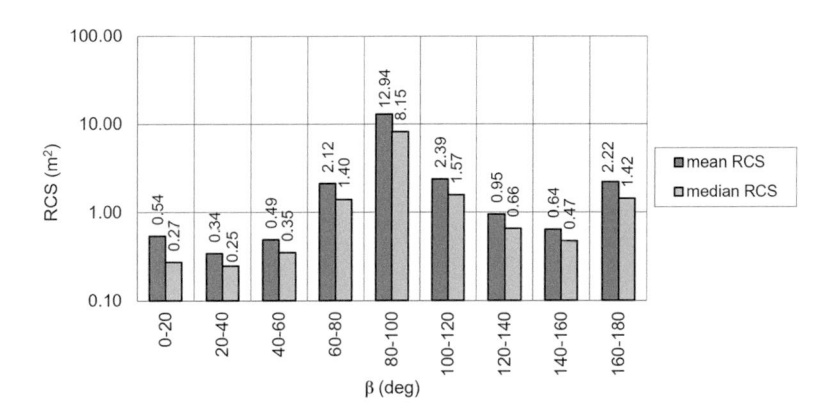

FIGURE 2.590 Diagrams of mean and median RCS of UAV Bayraktar TB2 model in 20-degree sectors of azimuth aspect given its radar observation at horizontal polarization and a carrier frequency of 3 GHz (10 cm wavelength).

extreme value distribution

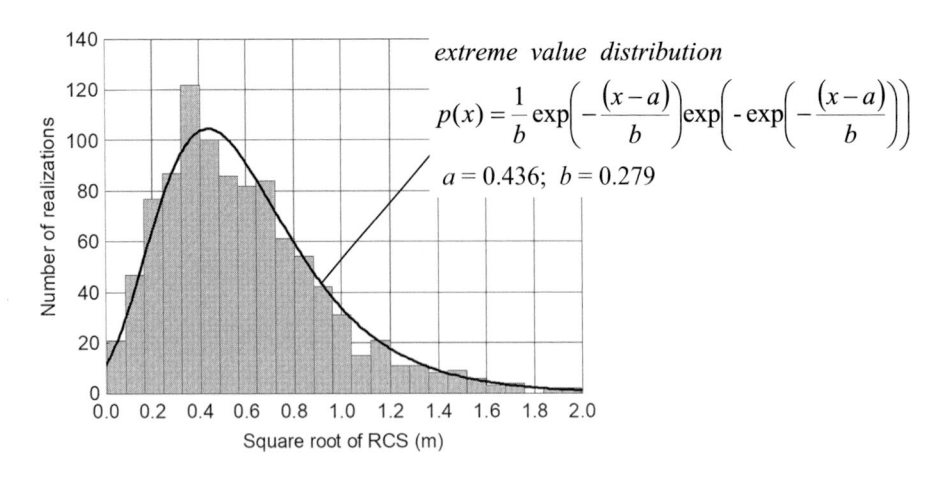

$$p(x) = \frac{1}{b}\exp\left(-\frac{(x-a)}{b}\right)\exp\left(-\exp\left(-\frac{(x-a)}{b}\right)\right)$$

$a = 0.436;\ b = 0.279$

FIGURE 2.591 Amplitude distribution of echo signal of UAV Bayraktar TB2 model at a carrier frequency of 3 GHz given its horizontal polarization.

Scattering characteristics of UAV Bayraktar TB2 model for sounding frequency 1 GHz (wavelength 30 cm).

Figure 2.592 shows the RCS circular diagram of the Bayraktar TB2. The noncoherent RCS circular diagram of the Bayraktar TB2 is shown in Figure 2.593.

The circular mean RCS of the UAV Bayraktar TB2 for horizontal polarization is 1.298 m². The circular median RCS for horizontal polarization is 0.467 m².

Figures 2.594 and 2.595 show the mean and median RCS for the main ranges of sounding azimuths (nose, side, and tail) and for ranges of 20°.

The expressions and parameters of probability distributions that are most consistent with the empirical distributions of the square root of the RCS for different ranges of sounding azimuths and polarizations are given in the electronic appendix to this book. The expressions and parameters of the probability distributions that are most consistent with the empirical distributions of the RCS (energy characteristic) for different ranges of sounding azimuths and polarizations are also given there.

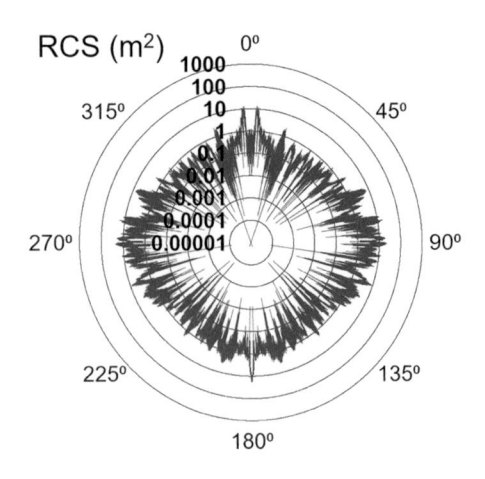

FIGURE 2.592 Circular diagram of RCS given radar observation of UAV Bayraktar TB2 model at a carrier frequency of 1 GHz (30 cm wavelength).

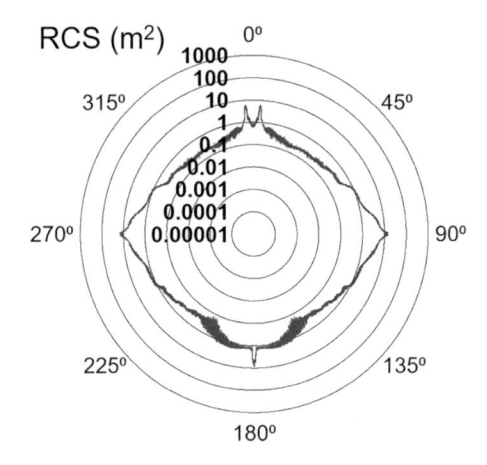

FIGURE 2.593 Circular diagram of noncoherent RCS given radar observation of UAV Bayraktar TB2 model at a carrier frequency of 1 GHz (30 cm wavelength).

Scattering Characteristics of Aerial Objects

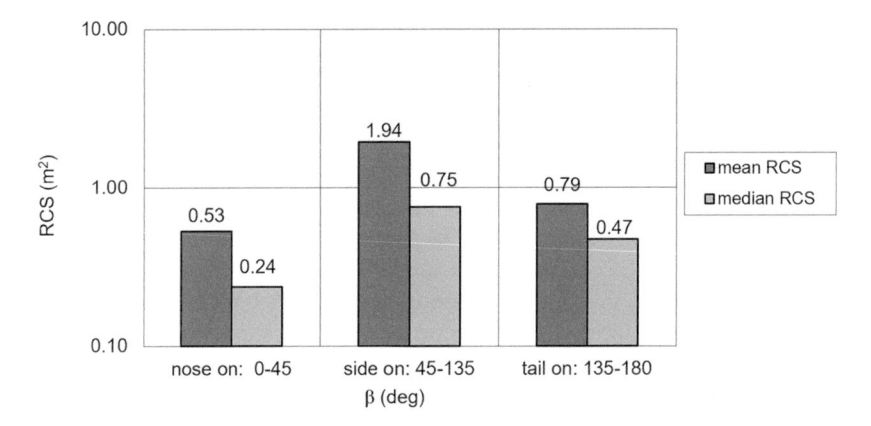

FIGURE 2.594 Diagrams of mean and median RCS of UAV Bayraktar TB2 model in three sectors of azimuth aspect given its radar observation at horizontal polarization and a carrier frequency of 1 GHz (30 cm wavelength).

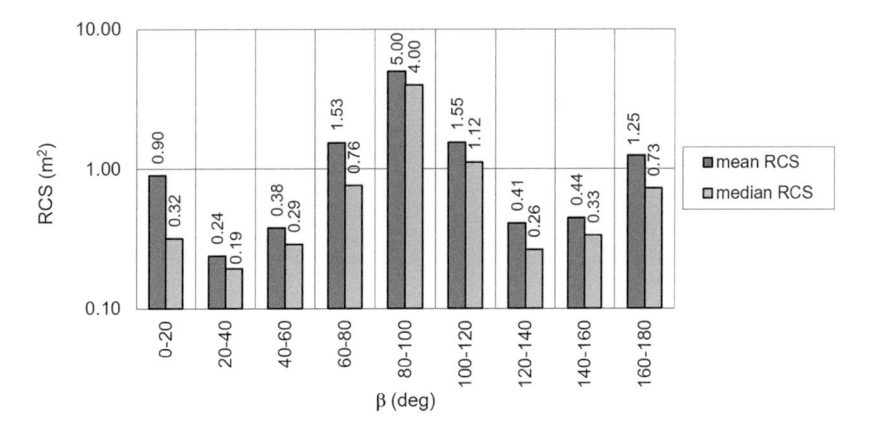

FIGURE 2.595 Diagrams of mean and median RCS of UAV Bayraktar TB2 model in 20-degree sectors of azimuth aspect given its radar observation at horizontal polarization and a carrier frequency of 1 GHz (30 cm wavelength).

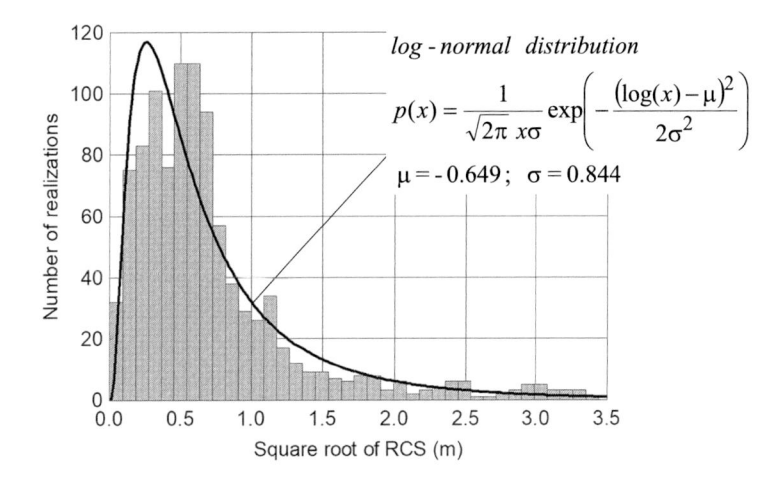

$$\text{log-normal distribution}$$
$$p(x) = \frac{1}{\sqrt{2\pi}\, x\sigma} \exp\left(-\frac{(\log(x)-\mu)^2}{2\sigma^2}\right)$$
$$\mu = -0.649;\quad \sigma = 0.844$$

FIGURE 2.596 Amplitude distribution of echo signal of UAV Bayraktar TB2 model at a carrier frequency of 1 GHz given its horizontal polarization.

2.36 UNMANNED AERIAL VEHICLE MOHAJER-6

Mohajer-6 (Figure 2.597) is a multirole unmanned aerial vehicle. The first flight took place in 2016.

General characteristics of Mohajer-6 [47]: wingspan –10.08 m, length – 5.73 m, weight – 600–670 kg, powerplant – 1 Rotax 912 or Rotax 914 similar internal combustion engine, speed – 130–200 km/h, range – 2 000 km, and service ceiling – 7 600 m.

In accordance with the design of the Mohajer-6, a model of its surface was created to obtaining scattering characteristics (in particular, RCS). The model is shown in Figure 2.598. The model surface was assumed to be perfectly conducting, except for the three-blade propeller, which is made of wood. The smooth parts of the perfectly conducting model surface were approximated by parts of 70 triaxial ellipsoids. The surface breaks were modeled using 33 straight edge scattering parts. The propeller was approximated by parts of 9 triaxial ellipsoids

FIGURE 2.597 Unmanned aerial vehicle Mohajer-6.

FIGURE 2.598 Surface model of unmanned aerial vehicle Mohajer-6.

Below are some scattering characteristics of the Mohajer-6 model at sounding frequencies of 10, 3, and 1 GHz (wavelengths of 3, 10, and 30 cm, respectively) for horizontal polarization of the probing signal. The scattering characteristics for this model at two polarizations, as well as scattering characteristics at sounding frequencies of 5 and 1.3 GHz (wavelengths of 6 and 23 cm, respectively), are given in the electronic appendix of this book.

Sounding parameters: The elevation angle is to be a random value distributed uniformly in the range −3° ± 4° with respect to the wing plane (elevation angle of −3° corresponds to the radar

Scattering Characteristics of Aerial Objects 223

observation from the lower hemisphere), azimuth aspect increment was 0.02°, the azimuth being counted off from the nose-on aspect (0° corresponds to the nose-on radar observation, 180° corresponds to the tail-on observation).

Scattering characteristics of UAV Mohajer-6 model for sounding frequency 10 GHz (wavelength 3 cm).

Figure 2.599 shows the RCS circular diagram of the Mohajer-6 model. The noncoherent RCS circular diagram of the Mohajer-6 is shown in Figure 2.600.

The circular mean RCS of the UAV Mohajer-6 model for horizontal polarization is 12.602 m^2. The circular median RCS (the RCS value used to calculate the detection range of an object with a probability of 0.5) for horizontal polarization is 0.940 m^2.

Figures 2.601 and 2.602 show the mean and median RCS for the main ranges of sounding azimuths (nose, side, and tail) and for ranges of 20°.

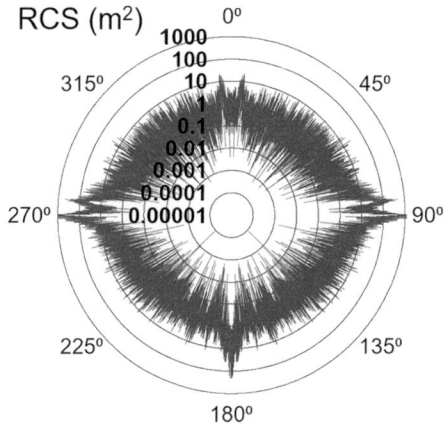

FIGURE 2.599 Circular diagram of RCS given radar observation of UAV Mohajer-6 model at a carrier frequency of 10 GHz (3 cm wavelength).

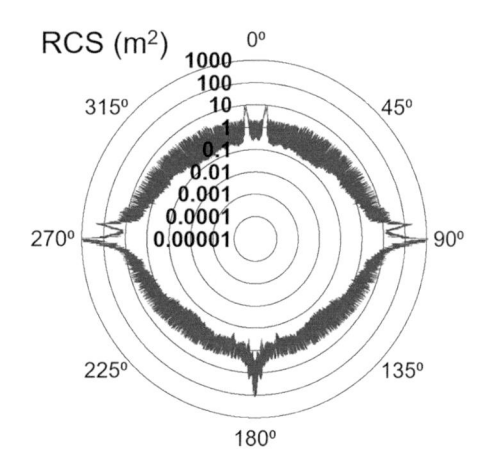

FIGURE 2.600 Circular diagram of noncoherent RCS given radar observation of UAV Mohajer-6 model at a carrier frequency of 10 GHz (3 cm wavelength).

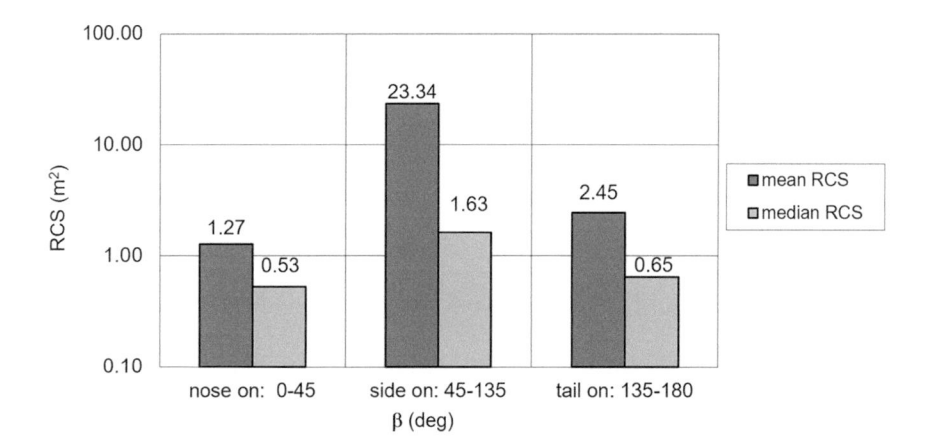

FIGURE 2.601 Diagrams of mean and median RCS of UAV Mohajer-6 model in three sectors of azimuth aspect given its radar observation at horizontal polarization and a carrier frequency of 10 GHz (3 cm wavelength).

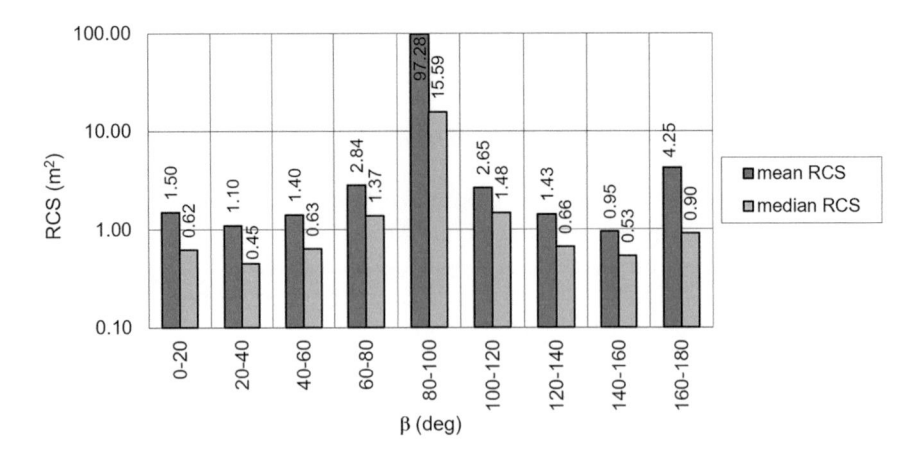

FIGURE 2.602 Diagrams of mean and median RCS of UAV Mohajer-6 model in 20-degree sectors of azimuth aspect given its radar observation at horizontal polarization and a carrier frequency of 10 GHz (3 cm wavelength).

Figures 2.603, 2.608, and 2.613 show histograms of the scattered signal amplitude (square root of the RCS) for the range of sounding azimuths −20° to +20° (sounding from the nose). The bold line shows the probability density functions of the distribution, which can be used to approximate the histogram of the signal amplitude.

$$p(x) = \frac{1}{\sqrt{2\pi}\, x\sigma} \exp\left(-\frac{(\log(x) - \mu)^2}{2\sigma^2}\right)$$

$\mu = -0.280; \quad \sigma = 0.755$

log - normal distribution

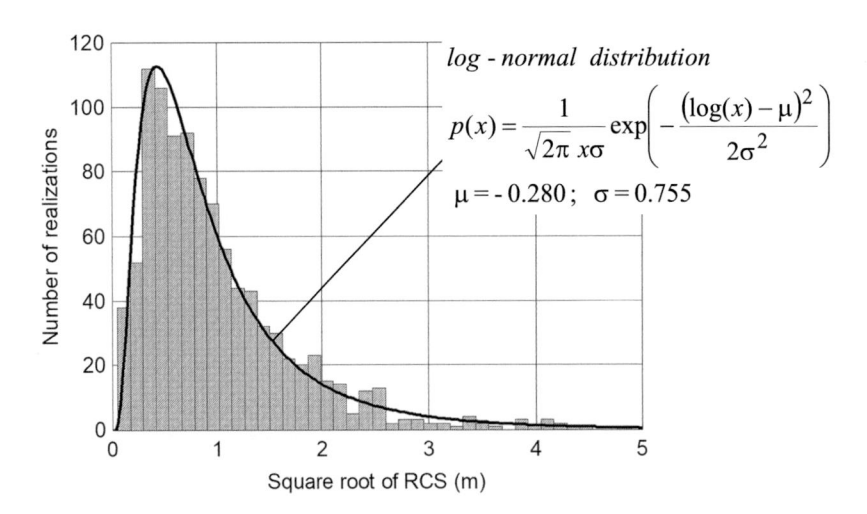

FIGURE 2.603 Amplitude distribution of echo signal of UAV Mohajer-6 model at a carrier frequency of 10 GHz given its horizontal polarization.

Scattering characteristics of UAV Mohajer-6 model for sounding frequency 3 GHz (wavelength 10 cm).

Figure 2.604 shows the RCS circular diagram of the Mohajer-6. The noncoherent RCS circular diagram of the Mohajer-6 is shown in Figure 2.605.

The circular mean RCS of the UAV Mohajer-6 for horizontal polarization is 14.818 m². The circular median RCS for horizontal polarization is 1.236 m².

Figures 2.606 and 2.607 show the mean and median RCS for the main ranges of sounding azimuths (nose, side, and tail) and for ranges of 20°.

Scattering Characteristics of Aerial Objects

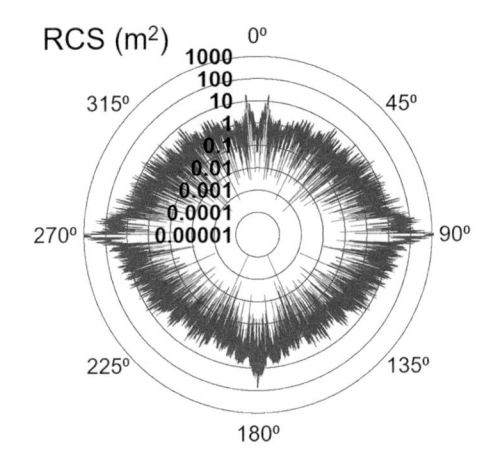

FIGURE 2.604 Circular diagram of RCS given radar observation of UAV Mohajer-6 model at a carrier frequency of 3 GHz (10 cm wavelength).

FIGURE 2.605 Circular diagram of noncoherent RCS given radar observation of UAV Mohajer-6 model at a carrier frequency of 3 GHz (10 cm wavelength).

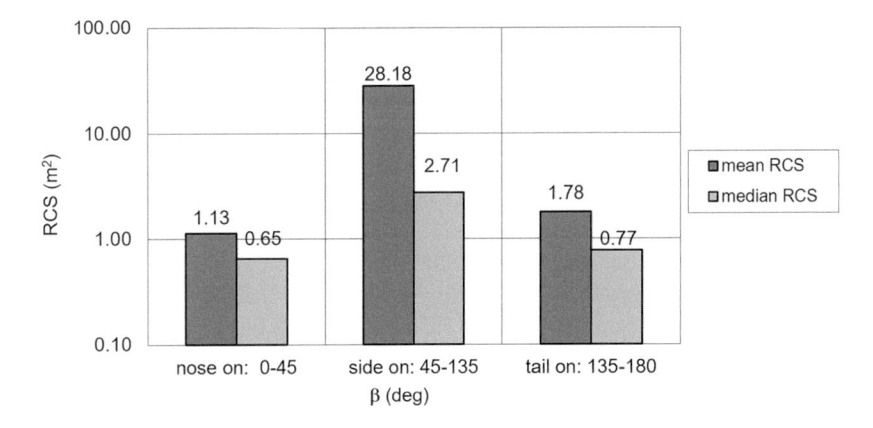

FIGURE 2.606 Diagrams of mean and median RCS of UAV Mohajer-6 model in three sectors of azimuth aspect given its radar observation at horizontal polarization and a carrier frequency of 3 GHz (10 cm wavelength).

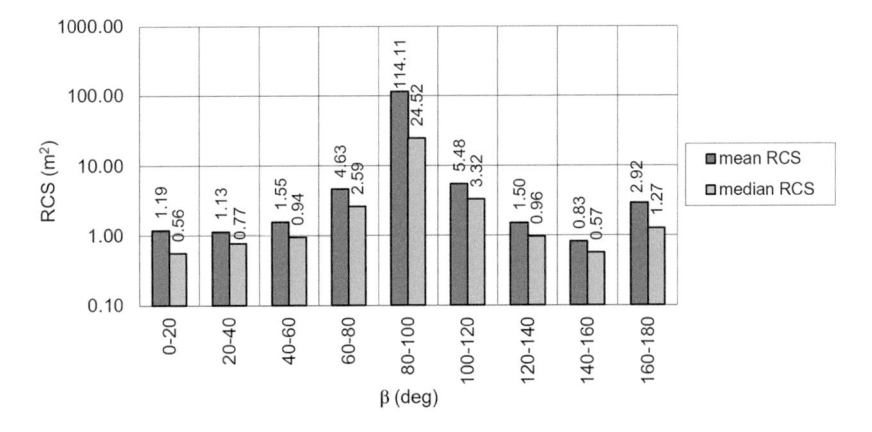

FIGURE 2.607 Diagrams of mean and median RCS of UAV Mohajer-6 model in 20-degree sectors of azimuth aspect given its radar observation at horizontal polarization and a carrier frequency of 3 GHz (10 cm wavelength).

$$p(x) = \frac{1}{\sqrt{2\pi}\, x\sigma} \exp\left(-\frac{(\log(x)-\mu)^2}{2\sigma^2}\right)$$

log - normal distribution

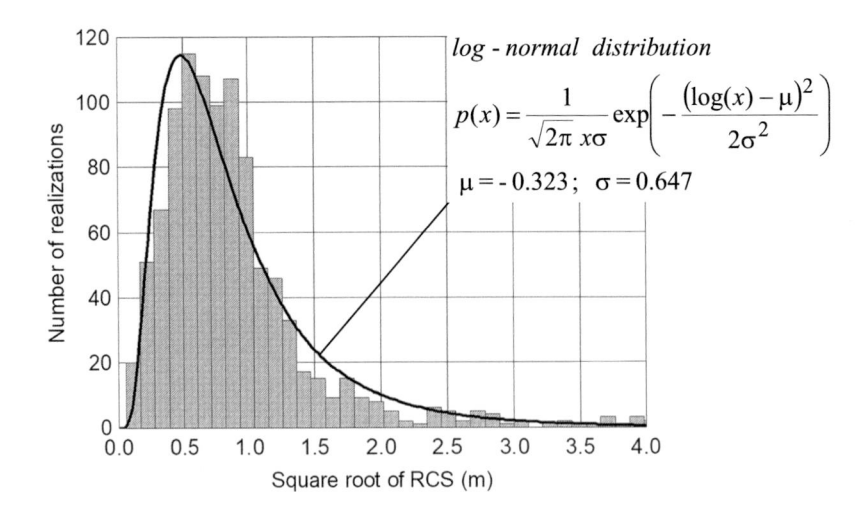

$\mu = -0.323;\ \sigma = 0.647$

FIGURE 2.608 Amplitude distribution of echo signal of UAV Mohajer-6 model at a carrier frequency of 3 GHz given its horizontal polarization.

Scattering characteristics of UAV Mohajer-6 model for sounding frequency 1 GHz (wavelength 30 cm).

Figure 2.609 shows the RCS circular diagram of the Mohajer-6. The noncoherent RCS circular diagram of the Mohajer-6 is shown in Figure 2.610.

The circular mean RCS of the UAV Mohajer-6 for horizontal polarization is 7.734 m². The circular median RCS for horizontal polarization is 0.835 m².

Figures 2.611 and 2.612 show the mean and median RCS for the main ranges of sounding azimuths (nose, side, and tail) and for ranges of 20°.

The expressions and parameters of probability distributions that are most consistent with the empirical distributions of the square root of the RCS for different ranges of sounding azimuths and polarizations are given in the electronic appendix to this book. The expressions and parameters of the probability distributions that are most consistent with the empirical distributions of the RCS (energy characteristic) for different ranges of sounding azimuths and polarizations are also given there.

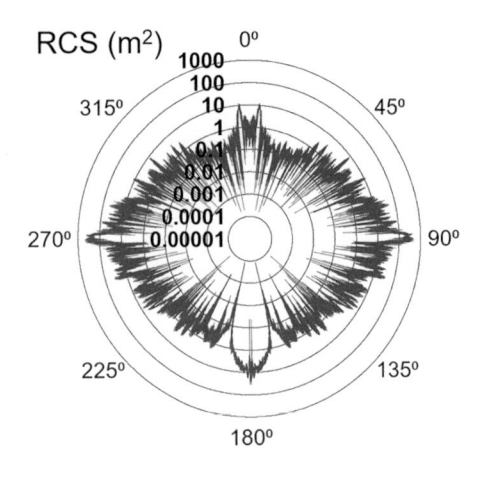

FIGURE 2.609 Circular diagram of RCS given radar observation of UAV Mohajer-6 model at a carrier frequency of 1 GHz (30 cm wavelength).

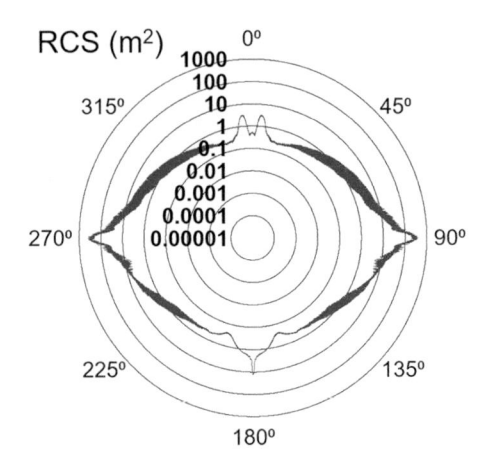

FIGURE 2.610 Circular diagram of noncoherent RCS given radar observation of UAV Mohajer-6 model at a carrier frequency of 1 GHz (30 cm wavelength).

Scattering Characteristics of Aerial Objects

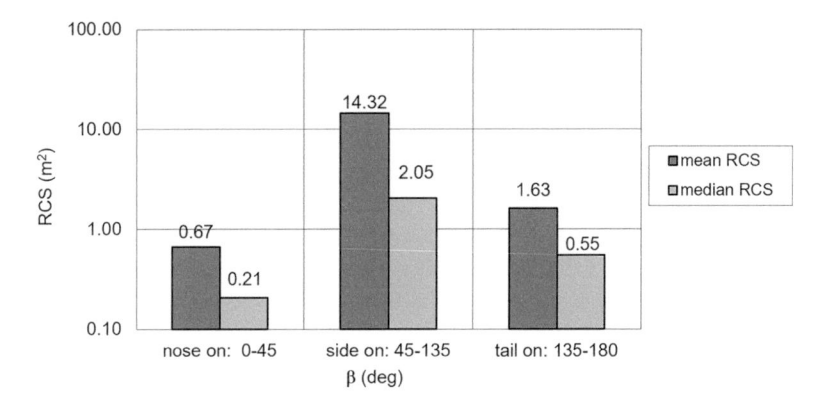

FIGURE 2.611 Diagrams of mean and median RCS of UAV Mohajer-6 model in three sectors of azimuth aspect given its radar observation at horizontal polarization and a carrier frequency of 1 GHz (30 cm wavelength).

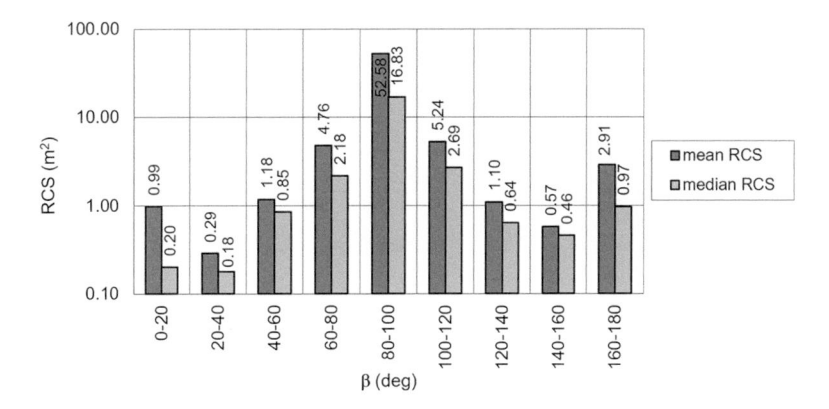

FIGURE 2.612 Diagrams of mean and median RCS of UAV Mohajer-6 model in 20-degree sectors of azimuth aspect given its radar observation at horizontal polarization and a carrier frequency of 1 GHz (30 cm wavelength).

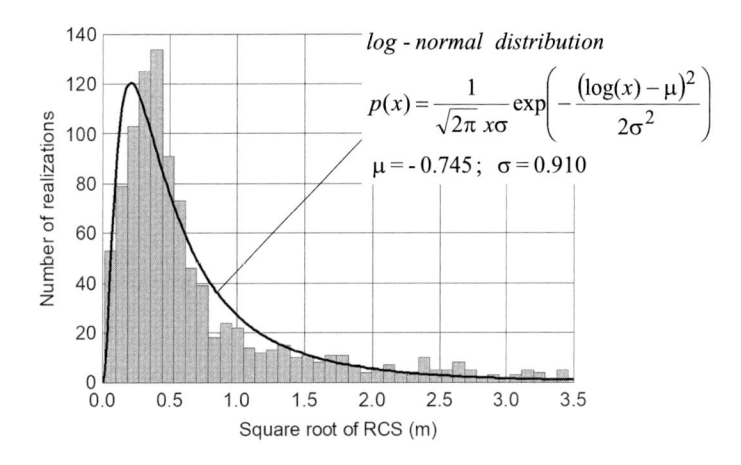

$$p(x) = \frac{1}{\sqrt{2\pi}\, x\sigma} \exp\left(-\frac{(\log(x) - \mu)^2}{2\sigma^2}\right)$$

$\mu = -0.745; \quad \sigma = 0.910$

FIGURE 2.613 Amplitude distribution of echo signal of UAV Mohajer-6 model at a carrier frequency of 1 GHz given its horizontal polarization.

2.37 UNMANNED AERIAL VEHICLE IAI SEARCHER II (FORPOST)

IAI Searcher II (Forpost) (Figure 2.614) is a reconnaissance unmanned aerial vehicle. The first flight took place in 1998.

General characteristics of IAI Searcher II (in round brackets – for Forpost) [48]: wingspan –8.55 m, length – 5.85 m, weight – 436 (500) kg, powerplant – 1 Limbach L 550 (APD85) engine, speed – 200 km/h, range – 300 (350) km, and service ceiling – 7 000 (5 500) m.

In accordance with the design of the IAI Searcher II (hereinafter - Forpost), a model of its surface was created to obtaining scattering characteristics (in particular, RCS). The model is shown in Figure 2.615. In the modeling, the plastic shells and propeller were approximated by parts of 23 triaxial ellipsoids. The smooth parts of the inner equipment surfaces and metal elements were approximated by parts of 90 triaxial ellipsoids. The surface breaks of the equipment were modeled using 124 straight edge scattering parts.

FIGURE 2.614 Unmanned aerial vehicle IAI Searcher II (Forpost).

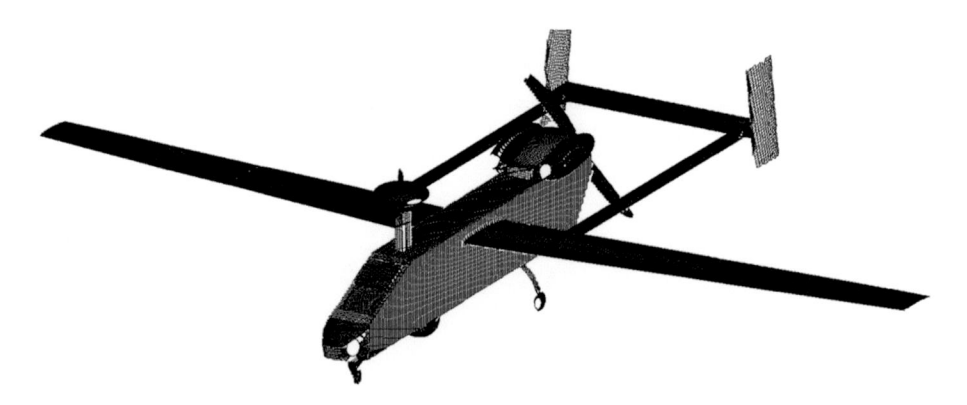

FIGURE 2.615 Surface model of unmanned aerial vehicle IAI Searcher II (Forpost).

Below are some scattering characteristics of the Forpost model at sounding frequencies of 10, 3, and 1 GHz (wavelengths of 3, 10, and 30 cm, respectively) for horizontal polarization of the probing signal. The scattering characteristics for this model at two polarizations, as well as scattering characteristics at sounding frequencies of 5 and 1.3 GHz (wavelengths of 6 and 23 cm, respectively), are given in the electronic appendix of this book.

Sounding parameters: The elevation angle is to be a random value distributed uniformly in the range $-3° \pm 4°$ with respect to the wing plane (elevation angle of $-3°$ corresponds to the radar observation from the lower hemisphere), azimuth aspect increment was $0.02°$, the azimuth being counted

Scattering Characteristics of Aerial Objects 229

off from the nose-on aspect (0° corresponds to the nose-on radar observation, 180° corresponds to the tail-on observation).

Scattering characteristics of UAV Forpost model for sounding frequency 10 GHz (wavelength 3 cm).

Figure 2.616 shows the RCS circular diagram of the Forpost model. The noncoherent RCS circular diagram of the Forpost is shown in Figure 2.617.

The circular mean RCS of the UAV Forpost model for horizontal polarization is 4.375 m². The circular median RCS (the RCS value used to calculate the detection range of an object with a probability of 0.5) for horizontal polarization is 0.739 m².

Figures 2.618 and 2.619 show the mean and median RCS for the main ranges of sounding azimuths (nose, side, and tail) and for ranges of 20°.

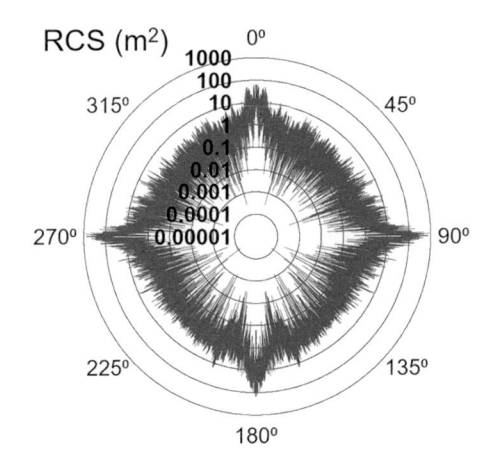

FIGURE 2.616 Circular diagram of RCS given radar observation of UAV Forpost model at a carrier frequency of 10 GHz (3 cm wavelength).

FIGURE 2.617 Circular diagram of noncoherent RCS given radar observation of UAV Forpost model at a carrier frequency of 10 GHz (3 cm wavelength).

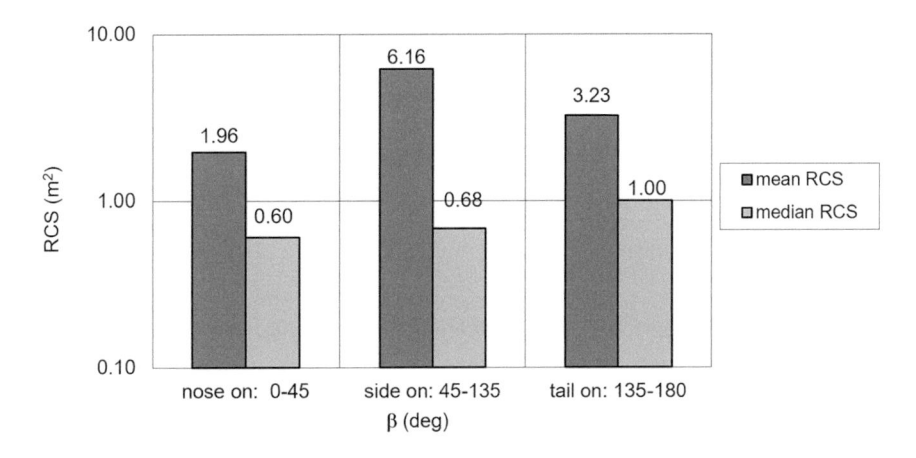

FIGURE 2.618 Diagrams of mean and median RCS of UAV Forpost model in three sectors of azimuth aspect given its radar observation at horizontal polarization and a carrier frequency of 10 GHz (3 cm wavelength).

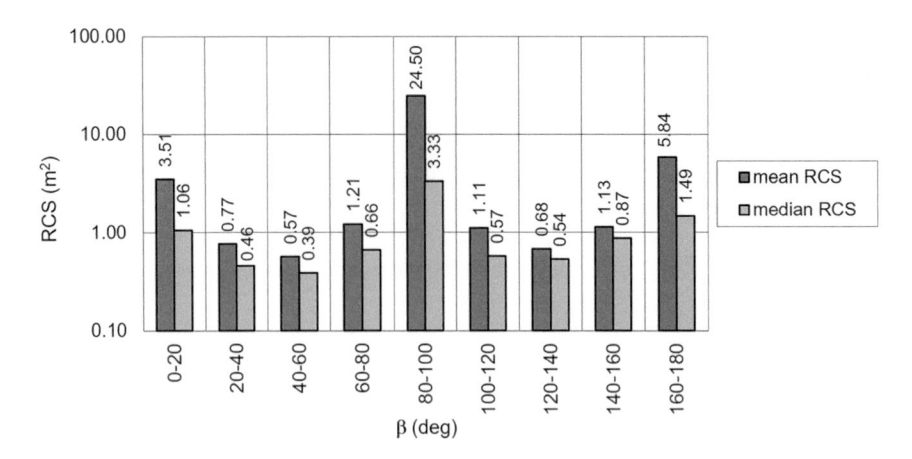

FIGURE 2.619 Diagrams of mean and median RCS of UAV Forpost model in 20-degree sectors of azimuth aspect given its radar observation at horizontal polarization and a carrier frequency of 10 GHz (3 cm wavelength).

Figures 2.620, 2.625, and 2.630 show histograms of the scattered signal amplitude (square root of the RCS) for the range of sounding azimuths −20° to +20° (sounding from the nose). The bold line shows the probability density functions of the distribution, which can be used to approximate the histogram of the signal amplitude.

$$p(x) = \frac{1}{\sqrt{2\pi}\, x\sigma} \exp\left(-\frac{(\log(x)-\mu)^2}{2\sigma^2}\right)$$

log - normal distribution

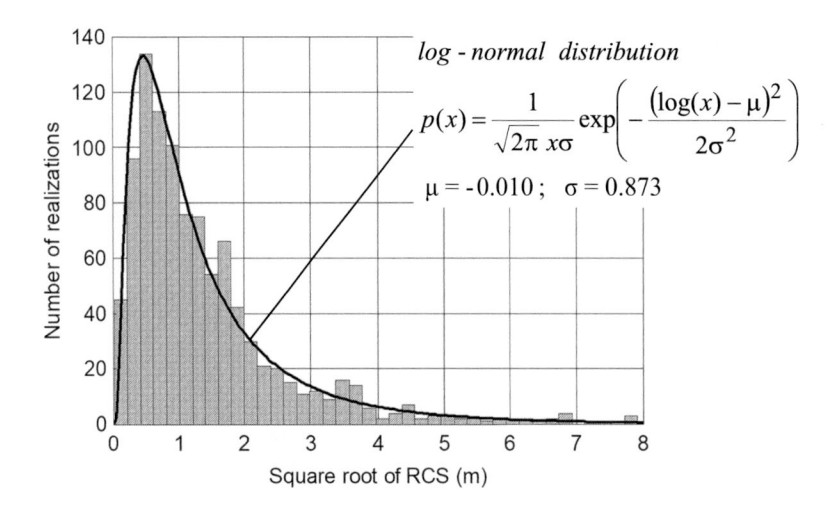

$\mu = -0.010$; $\sigma = 0.873$

FIGURE 2.620 Amplitude distribution of echo signal of UAV Forpost model at a carrier frequency of 10 GHz given its horizontal polarization.

Scattering characteristics of UAV Forpost model for sounding frequency 3 GHz (wavelength 10 cm).

Figure 2.621 shows the RCS circular diagram of the Forpost. The noncoherent RCS circular diagram of the Forpost is shown in Figure 2.622.

The circular mean RCS of the UAV Forpost for horizontal polarization is 5.142 m². The circular median RCS for horizontal polarization is 0.920 m².

Figures 2.623 and 2.624 show the mean and median RCS for the main ranges of sounding azimuths (nose, side, and tail) and for ranges of 20°.

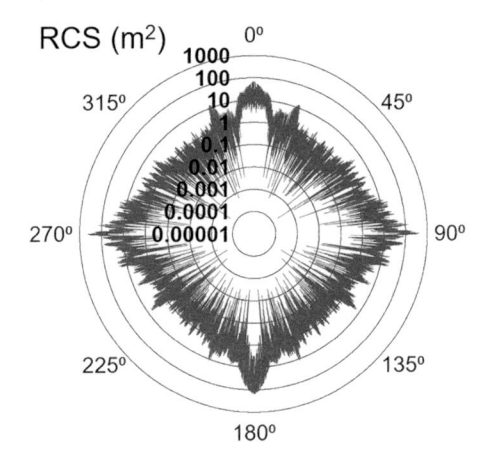

Scattering Characteristics of Aerial Objects

FIGURE 2.621 Circular diagram of RCS given radar observation of UAV Forpost model at a carrier frequency of 3 GHz (10 cm wavelength).

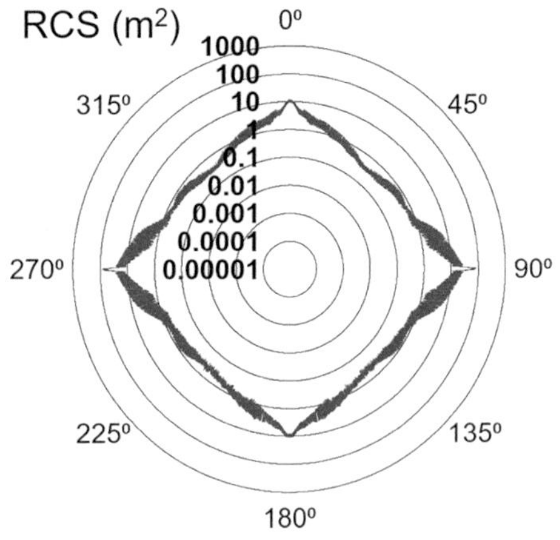

FIGURE 2.622 Circular diagram of noncoherent RCS given radar observation of UAV Forpost model at a carrier frequency of 3 GHz (10 cm wavelength).

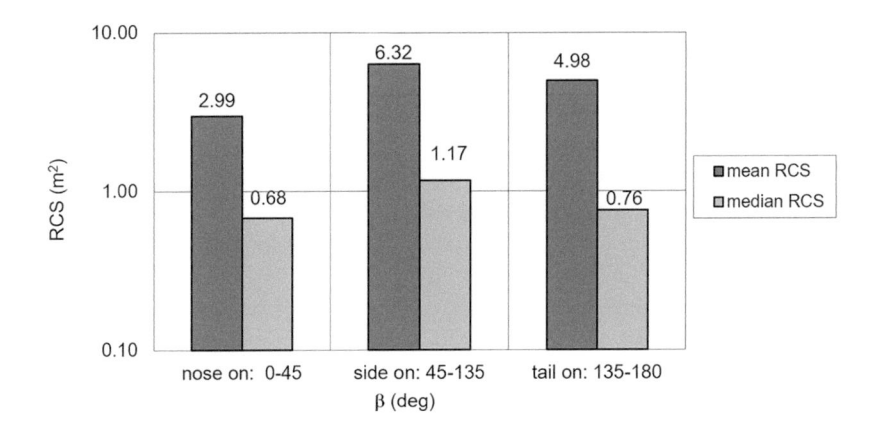

FIGURE 2.623 Diagrams of mean and median RCS of UAV Forpost model in three sectors of azimuth aspect given its radar observation at horizontal polarization and a carrier frequency of 3 GHz (10 cm wavelength).

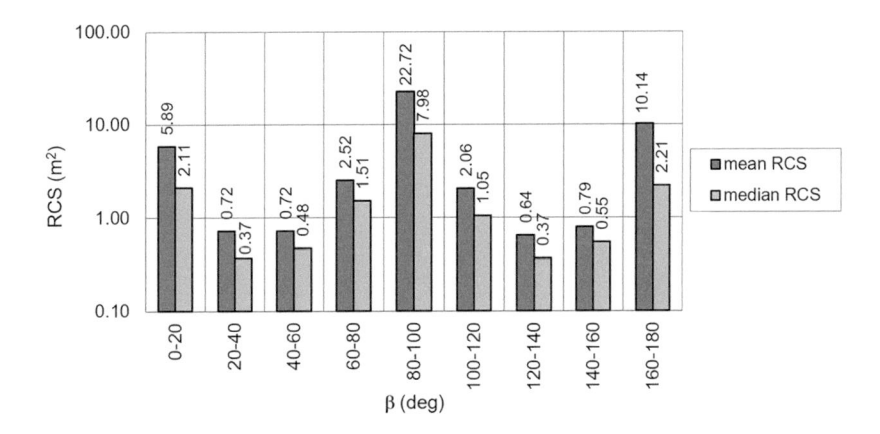

FIGURE 2.624 Diagrams of mean and median RCS of UAV Forpost model in 20-degree sectors of azimuth aspect given its radar observation at horizontal polarization and a carrier frequency of 3 GHz (10 cm wavelength).

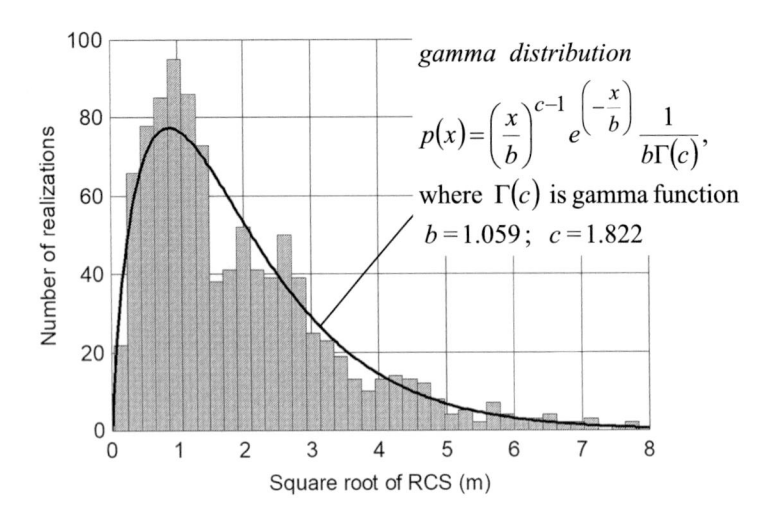

$$p(x) = \left(\frac{x}{b}\right)^{c-1} e^{\left(-\frac{x}{b}\right)} \frac{1}{b\Gamma(c)},$$

gamma distribution, where $\Gamma(c)$ is gamma function
$b = 1.059; \quad c = 1.822$

FIGURE 2.625 Amplitude distribution of echo signal of UAV Forpost model at a carrier frequency of 3 GHz given its horizontal polarization.

Scattering characteristics of UAV Forpost model for sounding frequency 1 GHz (wavelength 30 cm).

Figure 2.626 shows the RCS circular diagram of the Forpost. The noncoherent RCS circular diagram of the Forpost is shown in Figure 2.627.

The circular mean RCS of the UAV Forpost for horizontal polarization is 4.721 m². The circular median RCS for horizontal polarization is 1.798 m².

Figures 2.628 and 2.629 show the mean and median RCS for the main ranges of sounding azimuths (nose, side, and tail) and for ranges of 20°.

The expressions and parameters of probability distributions that are most consistent with the empirical distributions of the square root of the RCS for different ranges of sounding azimuths and polarizations are given in the electronic appendix to this book. The expressions and parameters of the probability distributions that are most consistent with the empirical distributions of the RCS (energy characteristic) for different ranges of sounding azimuths and polarizations are also given there.

FIGURE 2.626 Circular diagram of RCS given radar observation of UAV Forpost model at a carrier frequency of 1 GHz (30 cm wavelength).

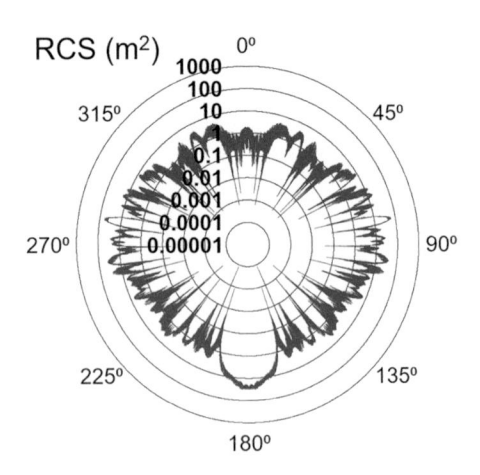

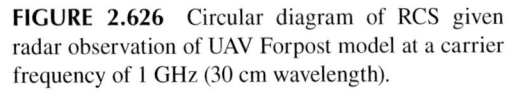

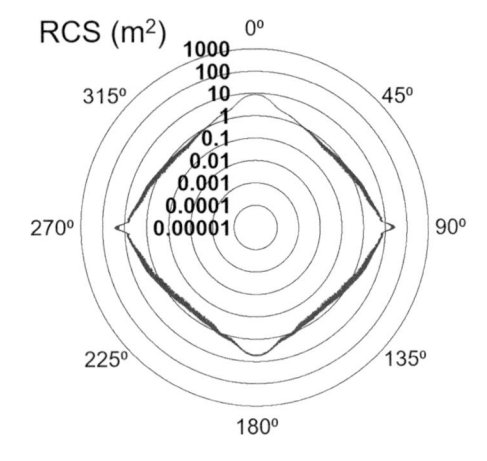

FIGURE 2.627 Circular diagram of noncoherent RCS given radar observation of UAV Forpost model at a carrier frequency of 1 GHz (30 cm wavelength).

Scattering Characteristics of Aerial Objects 233

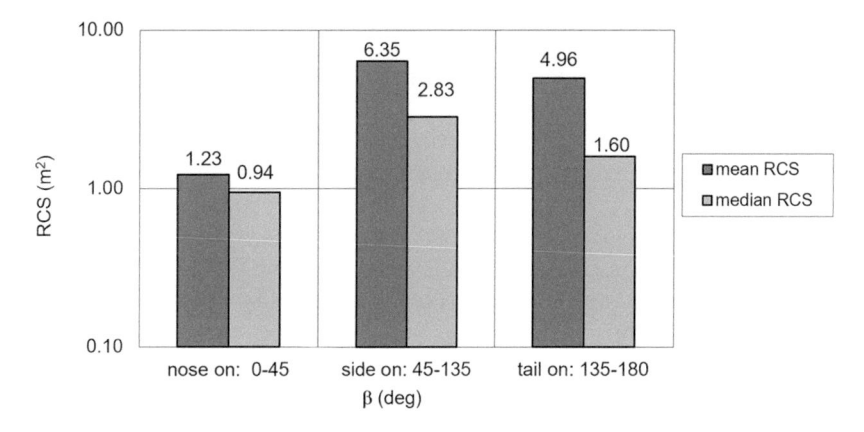

FIGURE 2.628 Diagrams of mean and median RCS of UAV Forpost model in three sectors of azimuth aspect given its radar observation at horizontal polarization and a carrier frequency of 1 GHz (30 cm wavelength).

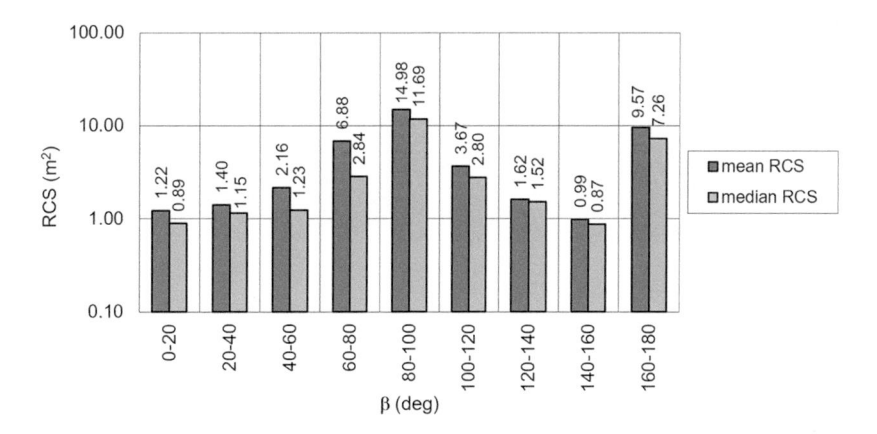

FIGURE 2.629 Diagrams of mean and median RCS of UAV Forpost model in 20-degree sectors of azimuth aspect given its radar observation at horizontal polarization and a carrier frequency of 1 GHz (30 cm wavelength).

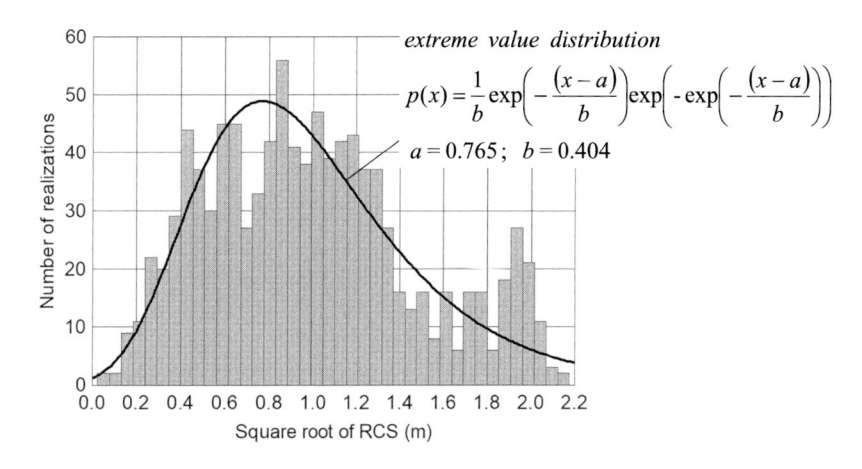

extreme value distribution

$$p(x) = \frac{1}{b}\exp\left(-\frac{(x-a)}{b}\right)\exp\left(-\exp\left(-\frac{(x-a)}{b}\right)\right)$$

$a = 0.765;\ b = 0.404$

FIGURE 2.630 Amplitude distribution of echo signal of UAV Forpost model at a carrier frequency of 1 GHz given its horizontal polarization.

2.38 UNMANNED AERIAL VEHICLE DOZOR-600

Dozor-600 (Figure 2.631) is a long-flight medium-heavy unmanned aerial vehicle. The first flight took place in 2009–2011.

General characteristics of Dozor-600 [49]: wingspan –12.0 m, length – 7.0 m, weight – 360–720 kg, powerplant – 1 engine with a pusher propeller, speed – 130–150 km/h, range – 3 000 km, and service ceiling – 7 100 m.

In accordance with the design of the Dozor-600, a model of its surface was created to obtaining scattering characteristics (in particular, RCS). The model is shown in Figure 2.632. The model surface was assumed to be perfectly conducting, except of the nose antenna radome, which is made of radar transparent plastic. The smooth parts of the perfectly conducting model surface were approximated by parts of 66 triaxial ellipsoids. The surface breaks were modeled using 12 straight edge scattering parts. The nose antenna radome was approximated by parts of 8 triaxial ellipsoids.

FIGURE 2.631 Unmanned aerial vehicle Dozor-600.

FIGURE 2.632 Surface model of unmanned aerial vehicle Dozor-600.

Below are some scattering characteristics of the Dozor-600 model at sounding frequencies of 10, 3, and 1 GHz (wavelengths of 3, 10, and 30 cm, respectively) for horizontal polarization of the probing signal. The scattering characteristics for this model at two polarizations, as well as scattering characteristics at sounding frequencies of 5 and 1.3 GHz (wavelengths of 6 and 23 cm, respectively), are given in the electronic appendix of this book.

Sounding parameters: The elevation angle is to be a random value distributed uniformly in the range $-3° \pm 4°$ with respect to the wing plane (elevation angle of $-3°$ corresponds to the radar observation from the lower hemisphere), azimuth aspect increment was $0.02°$, the azimuth being counted off from the nose-on aspect ($0°$ corresponds to the nose-on radar observation, $180°$ corresponds to the tail-on observation).

Scattering Characteristics of Aerial Objects   235

Scattering characteristics of UAV Dozor-600 model for sounding frequency 10 GHz (wavelength 3 cm).

Figure 2.633 shows the RCS circular diagram of the Dozor-600 model. The noncoherent RCS circular diagram of the Dozor-600 is shown in Figure 2.634.

The circular mean RCS of the UAV Dozor-600 model for horizontal polarization is 9.992 m². The circular median RCS (the RCS value used to calculate the detection range of an object with a probability of 0.5) for horizontal polarization is 0.538 m².

Figures 2.635 and 2.636 show the mean and median RCS for the main ranges of sounding azimuths (nose, side, and tail) and for ranges of 20°.

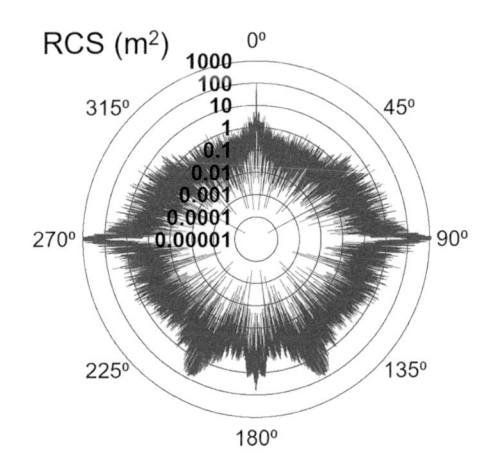

FIGURE 2.633 Circular diagram of RCS given radar observation of UAV Dozor-600 model at a carrier frequency of 10 GHz (3 cm wavelength).

FIGURE 2.634 Circular diagram of noncoherent RCS given radar observation of UAV Dozor-600 model at a carrier frequency of 10 GHz (3 cm wavelength).

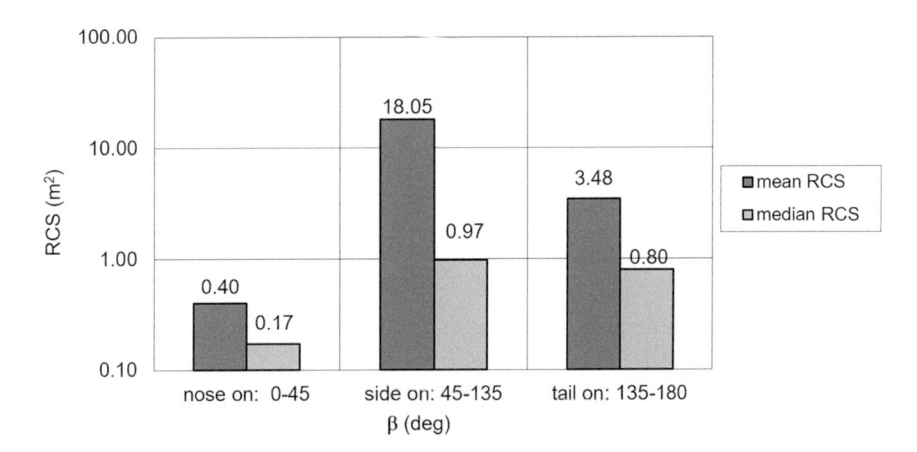

FIGURE 2.635 Diagrams of mean and median RCS of UAV Dozor-600 model in three sectors of azimuth aspect given its radar observation at horizontal polarization and a carrier frequency of 10 GHz (3 cm wavelength).

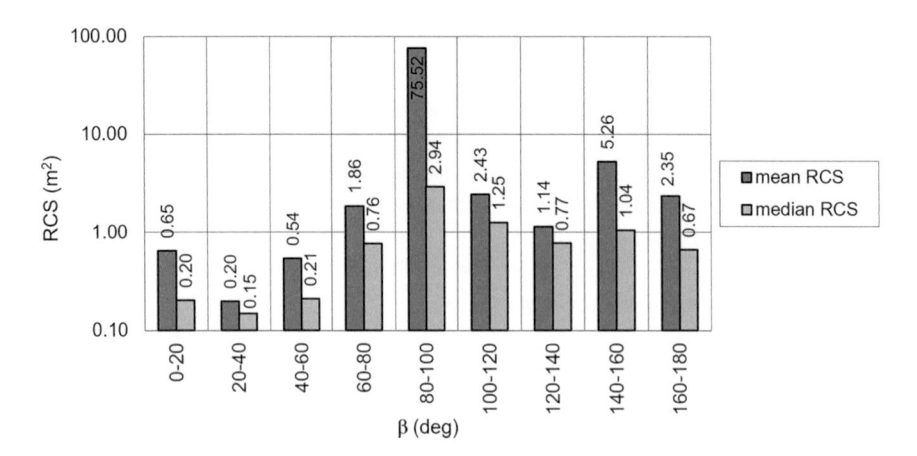

FIGURE 2.636 Diagrams of mean and median RCS of UAV Dozor-600 model in 20-degree sectors of azimuth aspect given its radar observation at horizontal polarization and a carrier frequency of 10 GHz (3 cm wavelength).

Figures 2.637, 2.642, and 2.647 show histograms of the scattered signal amplitude (square root of the RCS) for the range of sounding azimuths −20° to +20° (sounding from the nose). The bold line shows the probability density functions of the distribution, which can be used to approximate the histogram of the signal amplitude.

extreme value distribution

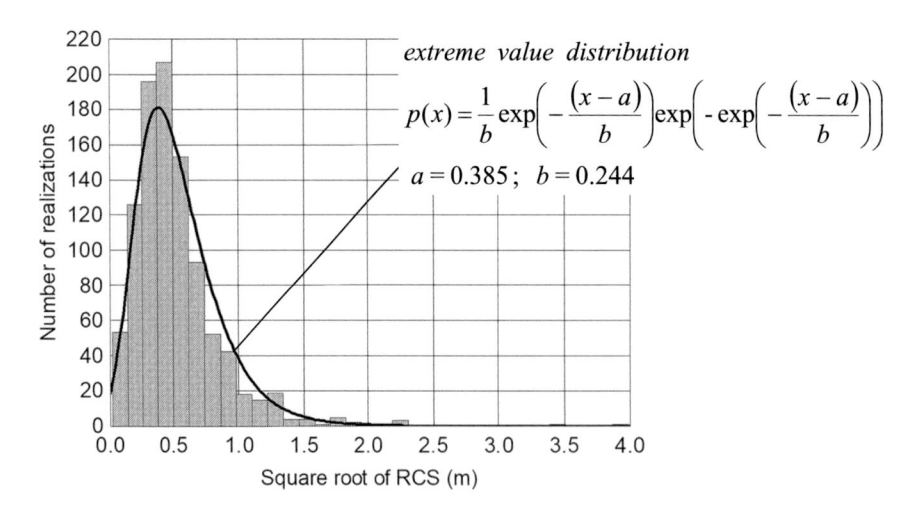

$$p(x) = \frac{1}{b}\exp\left(-\frac{(x-a)}{b}\right)\exp\left(-\exp\left(-\frac{(x-a)}{b}\right)\right)$$

$a = 0.385; \quad b = 0.244$

FIGURE 2.637 Amplitude distribution of echo signal of UAV Dozor-600 model at a carrier frequency of 10 GHz given its horizontal polarization.

Scattering characteristics of UAV Dozor-600 model for sounding frequency 3 GHz (wavelength 10 cm).

Figure 2.638 shows the RCS circular diagram of the Dozor-600. The noncoherent RCS circular diagram of the Dozor-600 is shown in Figure 2.639.

The circular mean RCS of the UAV Dozor-600 for horizontal polarization is 13.060 m². The circular median RCS for horizontal polarization is 0.906 m².

Figures 2.640 and 2.641 show the mean and median RCS for the main ranges of sounding azimuths (nose, side, and tail) and for ranges of 20°.

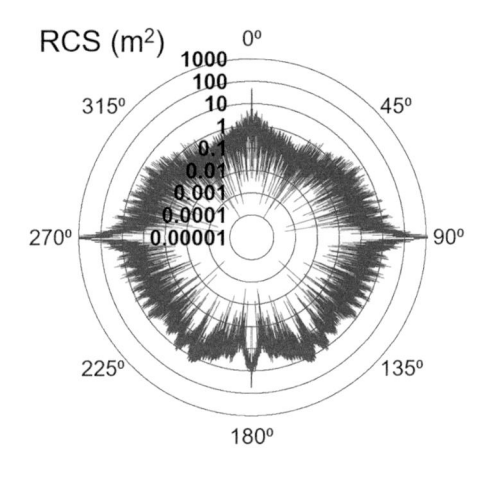

Scattering Characteristics of Aerial Objects

FIGURE 2.638 Circular diagram of RCS given radar observation of UAV Dozor-600 model at a carrier frequency of 3 GHz (10 cm wavelength).

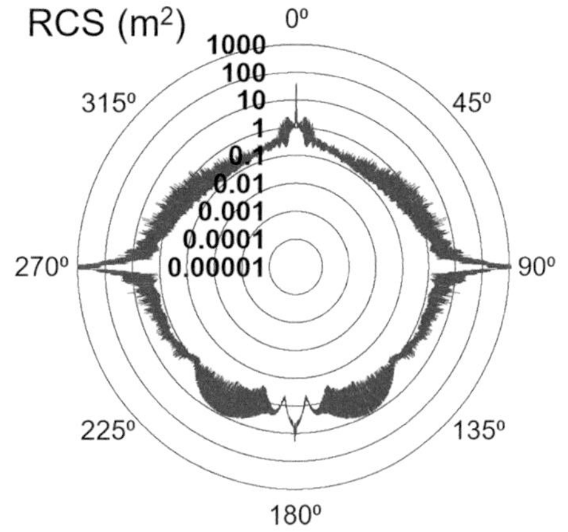

FIGURE 2.639 Circular diagram of noncoherent RCS given radar observation of UAV Dozor-600 model at a carrier frequency of 3 GHz (10 cm wavelength).

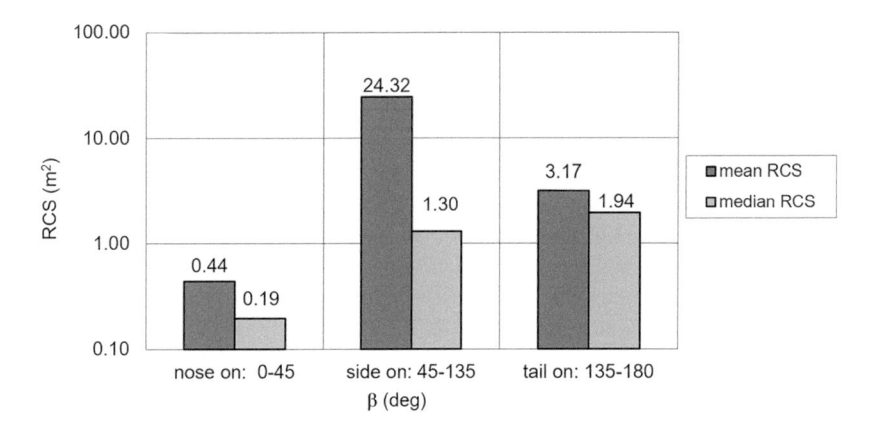

FIGURE 2.640 Diagrams of mean and median RCS of UAV Dozor-600 model in three sectors of azimuth aspect given its radar observation at horizontal polarization and a carrier frequency of 3 GHz (10 cm wavelength).

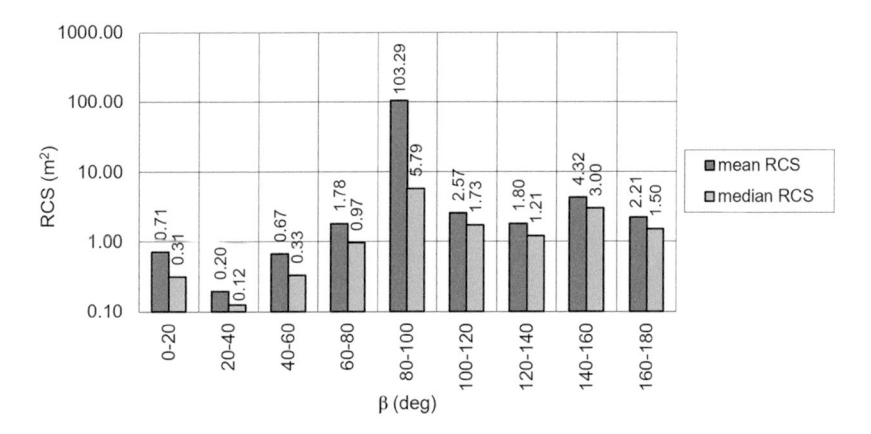

FIGURE 2.641 Diagrams of mean and median RCS of UAV Dozor-600 model in 20-degree sectors of azimuth aspect given its radar observation at horizontal polarization and a carrier frequency of 3 GHz (10 cm wavelength).

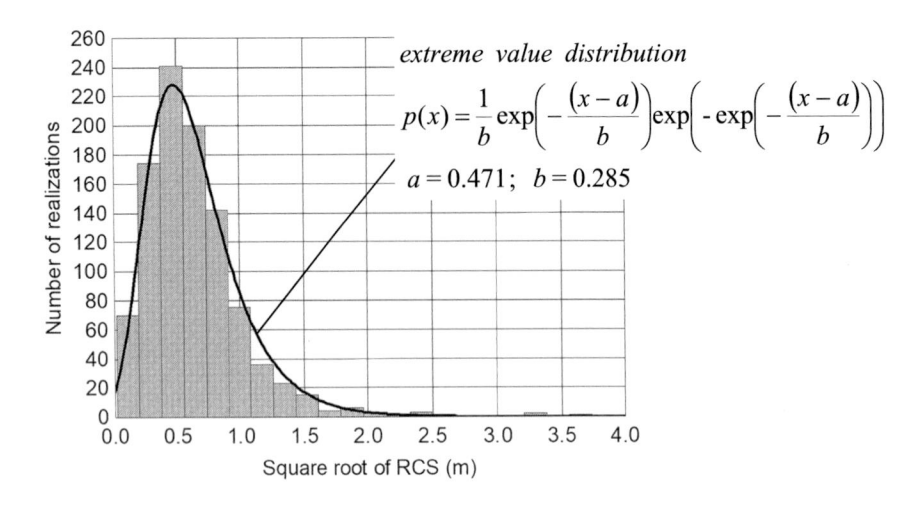

FIGURE 2.642 Amplitude distribution of echo signal of UAV Dozor-600 model at a carrier frequency of 3 GHz given its horizontal polarization.

extreme value distribution

$$p(x) = \frac{1}{b} \exp\left(-\frac{(x-a)}{b}\right) \exp\left(-\exp\left(-\frac{(x-a)}{b}\right)\right)$$

$a = 0.471; \quad b = 0.285$

Scattering characteristics of UAV Dozor-600 model for sounding frequency 1 GHz (wavelength 30 cm).

Figure 2.643 shows the RCS circular diagram of the Dozor-600. The noncoherent RCS circular diagram of the Dozor-600 is shown in Figure 2.644.

The circular mean RCS of the UAV Dozor-600 for horizontal polarization is 9.245 m². The circular median RCS for horizontal polarization is 1.269 m².

Figures 2.645 and 2.646 show the mean and median RCS for the main ranges of sounding azimuths (nose, side, and tail) and for ranges of 20°.

The expressions and parameters of probability distributions that are most consistent with the empirical distributions of the square root of the RCS for different ranges of sounding azimuths and polarizations are given in the electronic appendix to this book. The expressions and parameters of the probability distributions that are most consistent with the empirical distributions of the RCS (energy characteristic) for different ranges of sounding azimuths and polarizations are also given there.

FIGURE 2.643 Circular diagram of RCS given radar observation of UAV Dozor-600 model at a carrier frequency of 1 GHz (30 cm wavelength).

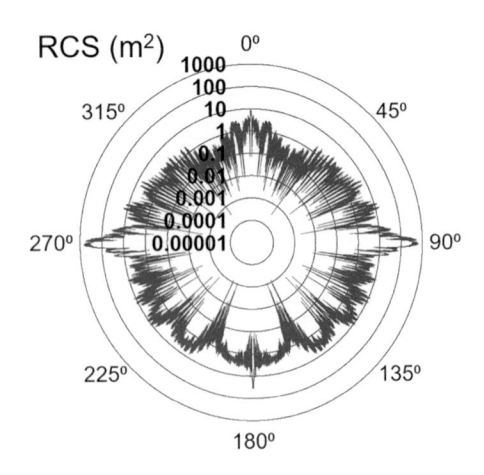

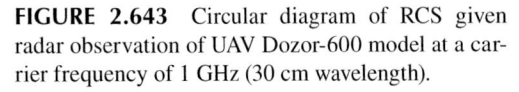

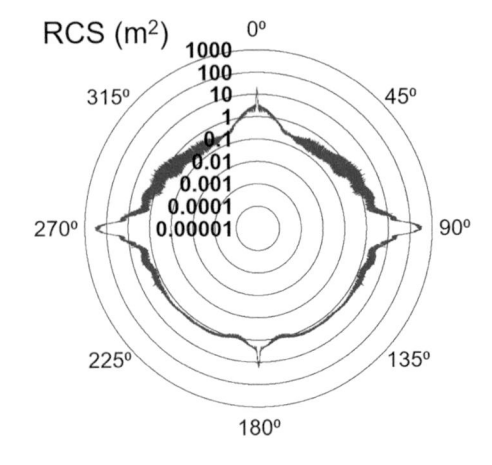

FIGURE 2.644 Circular diagram of noncoherent RCS given radar observation of UAV Dozor-600 model at a carrier frequency of 1 GHz (30 cm wavelength).

Scattering Characteristics of Aerial Objects

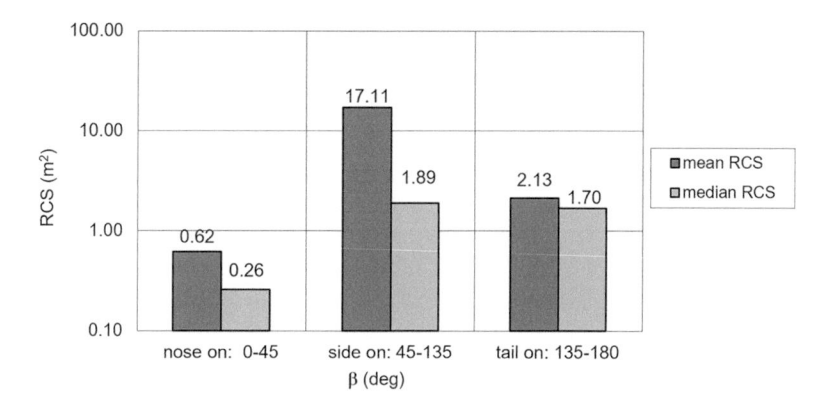

FIGURE 2.645 Diagrams of mean and median RCS of UAV Dozor-600 model in three sectors of azimuth aspect given its radar observation at horizontal polarization and a carrier frequency of 1 GHz (30 cm wavelength).

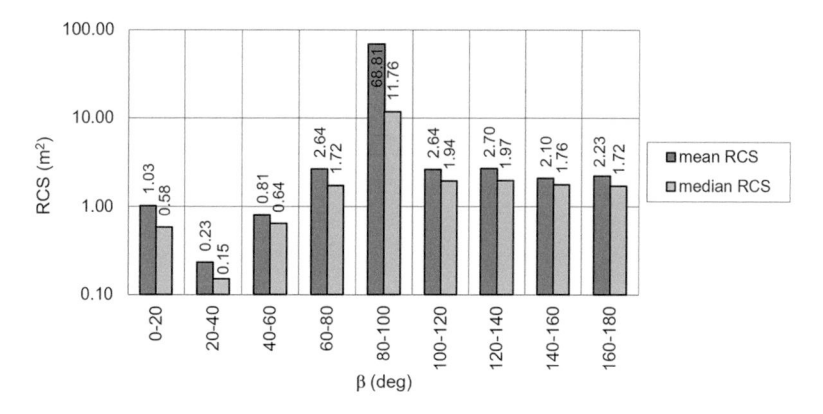

FIGURE 2.646 Diagrams of mean and median RCS of UAV Dozor-600 model in 20-degree sectors of azimuth aspect given its radar observation at horizontal polarization and a carrier frequency of 1 GHz (30 cm wavelength).

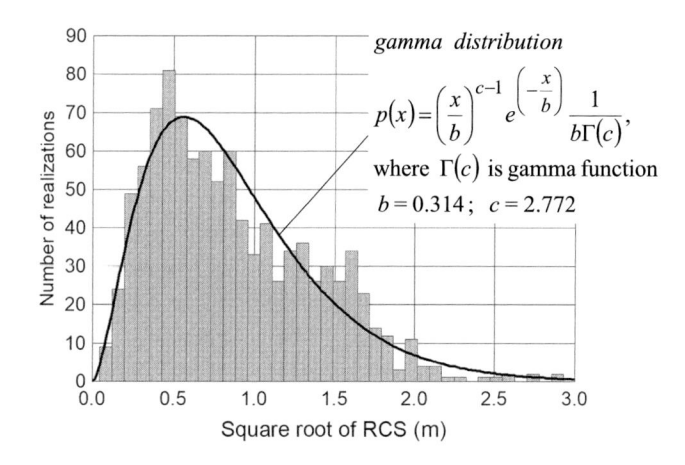

gamma distribution

$$p(x) = \left(\frac{x}{b}\right)^{c-1} e^{\left(-\frac{x}{b}\right)} \frac{1}{b\Gamma(c)},$$

where $\Gamma(c)$ is gamma function
$b = 0.314; \quad c = 2.772$

FIGURE 2.647 Amplitude distribution of echo signal of UAV Dozor-600 model at a carrier frequency of 1 GHz given its horizontal polarization.

2.39 UNMANNED AERIAL VEHICLE KRONSHTADT ORION

Kronshtadt Orion (Figure 2.648) is an unmanned surveillance and reconnaissance aerial vehicle (reconnaissance variant) and unmanned combat aerial vehicle (armed variant). The first flight took place in 2016.

General characteristics of Kronshtadt Orion [50]: wingspan –16.0 m, length – 8.0 m, height – 2.0 m, weight – 500–1 150 kg, powerplant – 1 conventional engine with a pusher propeller, speed – 120–200 km/h, range – 700–1 400 km, and service ceiling – 7 500 m.

In accordance with the design of the Orion, a model of its surface was created to obtaining scattering characteristics (in particular, RCS). The model is shown in Figure 2.649. The model surface was assumed to be perfectly conducting. The smooth parts of the model surface were approximated by parts of 54 triaxial ellipsoids. The surface breaks were modeled using 8 straight edge scattering parts.

FIGURE 2.648 Unmanned aerial vehicle Kronshtadt Orion.

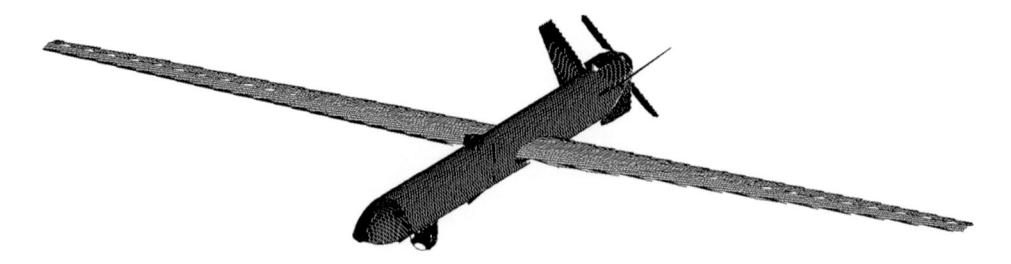

FIGURE 2.649 Surface model of unmanned aerial vehicle Kronshtadt Orion.

Below are some scattering characteristics of the Orion model at sounding frequencies of 10, 3, and 1 GHz (wavelengths of 3, 10, and 30 cm, respectively) for horizontal polarization of the probing signal. The scattering characteristics for this model at two polarizations, as well as scattering characteristics at sounding frequencies of 5 and 1.3 GHz (wavelengths of 6 and 23 cm, respectively), are given in the electronic appendix of this book.

Sounding parameters: The elevation angle is to be a random value distributed uniformly in the range −3° ± 4° with respect to the wing plane (elevation angle of −3° corresponds to the radar observation from the lower hemisphere), azimuth aspect increment was 0.02°, the azimuth being counted

Scattering Characteristics of Aerial Objects

off from the nose-on aspect (0° corresponds to the nose-on radar observation, 180° corresponds to the tail-on observation).

Scattering characteristics of UAV Orion model for sounding frequency 10 GHz (wavelength 3 cm).

Figure 2.650 shows the RCS circular diagram of the Orion model. The noncoherent RCS circular diagram of the Orion is shown in Figure 2.651.

The circular mean RCS of the UAV Orion model for horizontal polarization is 4.439 m². The circular median RCS (the RCS value used to calculate the detection range of an object with a probability of 0.5) for horizontal polarization is 0.628 m².

Figures 2.652 and 2.653 show the mean and median RCS for the main ranges of sounding azimuths (nose, side, and tail) and for ranges of 20°.

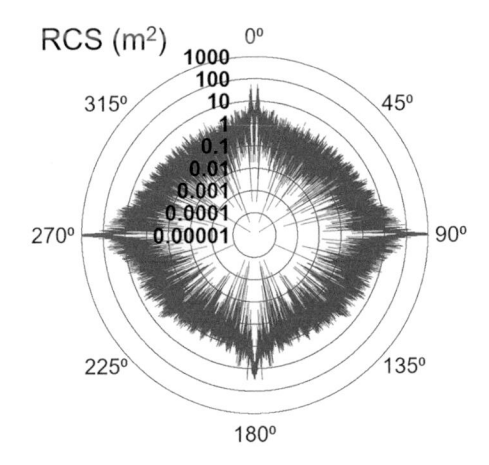

FIGURE 2.650 Circular diagram of RCS given radar observation of UAV Orion model at a carrier frequency of 10 GHz (3 cm wavelength).

FIGURE 2.651 Circular diagram of noncoherent RCS given radar observation of UAV Orion model at a carrier frequency of 10 GHz (3 cm wavelength).

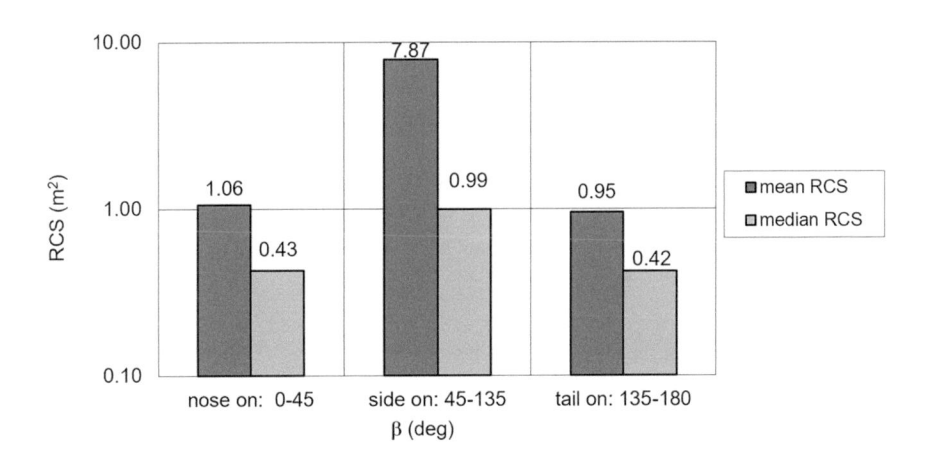

FIGURE 2.652 Diagrams of mean and median RCS of UAV Orion model in three sectors of azimuth aspect given its radar observation at horizontal polarization and a carrier frequency of 10 GHz (3 cm wavelength).

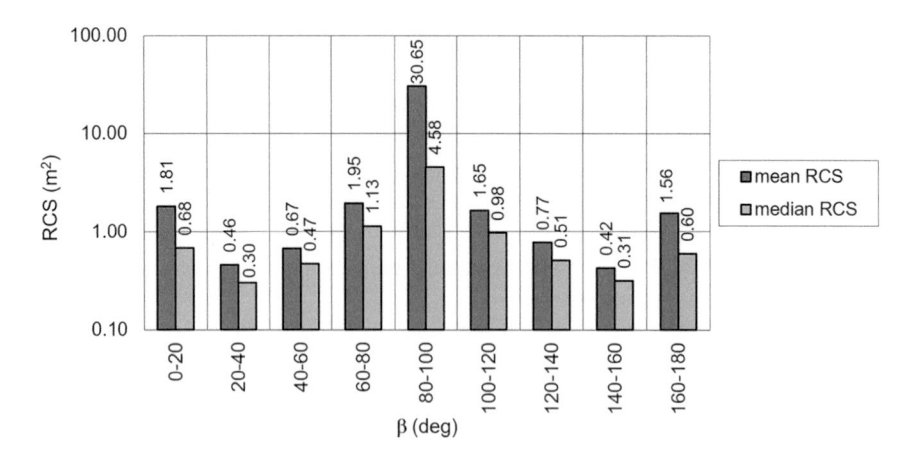

FIGURE 2.653 Diagrams of mean and median RCS of UAV Orion model in 20-degree sectors of azimuth aspect given its radar observation at horizontal polarization and a carrier frequency of 10 GHz (3 cm wavelength).

Figures 2.654, 2.659, and 2.664 show histograms of the scattered signal amplitude (square root of the RCS) for the range of sounding azimuths −20° to +20° (sounding from the nose). The bold line shows the probability density functions of the distribution, which can be used to approximate the histogram of the signal amplitude.

log - normal distribution

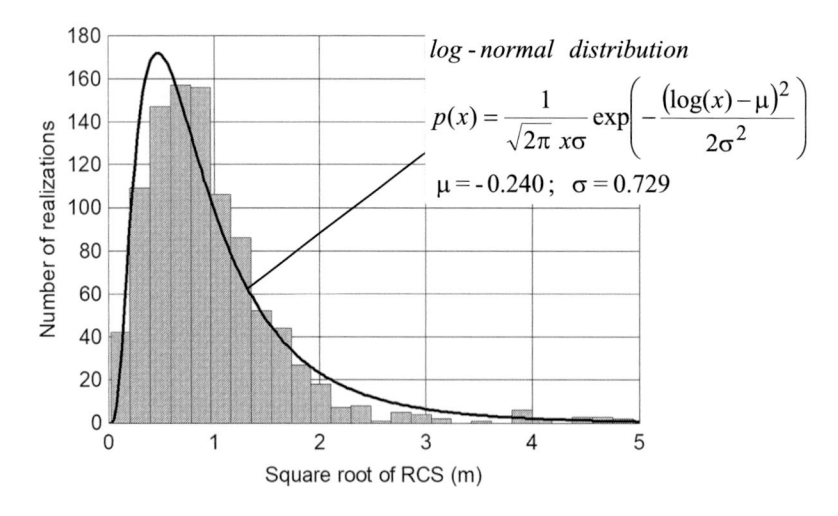

$$p(x) = \frac{1}{\sqrt{2\pi}\, x\sigma} \exp\left(-\frac{(\log(x)-\mu)^2}{2\sigma^2}\right)$$

$\mu = -0.240; \quad \sigma = 0.729$

FIGURE 2.654 Amplitude distribution of echo signal of UAV Orion model at a carrier frequency of 10 GHz given its horizontal polarization.

Scattering characteristics of UAV Orion model for sounding frequency 3 GHz (wavelength 10 cm).

Figure 2.655 shows the RCS circular diagram of the Orion. The noncoherent RCS circular diagram of the Orion is shown in Figure 2.656.

The circular mean RCS of the UAV Orion for horizontal polarization is 3.284 m². The circular median RCS for horizontal polarization is 0.586 m².

Figures 2.657 and 2.658 show the mean and median RCS for the main ranges of sounding azimuths (nose, side, and tail) and for ranges of 20°.

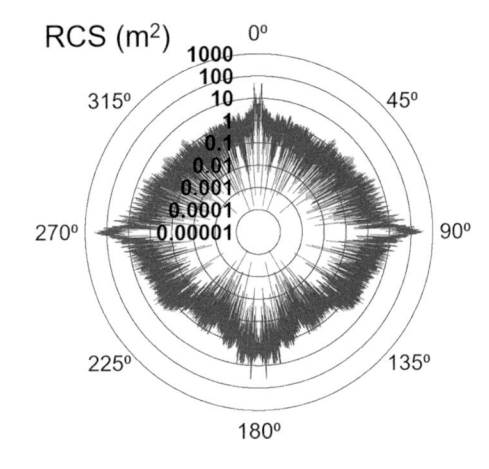

Scattering Characteristics of Aerial Objects

FIGURE 2.655 Circular diagram of RCS given radar observation of UAV Orion model at a carrier frequency of 3 GHz (10 cm wavelength).

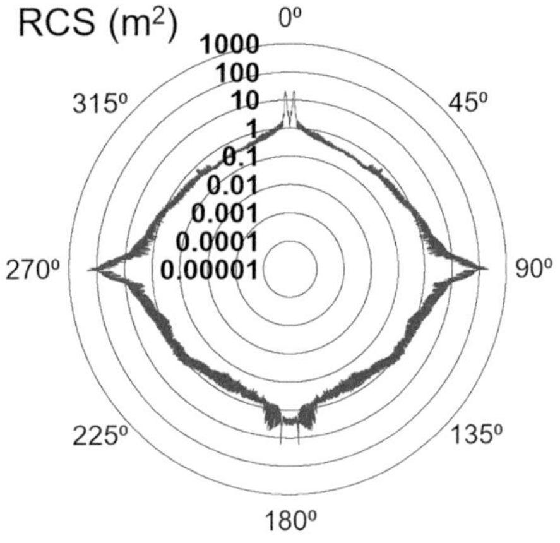

FIGURE 2.656 Circular diagram of noncoherent RCS given radar observation of UAV Orion model at a carrier frequency of 3 GHz (10 cm wavelength).

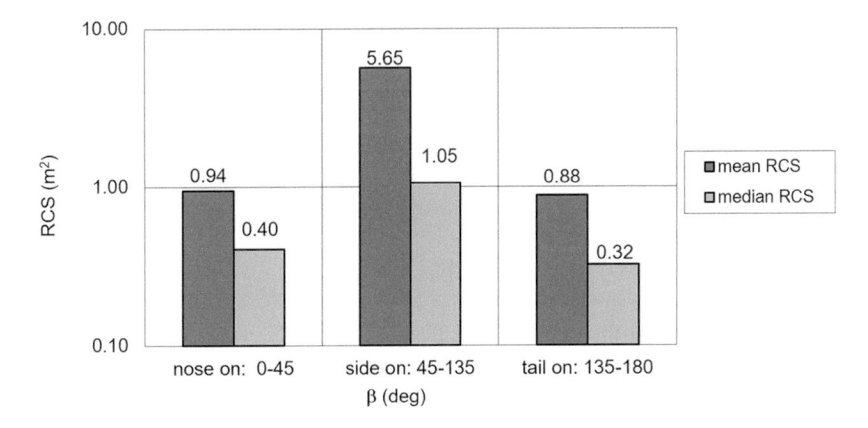

FIGURE 2.657 Diagrams of mean and median RCS of UAV Orion model in three sectors of azimuth aspect given its radar observation at horizontal polarization and a carrier frequency of 3 GHz (10 cm wavelength).

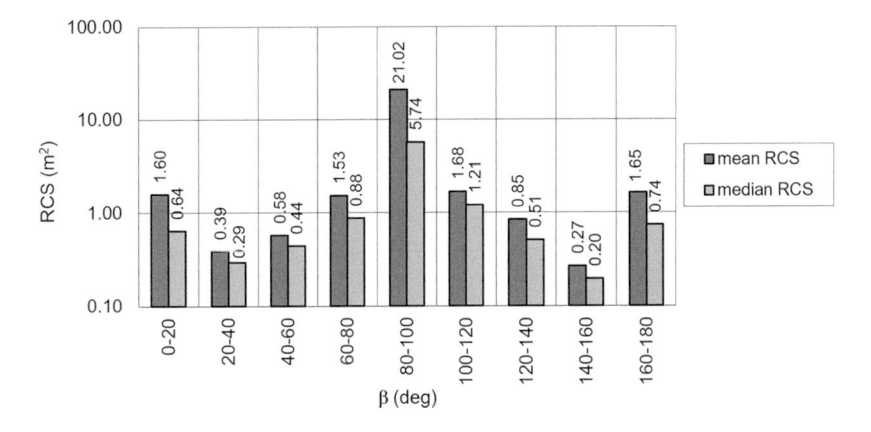

FIGURE 2.658 Diagrams of mean and median RCS of UAV Orion model in 20-degree sectors of azimuth aspect given its radar observation at horizontal polarization and a carrier frequency of 3 GHz (10 cm wavelength).

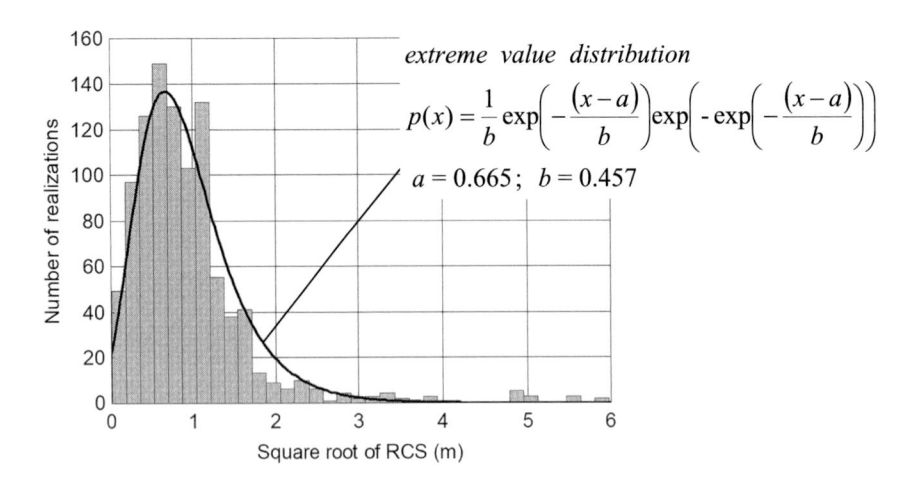

FIGURE 2.659 Amplitude distribution of echo signal of UAV Orion model at a carrier frequency of 3 GHz given its horizontal polarization.

extreme value distribution

$$p(x) = \frac{1}{b} \exp\left(-\frac{(x-a)}{b}\right) \exp\left(-\exp\left(-\frac{(x-a)}{b}\right)\right)$$

$a = 0.665;\ b = 0.457$

Scattering characteristics of UAV Orion model for sounding frequency 1 GHz (wavelength 30 cm).

Figure 2.660 shows the RCS circular diagram of the Orion. The noncoherent RCS circular diagram of the Orion is shown in Figure 2.661.

The circular mean RCS of the UAV Orion for horizontal polarization is 2.193 m². The circular median RCS for horizontal polarization is 0.784 m².

Figures 2.662 and 2.663 show the mean and median RCS for the main ranges of sounding azimuths (nose, side, and tail) and for ranges of 20°.

The expressions and parameters of probability distributions that are most consistent with the empirical distributions of the square root of the RCS for different ranges of sounding azimuths and polarizations are given in the electronic appendix to this book. The expressions and parameters of the probability distributions that are most consistent with the empirical distributions of the RCS (energy characteristic) for different ranges of sounding azimuths and polarizations are also given there.

FIGURE 2.660 Circular diagram of RCS given radar observation of UAV Orion model at a carrier frequency of 1 GHz (30 cm wavelength).

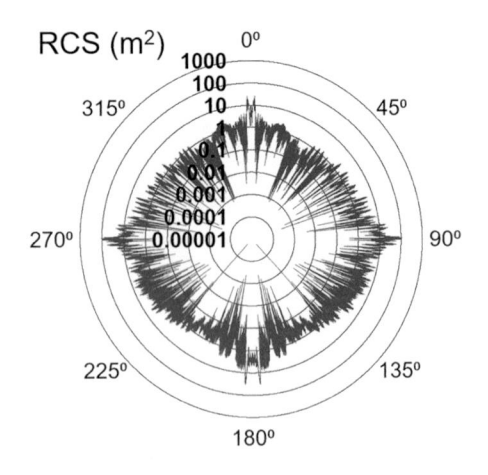

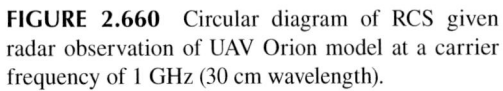

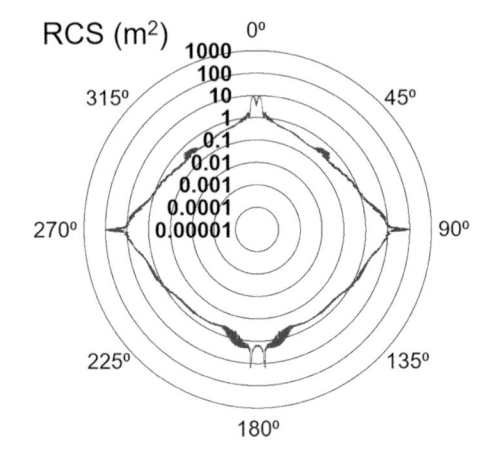

FIGURE 2.661 Circular diagram of noncoherent RCS given radar observation of UAV Orion model at a carrier frequency of 1 GHz (30 cm wavelength).

Scattering Characteristics of Aerial Objects

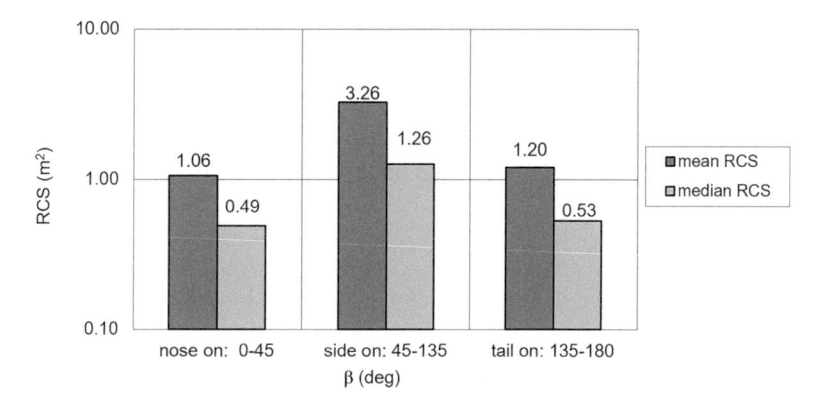

FIGURE 2.662 Diagrams of mean and median RCS of UAV Orion model in three sectors of azimuth aspect given its radar observation at horizontal polarization and a carrier frequency of 1 GHz (30 cm wavelength).

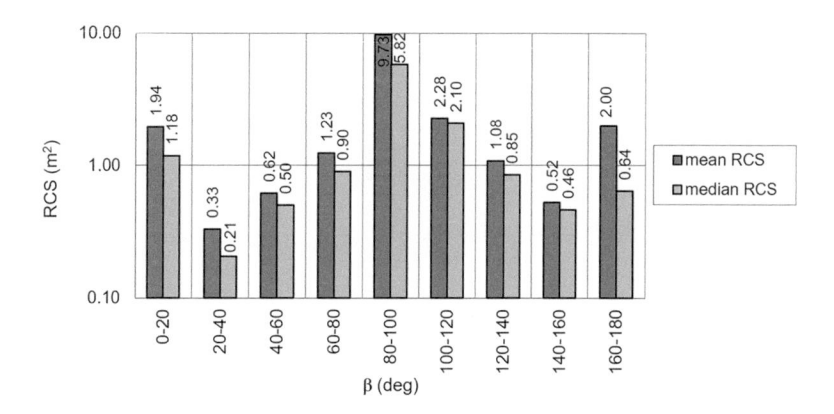

FIGURE 2.663 Diagrams of mean and median RCS of UAV Orion model in 20-degree sectors of azimuth aspect given its radar observation at horizontal polarization and a carrier frequency of 1 GHz (30 cm wavelength).

$$p(x) = \frac{1}{\sqrt{2\pi}\, x\sigma} \exp\left(-\frac{(\log(x)-\mu)^2}{2\sigma^2}\right)$$

log-normal distribution

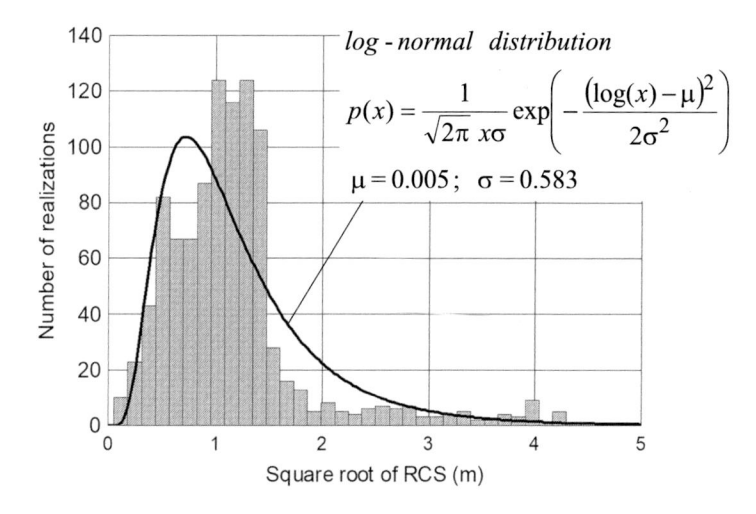

$\mu = 0.005;\ \sigma = 0.583$

FIGURE 2.664 Amplitude distribution of echo signal of UAV Orion model at a carrier frequency of 1 GHz given its horizontal polarization.

2.40 LOITERING MUNITION SHAHED 136

Shahed 136 (Figure 2.665) is loitering munition in the form of an autonomous pusher-prop drone. The first flight took place in 2019.

General characteristics of Shahed 136 [51]: wingspan –3.0 m, length – 3.25 m, weight – 200 kg, powerplant – 1 MADO MD-550 piston engine (Limbach L550E clone), speed – 180 km/h, range – 1 000–2 000 km, and service ceiling – 4 000 m.

In accordance with the design of the Shahed 136, a model of its surface was created to obtaining scattering characteristics (in particular, RCS). The model is shown in Figure 2.666. The model surface was assumed to be perfectly conducting, except of wood propeller and the vertical planes at the ends of the wing, which is made of plastic. The smooth parts of the perfectly conducting model surface were approximated by parts of 112 triaxial ellipsoids. The surface breaks were modeled using 14 straight edge scattering parts. The propeller and vertical plastic planes were approximated by parts of 8 triaxial ellipsoids.

FIGURE 2.665 Loitering munition Shahed 136.

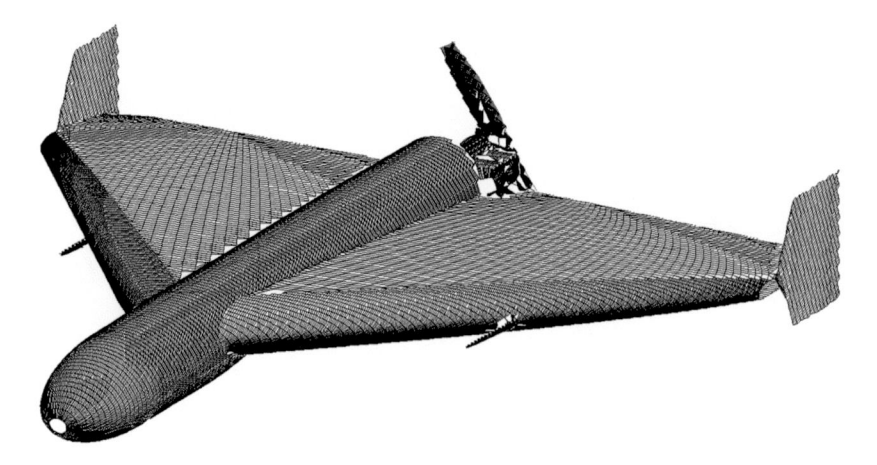

FIGURE 2.666 Surface model of loitering munition Shahed 136.

Below are some scattering characteristics of the Shahed 136 model at sounding frequencies of 10, 3, and 1 GHz (wavelengths of 3, 10, and 30 cm, respectively) for horizontal polarization of the probing signal. The scattering characteristics for this model at two polarizations, as well as scattering characteristics at sounding frequencies of 5 and 1.3 GHz (wavelengths of 6 and 23 cm, respectively), are given in the electronic appendix of this book.

Scattering Characteristics of Aerial Objects   247

Sounding parameters: The elevation angle is to be a random value distributed uniformly in the range $-3° \pm 4°$ with respect to the wing plane (elevation angle of $-3°$ corresponds to the radar observation from the lower hemisphere), azimuth aspect increment was $0.02°$, the azimuth being counted off from the nose-on aspect ($0°$ corresponds to the nose-on radar observation, $180°$ corresponds to the tail-on observation).

Scattering characteristics of Shahed 136 model for sounding frequency 10 GHz (wavelength 3 cm).

Figure 2.667 shows the RCS circular diagram of the Shahed 136 model. The noncoherent RCS circular diagram of the Shahed 136 is shown in Figure 2.668.

The circular mean RCS of the Shahed 136 model for horizontal polarization is 0.627 m². The circular median RCS (the RCS value used to calculate the detection range of an object with a probability of 0.5) for horizontal polarization is 0.200 m².

Figures 2.669 and 2.670 show the mean and median RCS for the main ranges of sounding azimuths (nose, side, and tail) and for ranges of $20°$.

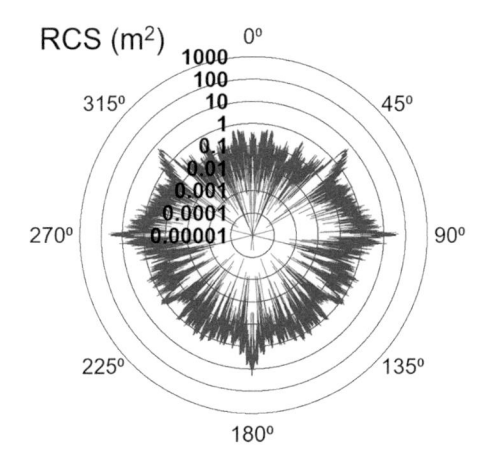

FIGURE 2.667 Circular diagram of RCS given radar observation of Shahed 136 model at a carrier frequency of 10 GHz (3 cm wavelength).

FIGURE 2.668 Circular diagram of noncoherent RCS given radar observation of Shahed 136 model at a carrier frequency of 10 GHz (3 cm wavelength).

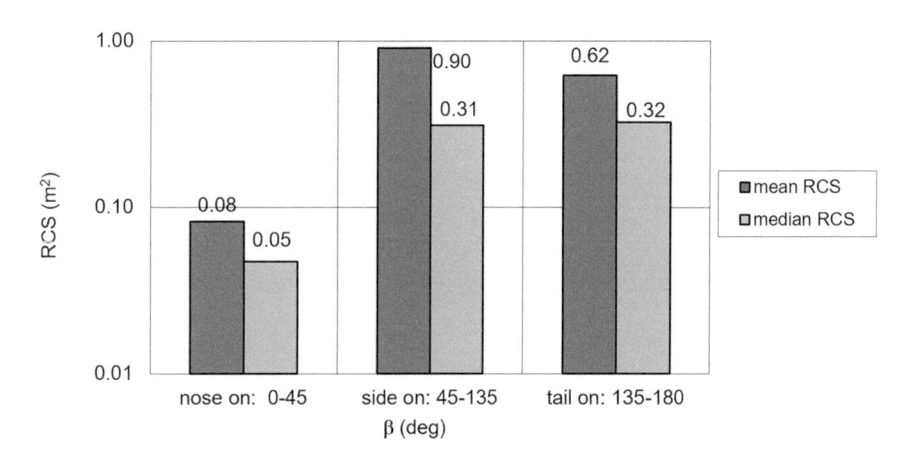

FIGURE 2.669 Diagrams of mean and median RCS of Shahed 136 model in three sectors of azimuth aspect given its radar observation at horizontal polarization and a carrier frequency of 10 GHz (3 cm wavelength).

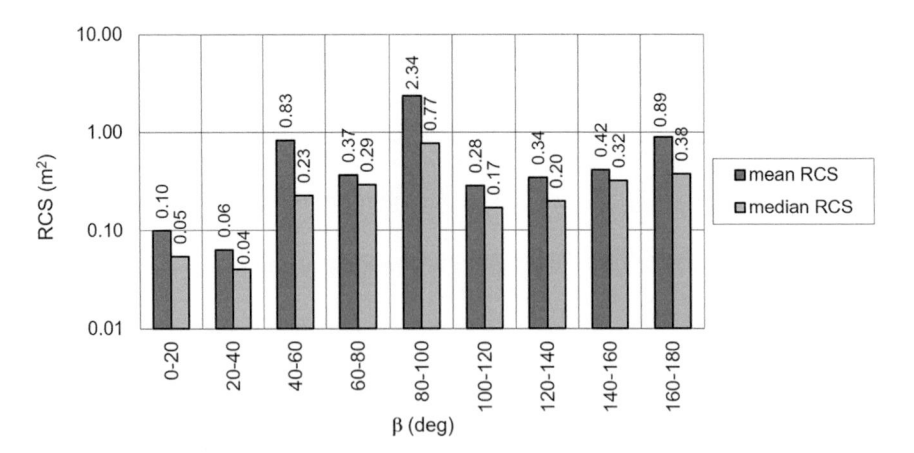

FIGURE 2.670 Diagrams of mean and median RCS of Shahed 136 model in 20-degree sectors of azimuth aspect given its radar observation at horizontal polarization and a carrier frequency of 10 GHz (3 cm wavelength).

Figures 2.671, 2.676, and 2.681 show histograms of the scattered signal amplitude (square root of the RCS) for the range of sounding azimuths −20° to +20° (sounding from the nose). The bold line shows the probability density functions of the distribution, which can be used to approximate the histogram of the signal amplitude.

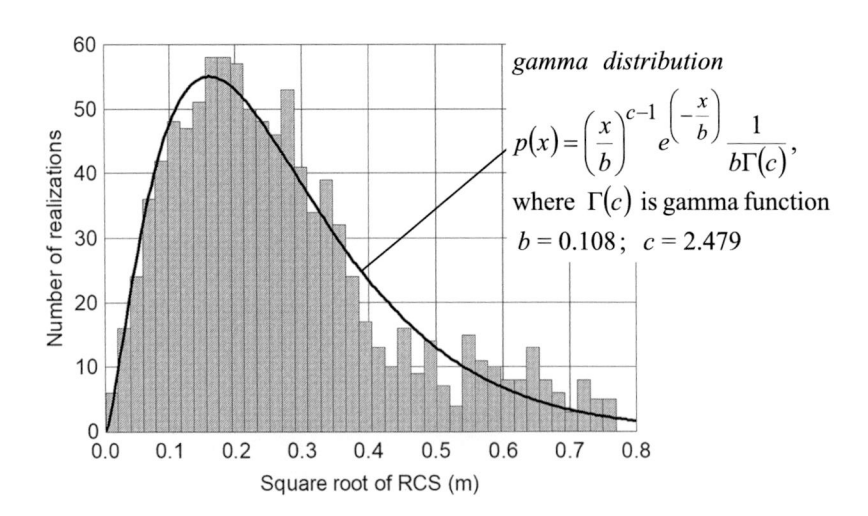

gamma distribution

$$p(x) = \left(\frac{x}{b}\right)^{c-1} e^{\left(-\frac{x}{b}\right)} \frac{1}{b\Gamma(c)},$$

where $\Gamma(c)$ is gamma function
$b = 0.108; \; c = 2.479$

FIGURE 2.671 Amplitude distribution of echo signal of Shahed 136 model at a carrier frequency of 10 GHz given its horizontal polarization.

Scattering characteristics of Shahed 136 model for sounding frequency 3 GHz (wavelength 10 cm).

Figure 2.672 shows the RCS circular diagram of the Shahed 136. The noncoherent RCS circular diagram of the Shahed 136 is shown in Figure 2.673.

The circular mean RCS of the Shahed 136 for horizontal polarization is 0.686 m². The circular median RCS for horizontal polarization is 0.194 m².

Figures 2.674 and 2.674 show the mean and median RCS for the main ranges of sounding azimuths (nose, side, and tail) and for ranges of 20°.

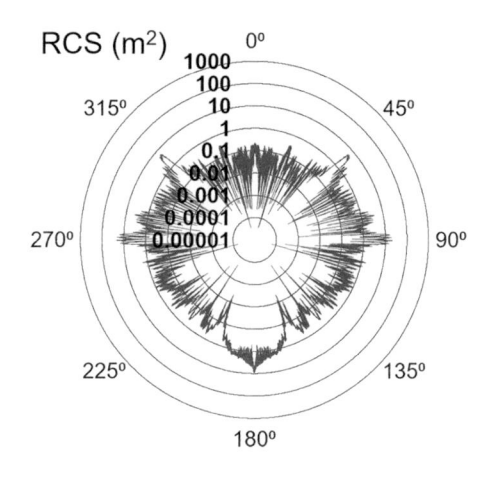

Scattering Characteristics of Aerial Objects

FIGURE 2.672 Circular diagram of RCS given radar observation of Shahed 136 model at a carrier frequency of 3 GHz (10 cm wavelength).

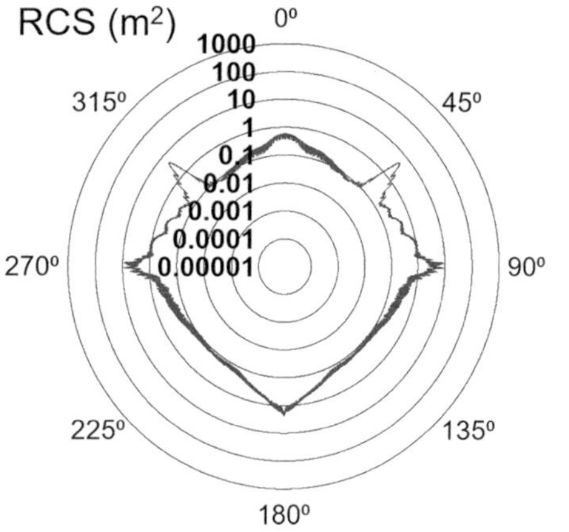

FIGURE 2.673 Circular diagram of noncoherent RCS given radar observation of Shahed 136 model at a carrier frequency of 3 GHz (10 cm wavelength).

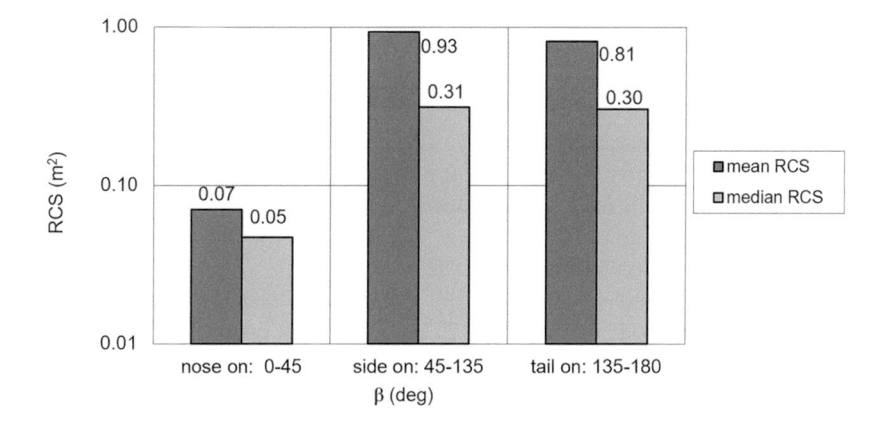

FIGURE 2.674 Diagrams of mean and median RCS of Shahed 136 model in three sectors of azimuth aspect given its radar observation at horizontal polarization and a carrier frequency of 3 GHz (10 cm wavelength).

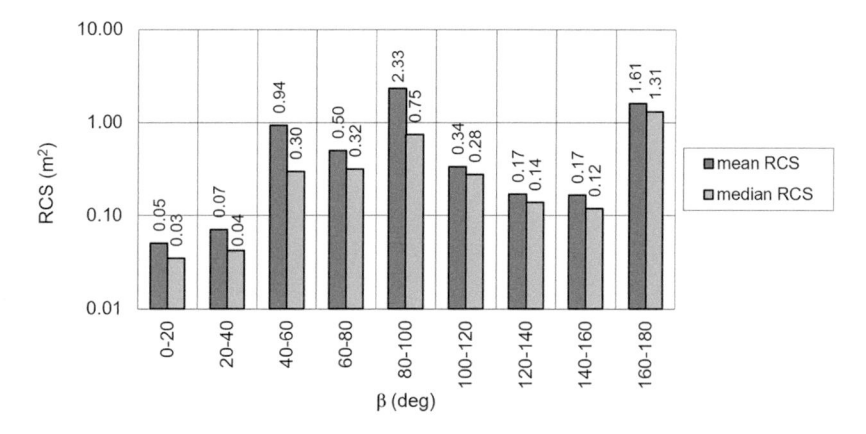

FIGURE 2.675 Diagrams of mean and median RCS of Shahed 136 model in 20-degree sectors of azimuth aspect given its radar observation at horizontal polarization and a carrier frequency of 3 GHz (10 cm wavelength).

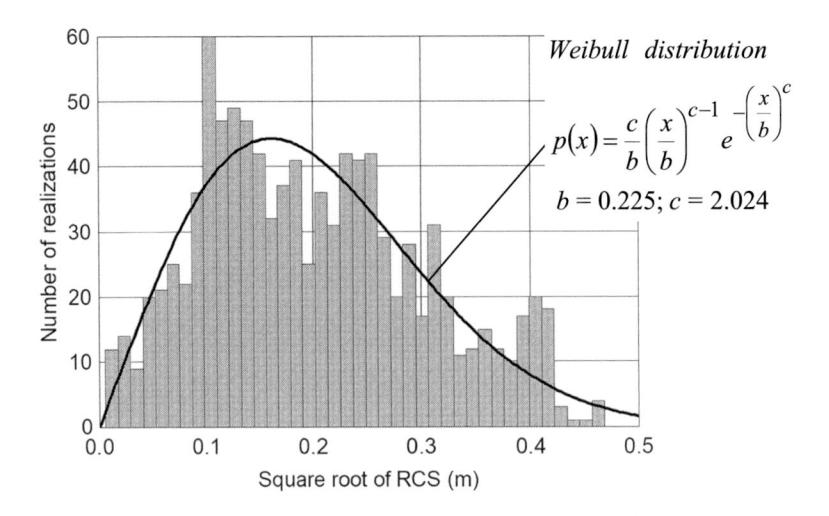

FIGURE 2.676 Amplitude distribution of echo signal of Shahed 136 model at a carrier frequency of 3 GHz given its horizontal polarization.

Scattering characteristics of Shahed 136 model for sounding frequency 1 GHz (wavelength 30 cm).

Figure 2.677 shows the RCS circular diagram of the Shahed 136. The noncoherent RCS circular diagram of the Shahed 136 is shown in Figure 2.678.

The circular mean RCS of the Shahed 136 for horizontal polarization is 0.793 m². The circular median RCS for horizontal polarization is 0.234 m².

Figures 2.679 and 2.680 show the mean and median RCS for the main ranges of sounding azimuths (nose, side, and tail) and for ranges of 20°.

The expressions and parameters of probability distributions that are most consistent with the empirical distributions of the square root of the RCS for different ranges of sounding azimuths and polarizations are given in the electronic appendix to this book. The expressions and parameters of the probability distributions that are most consistent with the empirical distributions of the RCS (energy characteristic) for different ranges of sounding azimuths and polarizations are also given there.

FIGURE 2.677 Circular diagram of RCS given radar observation of Shahed 136 model at a carrier frequency of 1 GHz (30 cm wavelength).

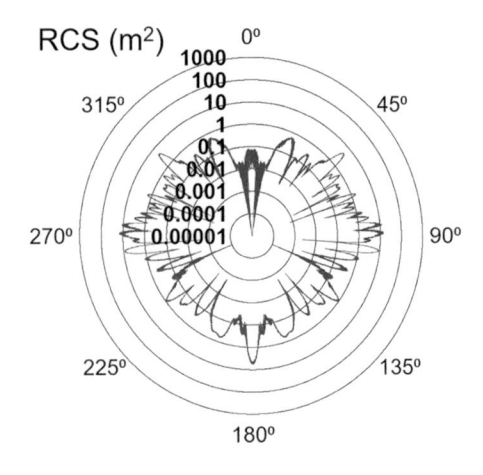

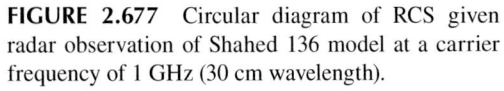

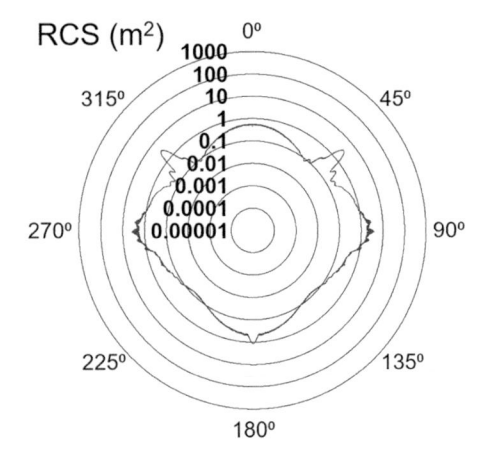

FIGURE 2.678 Circular diagram of noncoherent RCS given radar observation of Shahed 136 model at a carrier frequency of 1 GHz (30 cm wavelength).

Scattering Characteristics of Aerial Objects

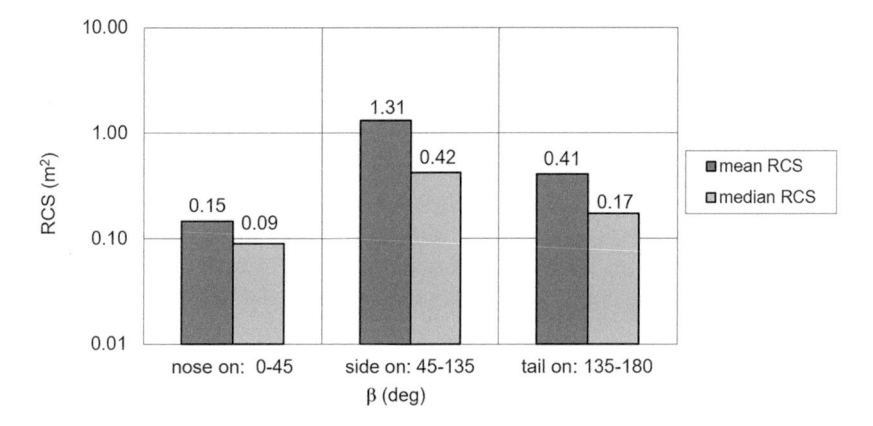

FIGURE 2.679 Diagrams of mean and median RCS of Shahed 136 model in three sectors of azimuth aspect given its radar observation at horizontal polarization and a carrier frequency of 1 GHz (30 cm wavelength).

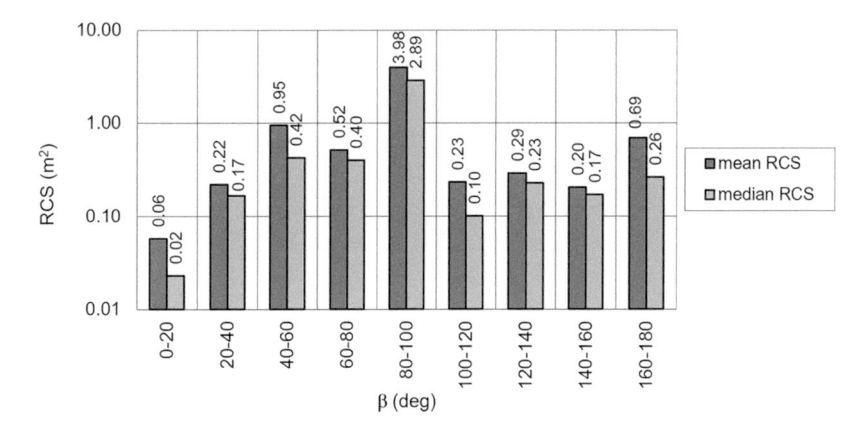

FIGURE 2.680 Diagrams of mean and median RCS of Shahed 136 model in 20-degree sectors of azimuth aspect given its radar observation at horizontal polarization and a carrier frequency of 1 GHz (30 cm wavelength).

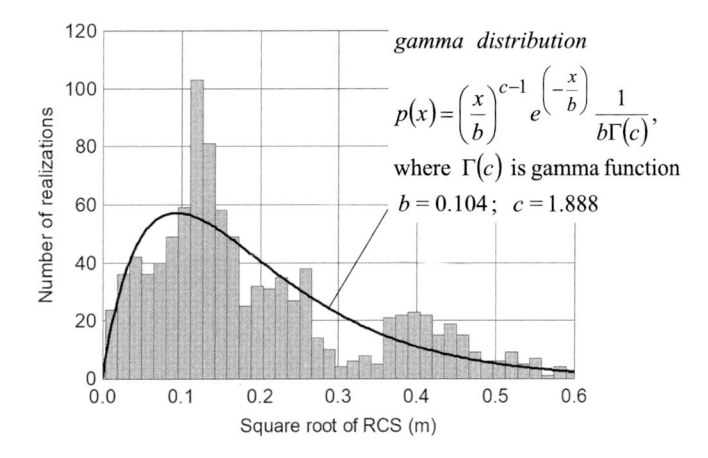

gamma distribution

$$p(x) = \left(\frac{x}{b}\right)^{c-1} e^{\left(-\frac{x}{b}\right)} \frac{1}{b\Gamma(c)},$$

where $\Gamma(c)$ is gamma function

$b = 0.104; \quad c = 1.888$

FIGURE 2.681 Amplitude distribution of echo signal of Shahed 136 model at a carrier frequency of 1 GHz given its horizontal polarization.

2.41 CRUISE MISSILE Kh-555

Kh-555 (Figure 2.682) is an air-launched subsonic cruise missile. The first flight took place in 1999.

General characteristics of Kh-555 [52]: wingspan –3.1 m, length – 5.88–6.04 m, fuselage diameter – 0.51–0.77 m, weight – 1 185–1 465 kg, powerplant – 1 R95TP-300 turbofan, speed – 260 m/s, and range – 2 500–3 500 km.

In accordance with the design of the Kh-555, a model of its surface was created to obtaining scattering characteristics (in particular, RCS). The model is shown in Figure 2.683. In the modeling, the cruise missile metal surfaces were approximated by parts of 33 triaxial ellipsoids. The cruise missile dielectric surfaces were approximated by parts of 11 triaxial ellipsoids.

FIGURE 2.682 Cruise missile Kh-555.

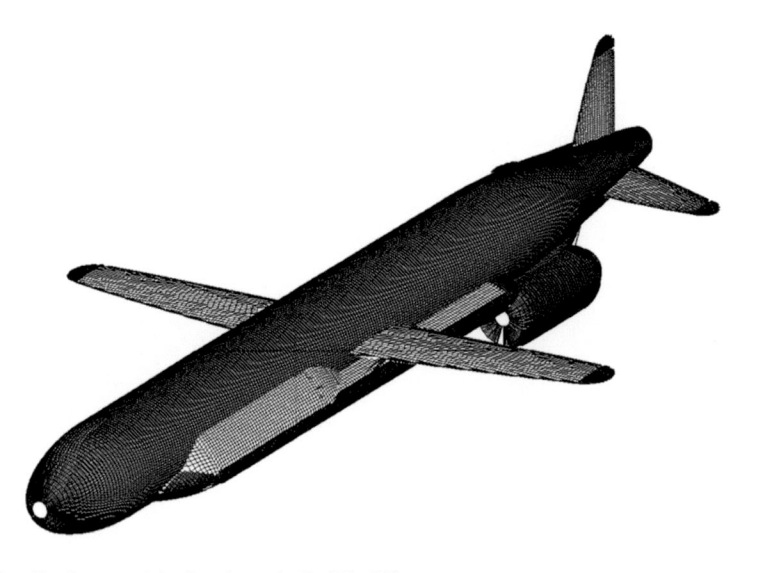

FIGURE 2.683 Surface model of cruise missile Kh-555.

Below are some scattering characteristics of the Kh-555 model at sounding frequencies of 10, 3, and 1 GHz (wavelengths of 3, 10, and 30 cm, respectively) for horizontal polarization of the probing signal. The scattering characteristics for this model at two polarizations, as well as scattering characteristics at sounding frequencies of 5 and 1.3 GHz (wavelengths of 6 and 23 cm, respectively), are given in the electronic appendix of this book.

Scattering Characteristics of Aerial Objects   253

Sounding parameters: The elevation angle is to be a random value distributed uniformly in the range −3° ± 4° with respect to the wing plane (elevation angle of −3° corresponds to the radar observation from the lower hemisphere), azimuth aspect increment was 0.02°, the azimuth being counted off from the nose-on aspect (0° corresponds to the nose-on radar observation, 180° corresponds to the tail-on observation).

Scattering characteristics of Kh-555 model for sounding frequency 10 GHz (wavelength 3 cm).

Figure 2.684 shows the RCS circular diagram of the Kh-555 model. The noncoherent RCS circular diagram of the Kh-555 is shown in Figure 2.685.

The circular mean RCS of the Kh-555 model for horizontal polarization is 1.743 m^2. The circular median RCS (the RCS value used to calculate the detection range of an object with a probability of 0.5) for horizontal polarization is 0.144 m^2.

Figures 2.686 and 2.687 show the mean and median RCS for the main ranges of sounding azimuths (nose, side, and tail) and for ranges of 20°.

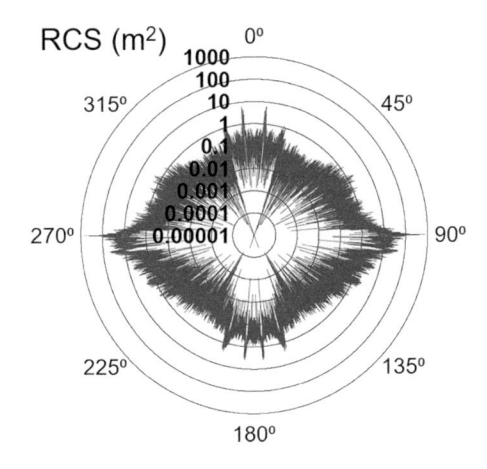

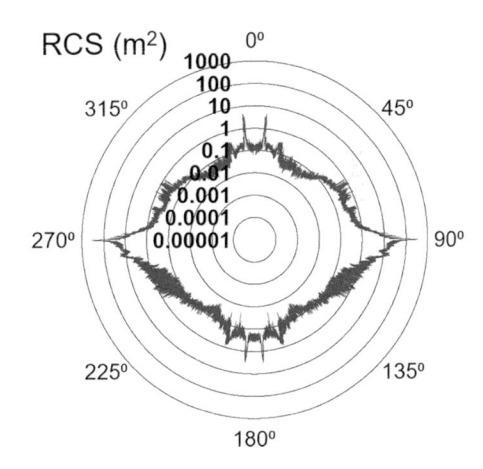

FIGURE 2.684 Circular diagram of RCS given radar observation of Kh-555 model at a carrier frequency of 10 GHz (3 cm wavelength).

FIGURE 2.685 Circular diagram of noncoherent RCS given radar observation of Kh-555 model at a carrier frequency of 10 GHz (3 cm wavelength).

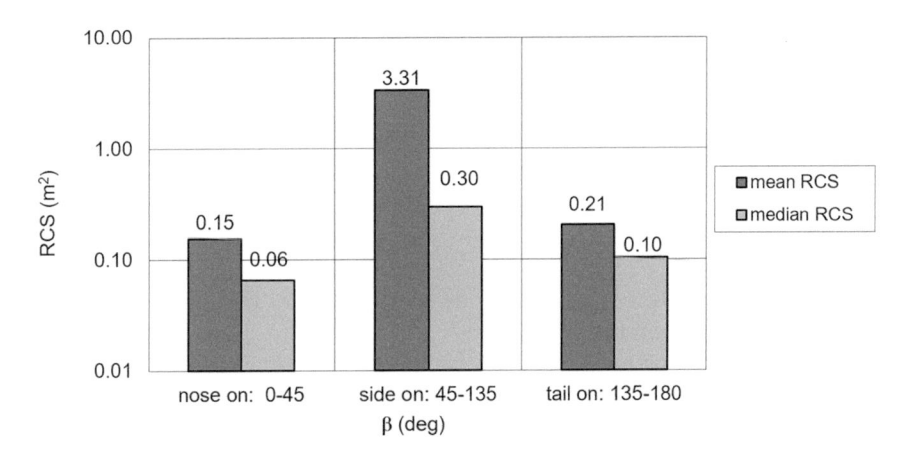

FIGURE 2.686 Diagrams of mean and median RCS of Kh-555 model in three sectors of azimuth aspect given its radar observation at horizontal polarization and a carrier frequency of 10 GHz (3 cm wavelength).

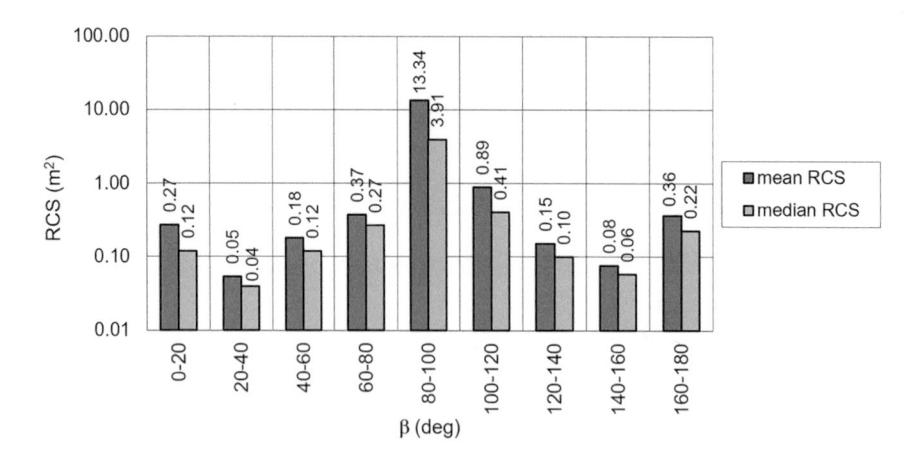

FIGURE 2.687 Diagrams of mean and median RCS of Kh-555 model in 20-degree sectors of azimuth aspect given its radar observation at horizontal polarization and a carrier frequency of 10 GHz (3 cm wavelength).

Figures 2.688, 2.693, and 2.698 show histograms of the scattered signal amplitude (square root of the RCS) for the range of sounding azimuths −20° to +20° (sounding from the nose). The bold line shows the probability density functions of the distribution, which can be used to approximate the histogram of the signal amplitude.

extreme value distribution

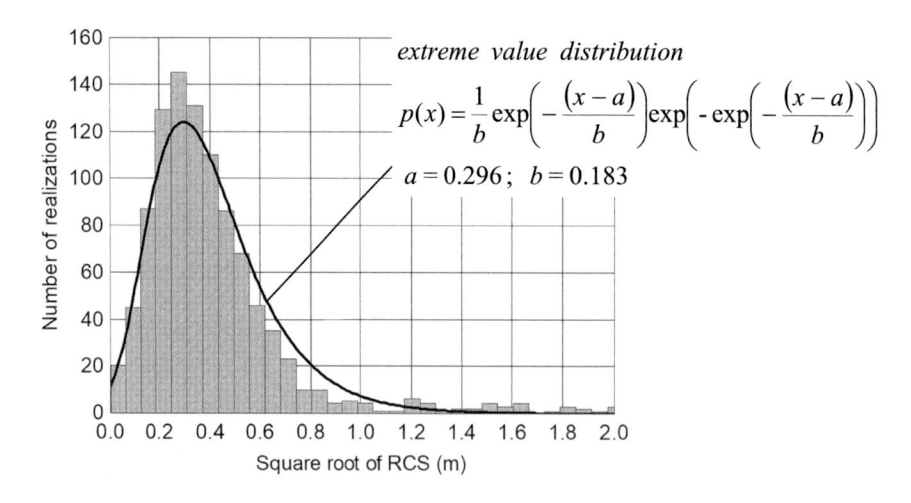

$$p(x) = \frac{1}{b}\exp\left(-\frac{(x-a)}{b}\right)\exp\left(-\exp\left(-\frac{(x-a)}{b}\right)\right)$$

$a = 0.296; \quad b = 0.183$

FIGURE 2.688 Amplitude distribution of echo signal of Kh-555 model at a carrier frequency of 10 GHz given its horizontal polarization.

Scattering characteristics of Kh-555 model for sounding frequency 3 GHz (wavelength 10 cm).

Figure 2.689 shows the RCS circular diagram of the Kh-555. The noncoherent RCS circular diagram of the Kh-555 is shown in Figure 2.690.

The circular mean RCS of the Kh-555 for horizontal polarization is 1.205 m². The circular median RCS for horizontal polarization is 0.101 m².

Figures 2.691 and 2.692 show the mean and median RCS for the main ranges of sounding azimuths (nose, side, and tail) and for ranges of 20°.

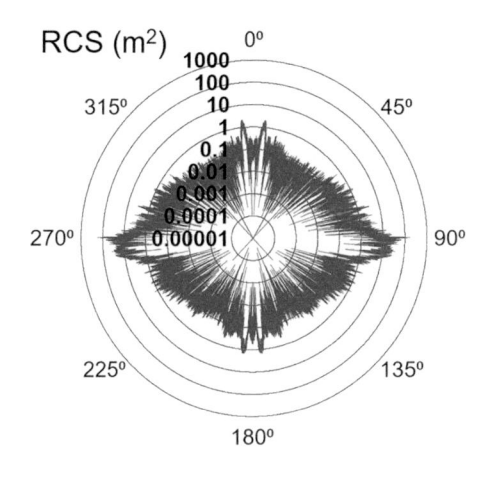

Scattering Characteristics of Aerial Objects

FIGURE 2.689 Circular diagram of RCS given radar observation of Kh-555 model at a carrier frequency of 3 GHz (10 cm wavelength).

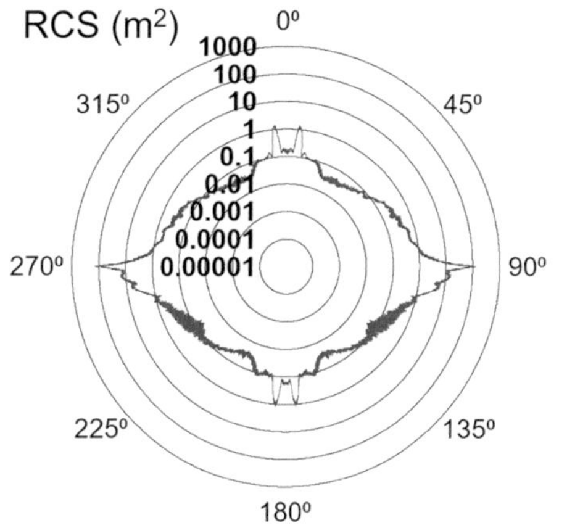

FIGURE 2.690 Circular diagram of noncoherent RCS given radar observation of Kh-555 model at a carrier frequency of 3 GHz (10 cm wavelength).

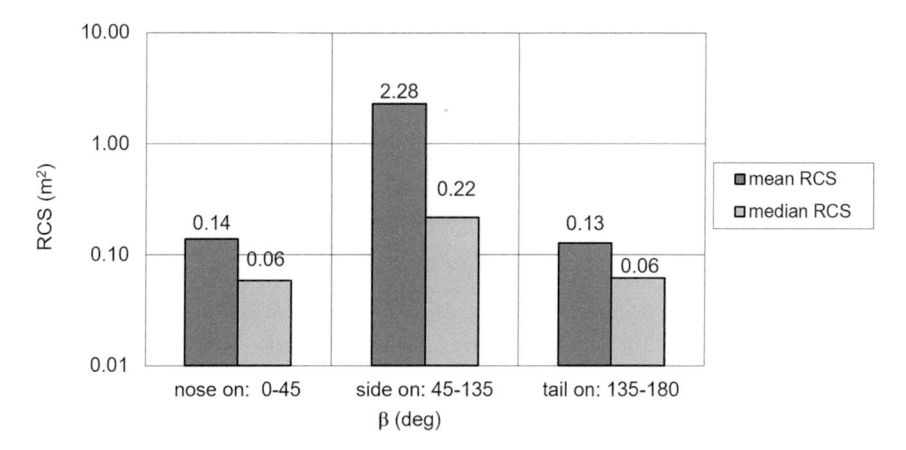

FIGURE 2.691 Diagrams of mean and median RCS of Kh-555 model in three sectors of azimuth aspect given its radar observation at horizontal polarization and a carrier frequency of 3 GHz (10 cm wavelength).

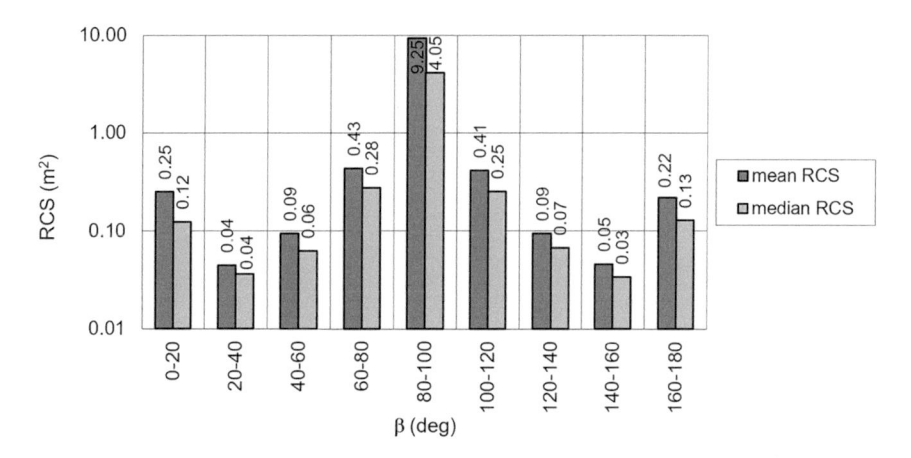

FIGURE 2.692 Diagrams of mean and median RCS of Kh-555 model in 20-degree sectors of azimuth aspect given its radar observation at horizontal polarization and a carrier frequency of 3 GHz (10 cm wavelength).

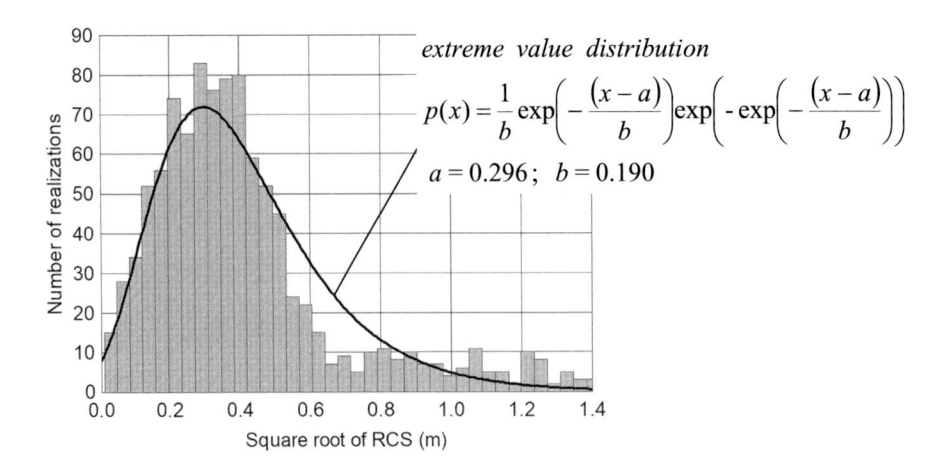

FIGURE 2.693 Amplitude distribution of echo signal of Kh-555 model at a carrier frequency of 3 GHz given its horizontal polarization.

extreme value distribution

$$p(x) = \frac{1}{b} \exp\left(-\frac{(x-a)}{b}\right) \exp\left(-\exp\left(-\frac{(x-a)}{b}\right)\right)$$

$a = 0.296; \quad b = 0.190$

Scattering characteristics of Kh-555 model for sounding frequency 1 GHz (wavelength 30 cm).

Figure 2.694 shows the RCS circular diagram of the Kh-555. The noncoherent RCS circular diagram of the Kh-555 is shown in Figure 2.695.

The circular mean RCS of the Kh-555 for horizontal polarization is 1.124 m². The circular median RCS for horizontal polarization is 0.088 m².

Figures 2.696 and 2.697 show the mean and median RCS for the main ranges of sounding azimuths (nose, side, and tail) and for ranges of 20°.

The expressions and parameters of probability distributions that are most consistent with the empirical distributions of the square root of the RCS for different ranges of sounding azimuths and polarizations are given in the electronic appendix to this book. The expressions and parameters of the probability distributions that are most consistent with the empirical distributions of the RCS (energy characteristic) for different ranges of sounding azimuths and polarizations are also given there.

FIGURE 2.694 Circular diagram of RCS given radar observation of Kh-555 model at a carrier frequency of 1 GHz (30 cm wavelength).

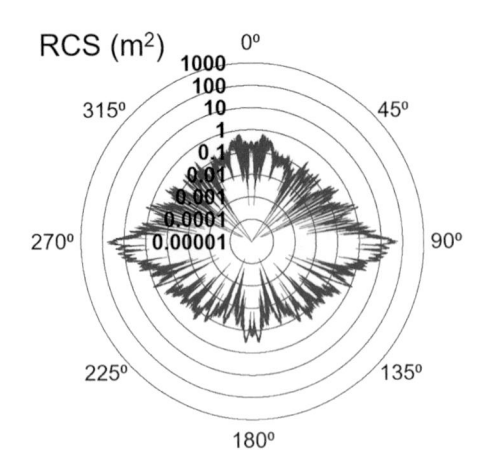

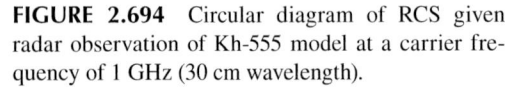

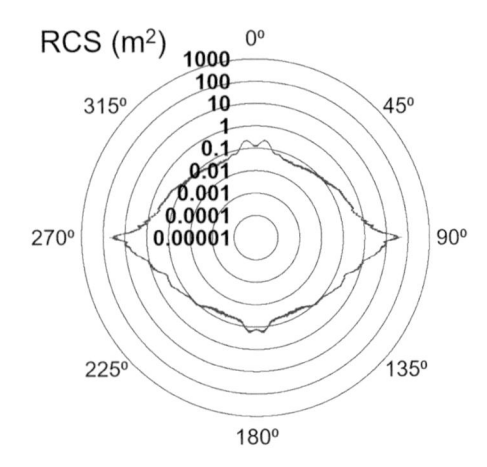

FIGURE 2.695 Circular diagram of noncoherent RCS given radar observation of Kh-555 model at a carrier frequency of 1 GHz (30 cm wavelength).

Scattering Characteristics of Aerial Objects

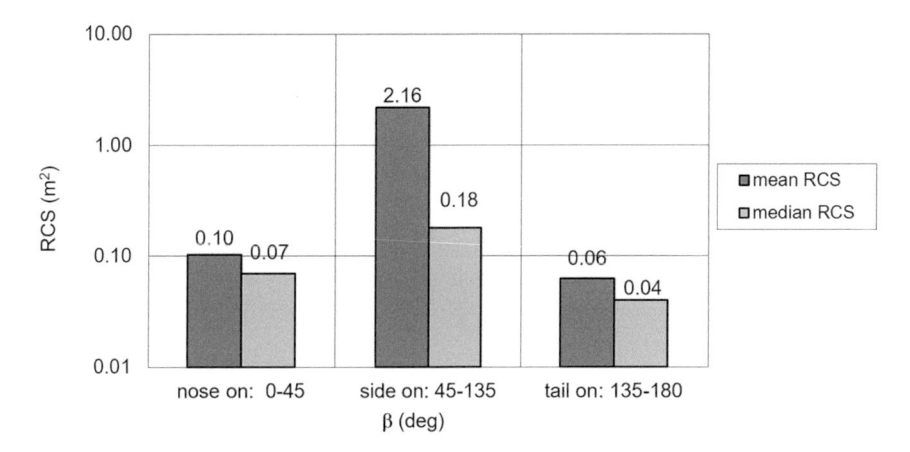

FIGURE 2.696 Diagrams of mean and median RCS of Kh-555 model in three sectors of azimuth aspect given its radar observation at horizontal polarization and a carrier frequency of 1 GHz (30 cm wavelength).

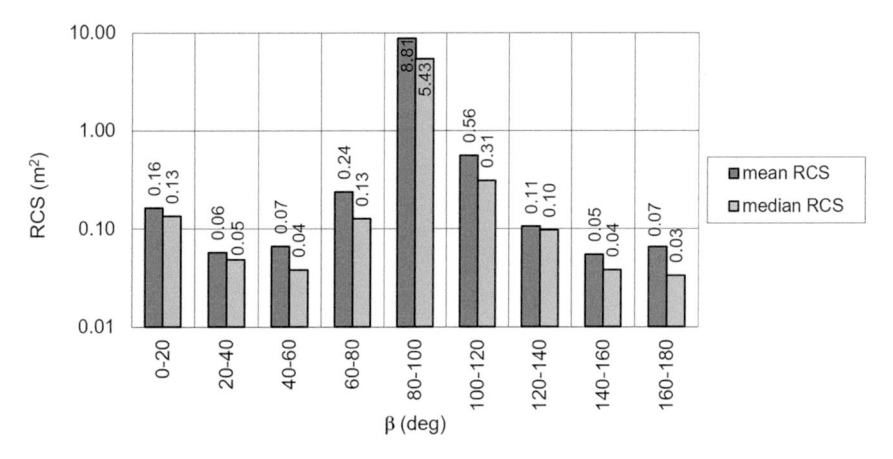

FIGURE 2.697 Diagrams of mean and median RCS of Kh-555 model in 20-degree sectors of azimuth aspect given its radar observation at horizontal polarization and a carrier frequency of 1 GHz (30 cm wavelength).

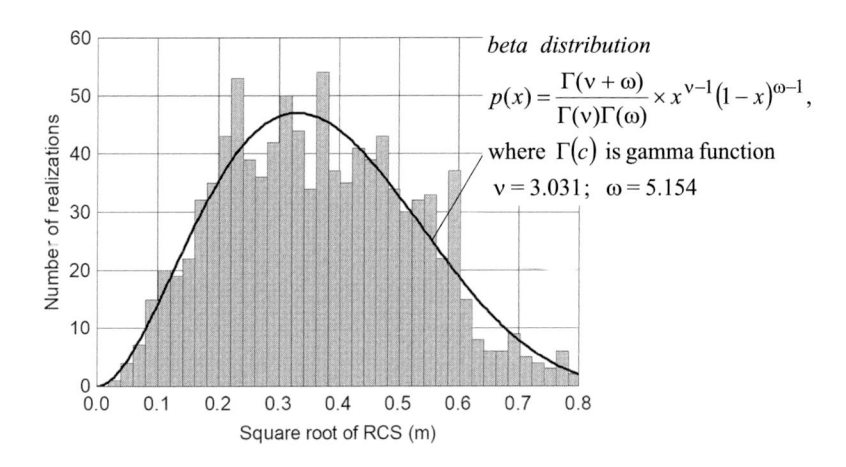

$$p(x) = \frac{\Gamma(\nu + \omega)}{\Gamma(\nu)\Gamma(\omega)} \times x^{\nu-1}(1-x)^{\omega-1},$$

where $\Gamma(c)$ is gamma function

$\nu = 3.031;\ \omega = 5.154$

FIGURE 2.698 Amplitude distribution of echo signal of Kh-555 model at a carrier frequency of 1 GHz given its horizontal polarization.

2.42 CRUISE MISSILE Kh-101

Kh-101 (Figure 2.699) is an air-launched subsonic cruise missile. The first flight took place in 1998.

General characteristics of Kh-101 [53]: wingspan –3 m, length – 7.45 m, fuselage diameter – 0.742 m, weight – 2 200–2 400 kg, powerplant – 1 TRDD-50A or 36MTturbofan, speed – 190–270 m/s, and range – 4 500–5 500 km.

In accordance with the design of the Kh-101, a model of its surface was created to obtaining scattering characteristics (in particular, RCS). The model is shown in Figure 2.700. In the modeling, the cruise missile metal surfaces were approximated by parts of 28 triaxial ellipsoids. The cruise missile dielectric surfaces were approximated by parts of 9 triaxial ellipsoids.

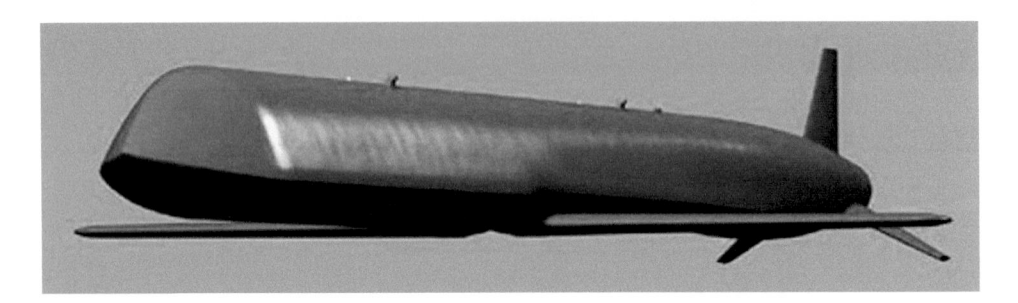

FIGURE 2.699 Cruise missile Kh-101.

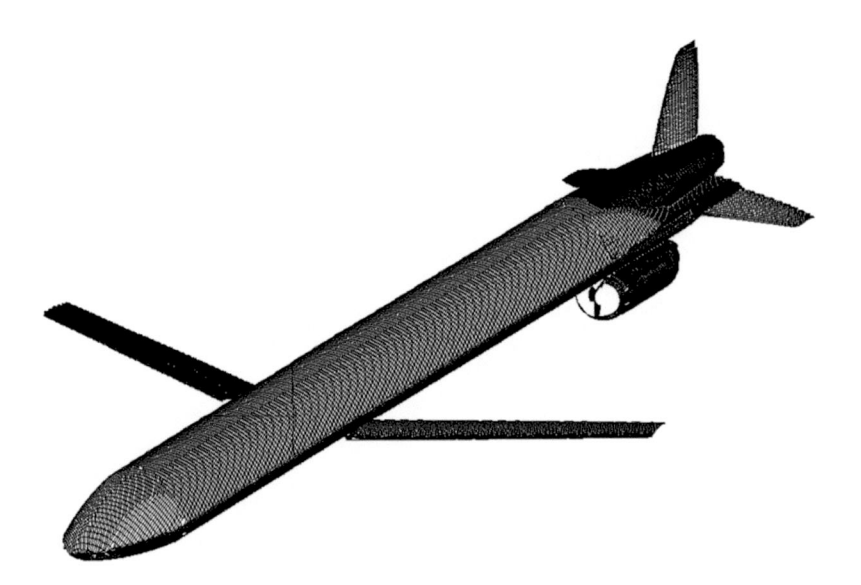

FIGURE 2.700 Surface model of cruise missile Kh-101.

Below are some scattering characteristics of the Kh-101 model at sounding frequencies of 10, 3, and 1 GHz (wavelengths of 3, 10, and 30 cm, respectively) for horizontal polarization of the probing signal. The scattering characteristics for this model at two polarizations, as well as scattering characteristics at sounding frequencies of 5 and 1.3 GHz (wavelengths of 6 and 23 cm, respectively), are given in the electronic appendix of this book.

Sounding parameters: The elevation angle is to be a random value distributed uniformly in the range −3° ± 4° with respect to the wing plane (elevation angle of −3° corresponds to the radar observation from the lower hemisphere), azimuth aspect increment was 0.02°, the azimuth being counted

Scattering Characteristics of Aerial Objects 259

off from the nose-on aspect (0° corresponds to the nose-on radar observation, 180° corresponds to the tail-on observation).

Scattering characteristics of Kh-101 model for sounding frequency 10 GHz (wavelength 3 cm).

Figure 2.701 shows the RCS circular diagram of the Kh-101 model. The noncoherent RCS circular diagram of the Kh-101 is shown in Figure 2.702.

The circular mean RCS of the Kh-101 model for horizontal polarization is 1.757 m^2. The circular median RCS (the RCS value used to calculate the detection range of an object with a probability of 0.5) for horizontal polarization is 0.081 m^2.

Figures 2.703 and 2.704 show the mean and median RCS for the main ranges of sounding azimuths (nose, side, and tail) and for ranges of 20°.

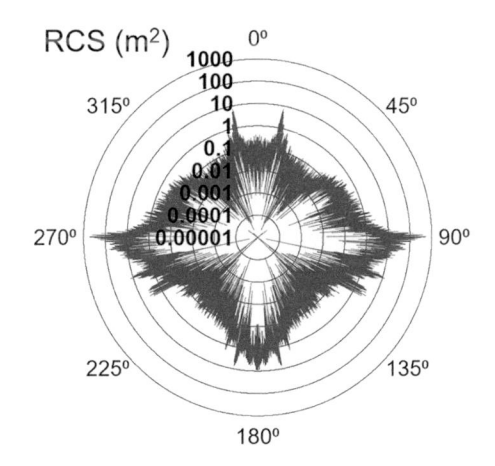

FIGURE 2.701 Circular diagram of RCS given radar observation of Kh-101 model at a carrier frequency of 10 GHz (3 cm wavelength).

FIGURE 2.702 Circular diagram of noncoherent RCS given radar observation of Kh-101 model at a carrier frequency of 10 GHz (3 cm wavelength).

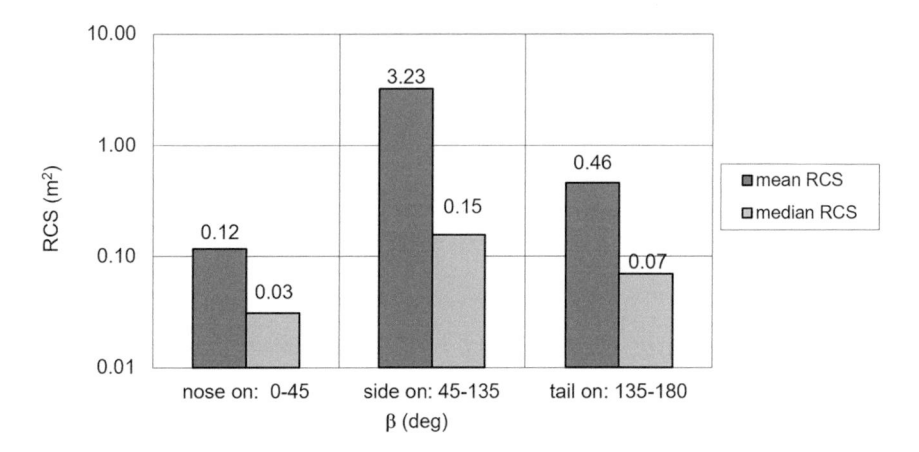

FIGURE 2.703 Diagrams of mean and median RCS of Kh-101 model in three sectors of azimuth aspect given its radar observation at horizontal polarization and a carrier frequency of 10 GHz (3 cm wavelength).

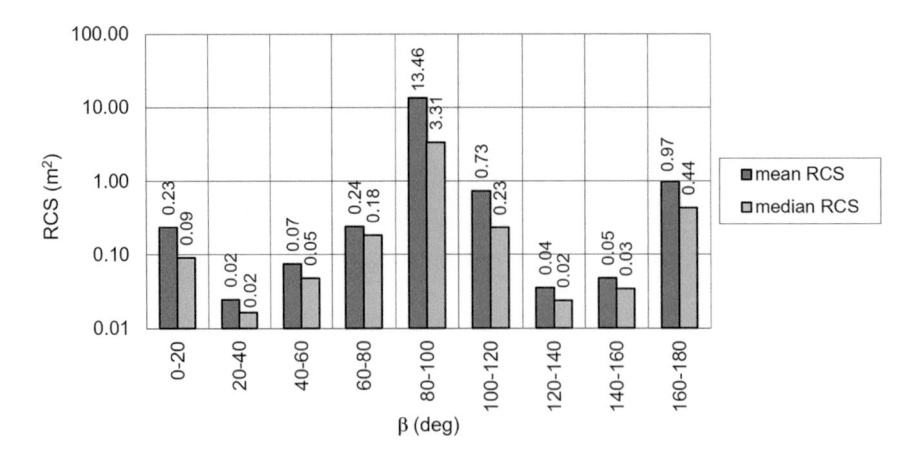

FIGURE 2.704 Diagrams of mean and median RCS of Kh-101 model in 20-degree sectors of azimuth aspect given its radar observation at horizontal polarization and a carrier frequency of 10 GHz (3 cm wavelength).

Figures 2.705, 2.710, and 2.710 show histograms of the scattered signal amplitude (square root of the RCS) for the range of sounding azimuths −20° to +20° (sounding from the nose). The bold line shows the probability density functions of the distribution, which can be used to approximate the histogram of the signal amplitude.

$$p(x) = \frac{1}{\sqrt{2\pi}\, x\sigma} \exp\left(-\frac{(\log(x) - \mu)^2}{2\sigma^2}\right)$$

$\mu = -1.256; \quad \sigma = 0.734$

log - normal distribution

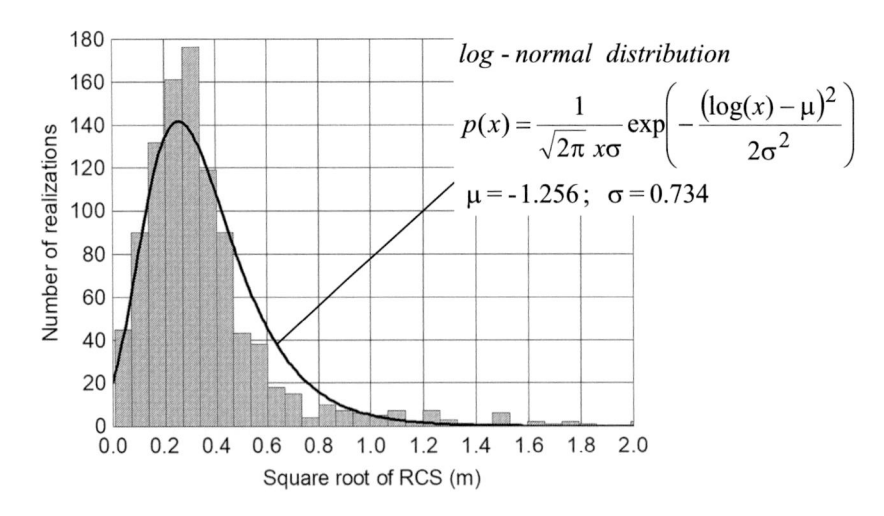

FIGURE 2.705 Amplitude distribution of echo signal of Kh-101 model at a carrier frequency of 10 GHz given its horizontal polarization.

Scattering characteristics of Kh-101 model for sounding frequency 3 GHz (wavelength 10 cm).

Figure 2.706 shows the RCS circular diagram of the Kh-101. The noncoherent RCS circular diagram of the Kh-101 is shown in Figure 2.707.

The circular mean RCS of the Kh-101 for horizontal polarization is 0.863 m². The circular median RCS for horizontal polarization is 0.073 m².

Figures 2.708 and 2.709 show the mean and median RCS for the main ranges of sounding azimuths (nose, side, and tail) and for ranges of 20°.

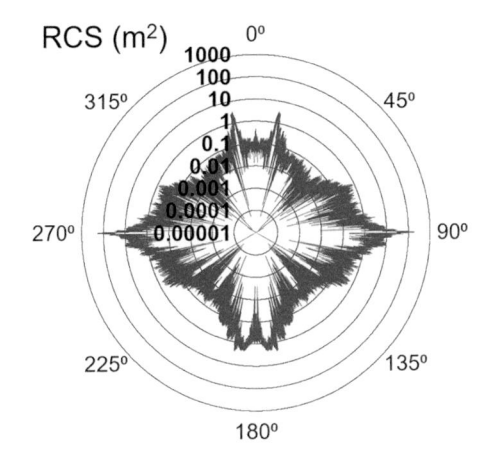

Scattering Characteristics of Aerial Objects

FIGURE 2.706 Circular diagram of RCS given radar observation of Kh-101 model at a carrier frequency of 3 GHz (10 cm wavelength).

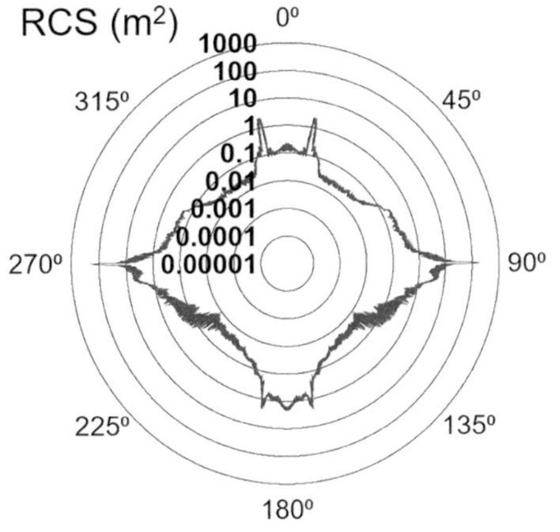

FIGURE 2.707 Circular diagram of noncoherent RCS given radar observation of Kh-101 model at a carrier frequency of 3 GHz (10 cm wavelength).

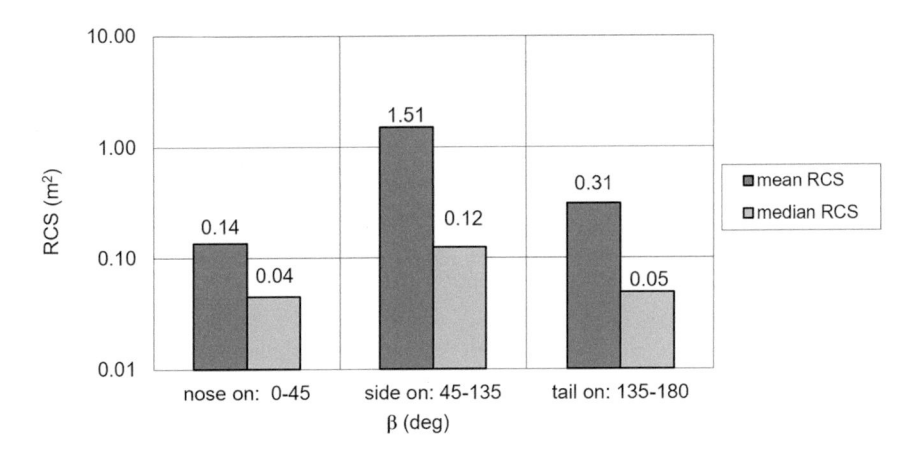

FIGURE 2.708 Diagrams of mean and median RCS of Kh-101 model in three sectors of azimuth aspect given its radar observation at horizontal polarization and a carrier frequency of 3 GHz (10 cm wavelength).

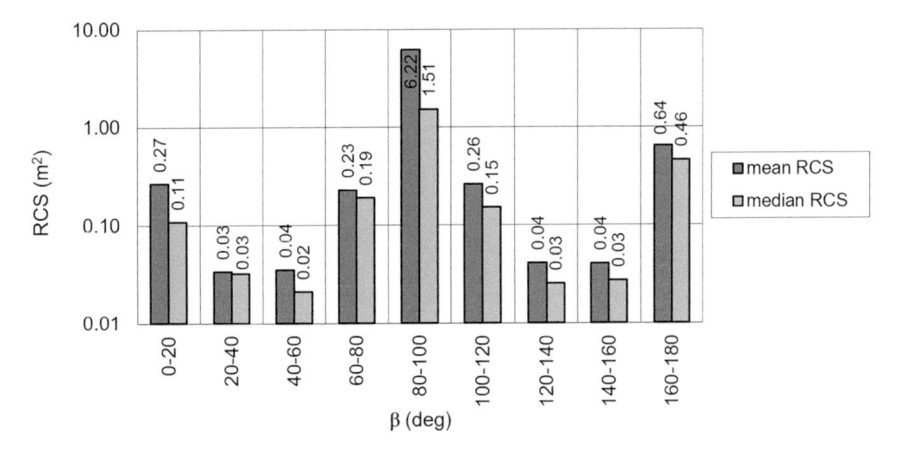

FIGURE 2.709 Diagrams of mean and median RCS of Kh-101 model in 20-degree sectors of azimuth aspect given its radar observation at horizontal polarization and a carrier frequency of 3 GHz (10 cm wavelength).

$$p(x) = \frac{1}{\sqrt{2\pi}\, x\sigma} \exp\left(-\frac{(\log(x) - \mu)^2}{2\sigma^2}\right)$$

log - normal distribution

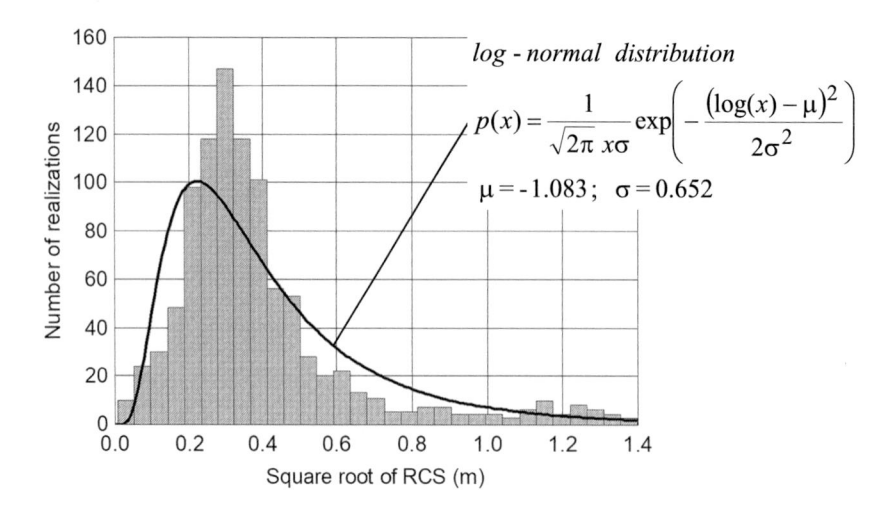

$\mu = -1.083$; $\sigma = 0.652$

FIGURE 2.710 Amplitude distribution of echo signal of Kh-101 model at a carrier frequency of 3 GHz given its horizontal polarization.

Scattering characteristics of Kh-101 model for sounding frequency 1 GHz (wavelength 30 cm).

Figure 2.711 shows the RCS circular diagram of the Kh-101. The noncoherent RCS circular diagram of the Kh-101 is shown in Figure 2.712.

The circular mean RCS of the Kh-101 for horizontal polarization is 0.979 m². The circular median RCS for horizontal polarization is 0.094 m².

Figures 2.713 and 2.714 show the mean and median RCS for the main ranges of sounding azimuths (nose, side, and tail) and for ranges of 20°.

The expressions and parameters of probability distributions that are most consistent with the empirical distributions of the square root of the RCS for different ranges of sounding azimuths and polarizations are given in the electronic appendix to this book. The expressions and parameters of the probability distributions that are most consistent with the empirical distributions of the RCS (energy characteristic) for different ranges of sounding azimuths and polarizations are also given there.

FIGURE 2.711 Circular diagram of RCS given radar observation of Kh-101 model at a carrier frequency of 1 GHz (30 cm wavelength).

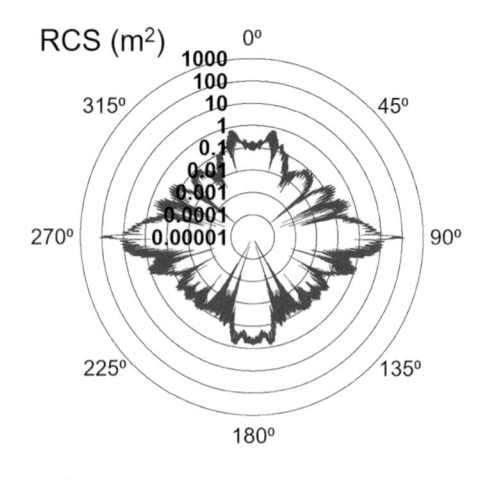

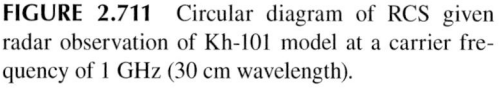

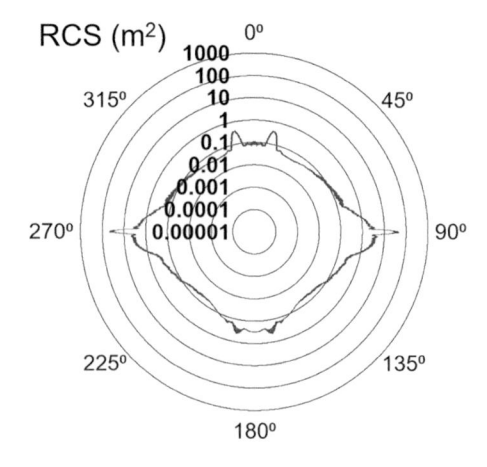

FIGURE 2.712 Circular diagram of noncoherent RCS given radar observation of Kh-101 model at a carrier frequency of 1 GHz (30 cm wavelength).

Scattering Characteristics of Aerial Objects

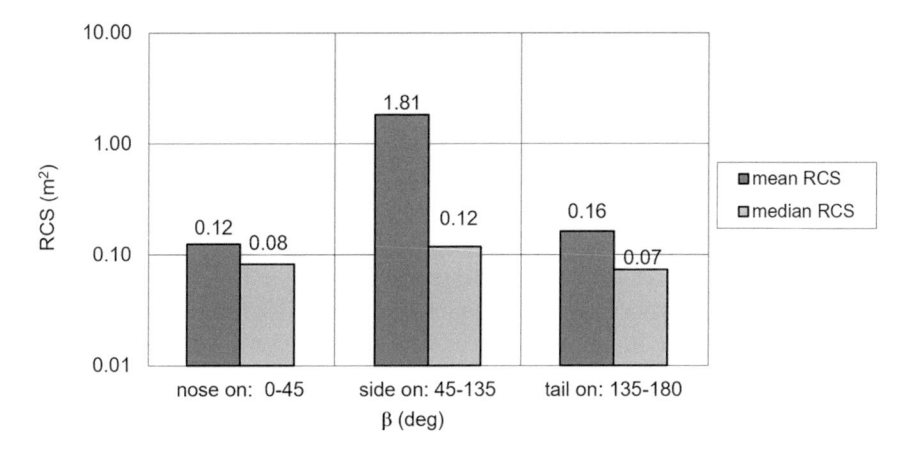

FIGURE 2.713 Diagrams of mean and median RCS of Kh-101 model in three sectors of azimuth aspect given its radar observation at horizontal polarization and a carrier frequency of 1 GHz (30 cm wavelength).

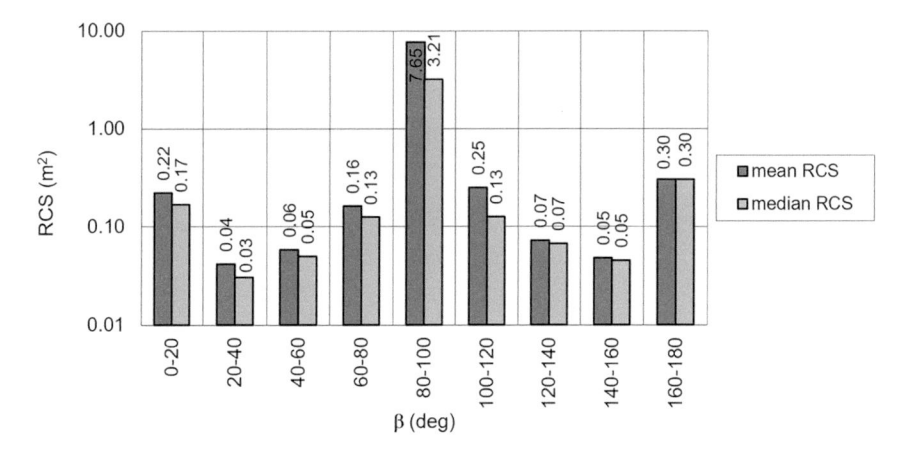

FIGURE 2.714 Diagrams of mean and median RCS of Kh-101 model in 20-degree sectors of azimuth aspect given its radar observation at horizontal polarization and a carrier frequency of 1 GHz (30 cm wavelength).

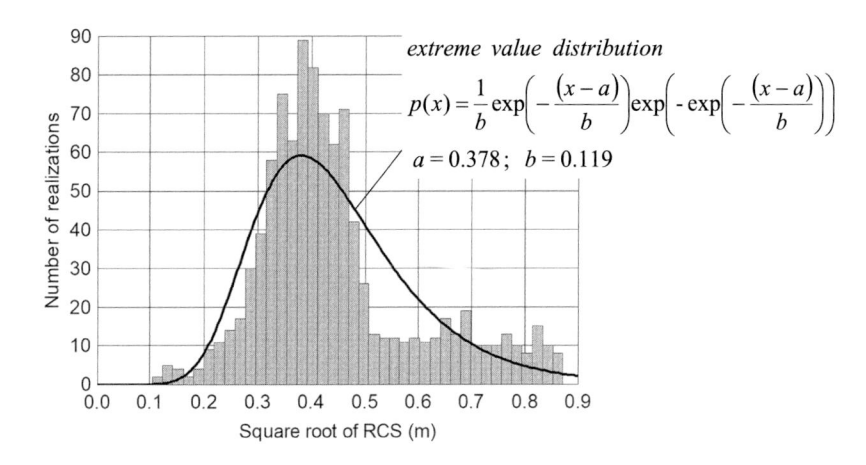

extreme value distribution

$$p(x) = \frac{1}{b}\exp\left(-\frac{(x-a)}{b}\right)\exp\left(-\exp\left(-\frac{(x-a)}{b}\right)\right)$$

$a = 0.378; \quad b = 0.119$

FIGURE 2.715 Amplitude distribution of echo signal of Kh-101 model at a carrier frequency of 1 GHz given its horizontal polarization.

2.43 CRUISE MISSILE P-700 GRANIT

P-700 Granit - SS-N-19 Shipwreck - (Figure 2.716) is a naval anti-ship cruise missile. The first flight took place in 1975–1979.

General characteristics of P-700 Granit [54]: wingspan –2.6 m, length – 10 m, fuselage diameter – 1.14 m, weight – 7 000 kg, powerplant – 1 KR-21–300 turbofan, speed – 1.6–2.5 M, and range – 600 km.

In accordance with the design of the P-700 Granit, a model of its surface was created to obtaining scattering characteristics (in particular, RCS). The model is shown in Figure 2.717. The model surface was assumed to be perfectly conducting. The smooth parts of the model surface were approximated by parts of 23 triaxial ellipsoids. The surface breaks were modeled using 18 straight edge scattering parts.

FIGURE 2.716 Cruise missile P-700 Granit.

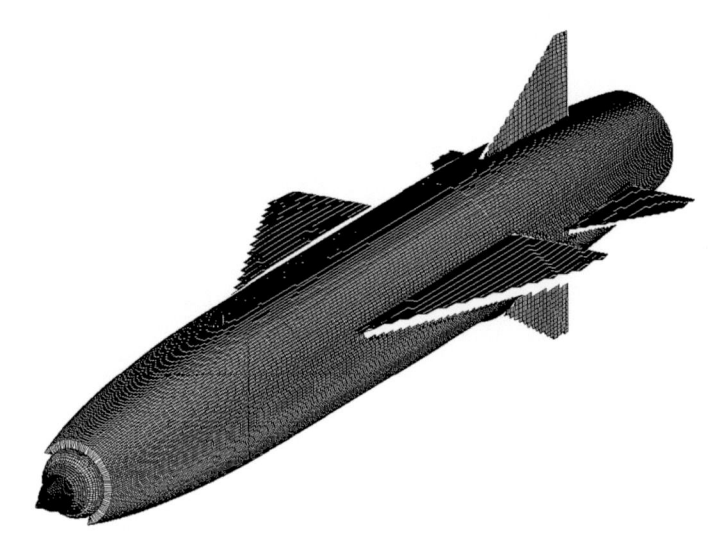

FIGURE 2.717 Surface model of cruise missile P-700 Granit.

Below are some scattering characteristics of the P-700 model at sounding frequencies of 10, 3, and 1 GHz (wavelengths of 3, 10, and 30 cm, respectively) for horizontal polarization of the probing signal. The scattering characteristics for this model at two polarizations, as well as scattering characteristics at sounding frequencies of 5 and 1.3 GHz (wavelengths of 6 and 23 cm, respectively), are given in the electronic appendix of this book.

Scattering Characteristics of Aerial Objects

Sounding parameters: The elevation angle is to be a random value distributed uniformly in the range −3° ± 4° with respect to the wing plane (elevation angle of −3° corresponds to the radar observation from the lower hemisphere), azimuth aspect increment was 0.02°, the azimuth being counted off from the nose-on aspect (0° corresponds to the nose-on radar observation, 180° corresponds to the tail-on observation).

Scattering characteristics of P-700 model for sounding frequency 10 GHz (wavelength 3 cm).
Figure 2.718 shows the RCS circular diagram of the P-700 model. The noncoherent RCS circular diagram of the P-700 is shown in Figure 2.719.

The circular mean RCS of the P-700 model for horizontal polarization is 8.781 m². The circular median RCS (the RCS value used to calculate the detection range of an object with a probability of 0.5) for horizontal polarization is 0.082 m².

Figures 2.720 and 2.721 show the mean and median RCS for the main ranges of sounding azimuths (nose, side, and tail) and for ranges of 20°.

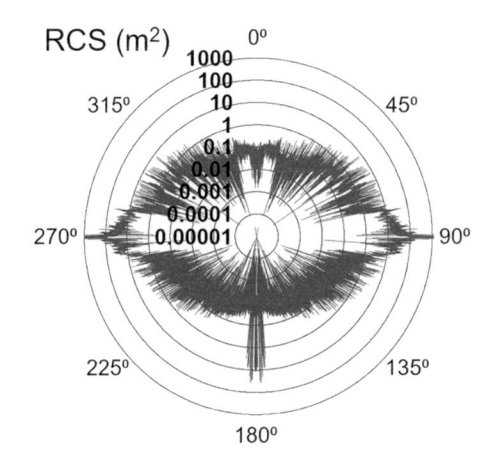

FIGURE 2.718 Circular diagram of RCS given radar observation of P-700 model at a carrier frequency of 10 GHz (3 cm wavelength).

FIGURE 2.719 Circular diagram of noncoherent RCS given radar observation of P-700 model at a carrier frequency of 10 GHz (3 cm wavelength).

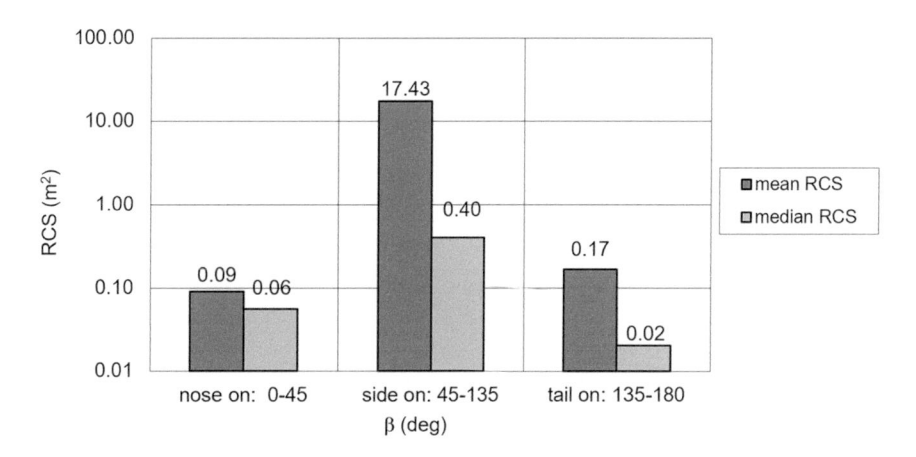

FIGURE 2.720 Diagrams of mean and median RCS of P-700 model in three sectors of azimuth aspect given its radar observation at horizontal polarization and a carrier frequency of 10 GHz (3 cm wavelength).

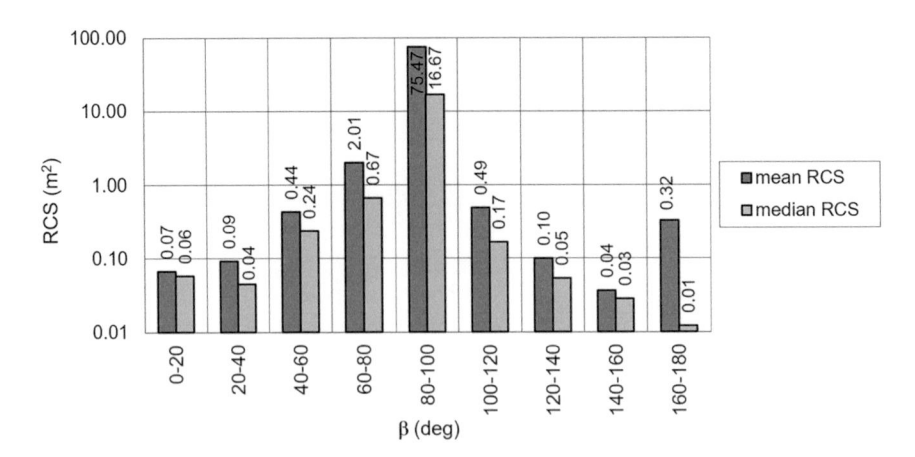

FIGURE 2.721 Diagrams of mean and median RCS of P-700 model in 20-degree sectors of azimuth aspect given its radar observation at horizontal polarization and a carrier frequency of 10 GHz (3 cm wavelength).

Figures 2.722, 2.727, and 2.732 show histograms of the scattered signal amplitude (square root of the RCS) for the range of sounding azimuths −20° to +20° (sounding from the nose). The bold line shows the probability density functions of the distribution, which can be used to approximate the histogram of the signal amplitude.

Weibull distribution

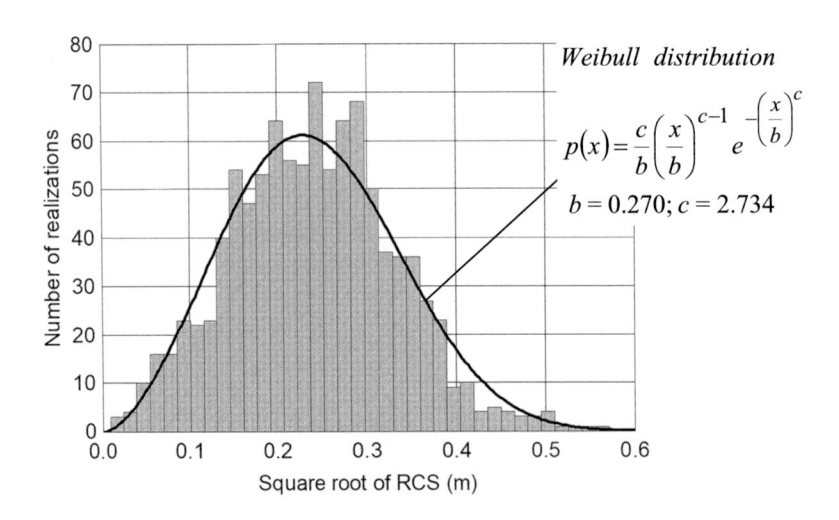

$$p(x) = \frac{c}{b}\left(\frac{x}{b}\right)^{c-1} e^{-\left(\frac{x}{b}\right)^c}$$

$b = 0.270; c = 2.734$

FIGURE 2.722 Amplitude distribution of echo signal of P-700 model at a carrier frequency of 10 GHz given its horizontal polarization.

Scattering characteristics of P-700 model for sounding frequency 3 GHz (wavelength 10 cm).

Figure 2.723 shows the RCS circular diagram of the P-700. The noncoherent RCS circular diagram of the P-700 is shown in Figure 2.724.

The circular mean RCS of the P-700 for horizontal polarization is 8.187 m². The circular median RCS for horizontal polarization is 0.128 m².

Figures 2.725 and 2.726 show the mean and median RCS for the main ranges of sounding azimuths (nose, side, and tail) and for ranges of 20°.

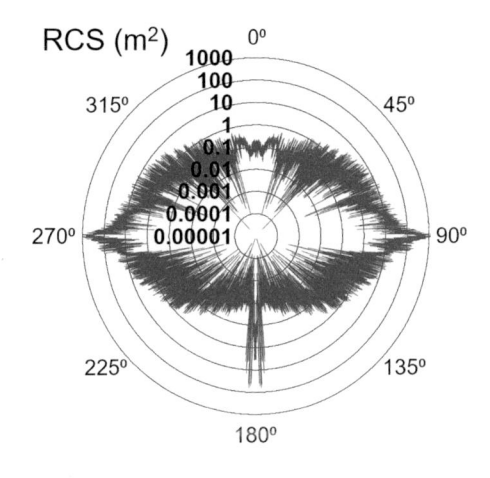

Scattering Characteristics of Aerial Objects

FIGURE 2.723 Circular diagram of RCS given radar observation of P-700 model at a carrier frequency of 3 GHz (10 cm wavelength).

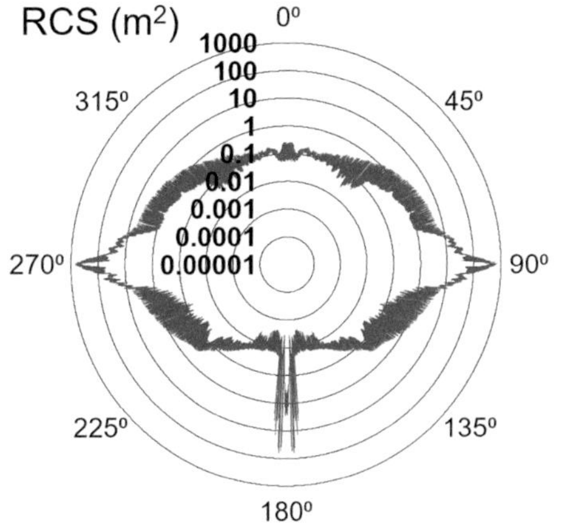

FIGURE 2.724 Circular diagram of noncoherent RCS given radar observation of P-700 model at a carrier frequency of 3 GHz (10 cm wavelength).

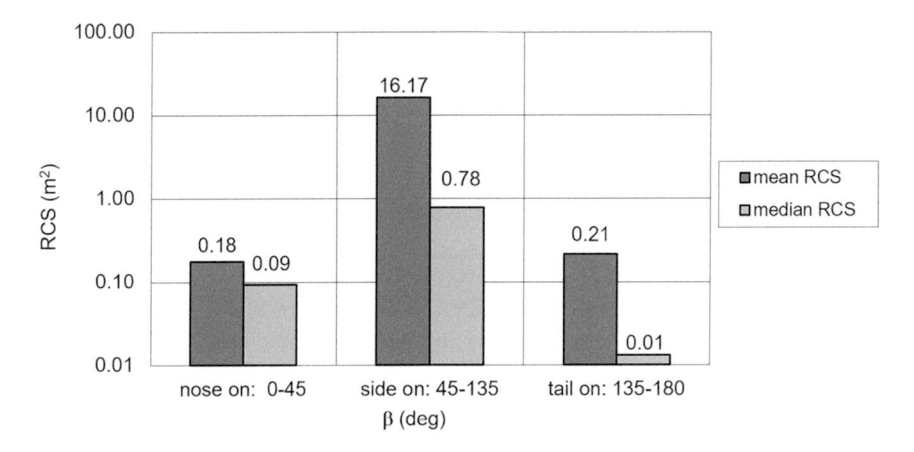

FIGURE 2.725 Diagrams of mean and median RCS of P-700 model in three sectors of azimuth aspect given its radar observation at horizontal polarization and a carrier frequency of 3 GHz (10 cm wavelength).

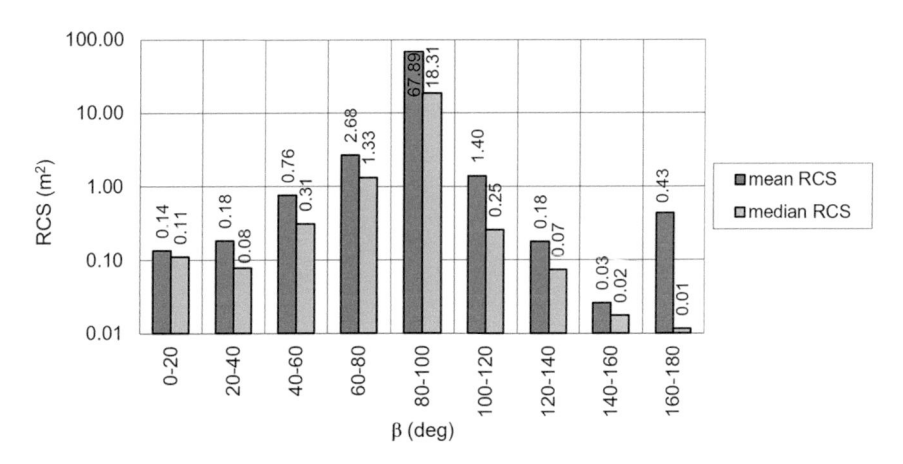

FIGURE 2.726 Diagrams of mean and median RCS of P-700 model in 20-degree sectors of azimuth aspect given its radar observation at horizontal polarization and a carrier frequency of 3 GHz (10 cm wavelength).

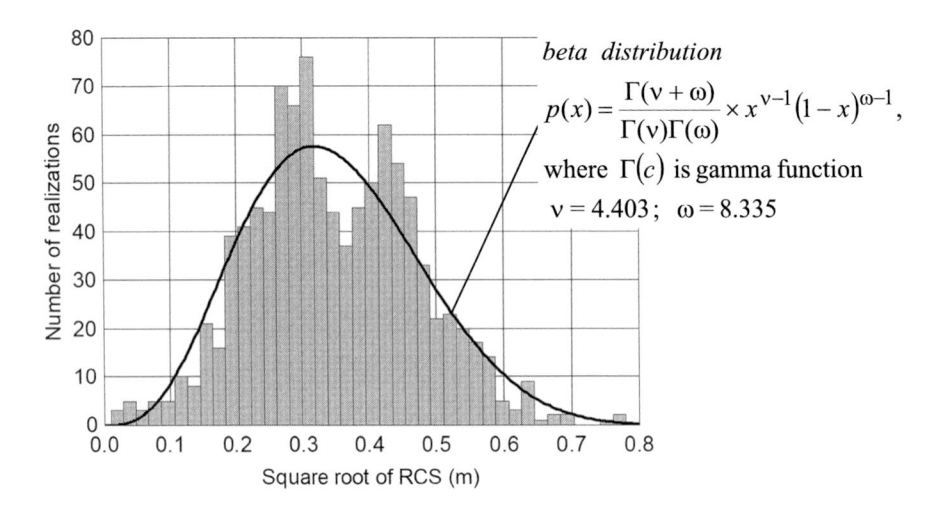

FIGURE 2.727 Amplitude distribution of echo signal of P-700 model at a carrier frequency of 3 GHz given its horizontal polarization.

$$p(x) = \frac{\Gamma(\nu + \omega)}{\Gamma(\nu)\Gamma(\omega)} \times x^{\nu-1}(1-x)^{\omega-1},$$

where $\Gamma(c)$ is gamma function
$\nu = 4.403$; $\omega = 8.335$

Scattering characteristics of P-700 model for sounding frequency 1 GHz (wavelength 30 cm).
Figure 2.728 shows the RCS circular diagram of the P-700. The noncoherent RCS circular diagram of the P-700 is shown in Figure 2.729.

The circular mean RCS of the P-700 for horizontal polarization is 9.760 m². The circular median RCS for horizontal polarization is 0.255 m².

Figures 2.730 and 2.731 show the mean and median RCS for the main ranges of sounding azimuths (nose, side, and tail) and for ranges of 20°.

The expressions and parameters of probability distributions that are most consistent with the empirical distributions of the square root of the RCS for different ranges of sounding azimuths and polarizations are given in the electronic appendix to this book. The expressions and parameters of the probability distributions that are most consistent with the empirical distributions of the RCS (energy characteristic) for different ranges of sounding azimuths and polarizations are also given there.

FIGURE 2.728 Circular diagram of RCS given radar observation of P-700 model at a carrier frequency of 1 GHz (30 cm wavelength).

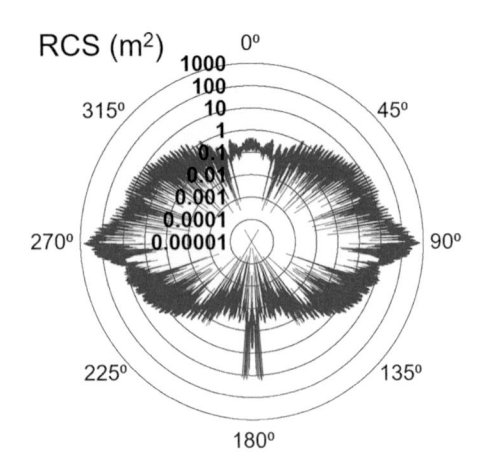

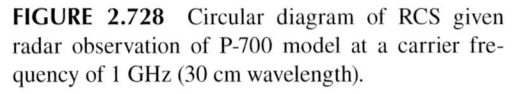

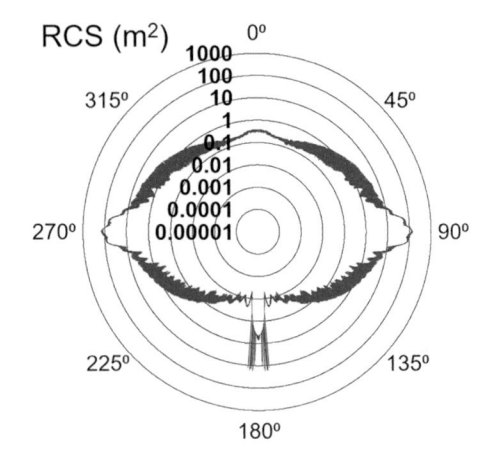

FIGURE 2.729 Circular diagram of noncoherent RCS given radar observation of P-700 model at a carrier frequency of 1 GHz (30 cm wavelength).

Scattering Characteristics of Aerial Objects

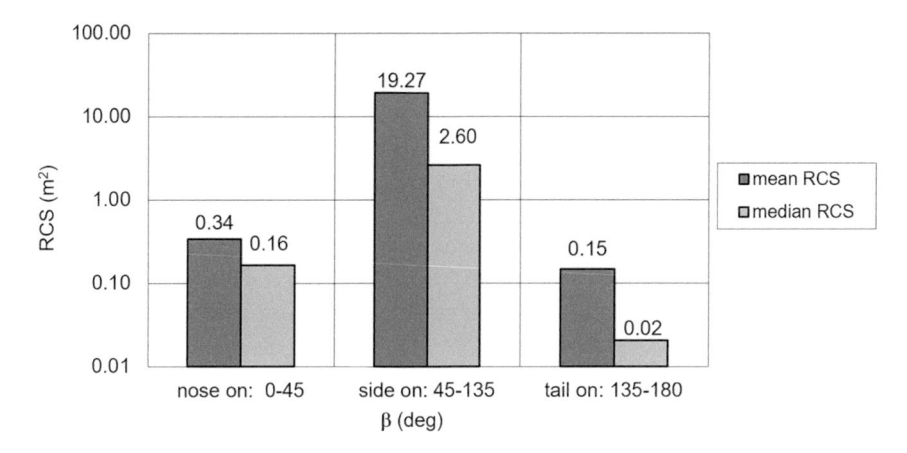

FIGURE 2.730 Diagrams of mean and median RCS of P-700 model in three sectors of azimuth aspect given its radar observation at horizontal polarization and a carrier frequency of 1 GHz (30 cm wavelength).

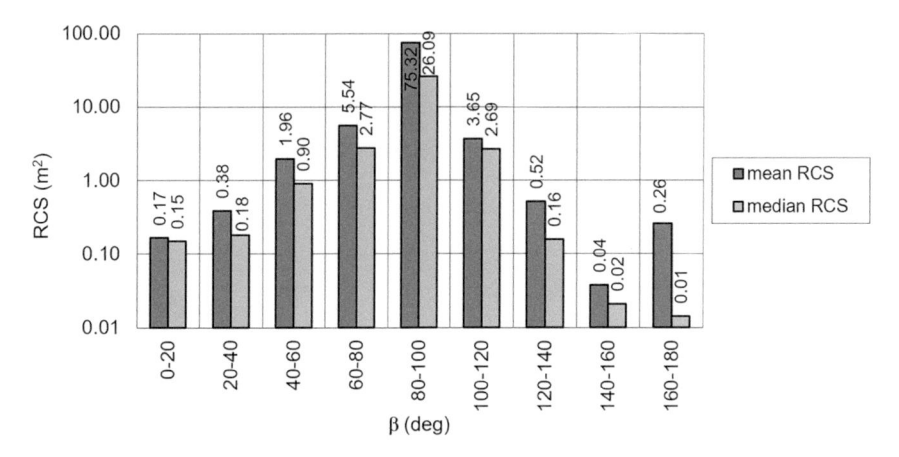

FIGURE 2.731 Diagrams of mean and median RCS of P-700 model in 20-degree sectors of azimuth aspect given its radar observation at horizontal polarization and a carrier frequency of 1 GHz (30 cm wavelength).

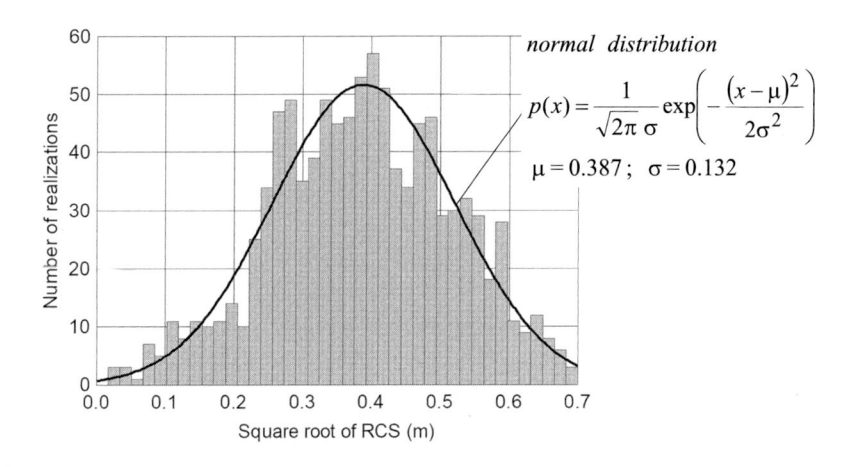

$$p(x) = \frac{1}{\sqrt{2\pi}\,\sigma} \exp\left(-\frac{(x-\mu)^2}{2\sigma^2}\right)$$

$\mu = 0.387$; $\sigma = 0.132$

FIGURE 2.732 Amplitude distribution of echo signal of P-700 model at a carrier frequency of 1 GHz given its horizontal polarization.

2.44 CRUISE MISSILE P-800 ONIKS

P-800 Oniks (Yahont) - SS-N-26 Strobile - (Figure 2.733) is a supersonic anti-ship cruise missile. The first flight took place in 1993.

General characteristics of P-800 Oniks [55]: wingspan –1.7 m, length – 8 m (ship-based variant), fuselage diameter – 0.67 m, weight – 2 500–3 000 kg, powerplant – 1 Ramjet, speed – 2.0–2.6 M, range – 120–300 (600- Oniks version for Russia) km.

In accordance with the design of the P-800 Oniks, a model of its surface was created to obtaining scattering characteristics (in particular, RCS). The model is shown in Figure 2.734. The model surface was assumed to be perfectly conducting. The smooth parts of the model surface were approximated by parts of 21 triaxial ellipsoids. The surface breaks were modeled using 24 straight edge scattering parts.

FIGURE 2.733 Cruise missile P-800 Oniks.

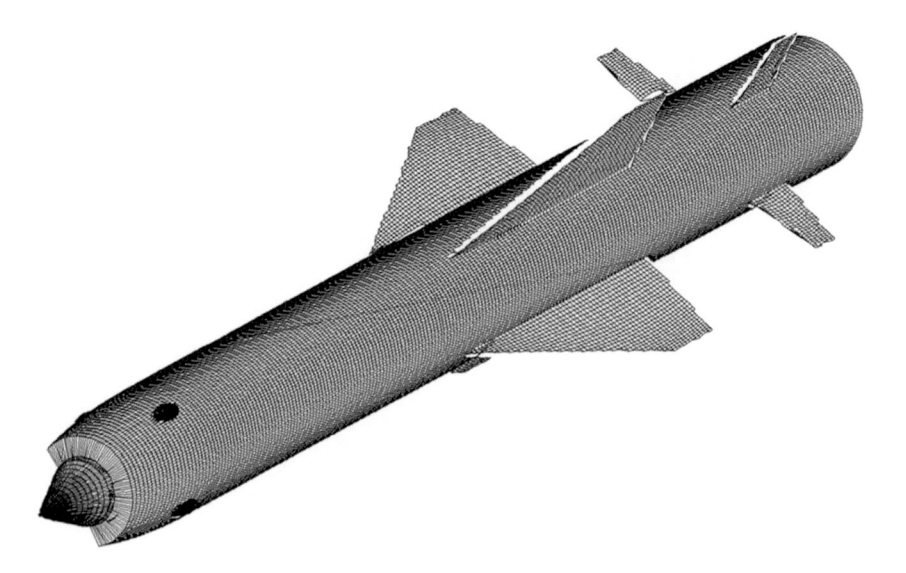

FIGURE 2.734 Surface model of cruise missile P-800 Oniks.

Below are some scattering characteristics of the P-800 model at sounding frequencies of 10, 3, and 1 GHz (wavelengths of 3, 10, and 30 cm, respectively) for horizontal polarization of the probing signal. The scattering characteristics for this model at two polarizations, as well as scattering characteristics at sounding frequencies of 5 and 1.3 GHz (wavelengths of 6 and 23 cm, respectively), are given in the electronic appendix of this book.

Sounding parameters: The elevation angle is to be a random value distributed uniformly in the range $-3° \pm 4°$ with respect to the wing plane (elevation angle of $-3°$ corresponds to the radar

Scattering Characteristics of Aerial Objects

observation from the lower hemisphere), azimuth aspect increment was 0.02°, the azimuth being counted off from the nose-on aspect (0° corresponds to the nose-on radar observation, 180° corresponds to the tail-on observation).

Scattering characteristics of P-800 model for sounding frequency 10 GHz (wavelength 3 cm).
Figure 2.735 shows the RCS circular diagram of the P-800 model. The noncoherent RCS circular diagram of the P-800 is shown in Figure 2.736.

The circular mean RCS of the P-800 model for horizontal polarization is 8.895 m². The circular median RCS (the RCS value used to calculate the detection range of an object with a probability of 0.5) for horizontal polarization is 0.015 m².

Figures 2.737 and 2.738 show the mean and median RCS for the main ranges of sounding azimuths (nose, side, and tail) and for ranges of 20°.

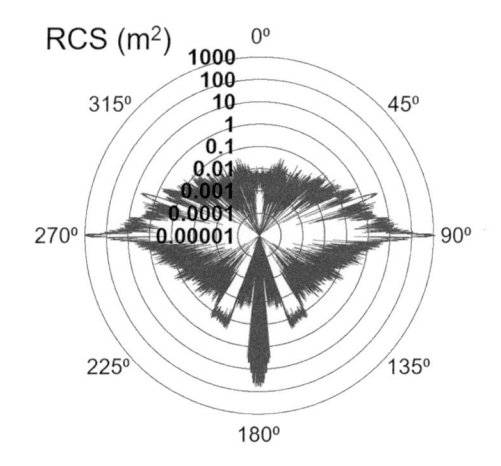

FIGURE 2.735 Circular diagram of RCS given radar observation of P-800 model at a carrier frequency of 10 GHz (3 cm wavelength).

FIGURE 2.736 Circular diagram of noncoherent RCS given radar observation of P-800 model at a carrier frequency of 10 GHz (3 cm wavelength).

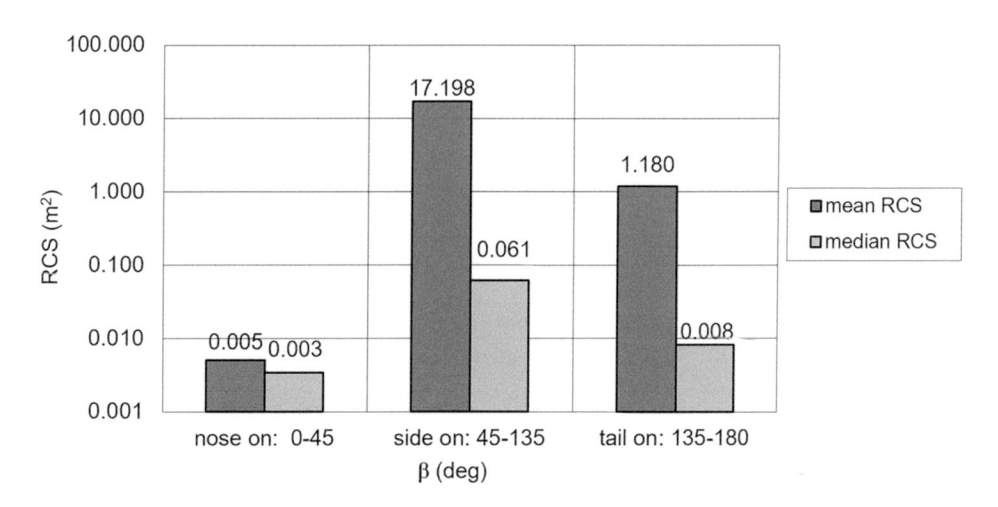

FIGURE 2.737 Diagrams of mean and median RCS of P-800 model in three sectors of azimuth aspect given its radar observation at horizontal polarization and a carrier frequency of 10 GHz (3 cm wavelength).

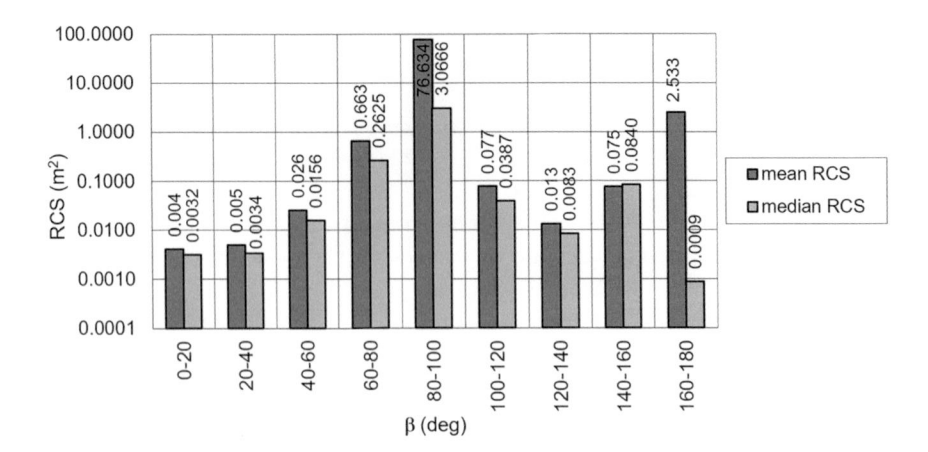

FIGURE 2.738 Diagrams of mean and median RCS of P-800 model in 20-degree sectors of azimuth aspect given its radar observation at horizontal polarization and a carrier frequency of 10 GHz (3 cm wavelength).

Figures 2.739, 2.744, and 2.749 show histograms of the scattered signal amplitude (square root of the RCS) for the range of sounding azimuths −20° to +20° (sounding from the nose). The bold line shows the probability density functions of the distribution, which can be used to approximate the histogram of the signal amplitude.

extreme value distribution

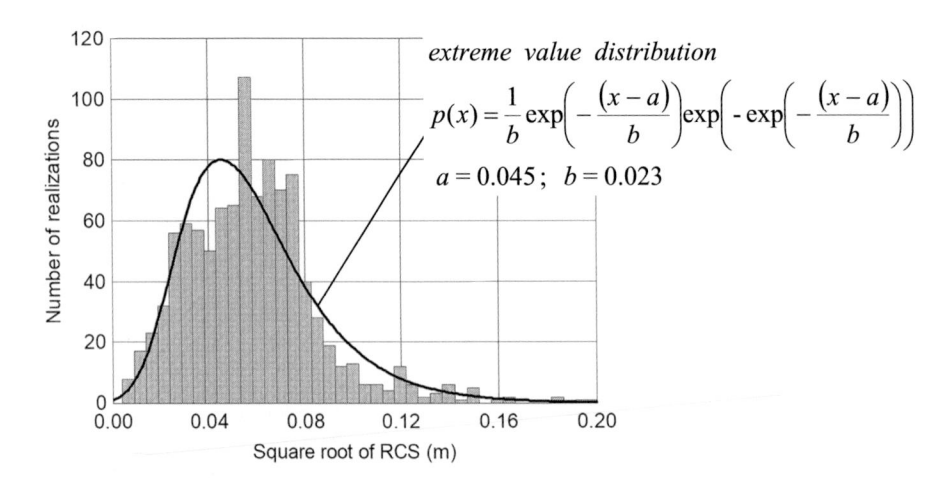

$$p(x) = \frac{1}{b} \exp\left(-\frac{(x-a)}{b}\right) \exp\left(-\exp\left(-\frac{(x-a)}{b}\right)\right)$$

$a = 0.045; \; b = 0.023$

FIGURE 2.739 Amplitude distribution of echo signal of P-800 model at a carrier frequency of 10 GHz given its horizontal polarization.

Scattering characteristics of P-800 model for sounding frequency 3 GHz (wavelength 10 cm).

Figure 2.740 shows the RCS circular diagram of the P-800. The noncoherent RCS circular diagram of the P-800 is shown in Figure 2.741.

The circular mean RCS of the P-800 for horizontal polarization is 4.368 m². The circular median RCS for horizontal polarization is 0.022 m².

Figures 2.742 and 2.743 show the mean and median RCS for the main ranges of sounding azimuths (nose, side, and tail) and for ranges of 20°.

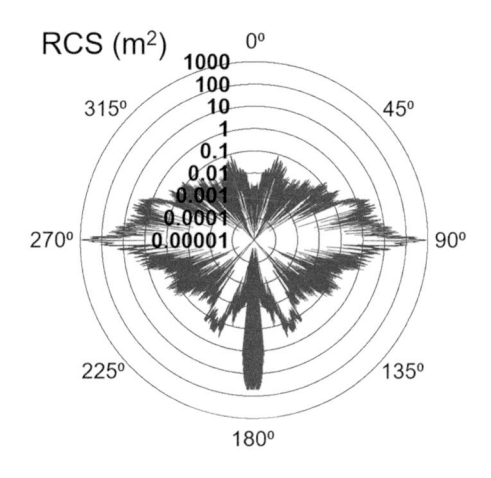

Scattering Characteristics of Aerial Objects 273

FIGURE 2.740 Circular diagram of RCS given radar observation of P-800 model at a carrier frequency of 3 GHz (10 cm wavelength).

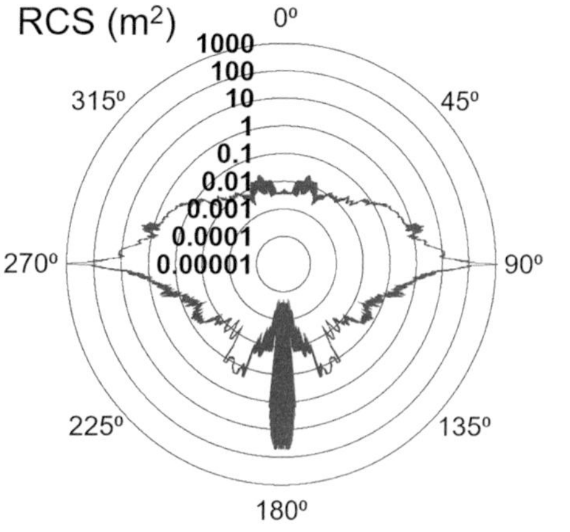

FIGURE 2.741 Circular diagram of noncoherent RCS given radar observation of P-800 model at a carrier frequency of 3 GHz (10 cm wavelength).

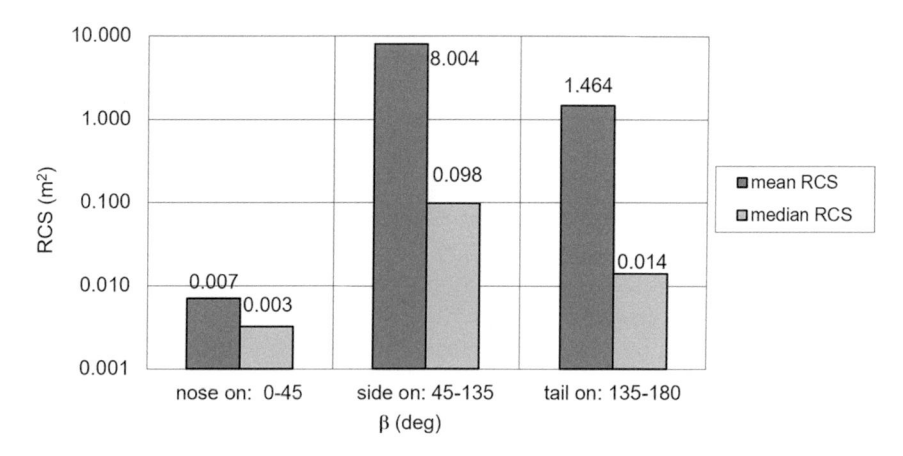

FIGURE 2.742 Diagrams of mean and median RCS of P-800 model in three sectors of azimuth aspect given its radar observation at horizontal polarization and a carrier frequency of 3 GHz (10 cm wavelength).

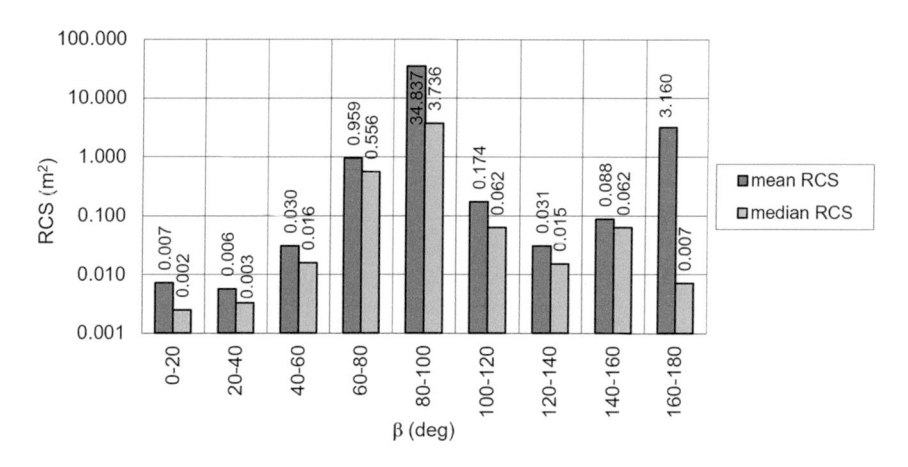

FIGURE 2.743 Diagrams of mean and median RCS of P-800 model in 20-degree sectors of azimuth aspect given its radar observation at horizontal polarization and a carrier frequency of 3 GHz (10 cm wavelength).

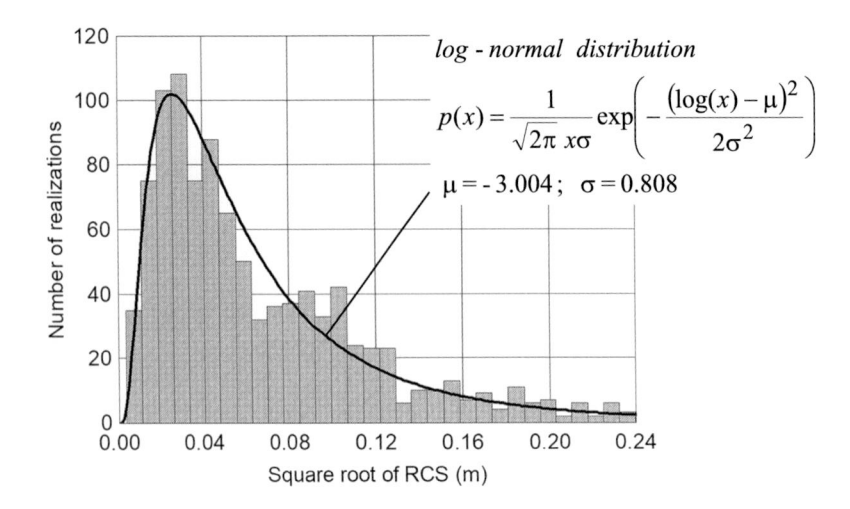

FIGURE 2.744 Amplitude distribution of echo signal of P-800 model at a carrier frequency of 3 GHz given its horizontal polarization.

Scattering characteristics of P-800 model for sounding frequency 1 GHz (wavelength 30 cm).
Figure 2.745 shows the RCS circular diagram of the P-800. The noncoherent RCS circular diagram of the P-800 is in Figure 2.746.

The circular mean RCS of the P-800 for horizontal polarization is 3.322 m². The circular median RCS for horizontal polarization is 0.070 m².

Figures 2.747 and 2.748 show the mean and median RCS for the main ranges of sounding azimuths (nose, side, and tail) and for ranges of 20°.

The expressions and parameters of probability distributions that are most consistent with the empirical distributions of the square root of the RCS for different ranges of sounding azimuths and polarizations are given in the electronic appendix to this book. The expressions and parameters of the probability distributions that are most consistent with the empirical distributions of the RCS (energy characteristic) for different ranges of sounding azimuths and polarizations are also given there.

Within the histogram figure:

log - normal distribution

$$p(x) = \frac{1}{\sqrt{2\pi}\, x\sigma} \exp\left(-\frac{(\log(x)-\mu)^2}{2\sigma^2}\right)$$

$\mu = -3.004;\ \sigma = 0.808$

FIGURE 2.745 Circular diagram of RCS given radar observation of P-800 model at a carrier frequency of 1 GHz (30 cm wavelength).

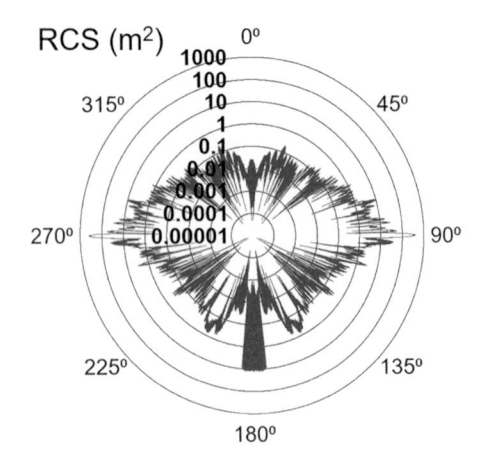

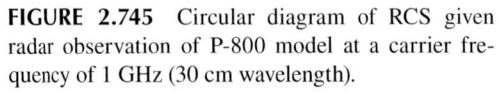

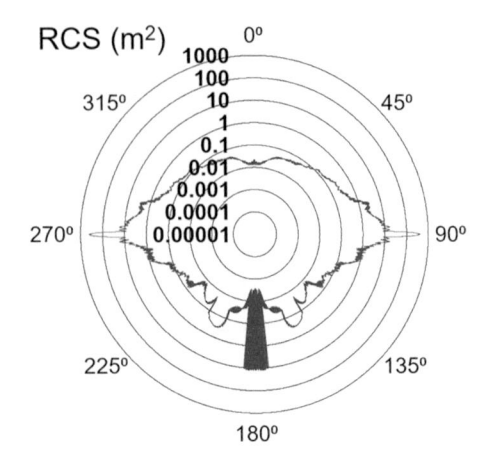

FIGURE 2.746 Circular diagram of noncoherent RCS given radar observation of P-800 model at a carrier frequency of 1 GHz (30 cm wavelength).

Scattering Characteristics of Aerial Objects

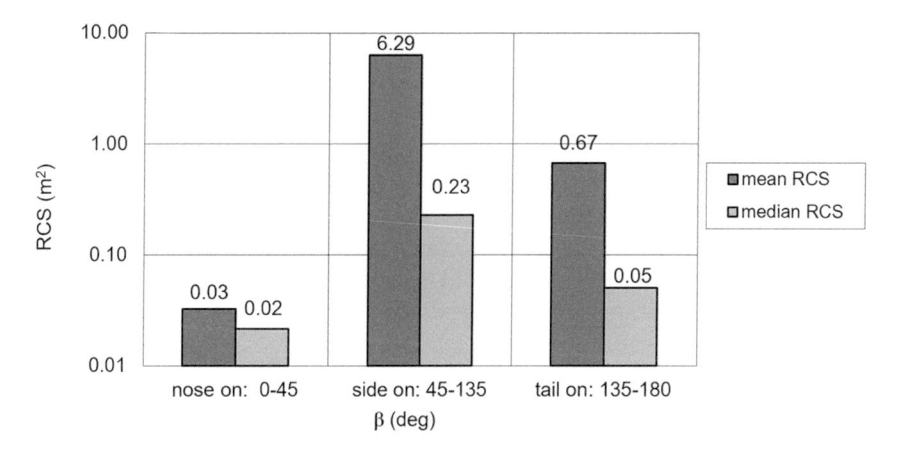

FIGURE 2.747 Diagrams of mean and median RCS of P-800 model in three sectors of azimuth aspect given its radar observation at horizontal polarization and a carrier frequency of 1 GHz (30 cm wavelength).

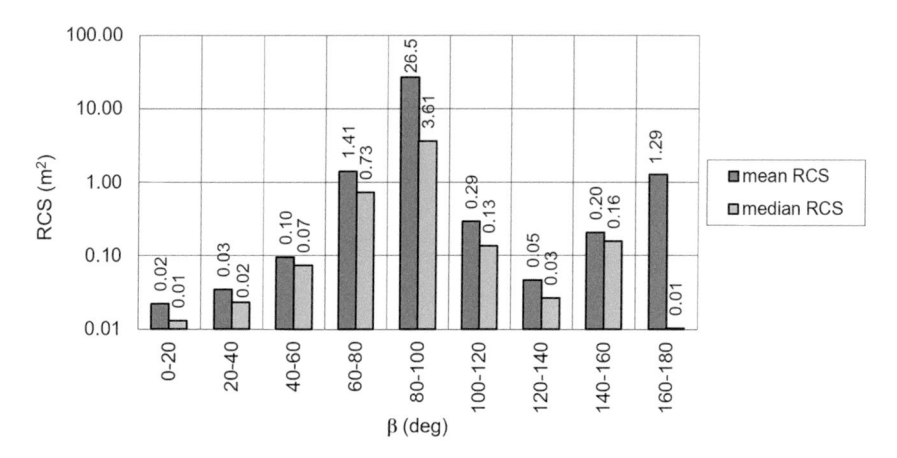

FIGURE 2.748 Diagrams of mean and median RCS of P-800 model in 20-degree sectors of azimuth aspect given its radar observation at horizontal polarization and a carrier frequency of 1 GHz (30 cm wavelength).

Weibull distribution

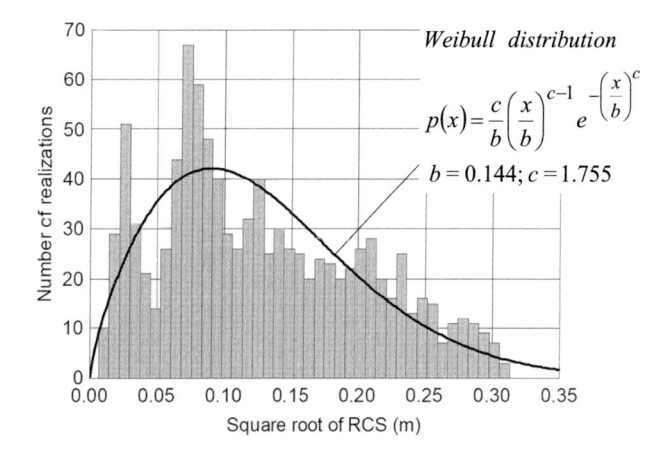

$$p(x) = \frac{c}{b}\left(\frac{x}{b}\right)^{c-1} e^{-\left(\frac{x}{b}\right)^c}$$

$b = 0.144; c = 1.755$

FIGURE 2.749 Amplitude distribution of echo signal of P-800 model at a carrier frequency of 1 GHz given its horizontal polarization.

2.45 CRUISE MISSILE 3M-14 KALIBR

3M-14 Kalibr - SS-N-30A - (Figure 2.750) is land attack cruise missile. The first flight took place in 1993.

General characteristics of 3M-14 Kalibr [56]: wingspan –2.7 m, length – 6.2 m, fuselage diameter – 0.533 m, weight – 1 770 kg, powerplant – 1 37-01E turbojet engine, speed – 180–240 m/s, range – 1 500–2 500 km.

In accordance with the design of the 3M-14 Kalibr, a model of its surface was created to obtaining scattering characteristics (in particular, RCS). The model is shown in Figure 2.751. The smooth parts of the model surface were approximated by parts of 16 triaxial ellipsoids. The surface breaks were modeled using 18 straight edge scattering parts.

FIGURE 2.750 Cruise missile 3M-14 Kalibr.

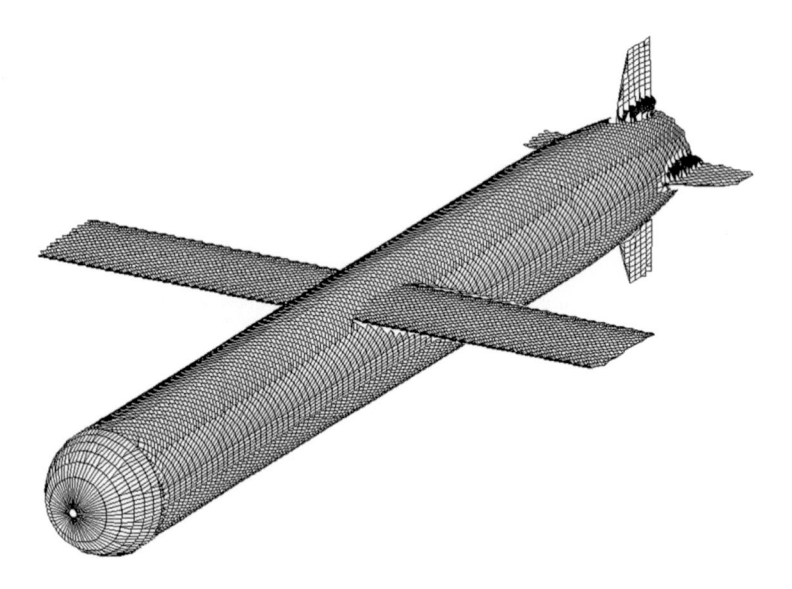

FIGURE 2.751 Surface model of cruise missile 3M-14 Kalibr.

Below are some scattering characteristics of the 3M-14 model at sounding frequencies of 10, 3, and 1 GHz (wavelengths of 3, 10, and 30 cm, respectively) for horizontal polarization of the probing signal. The scattering characteristics for this model at two polarizations, as well as scattering characteristics at sounding frequencies of 5 and 1.3 GHz (wavelengths of 6 and 23 cm, respectively), are given in the electronic appendix of this book.

Sounding parameters: The elevation angle is to be a random value distributed uniformly in the range −3° ± 4° with respect to the wing plane (elevation angle of −3° corresponds to the radar

observation from the lower hemisphere), azimuth aspect increment was 0.02°, the azimuth being counted off from the nose-on aspect (0° corresponds to the nose-on radar observation, 180° corresponds to the tail-on observation).

Scattering characteristics of 3M-14 model for sounding frequency 10 GHz (wavelength 3 cm).

Figure 2.752 shows the RCS circular diagram of the 3M-14 model. The noncoherent RCS circular diagram of the 3M-14 is shown in Figure 2.753.

The circular mean RCS of the 3M-14 model for horizontal polarization is 1.775 m². The circular median RCS (the RCS value used to calculate the detection range of an object with a probability of 0.5) for horizontal polarization is 0.049 m².

Figures 2.754 and 2.755 show the mean and median RCS for the main ranges of sounding azimuths (nose, side, and tail) and for ranges of 20°.

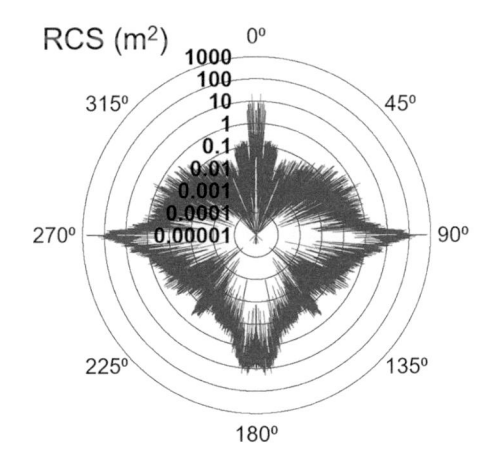

FIGURE 2.752 Circular diagram of RCS given radar observation of 3M-14 model at a carrier frequency of 10 GHz (3 cm wavelength).

FIGURE 2.753 Circular diagram of noncoherent RCS given radar observation of 3M-14 model at a carrier frequency of 10 GHz (3 cm wavelength).

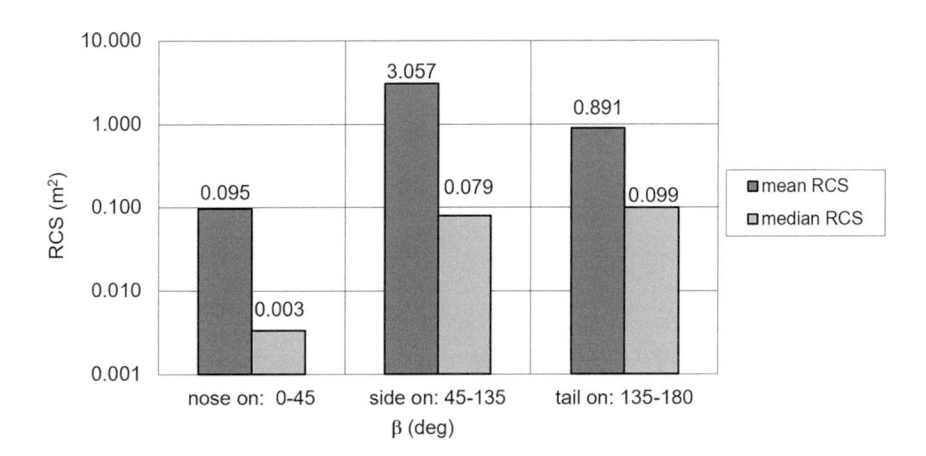

FIGURE 2.754 Diagrams of mean and median RCS of 3M-14 model in three sectors of azimuth aspect given its radar observation at horizontal polarization and a carrier frequency of 10 GHz (3 cm wavelength).

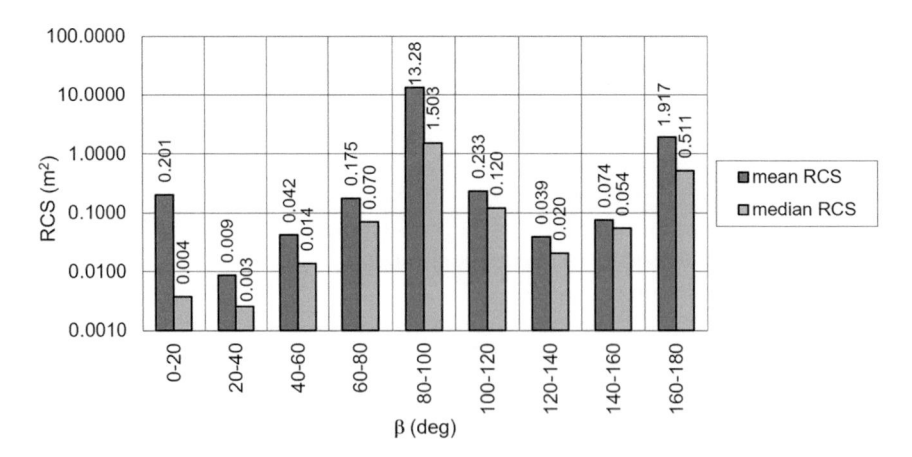

FIGURE 2.755 Diagrams of mean and median RCS of 3M-14 model in 20-degree sectors of azimuth aspect given its radar observation at horizontal polarization and a carrier frequency of 10 GHz (3 cm wavelength).

Figures 2.756, 2.761, and 2.766 show histograms of the scattered signal amplitude (square root of the RCS) for the range of sounding azimuths −20° to +20° (sounding from the nose). The bold line shows the probability density functions of the distribution, which can be used to approximate the histogram of the signal amplitude.

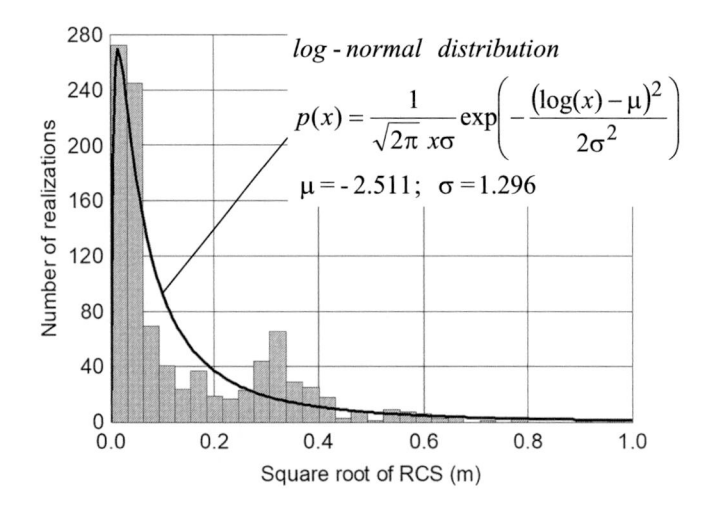

$$p(x) = \frac{1}{\sqrt{2\pi}\, x\sigma} \exp\left(-\frac{(\log(x) - \mu)^2}{2\sigma^2}\right)$$

$\mu = -2.511; \quad \sigma = 1.296$

FIGURE 2.756 Amplitude distribution of echo signal of 3M-14 model at a carrier frequency of 10 GHz given its horizontal polarization.

Scattering characteristics of 3M-14 model for sounding frequency 3 GHz (wavelength 10 cm).

Figure 2.757 shows the RCS circular diagram of the 3M-14. The noncoherent RCS circular diagram of the 3M-14 is shown in Figure 2.758.

The circular mean RCS of the 3M-14 for horizontal polarization is 2.094 m². The circular median RCS for horizontal polarization is 0.123 m².

Figures 2.759 and 2.760 show the mean and median RCS for the main ranges of sounding azimuths (nose, side, and tail) and for ranges of 20°.

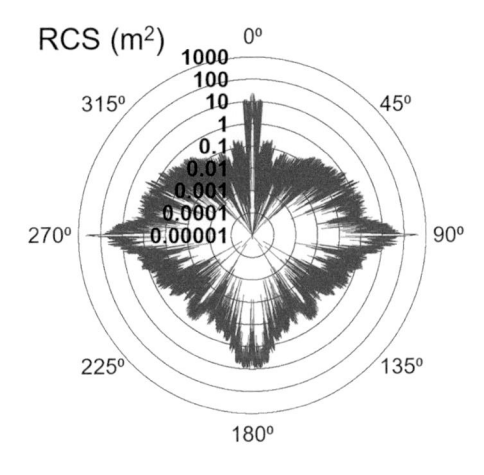

Scattering Characteristics of Aerial Objects

FIGURE 2.757 Circular diagram of RCS given radar observation of 3M-14 model at a carrier frequency of 3 GHz (10 cm wavelength).

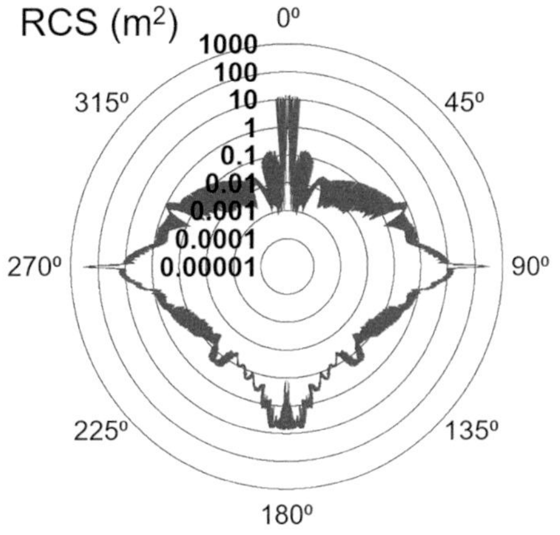

FIGURE 2.758 Circular diagram of noncoherent RCS given radar observation of 3M-14 model at a carrier frequency of 3 GHz (10 cm wavelength).

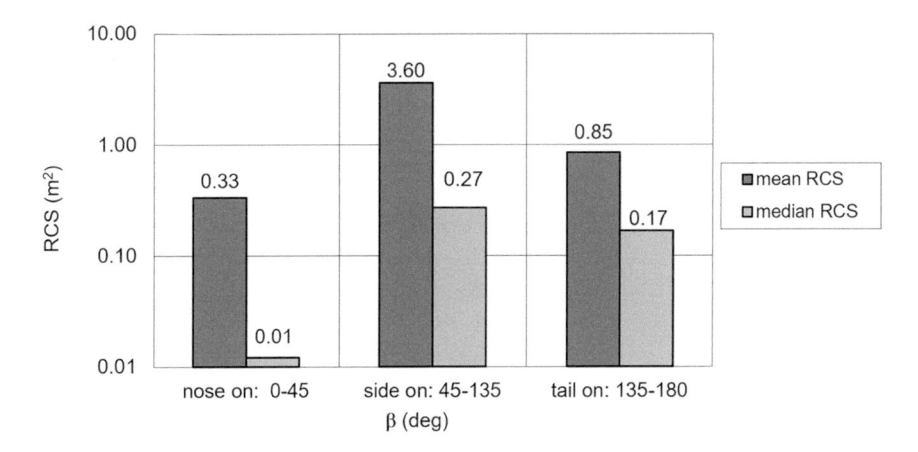

FIGURE 2.759 Diagrams of mean and median RCS of 3M-14 model in three sectors of azimuth aspect given its radar observation at horizontal polarization and a carrier frequency of 3 GHz (10 cm wavelength).

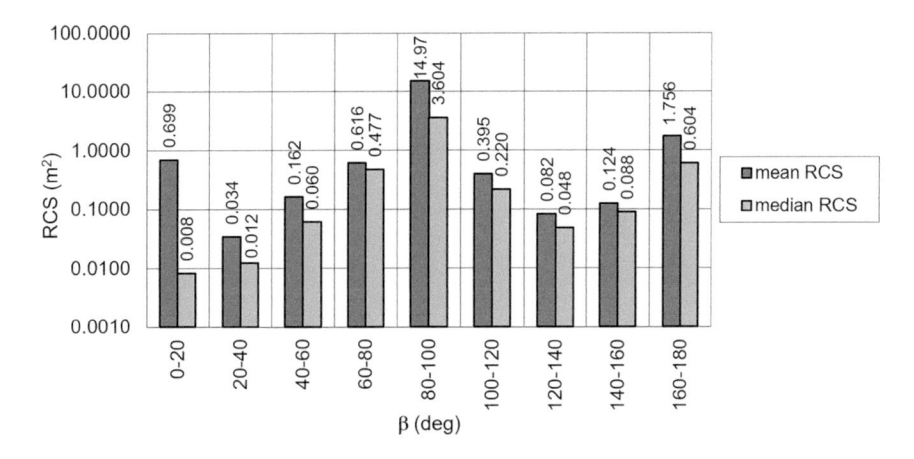

FIGURE 2.760 Diagrams of mean and median RCS of 3M-14 model in 20-degree sectors of azimuth aspect given its radar observation at horizontal polarization and a carrier frequency of 3 GHz (10 cm wavelength).

$$p(x) = \frac{1}{\sqrt{2\pi}\, x\sigma} \exp\left(-\frac{(\log(x)-\mu)^2}{2\sigma^2}\right)$$

log-normal distribution

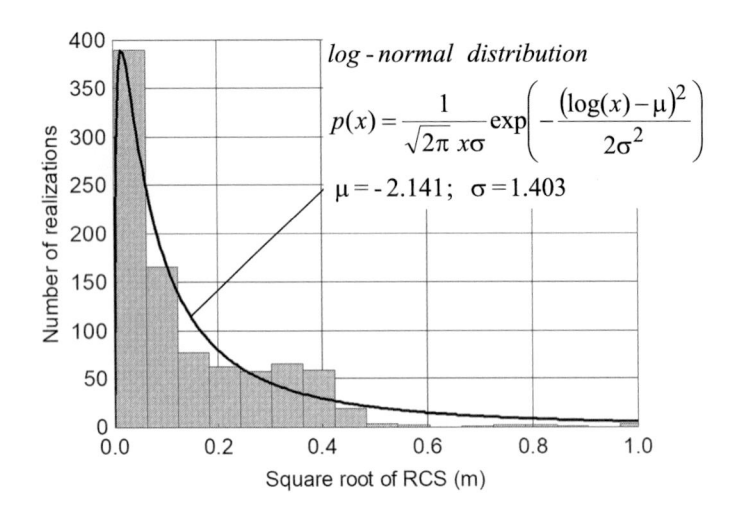

$\mu = -2.141; \quad \sigma = 1.403$

FIGURE 2.761 Amplitude distribution of echo signal of 3M-14 model at a carrier frequency of 3 GHz given its horizontal polarization.

Scattering characteristics of 3M-14 model for sounding frequency 1 GHz (wavelength 30 cm).

Figure 2.762 shows the RCS circular diagram of the 3M-14. The noncoherent RCS circular diagram of the 3M-14 is shown in Figure 2.763.

The circular mean RCS of the 3M-14 for horizontal polarization is 2.459 m². The circular median RCS for horizontal polarization is 0.332 m².

Figures 2.764 and 2.765 show the mean and median RCS for the main ranges of sounding azimuths (nose, side, and tail) and for ranges of 20°.

The expressions and parameters of probability distributions that are most consistent with the empirical distributions of the square root of the RCS for different ranges of sounding azimuths and polarizations are given in the electronic appendix to this book. The expressions and parameters of the probability distributions that are most consistent with the empirical distributions of the RCS (energy characteristic) for different ranges of sounding azimuths and polarizations are also given there.

FIGURE 2.762 Circular diagram of RCS given radar observation of 3M-14 model at a carrier frequency of 1 GHz (30 cm wavelength).

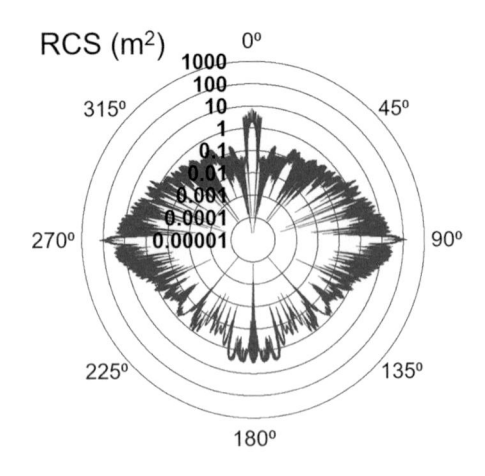

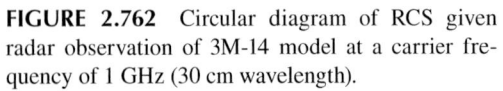

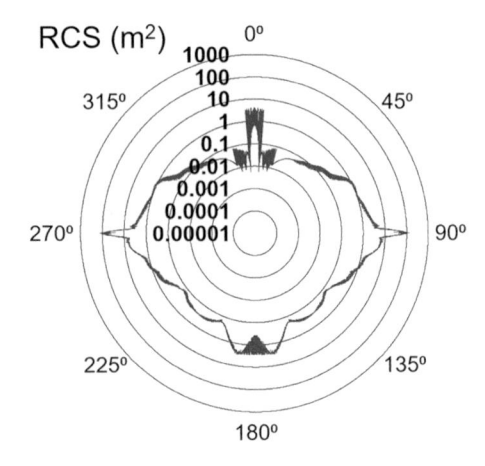

FIGURE 2.763 Circular diagram of noncoherent RCS given radar observation of 3M-14 model at a carrier frequency of 1 GHz (30 cm wavelength).

Scattering Characteristics of Aerial Objects

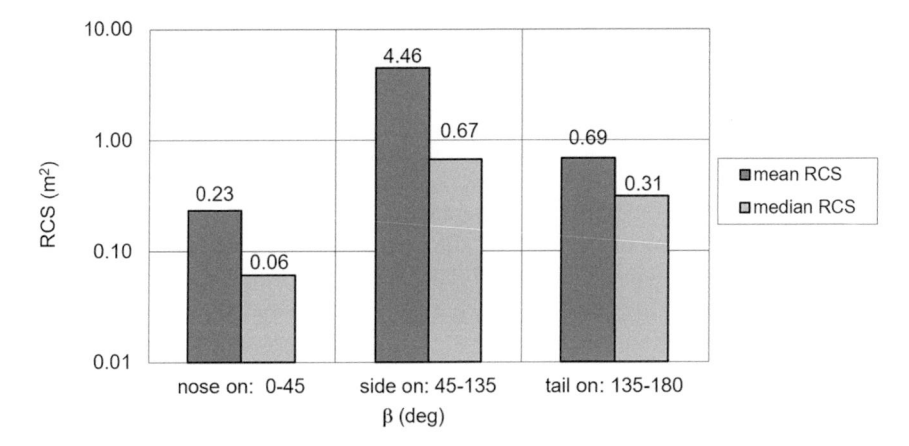

FIGURE 2.764 Diagrams of mean and median RCS of 3M-14 model in three sectors of azimuth aspect given its radar observation at horizontal polarization and a carrier frequency of 1 GHz (30 cm wavelength).

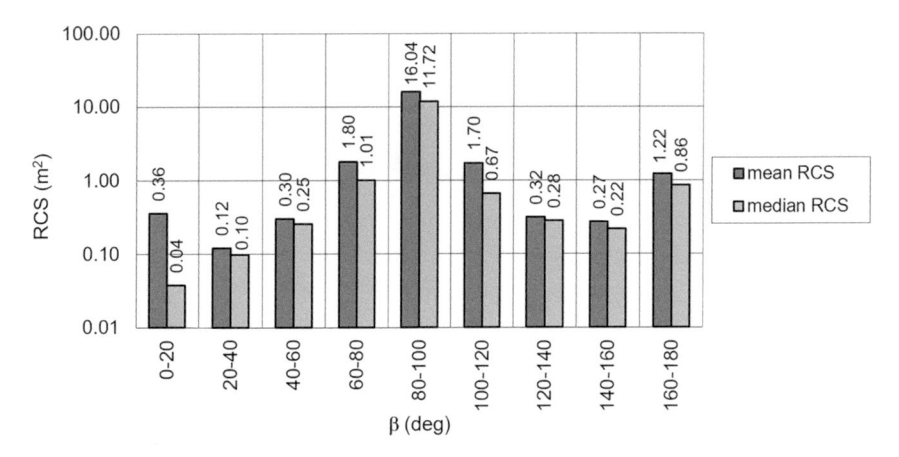

FIGURE 2.765 Diagrams of mean and median RCS of 3M-14 model in 20-degree sectors of azimuth aspect given its radar observation at horizontal polarization and a carrier frequency of 1 GHz (30 cm wavelength).

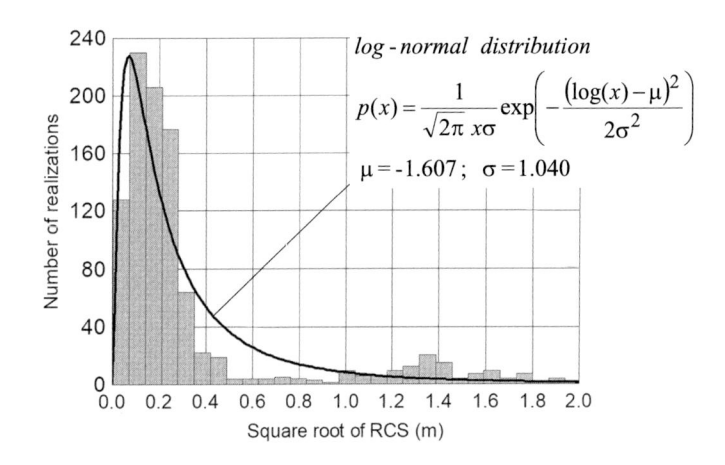

$$p(x) = \frac{1}{\sqrt{2\pi}\, x\sigma} \exp\left(-\frac{(\log(x)-\mu)^2}{2\sigma^2}\right)$$

$\mu = -1.607;\ \sigma = 1.040$

FIGURE 2.766 Amplitude distribution of echo signal of 3M-14 model at a carrier frequency of 1 GHz given its horizontal polarization.

2.46 CRUISE MISSILE 3M-54 KALIBR

3M-54 Kalibr - SS-N-27- (Figure 2.767) is anti-ship cruise missile. The first flight took place in 1987.

General characteristics of 3M-54 Kalibr [57]: wingspan –2.7 m, length – 8.22 m, fuselage diameter – 0.533–0.645 m, weight – 1 950 kg, powerplant – 1 R95–300 turbojet engine, speed – M = 0.6–0.8 (up to 3 at the final section), range – 220–300 km.

In accordance with the design of the 3M-54 Kalibr, a model of its surface was created to obtaining scattering characteristics (in particular, RCS). The model is shown in Figure 2.768. The smooth parts of the model surface were approximated by parts of 25 triaxial ellipsoids. The surface breaks were modeled using 22 straight edge scattering parts.

FIGURE 2.767 Cruise missile 3M-54 Kalibr.

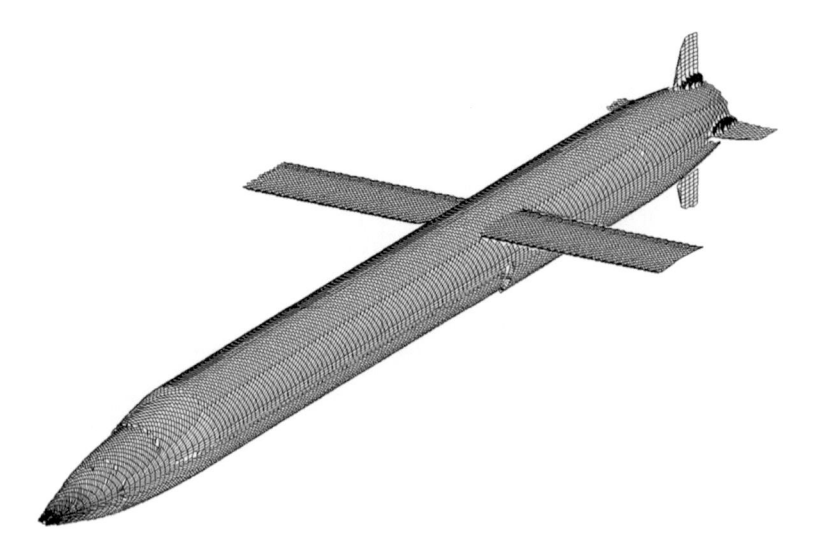

FIGURE 2.768 Surface model of cruise missile 3M-54 Kalibr.

Below are some scattering characteristics of the 3M-54 model at sounding frequencies of 10, 3, and 1 GHz (wavelengths of 3, 10, and 30 cm, respectively) for horizontal polarization of the probing signal. The scattering characteristics for this model at two polarizations, as well as scattering characteristics at sounding frequencies of 5 and 1.3 GHz (wavelengths of 6 and 23 cm, respectively), are given in the electronic appendix of this book.

Sounding parameters: The elevation angle is to be a random value distributed uniformly in the range −3° ± 4° with respect to the wing plane (elevation angle of −3° corresponds to the radar

Scattering Characteristics of Aerial Objects 283

observation from the lower hemisphere), azimuth aspect increment was 0.02°, the azimuth being counted off from the nose-on aspect (0° corresponds to the nose-on radar observation, 180° corresponds to the tail-on observation).

Scattering characteristics of 3M-54 model for sounding frequency 10 GHz (wavelength 3 cm).
Figure 2.769 shows the RCS circular diagram of the 3M-54 model. The noncoherent RCS circular diagram of the 3M-54 is shown in Figure 2.770.

The circular mean RCS of the 3M-54 model for horizontal polarization is 1.837 m^2. The circular median RCS (the RCS value used to calculate the detection range of an object with a probability of 0.5) for horizontal polarization is 0.075 m^2.

Figures 2.771 and 2.772 show the mean and median RCS for the main ranges of sounding azimuths (nose, side, and tail) and for ranges of 20°.

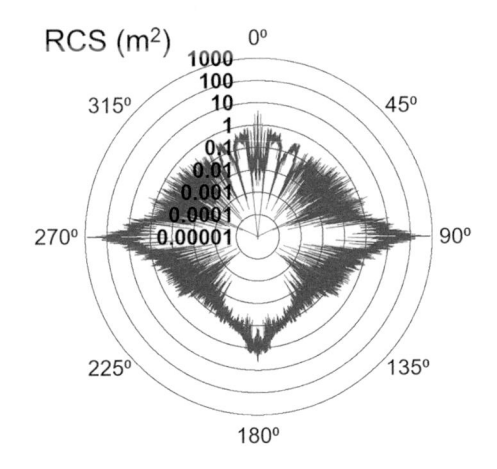

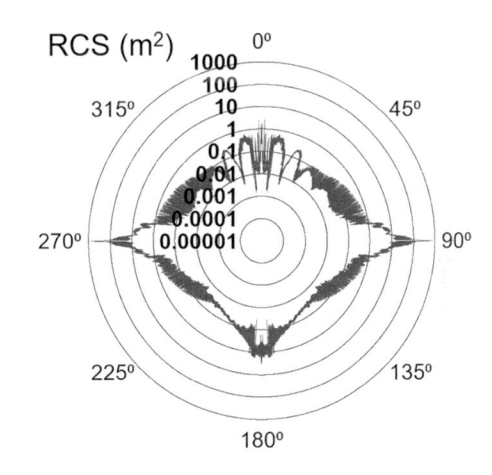

FIGURE 2.769 Circular diagram of RCS given radar observation of 3M-54 model at a carrier frequency of 10 GHz (3 cm wavelength).

FIGURE 2.770 Circular diagram of noncoherent RCS given radar observation of 3M-54 model at a carrier frequency of 10 GHz (3 cm wavelength).

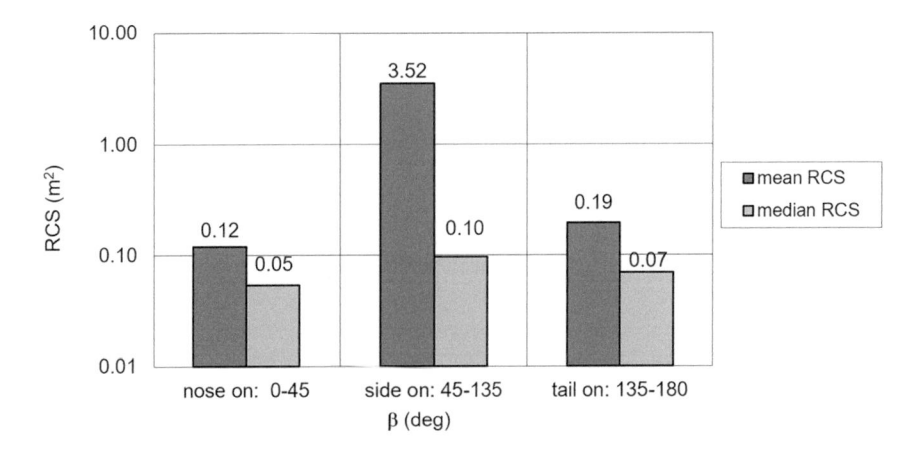

FIGURE 2.771 Diagrams of mean and median RCS of 3M-54 model in three sectors of azimuth aspect given its radar observation at horizontal polarization and a carrier frequency of 10 GHz (3 cm wavelength).

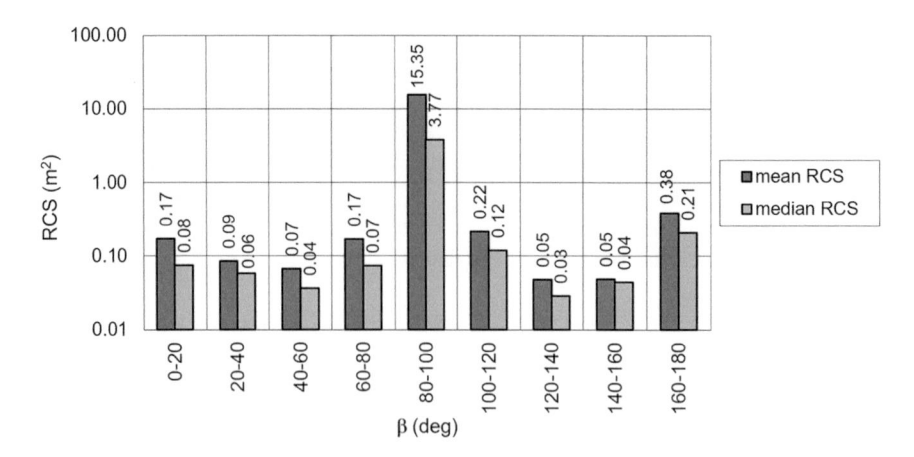

FIGURE 2.772 Diagrams of mean and median RCS of 3M-54 model in 20-degree sectors of azimuth aspect given its radar observation at horizontal polarization and a carrier frequency of 10 GHz (3 cm wavelength).

Figures 2.773, 2.778, and 2.783 show histograms of the scattered signal amplitude (square root of the RCS) for the range of sounding azimuths −20° to +20° (sounding from the nose). The bold line shows the probability density functions of the distribution, which can be used to approximate the histogram of the signal amplitude.

Weibull distribution

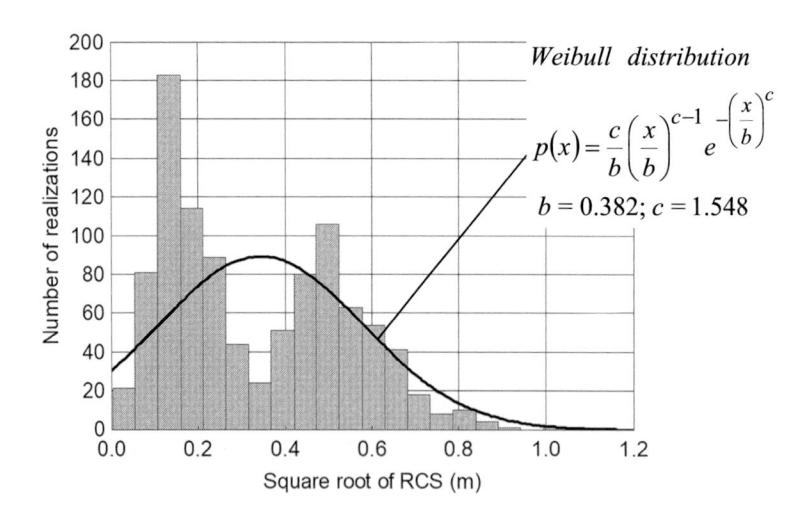

$$p(x) = \frac{c}{b}\left(\frac{x}{b}\right)^{c-1} e^{-\left(\frac{x}{b}\right)^c}$$

$b = 0.382; c = 1.548$

FIGURE 2.773 Amplitude distribution of echo signal of 3M-54 model at a carrier frequency of 10 GHz given its horizontal polarization.

Scattering characteristics of 3M-54 model for sounding frequency 3 GHz (wavelength 10 cm).
Figure 2.774 shows the RCS circular diagram of the 3M-54. The noncoherent RCS circular diagram of the 3M-54 is shown in Figure 2.775.

The circular mean RCS of the 3M-54 for horizontal polarization is 2.011 m². The circular median RCS for horizontal polarization is 0.177 m².

Figures 2.776 and 2.777 show the mean and median RCS for the main ranges of sounding azimuths (nose, side, and tail) and for ranges of 20°.

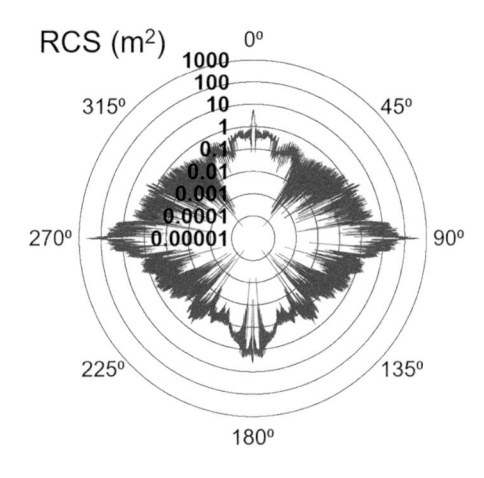

Scattering Characteristics of Aerial Objects   285

FIGURE 2.774 Circular diagram of RCS given radar observation of 3M-54 model at a carrier frequency of 3 GHz (10 cm wavelength).

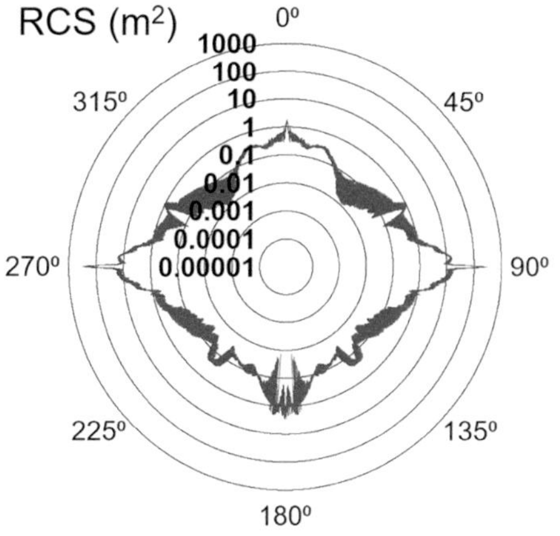

FIGURE 2.775 Circular diagram of noncoherent RCS given radar observation of 3M-54 model at a carrier frequency of 3 GHz (10 cm wavelength).

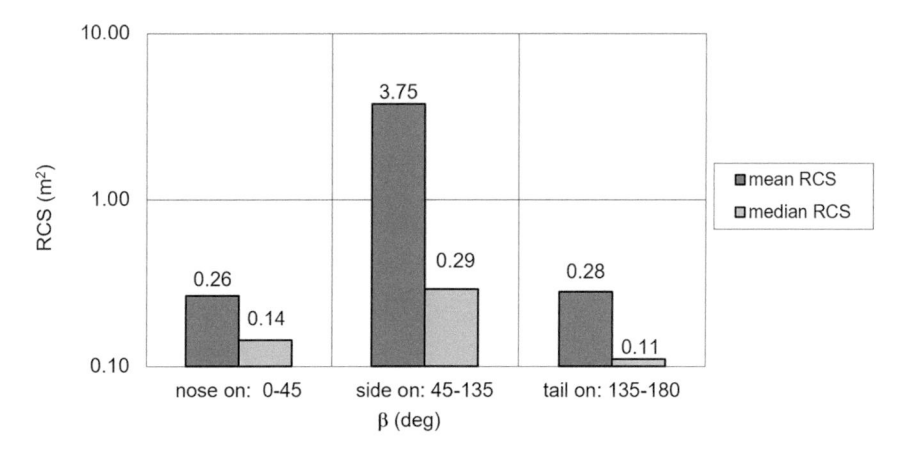

FIGURE 2.776 Diagrams of mean and median RCS of 3M-54 model in three sectors of azimuth aspect given its radar observation at horizontal polarization and a carrier frequency of 3 GHz (10 cm wavelength).

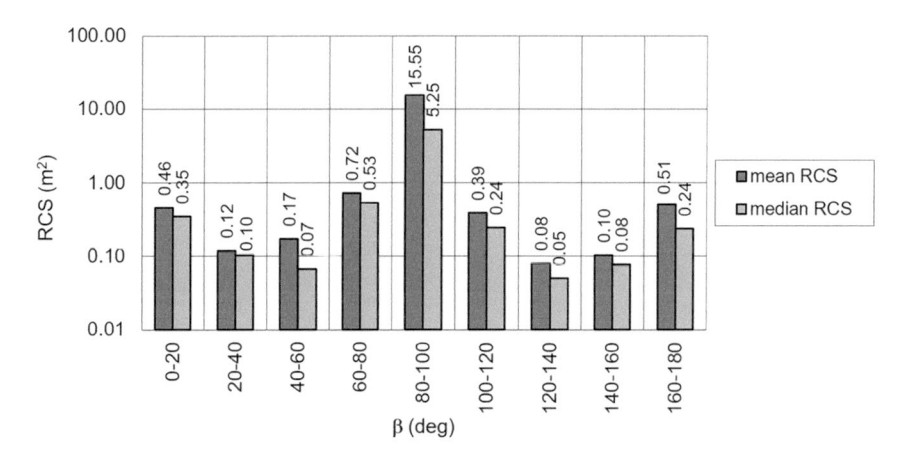

FIGURE 2.777 Diagrams of mean and median RCS of 3M-54 model in 20-degree sectors of azimuth aspect given its radar observation at horizontal polarization and a carrier frequency of 3 GHz (10 cm wavelength).

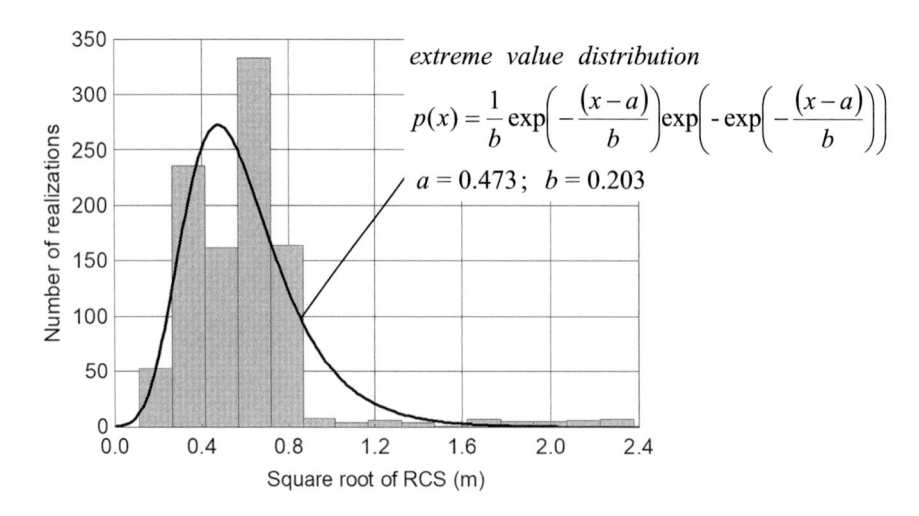

FIGURE 2.778 Amplitude distribution of echo signal of 3M-54 model at a carrier frequency of 3 GHz given its horizontal polarization.

$$p(x) = \frac{1}{b} \exp\left(-\frac{(x-a)}{b}\right) \exp\left(-\exp\left(-\frac{(x-a)}{b}\right)\right)$$

extreme value distribution; $a = 0.473$; $b = 0.203$

Scattering characteristics of 3M-54 model for sounding frequency 1 GHz (wavelength 30 cm).
Figure 2.779 shows the RCS circular diagram of the 3M-54. The noncoherent RCS circular diagram of the 3M-54 is shown in Figure 2.780.

The circular mean RCS of the 3M-54 for horizontal polarization is 2.515 m². The circular median RCS for horizontal polarization is 0.272 m².

Figures 2.781 and 2.782 show the mean and median RCS for the main ranges of sounding azimuths (nose, side, and tail) and for ranges of 20°.

The expressions and parameters of probability distributions that are most consistent with the empirical distributions of the square root of the RCS for different ranges of sounding azimuths and polarizations are given in the electronic appendix to this book. The expressions and parameters of the probability distributions that are most consistent with the empirical distributions of the RCS (energy characteristic) for different ranges of sounding azimuths and polarizations are also given there.

FIGURE 2.779 Circular diagram of RCS given radar observation of 3M-54 model at a carrier frequency of 1 GHz (30 cm wavelength).

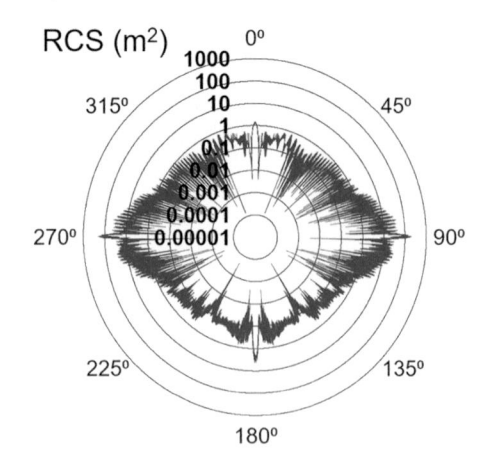

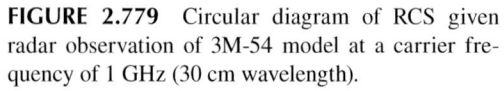

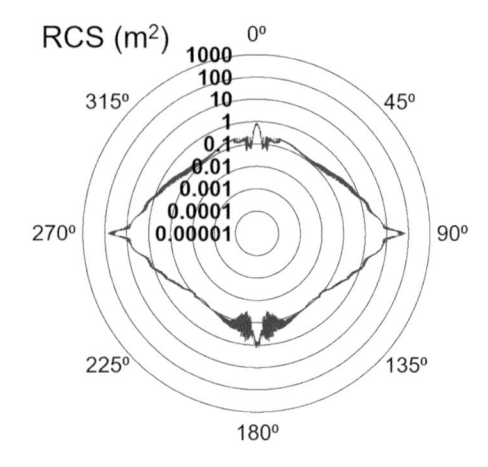

FIGURE 2.780 Circular diagram of noncoherent RCS given radar observation of 3M-54 model at a carrier frequency of 1 GHz (30 cm wavelength).

Scattering Characteristics of Aerial Objects

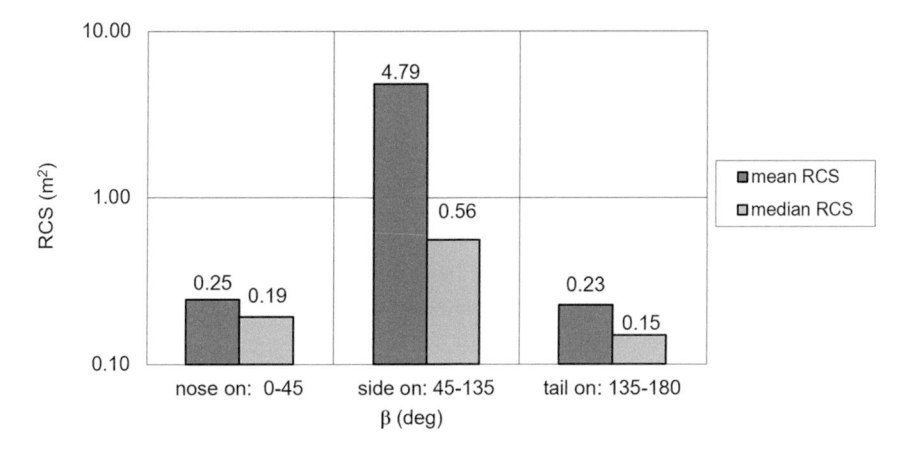

FIGURE 2.781 Diagrams of mean and median RCS of 3M-54 model in three sectors of azimuth aspect given its radar observation at horizontal polarization and a carrier frequency of 1 GHz (30 cm wavelength).

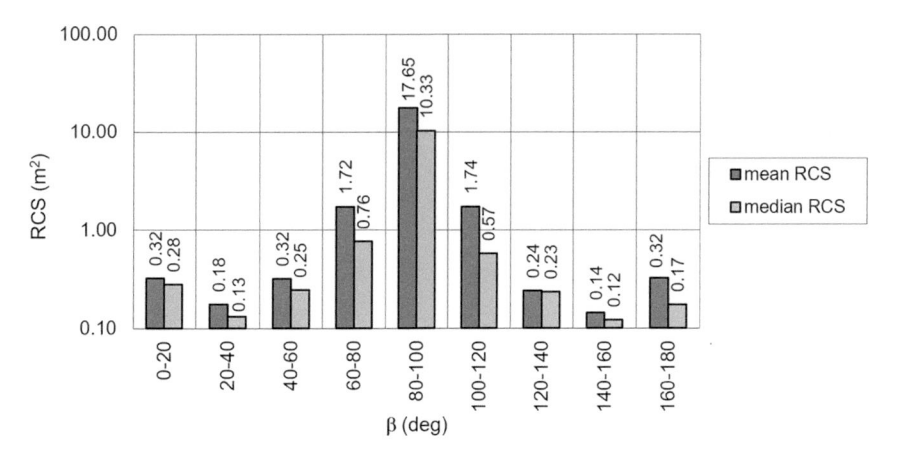

FIGURE 2.782 Diagrams of mean and median RCS of 3M-54 model in 20-degree sectors of azimuth aspect given its radar observation at horizontal polarization and a carrier frequency of 1 GHz (30 cm wavelength).

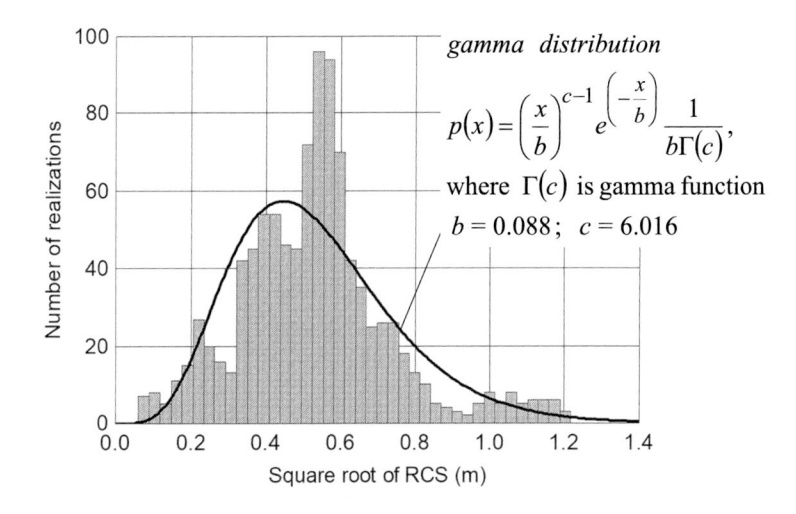

gamma distribution

$$p(x) = \left(\frac{x}{b}\right)^{c-1} e^{\left(-\frac{x}{b}\right)} \frac{1}{b\Gamma(c)},$$

where $\Gamma(c)$ is gamma function

$b = 0.088$; $c = 6.016$

FIGURE 2.783 Amplitude distribution of echo signal of 3M-54 model at a carrier frequency of 1 GHz given its horizontal polarization.

2.47 CRUISE MISSILE AGM-86C CALCM

AGM-86C CALCM - (Figure 2.784) is an air-to-ground strategic cruise missile. The first flight took place in 1986.

General characteristics of AGM-86C CALCM [58]: wingspan –3.66 m, length – 6.32 m, fuselage diameter – 0.622 m, weight – 1 450 kg, powerplant – 1 F-107-WR-101 turbofan engine, speed – 800 km/h, range – 1200 km.

In accordance with the design of the AGM-86C CALCM a model of its surface was created to obtaining scattering characteristics (in particular, RCS). The model is shown in Figure 2.785. The smooth parts of the model surface were approximated by parts of 12 triaxial ellipsoids. The surface breaks were modeled using 15 straight edge scattering parts.

FIGURE 2.784 Cruise missile AGM-86C CALCM.

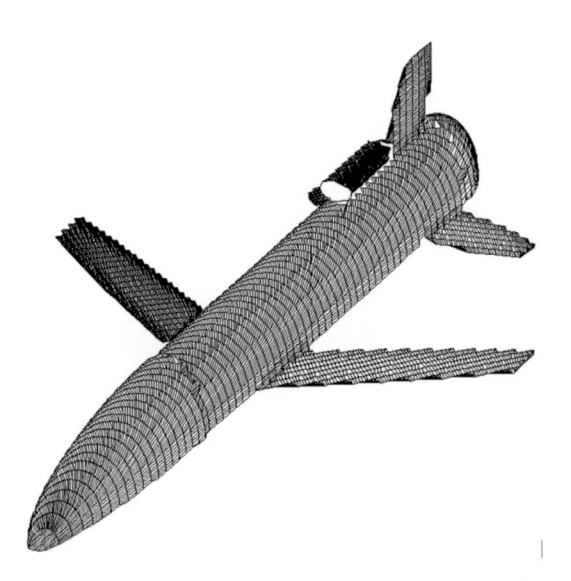

FIGURE 2.785 Surface model of cruise missile AGM-86C CALCM.

Below are some scattering characteristics of the AGM-86C model at sounding frequencies of 10, 3, and 1 GHz (wavelengths of 3, 10, and 30 cm, respectively) for horizontal polarization of the probing signal. The scattering characteristics for this model at two polarizations, as well as scattering characteristics at sounding frequencies of 5 and 1.3 GHz (wavelengths of 6 and 23 cm, respectively), are given in the electronic appendix of this book.

Scattering Characteristics of Aerial Objects

Sounding parameters: The elevation angle is to be a random value distributed uniformly in the range −3° ± 4° with respect to the wing plane (elevation angle of −3° corresponds to the radar observation from the lower hemisphere), azimuth aspect increment was 0.02°, the azimuth being counted off from the nose-on aspect (0° corresponds to the nose-on radar observation, 180° corresponds to the tail-on observation).

Scattering characteristics of AGM-86C model for sounding frequency 10 GHz (wavelength 3 cm).

Figure 2.786 shows the RCS circular diagram of the AGM-86C model. The noncoherent RCS circular diagram of the AGM-86C is shown in Figure 2.787.

The circular mean RCS of the AGM-86C model for horizontal polarization is 2.674 m^2. The circular median RCS (the RCS value used to calculate the detection range of an object with a probability of 0.5) for horizontal polarization is 0.065 m^2.

Figures 2.788 and 2.789 show the mean and median RCS for the main ranges of sounding azimuths (nose, side, and tail) and for ranges of 20°.

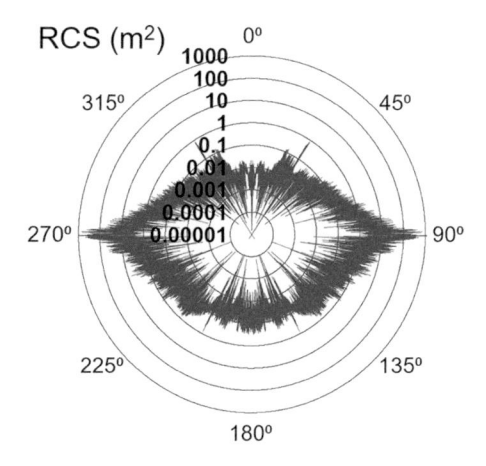

FIGURE 2.786 Circular diagram of RCS given radar observation of AGM-86C model at a carrier frequency of 10 GHz (3 cm wavelength).

FIGURE 2.787 Circular diagram of noncoherent RCS given radar observation of AGM-86C model at a carrier frequency of 10 GHz (3 cm wavelength).

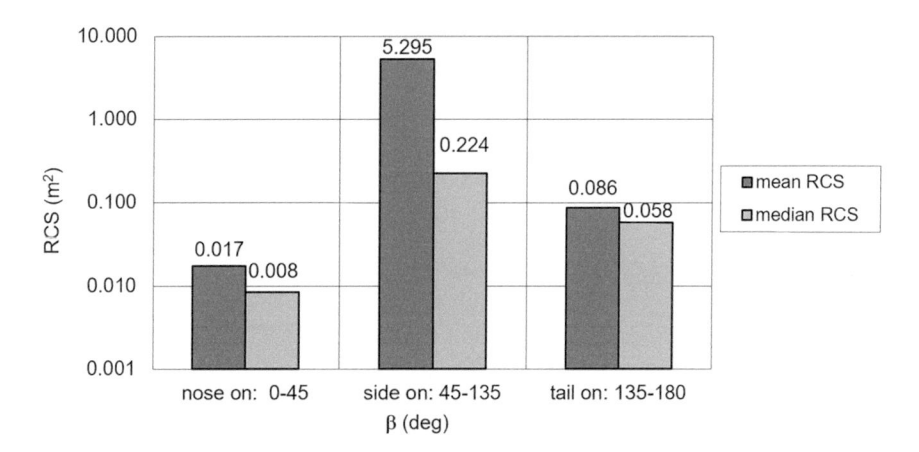

FIGURE 2.788 Diagrams of mean and median RCS of AGM-86C model in three sectors of azimuth aspect given its radar observation at horizontal polarization and a carrier frequency of 10 GHz (3 cm wavelength).

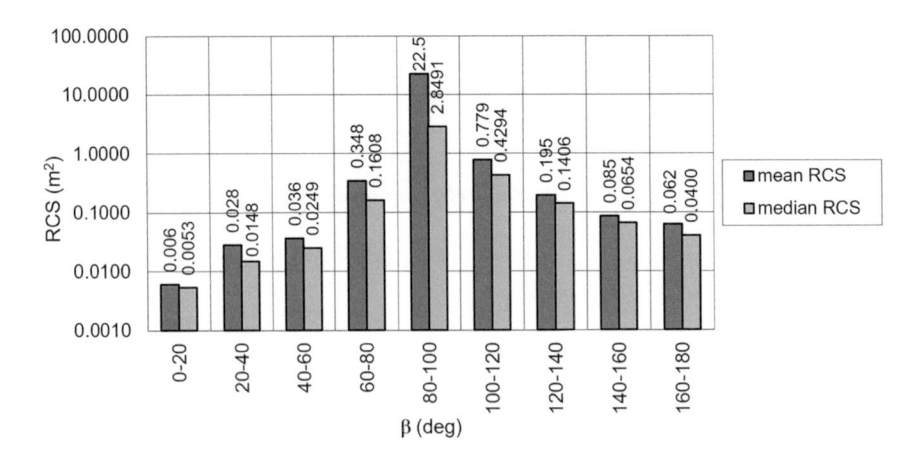

FIGURE 2.789 Diagrams of mean and median RCS of AGM-86C model in 20-degree sectors of azimuth aspect given its radar observation at horizontal polarization and a carrier frequency of 10 GHz (3 cm wavelength).

Figures 2.790, 2.795, and 2.800 show histograms of the scattered signal amplitude (square root of the RCS) for the range of sounding azimuths −20° to +20° (sounding from the nose). The bold line shows the probability density functions of the distribution, which can be used to approximate the histogram of the signal amplitude.

$$p(x) = \frac{c}{b}\left(\frac{x}{b}\right)^{c-1} e^{-\left(\frac{x}{b}\right)^c}$$

$b = 0.080; c = 2.734$

Weibull distribution

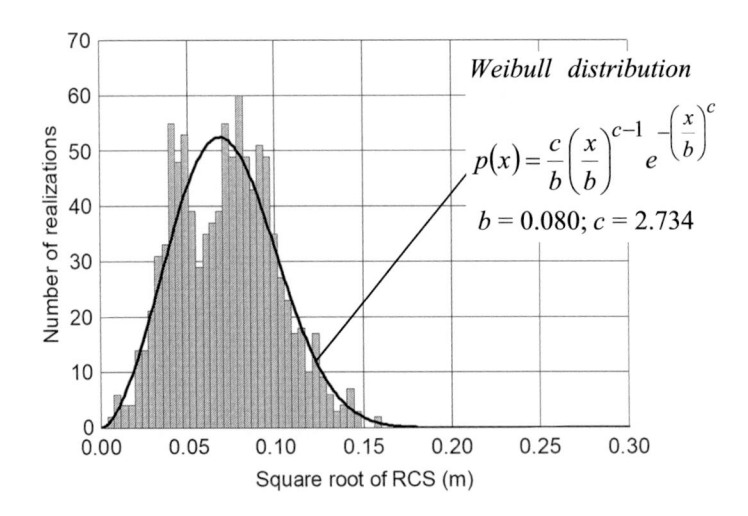

FIGURE 2.790 Amplitude distribution of echo signal of AGM-86C model at a carrier frequency of 10 GHz given its horizontal polarization.

Scattering characteristics of AGM-86C model for sounding frequency 3 GHz (wavelength 10 cm).

Figure 2.791 shows the RCS circular diagram of the AGM-86C. The noncoherent RCS circular diagram of the AGM-86C is shown in Figure 2.792.

The circular mean RCS of the AGM-86C for horizontal polarization is 2.074 m². The circular median RCS for horizontal polarization is 0.097 m².

Figures 2.793 and 2.794 show the mean and median RCS for the main ranges of sounding azimuths (nose, side, and tail) and for ranges of 20°.

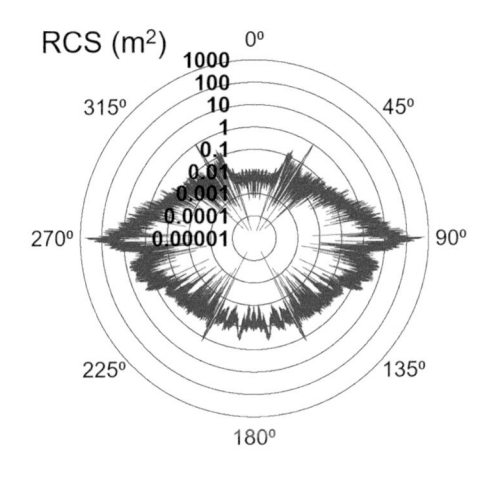

Scattering Characteristics of Aerial Objects 291

FIGURE 2.791 Circular diagram of RCS given radar observation of AGM-86C model at a carrier frequency of 3 GHz (10 cm wavelength).

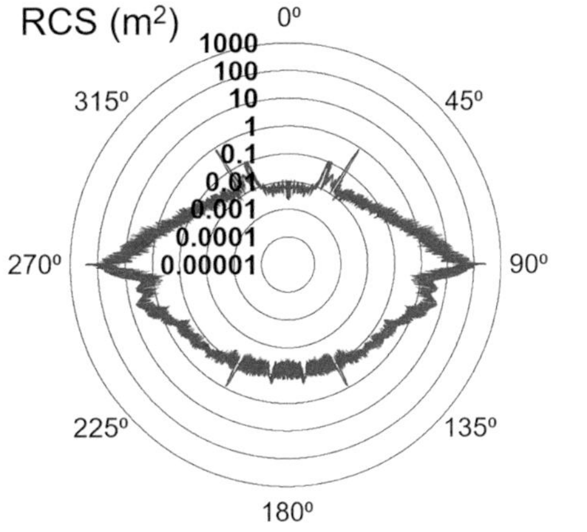

FIGURE 2.792 Circular diagram of noncoherent RCS given radar observation of AGM-86C model at a carrier frequency of 3 GHz (10 cm wavelength).

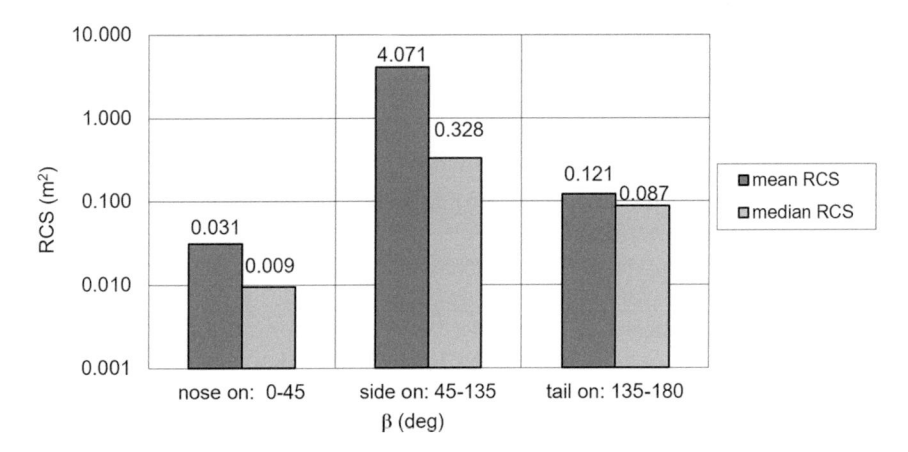

FIGURE 2.793 Diagrams of mean and median RCS of AGM-86C model in three sectors of azimuth aspect given its radar observation at horizontal polarization and a carrier frequency of 3 GHz (10 cm wavelength).

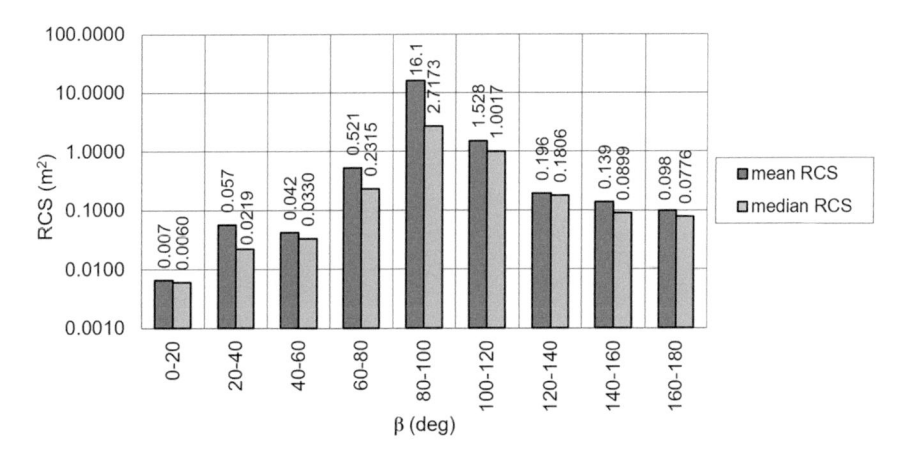

FIGURE 2.794 Diagrams of mean and median RCS of AGM-86C model in 20-degree sectors of azimuth aspect given its radar observation at horizontal polarization and a carrier frequency of 3 GHz (10 cm wavelength).

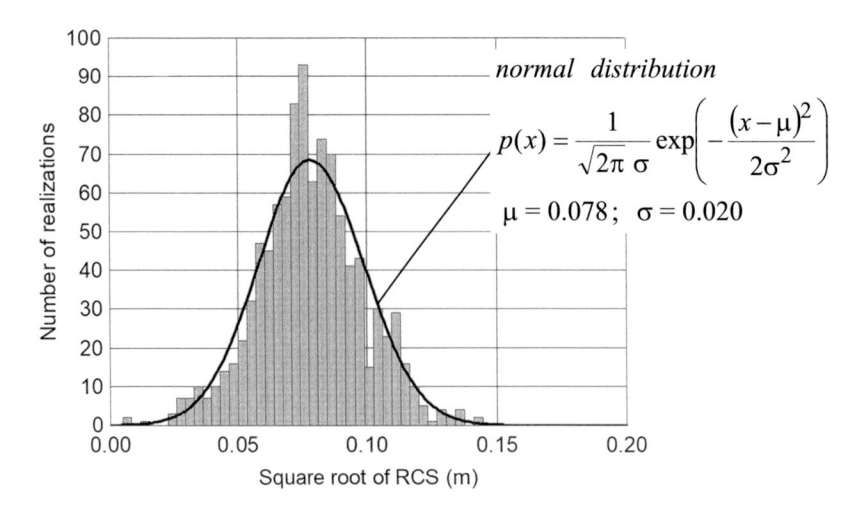

FIGURE 2.795 Amplitude distribution of echo signal of AGM-86C model at a carrier frequency of 3 GHz given its horizontal polarization.

$$p(x) = \frac{1}{\sqrt{2\pi}\,\sigma} \exp\left(-\frac{(x-\mu)^2}{2\sigma^2}\right)$$

$\mu = 0.078;\ \sigma = 0.020$

Scattering characteristics of AGM-86C model for sounding frequency 1 GHz (wavelength 30 cm).

Figure 2.796 shows the RCS circular diagram of the AGM-86C. The noncoherent RCS circular diagram of the AGM-86C is shown in Figure 2.797.

The circular mean RCS of the AGM-86C for horizontal polarization is 1.613 m². The circular median RCS for horizontal polarization is 0.135 m².

Figures 2.798 and 2.799 show the mean and median RCS for the main ranges of sounding azimuths (nose, side, and tail) and for ranges of 20°.

The expressions and parameters of probability distributions that are most consistent with the empirical distributions of the square root of the RCS for different ranges of sounding azimuths and polarizations are given in the electronic appendix to this book. The expressions and parameters of the probability distributions that are most consistent with the empirical distributions of the RCS (energy characteristic) for different ranges of sounding azimuths and polarizations are also given there.

FIGURE 2.796 Circular diagram of RCS given radar observation of AGM-86C model at a carrier frequency of 1 GHz (30 cm wavelength).

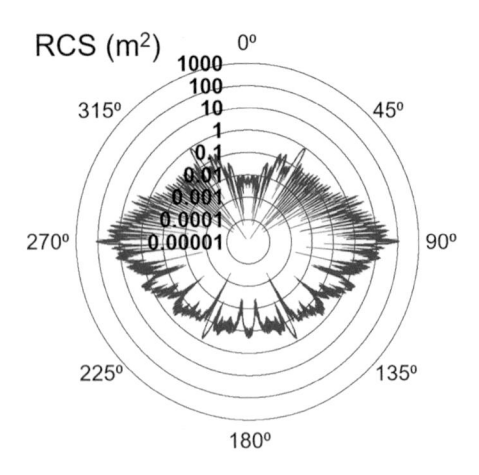

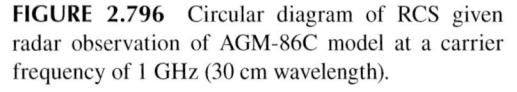

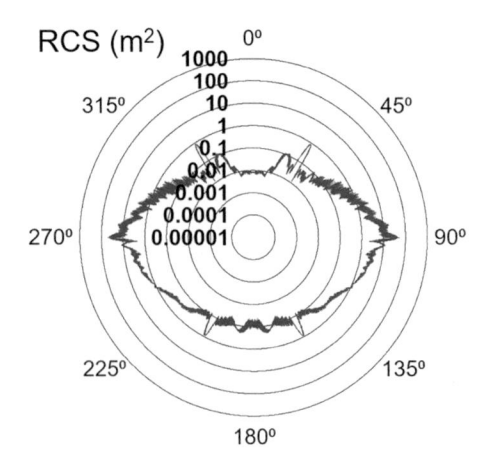

FIGURE 2.797 Circular diagram of noncoherent RCS given radar observation of AGM-86C model at a carrier frequency of 1 GHz (30 cm wavelength).

Scattering Characteristics of Aerial Objects

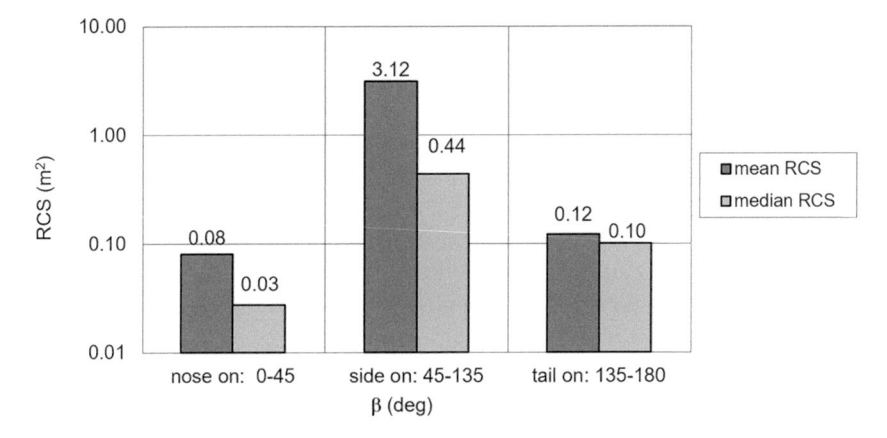

FIGURE 2.798 Diagrams of mean and median RCS of AGM-86C model in three sectors of azimuth aspect given its radar observation at horizontal polarization and a carrier frequency of 1 GHz (30 cm wavelength).

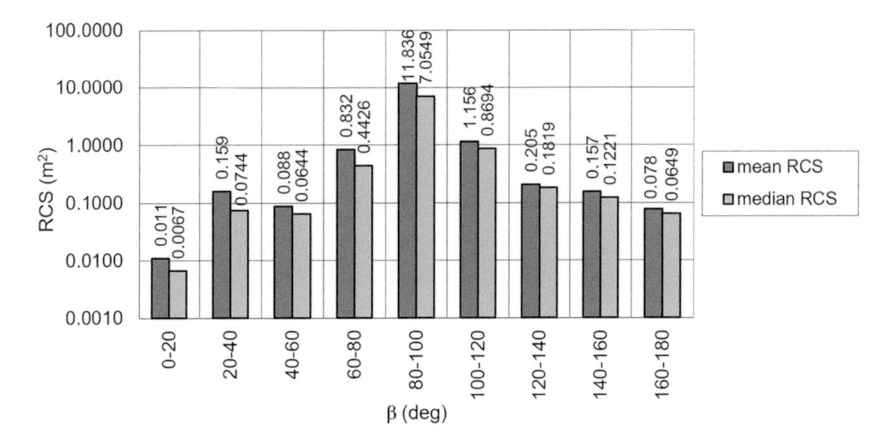

FIGURE 2.799 Diagrams of mean and median RCS of AGM-86C model in 20-degree sectors of azimuth aspect given its radar observation at horizontal polarization and a carrier frequency of 1 GHz (30 cm wavelength).

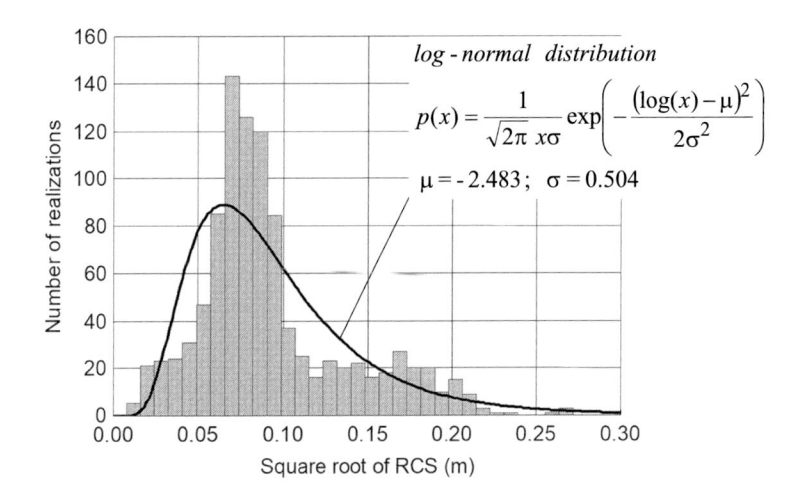

$$p(x) = \frac{1}{\sqrt{2\pi}\, x\sigma} \exp\left(-\frac{(\log(x)-\mu)^2}{2\sigma^2}\right)$$

$\mu = -2.483;\ \sigma = 0.504$

FIGURE 2.800 Amplitude distribution of echo signal of AGM-86C model at a carrier frequency of 1 GHz given its horizontal polarization.

2.48 HYPERSONIC MISSILE Kh-47M2 KINZHAL

Kh-47M2 Kinzhal - AS-24 Killjoy - (Figure 2.801) is a hypersonic air-launched ballistic missile. The first flight took place in 2017.

General characteristics of Kh-47M2 Kinzhal [59]: wingspan –1.48 m, length – 7.00 m, fuselage diameter – 0.92 m, weight – 3 990 kg, powerplant – 1 solid-propellant rocket motor, speed – 10 M, range – 1 000–2 000 km.

In accordance with the design of the Kh-47M2 Kinzhal a model of its surface was created to obtaining scattering characteristics (in particular, RCS). The model is shown in Figure 2.802. The smooth parts of the model surface were approximated by parts of 46 triaxial ellipsoids. The surface breaks were modeled using 12 straight edge scattering parts.

FIGURE 2.801 Hypersonic missile Kh-47M2 Kinzhal.

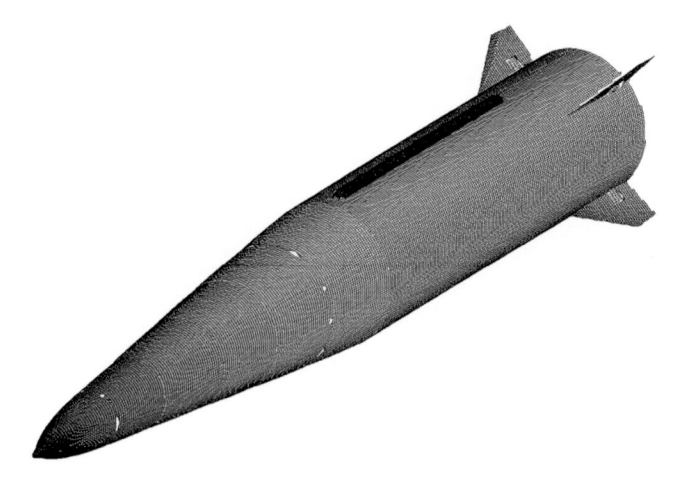

FIGURE 2.802 Surface model of hypersonic missile Kh-47M2 Kinzhal.

Below are some scattering characteristics of the Kh-47M2 Kinzhal model at sounding frequencies of 10, 3, and 1 GHz (wavelengths of 3, 10, and 30 cm, respectively) for horizontal polarization of the probing signal. The scattering characteristics for this model at two polarizations, as well as scattering characteristics at sounding frequencies of 5 and 1.3 GHz (wavelengths of 6 and 23 cm, respectively), are given in the electronic appendix of this book.

Sounding parameters: The elevation angle is to be a random value distributed uniformly in the range −3° ± 4° with respect to the wing plane (elevation angle of −3° corresponds to the radar observation from the lower hemisphere), azimuth aspect increment was 0.02°, the azimuth being counted off from the nose-on aspect (0° corresponds to the nose-on radar observation, 180° corresponds to the tail-on observation).

Scattering characteristics of Kh-47M2 Kinzhal model for sounding frequency 10 GHz (wavelength 3 cm).

Figure 2.803 shows the RCS circular diagram of the Kh-47M2 Kinzhal model. The noncoherent RCS circular diagram of the Kh-47M2 Kinzhal is shown in Figure 2.804.

The circular mean RCS of the Kh-47M2 Kinzhal model for horizontal polarization is 3.435 m². The circular median RCS (the RCS value used to calculate the detection range of an object with a probability of 0.5) for horizontal polarization is 0.139 m².

Figures 2.805 and 2.806 show the mean and median RCS for the main ranges of sounding azimuths (nose, side, and tail) and for ranges of 20°.

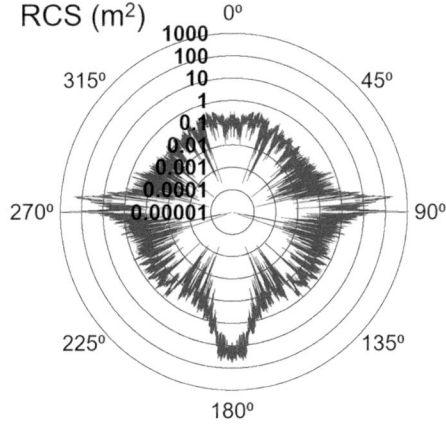

FIGURE 2.803 Circular diagram of RCS given radar observation of Kh-47M2 Kinzhal model at a carrier frequency of 10 GHz (3 cm wavelength).

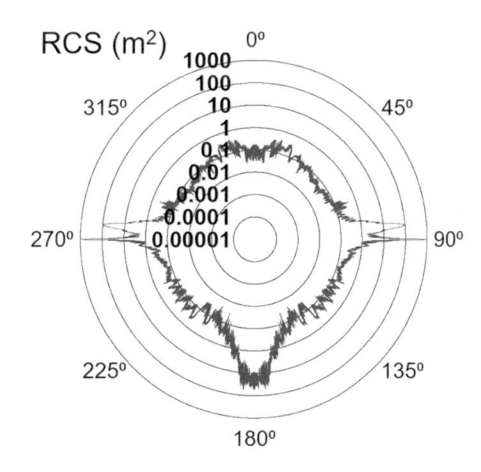

FIGURE 2.804 Circular diagram of noncoherent RCS given radar observation of Kh-47M2 Kinzhal model at a carrier frequency of 10 GHz (3 cm wavelength).

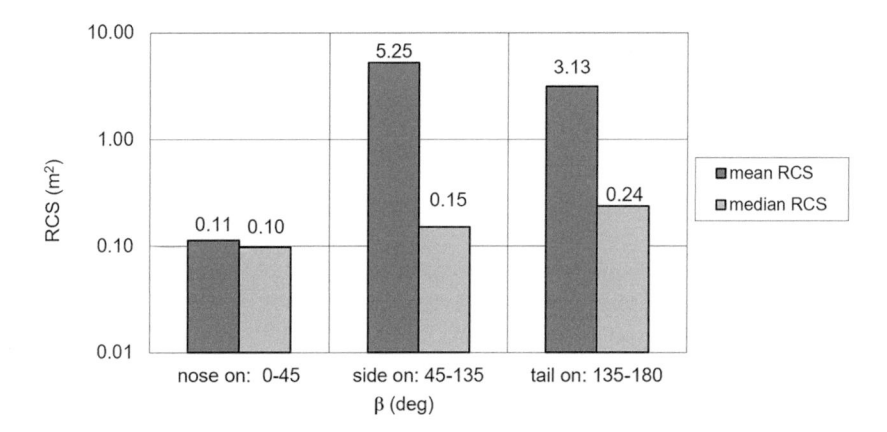

FIGURE 2.805 Diagrams of mean and median RCS of Kh-47M2 Kinzhal model in three sectors of azimuth aspect given its radar observation at horizontal polarization and a carrier frequency of 10 GHz (3 cm wavelength).

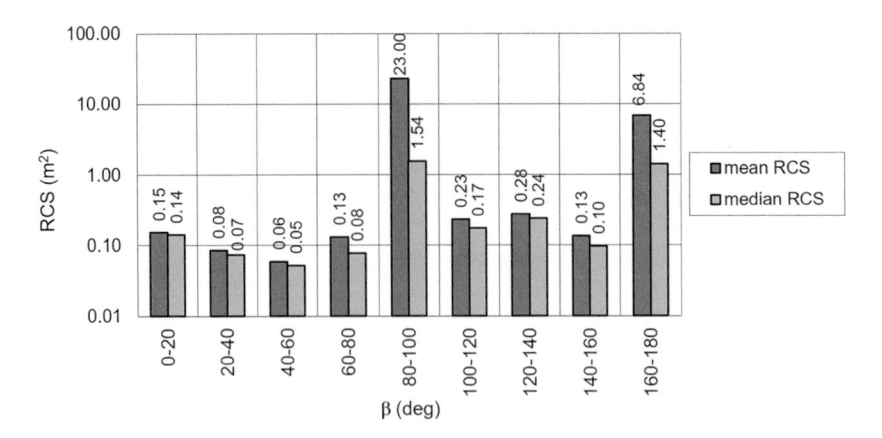

FIGURE 2.806 Diagrams of mean and median RCS of Kh-47M2 Kinzhal model in 20-degree sectors of azimuth aspect given its radar observation at horizontal polarization and a carrier frequency of 10 GHz (3 cm wavelength).

Figures 2.807, 2.812, and 2.817 show histograms of the scattered signal amplitude (square root of the RCS) for the range of sounding azimuths −20° to +20° (sounding from the nose). The bold line shows the probability density functions of the distribution, which can be used to approximate the histogram of the signal amplitude.

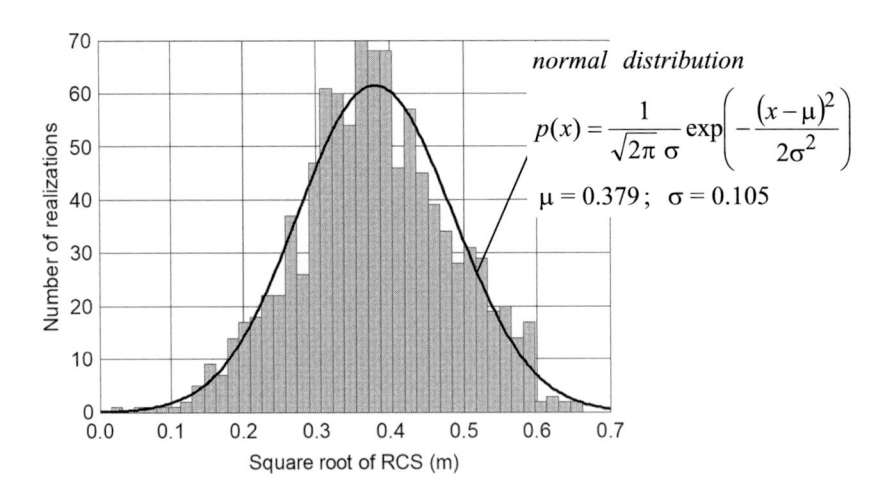

$$p(x) = \frac{1}{\sqrt{2\pi}\,\sigma} \exp\left(-\frac{(x-\mu)^2}{2\sigma^2}\right)$$

$\mu = 0.379; \quad \sigma = 0.105$

FIGURE 2.807 Amplitude distribution of echo signal of Kh-47M2 Kinzhal model at a carrier frequency of 10 GHz given its horizontal polarization.

Scattering characteristics of Kh-47M2 Kinzhal model for sounding frequency 3 GHz (wavelength 10 cm).

Figure 2.808 shows the RCS circular diagram of the Kh-47M2 Kinzhal. The noncoherent RCS circular diagram of the Kh-47M2 Kinzhal is shown in Figure 2.809.

The circular mean RCS of the Kh-47M2 Kinzhal for horizontal polarization is 3.213 m². The circular median RCS for horizontal polarization is 0.148 m².

Figures 2.810 and 2.811 show the mean and median RCS for the main ranges of sounding azimuths (nose, side, and tail) and for ranges of 20°.

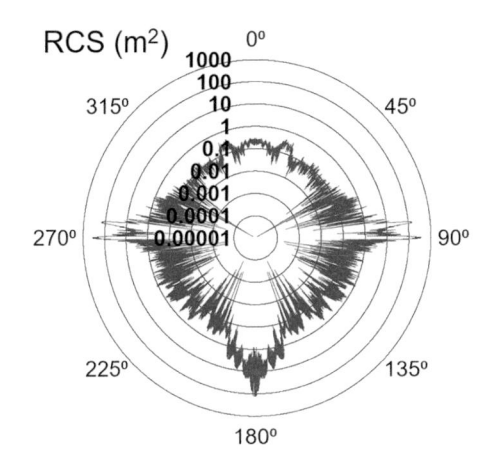

Scattering Characteristics of Aerial Objects

FIGURE 2.808 Circular diagram of RCS given radar observation of Kh-47M2 Kinzhal model at a carrier frequency of 3 GHz (10 cm wavelength).

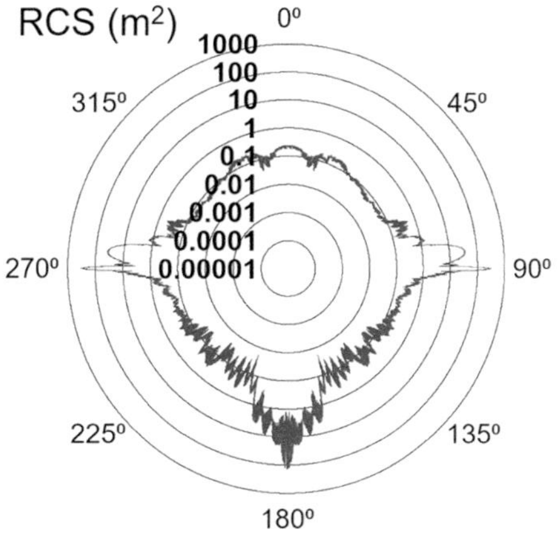

FIGURE 2.809 Circular diagram of noncoherent RCS given radar observation of Kh-47M2 Kinzhal model at a carrier frequency of 3 GHz (10 cm wavelength).

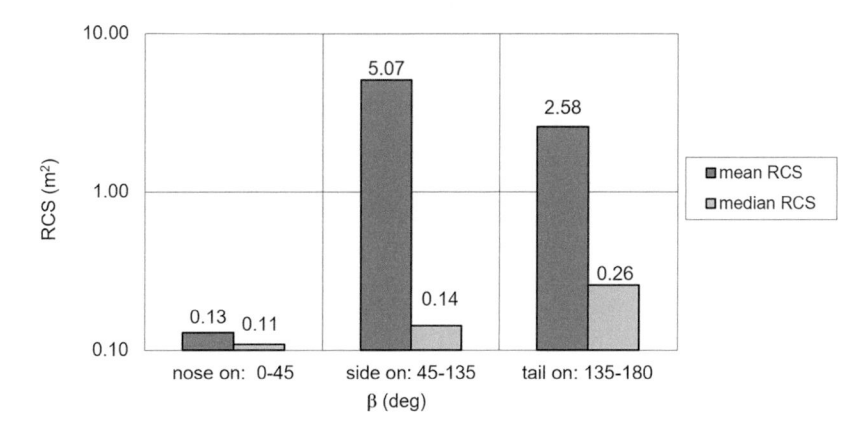

FIGURE 2.810 Diagrams of mean and median RCS of Kh-47M2 Kinzhal model in three sectors of azimuth aspect given its radar observation at horizontal polarization and a carrier frequency of 3 GHz (10 cm wavelength).

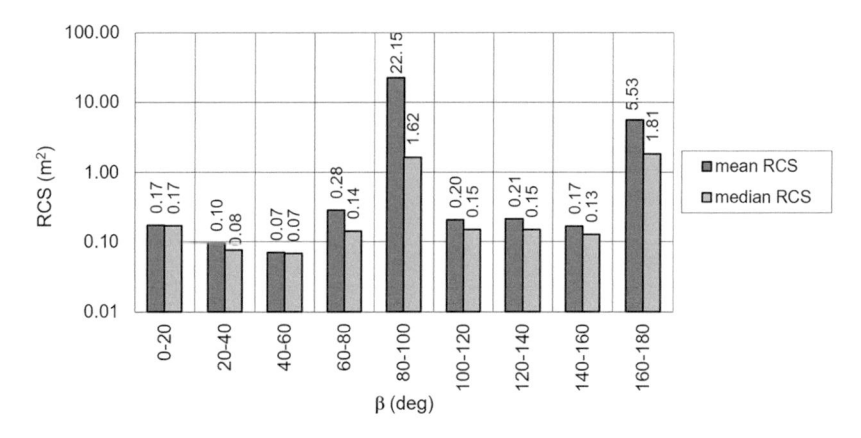

FIGURE 2.811 Diagrams of mean and median RCS of Kh-47M2 Kinzhal model in 20-degree sectors of azimuth aspect given its radar observation at horizontal polarization and a carrier frequency of 3 GHz (10 cm wavelength).

Weibull distribution

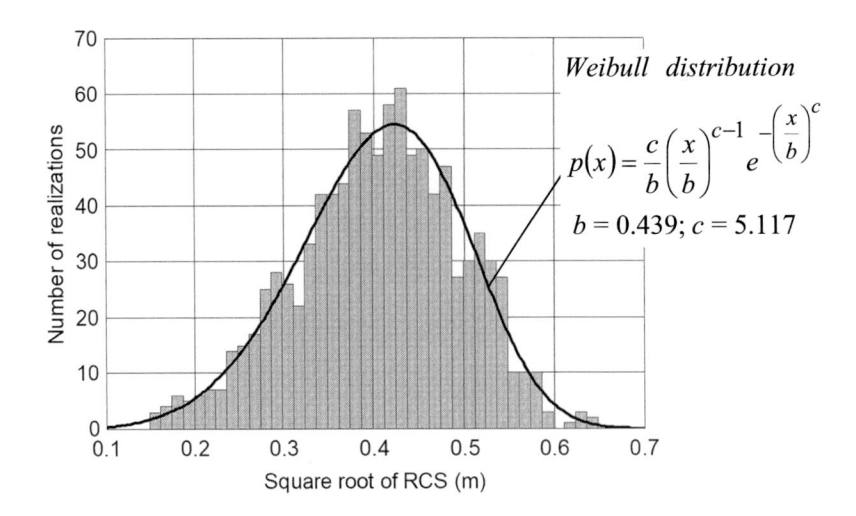

$$p(x) = \frac{c}{b}\left(\frac{x}{b}\right)^{c-1} e^{-\left(\frac{x}{b}\right)^c}$$

$b = 0.439; c = 5.117$

FIGURE 2.812 Amplitude distribution of echo signal of Kh-47M2 Kinzhal model at a carrier frequency of 3 GHz given its horizontal polarization.

Scattering characteristics of Kh-47M2 Kinzhal model for sounding frequency 1 GHz (wavelength 30 cm).

Figure 2.813 shows the RCS circular diagram of the Kh-47M2 Kinzhal. The noncoherent RCS circular diagram of the Kh-47M2 Kinzhal is shown in Figure 2.814.

The circular mean RCS of the Kh-47M2 Kinzhal model for horizontal polarization is 2.862 m². The circular median RCS for horizontal polarization is 0.265 m².

Figures 2.815 and 2.816 show the mean and median RCS for the main ranges of sounding azimuths (nose, side, and tail) and for ranges of 20°.

The expressions and parameters of probability distributions that are most consistent with the empirical distributions of the square root of the RCS for different ranges of sounding azimuths and polarizations are given in the electronic appendix to this book. The expressions and parameters of the probability distributions that are most consistent with the empirical distributions of the RCS (energy characteristic) for different ranges of sounding azimuths and polarizations are also given there.

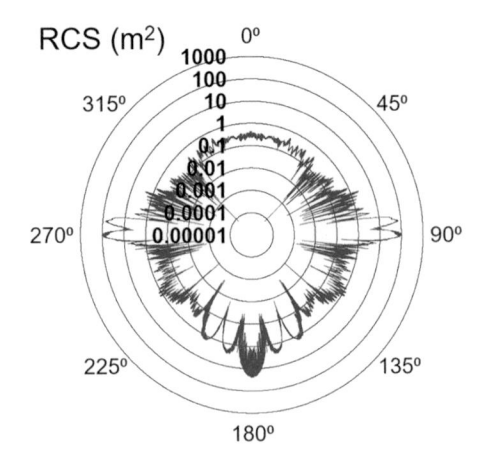

FIGURE 2.813 Circular diagram of RCS given radar observation of Kh-47M2 Kinzhal model at a carrier frequency of 1 GHz (30 cm wavelength).

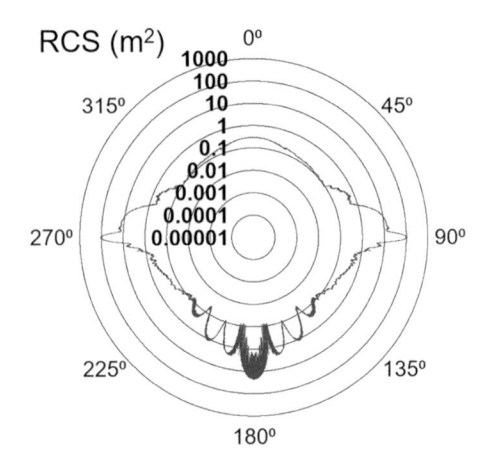

FIGURE 2.814 Circular diagram of noncoherent RCS given radar observation of Kh-47M2 Kinzhal model at a carrier frequency of 1 GHz (30 cm wavelength).

Scattering Characteristics of Aerial Objects

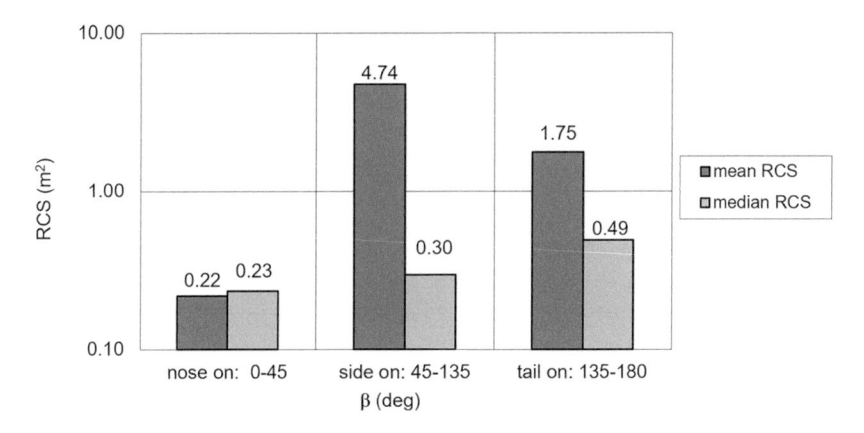

FIGURE 2.815 Diagrams of mean and median RCS of Kh-47M2 Kinzhal model in three sectors of azimuth aspect given its radar observation at horizontal polarization and a carrier frequency of 1 GHz (30 cm wavelength).

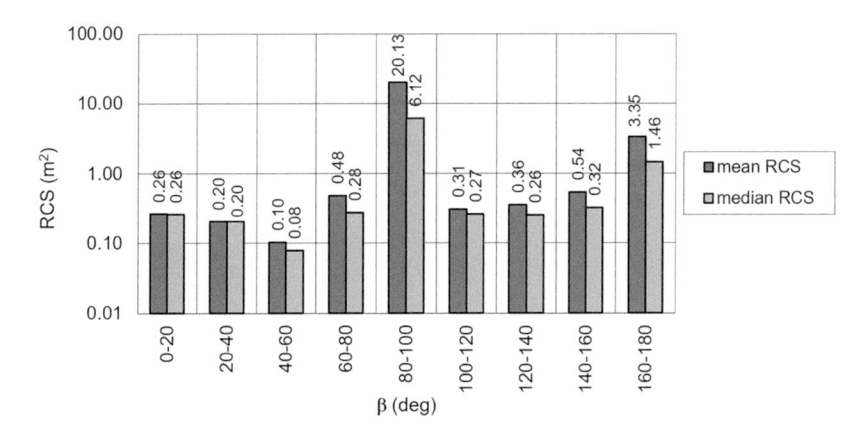

FIGURE 2.816 Diagrams of mean and median RCS of Kh-47M2 Kinzhal model in 20-degree sectors of azimuth aspect given its radar observation at horizontal polarization and a carrier frequency of 1 GHz (30 cm wavelength).

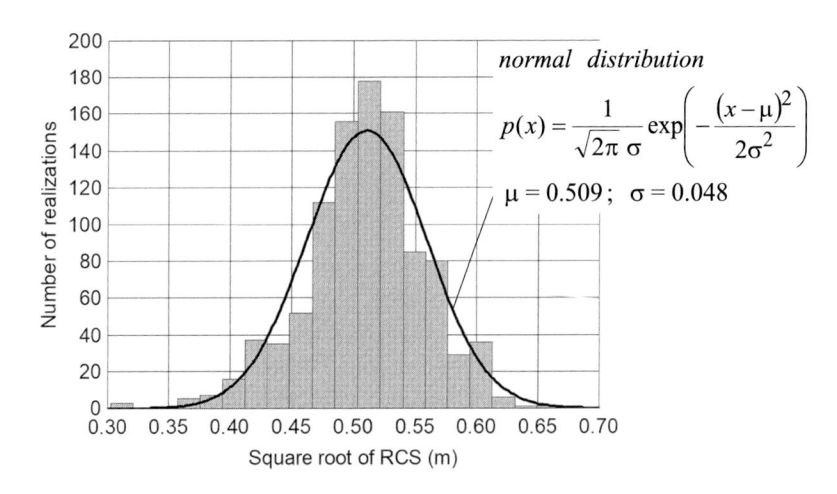

$$p(x) = \frac{1}{\sqrt{2\pi}\,\sigma} \exp\left(-\frac{(x-\mu)^2}{2\sigma^2}\right)$$

$\mu = 0.509;\ \sigma = 0.048$

FIGURE 2.817 Amplitude distribution of echo signal of Kh-47M2 Kinzhal model at a carrier frequency of 1 GHz given its horizontal polarization.

2.49 ANTI-RADIATION MISSILE Kh-25MPU

Kh-25MPU - AS-12 Kegler - (Figure 2.818) is an air-launched anti-radiation missile. The first flight took place in 1974.

General characteristics of Kh-25MPU [60]: wingspan – 0.755–0,820 m, length – 4.30 m, fuselage diameter – 0.276 m, weight – 310–320 kg, powerplant – 1 solid-propellant PRD-276, speed – 400–900 m/s, range – 3–40 km.

In accordance with the design of the Kh-25MPU a model of its surface was created to obtaining scattering characteristics (in particular, RCS). The model is shown in Figure 2.819. The perfectly conducting smooth parts of the model surface were approximated by parts of 35 triaxial ellipsoids. Dielectric parts of the model surface were approximated by parts of 4 triaxial ellipsoids The surface breaks were modeled using 36 straight edge scattering parts.

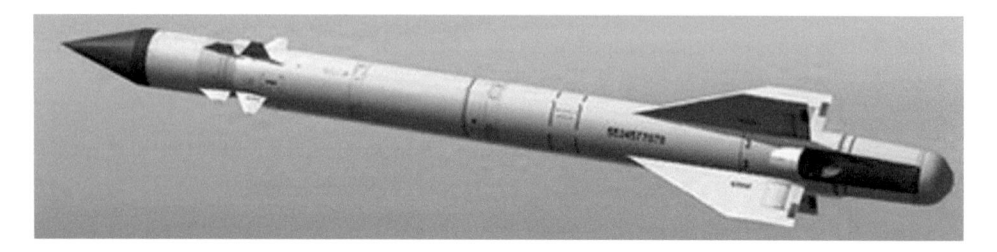

FIGURE 2.818 Anti-radiation missile Kh25MPU.

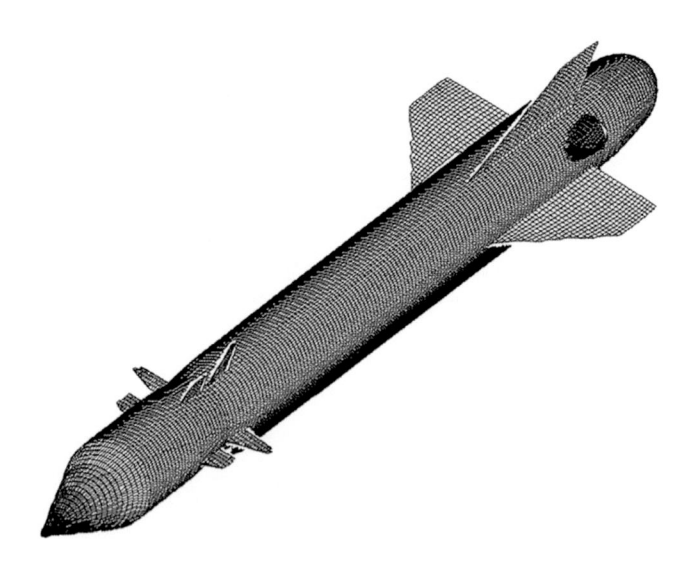

FIGURE 2.819 Surface model of anti-radiation missile Kh-25MPU.

Below are some scattering characteristics of the Kh-25MPU model at sounding frequencies of 10, 3, and 1 GHz (wavelengths of 3, 10, and 30 cm, respectively) for horizontal polarization of the probing signal. The scattering characteristics for this model at two polarizations, as well as scattering characteristics at sounding frequencies of 5 and 1.3 GHz (wavelengths of 6 and 23 cm, respectively), are given in the electronic appendix of this book.

Sounding parameters: The elevation angle is to be a random value distributed uniformly in the range $-3° \pm 4°$ with respect to the wing plane (elevation angle of $-3°$ corresponds to the radar observation from the lower hemisphere), azimuth aspect increment was $0.02°$, the azimuth being counted

off from the nose-on aspect (0° corresponds to the nose-on radar observation, 180° corresponds to the tail-on observation).

Scattering characteristics of Kh-25MPU model for sounding frequency 10 GHz (wavelength 3 cm).

Figure 2.820 shows the RCS circular diagram of the Kh-25MPU model. The noncoherent RCS circular diagram of the Kh-25MPU is shown in Figure 2.821.

The circular mean RCS of the Kh-25MPU model for horizontal polarization is 1.655 m². The circular median RCS (the RCS value used to calculate the detection range of an object with a probability of 0.5) for horizontal polarization is 0.032 m².

Figures 2.822 and 2.823 show the mean and median RCS for the main ranges of sounding azimuths (nose, side, and tail) and for ranges of 20°.

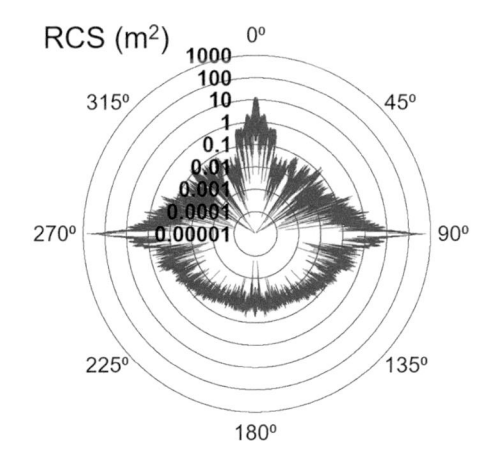

FIGURE 2.820 Circular diagram of RCS given radar observation of Kh-25MPU model at a carrier frequency of 10 GHz (3 cm wavelength).

FIGURE 2.821 Circular diagram of noncoherent RCS given radar observation of Kh-25MPU model at a carrier frequency of 10 GHz (3 cm wavelength).

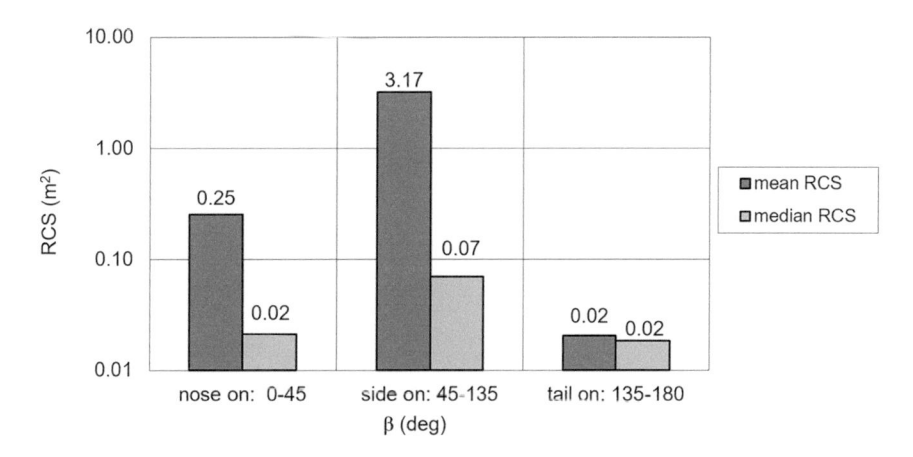

FIGURE 2.822 Diagrams of mean and median RCS of Kh-25MPU model in three sectors of azimuth aspect given its radar observation at horizontal polarization and a carrier frequency of 10 GHz (3 cm wavelength).

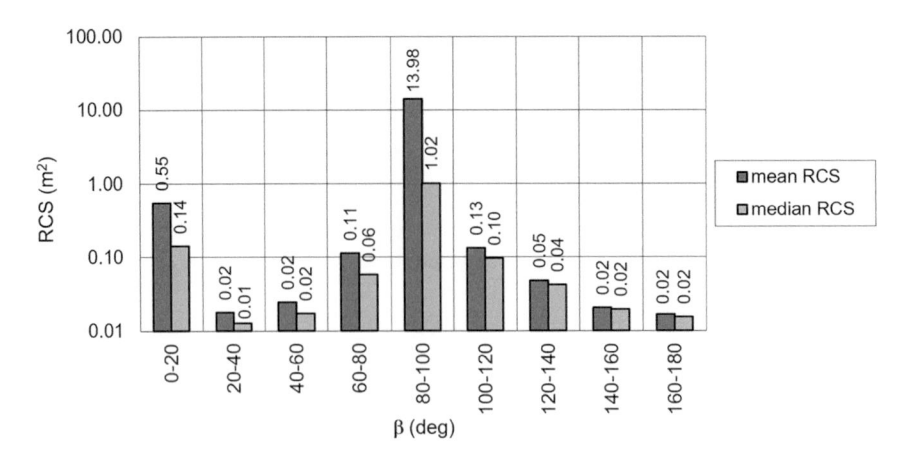

FIGURE 2.823 Diagrams of mean and median RCS of Kh-25MPU model in 20-degree sectors of azimuth aspect given its radar observation at horizontal polarization and a carrier frequency of 10 GHz (3 cm wavelength).

Figures 2.824, 2.829, and 2.834 show histograms of the scattered signal amplitude (square root of the RCS) for the range of sounding azimuths −20° to +20° (sounding from the nose). The bold line shows the probability density functions of the distribution, which can be used to approximate the histogram of the signal amplitude.

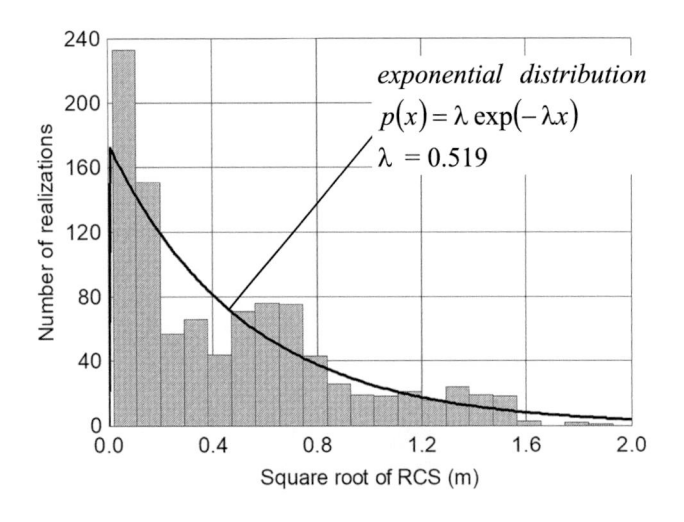

FIGURE 2.824 Amplitude distribution of echo signal of Kh-25MPU model at a carrier frequency of 10 GHz given its horizontal polarization.

Scattering characteristics of Kh-25MPU model for sounding frequency 3 GHz (wavelength 10 cm).

Figure 2.825 shows the RCS circular diagram of the Kh-25MPU. The noncoherent RCS circular diagram of the Kh-25MPU is shown in Figure 2.826.

The circular mean RCS of the Kh-25MPU for horizontal polarization is 1.201 m². The circular median RCS for horizontal polarization is 0.048 m².

Figures 2.827 and 2.828 show the mean and median RCS for the main ranges of sounding azimuths (nose, side, and tail) and for ranges of 20°.

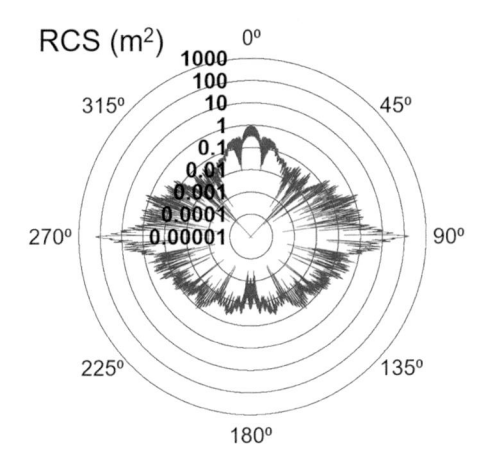

Scattering Characteristics of Aerial Objects

FIGURE 2.825 Circular diagram of RCS given radar observation of Kh-25MPU model at a carrier frequency of 3 GHz (10 cm wavelength).

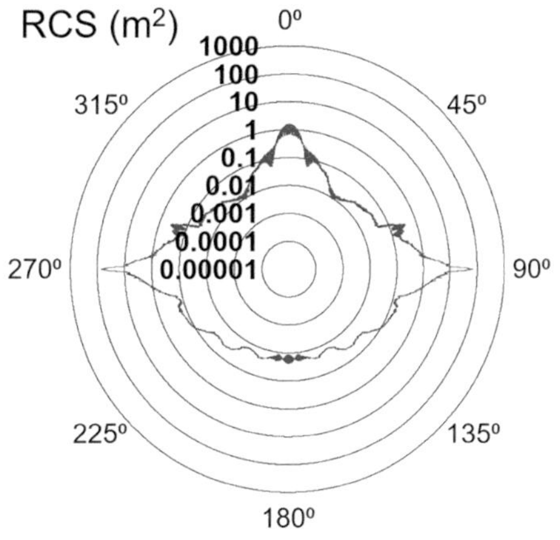

FIGURE 2.826 Circular diagram of noncoherent RCS given radar observation of Kh-25MPU model at a carrier frequency of 3 GHz (10 cm wavelength).

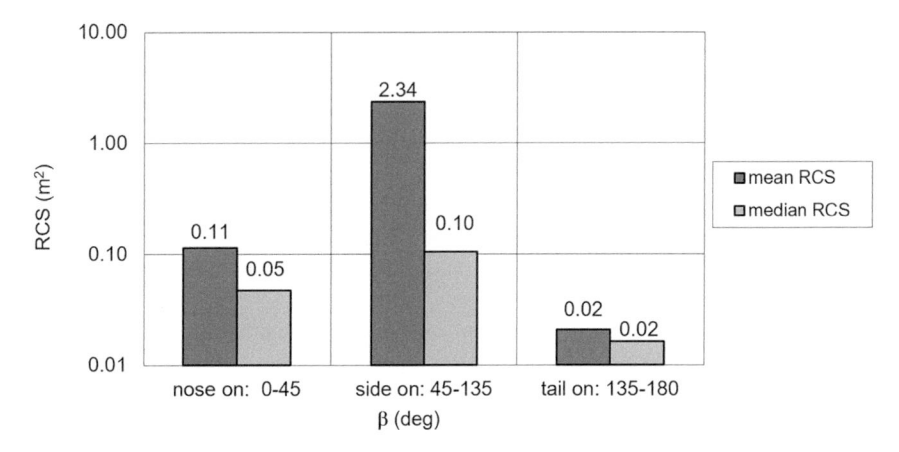

FIGURE 2.827 Diagrams of mean and median RCS of Kh-25MPU model in three sectors of azimuth aspect given its radar observation at horizontal polarization and a carrier frequency of 3 GHz (10 cm wavelength).

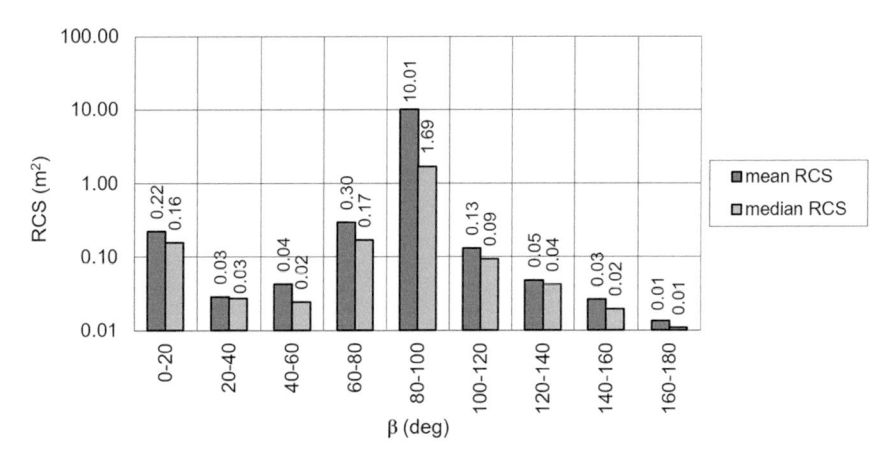

FIGURE 2.828 Diagrams of mean and median RCS of Kh-25MPU model in 20-degree sectors of azimuth aspect given its radar observation at horizontal polarization and a carrier frequency of 3 GHz (10 cm wavelength).

$$p(x) = \frac{1}{\sqrt{2\pi}\, x\sigma} \exp\left(-\frac{(\log(x)-\mu)^2}{2\sigma^2}\right)$$

log - normal distribution

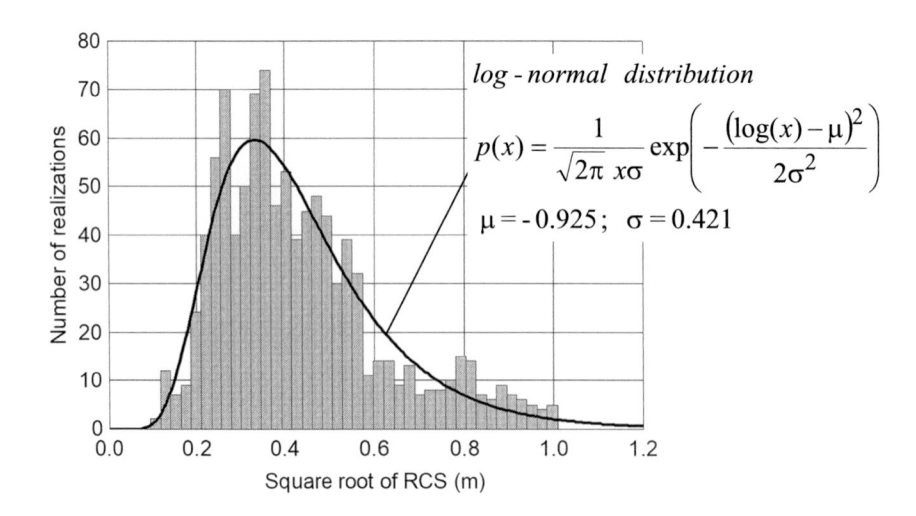

$\mu = -0.925;\ \sigma = 0.421$

FIGURE 2.829 Amplitude distribution of echo signal of Kh-25MPU model at a carrier frequency of 3 GHz given its horizontal polarization.

Scattering characteristics of Kh-25MPU model for sounding frequency 1 GHz (wavelength 30 cm).

Figure 2.830 shows the RCS circular diagram of the Kh-25MPU. The noncoherent RCS circular diagram of the Kh-25MPU is shown in Figure 2.831.

The circular mean RCS of the Kh-25MPU model for horizontal polarization is 0.619 m². The circular median RCS for horizontal polarization is 0.170 m².

Figures 2.832 and 2.833 show the mean and median RCS for the main ranges of sounding azimuths (nose, side, and tail) and for ranges of 20°.

The expressions and parameters of probability distributions that are most consistent with the empirical distributions of the square root of the RCS for different ranges of sounding azimuths and polarizations are given in the electronic appendix to this book. The expressions and parameters of the probability distributions that are most consistent with the empirical distributions of the RCS (energy characteristic) for different ranges of sounding azimuths and polarizations are also given there.

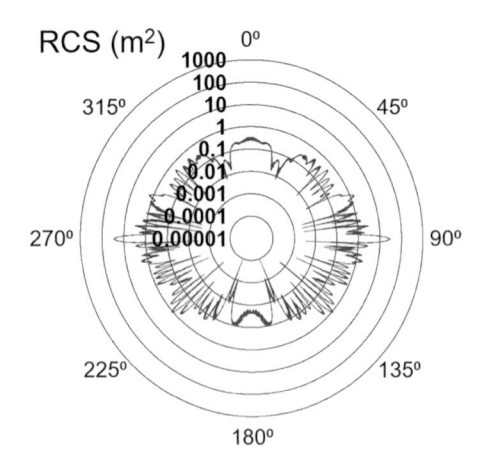

FIGURE 2.830 Circular diagram of RCS given radar observation of Kh-25MPU model at a carrier frequency of 1 GHz (30 cm wavelength).

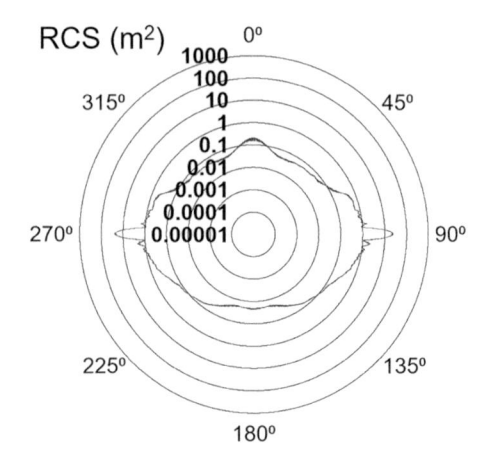

FIGURE 2.831 Circular diagram of noncoherent RCS given radar observation of Kh-25MPU model at a carrier frequency of 1 GHz (30 cm wavelength).

Scattering Characteristics of Aerial Objects

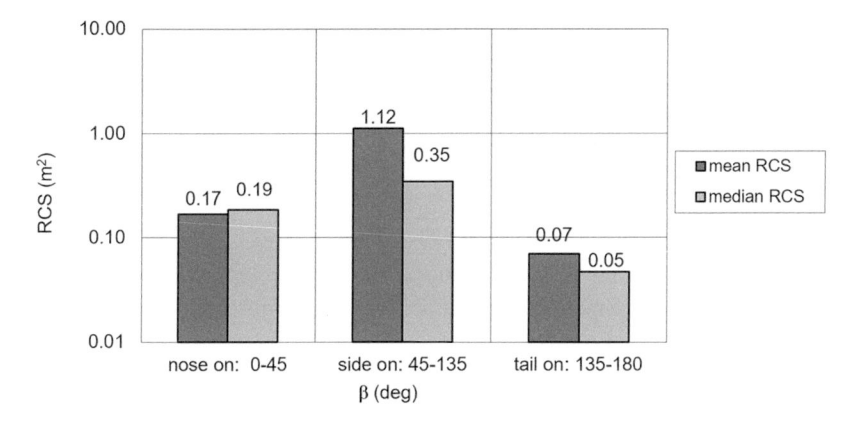

FIGURE 2.832 Diagrams of mean and median RCS of Kh-25MPU model in three sectors of azimuth aspect given its radar observation at horizontal polarization and a carrier frequency of 1 GHz (30 cm wavelength).

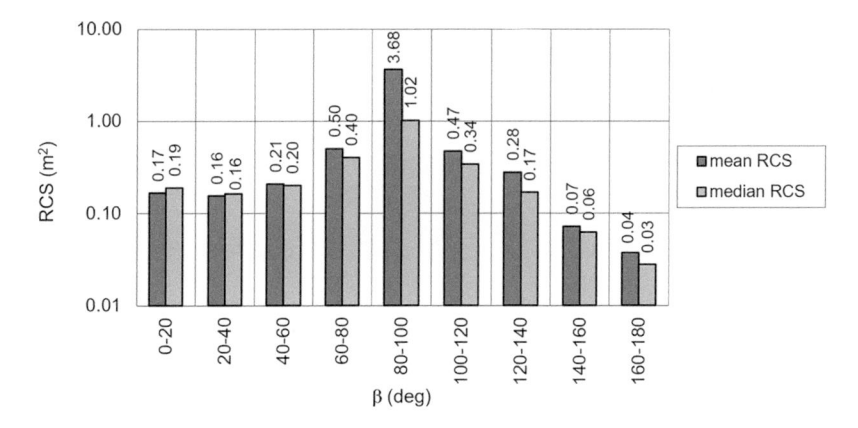

FIGURE 2.833 Diagrams of mean and median RCS of Kh-25MPU model in 20-degree sectors of azimuth aspect given its radar observation at horizontal polarization and a carrier frequency of 1 GHz (30 cm wavelength).

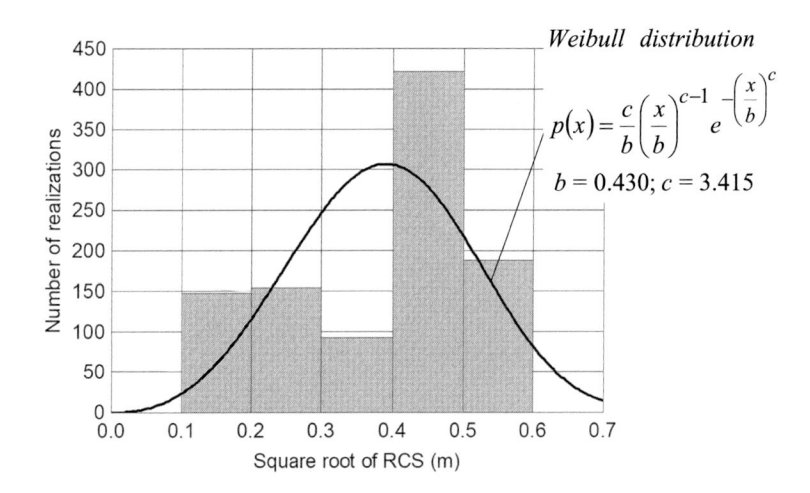

$$p(x) = \frac{c}{b}\left(\frac{x}{b}\right)^{c-1} e^{-\left(\frac{x}{b}\right)^c}$$

$b = 0.430; c = 3.415$

FIGURE 2.834 Amplitude distribution of echo signal of Kh-25MPU model at a carrier frequency of 1 GHz given its horizontal polarization.

2.50 AIR-TO-SURFACE MISSILE Kh-29T

Kh-29T - AS-14 Kedge-B - (Figure 2.835) is a TV-guided air-to-ground missile. The first flight took place in 1980.

General characteristics of Kh-29T [61]: wingspan – 1.1 m, length – 4.30 m, fuselage diameter – 0.38–0.40 m, weight – 660–680 kg, powerplant – 1 fixed thrust solid-fuel rocket PRD-280 engine, speed – 250–450 m/s, and range – 3–12 km.

In accordance with the design of the Kh-29T a model of its surface was created to obtaining scattering characteristics (in particular, RCS). The model is shown in Figure 2.836. The smooth parts of the model surface were approximated by parts of 35 triaxial ellipsoids. The surface breaks were modeled using 40 straight edge scattering parts.

FIGURE 2.835 Air-to-surface missile Kh29T.

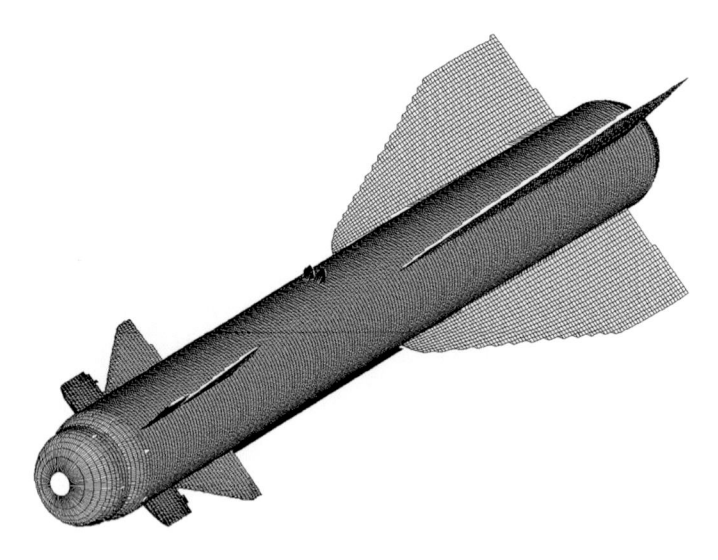

FIGURE 2.836 Surface model of air-to-surface missile Kh-29T.

Below are some scattering characteristics of the Kh-29T model at sounding frequencies of 10, 3, and 1 GHz (wavelengths of 3, 10, and 30 cm, respectively) for horizontal polarization of the probing signal. The scattering characteristics for this model at two polarizations, as well as scattering characteristics at sounding frequencies of 5 and 1.3 GHz (wavelengths of 6 and 23 cm, respectively), are given in the electronic appendix of this book.

Scattering Characteristics of Aerial Objects 307

Sounding parameters: The elevation angle is to be a random value distributed uniformly in the range −3° ± 4° with respect to the wing plane (elevation angle of −3° corresponds to the radar observation from the lower hemisphere), azimuth aspect increment was 0.02°, the azimuth being counted off from the nose-on aspect (0° corresponds to the nose-on radar observation, 180° corresponds to the tail-on observation).

Scattering characteristics of Kh-29T model for sounding frequency 10 GHz (wavelength 3 cm).

Figure 2.837 shows the RCS circular diagram of the Kh-29T model. The noncoherent RCS circular diagram of the Kh-29T is shown in Figure 2.838.

The circular mean RCS of the Kh-29T model for horizontal polarization is 4.642 m². The circular median RCS (the RCS value used to calculate the detection range of an object with a probability of 0.5) for horizontal polarization is 0.052 m².

Figures 2.839 and 2.840 show the mean and median RCS for the main ranges of sounding azimuths (nose, side, and tail) and for ranges of 20°.

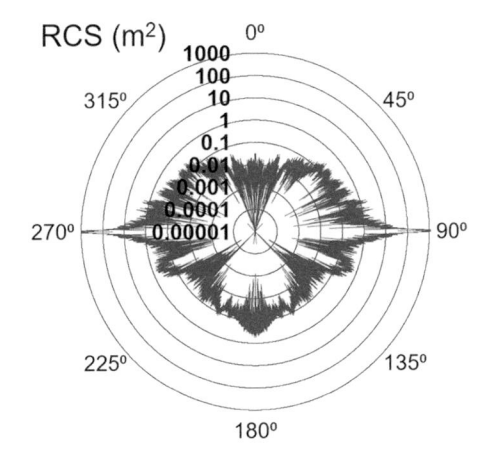

FIGURE 2.837 Circular diagram of RCS given radar observation of Kh-29T model at a carrier frequency of 10 GHz (3 cm wavelength).

FIGURE 2.838 Circular diagram of noncoherent RCS given radar observation of Kh-29T model at a carrier frequency of 10 GHz (3 cm wavelength).

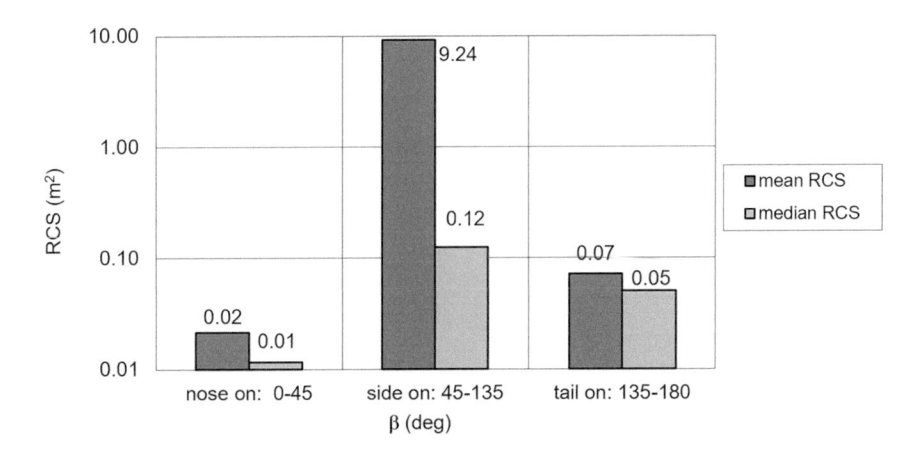

FIGURE 2.839 Diagrams of mean and median RCS of Kh-29T model in three sectors of azimuth aspect given its radar observation at horizontal polarization and a carrier frequency of 10 GHz (3 cm wavelength).

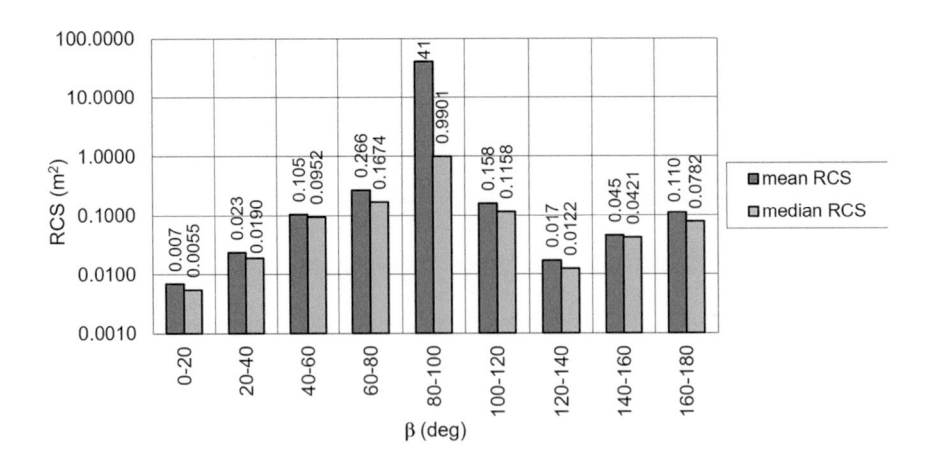

FIGURE 2.840 Diagrams of mean and median RCS of Kh-29T model in 20-degree sectors of azimuth aspect given its radar observation at horizontal polarization and a carrier frequency of 10 GHz (3 cm wavelength).

Figures 2.841, 2.846, and 2.851 show histograms of the scattered signal amplitude (square root of the RCS) for the range of sounding azimuths −20° to +20° (sounding from the nose). The bold line shows the probability density functions of the distribution, which can be used to approximate the histogram of the signal amplitude.

normal distribution

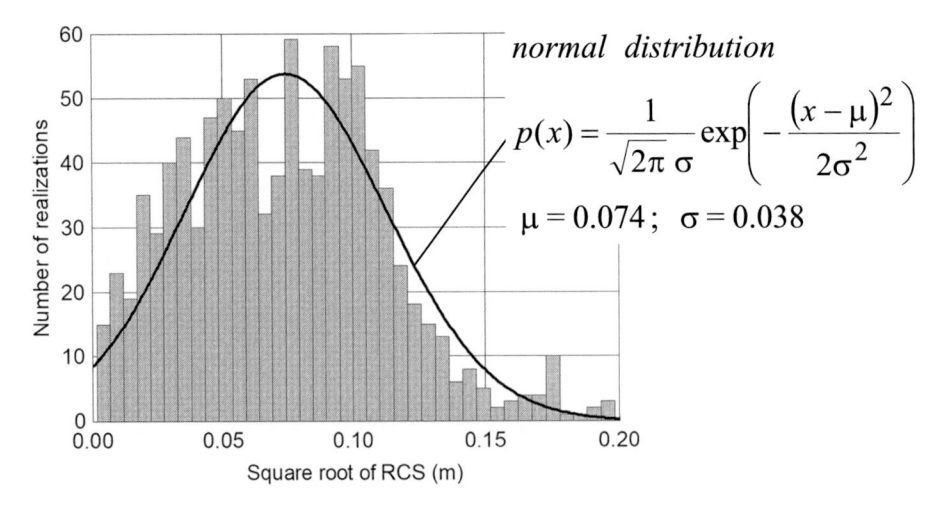

$$p(x) = \frac{1}{\sqrt{2\pi}\,\sigma} \exp\left(-\frac{(x-\mu)^2}{2\sigma^2}\right)$$

$\mu = 0.074;\ \sigma = 0.038$

FIGURE 2.841 Amplitude distribution of echo signal of Kh-29T model at a carrier frequency of 10 GHz given its horizontal polarization.

Scattering characteristics of Kh-29T model for sounding frequency 3 GHz (wavelength 10 cm).

Figure 2.842 shows the RCS circular diagram of the Kh-29T. The noncoherent RCS circular diagram of the Kh-29T is shown in Figure 2.843.

The circular mean RCS of the Kh-29T for horizontal polarization is 2.870 m². The circular median RCS for horizontal polarization is 0.066 m².

Figures 2.844 and 2.845 show the mean and median RCS for the main ranges of sounding azimuths (nose, side, and tail) and for ranges of 20°.

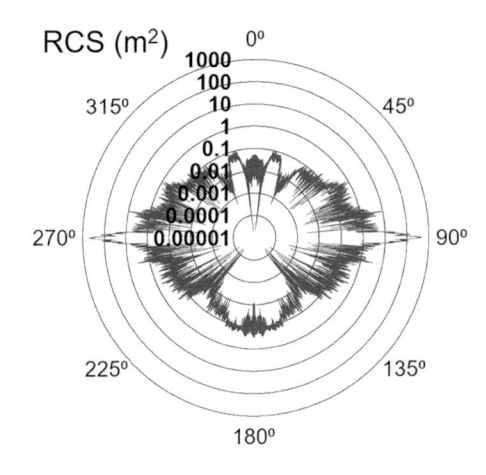

Scattering Characteristics of Aerial Objects 309

FIGURE 2.842 Circular diagram of RCS given radar observation of Kh-29T model at a carrier frequency of 3 GHz (10 cm wavelength).

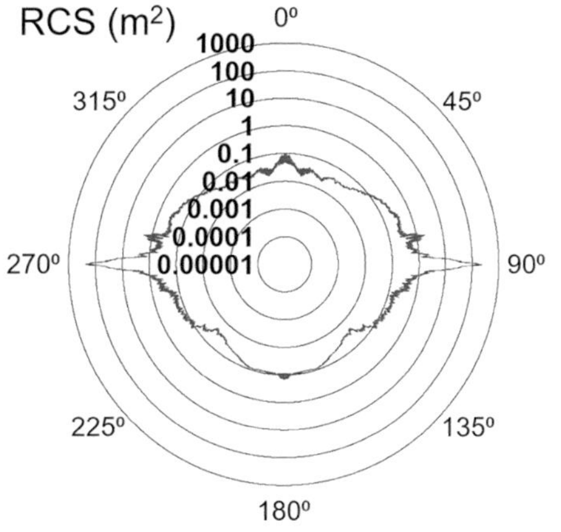

FIGURE 2.843 Circular diagram of noncoherent RCS given radar observation of Kh-29T model at a carrier frequency of 3 GHz (10 cm wavelength).

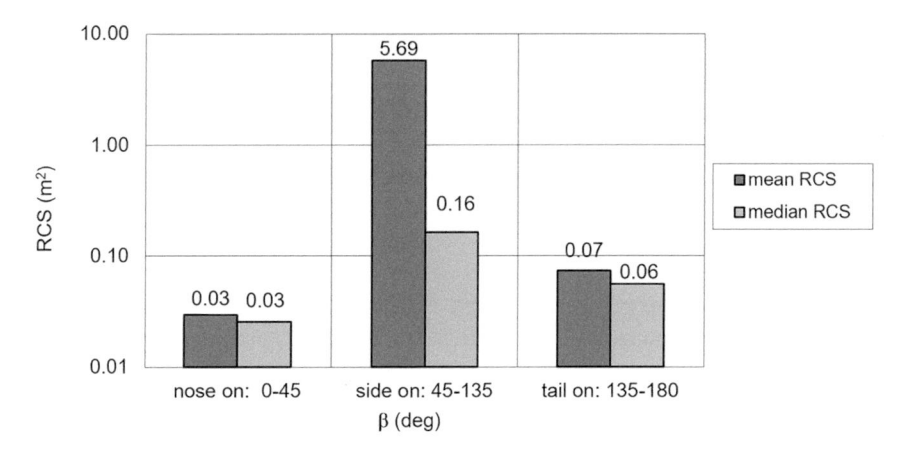

FIGURE 2.844 Diagrams of mean and median RCS of Kh-29T model in three sectors of azimuth aspect given its radar observation at horizontal polarization and a carrier frequency of 3 GHz (10 cm wavelength).

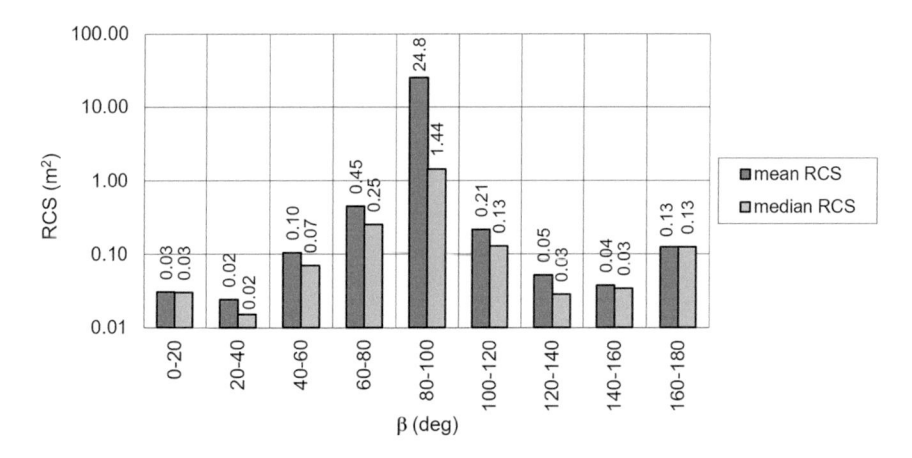

FIGURE 2.845 Diagrams of mean and median RCS of Kh-29T model in 20-degree sectors of azimuth aspect given its radar observation at horizontal polarization and a carrier frequency of 3 GHz (10 cm wavelength).

$$p(x) = \frac{1}{\sqrt{2\pi}\,\sigma} \exp\left(-\frac{(x-\mu)^2}{2\sigma^2}\right)$$

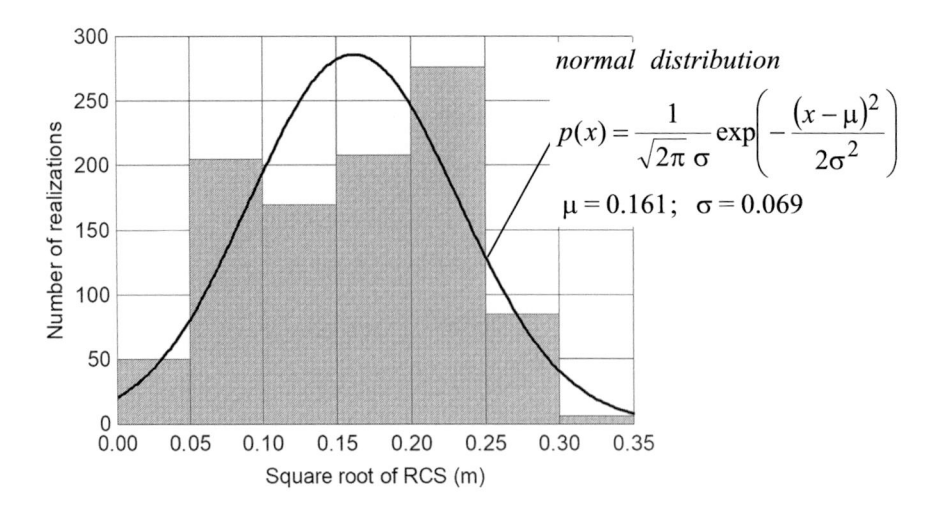

$\mu = 0.161;\quad \sigma = 0.069$

FIGURE 2.846 Amplitude distribution of echo signal of Kh-29T model at a carrier frequency of 3 GHz given its horizontal polarization.

Scattering characteristics of Kh-29T model for sounding frequency 1 GHz (wavelength 30 cm).

Figure 2.847 shows the RCS circular diagram of the Kh-29T. The noncoherent RCS circular diagram of the Kh-29T is shown in Figure 2.848.

The circular mean RCS of the Kh-29T model for horizontal polarization is 2.044 m². The circular median RCS for horizontal polarization is 0.132 m².

Figures 2.849 and 2.850 show the mean and median RCS for the main ranges of sounding azimuths (nose, side, and tail) and for ranges of 20°.

The expressions and parameters of probability distributions that are most consistent with the empirical distributions of the square root of the RCS for different ranges of sounding azimuths and polarizations are given in the electronic appendix to this book. The expressions and parameters of the probability distributions that are most consistent with the empirical distributions of the RCS (energy characteristic) for different ranges of sounding azimuths and polarizations are also given there.

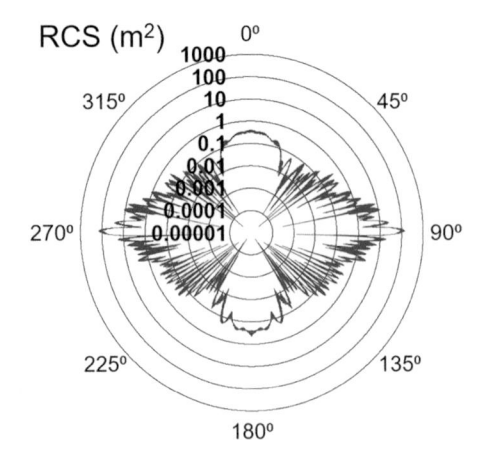

FIGURE 2.847 Circular diagram of RCS given radar observation of Kh-29T model at a carrier frequency of 1 GHz (30 cm wavelength).

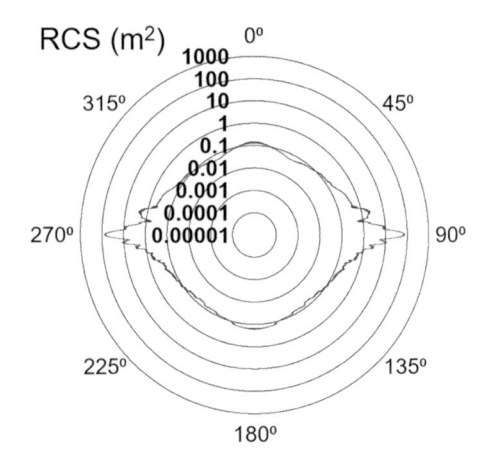

FIGURE 2.848 Circular diagram of noncoherent RCS given radar observation of Kh-29T model at a carrier frequency of 1 GHz (30 cm wavelength).

Scattering Characteristics of Aerial Objects

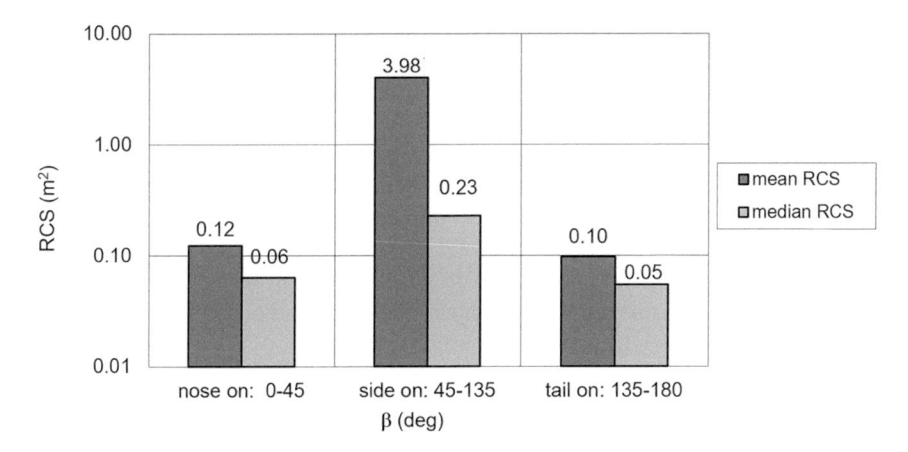

FIGURE 2.849 Diagrams of mean and median RCS of Kh-29T model in three sectors of azimuth aspect given its radar observation at horizontal polarization and a carrier frequency of 1 GHz (30 cm wavelength).

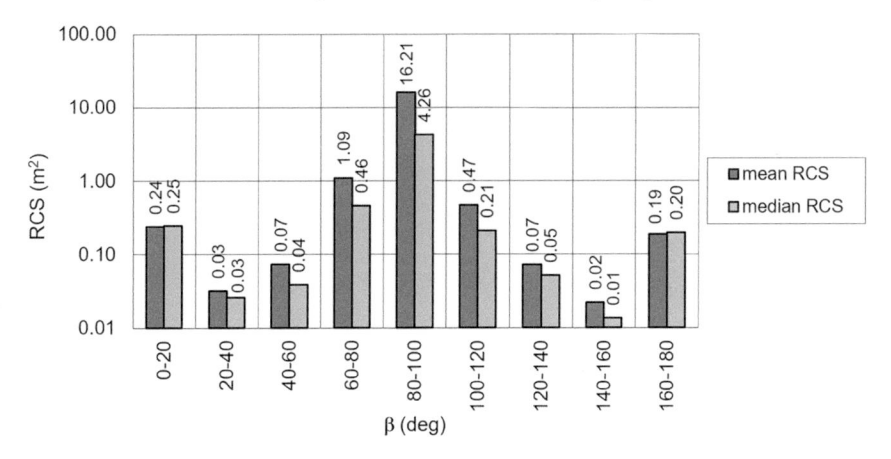

FIGURE 2.850 Diagrams of mean and median RCS of Kh-29T model in 20-degree sectors of azimuth aspect given its radar observation at horizontal polarization and a carrier frequency of 1 GHz (30 cm wavelength).

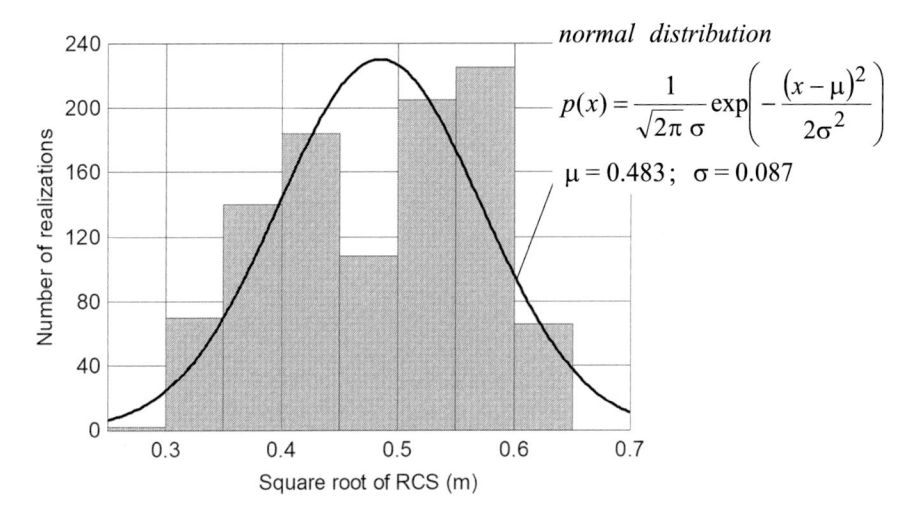

$$p(x) = \frac{1}{\sqrt{2\pi}\,\sigma}\exp\left(-\frac{(x-\mu)^2}{2\sigma^2}\right)$$

$\mu = 0.483;\ \sigma = 0.087$

FIGURE 2.851 Amplitude distribution of echo signal of Kh-29T model at a carrier frequency of 1 GHz given its horizontal polarization.

2.51 ANTI-RADIATION MISSILE Kh-31PD

Kh-31PD - AS-17 Krypton Mod 2- (Figure 2.852) is a supersonic anti-radiation missile. The first flight took place in 2010–2012.

General characteristics of Kh-31PD [62]: wingspan – 0.954 m, length – 5.34 m, fuselage diameter – 0.36 m, weight – 715 kg, powerplant – solid-fuel rocket in initial stage, ramjet for rest of trajectory, speed – 600–1 000 m/s, range – 15–250 km.

In accordance with the design of the Kh-31PD, a model of its surface was created to obtaining scattering characteristics (in particular, RCS). The model is shown in Figure 2.853. The smooth parts of the model surface were approximated by parts of 46 triaxial ellipsoids. The surface breaks were modeled using 24 straight edge scattering parts.

FIGURE 2.852 Anti-radiation missile Kh-31PD.

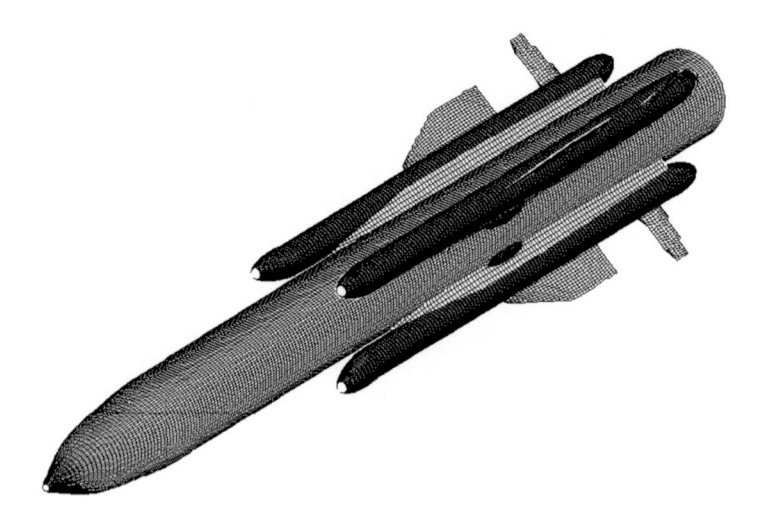

FIGURE 2.853 Surface model of anti-radiation missile Kh-31PD.

Below are some scattering characteristics of the Kh-31PD model at sounding frequencies of 10, 3, and 1 GHz (wavelengths of 3, 10, and 30 cm, respectively) for horizontal polarization of the probing signal. The scattering characteristics for this model at two polarizations, as well as scattering characteristics at sounding frequencies of 5 and 1.3 GHz (wavelengths of 6 and 23 cm, respectively), are given in the electronic appendix of this book.

Scattering Characteristics of Aerial Objects 313

Sounding parameters: The elevation angle is to be a random value distributed uniformly in the range −3° ± 4° with respect to the wing plane (elevation angle of −3° corresponds to the radar observation from the lower hemisphere), azimuth aspect increment was 0.02°, the azimuth being counted off from the nose-on aspect (0° corresponds to the nose-on radar observation, 180° corresponds to the tail-on observation).

Scattering characteristics of Kh-31PD model for sounding frequency 10 GHz (wavelength 3 cm).

Figure 2.854 shows the RCS circular diagram of the Kh-31PD model. The noncoherent RCS circular diagram of the Kh-31PD is shown in Figure 2.855.

The circular mean RCS of the Kh-31PD model for horizontal polarization is 1.660 m². The circular median RCS (the RCS value used to calculate the detection range of an object with a probability of 0.5) for horizontal polarization is 0.067 m².

Figures 2.856 and 2.857 show the mean and median RCS for the main ranges of sounding azimuths (nose, side, and tail) and for ranges of 20°.

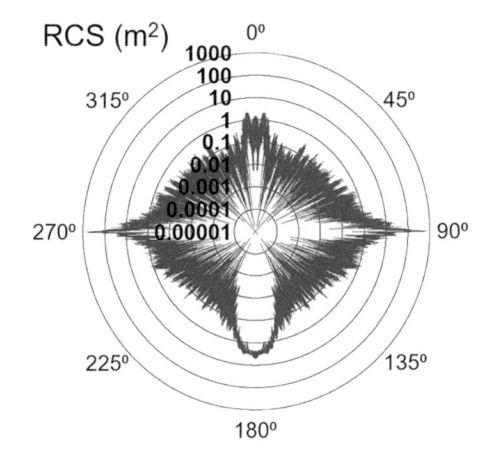

FIGURE 2.854 Circular diagram of RCS given radar observation of Kh-31PD model at a carrier frequency of 10 GHz (3 cm wavelength).

FIGURE 2.855 Circular diagram of noncoherent RCS given radar observation of Kh-31PD model at a carrier frequency of 10 GHz (3 cm wavelength).

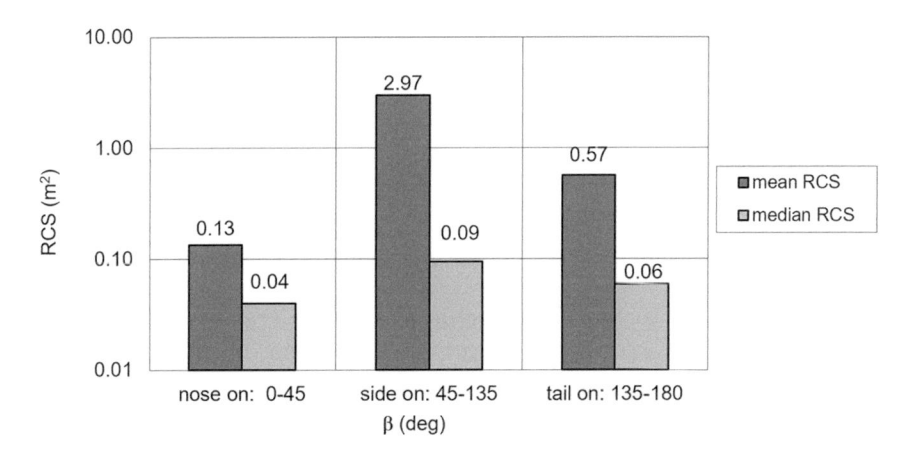

FIGURE 2.856 Diagrams of mean and median RCS of Kh-31PD model in three sectors of azimuth aspect given its radar observation at horizontal polarization and a carrier frequency of 10 GHz (3 cm wavelength).

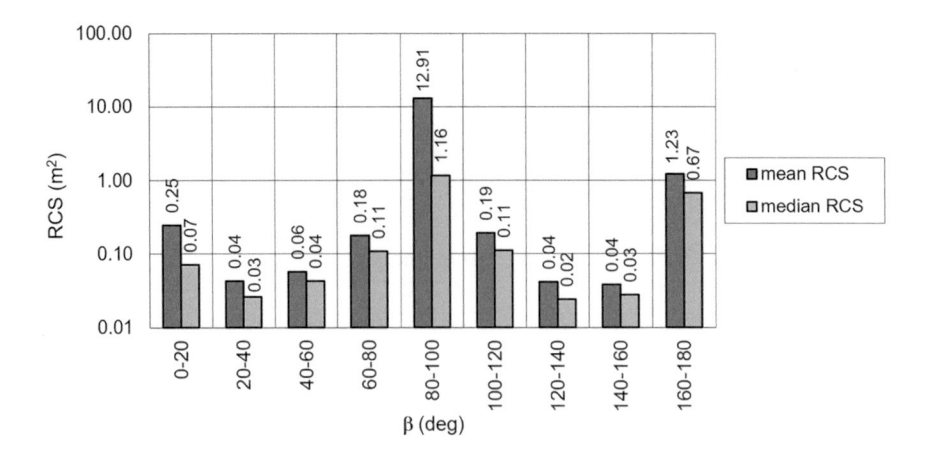

FIGURE 2.857 Diagrams of mean and median RCS of Kh-31PD model in 20-degree sectors of azimuth aspect given its radar observation at horizontal polarization and a carrier frequency of 10 GHz (3 cm wavelength).

Figures 2.858, 2.863, and 2.868 show histograms of the scattered signal amplitude (square root of the RCS) for the range of sounding azimuths −20° to +20° (sounding from the nose). The bold line shows the probability density functions of the distribution, which can be used to approximate the histogram of the signal amplitude.

log - normal distribution

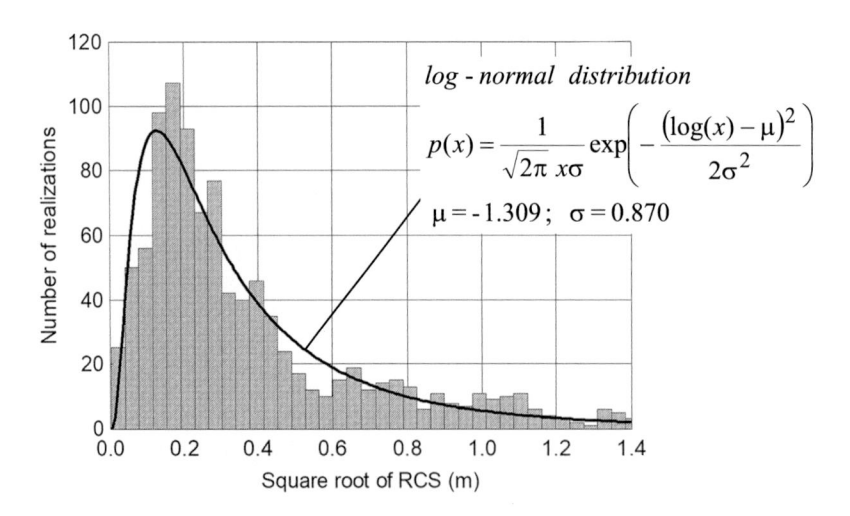

$$p(x) = \frac{1}{\sqrt{2\pi}\, x\sigma} \exp\left(-\frac{(\log(x) - \mu)^2}{2\sigma^2}\right)$$

$\mu = -1.309;\ \sigma = 0.870$

FIGURE 2.858 Amplitude distribution of echo signal of Kh-31PD model at a carrier frequency of 10 GHz given its horizontal polarization.

Scattering characteristics of Kh-31PD model for sounding frequency 3 GHz (wavelength 10 cm).

Figure 2.859 shows the RCS circular diagram of the Kh-31PD. The noncoherent RCS circular diagram of the Kh-31PD is shown in Figure 2.860.

The circular mean RCS of the Kh-31PD for horizontal polarization is 2.123 m². The circular median RCS for horizontal polarization is 0.089 m².

Figures 2.861 and 2.862 show the mean and median RCS for the main ranges of sounding azimuths (nose, side, and tail) and for ranges of 20°.

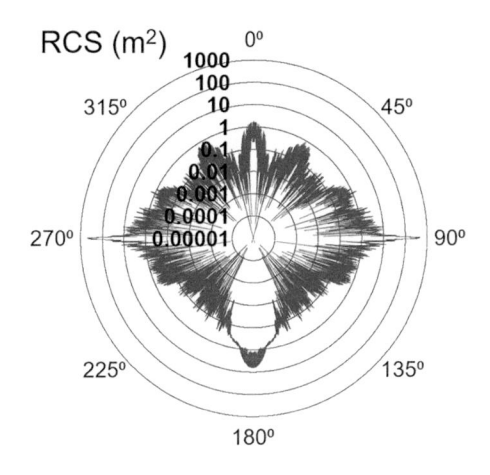

FIGURE 2.859 Circular diagram of RCS given radar observation of Kh-31PD model at a carrier frequency of 3 GHz (10 cm wavelength).

FIGURE 2.860 Circular diagram of noncoherent RCS given radar observation of Kh-31PD model at a carrier frequency of 3 GHz (10 cm wavelength).

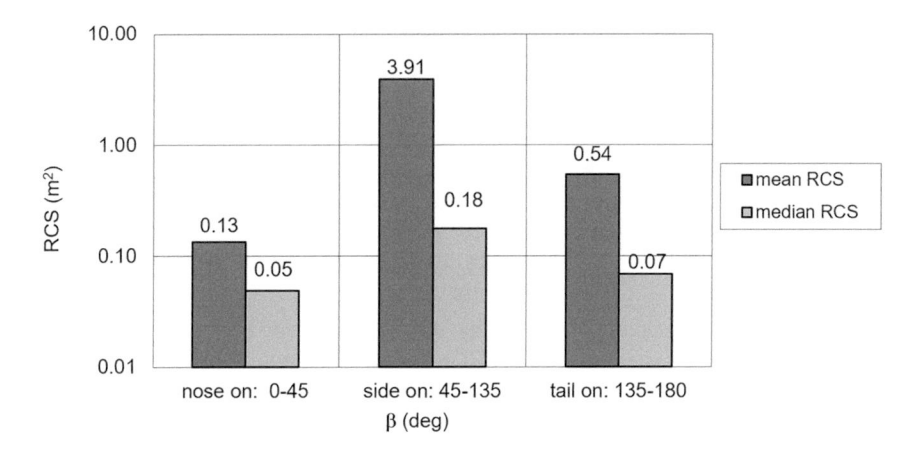

FIGURE 2.861 Diagrams of mean and median RCS of Kh-31PD model in three sectors of azimuth aspect given its radar observation at horizontal polarization and a carrier frequency of 3 GHz (10 cm wavelength).

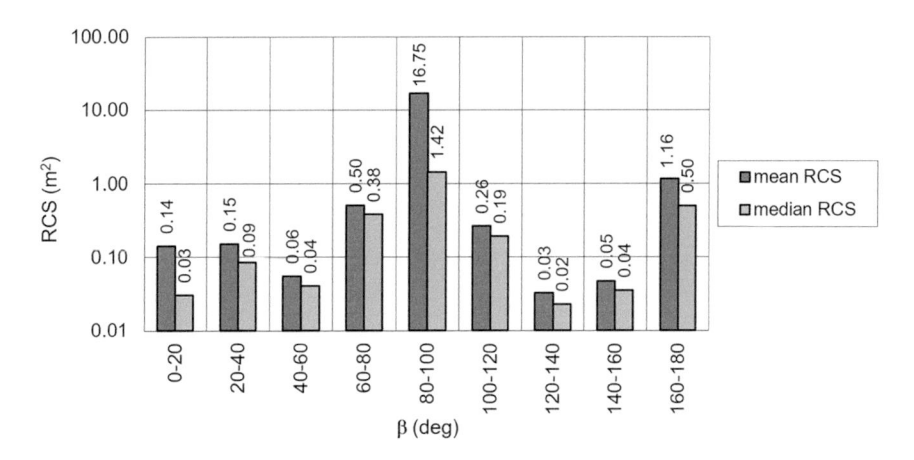

FIGURE 2.862 Diagrams of mean and median RCS of Kh-31PD model in 20-degree sectors of azimuth aspect given its radar observation at horizontal polarization and a carrier frequency of 3 GHz (10 cm wavelength).

$$p(x) = \frac{1}{\sqrt{2\pi}\, x\sigma} \exp\left(-\frac{(\log(x) - \mu)^2}{2\sigma^2}\right)$$

log - normal distribution

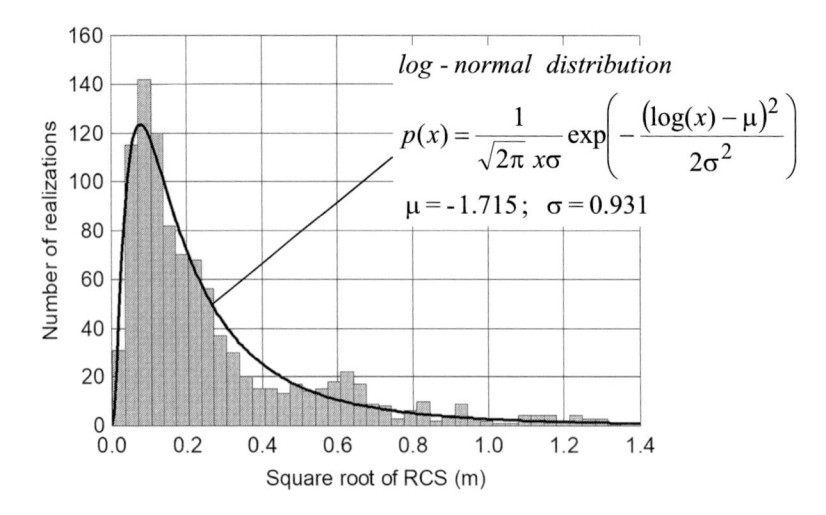

$\mu = -1.715;\ \sigma = 0.931$

FIGURE 2.863 Amplitude distribution of echo signal of Kh-31PD model at a carrier frequency of 3 GHz given its horizontal polarization.

Scattering characteristics of Kh-31PD model for sounding frequency 1 GHz (wavelength 30 cm).

Figure 2.864 shows the RCS circular diagram of the Kh-31PD. The noncoherent RCS circular diagram of the Kh-31PD is shown in Figure 2.865.

The circular mean RCS of the Kh-31PD model for horizontal polarization is 0.850 m². The circular median RCS for horizontal polarization is 0.138 m².

Figures 2.866 and 2.867 show the mean and median RCS for the main ranges of sounding azimuths (nose, side, and tail) and for ranges of 20°.

The expressions and parameters of probability distributions that are most consistent with the empirical distributions of the square root of the RCS for different ranges of sounding azimuths and polarizations are given in the electronic appendix to this book. The expressions and parameters of the probability distributions that are most consistent with the empirical distributions of the RCS (energy characteristic) for different ranges of sounding azimuths and polarizations are also given there.

FIGURE 2.864 Circular diagram of RCS given radar observation of Kh-31PD model at a carrier frequency of 1 GHz (30 cm wavelength).

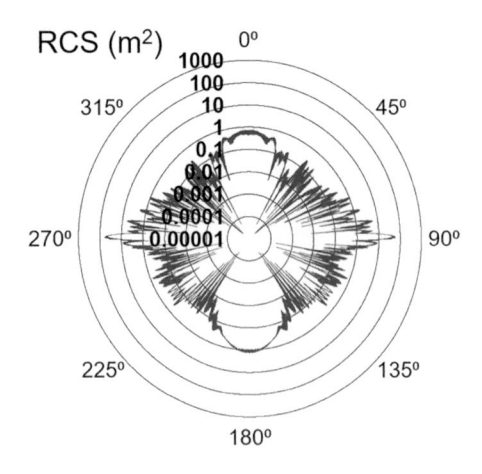

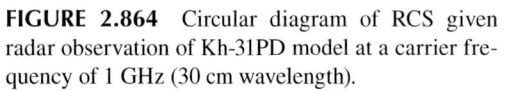

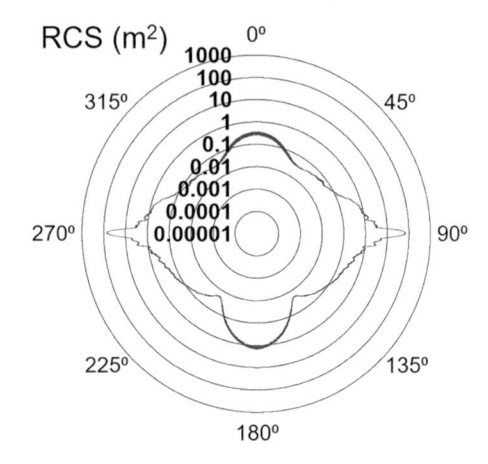

FIGURE 2.865 Circular diagram of noncoherent RCS given radar observation of Kh-31PD model at a carrier frequency of 1 GHz (30 cm wavelength).

Scattering Characteristics of Aerial Objects

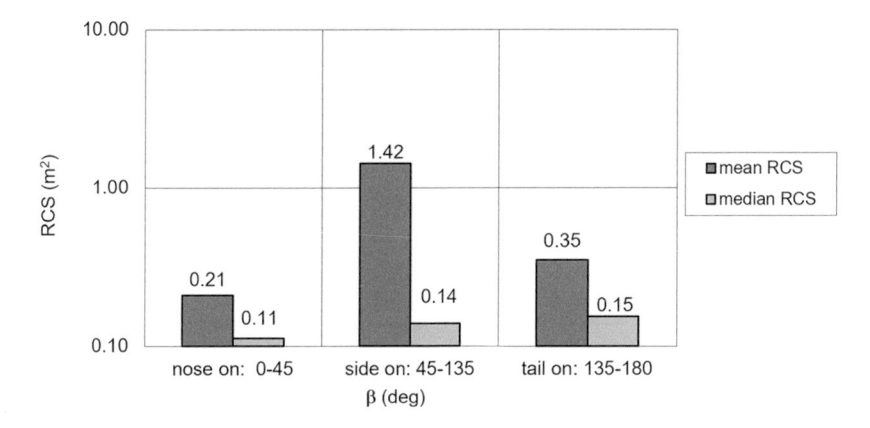

FIGURE 2.866 Diagrams of mean and median RCS of Kh-31PD model in three sectors of azimuth aspect given its radar observation at horizontal polarization and a carrier frequency of 1 GHz (30 cm wavelength).

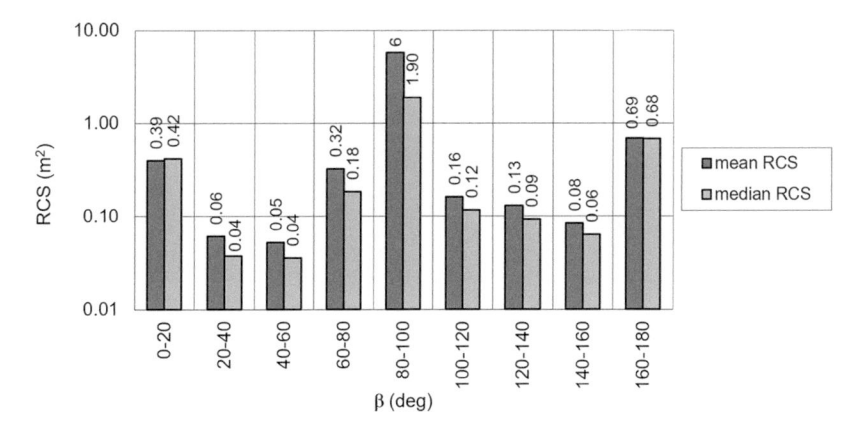

FIGURE 2.867 Diagrams of mean and median RCS of Kh-31PD model in 20-degree sectors of azimuth aspect given its radar observation at horizontal polarization and a carrier frequency of 1 GHz (30 cm wavelength).

Weibull distribution

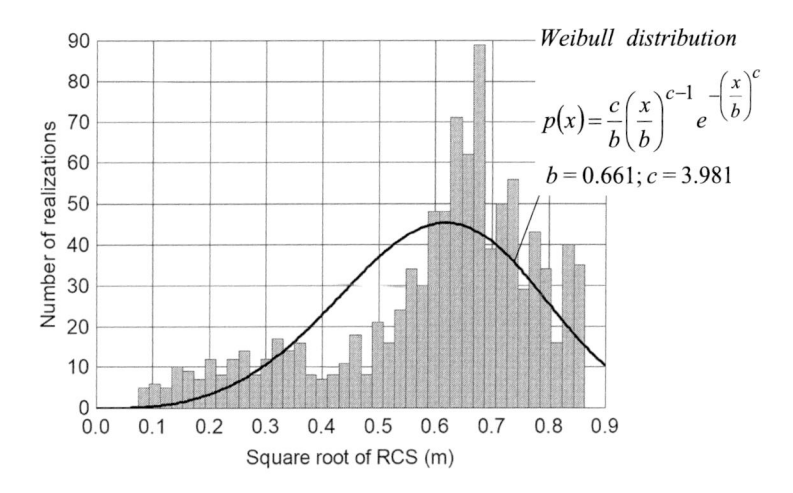

$$p(x) = \frac{c}{b}\left(\frac{x}{b}\right)^{c-1} e^{-\left(\frac{x}{b}\right)^c}$$

$b = 0.661; c = 3.981$

FIGURE 2.868 Amplitude distribution of echo signal of Kh-31PD model at a carrier frequency of 1 GHz given its horizontal polarization.

2.52 CRUISE MISSILE Kh-32

Kh-32 (Figure 2.869) is a supersonic air-launched cruise missile. The first flight took place in 1998.

General characteristics of Kh-32 [63]: wingspan – 2.9–3.0 m, length – 11.65 m, fuselage diameter – 0.9–0.92 m, weight – 5 780 kg, speed – 3.5–4.6 M, range – 600–1000 km.

In accordance with the design of the Kh-32, a model of its surface was created to obtaining scattering characteristics (in particular, RCS). The model is shown in Figure 2.870. The smooth parts of the model surface were approximated by parts of 24 triaxial ellipsoids. The surface breaks were modeled using 21 straight edge scattering parts.

FIGURE 2.869 Cruise missile Kh-32.

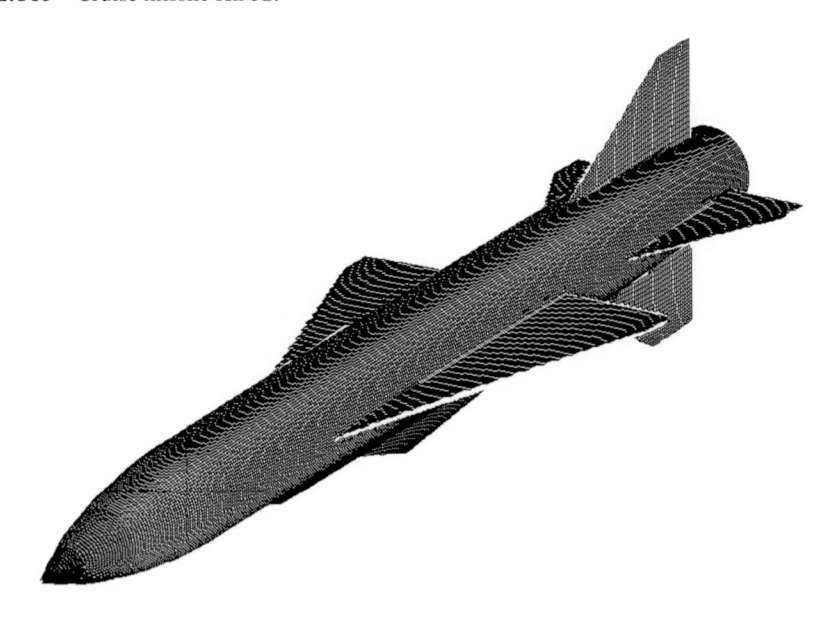

FIGURE 2.870 Surface model of cruise missile Kh-32.

Below are some scattering characteristics of the Kh-32 model at sounding frequencies of 10, 3, and 1 GHz (wavelengths of 3, 10, and 30 cm, respectively) for horizontal polarization of the probing signal. The scattering characteristics for this model at two polarizations, as well as scattering

Scattering Characteristics of Aerial Objects 319

characteristics at sounding frequencies of 5 and 1.3 GHz (wavelengths of 6 and 23 cm, respectively), are given in the electronic appendix of this book.

Sounding parameters: The elevation angle is to be a random value distributed uniformly in the range −3° ± 4° with respect to the wing plane (elevation angle of −3° corresponds to the radar observation from the lower hemisphere), azimuth aspect increment was 0.02°, the azimuth being counted off from the nose-on aspect (0° corresponds to the nose-on radar observation, 180° corresponds to the tail-on observation).

Scattering characteristics of Kh-32 model for sounding frequency 10 GHz (wavelength 3 cm).

Figure 2.871 shows the RCS circular diagram of the Kh-32 model. The noncoherent RCS circular diagram of the Kh-32 is shown in Figure 2.872.

The circular mean RCS of the Kh-32 model for horizontal polarization is 16.394 m². The circular median RCS (the RCS value used to calculate the detection range of an object with a probability of 0.5) for horizontal polarization is 0.200 m².

Figures 2.873 and 2.874 show the mean and median RCS for the main ranges of sounding azimuths (nose, side, and tail) and for ranges of 20°.

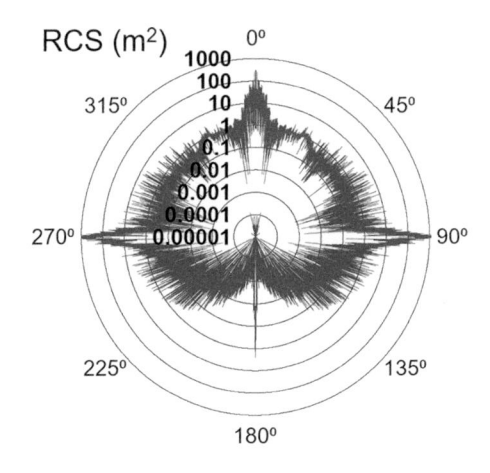

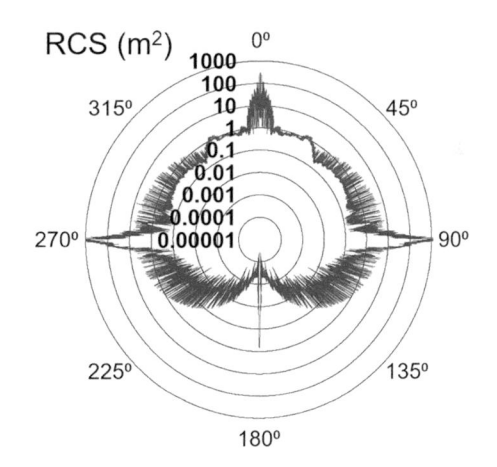

FIGURE 2.871 Circular diagram of RCS given radar observation of Kh-32 model at a carrier frequency of 10 GHz (3 cm wavelength).

FIGURE 2.872 Circular diagram of noncoherent RCS given radar observation of Kh-32 model at a carrier frequency of 10 GHz (3 cm wavelength).

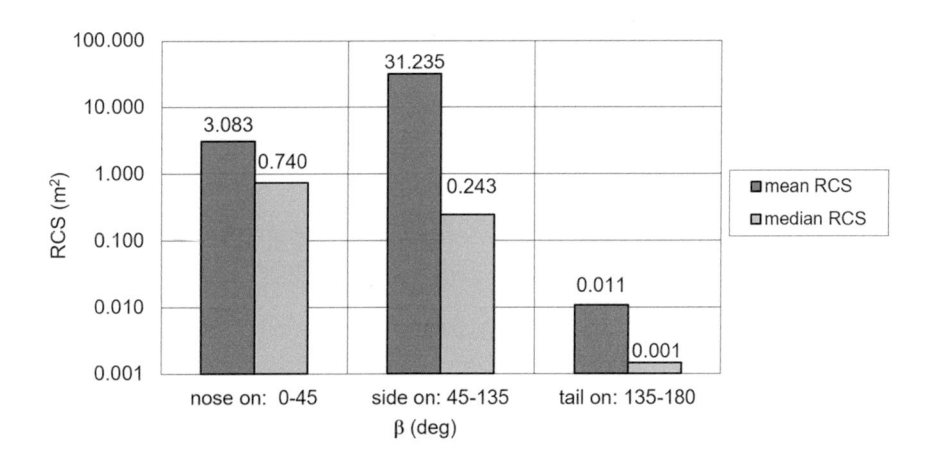

FIGURE 2.873 Diagrams of mean and median RCS of Kh-32 model in three sectors of azimuth aspect given its radar observation at horizontal polarization and a carrier frequency of 10 GHz (3 cm wavelength).

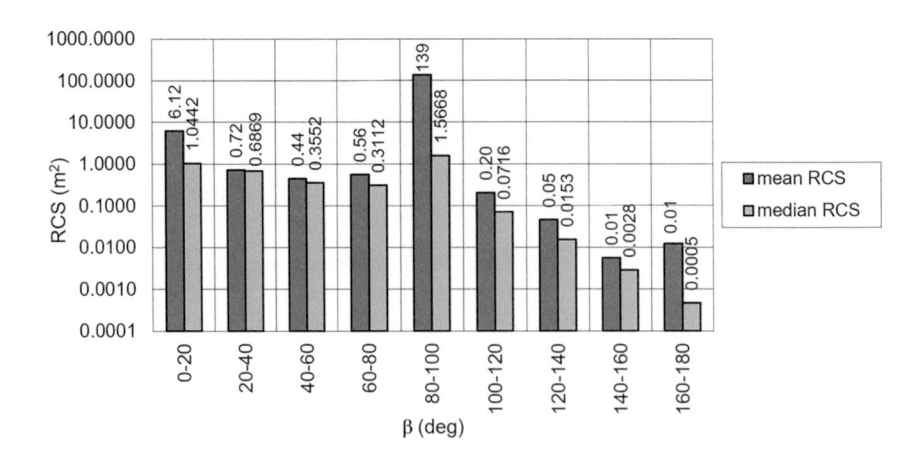

FIGURE 2.874 Diagrams of mean and median RCS of Kh-32 model in 20-degree sectors of azimuth aspect given its radar observation at horizontal polarization and a carrier frequency of 10 GHz (3 cm wavelength).

Figures 2.875, 2.880, and 2.885 show histograms of the scattered signal amplitude (square root of the RCS) for the range of sounding azimuths −20° to +20° (sounding from the nose). The bold line shows the probability density functions of the distribution, which can be used to approximate the histogram of the signal amplitude.

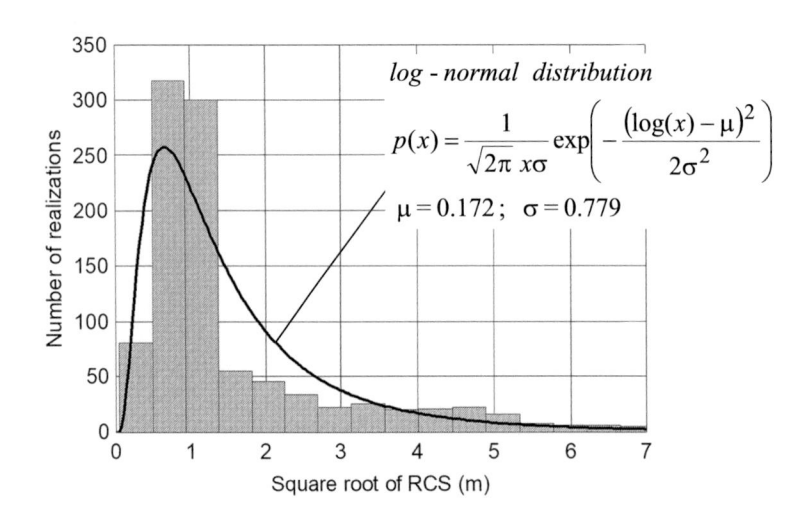

$$\text{log - normal distribution}$$
$$p(x) = \frac{1}{\sqrt{2\pi}\, x\sigma} \exp\left(-\frac{(\log(x) - \mu)^2}{2\sigma^2}\right)$$
$$\mu = 0.172; \quad \sigma = 0.779$$

FIGURE 2.875 Amplitude distribution of echo signal of Kh-32 model at a carrier frequency of 10 GHz given its horizontal polarization.

Scattering characteristics of Kh-32 model for sounding frequency 3 GHz (wavelength 10 cm).

Figure 2.876 shows the RCS circular diagram of the Kh-32. The noncoherent RCS circular diagram of the Kh-32 is shown in Figure 2.877.

The circular mean RCS of the Kh-32 for horizontal polarization is 14.883 m². The circular median RCS for horizontal polarization is 0.242 m².

Figures 2.878 and 2.879 show the mean and median RCS for the main ranges of sounding azimuths (nose, side, and tail) and for ranges of 20°.

Scattering Characteristics of Aerial Objects

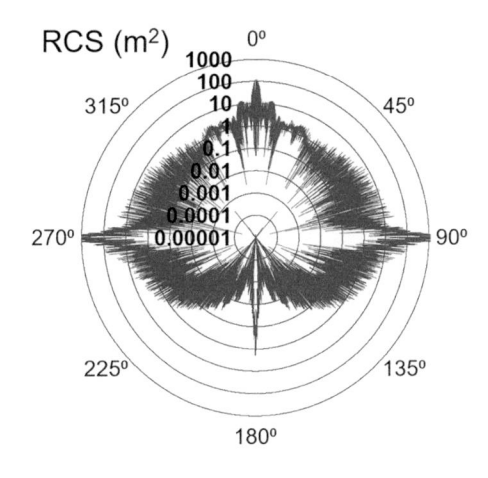

FIGURE 2.876 Circular diagram of RCS given radar observation of Kh-32 model at a carrier frequency of 3 GHz (10 cm wavelength).

FIGURE 2.877 Circular diagram of noncoherent RCS given radar observation of Kh-32 model at a carrier frequency of 3 GHz (10 cm wavelength).

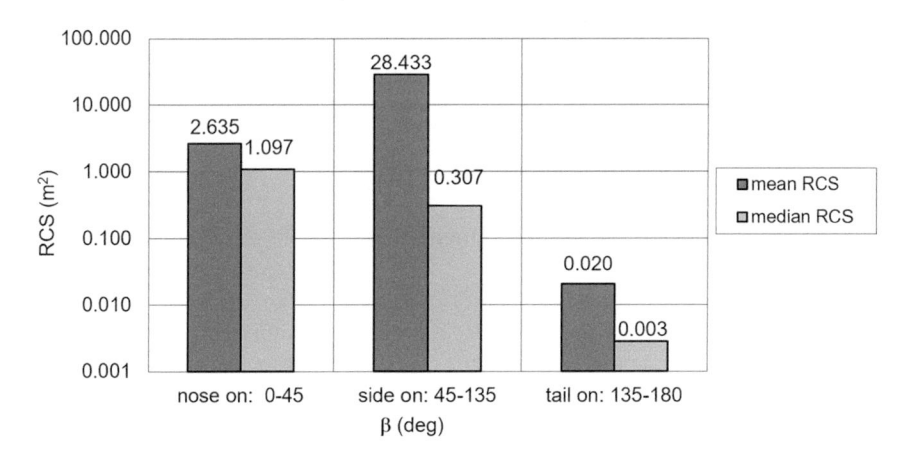

FIGURE 2.878 Diagrams of mean and median RCS of Kh-32 model in three sectors of azimuth aspect given its radar observation at horizontal polarization and a carrier frequency of 3 GHz (10 cm wavelength).

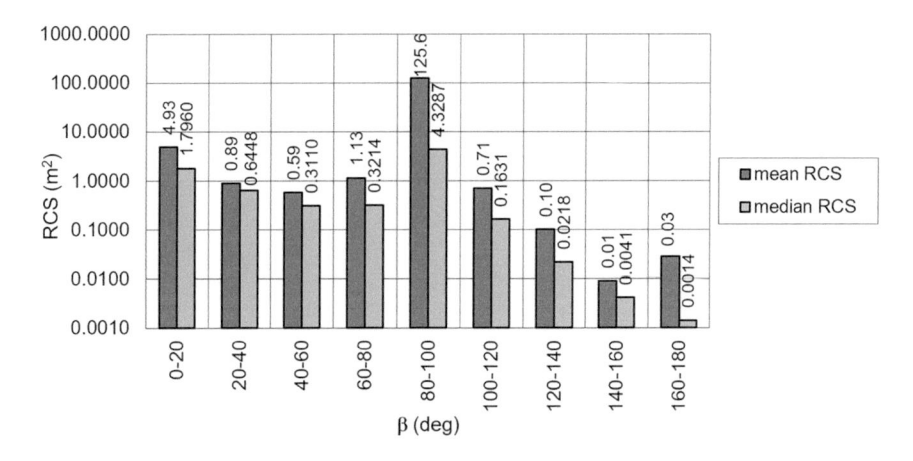

FIGURE 2.879 Diagrams of mean and median RCS of Kh-32 model in 20-degree sectors of azimuth aspect given its radar observation at horizontal polarization and a carrier frequency of 3 GHz (10 cm wavelength).

$$p(x) = \frac{1}{\sqrt{2\pi}\, x\sigma} \exp\left(-\frac{(\log(x)-\mu)^2}{2\sigma^2}\right)$$

log - normal distribution

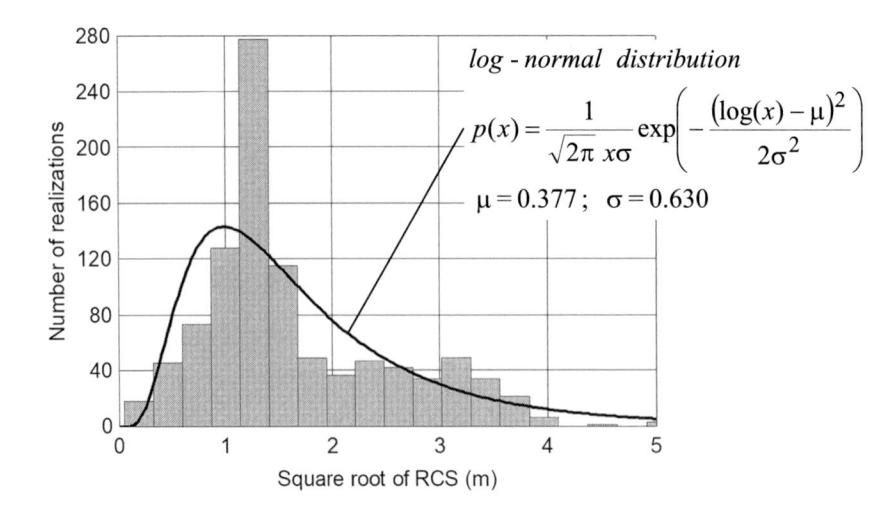

$\mu = 0.377;\ \sigma = 0.630$

FIGURE 2.880 Amplitude distribution of echo signal of Kh-32 model at a carrier frequency of 3 GHz given its horizontal polarization.

Scattering characteristics of Kh-32 model for sounding frequency 1 GHz (wavelength 30 cm).
Figure 2.881 shows the RCS circular diagram of the Kh-32. The noncoherent RCS circular diagram of the Kh-32 is shown in Figure 2.882.

The circular mean RCS of the Kh-32 model for horizontal polarization is 17.784 m². The circular median RCS for horizontal polarization is 0.513 m².

Figures 2.883 and 2.884 show the mean and median RCS for the main ranges of sounding azimuths (nose, side, and tail) and for ranges of 20°.

The expressions and parameters of probability distributions that are most consistent with the empirical distributions of the square root of the RCS for different ranges of sounding azimuths and polarizations are given in the electronic appendix to this book. The expressions and parameters of the probability distributions that are most consistent with the empirical distributions of the RCS (energy characteristic) for different ranges of sounding azimuths and polarizations are also given there.

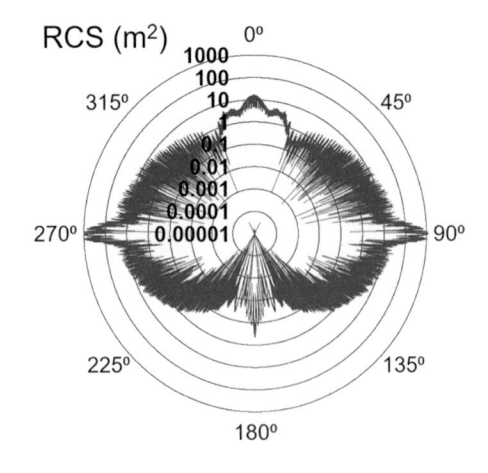

FIGURE 2.881 Circular diagram of RCS given radar observation of Kh-32 model at a carrier frequency of 1 GHz (30 cm wavelength).

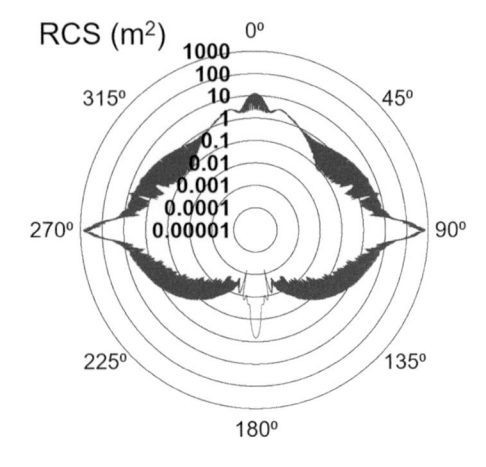

FIGURE 2.882 Circular diagram of noncoherent RCS given radar observation of Kh-32 model at a carrier frequency of 1 GHz (30 cm wavelength).

Scattering Characteristics of Aerial Objects

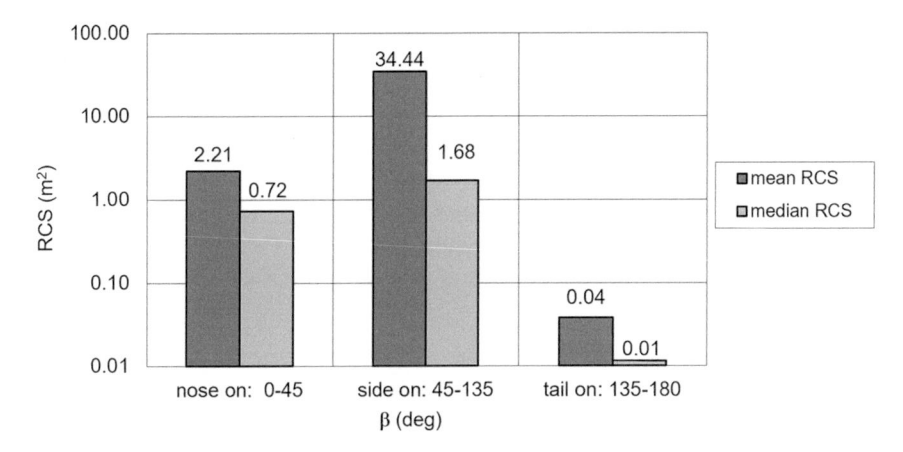

FIGURE 2.883 Diagrams of mean and median RCS of Kh-32 model in three sectors of azimuth aspect given its radar observation at horizontal polarization and a carrier frequency of 1 GHz (30 cm wavelength).

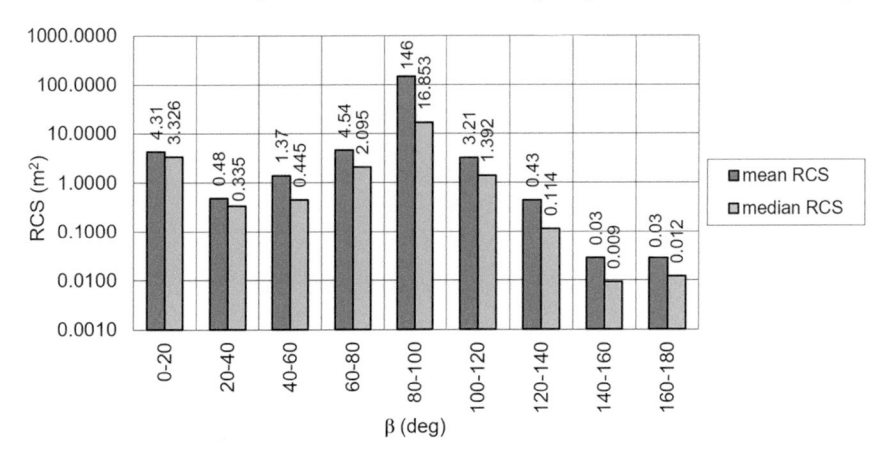

FIGURE 2.884 Diagrams of mean and median RCS of Kh-32 model in 20-degree sectors of azimuth aspect given its radar observation at horizontal polarization and a carrier frequency of 1 GHz (30 cm wavelength).

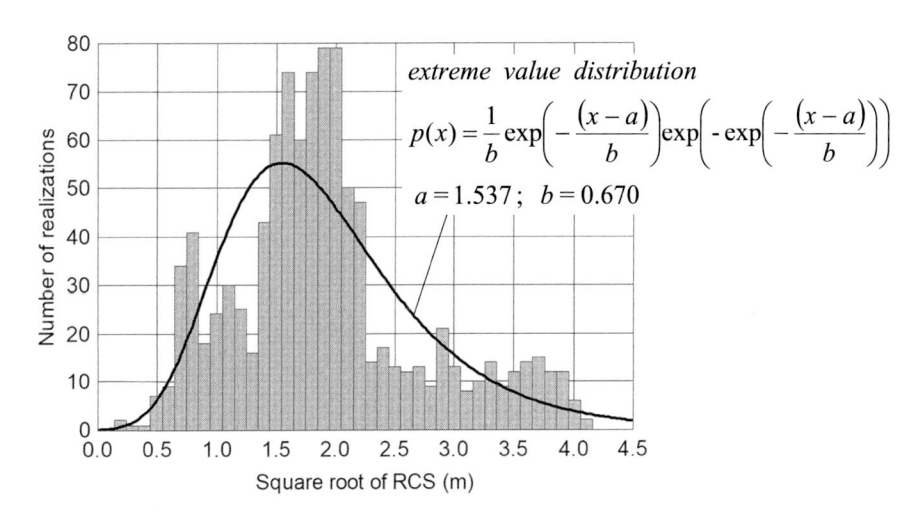

extreme value distribution

$$p(x) = \frac{1}{b}\exp\left(-\frac{(x-a)}{b}\right)\exp\left(-\exp\left(-\frac{(x-a)}{b}\right)\right)$$

$a = 1.537; \quad b = 0.670$

FIGURE 2.885 Amplitude distribution of echo signal of Kh-32 model at a carrier frequency of 1 GHz given its horizontal polarization.

2.53 CRUISE MISSILE Kh-35

Kh-35-AS-20 Kayak - (Figure 2.886) is a subsonic cruise anti-ship missile. The first flight took place in 1992.

General characteristics of Kh-35 [64]: wingspan – 1.33 m, length – 3.85 m (4.4 m with booster section), fuselage diameter – 0.42 m, weight – 520–610 kg, speed – 270–280 m/s, range – 7–130 km.

In accordance with the design of the Kh-35, a model of its surface was created to obtaining scattering characteristics (in particular, RCS). The model is shown in Figure 2.887. The smooth parts of the model surface were approximated by parts of 27 triaxial ellipsoids. The surface breaks were modeled using 35 straight edge scattering parts.

FIGURE 2.886   Cruise missile Kh-35.

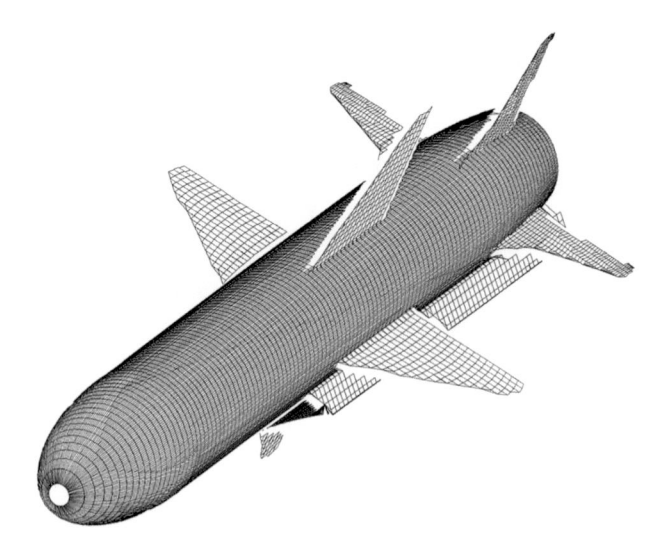

FIGURE 2.887   Surface model of cruise missile Kh-35.

Below are some scattering characteristics of the Kh-35 model at sounding frequencies of 10, 3, and 1 GHz (wavelengths of 3, 10, and 30 cm, respectively) for horizontal polarization of the probing

Scattering Characteristics of Aerial Objects

signal. The scattering characteristics for this model at two polarizations, as well as scattering characteristics at sounding frequencies of 5 and 1.3 GHz (wavelengths of 6 and 23 cm, respectively), are given in the electronic appendix of this book.

Sounding parameters: The elevation angle is to be a random value distributed uniformly in the range $-3° \pm 4°$ with respect to the wing plane (elevation angle of $-3°$ corresponds to the radar observation from the lower hemisphere), azimuth aspect increment was 0.02°, the azimuth being counted off from the nose-on aspect (0° corresponds to the nose-on radar observation, 180° corresponds to the tail-on observation).

Scattering characteristics of Kh-35 model for sounding frequency 10 GHz (wavelength 3 cm).

Figure 2.888 shows the RCS circular diagram of the Kh-35 model. The noncoherent RCS circular diagram of the Kh-35 is shown in Figure 2.889.

The circular mean RCS of the Kh-35 model for horizontal polarization is 3.508 m^2. The circular median RCS (the RCS value used to calculate the detection range of an object with a probability of 0.5) for horizontal polarization is 0.082 m^2.

Figure 2.890, 2.891 show the mean and median RCS for the main ranges of sounding azimuths (nose, side, and tail) and for ranges of 20°.

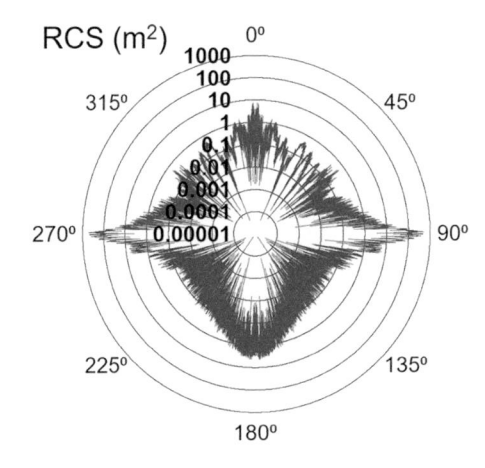

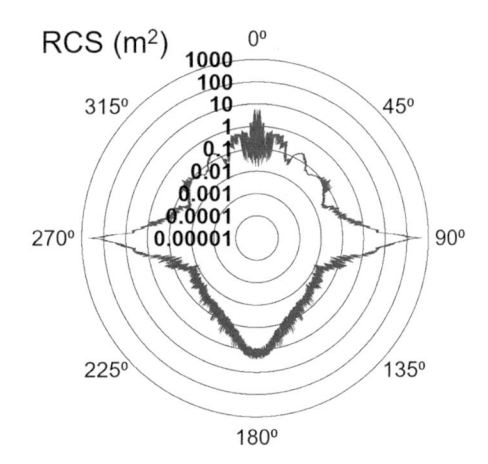

FIGURE 2.888 Circular diagram of RCS given radar observation of Kh-35 model at a carrier frequency of 10 GHz (3 cm wavelength).

FIGURE 2.889 Circular diagram of noncoherent RCS given radar observation of Kh-35 model at a carrier frequency of 10 GHz (3 cm wavelength).

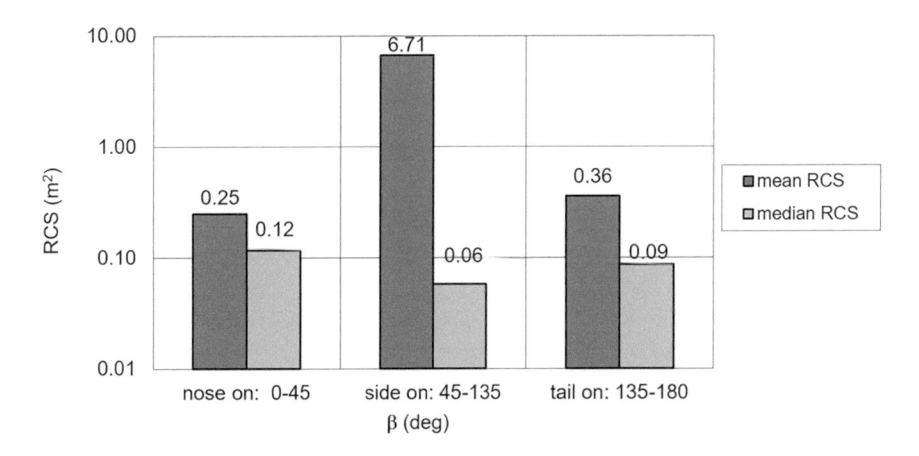

FIGURE 2.890 Diagrams of mean and median RCS of Kh-35 model in three sectors of azimuth aspect given its radar observation at horizontal polarization and a carrier frequency of 10 GHz (3 cm wavelength).

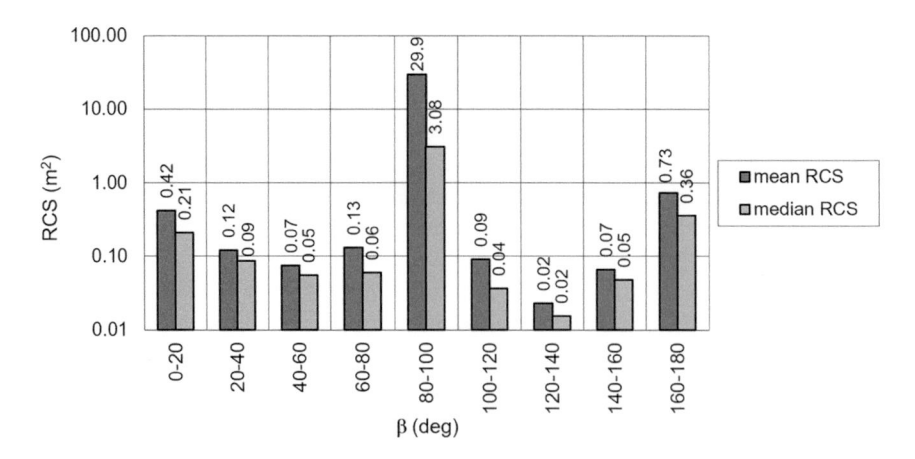

FIGURE 2.891 Diagrams of mean and median RCS of Kh-35 model in 20-degree sectors of azimuth aspect given its radar observation at horizontal polarization and a carrier frequency of 10 GHz (3 cm wavelength).

Figures 2.892, 2.897, and 2.902 show histograms of the scattered signal amplitude (square root of the RCS) for the range of sounding azimuths −20° to +20° (sounding from the nose). The bold line shows the probability density functions of the distribution, which can be used to approximate the histogram of the signal amplitude.

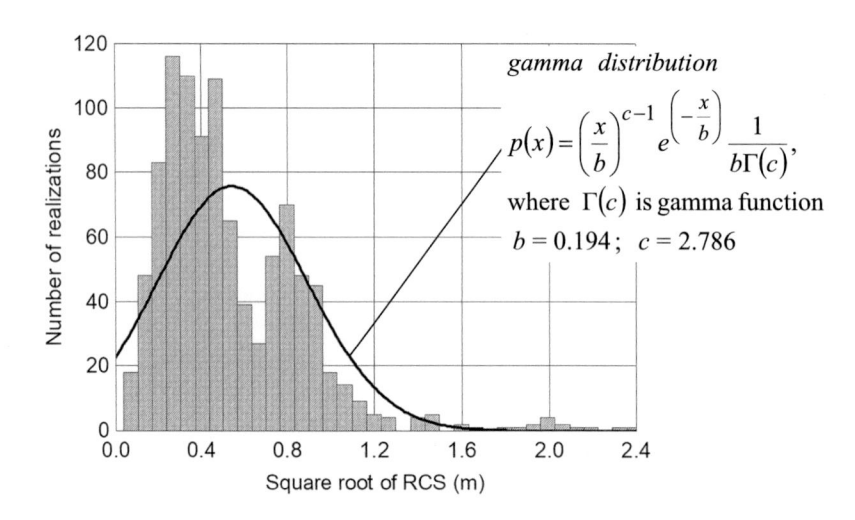

gamma distribution

$$p(x) = \left(\frac{x}{b}\right)^{c-1} e^{\left(-\frac{x}{b}\right)} \frac{1}{b\Gamma(c)},$$

where $\Gamma(c)$ is gamma function
$b = 0.194$; $c = 2.786$

FIGURE 2.892 Amplitude distribution of echo signal of Kh-35 model at a carrier frequency of 10 GHz given its horizontal polarization.

Scattering characteristics of Kh-35 model for sounding frequency 3 GHz (wavelength 10 cm).

Figure 2.893 shows the RCS circular diagram of the Kh-35. The noncoherent RCS circular diagram of the Kh-35 is shown in Figure 2.894.

The circular mean RCS of the Kh-35 for horizontal polarization is 1.551 m². The circular median RCS for horizontal polarization is 0.087 m².

Figures 2.895 and 2.896 show the mean and median RCS for the main ranges of sounding azimuths (nose, side, and tail) and for ranges of 20°.

Scattering Characteristics of Aerial Objects 327

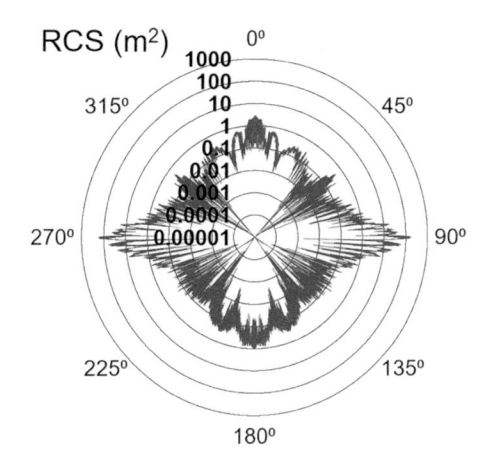

FIGURE 2.893 Circular diagram of RCS given radar observation of Kh-35 model at a carrier frequency of 3 GHz (10 cm wavelength).

FIGURE 2.894 Circular diagram of noncoherent RCS given radar observation of Kh-35 model at a carrier frequency of 3 GHz (10 cm wavelength).

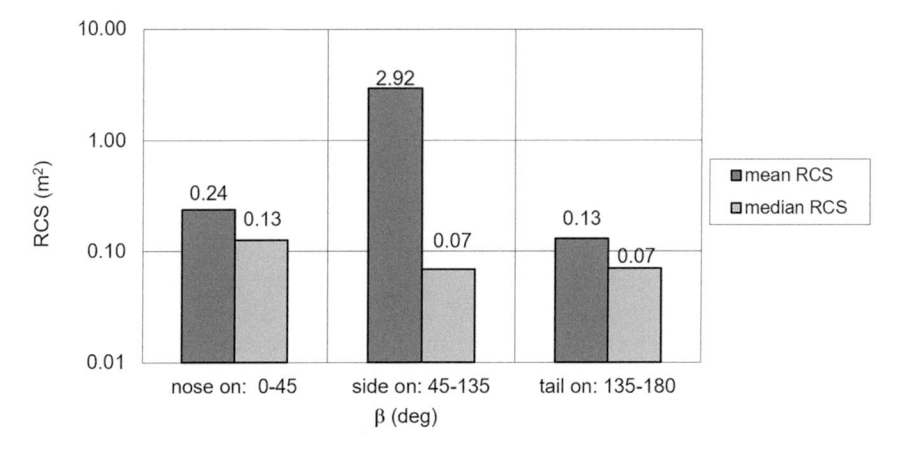

FIGURE 2.895 Diagrams of mean and median RCS of Kh-35 model in three sectors of azimuth aspect given its radar observation at horizontal polarization and a carrier frequency of 3 GHz (10 cm wavelength).

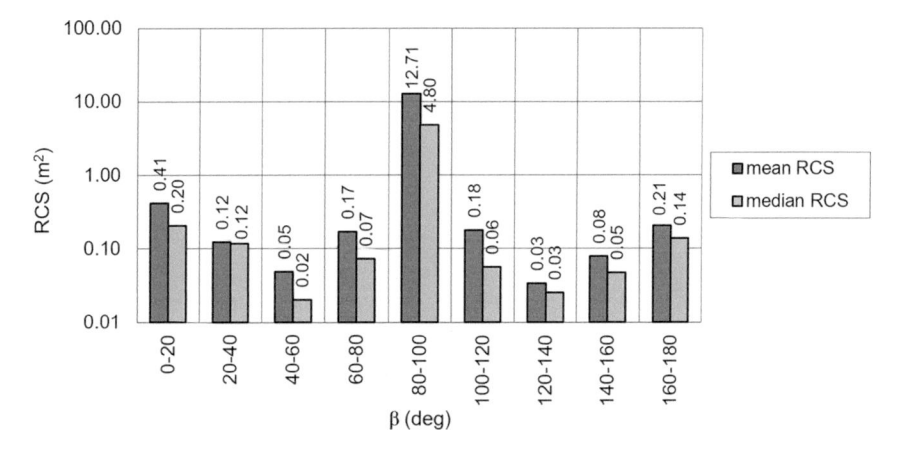

FIGURE 2.896 Diagrams of mean and median RCS of Kh-35 model in 20-degree sectors of azimuth aspect given its radar observation at horizontal polarization and a carrier frequency of 3 GHz (10 cm wavelength).

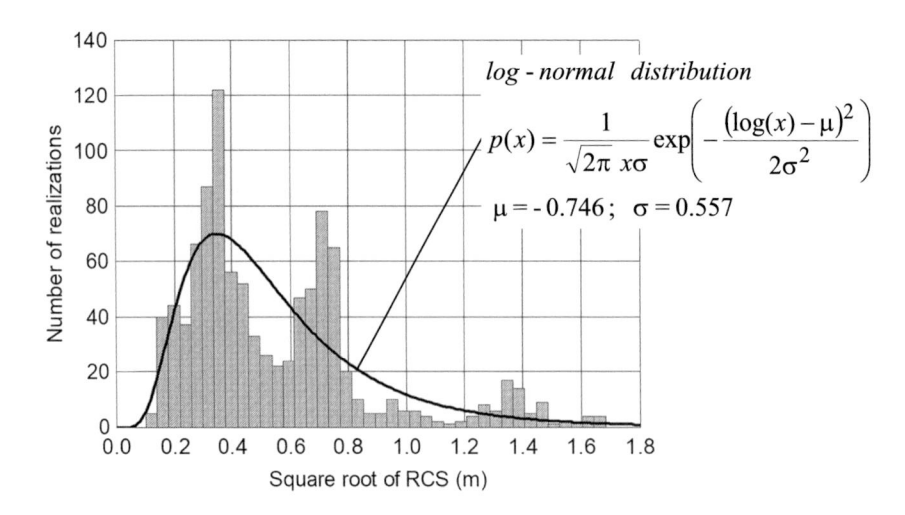

FIGURE 2.897 Amplitude distribution of echo signal of Kh-35 model at a carrier frequency of 3 GHz given its horizontal polarization.

$$p(x) = \frac{1}{\sqrt{2\pi}\, x\sigma} \exp\left(-\frac{(\log(x) - \mu)^2}{2\sigma^2}\right)$$

$\mu = -0.746$; $\sigma = 0.557$

Scattering characteristics of Kh-35 model for sounding frequency 1 GHz (wavelength 30 cm).
Figure 2.898 shows the RCS circular diagram of the Kh-35. The noncoherent RCS circular diagram of the Kh-35 is shown in Figure 2.899.

The circular mean RCS of the Kh-35 model for horizontal polarization is 1.341 m². The circular median RCS for horizontal polarization is 0.233 m².

Figures 2.900 and 2.901 show the mean and median RCS for the main ranges of sounding azimuths (nose, side, and tail) and for ranges of 20°.

The expressions and parameters of probability distributions that are most consistent with the empirical distributions of the square root of the RCS for different ranges of sounding azimuths and polarizations are given in the electronic appendix to this book. The expressions and parameters of the probability distributions that are most consistent with the empirical distributions of the RCS (energy characteristic) for different ranges of sounding azimuths and polarizations are also given there.

FIGURE 2.898 Circular diagram of RCS given radar observation of Kh-35 model at a carrier frequency of 1 GHz (30 cm wavelength).

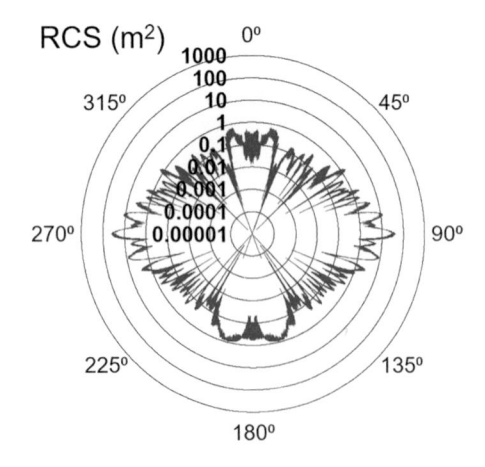

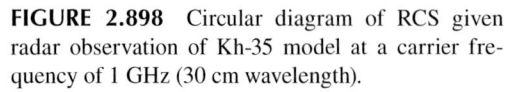

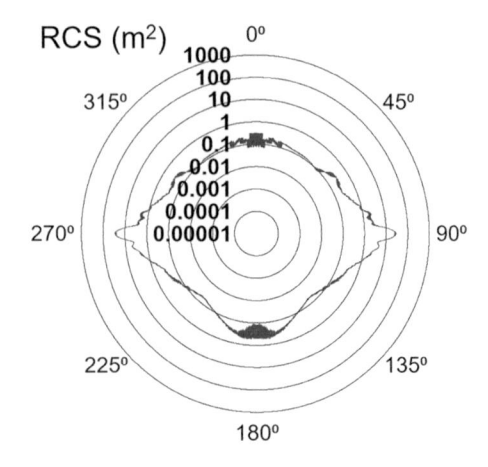

FIGURE 2.899 Circular diagram of noncoherent RCS given radar observation of Kh-35 model at a carrier frequency of 1 GHz (30 cm wavelength).

Scattering Characteristics of Aerial Objects

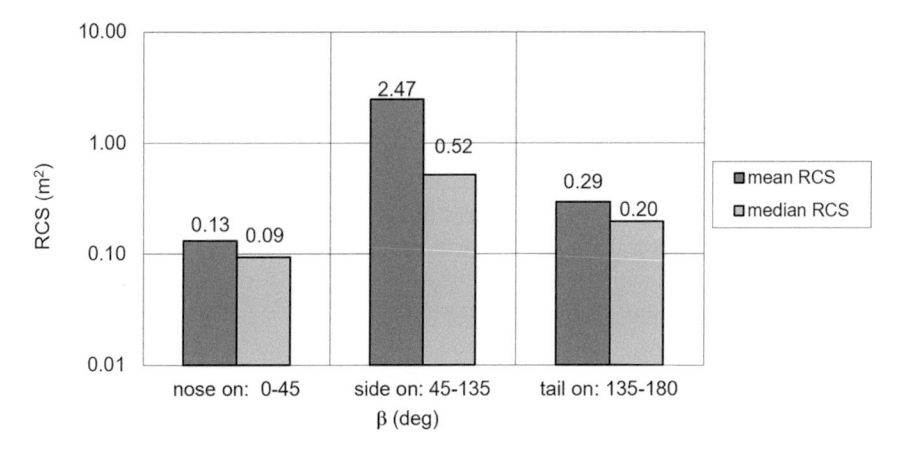

FIGURE 2.900 Diagrams of mean and median RCS of Kh-35 model in three sectors of azimuth aspect given its radar observation at horizontal polarization and a carrier frequency of 1 GHz (30 cm wavelength).

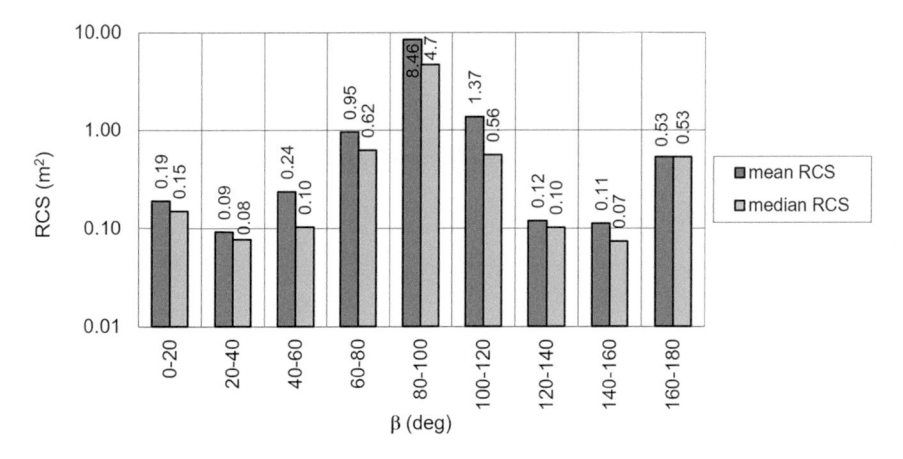

FIGURE 2.901 Diagrams of mean and median RCS of Kh-35 model in 20-degree sectors of azimuth aspect given its radar observation at horizontal polarization and a carrier frequency of 1 GHz (30 cm wavelength).

normal distribution

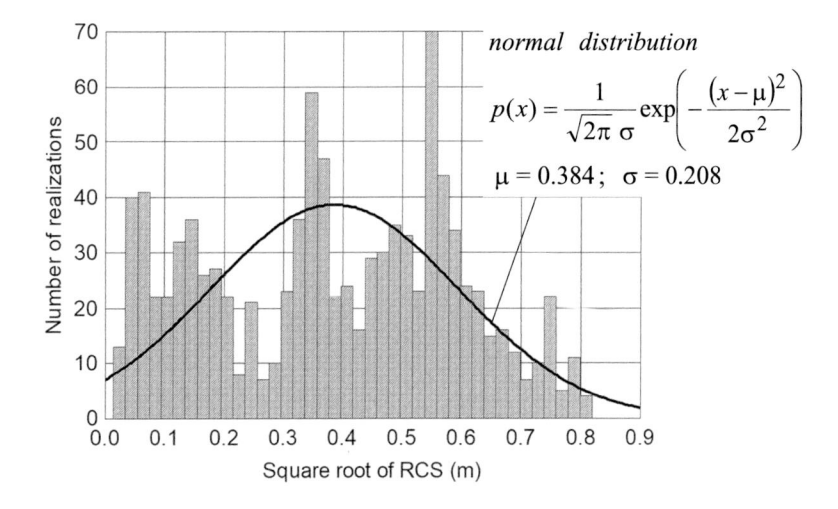

$$p(x) = \frac{1}{\sqrt{2\pi}\,\sigma} \exp\left(-\frac{(x-\mu)^2}{2\sigma^2}\right)$$

$\mu = 0.384; \quad \sigma = 0.208$

FIGURE 2.902 Amplitude distribution of echo signal of Kh-35 model at a carrier frequency of 1 GHz given its horizontal polarization.

2.54 AIR-TO-SURFACE MISSILE Kh-38ML

Kh-38ML (Figure 2.903) is a tactical air-to-surface missile with semi-active laser guidance. The first flight took place in 2010.

General characteristics of Kh-38ML [65]: wingspan – 1.14 m, length – 4.2 m, fuselage diameter – 0.31 m, weight – 520 kg, powerplant - two-stage solid rocket motor, speed – 2.2 M, range – 3–40 km. In accordance with the design of the Kh-38ML a ml of its surface was created to obtaining scattering characteristics (in particular, RCS). The model is shown in Figure 2.904. The smooth parts of the model surface were approximated by parts of 38 triaxial ellipsoids. The surface breaks were modeled using 36 straight edge scattering parts.

FIGURE 2.903 Air-to-surface missile Kh-38ML.

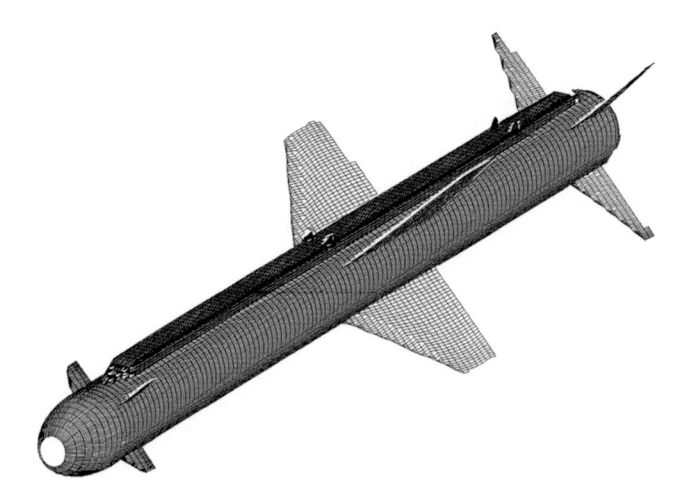

FIGURE 2.904 Surface model of air-to-surface missile Kh-38ML.

Below are some scattering characteristics of the Kh-38ML model at sounding frequencies of 10, 3, and 1 GHz (wavelengths of 3, 10, and 30 cm, respectively) for horizontal polarization of the probing signal. The scattering characteristics for this model at two polarizations, as well as scattering characteristics at sounding frequencies of 5 and 1.3 GHz (wavelengths of 6 and 23 cm, respectively), are given in the electronic appendix of this book.

Sounding parameters: The elevation angle is to be a random value distributed uniformly in the range $-3° \pm 4°$ with respect to the wing plane (elevation angle of $-3°$ corresponds to the radar observation from the lower hemisphere), azimuth aspect increment was $0.02°$, the azimuth being counted

off from the nose-on aspect (0° corresponds to the nose-on radar observation, 180° corresponds to the tail-on observation).

Scattering characteristics of Kh-38ML model for sounding frequency 10 GHz (wavelength 3 cm).

Figure 2.905 shows the RCS circular diagram of the Kh-38ML model. The noncoherent RCS circular diagram of the Kh-38ML is shown in Figure 2.906.

The circular mean RCS of the Kh-38ML model for horizontal polarization is 6.457 m². The circular median RCS (the RCS value used to calculate the detection range of an object with a probability of 0.5) for horizontal polarization is 0.044 m².

Figures 2.907 and 2.908 show the mean and median RCS for the main ranges of sounding azimuths (nose, side, and tail) and for ranges of 20°.

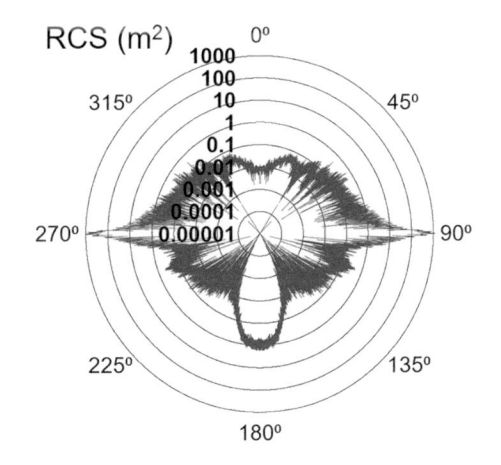

FIGURE 2.905 Circular diagram of RCS given radar observation of Kh-38ML model at a carrier frequency of 10 GHz (3 cm wavelength).

FIGURE 2.906 Circular diagram of noncoherent RCS given radar observation of Kh-38ML model at a carrier frequency of 10 GHz (3 cm wavelength).

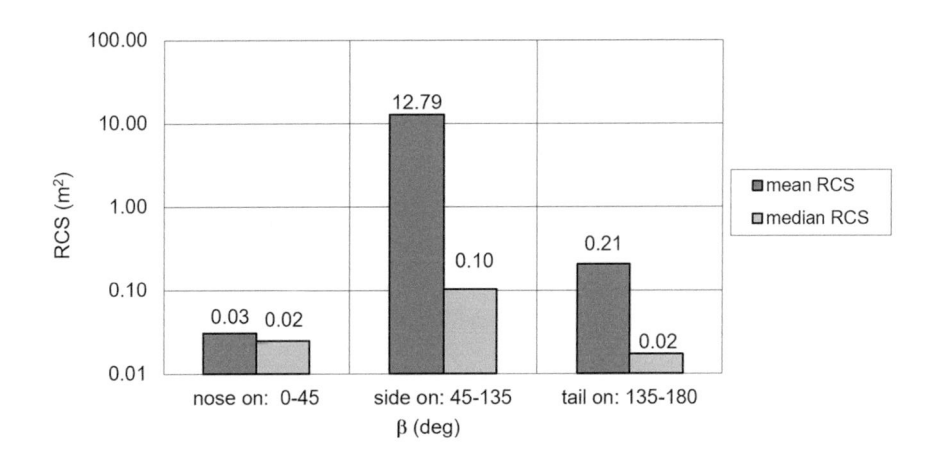

FIGURE 2.907 Diagrams of mean and median RCS of Kh-38ML model in three sectors of azimuth aspect given its radar observation at horizontal polarization and a carrier frequency of 10 GHz (3 cm wavelength).

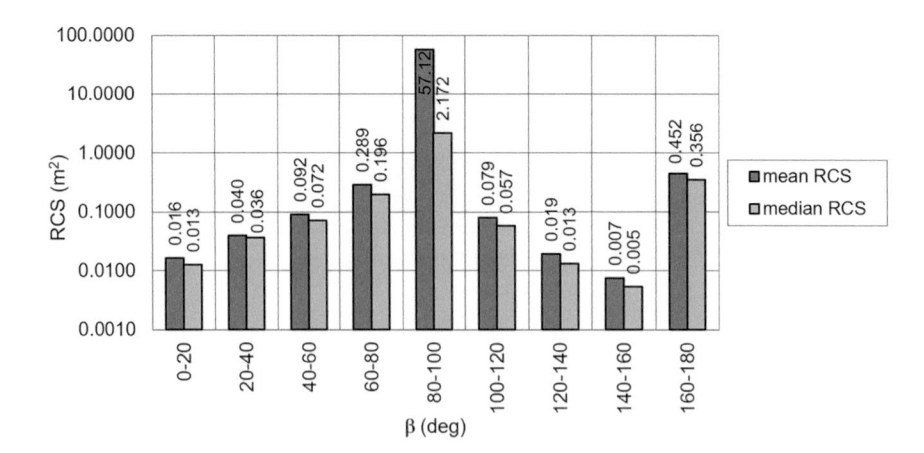

FIGURE 2.908 Diagrams of mean and median RCS of Kh-38ML model in 20-degree sectors of azimuth aspect given its radar observation at horizontal polarization and a carrier frequency of 10 GHz (3 cm wavelength).

Figures 2.909, 2.914, and 2.919 show histograms of the scattered signal amplitude (square root of the RCS) for the range of sounding azimuths −20° to +20° (sounding from the nose). The bold line shows the probability density functions of the distribution, which can be used to approximate the histogram of the signal amplitude.

extreme value distribution

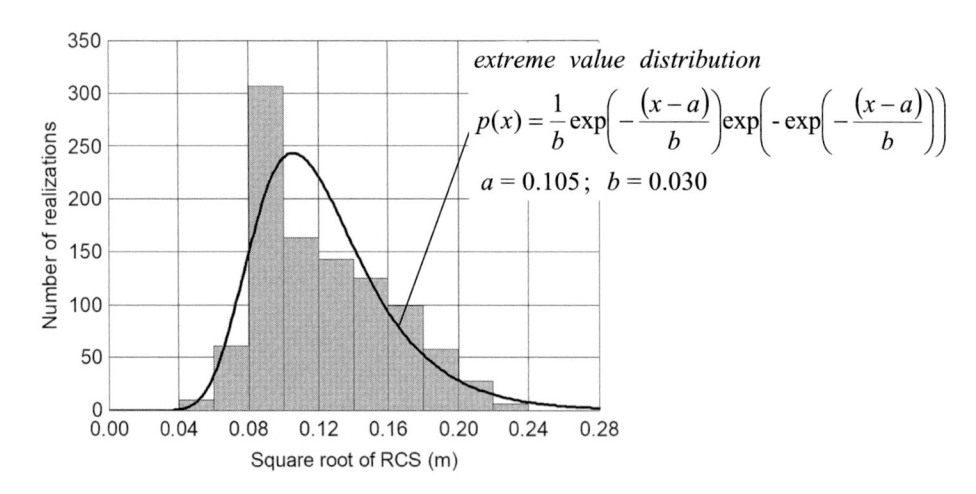

$$p(x) = \frac{1}{b}\exp\left(-\frac{(x-a)}{b}\right)\exp\left(-\exp\left(-\frac{(x-a)}{b}\right)\right)$$

$a = 0.105; \quad b = 0.030$

FIGURE 2.909 Amplitude distribution of echo signal of Kh-38ML model at a carrier frequency of 10 GHz given its horizontal polarization.

Scattering characteristics of Kh-38ML model for sounding frequency 3 GHz (wavelength 10 cm).

Figure 2.910 shows the RCS circular diagram of the Kh-38ML. The noncoherent RCS circular diagram of the Kh-38ML is shown in Figure 2.911.

The circular mean RCS of the Kh-38ML for horizontal polarization is 2.570 m². The circular median RCS for horizontal polarization is 0.048 m².

Figures 2.912 and 2.913 show the mean and median RCS for the main ranges of sounding azimuths (nose, side, and tail) and for ranges of 20°.

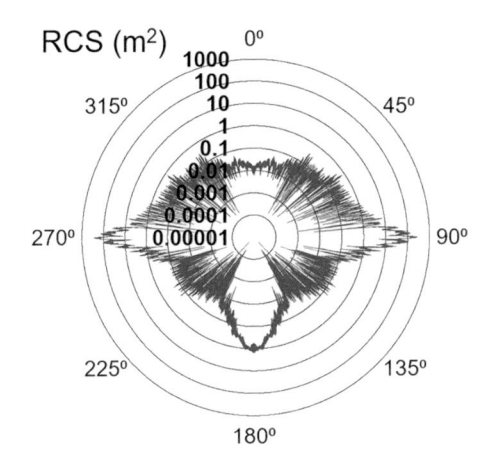

Scattering Characteristics of Aerial Objects

FIGURE 2.910 Circular diagram of RCS given radar observation of Kh-38ML model at a carrier frequency of 3 GHz (10 cm wavelength).

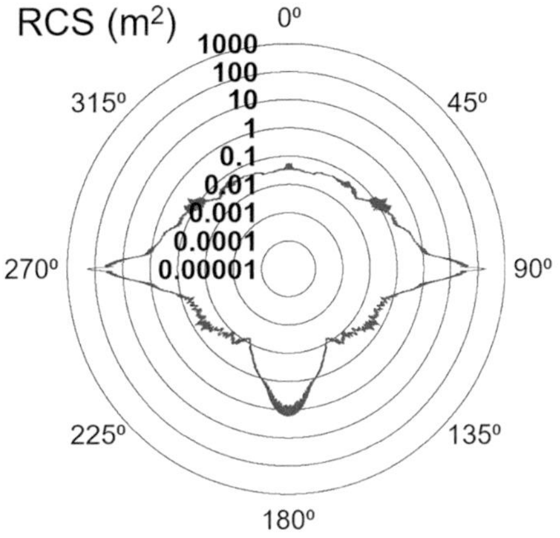

FIGURE 2.911 Circular diagram of noncoherent RCS given radar observation of Kh-38ML model at a carrier frequency of 3 GHz (10 cm wavelength).

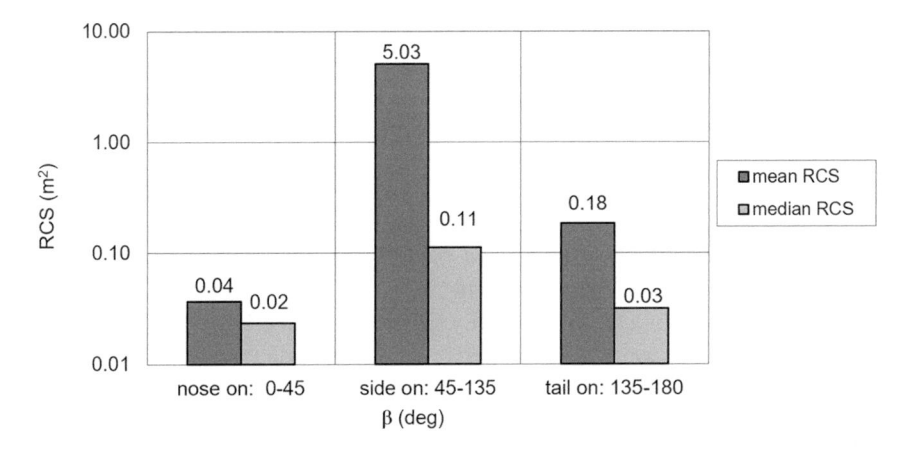

FIGURE 2.912 Diagrams of mean and median RCS of Kh-38ML model in three sectors of azimuth aspect given its radar observation at horizontal polarization and a carrier frequency of 3 GHz (10 cm wavelength).

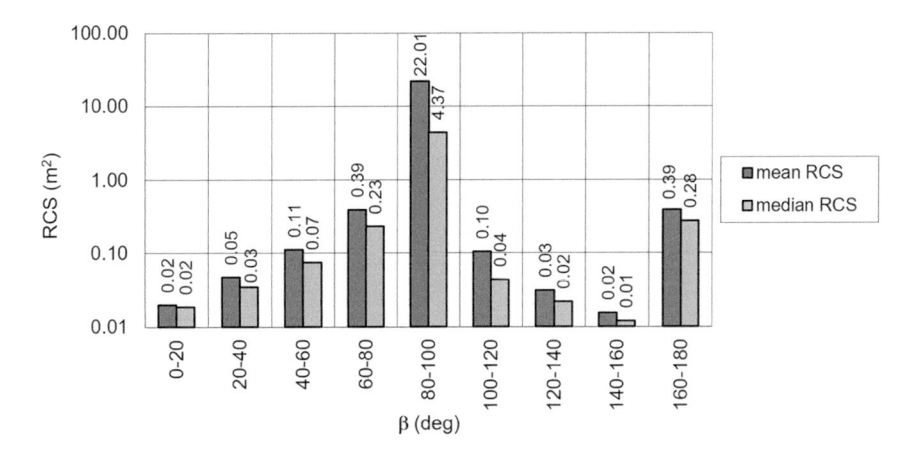

FIGURE 2.913 Diagrams of mean and median RCS of Kh-38ML model in 20-degree sectors of azimuth aspect given its radar observation at horizontal polarization and a carrier frequency of 3 GHz (10 cm wavelength).

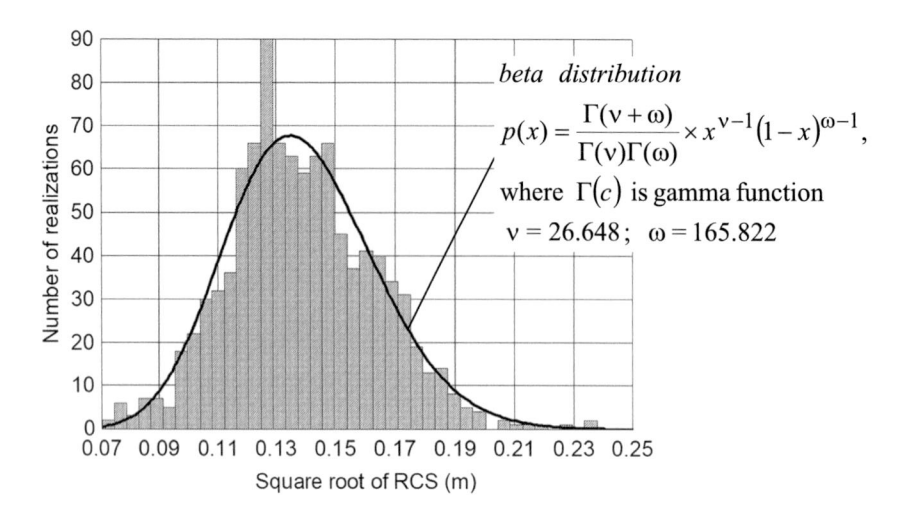

FIGURE 2.914 Amplitude distribution of echo signal of Kh-38ML model at a carrier frequency of 3 GHz given its horizontal polarization.

Scattering characteristics of Kh-38ML model for sounding frequency 1 GHz (wavelength 30 cm).

Figure 2.915 shows the RCS circular diagram of the Kh-38ML. The noncoherent RCS circular diagram of the Kh-38ML is shown in Figure 2.916.

The circular mean RCS of the Kh-38ML model for horizontal polarization is 0.648 m². The circular median RCS for horizontal polarization is 0.108 m².

Figures 2.917 and 2.918 show the mean and median RCS for the main ranges of sounding azimuths (nose, side, and tail) and for ranges of 20°.

The expressions and parameters of probability distributions that are most consistent with the empirical distributions of the square root of the RCS for different ranges of sounding azimuths and polarizations are given in the electronic appendix to this book. The expressions and parameters of the probability distributions that are most consistent with the empirical distributions of the RCS (energy characteristic) for different ranges of sounding azimuths and polarizations are also given there.

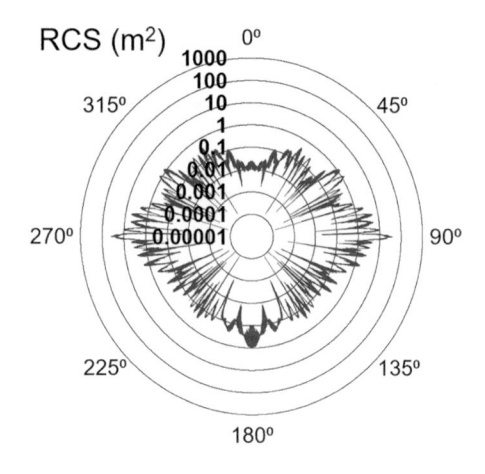

FIGURE 2.915 Circular diagram of RCS given radar observation of Kh-38ML model at a carrier frequency of 1 GHz (30 cm wavelength).

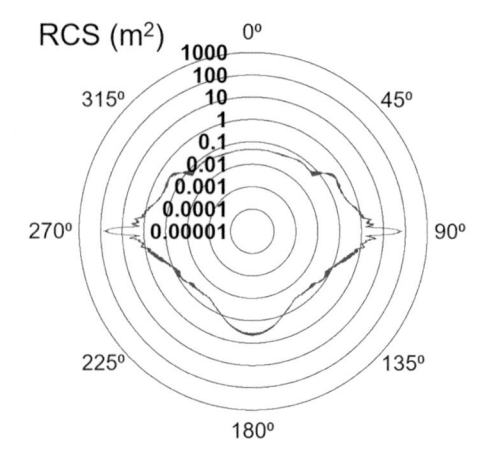

FIGURE 2.916 Circular diagram of noncoherent RCS given radar observation of Kh-38ML model at a carrier frequency of 1 GHz (30 cm wavelength).

The beta distribution shown in Figure 2.914:

$$p(x) = \frac{\Gamma(\nu+\omega)}{\Gamma(\nu)\Gamma(\omega)} \times x^{\nu-1}(1-x)^{\omega-1},$$

where $\Gamma(c)$ is gamma function, $\nu = 26.648$; $\omega = 165.822$.

Scattering Characteristics of Aerial Objects

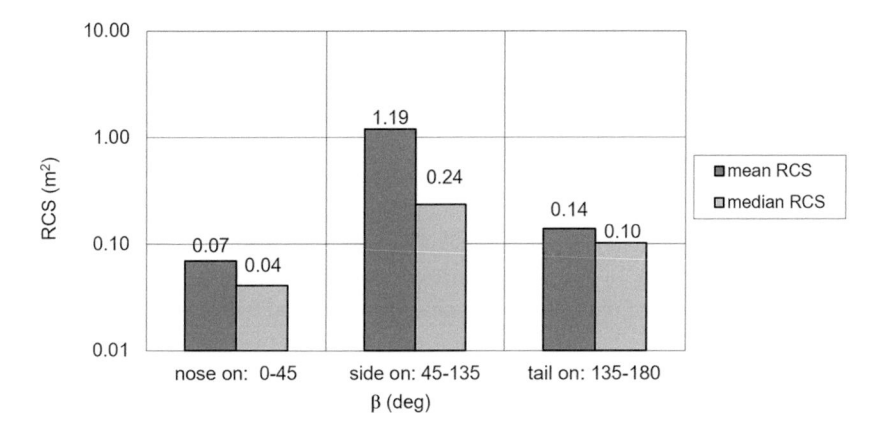

FIGURE 2.917 Diagrams of mean and median RCS of Kh-38ML model in three sectors of azimuth aspect given its radar observation at horizontal polarization and a carrier frequency of 1 GHz (30 cm wavelength).

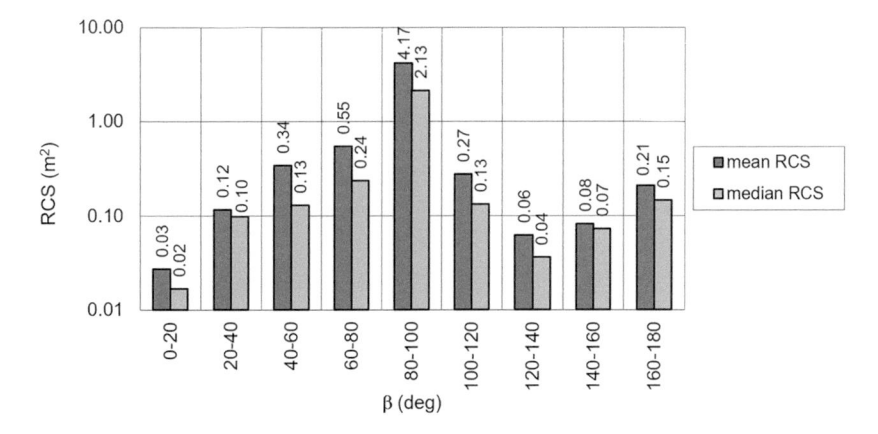

FIGURE 2.918 Diagrams of mean and median RCS of Kh-38ML model in 20-degree sectors of azimuth aspect given its radar observation at horizontal polarization and a carrier frequency of 1 GHz (30 cm wavelength).

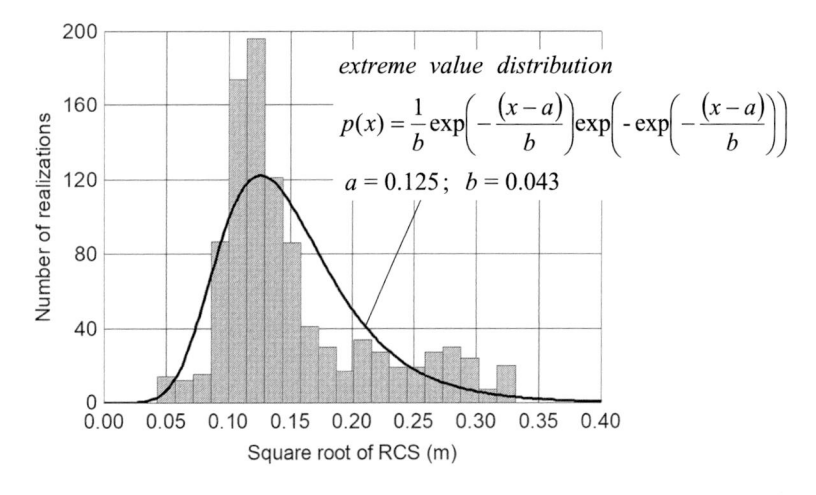

extreme value distribution

$$p(x) = \frac{1}{b}\exp\left(-\frac{(x-a)}{b}\right)\exp\left(-\exp\left(-\frac{(x-a)}{b}\right)\right)$$

$a = 0.125; \ b = 0.043$

FIGURE 2.919 Amplitude distribution of echo signal of Kh-38ML model at a carrier frequency of 1 GHz given its horizontal polarization.

2.55 ANTI-RADIATION MISSILE Kh-58UShKE

Kh-58UShKE (Figure 2.920) is an air-launched anti-radiation missile. The first flight took place in 2008.

General characteristics of Kh-58UShKE [66]: wingspan – 0.8 m, length – 4.19 m, fuselage diameter – 0.4 m, weight – 650 kg, powerplant - solid rocket motor, speed – 4200 km/h, range – 76–245 km.

In accordance with the design of the Kh-58UShKE, a model of its surface was created to obtaining scattering characteristics (in particular, RCS). The model is shown in Figure 2.921. The perfectly conducting smooth parts of the model surface were approximated by parts of 53 triaxial ellipsoids. Dielectric parts of the model surface were approximated by parts of 4 triaxial ellipsoids. The surface breaks were modeled using 24 straight edge scattering parts.

FIGURE 2.920 Anti-radiation missile Kh58UShKE.

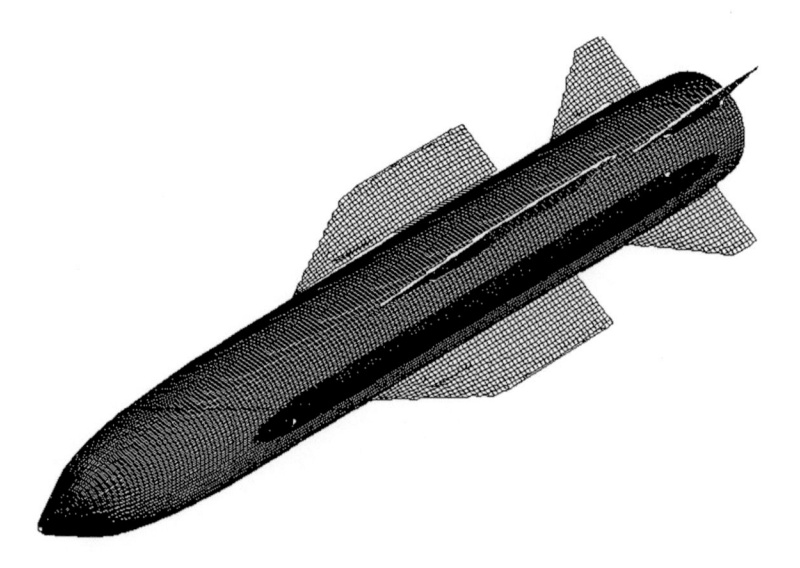

FIGURE 2.921 Surface model of anti-radiation missile Kh-58UShKE.

Below are some scattering characteristics of the Kh-58UShKE model at sounding frequencies of 10, 3, and 1 GHz (wavelengths of 3, 10, and 30 cm, respectively) for horizontal polarization of the probing signal. The scattering characteristics for this model at two polarizations, as well as scattering characteristics at sounding frequencies of 5 and 1.3 GHz (wavelengths of 6 and 23 cm, respectively), are given in the electronic appendix of this book.

Sounding parameters: The elevation angle is to be a random value distributed uniformly in the range −3° ± 4° with respect to the wing plane (elevation angle of −3° corresponds to the radar observation from the lower hemisphere), azimuth aspect increment was 0.02°, the azimuth being counted

Scattering Characteristics of Aerial Objects 337

off from the nose-on aspect (0° corresponds to the nose-on radar observation, 180° corresponds to the tail-on observation).

Scattering characteristics of Kh-58UShKE model for sounding frequency 10 GHz (wavelength 3 cm).

Figure 2.922 shows the RCS circular diagram of the Kh-58UShKE model. The noncoherent RCS circular diagram of the Kh-58UShKE is shown in Figure 2.923.

The circular mean RCS of the Kh-58UShKE model for horizontal polarization is 1.149 m². The circular median RCS (the RCS value used to calculate the detection range of an object with a probability of 0.5) for horizontal polarization is 0.036 m².

Figures 2.924 and 2.925 show the mean and median RCS for the main ranges of sounding azimuths (nose, side, and tail) and for ranges of 20°.

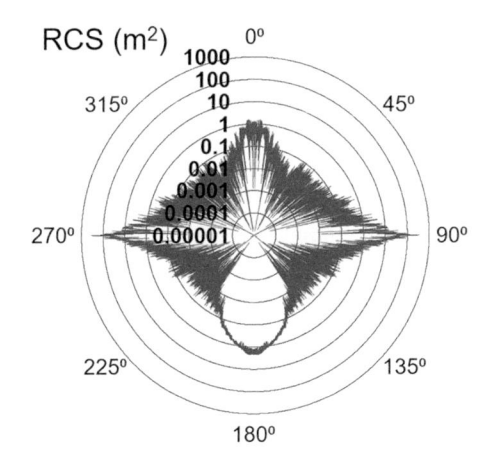

FIGURE 2.922 Circular diagram of RCS given radar observation of Kh-58UShKE model at a carrier frequency of 10 GHz (3 cm wavelength).

FIGURE 2.923 Circular diagram of noncoherent RCS given radar observation of Kh-58UShKE model at a carrier frequency of 10 GHz (3 cm wavelength).

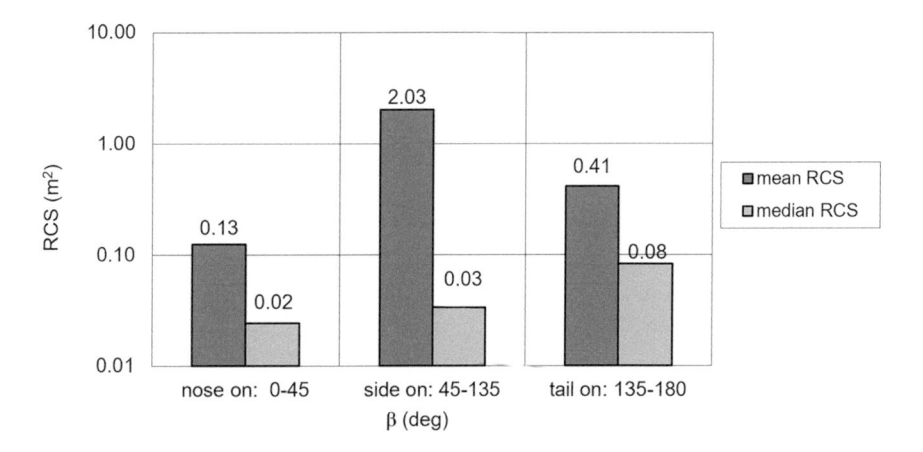

FIGURE 2.924 Diagrams of mean and median RCS of Kh-58UShKE model in three sectors of azimuth aspect given its radar observation at horizontal polarization and a carrier frequency of 10 GHz (3 cm wavelength).

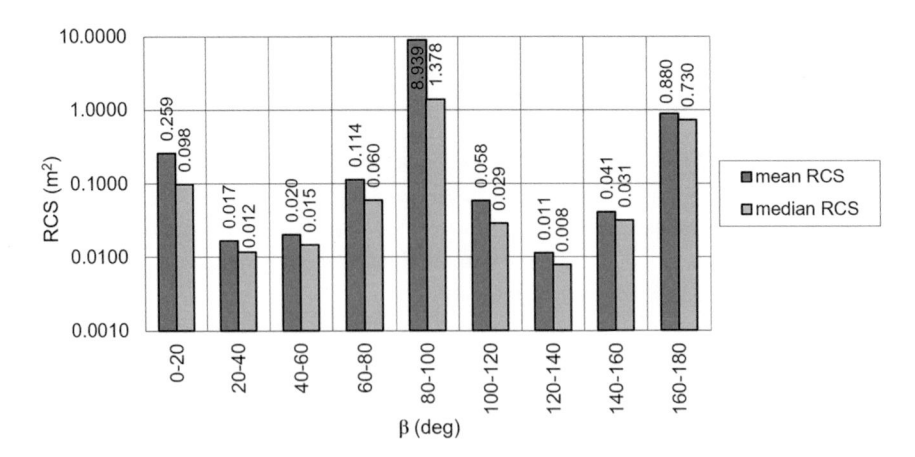

FIGURE 2.925 Diagrams of mean and median RCS of Kh-58UShKE model in 20-degree sectors of azimuth aspect given its radar observation at horizontal polarization and a carrier frequency of 10 GHz (3 cm wavelength).

Figures 2.926, 2.931, and 2.936 show histograms of the scattered signal amplitude (square root of the RCS) for the range of sounding azimuths −20° to +20° (sounding from the nose). The bold line shows the probability density functions of the distribution, which can be used to approximate the histogram of the signal amplitude.

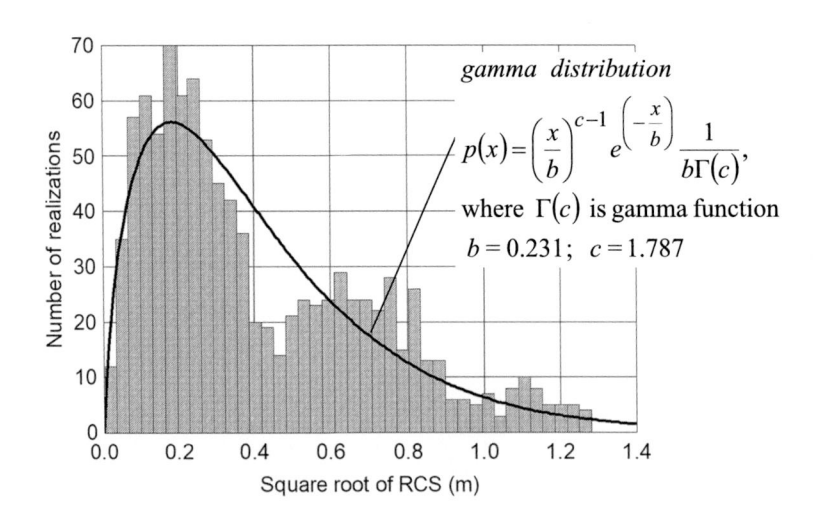

$$p(x) = \left(\frac{x}{b}\right)^{c-1} e^{\left(-\frac{x}{b}\right)} \frac{1}{b\Gamma(c)},$$

where $\Gamma(c)$ is gamma function
$b = 0.231; \quad c = 1.787$

FIGURE 2.926 Amplitude distribution of echo signal of Kh-58UShKE model at a carrier frequency of 10 GHz given its horizontal polarization.

Scattering characteristics of Kh-58UShKE model for sounding frequency 3 GHz (wavelength 10 cm).

Figure 2.927 shows the RCS circular diagram of the Kh-58UShKE. The noncoherent RCS circular diagram of the Kh-58UShKE is shown in Figure 2.928.

The circular mean RCS of the Kh-58UShKE for horizontal polarization is 0.541 m². The circular median RCS for horizontal polarization is 0.047 m².

Figures 2.929 and 2.930 show the mean and median RCS for the main ranges of sounding azimuths (nose, side, and tail) and for ranges of 20°.

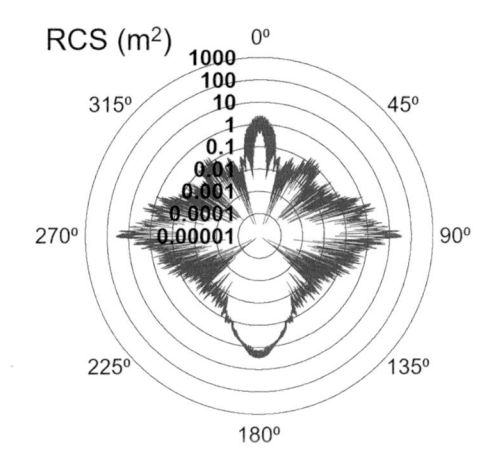

Scattering Characteristics of Aerial Objects

FIGURE 2.927 Circular diagram of RCS given radar observation of Kh-58UShKE model at a carrier frequency of 3 GHz (10 cm wavelength).

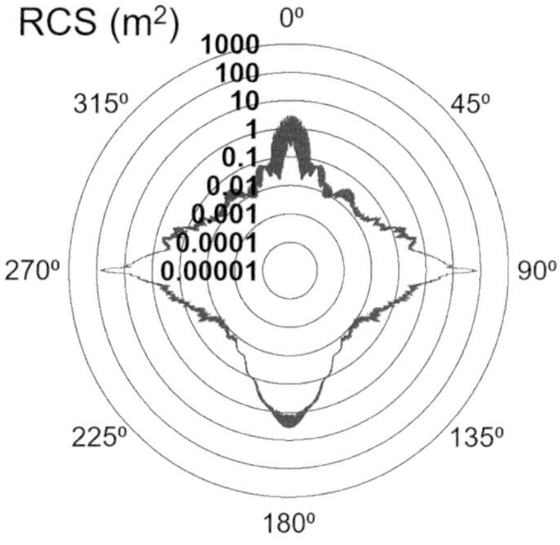

FIGURE 2.928 Circular diagram of noncoherent RCS given radar observation of Kh-58UShKE model at a carrier frequency of 3 GHz (10 cm wavelength).

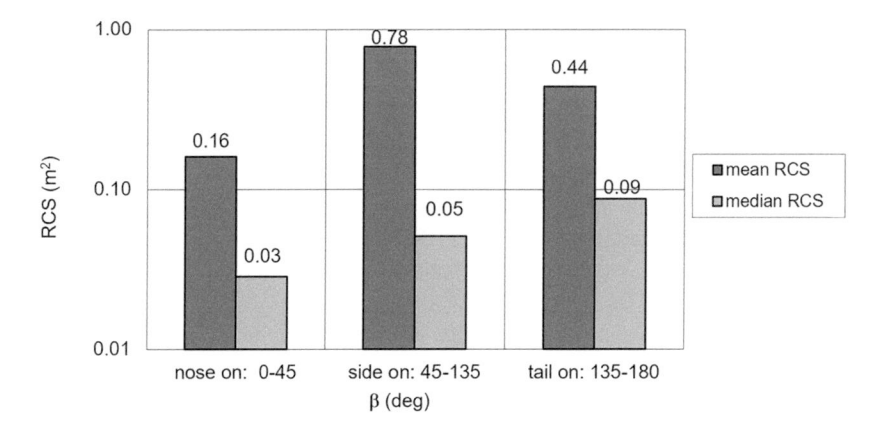

FIGURE 2.929 Diagrams of mean and median RCS of Kh-58UShKE model in three sectors of azimuth aspect given its radar observation at horizontal polarization and a carrier frequency of 3 GHz (10 cm wavelength).

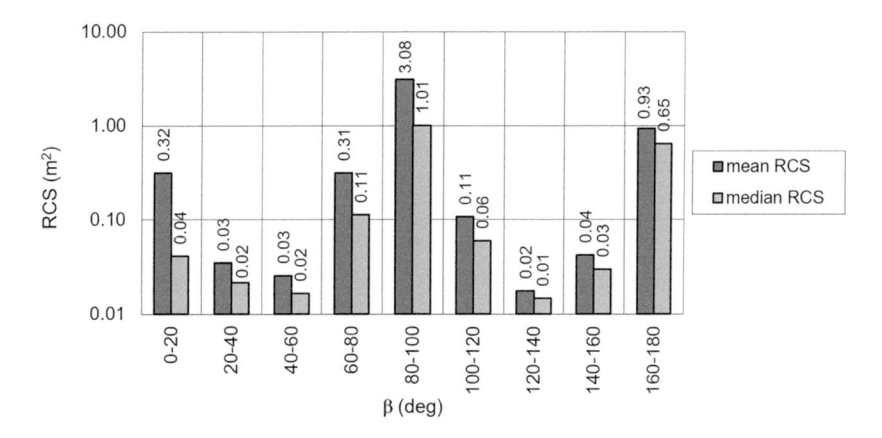

FIGURE 2.930 Diagrams of mean and median RCS of Kh-58UShKE model in 20-degree sectors of azimuth aspect given its radar observation at horizontal polarization and a carrier frequency of 3 GHz (10 cm wavelength).

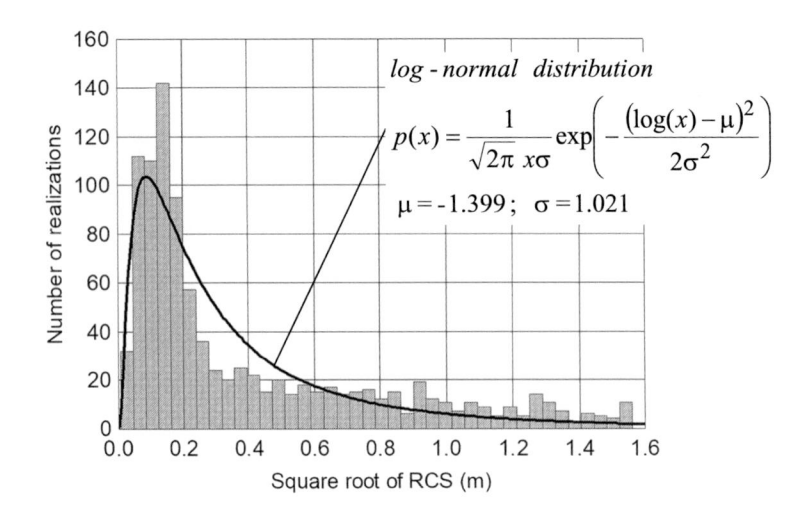

FIGURE 2.931 Amplitude distribution of echo signal of Kh-58UShKE model at a carrier frequency of 3 GHz given its horizontal polarization.

The log-normal distribution shown:
$$p(x) = \frac{1}{\sqrt{2\pi}\, x\sigma} \exp\left(-\frac{(\log(x)-\mu)^2}{2\sigma^2}\right)$$
$\mu = -1.399;\ \sigma = 1.021$

Scattering characteristics of Kh-58UShKE model for sounding frequency 1 GHz (wavelength 30 cm).

Figure 2.932 shows the RCS circular diagram of the Kh-58UShKE. The noncoherent RCS circular diagram of the Kh-58UShKE is shown in Figure 2.933.

The circular mean RCS of the Kh-58UShKE model for horizontal polarization is 1.517 m². The circular median RCS for horizontal polarization is 0.171 m².

Figures 2.934 and 2.935 show the mean and median RCS for the main ranges of sounding azimuths (nose, side, and tail) and for ranges of 20°.

The expressions and parameters of probability distributions that are most consistent with the empirical distributions of the square root of the RCS for different ranges of sounding azimuths and polarizations are given in the electronic appendix to this book. The expressions and parameters of the probability distributions that are most consistent with the empirical distributions of the RCS (energy characteristic) for different ranges of sounding azimuths and polarizations are also given there.

FIGURE 2.932 Circular diagram of RCS given radar observation of Kh-58UShKE model at a carrier frequency of 1 GHz (30 cm wavelength).

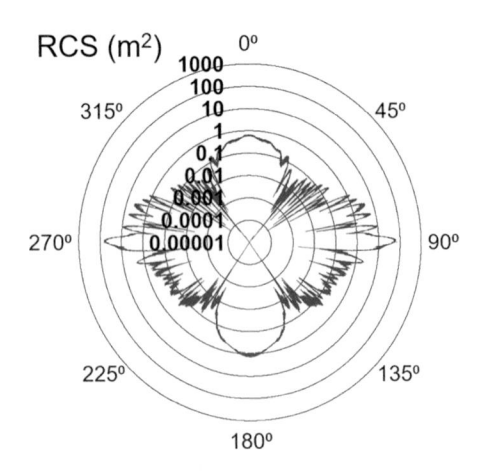

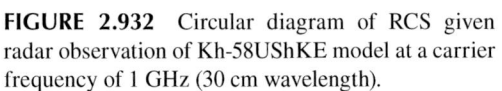

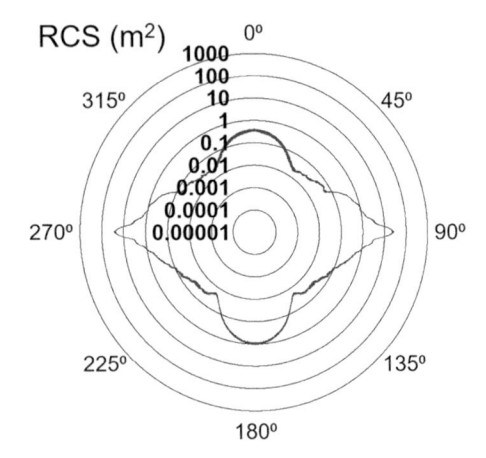

FIGURE 2.933 Circular diagram of noncoherent RCS given radar observation of Kh-58UShKE model at a carrier frequency of 1 GHz (30 cm wavelength).

Scattering Characteristics of Aerial Objects

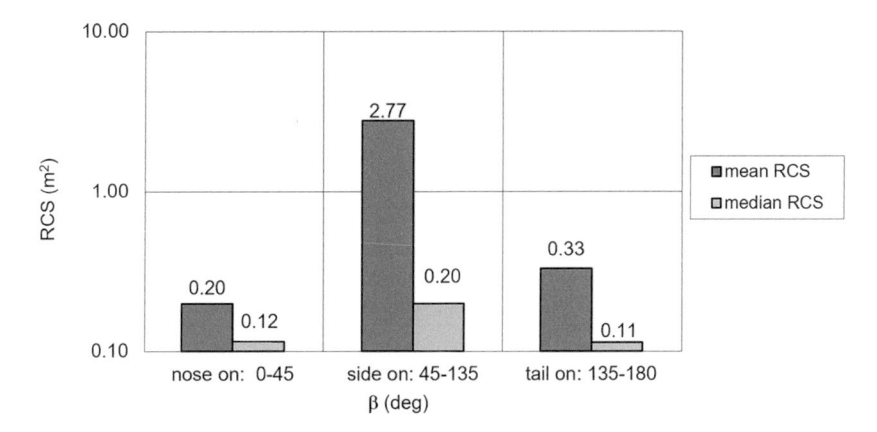

FIGURE 2.934 Diagrams of mean and median RCS of Kh-58UShKE model in three sectors of azimuth aspect given its radar observation at horizontal polarization and a carrier frequency of 1 GHz (30 cm wavelength).

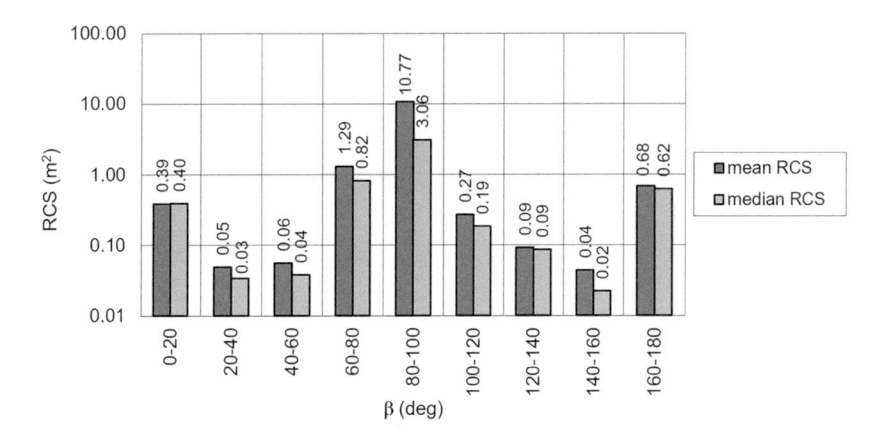

FIGURE 2.935 Diagrams of mean and median RCS of Kh-58UShKE model in 20-degree sectors of azimuth aspect given its radar observation at horizontal polarization and a carrier frequency of 1 GHz (30 cm wavelength).

Weibull distribution

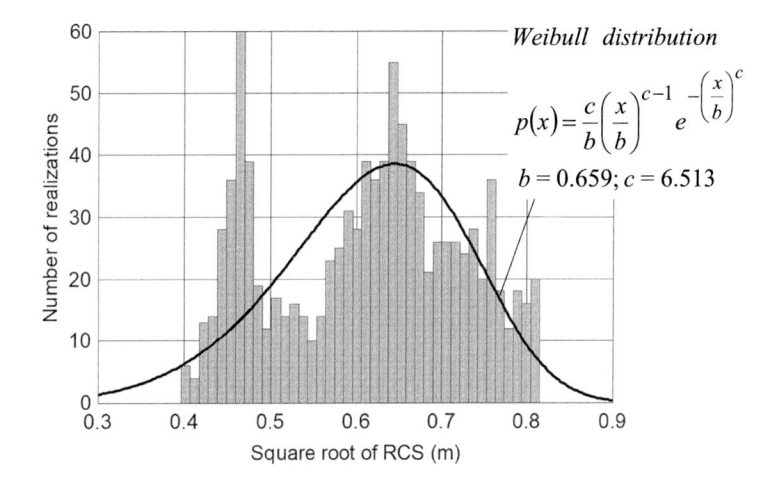

$$p(x) = \frac{c}{b}\left(\frac{x}{b}\right)^{c-1} e^{-\left(\frac{x}{b}\right)^c}$$

$b = 0.659; c = 6.513$

FIGURE 2.936 Amplitude distribution of echo signal of Kh-58UShKE model at a carrier frequency of 1 GHz given its horizontal polarization.

2.56 AIR-TO-SURFACE MISSILE Kh-59M

Kh-59M (Figure 2.937) is an air-launched air-to-surface missile. The first flight took place in 1990.

General characteristics of Kh-59M [67]: wingspan – 1.26–1.30 m, length – 5.69 m, fuselage diameter – 0.38 m, weight – 920 kg, powerplant – 1 RDK-300 turbojet engine, speed – 860–1 000 km/h, range – 10–115 km.

In accordance with the design of the Kh-59M, a model of its surface was created to obtaining scattering characteristics (in particular, RCS). The model is shown in Figure 2.938. The smooth parts of the model surface were approximated by parts of 34 triaxial ellipsoids. The surface breaks were modeled using 20 straight edge scattering parts.

FIGURE 2.937 Air-to-surface missile Kh-59M.

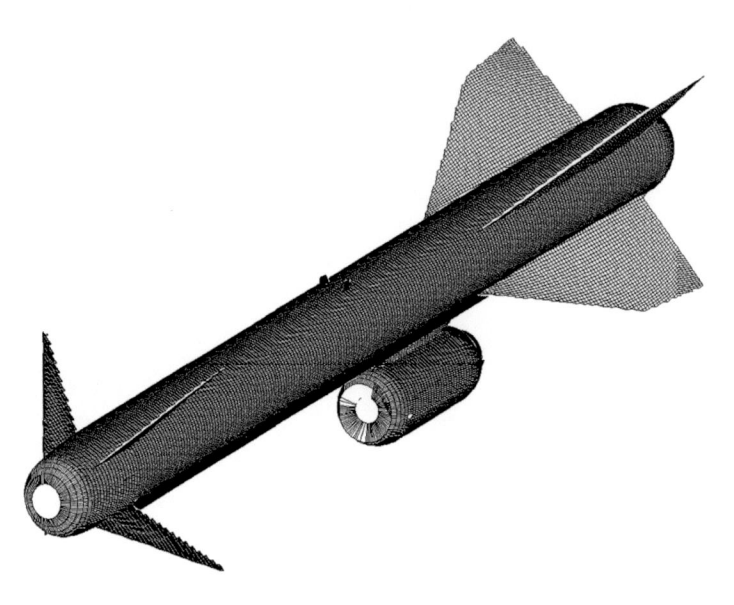

FIGURE 2.938 Surface model of air-to-surface missile Kh-59M.

Below are some scattering characteristics of the Kh-59M model at sounding frequencies of 10, 3, and 1 GHz (wavelengths of 3, 10, and 30 cm, respectively) for horizontal polarization of the probing signal. The scattering characteristics for this model at two polarizations, as well as scattering characteristics at sounding frequencies of 5 and 1.3 GHz (wavelengths of 6 and 23 cm, respectively), are given in the electronic appendix of this book.

Scattering Characteristics of Aerial Objects 343

Sounding parameters: The elevation angle is to be a random value distributed uniformly in the range −3° ± 4° with respect to the wing plane (elevation angle of −3° corresponds to the radar observation from the lower hemisphere), azimuth aspect increment was 0.02°, the azimuth being counted off from the nose-on aspect (0° corresponds to the nose-on radar observation, 180° corresponds to the tail-on observation).

Scattering characteristics of Kh-59M model for sounding frequency 10 GHz (wavelength 3 cm).

Figure 2.939 shows the RCS circular diagram of the Kh-59M model. The noncoherent RCS circular diagram of the Kh-59M is shown in Figure 2.940.

The circular mean RCS of the Kh-59M model for horizontal polarization is 7.714 m². The circular median RCS (the RCS value used to calculate the detection range of an object with a probability of 0.5) for horizontal polarization is 0.140 m².

Figures 2.941 and 2.942 show the mean and median RCS for the main ranges of sounding azimuths (nose, side, and tail) and for ranges of 20°.

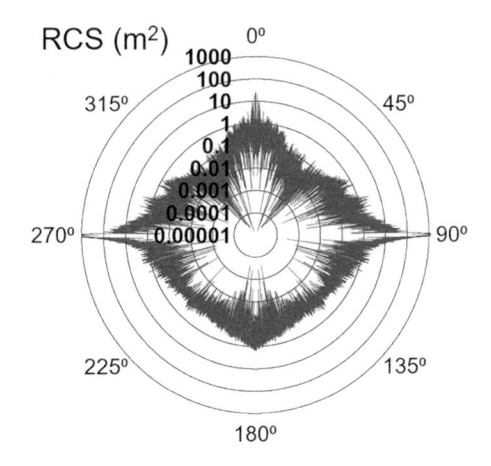

FIGURE 2.939 Circular diagram of RCS given radar observation of Kh-59M model at a carrier frequency of 10 GHz (3 cm wavelength).

FIGURE 2.940 Circular diagram of noncoherent RCS given radar observation of Kh-59M model at a carrier frequency of 10 GHz (3 cm wavelength).

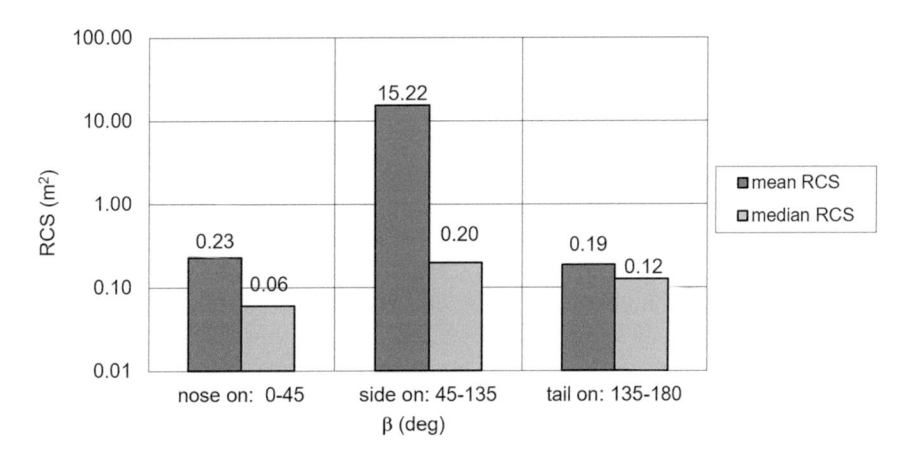

FIGURE 2.941 Diagrams of mean and median RCS of Kh-59M model in three sectors of azimuth aspect given its radar observation at horizontal polarization and a carrier frequency of 10 GHz (3 cm wavelength).

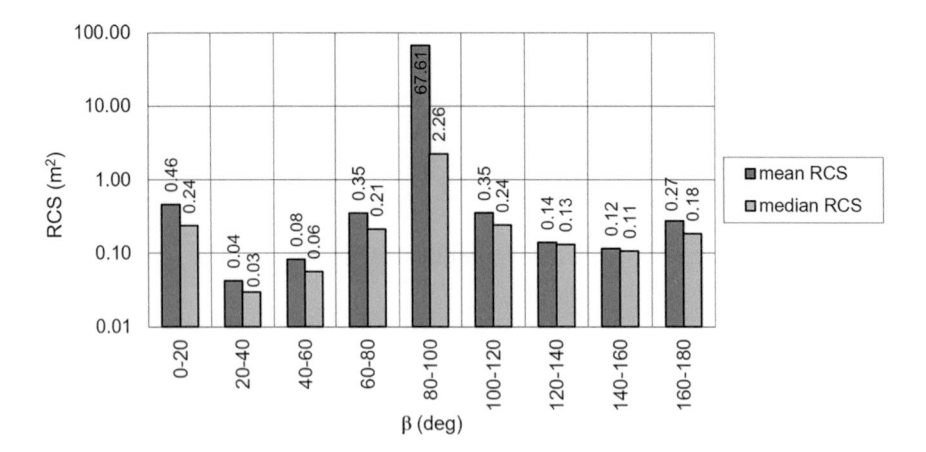

FIGURE 2.942 Diagrams of mean and median RCS of Kh-59M model in 20-degree sectors of azimuth aspect given its radar observation at horizontal polarization and a carrier frequency of 10 GHz (3 cm wavelength).

Figures 2.943, 2.948, and 2.953 show histograms of the scattered signal amplitude (square root of the RCS) for the range of sounding azimuths −20° to +20° (sounding from the nose). The bold line shows the probability density functions of the distribution, which can be used to approximate the histogram of the signal amplitude.

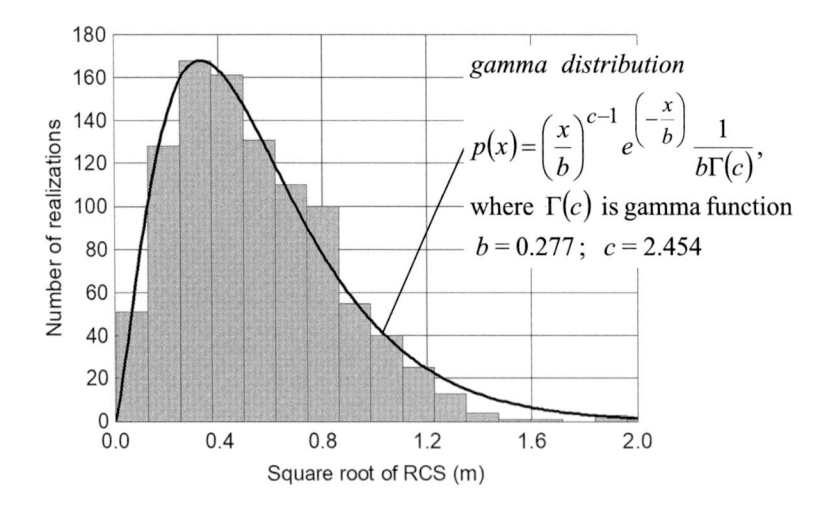

gamma distribution

$$p(x) = \left(\frac{x}{b}\right)^{c-1} e^{\left(-\frac{x}{b}\right)} \frac{1}{b\Gamma(c)},$$

where $\Gamma(c)$ is gamma function
$b = 0.277; \quad c = 2.454$

FIGURE 2.943 Amplitude distribution of echo signal of Kh-59M model at a carrier frequency of 10 GHz given its horizontal polarization.

Scattering characteristics of Kh-59M model for sounding frequency 3 GHz (wavelength 10 cm).

Figure 2.944 shows the RCS circular diagram of the Kh-59M. The noncoherent RCS circular diagram of the Kh-59M is shown in Figure 2.945.

The circular mean RCS of the Kh-59M for horizontal polarization is 4.193 m². The circular median RCS for horizontal polarization is 0.143 m².

Figures 2.946 and 2.947 show the mean and median RCS for the main ranges of sounding azimuths (nose, side, and tail) and for ranges of 20°.

Scattering Characteristics of Aerial Objects

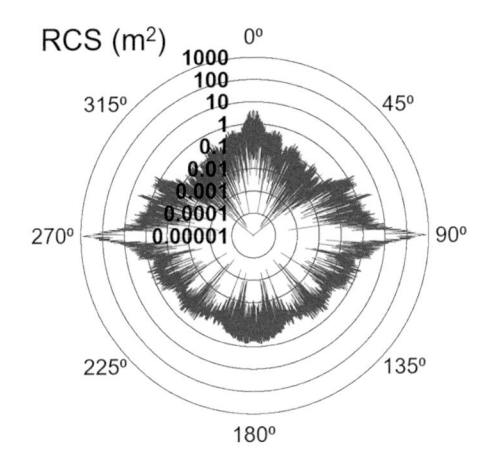

FIGURE 2.944 Circular diagram of RCS given radar observation of Kh-59M model at a carrier frequency of 3 GHz (10 cm wavelength).

FIGURE 2.945 Circular diagram of noncoherent RCS given radar observation of Kh-59M model at a carrier frequency of 3 GHz (10 cm wavelength).

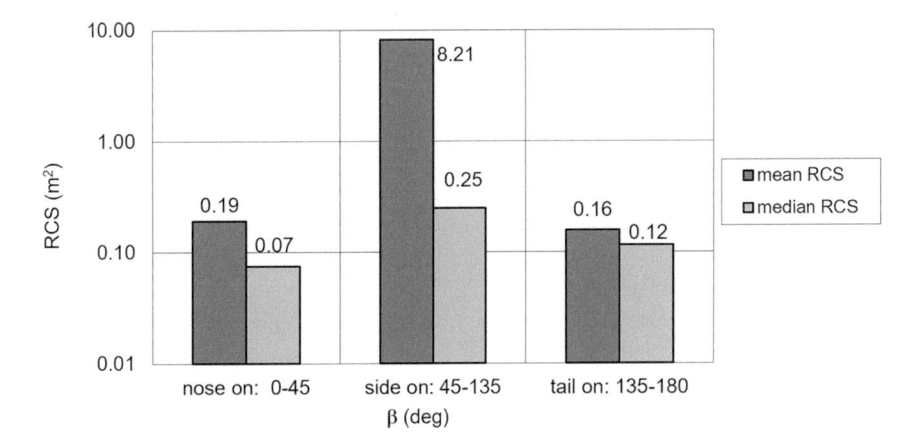

FIGURE 2.946 Diagrams of mean and median RCS of Kh-59M model in three sectors of azimuth aspect given its radar observation at horizontal polarization and a carrier frequency of 3 GHz (10 cm wavelength).

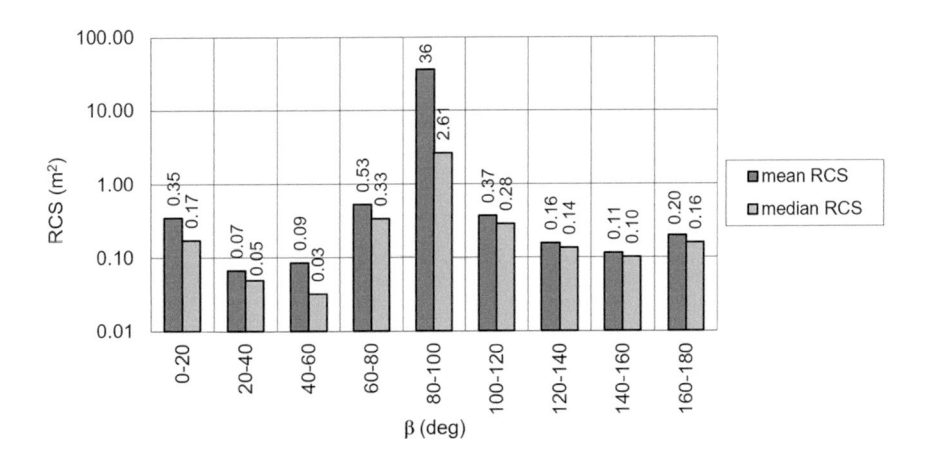

FIGURE 2.947 Diagrams of mean and median RCS of Kh-59M model in 20-degree sectors of azimuth aspect given its radar observation at horizontal polarization and a carrier frequency of 3 GHz (10 cm wavelength).

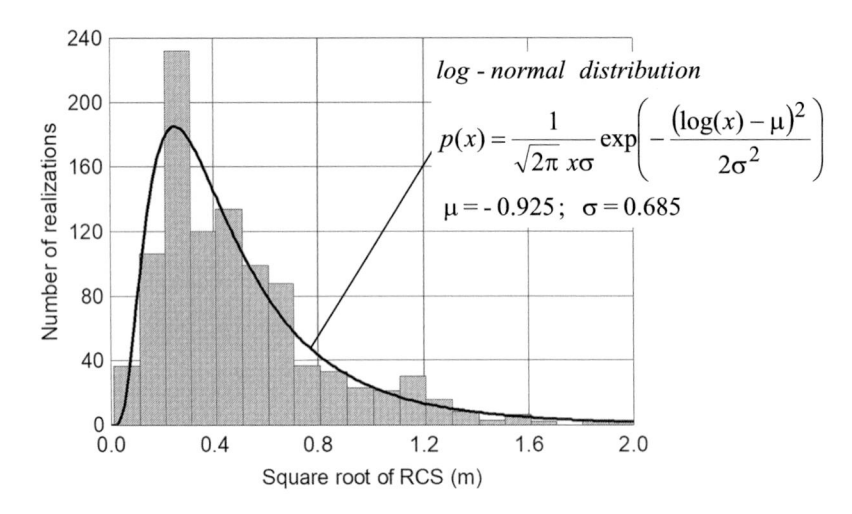

FIGURE 2.948 Amplitude distribution of echo signal of Kh-59M model at a carrier frequency of 3 GHz given its horizontal polarization.

Scattering characteristics of Kh-59M model for sounding frequency 1 GHz (wavelength 30 cm).

Figure 2.949 shows the RCS circular diagram of the Kh-59M. The noncoherent RCS circular diagram of the Kh-59M is shown in Figure 2.950.

The circular mean RCS of the Kh-59M model for horizontal polarization is 2.072 m². The circular median RCS for horizontal polarization is 0.193 m².

Figures 2.951 and 2.952 show the mean and median RCS for the main ranges of sounding azimuths (nose, side, and tail) and for ranges of 20°.

The expressions and parameters of probability distributions that are most consistent with the empirical distributions of the square root of the RCS for different ranges of sounding azimuths and polarizations are given in the electronic appendix to this book. The expressions and parameters of the probability distributions that are most consistent with the empirical distributions of the RCS (energy characteristic) for different ranges of sounding azimuths and polarizations are also given there.

FIGURE 2.949 Circular diagram of RCS given radar observation of Kh-59M model at a carrier frequency of 1 GHz (30 cm wavelength).

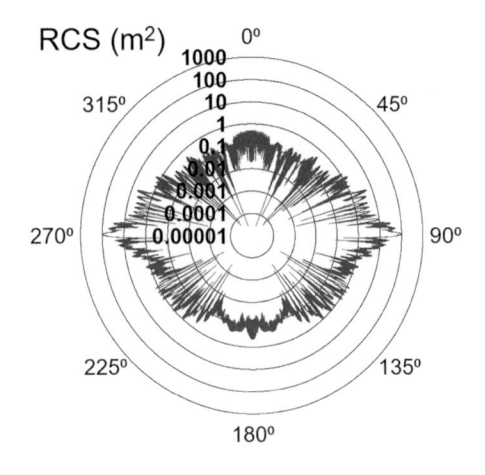

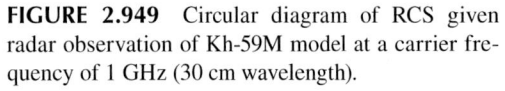

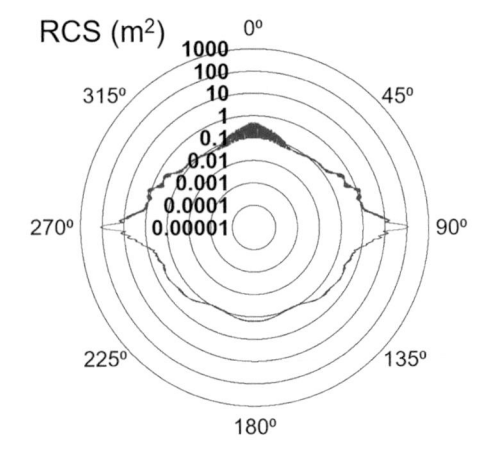

FIGURE 2.950 Circular diagram of noncoherent RCS given radar observation of Kh-59M model at a carrier frequency of 1 GHz (30 cm wavelength).

Scattering Characteristics of Aerial Objects

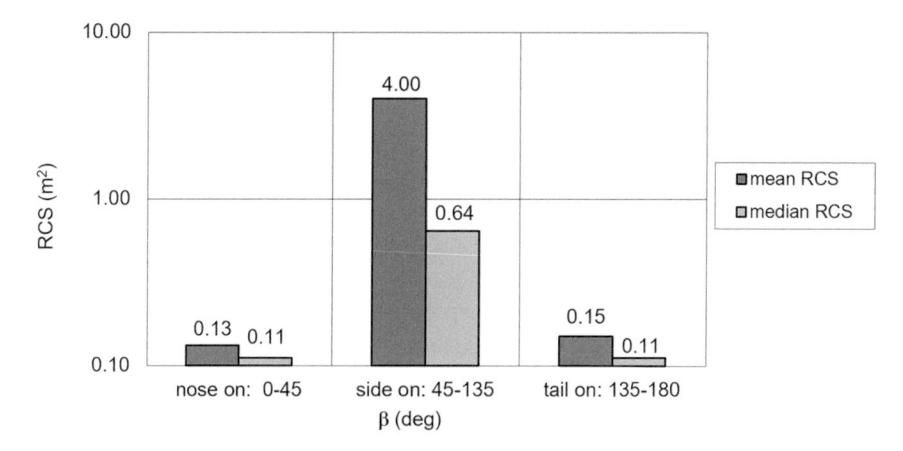

FIGURE 2.951 Diagrams of mean and median RCS of Kh-59M model in three sectors of azimuth aspect given its radar observation at horizontal polarization and a carrier frequency of 1 GHz (30 cm wavelength).

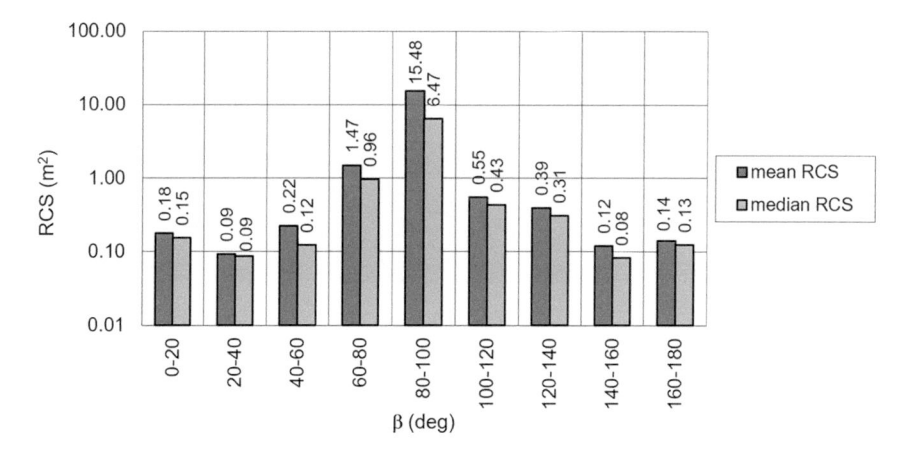

FIGURE 2.952 Diagrams of mean and median RCS of Kh-59M model in 20-degree sectors of azimuth aspect given its radar observation at horizontal polarization and a carrier frequency of 1 GHz (30 cm wavelength).

Weibull distribution

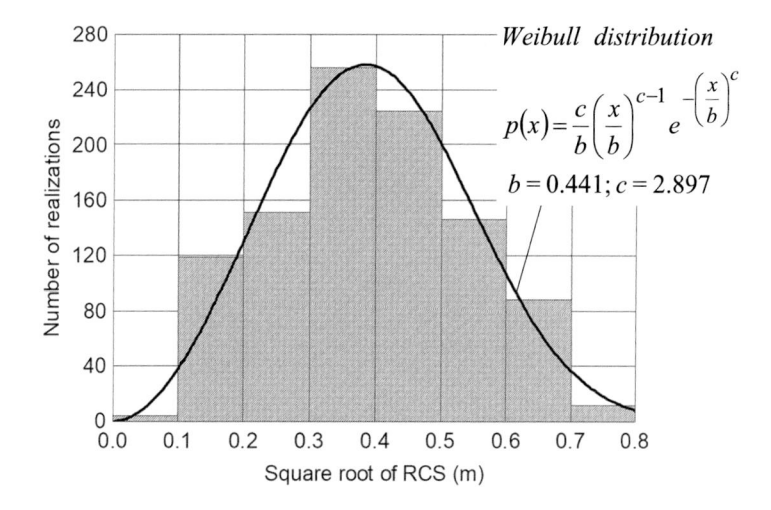

$$p(x) = \frac{c}{b}\left(\frac{x}{b}\right)^{c-1} e^{-\left(\frac{x}{b}\right)^c}$$

$b = 0.441; c = 2.897$

FIGURE 2.953 Amplitude distribution of echo signal of Kh-59M model at a carrier frequency of 1 GHz given its horizontal polarization.

2.57 AIR-TO-SURFACE MISSILE AGM-65 MAVERICK

AGM-65 Maverick (Figure 2.954) is a tactical, air-to-surface guided missile. The first flight took place in 1969.

General characteristics of AGM-65 [68]: wingspan – 0.711 m, length – 2.49 m, fuselage diameter – 0.304 m, weight – 218 kg, powerplant – Thiokol SR109-TC-1 or SR114-TC-1 or Aerojet SR115-AJ-1 solid-propellant rocket motor, speed –1 150 km/h, range – 27 km.

In accordance with the design of the AGM-65 a model of its surface was created to obtaining scattering characteristics (in particular, RCS). The model is shown in Figure 2.955. The smooth parts of the model surface were approximated by parts of 30 triaxial ellipsoids. The surface breaks were modeled using 20 straight edge scattering parts.

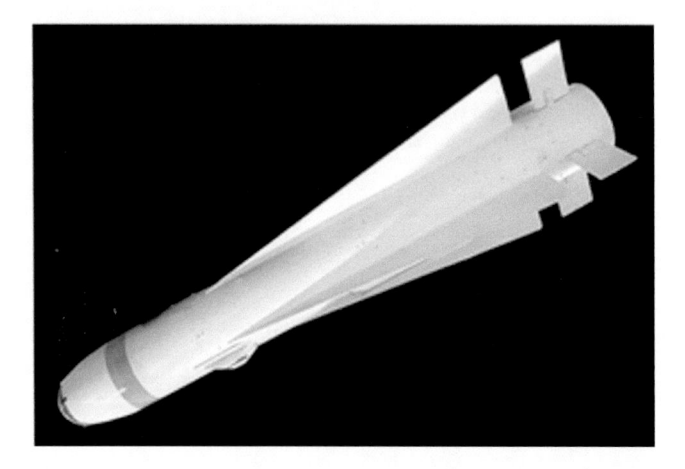

FIGURE 2.954 Air-to-surface missile AGM65 Maverick.

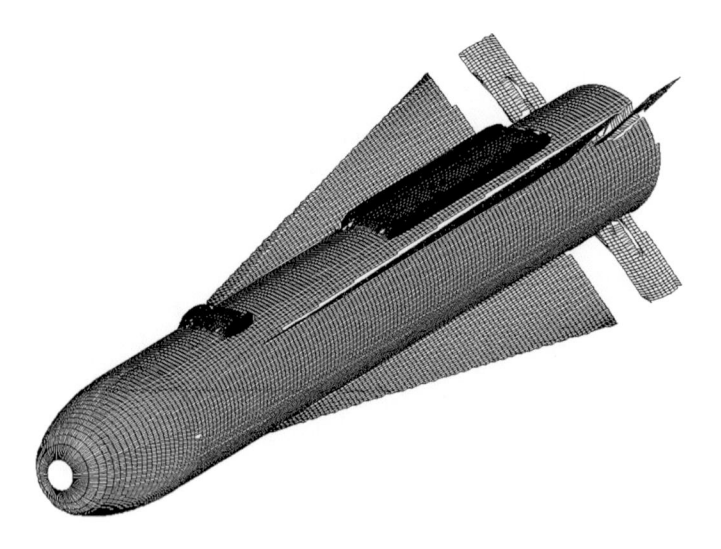

FIGURE 2.955 Surface model of air-to-surface missile AGM65 Maverick.

Below are some scattering characteristics of the AGM-65 model at sounding frequencies of 10, 3, and 1 GHz (wavelengths of 3, 10, and 30 cm, respectively) for horizontal polarization of the probing signal. The scattering characteristics for this model at two polarizations, as well as scattering characteristics at sounding frequencies of 5 and 1.3 GHz (wavelengths of 6 and 23 cm, respectively), are given in the electronic appendix of this book.

Scattering Characteristics of Aerial Objects

Sounding parameters: The elevation angle is to be a random value distributed uniformly in the range −3° ± 4° with respect to the wing plane (elevation angle of −3° corresponds to the radar observation from the lower hemisphere), azimuth aspect increment was 0.02°, the azimuth being counted off from the nose-on aspect (0° corresponds to the nose-on radar observation, 180° corresponds to the tail-on observation).

Scattering characteristics of AGM-65 model for sounding frequency 10 GHz (wavelength 3 cm).

Figure 2.956 shows the RCS circular diagram of the AGM-65 model. The noncoherent RCS circular diagram of the AGM-65 is shown in Figure 2.957.

The circular mean RCS of the AGM-65 model for horizontal polarization is 1.436 m². The circular median RCS (the RCS value used to calculate the detection range of an object with a probability of 0.5) for horizontal polarization is 0.040 m².

Figures 2.958 and 2.959 show the mean and median RCS for the main ranges of sounding azimuths (nose, side, and tail) and for ranges of 20°.

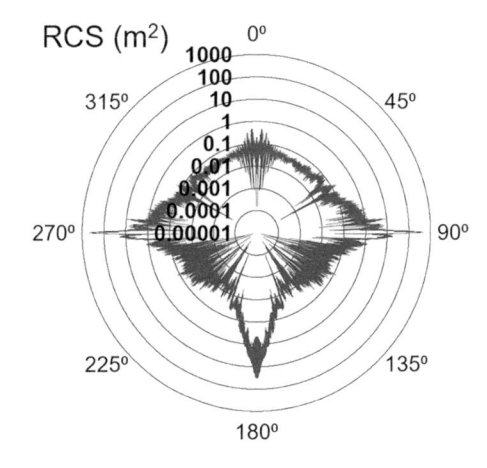

FIGURE 2.956 Circular diagram of RCS given radar observation of AGM-65 model at a carrier frequency of 10 GHz (3 cm wavelength).

FIGURE 2.957 Circular diagram of noncoherent RCS given radar observation of AGM-65 model at a carrier frequency of 10 GHz (3 cm wavelength).

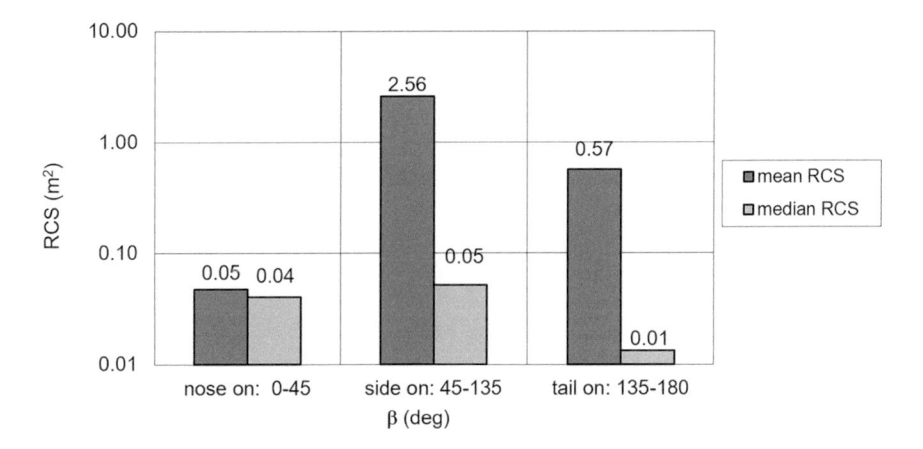

FIGURE 2.958 Diagrams of mean and median RCS of AGM-65 model in three sectors of azimuth aspect given its radar observation at horizontal polarization and a carrier frequency of 10 GHz (3 cm wavelength).

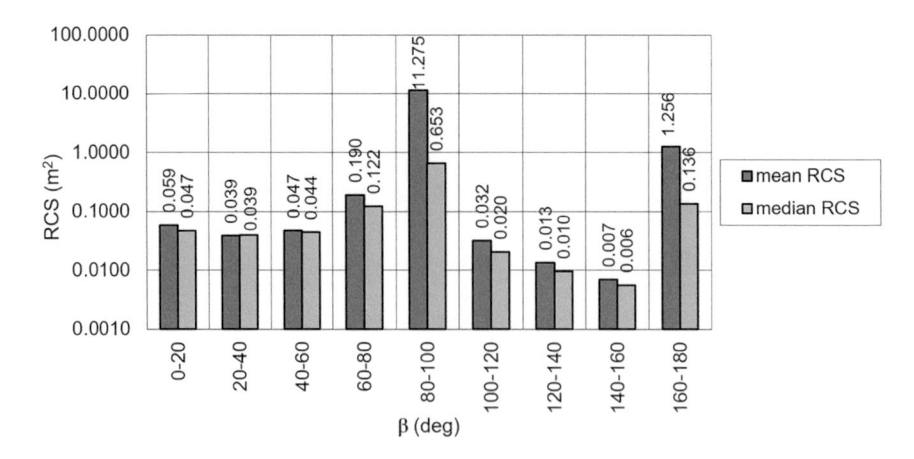

FIGURE 2.959 Diagrams of mean and median RCS of AGM-65 model in 20-degree sectors of azimuth aspect given its radar observation at horizontal polarization and a carrier frequency of 10 GHz (3 cm wavelength).

Figures 2.960, 2.965, and 2.970 show histograms of the scattered signal amplitude (square root of the RCS) for the range of sounding azimuths −20° to +20° (sounding from the nose). The bold line shows the probability density functions of the distribution, which can be used to approximate the histogram of the signal amplitude.

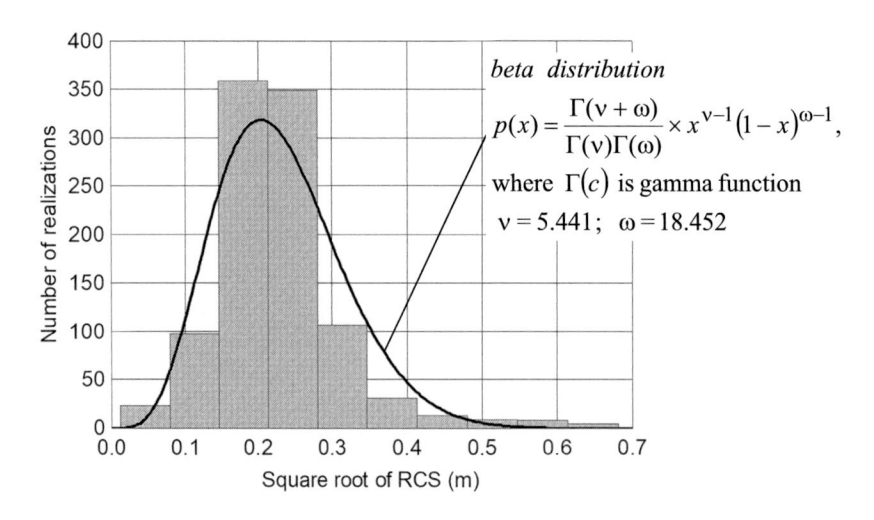

beta distribution

$$p(x) = \frac{\Gamma(\nu+\omega)}{\Gamma(\nu)\Gamma(\omega)} \times x^{\nu-1}(1-x)^{\omega-1},$$

where $\Gamma(c)$ is gamma function

$\nu = 5.441$; $\omega = 18.452$

FIGURE 2.960 Amplitude distribution of echo signal of AGM-65 model at a carrier frequency of 10 GHz given its horizontal polarization.

Scattering characteristics of AGM-65 model for sounding frequency 3 GHz (wavelength 10 cm).

Figure 2.961 shows the RCS circular diagram of the AGM-65. The noncoherent RCS circular diagram of the AGM-65 is shown in Figure 2.962.

The circular mean RCS of the AGM-65 for horizontal polarization is 0.888 m². The circular median RCS for horizontal polarization is 0.042 m².

Figures 2.963 and 2.964 show the mean and median RCS for the main ranges of sounding azimuths (nose, side, and tail) and for ranges of 20°.

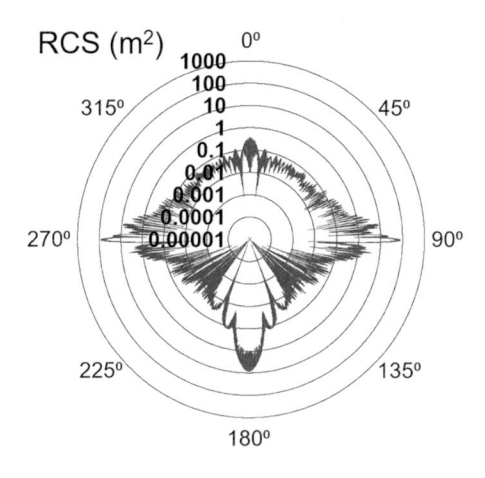

Scattering Characteristics of Aerial Objects

FIGURE 2.961 Circular diagram of RCS given radar observation of AGM-65 model at a carrier frequency of 3 GHz (10 cm wavelength).

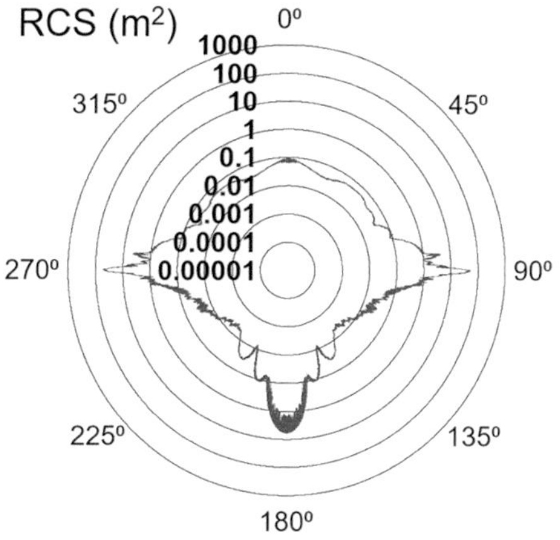

FIGURE 2.962 Circular diagram of noncoherent RCS given radar observation of AGM-65 model at a carrier frequency of 3 GHz (10 cm wavelength).

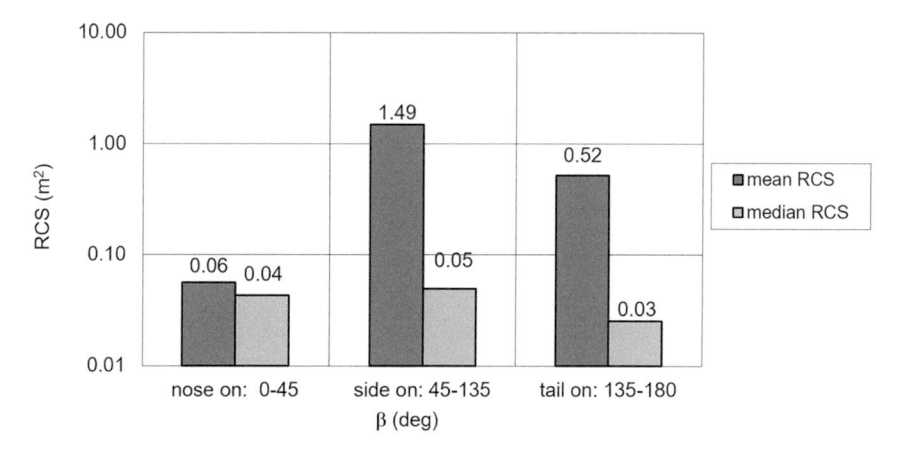

FIGURE 2.963 Diagrams of mean and median RCS of AGM-65 model in three sectors of azimuth aspect given its radar observation at horizontal polarization and a carrier frequency of 3 GHz (10 cm wavelength).

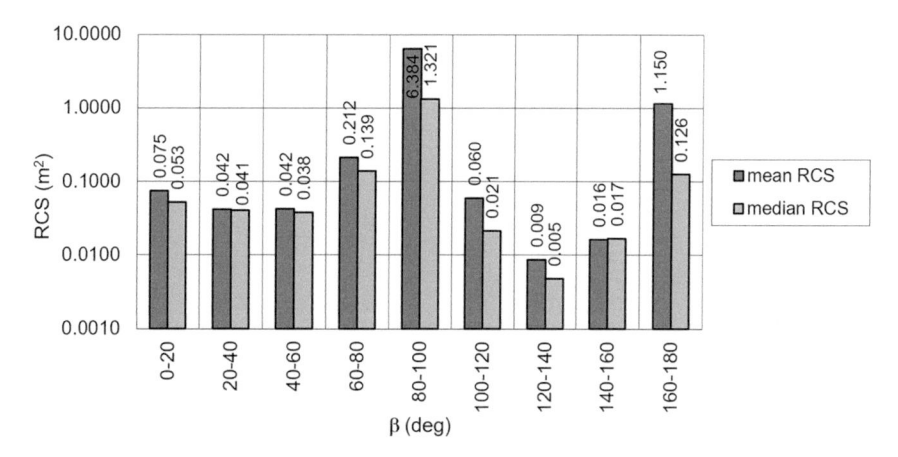

FIGURE 2.964 Diagrams of mean and median RCS of AGM-65 model in 20-degree sectors of azimuth aspect given its radar observation at horizontal polarization and a carrier frequency of 3 GHz (10 cm wavelength).

$$p(x) = \frac{1}{\sqrt{2\pi}\, x\sigma} \exp\left(-\frac{(\log(x) - \mu)^2}{2\sigma^2}\right)$$

log - normal distribution

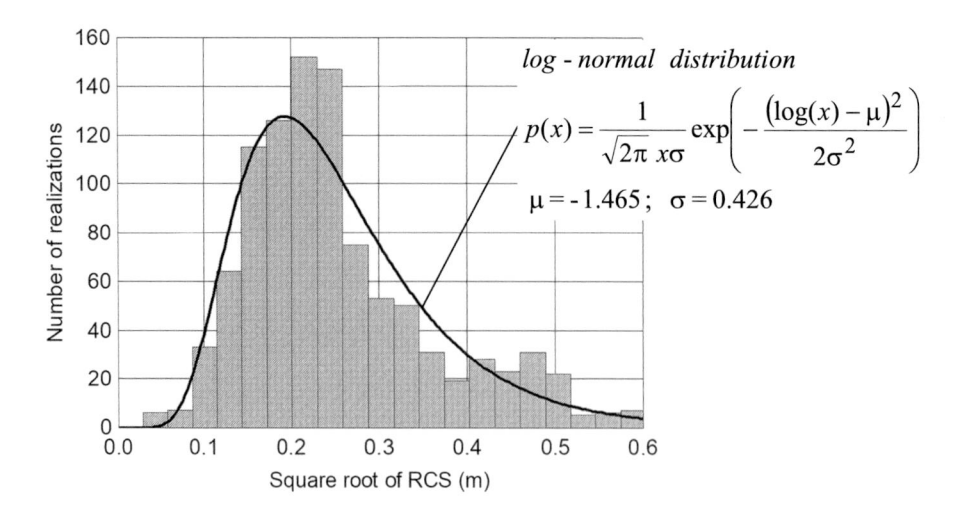

$\mu = -1.465;\ \sigma = 0.426$

FIGURE 2.965 Amplitude distribution of echo signal of AGM-65 model at a carrier frequency of 3 GHz given its horizontal polarization.

Scattering characteristics of AGM-65 model for sounding frequency 1 GHz (wavelength 30 cm).

Figure 2.966 shows the RCS circular diagram of the AGM-65. The noncoherent RCS circular diagram of the AGM-65 is shown in Figure 2.967.

The circular mean RCS of the AGM-65 model for horizontal polarization is 0.707 m². The circular median RCS for horizontal polarization is 0.101 m².

Figures 2.968 and 2.969 show the mean and median RCS for the main ranges of sounding azimuths (nose, side, and tail) and for ranges of 20°.

The expressions and parameters of probability distributions that are most consistent with the empirical distributions of the square root of the RCS for different ranges of sounding azimuths and polarizations are given in the electronic appendix to this book. The expressions and parameters of the probability distributions that are most consistent with the empirical distributions of the RCS (energy characteristic) for different ranges of sounding azimuths and polarizations are also given there.

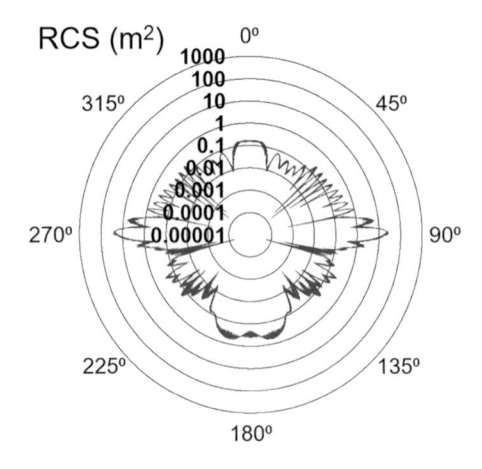

FIGURE 2.966 Circular diagram of RCS given radar observation of AGM-65 model at a carrier frequency of 1 GHz (30 cm wavelength).

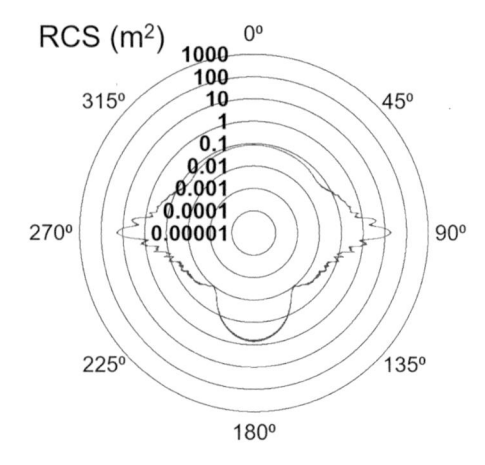

FIGURE 2.967 Circular diagram of noncoherent RCS given radar observation of AGM-65 model at a carrier frequency of 1 GHz (30 cm wavelength).

Scattering Characteristics of Aerial Objects

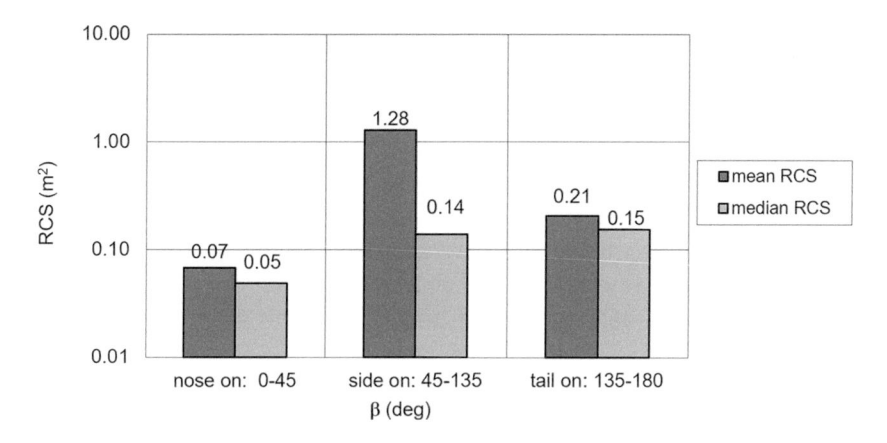

FIGURE 2.968 Diagrams of mean and median RCS of AGM-65 model in three sectors of azimuth aspect given its radar observation at horizontal polarization and a carrier frequency of 1 GHz (30 cm wavelength).

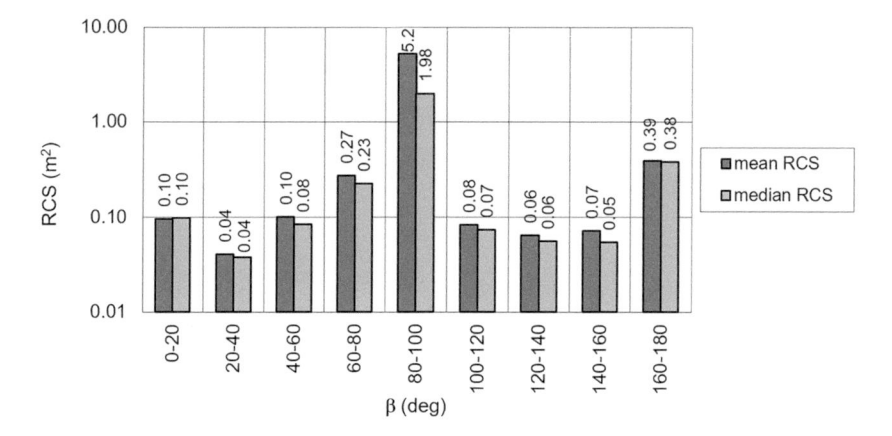

FIGURE 2.969 Diagrams of mean and median RCS of AGM-65 model in 20-degree sectors of azimuth aspect given its radar observation at horizontal polarization and a carrier frequency of 1 GHz (30 cm wavelength).

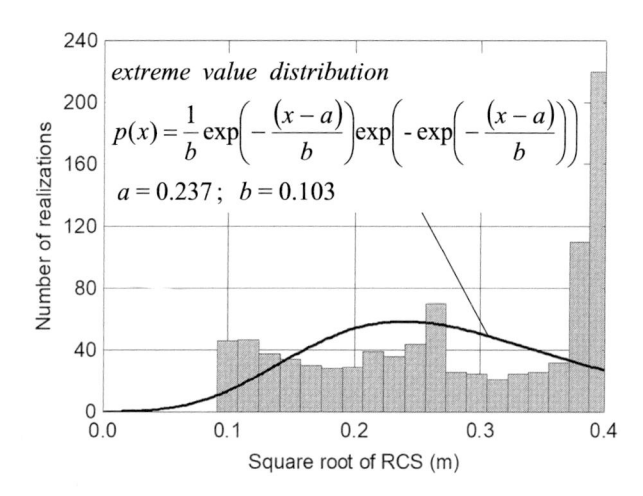

extreme value distribution

$$p(x) = \frac{1}{b}\exp\left(-\frac{(x-a)}{b}\right)\exp\left(-\exp\left(-\frac{(x-a)}{b}\right)\right)$$

$a = 0.237; \quad b = 0.103$

FIGURE 2.970 Amplitude distribution of echo signal of AGM-65 model at a carrier frequency of 1 GHz given its horizontal polarization.

2.58 AIR-TO-SURFACE MISSILE AGM-114 HELLFIRE

AGM-114 Hellfire (Figure 2.971) is an air-to-ground, laser-guided, subsonic missile. The first flight took place in 1978.

General characteristics of AGM-114 [69]: wingspan – 0.33 m, length – 1.8 m (AGM-114F Interim Hellfire), fuselage diameter – 0.178 m, weight – 48.6 kg, powerplant – Thiokol TX-657 solid-fuel rocket, speed –425 m/s, range – 11 km.

In accordance with the design of the AGM-114, a model of its surface was created to obtaining scattering characteristics (in particular, RCS). The model is shown in Figure 2.972. The smooth parts of the model surface were approximated by parts of 29 triaxial ellipsoids. The surface breaks were modeled using 28 straight edge scattering parts.

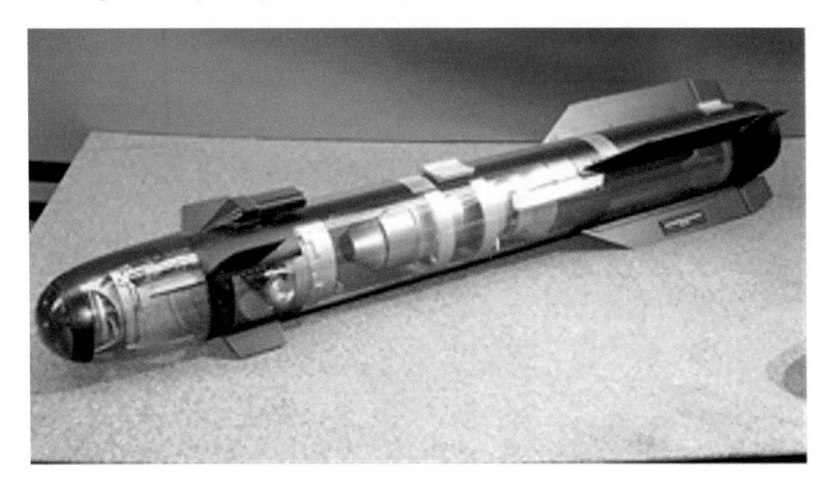

FIGURE 2.971 Air-to-surface missile AGM114 Hellfire.

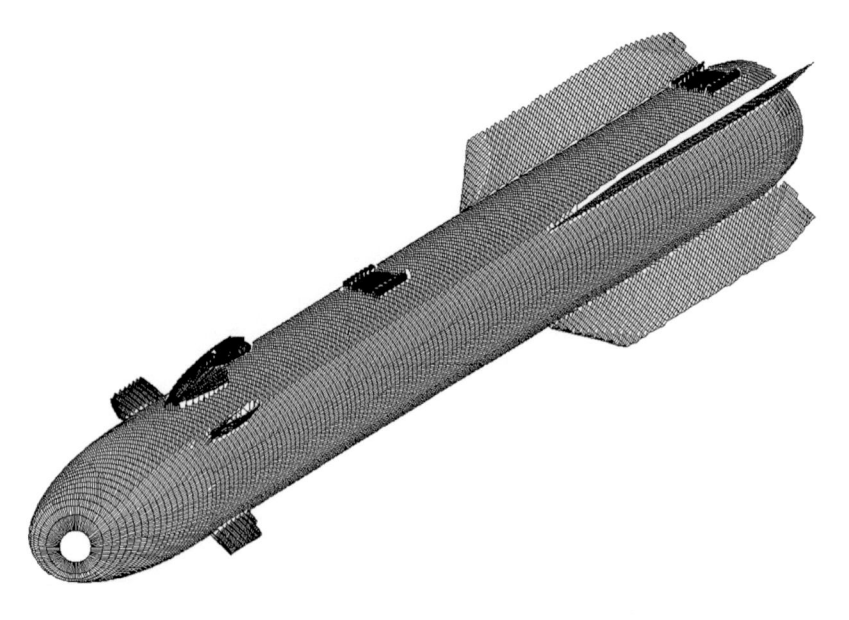

FIGURE 2.972 Surface model of air-to-surface missile AGM-114 Hellfire.

Below are some scattering characteristics of the AGM-114 model at sounding frequencies of 10, 3, and 1 GHz (wavelengths of 3, 10, and 30 cm, respectively) for horizontal polarization of the probing signal. The scattering characteristics for this model at two polarizations, as well as scattering characteristics at sounding frequencies of 5 and 1.3 GHz (wavelengths of 6 and 23 cm, respectively), are given in the electronic appendix of this book.

Sounding parameters: The elevation angle is to be a random value distributed uniformly in the range $-3° \pm 4°$ with respect to the wing plane (elevation angle of $-3°$ corresponds to the radar observation from the lower hemisphere), azimuth aspect increment was 0.02°, the azimuth being counted off from the nose-on aspect (0° corresponds to the nose-on radar observation, 180° corresponds to the tail-on observation).

Scattering characteristics of AGM-114 model for sounding frequency 10 GHz (wavelength 3 cm).

Figure 2.973 shows the RCS circular diagram of the AGM-114 model. The noncoherent RCS circular diagram of the AGM-114 is shown in Figure 2.974.

The circular mean RCS of the AGM-114 model for horizontal polarization is 0.388 m². The circular median RCS (the RCS value used to calculate the detection range of an object with a probability of 0.5) for horizontal polarization is 0.006 m².

Figures 2.975 and 2.976 show the mean and median RCS for the main ranges of sounding azimuths (nose, side, and tail) and for ranges of 20°.

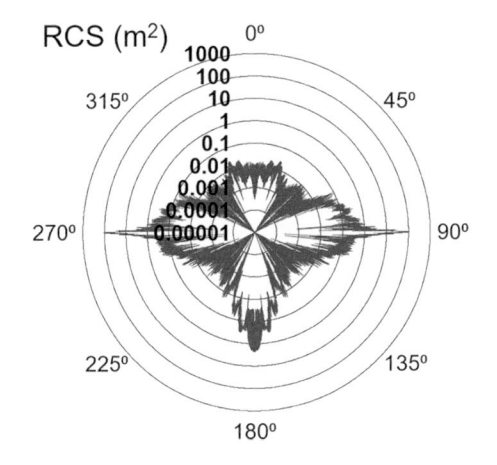

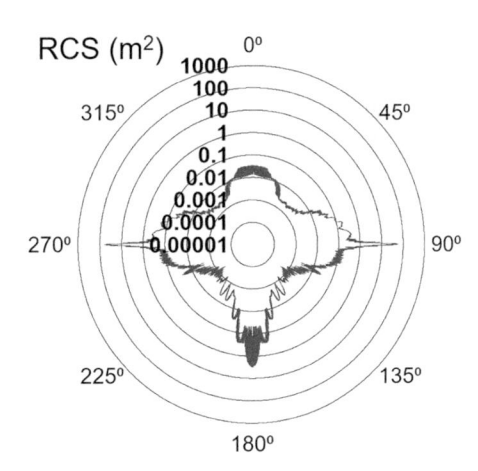

FIGURE 2.973 Circular diagram of RCS given radar observation of AGM-114 model at a carrier frequency of 10 GHz (3 cm wavelength).

FIGURE 2.974 Circular diagram of noncoherent RCS given radar observation of AGM-114 model at a carrier frequency of 10 GHz (3 cm wavelength).

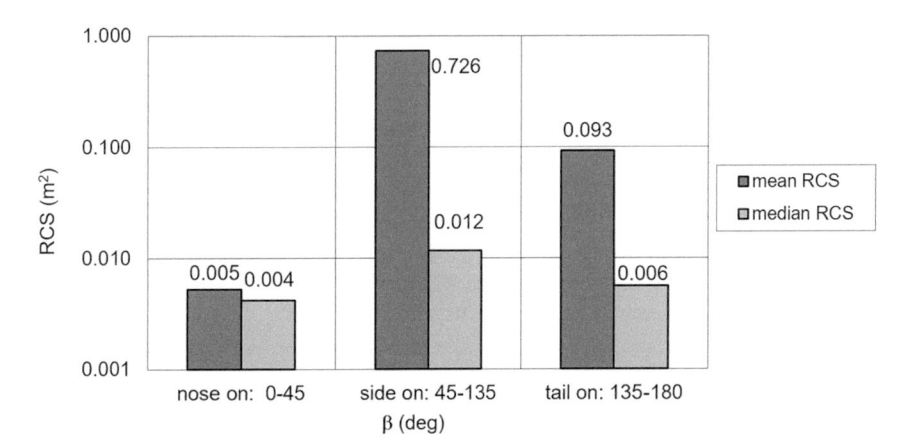

FIGURE 2.975 Diagrams of mean and median RCS of AGM-114 model in three sectors of azimuth aspect given its radar observation at horizontal polarization and a carrier frequency of 10 GHz (3 cm wavelength).

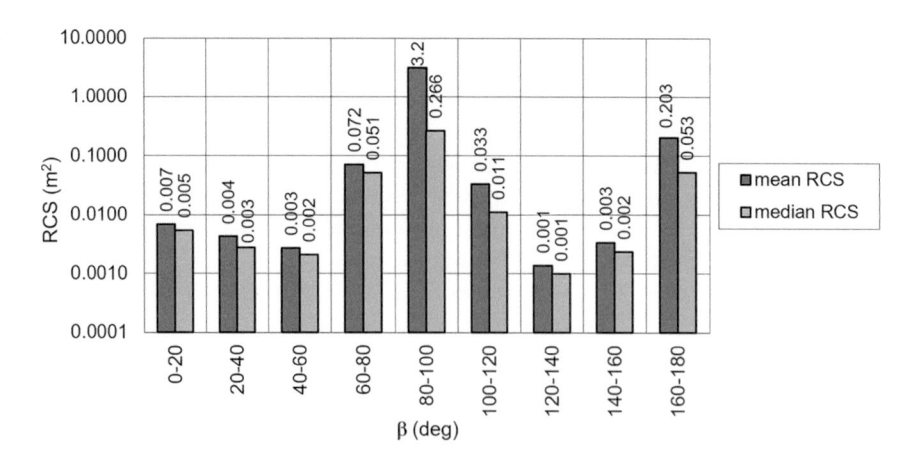

FIGURE 2.976 Diagrams of mean and median RCS of AGM-114 model in 20-degree sectors of azimuth aspect given its radar observation at horizontal polarization and a carrier frequency of 10 GHz (3 cm wavelength).

Figures 2.977, 2.982, and 2.987 show histograms of the scattered signal amplitude (square root of the RCS) for the range of sounding azimuths −20° to +20° (sounding from the nose). The bold line shows the probability density functions of the distribution, which can be used to approximate the histogram of the signal amplitude.

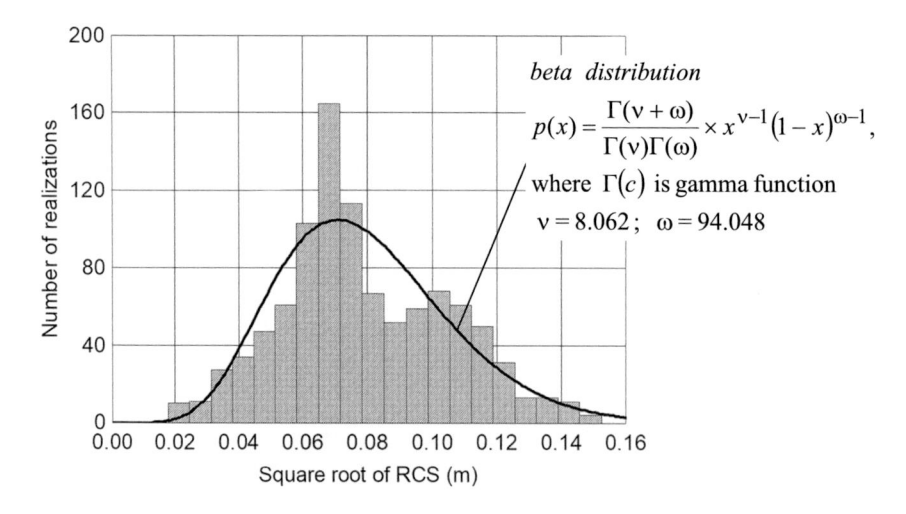

beta distribution

$$p(x) = \frac{\Gamma(\nu + \omega)}{\Gamma(\nu)\Gamma(\omega)} \times x^{\nu-1}(1-x)^{\omega-1},$$

where $\Gamma(c)$ is gamma function

$\nu = 8.062; \quad \omega = 94.048$

FIGURE 2.977 Amplitude distribution of echo signal of AGM-114 model at a carrier frequency of 10 GHz given its horizontal polarization.

Scattering characteristics of AGM-114 model for sounding frequency 3 GHz (wavelength 10 cm).

Figure 2.978 shows the RCS circular diagram of the AGM-114. The noncoherent RCS circular diagram of the AGM-114 is shown in Figure 2.979.

The circular mean RCS of the AGM-114 for horizontal polarization is 0.369 m². The circular median RCS for horizontal polarization is 0.016 m².

Figures 2.980 and 2.981 show the mean and median RCS for the main ranges of sounding azimuths (nose, side, and tail) and for ranges of 20°.

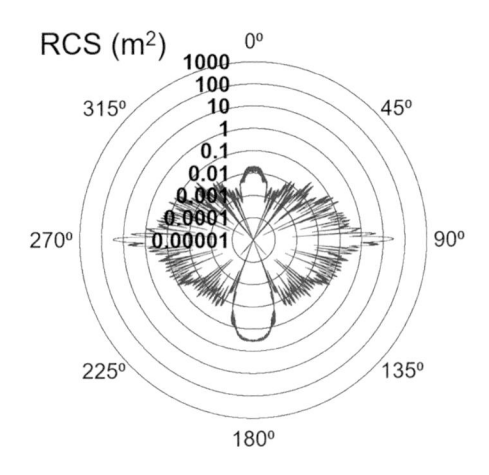

Scattering Characteristics of Aerial Objects 357

FIGURE 2.978 Circular diagram of RCS given radar observation of AGM-114 model at a carrier frequency of 3 GHz (10 cm wavelength).

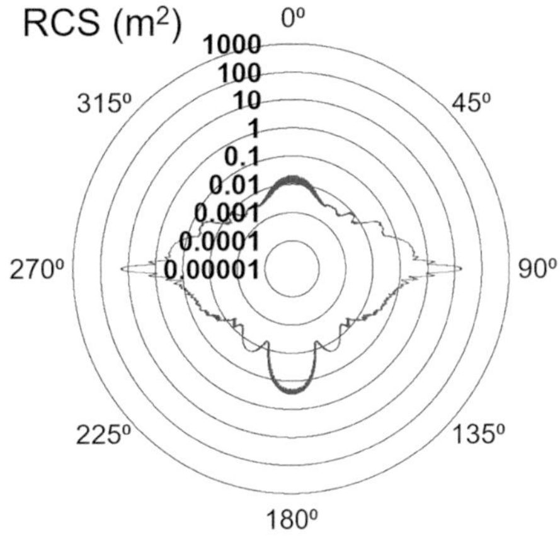

FIGURE 2.979 Circular diagram of noncoherent RCS given radar observation of AGM-114 model at a carrier frequency of 3 GHz (10 cm wavelength).

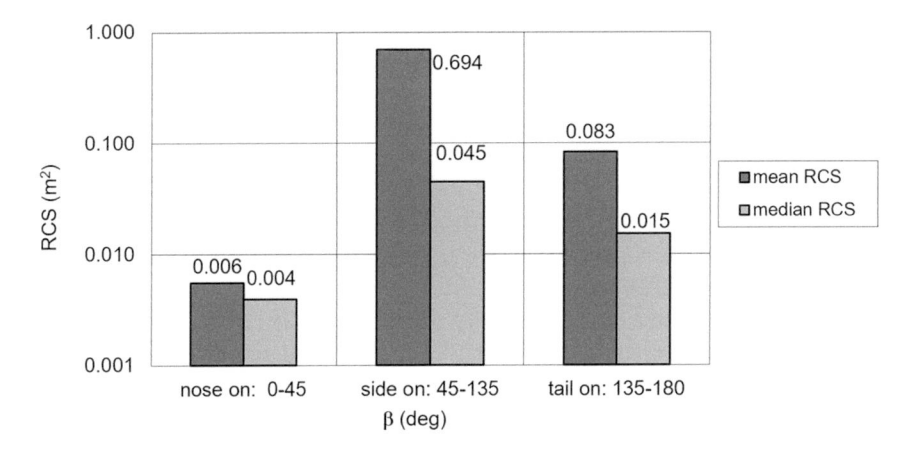

FIGURE 2.980 Diagrams of mean and median RCS of AGM-114 model in three sectors of azimuth aspect given its radar observation at horizontal polarization and a carrier frequency of 3 GHz (10 cm wavelength).

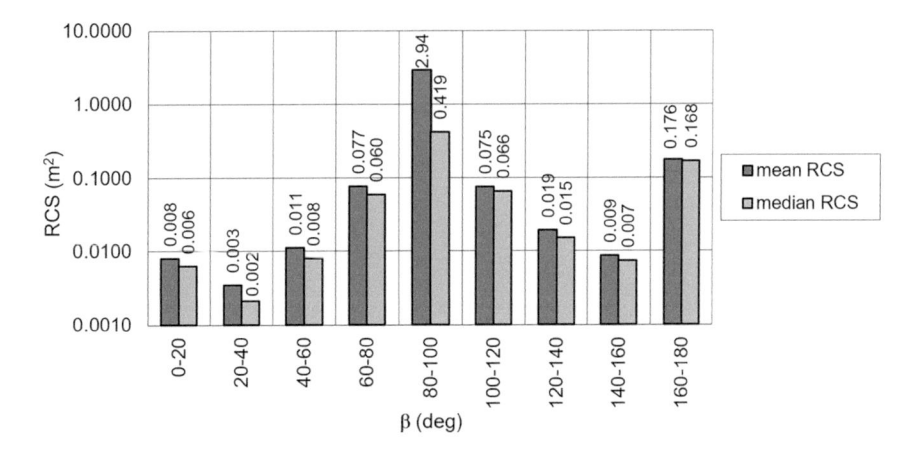

FIGURE 2.981 Diagrams of mean and median RCS of AGM-114 model in 20-degree sectors of azimuth aspect given its radar observation at horizontal polarization and a carrier frequency of 3 GHz (10 cm wavelength).

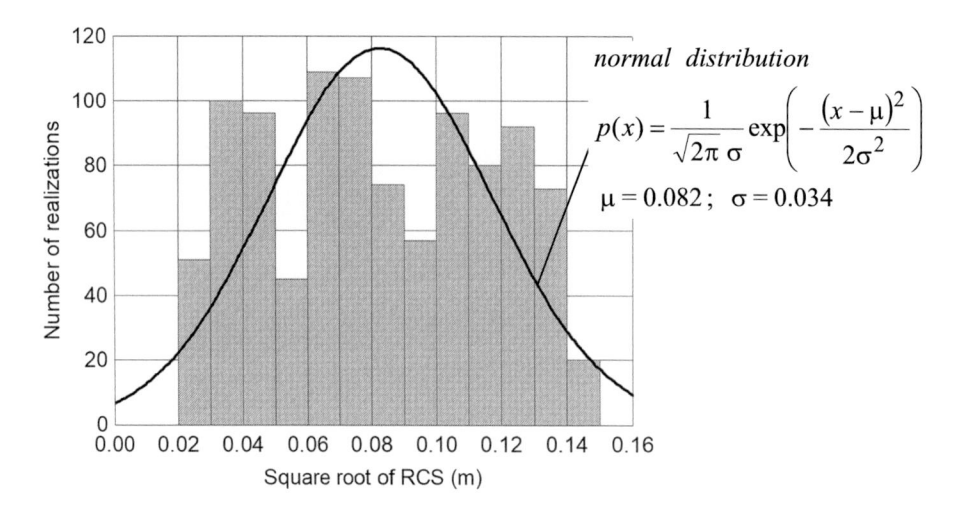

FIGURE 2.982 Amplitude distribution of echo signal of AGM-114 model at a carrier frequency of 3 GHz given its horizontal polarization.

Scattering characteristics of AGM-114 model for sounding frequency 1 GHz (wavelength 30 cm).

Figure 2.983 shows the RCS circular diagram of the AGM-114. The noncoherent RCS circular diagram of the AGM-114 is shown in Figure 2.984.

The circular mean RCS of the AGM-114 model for horizontal polarization is 0.540 m². The circular median RCS for horizontal polarization is 0.033 m².

Figures 2.985 and 2.986 show the mean and median RCS for the main ranges of sounding azimuths (nose, side, and tail) and for ranges of 20°.

The expressions and parameters of probability distributions that are most consistent with the empirical distributions of the square root of the RCS for different ranges of sounding azimuths and polarizations are given in the electronic appendix to this book. The expressions and parameters of the probability distributions that are most consistent with the empirical distributions of the RCS (energy characteristic) for different ranges of sounding azimuths and polarizations are also given there.

FIGURE 2.983 Circular diagram of RCS given radar observation of AGM-114 model at a carrier frequency of 1 GHz (30 cm wavelength).

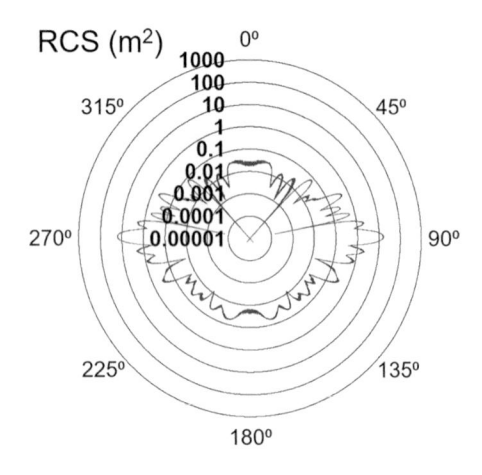

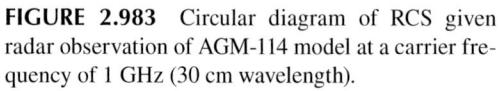

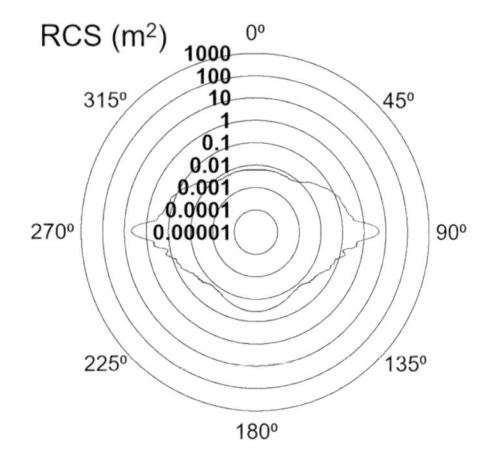

FIGURE 2.984 Circular diagram of noncoherent RCS given radar observation of AGM-114 model at a carrier frequency of 1 GHz (30 cm wavelength).

Scattering Characteristics of Aerial Objects

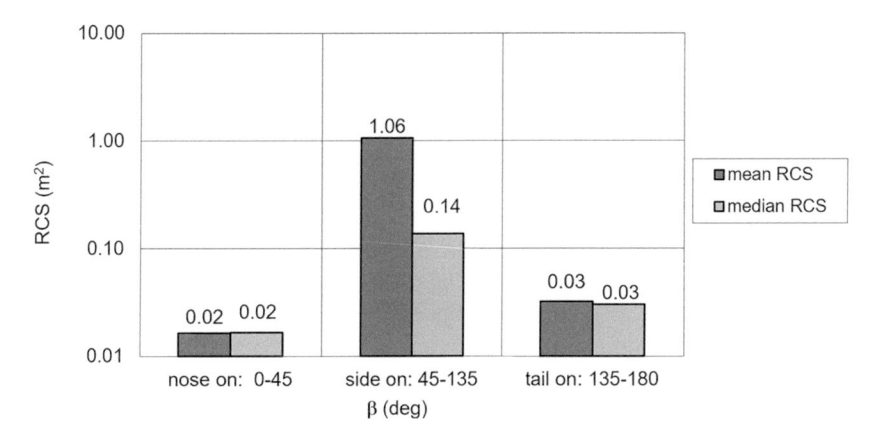

FIGURE 2.985 Diagrams of mean and median RCS of AGM-114 model in three sectors of azimuth aspect given its radar observation at horizontal polarization and a carrier frequency of 1 GHz (30 cm wavelength).

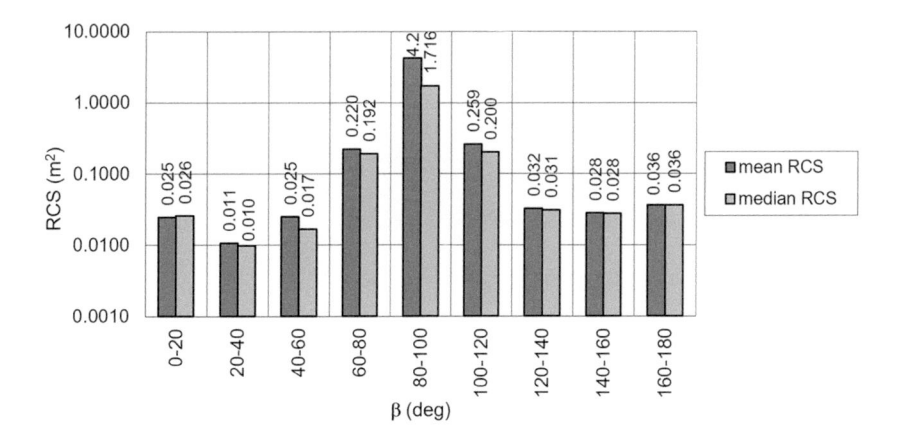

FIGURE 2.986 Diagrams of mean and median RCS of AGM-114 model in 20-degree sectors of azimuth aspect given its radar observation at horizontal polarization and a carrier frequency of 1 GHz (30 cm wavelength).

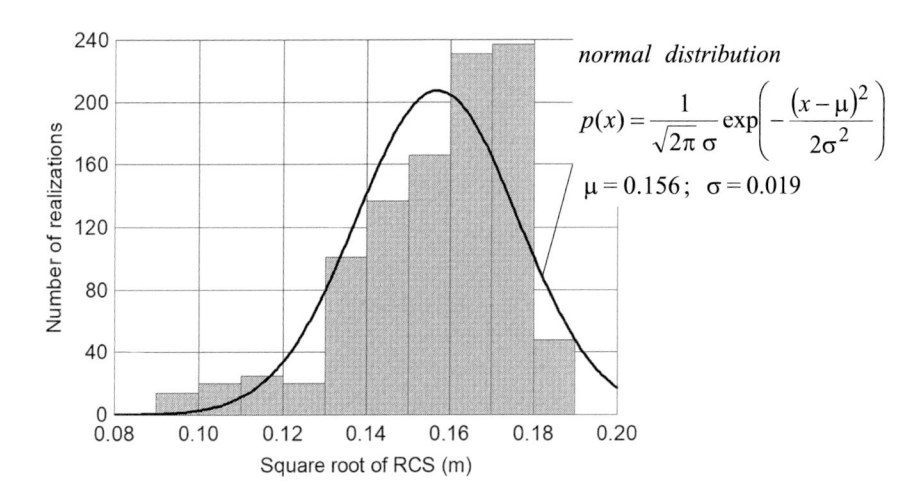

$$p(x) = \frac{1}{\sqrt{2\pi}\,\sigma}\exp\left(-\frac{(x-\mu)^2}{2\sigma^2}\right)$$

$\mu = 0.156;\ \sigma = 0.019$

FIGURE 2.987 Amplitude distribution of echo signal of AGM-114 model at a carrier frequency of 1 GHz given its horizontal polarization.

2.59 ANTI-RADIATION MISSILE AGM-88 HARM

AGM-88 HARM (Figure 2.988) is a tactical, air-to-surface anti-radiation missile. The first flight took place in 1986.

General characteristics of AGM-88 [70]: wingspan – 1.13 m, length – 4.2 m, fuselage diameter – 0.25 m, weight – 354 kg, powerplant – Thiokol SR113-TC-1 dual-thrust rocket engine, speed –2 280 km/h, range – 48 km.

In accordance with the design of the AGM-88, a model of its surface was created to obtaining scattering characteristics (in particular, RCS). The model is shown in Figure 2.989. The perfectly conducting smooth parts of the model surface were approximated by parts of 23 triaxial ellipsoids. Dielectric parts of the model surface were approximated by parts of 4 triaxial ellipsoids. The surface breaks were modeled using 28 straight edge scattering parts.

FIGURE 2.988 Anti-radiation missile AGM-88 HARM.

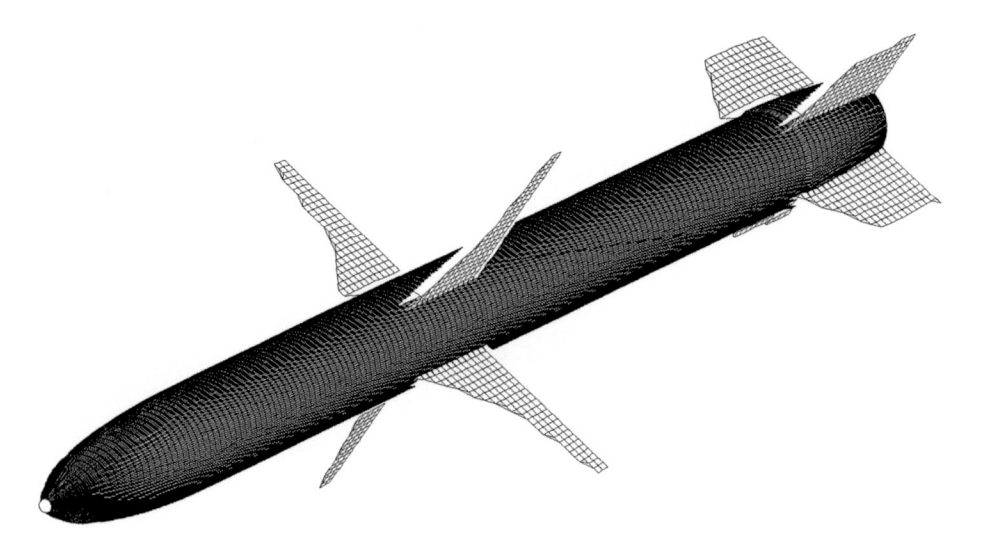

FIGURE 2.989 Surface model of anti-radiation missile AGM-88 HARM.

Below are some scattering characteristics of the AGM-88 model at sounding frequencies of 10, 3, and 1 GHz (wavelengths of 3, 10, and 30 cm, respectively) for horizontal polarization of the probing signal. The scattering characteristics for this model at two polarizations, as well as scattering characteristics at sounding frequencies of 5 and 1.3 GHz (wavelengths of 6 and 23 cm, respectively), are given in the electronic appendix of this book.

Scattering Characteristics of Aerial Objects 361

Sounding parameters: The elevation angle is to be a random value distributed uniformly in the range −3° ± 4° with respect to the wing plane (elevation angle of −3° corresponds to the radar observation from the lower hemisphere), azimuth aspect increment was 0.02°, the azimuth being counted off from the nose-on aspect (0° corresponds to the nose-on radar observation, 180° corresponds to the tail-on observation).

Scattering characteristics of AGM-88 model for sounding frequency 10 GHz (wavelength 3 cm).

Figure 2.990 shows the RCS circular diagram of the AGM-88 model. The noncoherent RCS circular diagram of the AGM-88 is shown in Figure 2.991.

The circular mean RCS of the AGM-88 model for horizontal polarization is 1.826 m². The circular median RCS (the RCS value used to calculate the detection range of an object with a probability of 0.5) for horizontal polarization is 0.042 m².

Figures 2.992 and 2.993 show the mean and median RCS for the main ranges of sounding azimuths (nose, side, and tail) and for ranges of 20°.

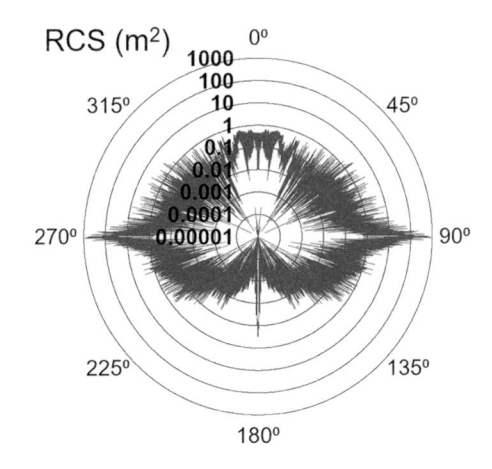

FIGURE 2.990 Circular diagram of RCS given radar observation of AGM-88 model at a carrier frequency of 10 GHz (3 cm wavelength).

FIGURE 2.991 Circular diagram of noncoherent RCS given radar observation of AGM-88 model at a carrier frequency of 10 GHz (3 cm wavelength).

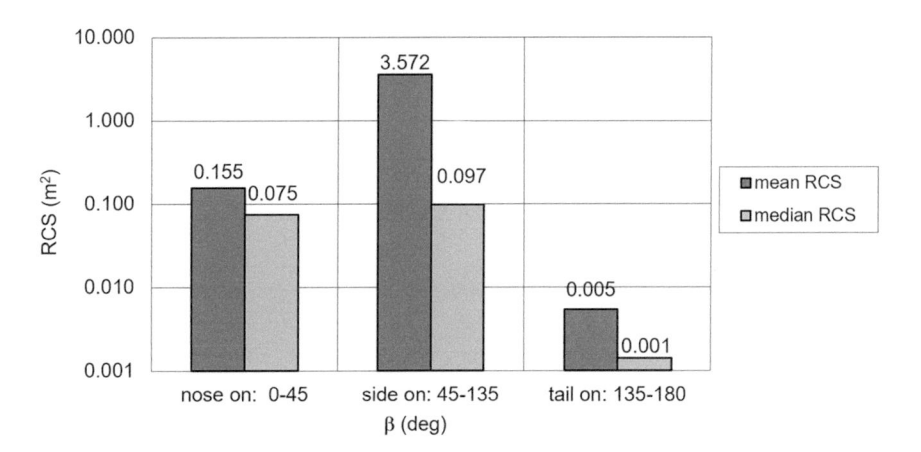

FIGURE 2.992 Diagrams of mean and median RCS of AGM-88 model in three sectors of azimuth aspect given its radar observation at horizontal polarization and a carrier frequency of 10 GHz (3 cm wavelength).

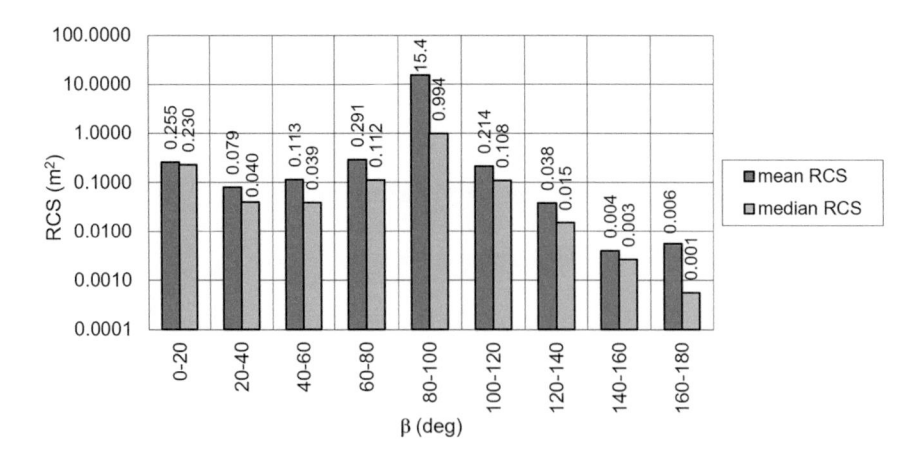

FIGURE 2.993 Diagrams of mean and median RCS of AGM-88 model in 20-degree sectors of azimuth aspect given its radar observation at horizontal polarization and a carrier frequency of 10 GHz (3 cm wavelength).

Figures 2.994, 2.999, and 2.1004 show histograms of the scattered signal amplitude (square root of the RCS) for the range of sounding azimuths −20° to +20° (sounding from the nose). The bold line shows the probability density functions of the distribution, which can be used to approximate the histogram of the signal amplitude.

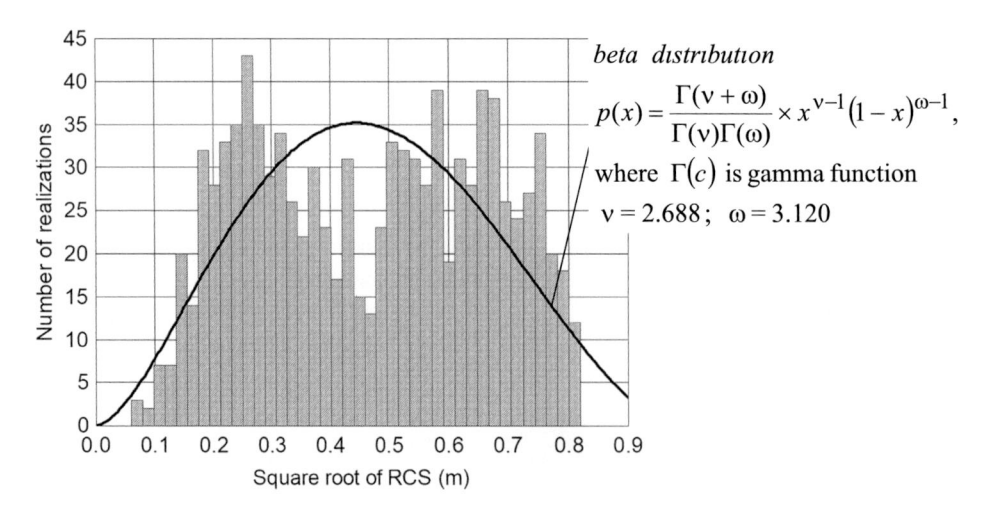

$$p(x) = \frac{\Gamma(\nu + \omega)}{\Gamma(\nu)\Gamma(\omega)} \times x^{\nu-1}(1-x)^{\omega-1},$$

where $\Gamma(c)$ is gamma function
$\nu = 2.688; \omega = 3.120$

beta distribution

FIGURE 2.994 Amplitude distribution of echo signal of AGM-88 model at a carrier frequency of 10 GHz given its horizontal polarization.

Scattering characteristics of AGM-88 model for sounding frequency 3 GHz (wavelength 10 cm).

Figure 2.995 shows the RCS circular diagram of the AGM-88. The noncoherent RCS circular diagram of the AGM-88 is shown in Figure 2.996.

The circular mean RCS of the AGM-88 for horizontal polarization is 2.080 m². The circular median RCS for horizontal polarization is 0.116 m².

Figures 2.997 and 2.998 show the mean and median RCS for the main ranges of sounding azimuths (nose, side, and tail) and for ranges of 20°.

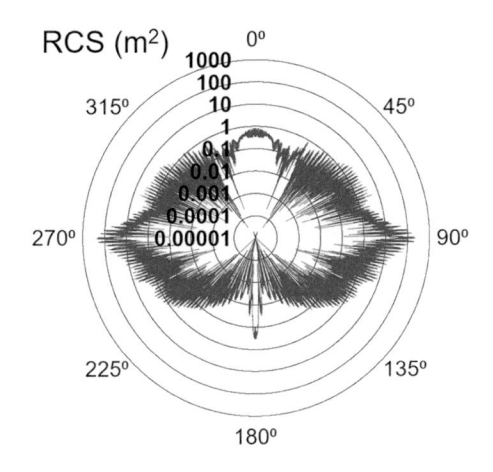

Scattering Characteristics of Aerial Objects

FIGURE 2.995 Circular diagram of RCS given radar observation of AGM-88 model at a carrier frequency of 3 GHz (10 cm wavelength).

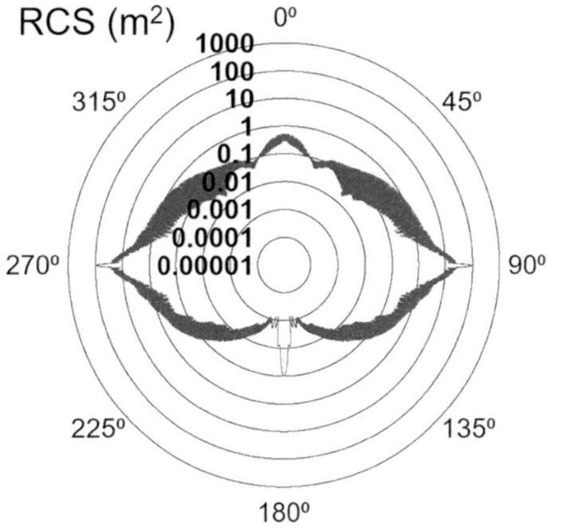

FIGURE 2.996 Circular diagram of noncoherent RCS given radar observation of AGM-88 model at a carrier frequency of 3 GHz (10 cm wavelength).

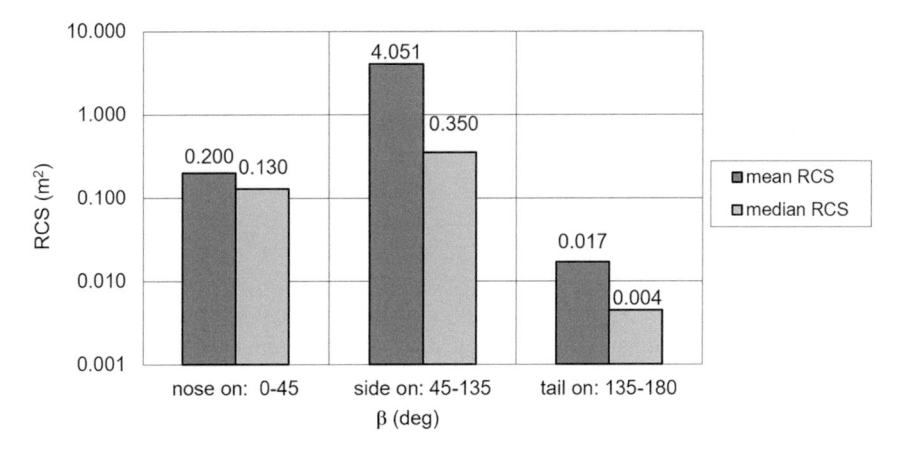

FIGURE 2.997 Diagrams of mean and median RCS of AGM-88 model in three sectors of azimuth aspect given its radar observation at horizontal polarization and a carrier frequency of 3 GHz (10 cm wavelength).

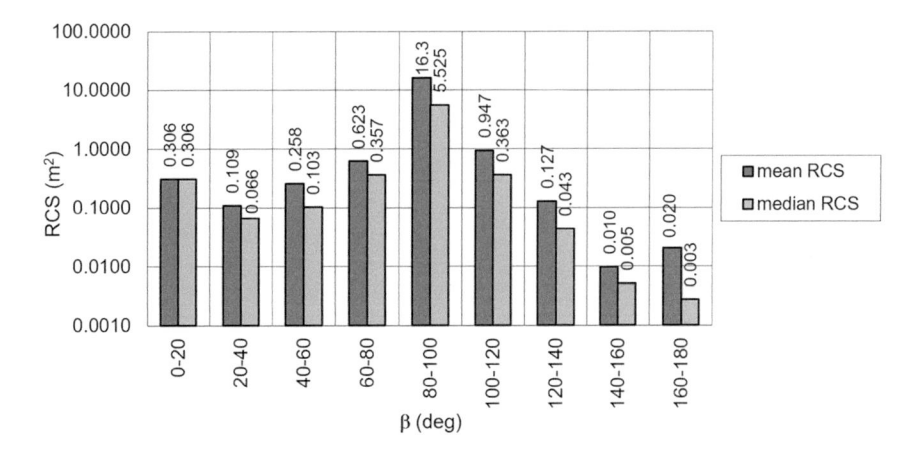

FIGURE 2.998 Diagrams of mean and median RCS of AGM-88 model in 20-degree sectors of azimuth aspect given its radar observation at horizontal polarization and a carrier frequency of 3 GHz (10 cm wavelength).

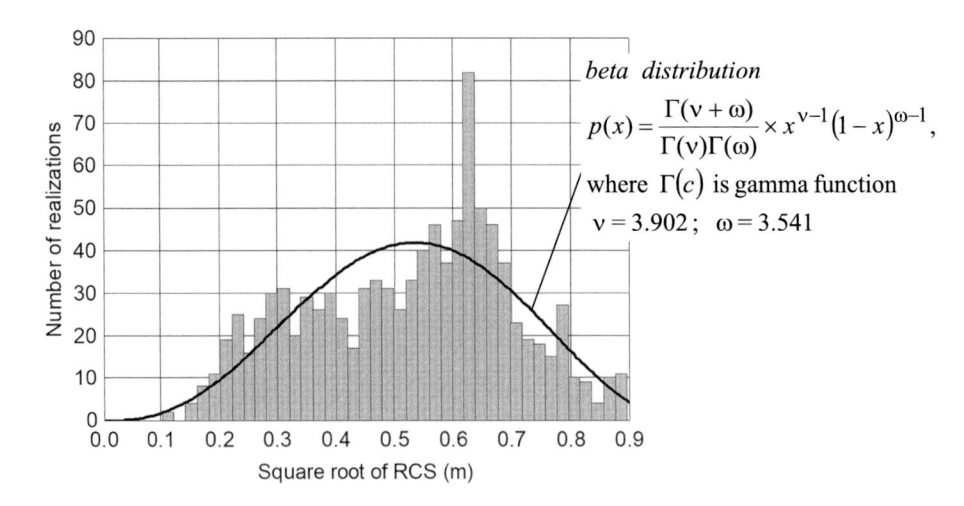

FIGURE 2.999 Amplitude distribution of echo signal of AGM-88 model at a carrier frequency of 3 GHz given its horizontal polarization.

beta distribution
$$p(x) = \frac{\Gamma(\nu+\omega)}{\Gamma(\nu)\Gamma(\omega)} \times x^{\nu-1}(1-x)^{\omega-1},$$
where $\Gamma(c)$ is gamma function
$\nu = 3.902$; $\omega = 3.541$

Scattering characteristics of AGM-88 model for sounding frequency 1 GHz (wavelength 30 cm).

Figure 2.1000 shows the RCS circular diagram of the AGM-88. The noncoherent RCS circular diagram of the AGM-88 is shown in Figure 2.1001.

The circular mean RCS of the AGM-88 model for horizontal polarization is 3.342 m². The circular median RCS for horizontal polarization is 0.253 m².

Figures 2.1002 and 2.1003 show the mean and median RCS for the main ranges of sounding azimuths (nose, side, and tail) and for ranges of 20°.

The expressions and parameters of probability distributions that are most consistent with the empirical distributions of the square root of the RCS for different ranges of sounding azimuths and polarizations are given in the electronic appendix to this book. The expressions and parameters of the probability distributions that are most consistent with the empirical distributions of the RCS (energy characteristic) for different ranges of sounding azimuths and polarizations are also given there.

FIGURE 2.1000 Circular diagram of RCS given radar observation of AGM-88 model at a carrier frequency of 1 GHz (30 cm wavelength).

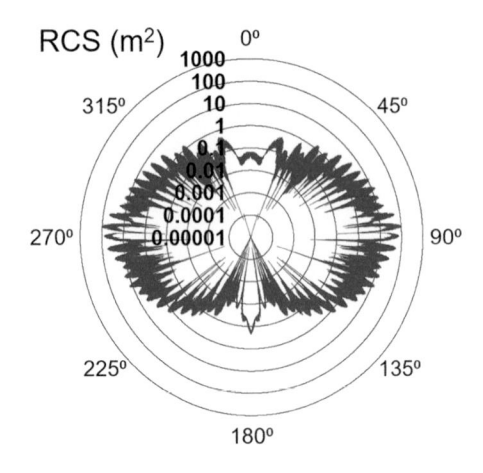

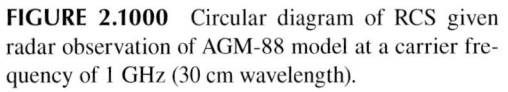

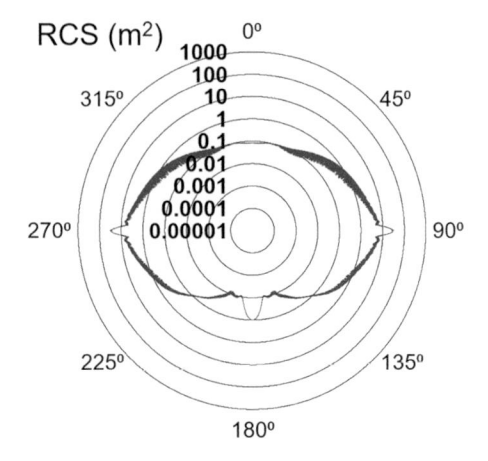

FIGURE 2.1001 Circular diagram of noncoherent RCS given radar observation of AGM-88 model at a carrier frequency of 1 GHz (30 cm wavelength).

Scattering Characteristics of Aerial Objects

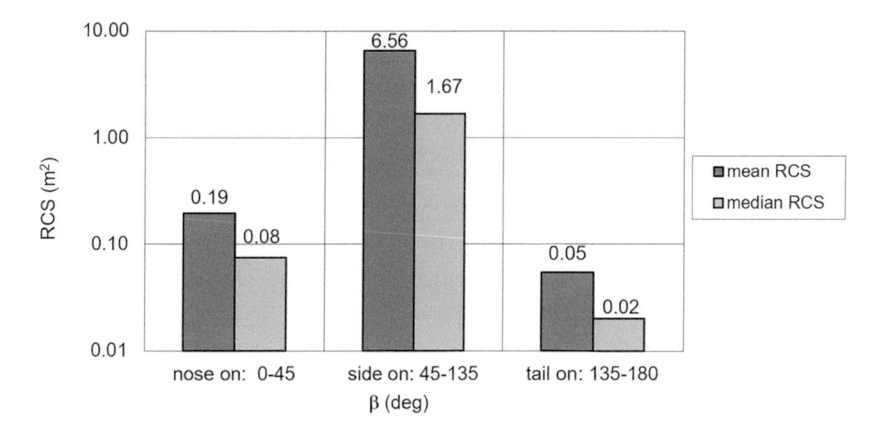

FIGURE 2.1002 Diagrams of mean and median RCS of AGM-88 model in three sectors of azimuth aspect given its radar observation at horizontal polarization and a carrier frequency of 1 GHz (30 cm wavelength).

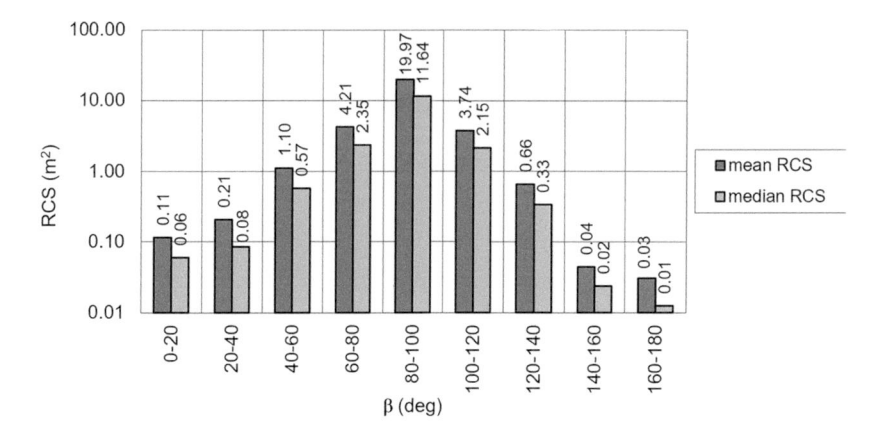

FIGURE 2.1003 Diagrams of mean and median RCS of AGM-88 model in 20-degree sectors of azimuth aspect given its radar observation at horizontal polarization and a carrier frequency of 1 GHz (30 cm wavelength).

log - normal distribution

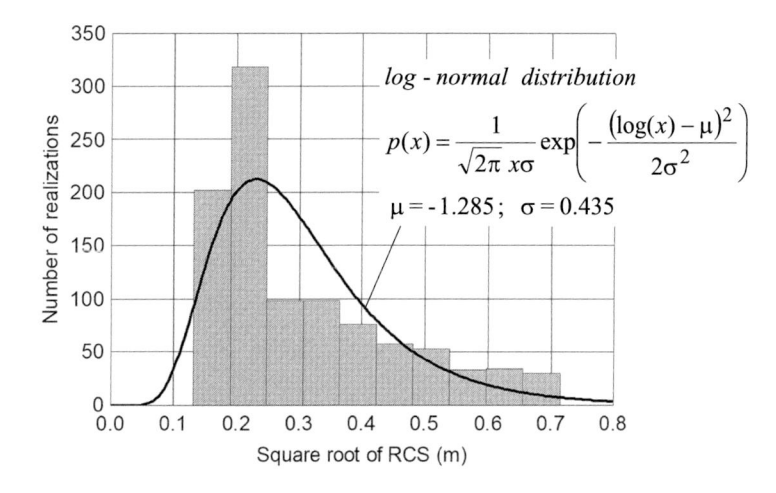

$$p(x) = \frac{1}{\sqrt{2\pi}\, x\sigma} \exp\left(-\frac{(\log(x) - \mu)^2}{2\sigma^2}\right)$$

$\mu = -1.285;\ \sigma = 0.435$

FIGURE 2.1004 Amplitude distribution of echo signal of AGM-88 model at a carrier frequency of 1 GHz given its horizontal polarization.

2.60 CRUISE MISSILE TAURUS KEPD 350

Taurus KEPD 350 (Figure 2.1005) is an air-launched cruise missile. The first flight took place in 2000.

General characteristics of Taurus KEPD 350 [71]: wingspan – 2.064 m, length – 5.100 m, fuselage height/width – 0.470/0.760 m, weight – 1 360 kg, powerplant – Williams International P8300-15, speed – 0.6–0.95 M, range – 300–500 km.

In accordance with the design of the Taurus KEPD 350, a model of its surface was created to obtaining scattering characteristics (in particular, RCS). The model is shown in Figure 2.1006. The smooth parts of the model surface were approximated by parts of 66 triaxial ellipsoids. The surface breaks were modeled using 18 straight edge scattering parts.

FIGURE 2.1005 Cruise missile Taurus KEPD 350.

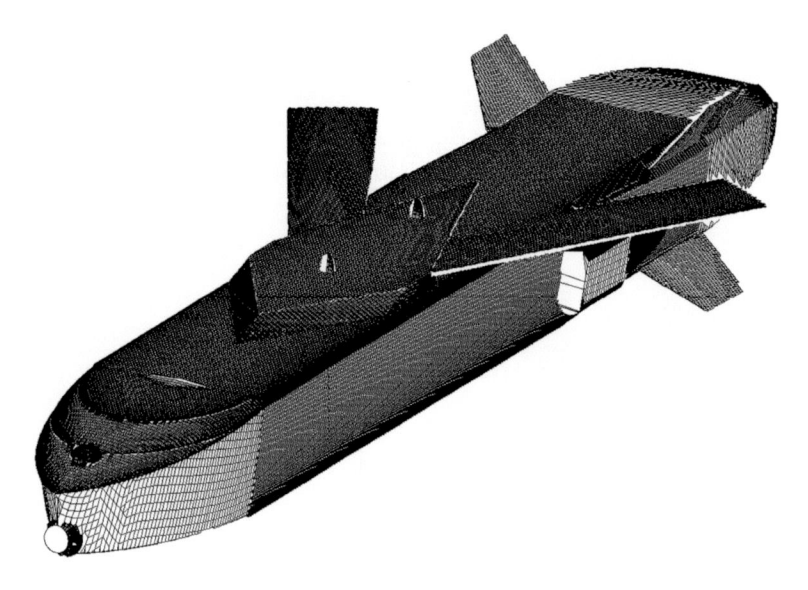

FIGURE 2.1006 Surface model of cruise missile Taurus KEPD 350.

Below are some scattering characteristics of the Taurus KEPD 350 model at sounding frequencies of 10, 3, and 1 GHz (wavelengths of 3, 10, and 30 cm, respectively) for horizontal polarization of the probing signal. The scattering characteristics for this model at two polarizations, as well as scattering characteristics at sounding frequencies of 5 and 1.3 GHz (wavelengths of 6 and 23 cm, respectively), are given in the electronic appendix of this book.

Scattering Characteristics of Aerial Objects 367

Sounding parameters: The elevation angle is to be a random value distributed uniformly in the range −3° ± 4° with respect to the wing plane (elevation angle of −3° corresponds to the radar observation from the lower hemisphere), azimuth aspect increment was 0.02°, the azimuth being counted off from the nose-on aspect (0° corresponds to the nose-on radar observation, 180° corresponds to the tail-on observation).

Scattering characteristics of Taurus KEPD 350 model for sounding frequency 10 GHz (wavelength 3 cm).

Figure 2.1007 shows the RCS circular diagram of the Taurus KEPD 350 model. The noncoherent RCS circular diagram of the Taurus KEPD 350 is shown in Figure 2.1008.

The circular mean RCS of the Taurus KEPD 350 model for horizontal polarization is 1.728 m^2. The circular median RCS (the RCS value used to calculate the detection range of an object with a probability of 0.5) for horizontal polarization is 0.128 m^2.

Figures 2.1009 and 2.1010 show the mean and median RCS for the main ranges of sounding azimuths (nose, side, and tail) and for ranges of 20°.

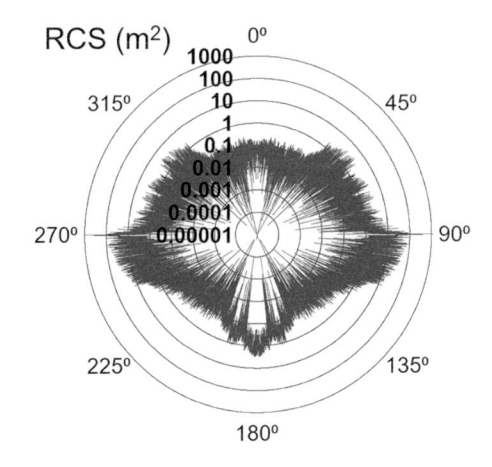

FIGURE 2.1007 Circular diagram of RCS given radar observation of Taurus KEPD 350 model at a carrier frequency of 10 GHz (3 cm wavelength).

FIGURE 2.1008 Circular diagram of noncoherent RCS given radar observation of Taurus KEPD 350 model at a carrier frequency of 10 GHz (3 cm wavelength).

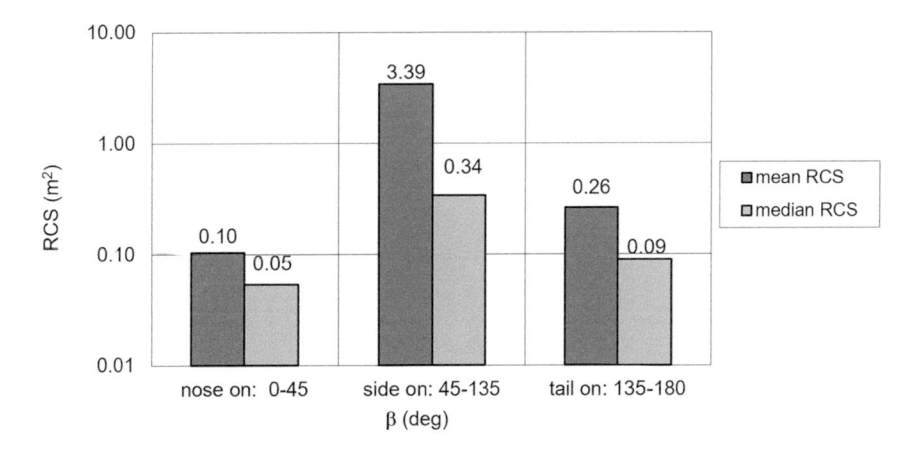

FIGURE 2.1009 Diagrams of mean and median RCS of Taurus KEPD 350 model in three sectors of azimuth aspect given its radar observation at horizontal polarization and a carrier frequency of 10 GHz (3 cm wavelength).

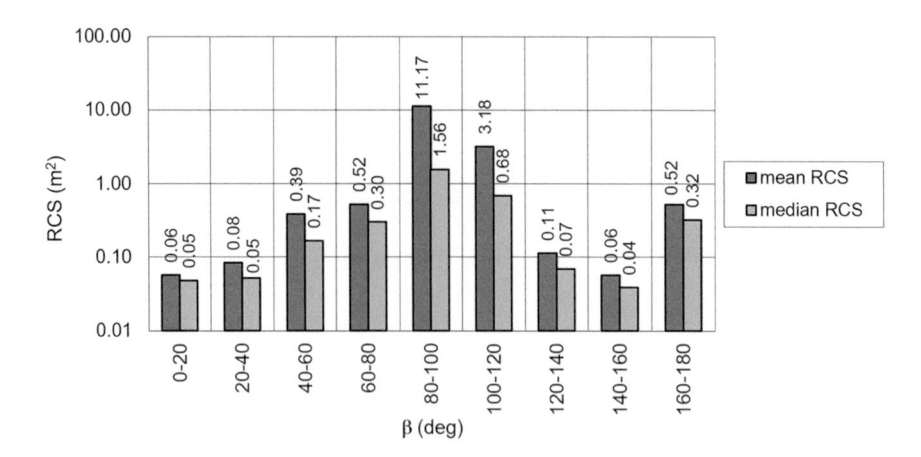

FIGURE 2.1010 Diagrams of mean and median RCS of Taurus KEPD 350 model in 20-degree sectors of azimuth aspect given its radar observation at horizontal polarization and a carrier frequency of 10 GHz (3 cm wavelength).

Figures 2.1011, 2.1016, and 2.1021 show histograms of the scattered signal amplitude (square root of the RCS) for the range of sounding azimuths −20° to +20° (sounding from the nose). The bold line shows the probability density functions of the distribution, which can be used to approximate the histogram of the signal amplitude.

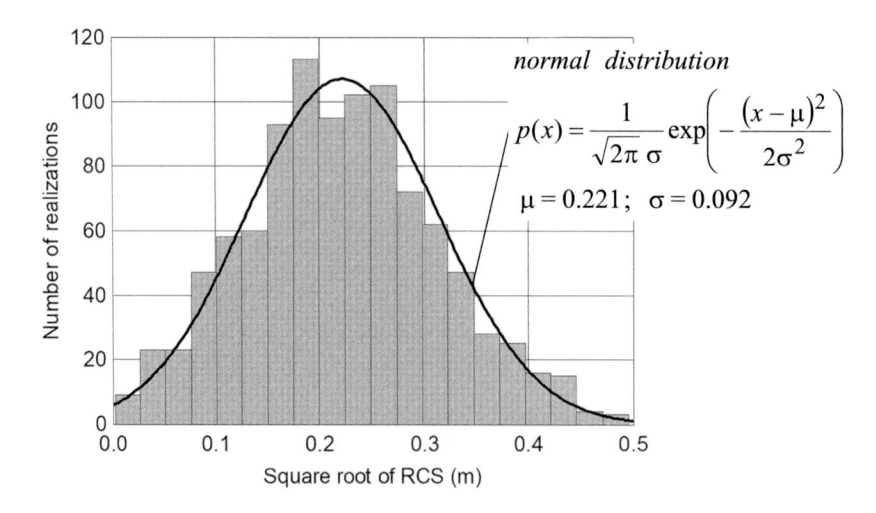

$$p(x) = \frac{1}{\sqrt{2\pi}\,\sigma} \exp\left(-\frac{(x-\mu)^2}{2\sigma^2}\right)$$

$\mu = 0.221;\ \sigma = 0.092$

FIGURE 2.1011 Amplitude distribution of echo signal of Taurus KEPD 350 model at a carrier frequency of 10 GHz given its horizontal polarization.

Scattering characteristics of Taurus KEPD 350 model for sounding frequency 3 GHz (wavelength 10 cm).

Figure 2.1012 shows the RCS circular diagram of the Taurus KEPD 350. The noncoherent RCS circular diagram of the Taurus KEPD 350 is shown in Figure 2.1013.

The circular mean RCS of the Taurus KEPD 350 for horizontal polarization is 3.256 m². The circular median RCS for horizontal polarization is 0.170 m².

Figures 2.1014 and 2.1015 show the mean and median RCS for the main ranges of sounding azimuths (nose, side, and tail) and for ranges of 20°.

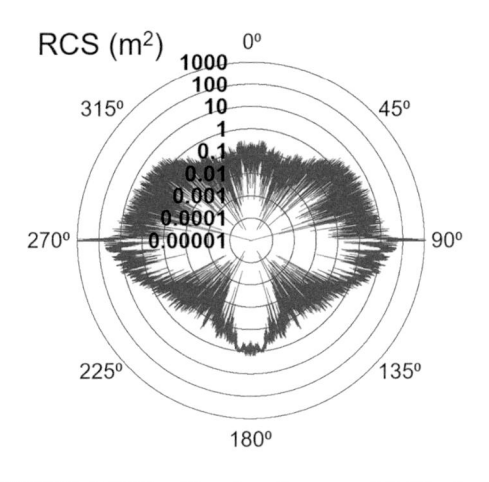

Scattering Characteristics of Aerial Objects 369

FIGURE 2.1012 Circular diagram of RCS given radar observation of Taurus KEPD 350 model at a carrier frequency of 3 GHz (10 cm wavelength).

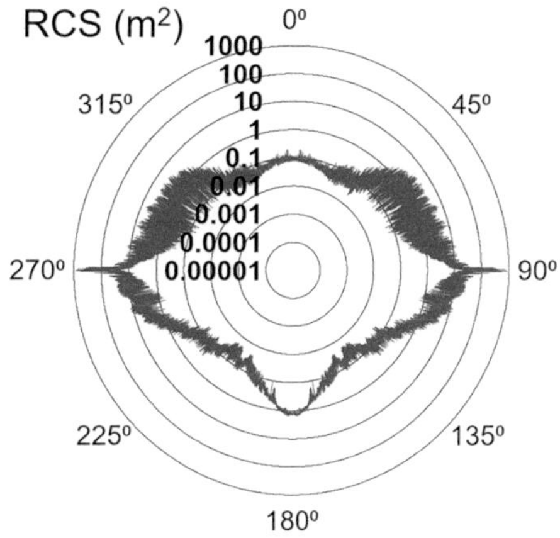

FIGURE 2.1013 Circular diagram of noncoherent RCS given radar observation of Taurus KEPD 350 model at a carrier frequency of 3 GHz (10 cm wavelength).

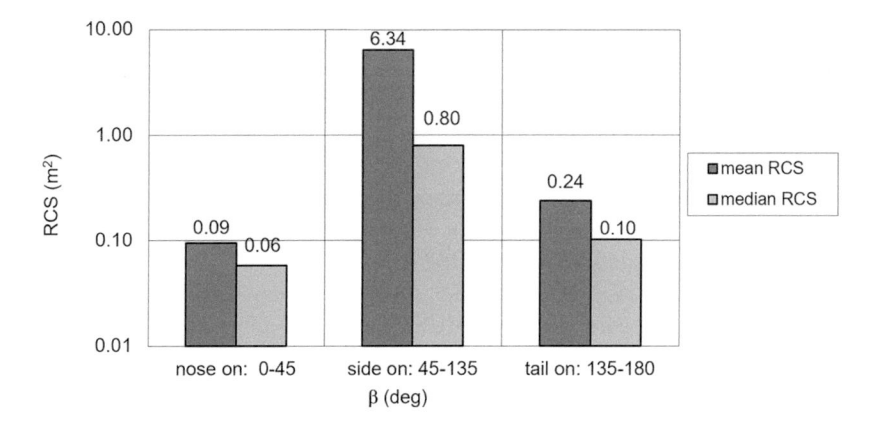

FIGURE 2.1014 Diagrams of mean and median RCS of Taurus KEPD 350 model in three sectors of azimuth aspect given its radar observation at horizontal polarization and a carrier frequency of 3 GHz (10 cm wavelength).

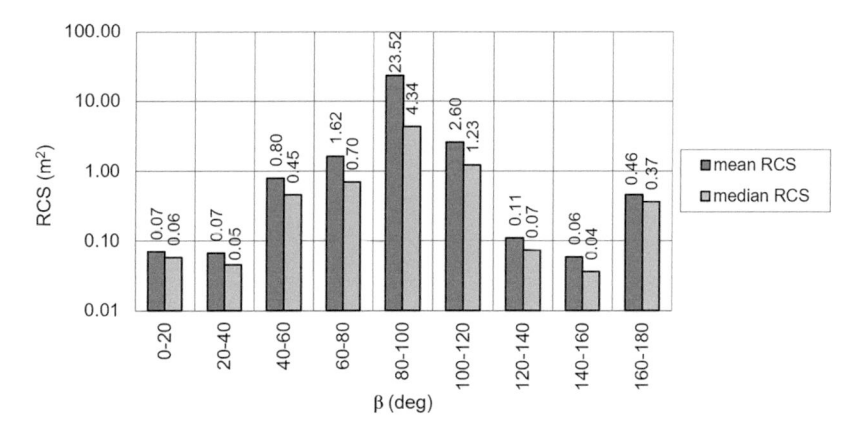

FIGURE 2.1015 Diagrams of mean and median RCS of Taurus KEPD 350 model in 20-degree sectors of azimuth aspect given its radar observation at horizontal polarization and a carrier frequency of 3 GHz (10 cm wavelength).

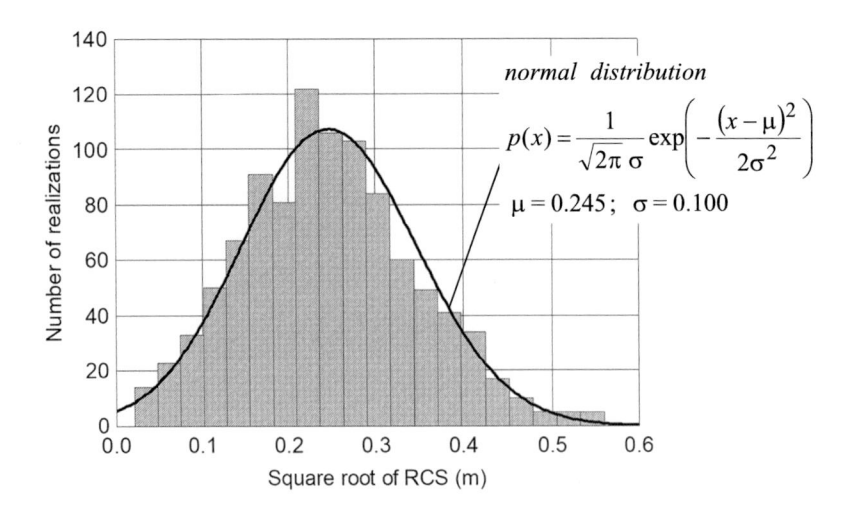

FIGURE 2.1016 Amplitude distribution of echo signal of Taurus KEPD 350 model at a carrier frequency of 3 GHz given its horizontal polarization.

Scattering characteristics of Taurus KEPD 350 model for sounding frequency 1 GHz (wavelength 30 cm).

Figure 2.1017 shows the RCS circular diagram of the Taurus KEPD 350. The noncoherent RCS circular diagram of the Taurus KEPD 350 is shown in Figure 2.1018.

The circular mean RCS of the Taurus KEPD 350 model for horizontal polarization is 4.019 m². The circular median RCS for horizontal polarization is 0.191 m².

Figures 2.1019 and 2.1020 show the mean and median RCS for the main ranges of sounding azimuths (nose, side, and tail) and for ranges of 20°.

The expressions and parameters of probability distributions that are most consistent with the empirical distributions of the square root of the RCS for different ranges of sounding azimuths and polarizations are given in the electronic appendix to this book. The expressions and parameters of the probability distributions that are most consistent with the empirical distributions of the RCS (energy characteristic) for different ranges of sounding azimuths and polarizations are also given there.

FIGURE 2.1017 Circular diagram of RCS given radar observation of Taurus KEPD 350 model at a carrier frequency of 1 GHz (30 cm wavelength).

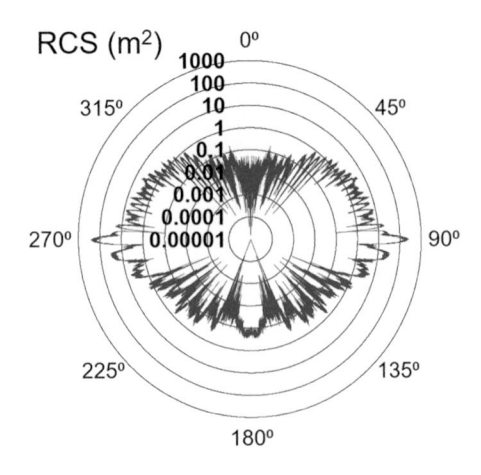

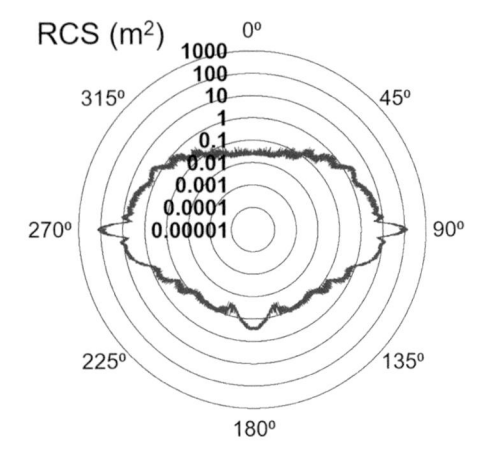

FIGURE 2.1018 Circular diagram of noncoherent RCS given radar observation of Taurus KEPD 350 model at a carrier frequency of 1 GHz (30 cm wavelength).

Scattering Characteristics of Aerial Objects

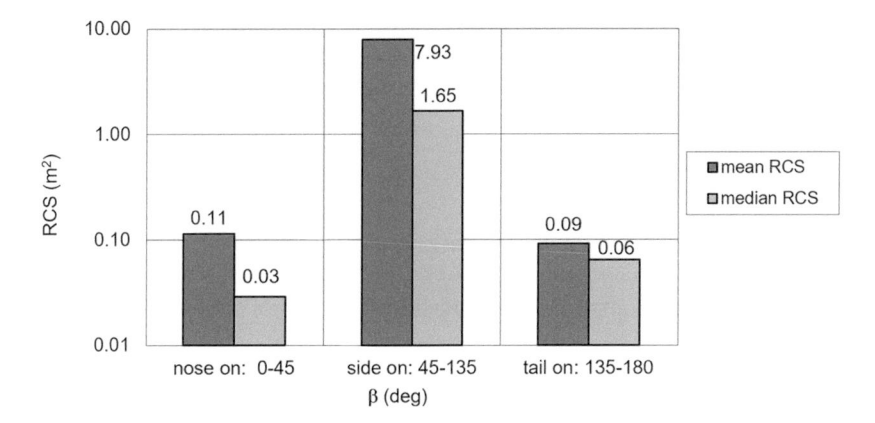

FIGURE 2.1019 Diagrams of mean and median RCS of Taurus KEPD 350 model in three sectors of azimuth aspect given its radar observation at horizontal polarization and a carrier frequency of 1 GHz (30 cm wavelength).

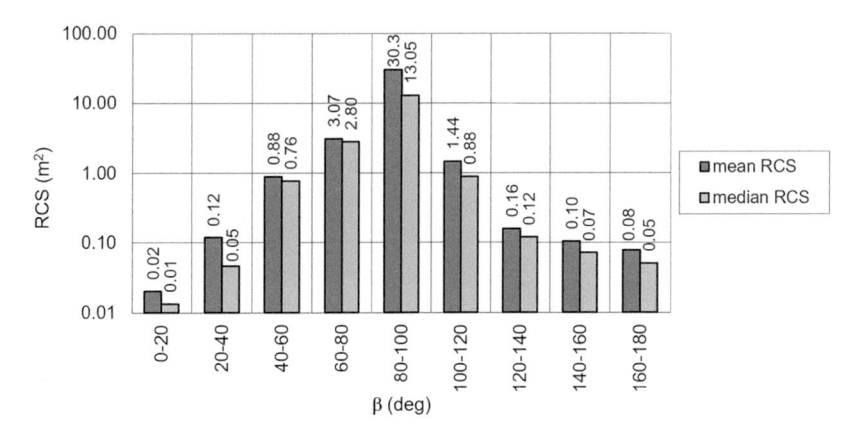

FIGURE 2.1020 Diagrams of mean and median RCS of Taurus KEPD 350 model in 20-degree sectors of azimuth aspect given its radar observation at horizontal polarization and a carrier frequency of 1 GHz (30 cm wavelength).

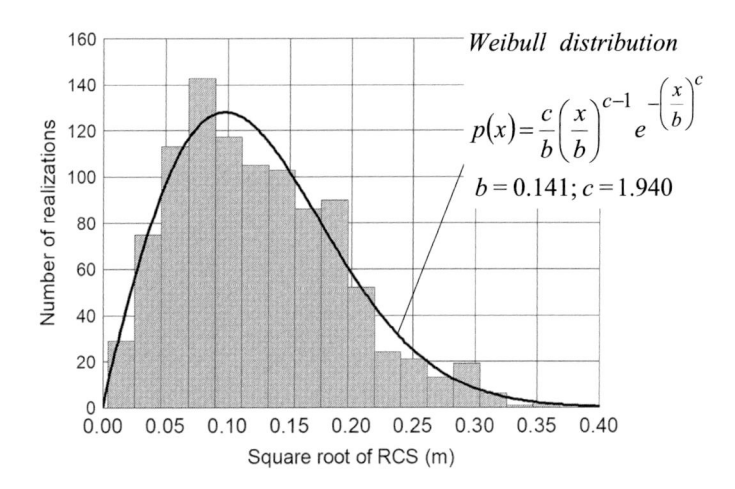

$$p(x) = \frac{c}{b}\left(\frac{x}{b}\right)^{c-1} e^{-\left(\frac{x}{b}\right)^c}$$

$b = 0.141; c = 1.940$

FIGURE 2.1021 Amplitude distribution of echo signal of Taurus KEPD 350 model at a carrier frequency of 1 GHz given its horizontal polarization.

2.61 CRUISE MISSILE STORM SHADOW

Storm Shadow (Figure 2.1022) is a long-range air-launched cruise missile. The first flight took place in 2000.

General characteristics of Storm Shadow [72]: wingspan – 2.840 m, length – 5.100 m, fuselage height/width – 0.510/0.420 m, weight – 1 300 kg, powerplant – Microturbo TRI 6030 Turbojet, speed – 0.8 M, range – 140–550 km.

In accordance with the design of the Storm Shadow, a model of its surface was created to obtaining scattering characteristics (in particular, RCS). The model is shown in Figure 2.1023. The smooth parts of the model surface were approximated by parts of 36 triaxial ellipsoids. The surface breaks were modeled using 24 straight edge scattering parts.

FIGURE 2.1022 Cruise missile Storm Shadow.

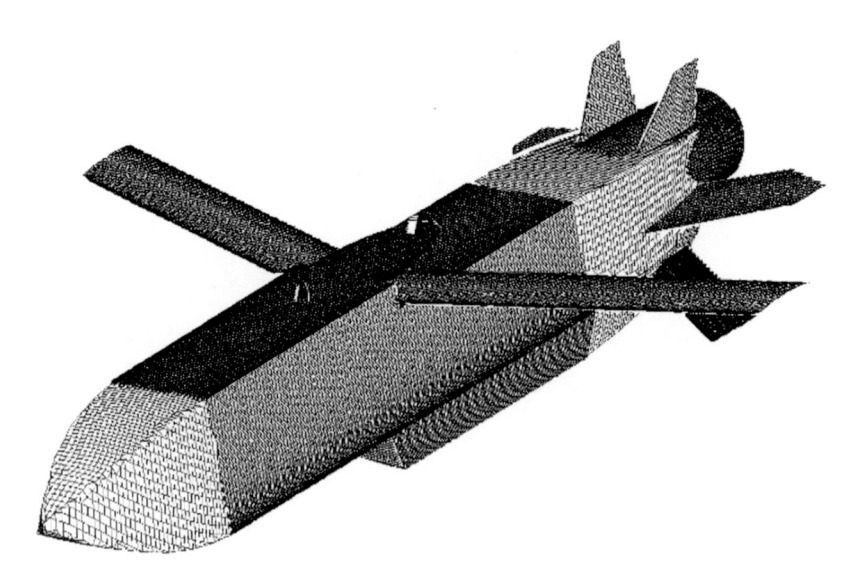

FIGURE 2.1023 Surface model of cruise missile Storm Shadow.

Below are some scattering characteristics of the Storm Shadow model at sounding frequencies of 10, 3, and 1 GHz (wavelengths of 3, 10, and 30 cm, respectively) for horizontal polarization of the probing signal. The scattering characteristics for this model at two polarizations, as well as scattering characteristics at sounding frequencies of 5 and 1.3 GHz (wavelengths of 6 and 23 cm, respectively), are given in the electronic appendix of this book.

Scattering Characteristics of Aerial Objects

Sounding parameters: The elevation angle is to be a random value distributed uniformly in the range $-3° \pm 4°$ with respect to the wing plane (elevation angle of $-3°$ corresponds to the radar observation from the lower hemisphere), azimuth aspect increment was $0.02°$, the azimuth being counted off from the nose-on aspect ($0°$ corresponds to the nose-on radar observation, $180°$ corresponds to the tail-on observation).

Scattering characteristics of Storm Shadow model for sounding frequency 10 GHz (wavelength 3 cm).

Figure 2.1024 shows the RCS circular diagram of the Storm Shadow model. The noncoherent RCS circular diagram of the Storm Shadow is shown in Figure 2.1025.

The circular mean RCS of the Storm Shadow model for horizontal polarization is 1.118 m^2. The circular median RCS (the RCS value used to calculate the detection range of an object with a probability of 0.5) for horizontal polarization is 0.062 m^2.

Figures 2.1026 and 2.1027 show the mean and median RCS for the main ranges of sounding azimuths (nose, side, and tail) and for ranges of $20°$.

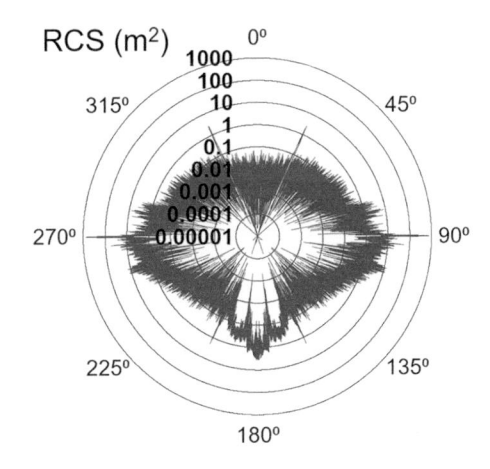

FIGURE 2.1024 Circular diagram of RCS given radar observation of Storm Shadow model at a carrier frequency of 10 GHz (3 cm wavelength).

FIGURE 2.1025 Circular diagram of noncoherent RCS given radar observation of Storm Shadow model at a carrier frequency of 10 GHz (3 cm wavelength).

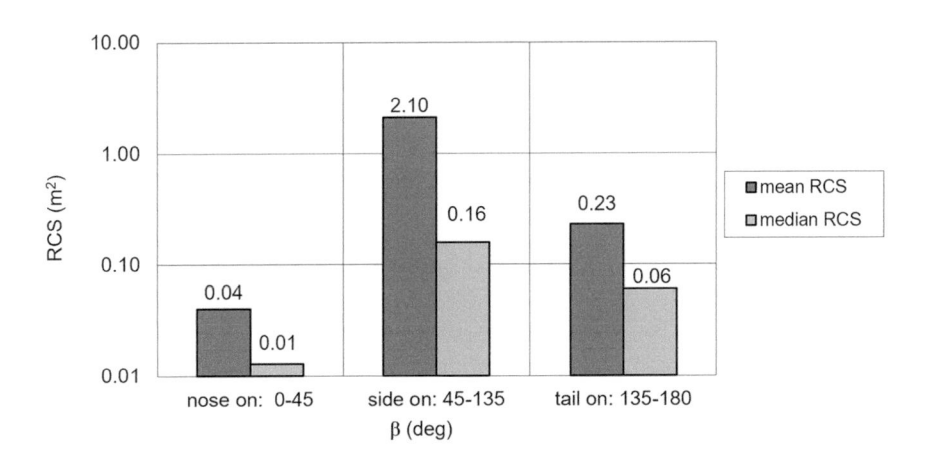

FIGURE 2.1026 Diagrams of mean and median RCS of Storm Shadow model in three sectors of azimuth aspect given its radar observation at horizontal polarization and a carrier frequency of 10 GHz (3 cm wavelength).

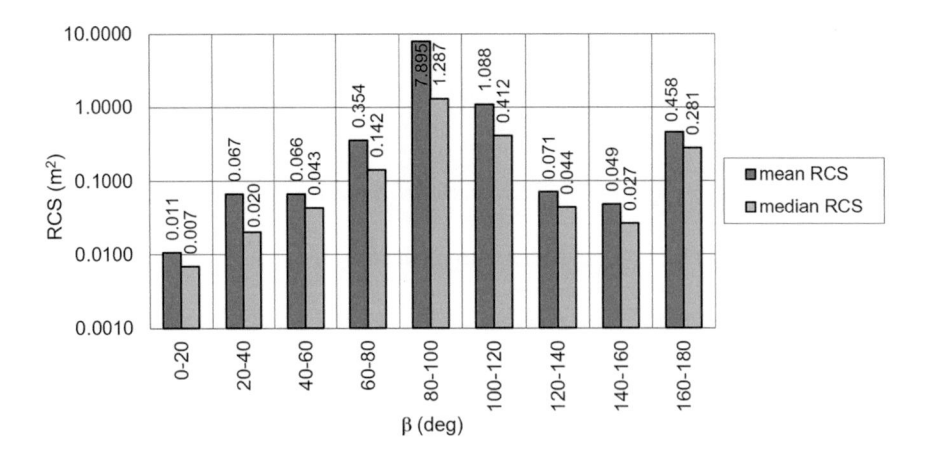

FIGURE 2.1027 Diagrams of mean and median RCS of Storm Shadow model in 20-degree sectors of azimuth aspect given its radar observation at horizontal polarization and a carrier frequency of 10 GHz (3 cm wavelength).

Figures 2.1028, 2.1033, and 2.1038 show histograms of the scattered signal amplitude (square root of the RCS) for the range of sounding azimuths −20° to +20° (sounding from the nose). The bold line shows the probability density functions of the distribution, which can be used to approximate the histogram of the signal amplitude.

Weibull distribution

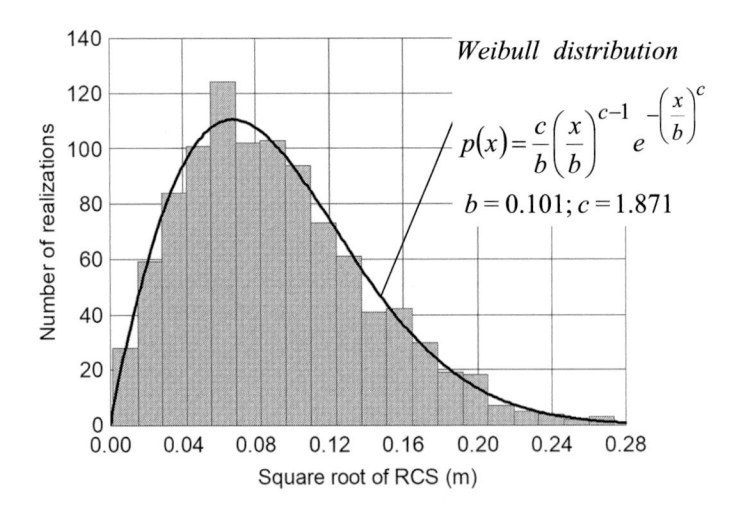

$$p(x) = \frac{c}{b}\left(\frac{x}{b}\right)^{c-1} e^{-\left(\frac{x}{b}\right)^c}$$

$b = 0.101;\ c = 1.871$

FIGURE 2.1028 Amplitude distribution of echo signal of Storm Shadow model at a carrier frequency of 10 GHz given its horizontal polarization.

Scattering characteristics of Storm Shadow model for sounding frequency 3 GHz (wavelength 10 cm).

Figure 2.1029 shows the RCS circular diagram of the Storm Shadow. The noncoherent RCS circular diagram of the Storm Shadow is shown in Figure 2.1030.

The circular mean RCS of the Storm Shadow for horizontal polarization is 1.381 m². The circular median RCS for horizontal polarization is 0.098 m².

Figures 2.1032 and 2.1032 show the mean and median RCS for the main ranges of sounding azimuths (nose, side, and tail) and for ranges of 20°.

Scattering Characteristics of Aerial Objects 375

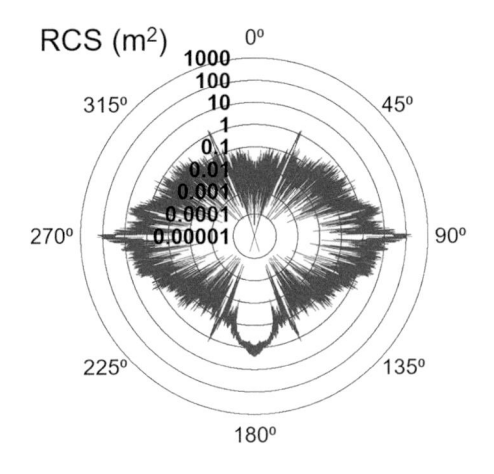

FIGURE 2.1029 Circular diagram of RCS given radar observation of Storm Shadow model at a carrier frequency of 3 GHz (10 cm wavelength).

FIGURE 2.1030 Circular diagram of noncoherent RCS given radar observation of Storm Shadow model at a carrier frequency of 3 GHz (10 cm wavelength).

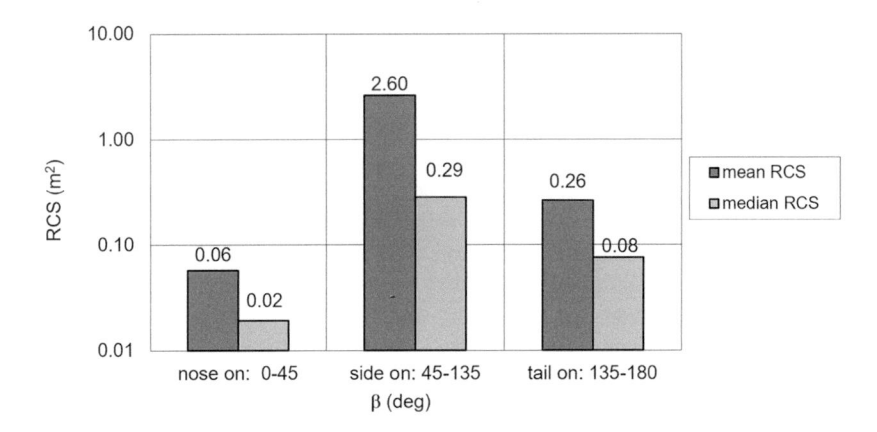

FIGURE 2.1031 Diagrams of mean and median RCS of Storm Shadow model in three sectors of azimuth aspect given its radar observation at horizontal polarization and a carrier frequency of 3 GHz (10 cm wavelength).

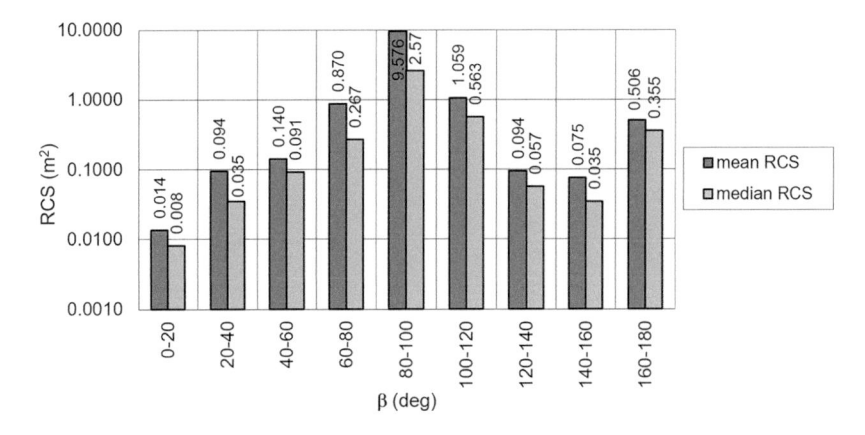

FIGURE 2.1032 Diagrams of mean and median RCS of Storm Shadow model in 20-degree sectors of azimuth aspect given its radar observation at horizontal polarization and a carrier frequency of 3 GHz (10 cm wavelength).

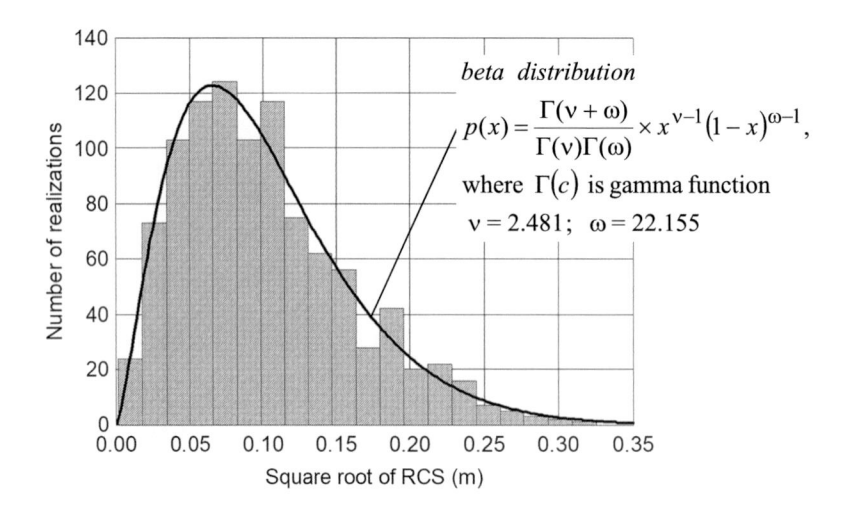

$$p(x) = \frac{\Gamma(\nu+\omega)}{\Gamma(\nu)\Gamma(\omega)} \times x^{\nu-1}(1-x)^{\omega-1},$$

where $\Gamma(c)$ is gamma function

$\nu = 2.481;\ \omega = 22.155$

FIGURE 2.1033 Amplitude distribution of echo signal of Storm Shadow model at a carrier frequency of 3 GHz given its horizontal polarization.

Scattering characteristics of Storm Shadow model for sounding frequency 1 GHz (wavelength 30 cm).

Figure 2.1034 shows the RCS circular diagram of the Storm Shadow. The noncoherent RCS circular diagram of the Storm Shadow is shown in Figure 2.1035.

The circular mean RCS of the Storm Shadow model for horizontal polarization is 1.402 m². The circular median RCS for horizontal polarization is 0.168 m².

Figures 2.1036 and 2.1037 show the mean and median RCS for the main ranges of sounding azimuths (nose, side, and tail) and for ranges of 20°.

The expressions and parameters of probability distributions that are most consistent with the empirical distributions of the square root of the RCS for different ranges of sounding azimuths and polarizations are given in the electronic appendix to this book. The expressions and parameters of the probability distributions that are most consistent with the empirical distributions of the RCS (energy characteristic) for different ranges of sounding azimuths and polarizations are also given there.

FIGURE 2.1034 Circular diagram of RCS given radar observation of Storm Shadow model at a carrier frequency of 1 GHz (30 cm wavelength).

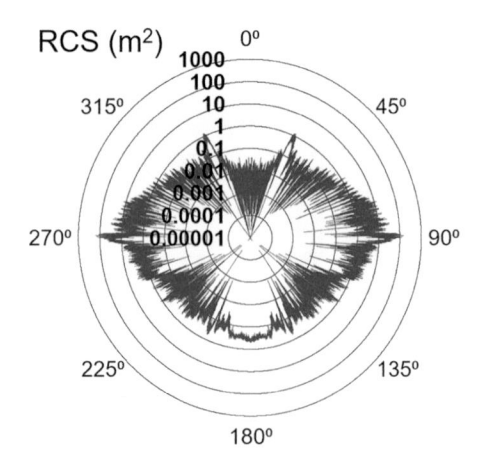

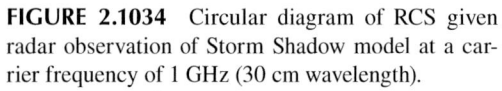

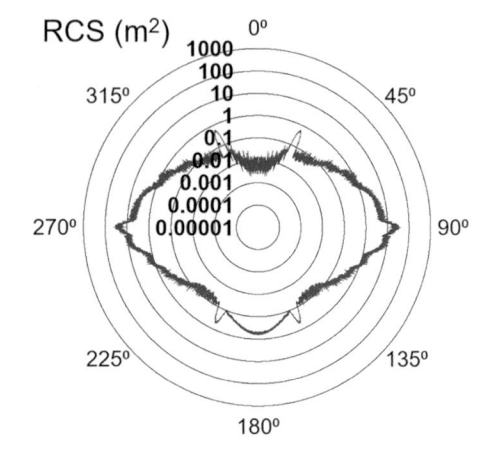

FIGURE 2.1035 Circular diagram of noncoherent RCS given radar observation of Storm Shadow model at a carrier frequency of 1 GHz (30 cm wavelength).

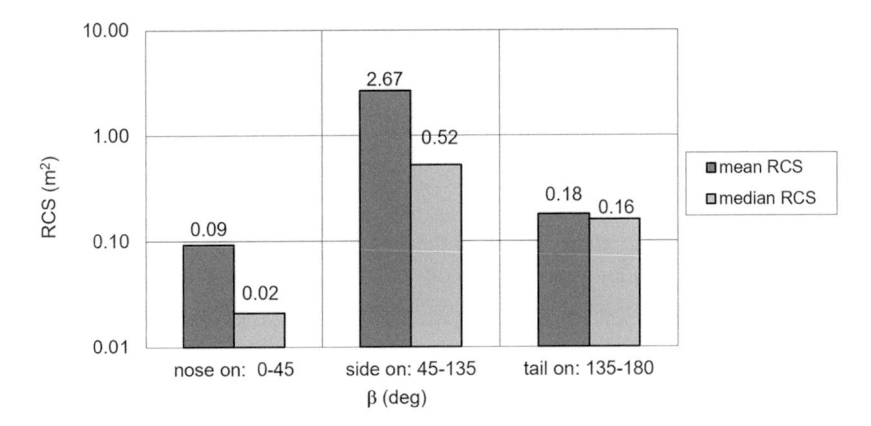

FIGURE 2.1036 Diagrams of mean and median RCS of Storm Shadow model in three sectors of azimuth aspect given its radar observation at horizontal polarization and a carrier frequency of 1 GHz (30 cm wavelength).

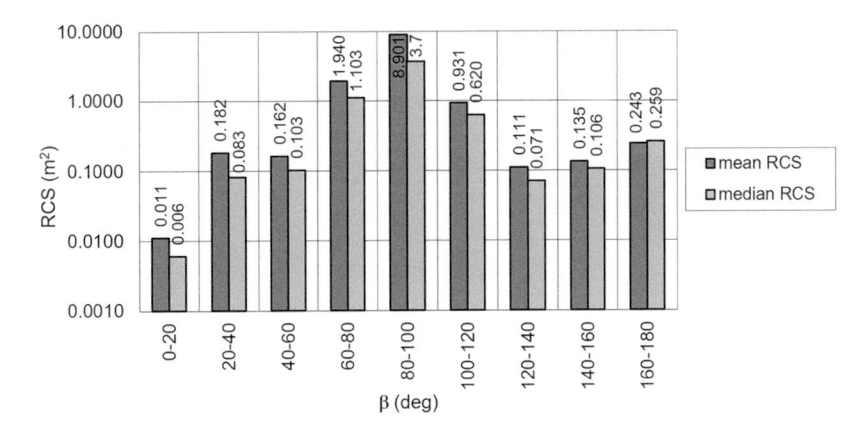

FIGURE 2.1037 Diagrams of mean and median RCS of Storm Shadow model in 20-degree sectors of azimuth aspect given its radar observation at horizontal polarization and a carrier frequency of 1 GHz (30 cm wavelength).

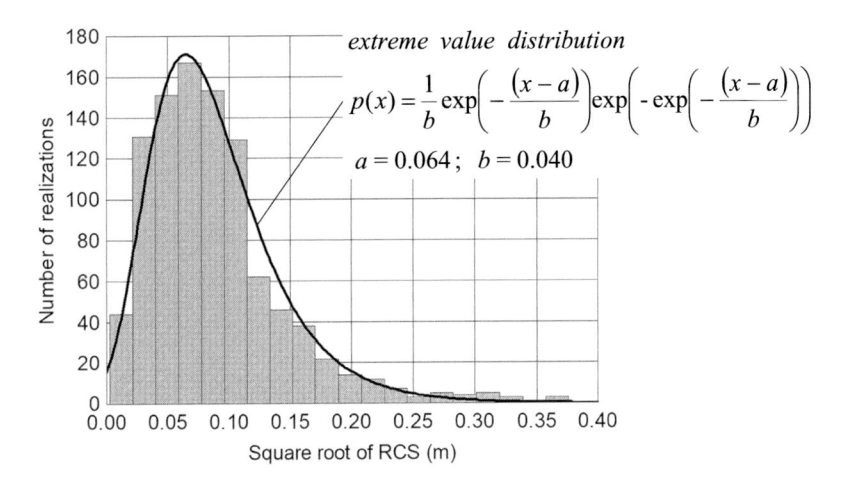

$$p(x) = \frac{1}{b}\exp\left(-\frac{(x-a)}{b}\right)\exp\left(-\exp\left(-\frac{(x-a)}{b}\right)\right)$$

$a = 0.064; \ b = 0.040$

FIGURE 2.1038 Amplitude distribution of echo signal of Storm Shadow model at a carrier frequency of 1 GHz given its horizontal polarization.

2.62 ANTI-AIRCRAFT MISSILE 5V55R

5V55R (Figure 2.1039) is the main missile for S-300 (SA-10 Grumble) family of surface-to-air missile systems. The missile can be used against ground targets. The first flight took place in 1973.

General characteristics of 5V55R [73]: wingspan – 1.124 m, length – 7.25 m, fuselage diameter – 0.508 m, weight – 1 665 kg, powerplant – 1 solid-fuel rocket, speed – 2 000 m/s, range – 5–75 km (against ground targets – 130 km).

In accordance with the design of the 5V55R, a model of its surface was created to obtaining scattering characteristics (in particular, RCS). The model is shown in Figure 2.1040. The smooth parts of the model surface were approximated by parts of 19 triaxial ellipsoids. The surface breaks were modeled using 12 straight edge scattering parts.

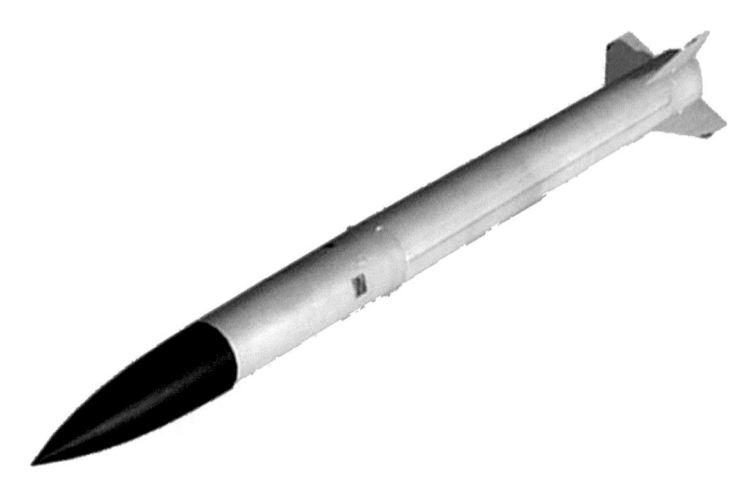

FIGURE 2.1039 Anti-aircraft missile 5V55R.

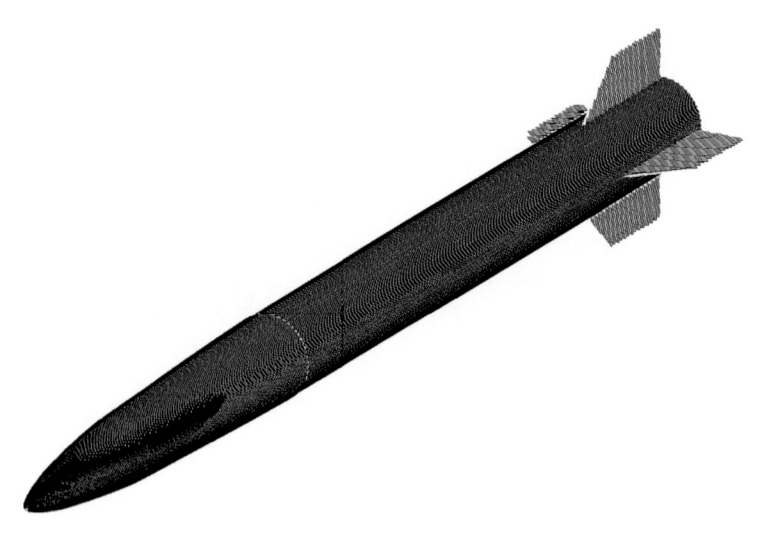

FIGURE 2.1040 Surface model of an anti-aircraft missile 5V55R.

Below are some scattering characteristics of the 5V55R model at sounding frequencies of 10, 3, and 1 GHz (wavelengths of 3, 10, and 30 cm, respectively) for horizontal polarization of the probing signal. The scattering characteristics for this model at two polarizations, as well as scattering characteristics at sounding frequencies of 5 and 1.3 GHz (wavelengths of 6 and 23 cm, respectively), are given in the electronic appendix of this book.

Scattering Characteristics of Aerial Objects 379

Sounding parameters: The elevation angle is to be a random value distributed uniformly in the range −3° ± 4° with respect to the wing plane (elevation angle of −3° corresponds to the radar observation from the lower hemisphere), azimuth aspect increment was 0.02°, the azimuth being counted off from the nose-on aspect (0° corresponds to the nose-on radar observation, 180° corresponds to the tail-on observation).

Scattering characteristics of the 5V55R model for sounding frequency 10 GHz (wavelength 3 cm).

Figure 2.1041 shows the RCS circular diagram of the 5V55R model. The noncoherent RCS circular diagram of the 5V55R is shown in Figure 2.1042.

The circular mean RCS of the 5V55R model for horizontal polarization is 3.897 m². The circular median RCS (the RCS value used to calculate the detection range of an object with a probability of 0.5) for horizontal polarization is 0.012 m².

Figures 2.1043 and 2.1044 show the mean and median RCS for the main ranges of sounding azimuths (nose, side, and tail) and for ranges of 20°.

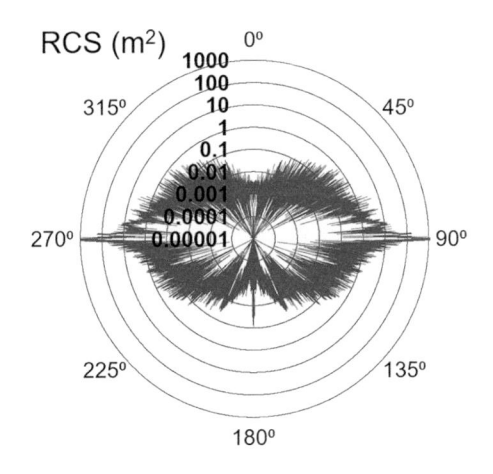

FIGURE 2.1041 Circular diagram of RCS given radar observation of 5V55R model at a carrier frequency of 10 GHz (3 cm wavelength).

FIGURE 2.1042 Circular diagram of noncoherent RCS given radar observation of 5V55R model at a carrier frequency of 10 GHz (3 cm wavelength).

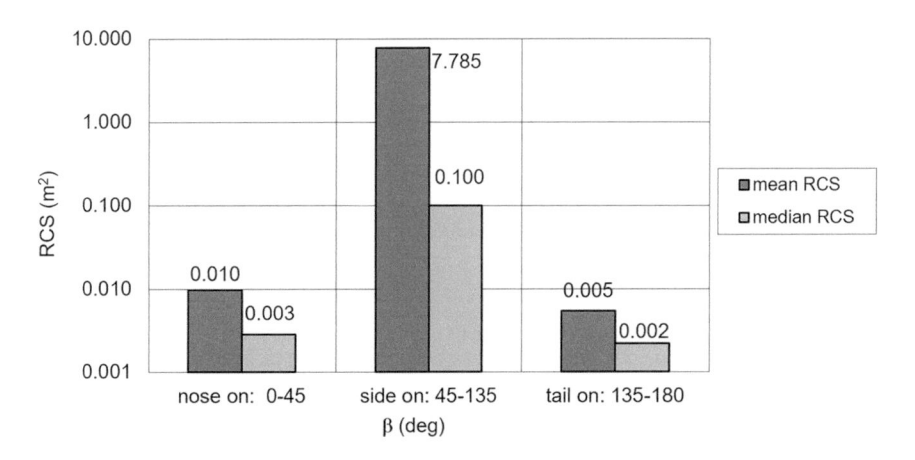

FIGURE 2.1043 Diagrams of mean and median RCS of 5V55R model in three sectors of azimuth aspect given its radar observation at horizontal polarization and a carrier frequency of 10 GHz (3 cm wavelength).

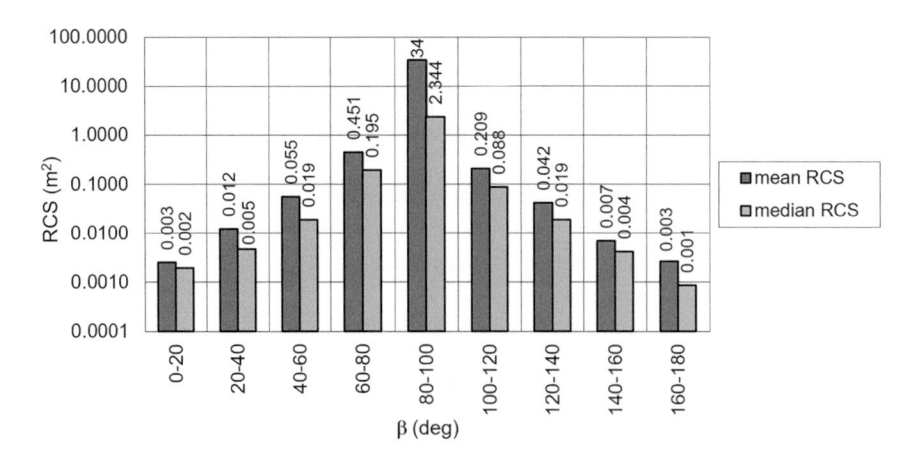

FIGURE 2.1044 Diagrams of mean and median RCS of 5V55R model in 20-degree sectors of azimuth aspect given its radar observation at horizontal polarization and a carrier frequency of 10 GHz (3 cm wavelength).

Figures 2.1045, 2.1050, and 2.1055 show histograms of the scattered signal amplitude (square root of the RCS) for the range of sounding azimuths −20° to +20° (sounding from the nose). The bold line shows the probability density functions of the distribution, which can be used to approximate the histogram of the signal amplitude.

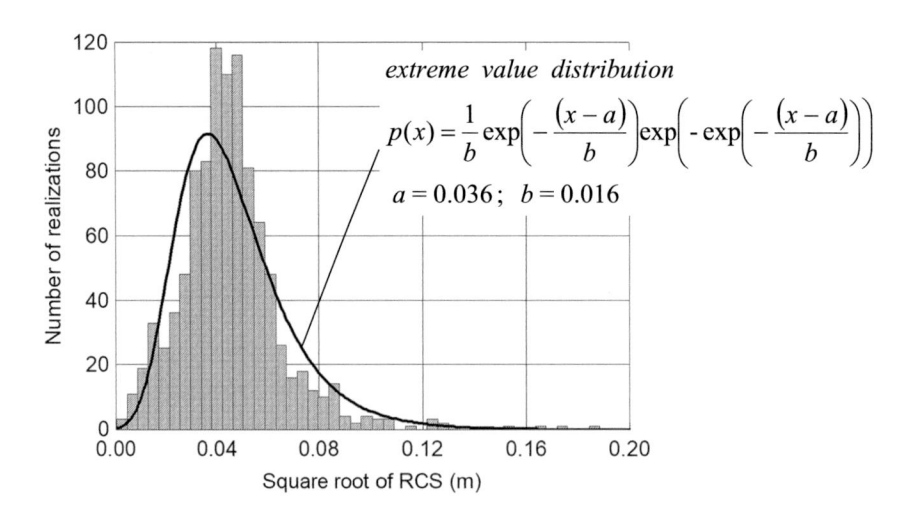

extreme value distribution

$$p(x) = \frac{1}{b}\exp\left(-\frac{(x-a)}{b}\right)\exp\left(-\exp\left(-\frac{(x-a)}{b}\right)\right)$$

$a = 0.036;\ b = 0.016$

FIGURE 2.1045 Amplitude distribution of echo signal of 5V55R model at a carrier frequency of 10 GHz given its horizontal polarization.

Scattering characteristics of the 5V55R model for sounding frequency 3 GHz (wavelength 10 cm).

Figure 2.1046 shows the RCS circular diagram of the 5V55R. The noncoherent RCS circular diagram of the 5V55R is shown in Figure 2.1047.

The circular mean RCS of the 5V55R for horizontal polarization is 3.955 m². The circular median RCS for horizontal polarization is 0.024 m².

Figures 2.1048 and 2.1049 show the mean and median RCS for the main ranges of sounding azimuths (nose, side, and tail) and for ranges of 20°.

Scattering Characteristics of Aerial Objects   381

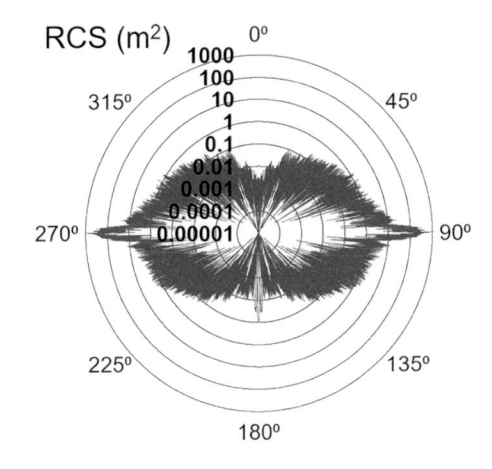

FIGURE 2.1046 Circular diagram of RCS given radar observation of 5V55R model at a carrier frequency of 3 GHz (10 cm wavelength).

FIGURE 2.1047 Circular diagram of noncoherent RCS given radar observation of 5V55R model at a carrier frequency of 3 GHz (10 cm wavelength).

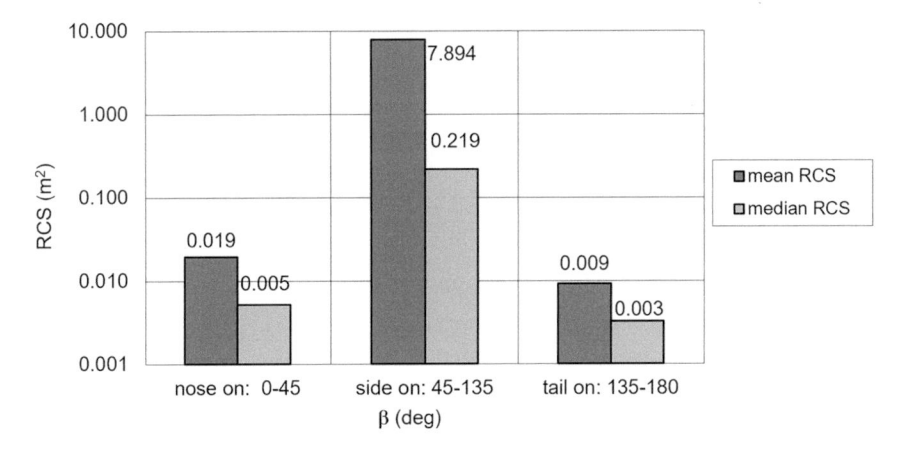

FIGURE 2.1048 Diagrams of mean and median RCS of 5V55R model in three sectors of azimuth aspect given its radar observation at horizontal polarization and a carrier frequency of 3 GHz (10 cm wavelength).

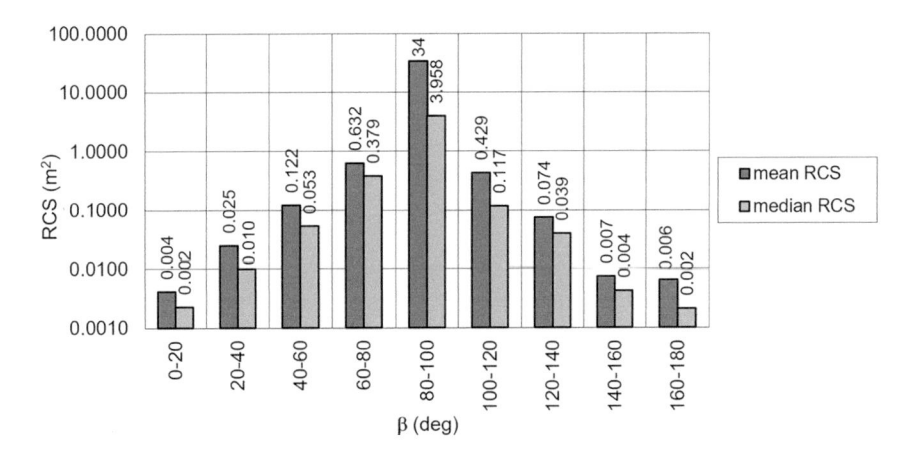

FIGURE 2.1049 Diagrams of mean and median RCS of 5V55R model in 20-degree sectors of azimuth aspect given its radar observation at horizontal polarization and a carrier frequency of 3 GHz (10 cm wavelength).

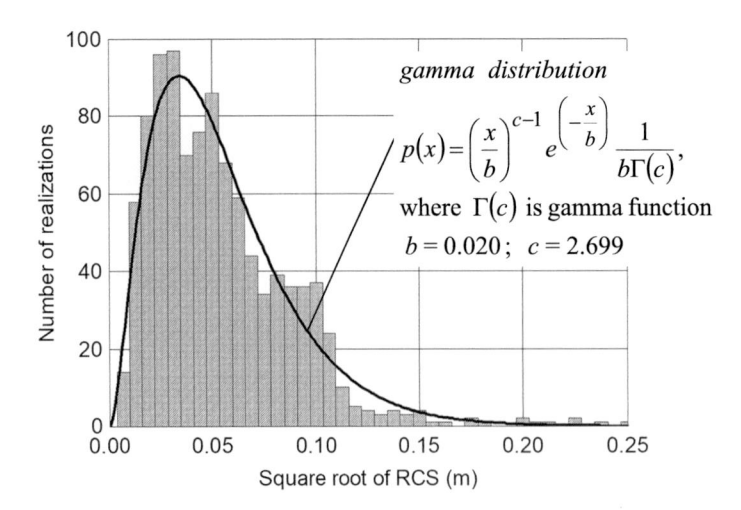

FIGURE 2.1050 Amplitude distribution of echo signal of 5V55R model at a carrier frequency of 3 GHz given its horizontal polarization.

gamma distribution

$$p(x) = \left(\frac{x}{b}\right)^{c-1} e^{\left(-\frac{x}{b}\right)} \frac{1}{b\Gamma(c)},$$

where $\Gamma(c)$ is gamma function
$b = 0.020; \quad c = 2.699$

Scattering characteristics of the 5V55R model for sounding frequency 1 GHz (wavelength 30 cm).

Figure 2.1051 shows the RCS circular diagram of the 5V55R. The noncoherent RCS circular diagram of the 5V55R is shown in Figure 2.1052.

The circular mean RCS of the 5V55R model for horizontal polarization is 3.040 m². The circular median RCS for horizontal polarization is 0.081 m².

Figures 2.1053 and 2.1054 show the mean and median RCS for the main ranges of sounding azimuths (nose, side, and tail) and for ranges of 20°.

The expressions and parameters of probability distributions that are most consistent with the empirical distributions of the square root of the RCS for different ranges of sounding azimuths and polarizations are given in the electronic appendix to this book. The expressions and parameters of the probability distributions that are most consistent with the empirical distributions of the RCS (energy characteristic) for different ranges of sounding azimuths and polarizations are also given there.

FIGURE 2.1051 Circular diagram of RCS given radar observation of 5V55R model at a carrier frequency of 1 GHz (30 cm wavelength).

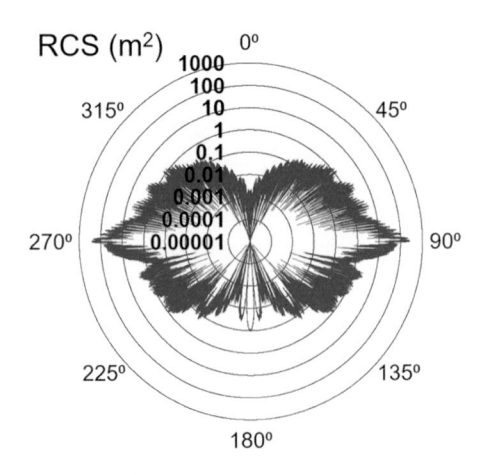

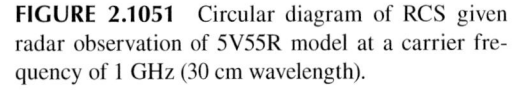

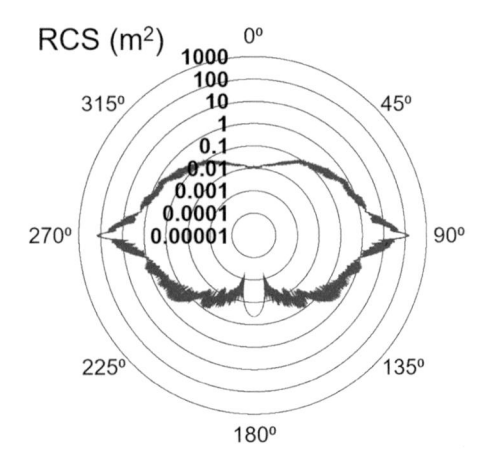

FIGURE 2.1052 Circular diagram of noncoherent RCS given radar observation of 5V55R model at a carrier frequency of 1 GHz (30 cm wavelength).

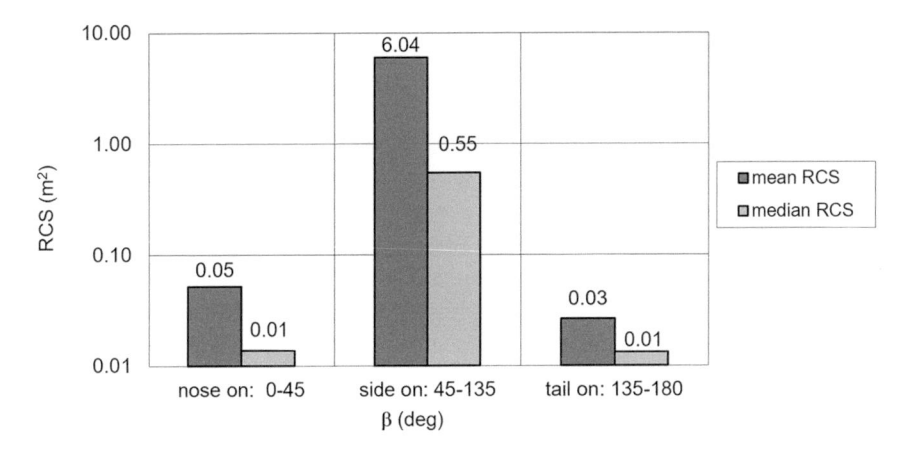

FIGURE 2.1053 Diagrams of mean and median RCS of 5V55R model in three sectors of azimuth aspect given its radar observation at horizontal polarization and a carrier frequency of 1 GHz (30 cm wavelength).

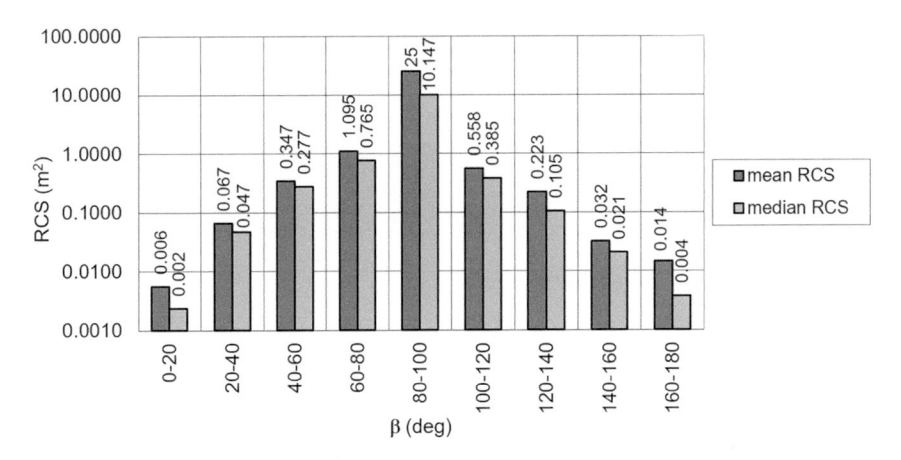

FIGURE 2.1054 Diagrams of mean and median RCS of 5V55R model in 20-degree sectors of azimuth aspect given its radar observation at horizontal polarization and a carrier frequency of 1 GHz (30 cm wavelength).

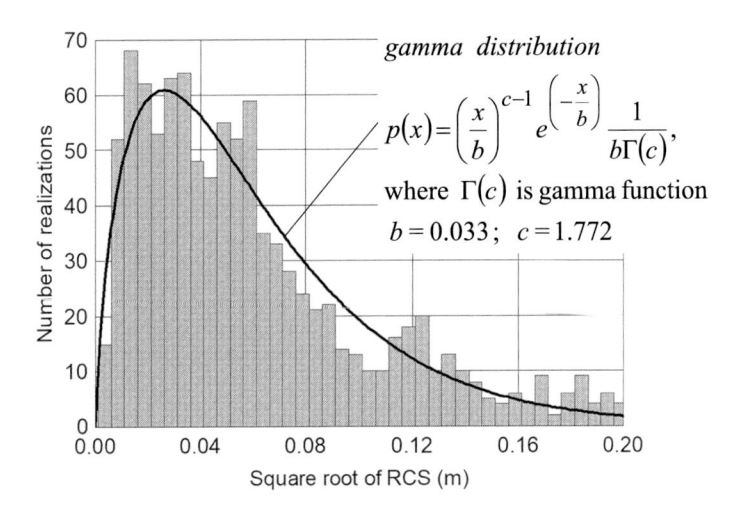

gamma distribution

$$p(x) = \left(\frac{x}{b}\right)^{c-1} e^{\left(-\frac{x}{b}\right)} \frac{1}{b\Gamma(c)},$$

where $\Gamma(c)$ is gamma function

$b = 0.033;\ c = 1.772$

FIGURE 2.1055 Amplitude distribution of echo signal of 5V55R model at a carrier frequency of 1 GHz given its horizontal polarization.

2.63 DECOY MISSILE ADM-141C iTALD

ADM-141C iTALD (Figure 2.1056) is a tactical air-launched decoy missile. The first flight took place in 1993.

General characteristics of ADM-141C iTALD [74]: wingspan – 1.55 m, length – 2.34 m, fuselage diameter – 0.533 m, weight – 180 kg, powerplant – 1 Teledyne CAE-312 turbofan engine, speed – 0.8 M, range – 300 km.

In accordance with the design of the ADM-141C, a model of its surface was created to obtaining scattering characteristics (in particular, RCS). The model is shown in Figure 2.1057. The perfectly conducting smooth parts of the model surface were approximated by parts of 29 triaxial ellipsoids. Dielectric parts of the model surface were approximated by parts of 5 triaxial ellipsoids. To increase the model RCS in a wide range of angles, a spherical reflector is placed under dielectric nose of the model fuselage. The surface breaks were modeled using 14 straight edge scattering parts.

FIGURE 2.1056 Decoy missile ADM-141C iTALD.

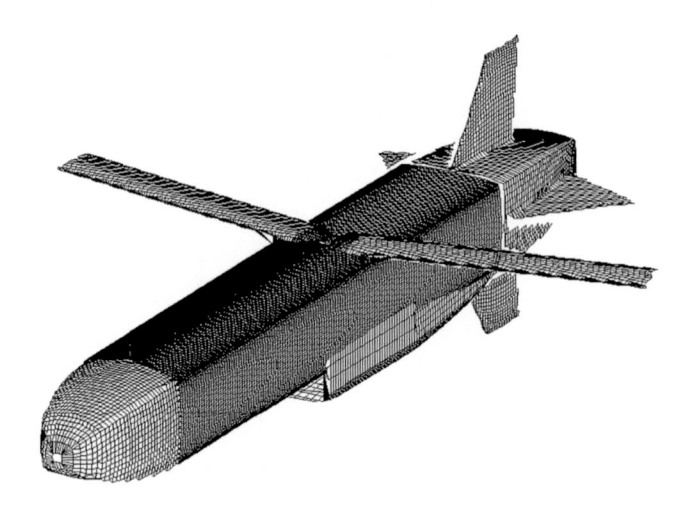

FIGURE 2.1057 Surface model of a decoy missile ADM-141C iTALD.

Below are some scattering characteristics of the ADM-141C model at sounding frequencies of 10, 3, and 1 GHz (wavelengths of 3, 10, and 30 cm, respectively) for horizontal polarization of the probing signal. The scattering characteristics for this model at two polarizations, as well as scattering characteristics at sounding frequencies of 5 and 1.3 GHz (wavelengths of 6 and 23 cm, respectively), are given in the electronic appendix of this book.

Sounding parameters: The elevation angle is to be a random value distributed uniformly in the range $-3° \pm 4°$ with respect to the wing plane (elevation angle of $-3°$ corresponds to the radar

Scattering Characteristics of Aerial Objects

observation from the lower hemisphere), azimuth aspect increment was 0.02°, the azimuth being counted off from the nose-on aspect (0° corresponds to the nose-on radar observation, 180° corresponds to the tail-on observation).

Scattering characteristics of ADM-141C model for sounding frequency 10 GHz (wavelength 3 cm).

Figure 2.1058 shows the RCS circular diagram of the ADM-141C model. The noncoherent RCS circular diagram of the ADM-141C is shown in Figure 2.1059.

The circular mean RCS of the ADM-141C model for horizontal polarization is 7.547 m². The circular median RCS (the RCS value used to calculate the detection range of an object with a probability of 0.5) for horizontal polarization is 2.053 m².

Figures 2.1060 and 2.1061 show the mean and median RCS for the main ranges of sounding azimuths (nose, side, and tail) and for ranges of 20°.

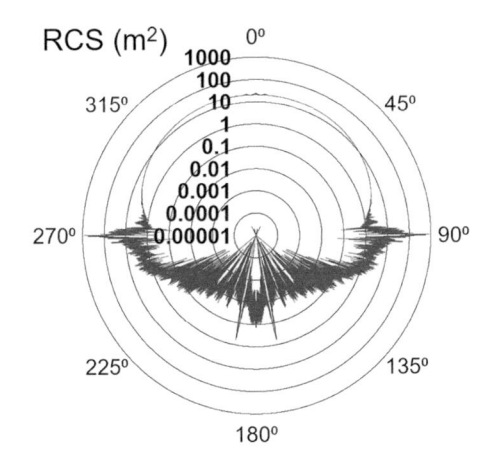

FIGURE 2.1058 Circular diagram of RCS given radar observation of ADM-141C model at a carrier frequency of 10 GHz (3 cm wavelength).

FIGURE 2.1059 Circular diagram of noncoherent RCS given radar observation of ADM-141C model at a carrier frequency of 10 GHz (3 cm wavelength).

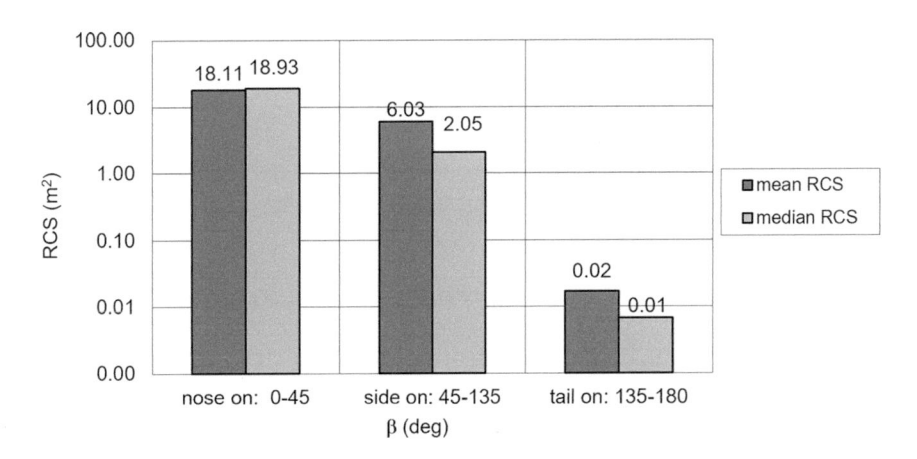

FIGURE 2.1060 Diagrams of mean and median RCS of ADM-141C model in three sectors of azimuth aspect given its radar observation at horizontal polarization and a carrier frequency of 10 GHz (3 cm wavelength).

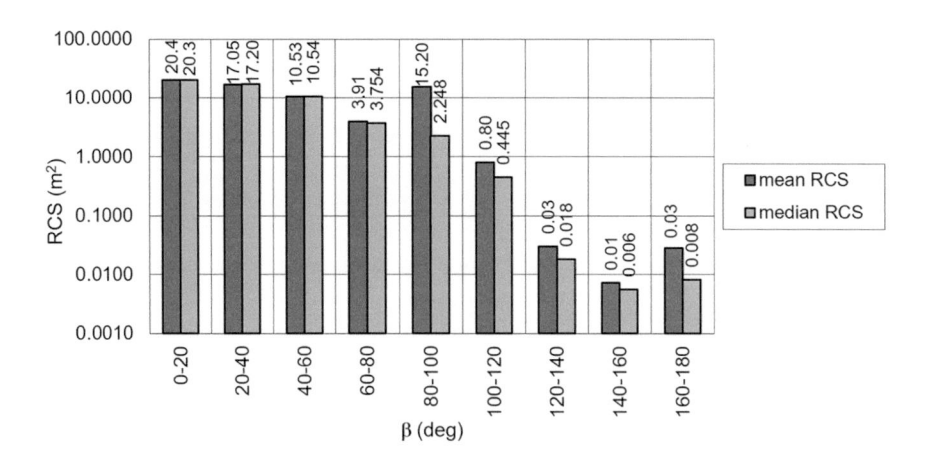

FIGURE 2.1061 Diagrams of mean and median RCS of ADM-141C model in 20-degree sectors of azimuth aspect given its radar observation at horizontal polarization and a carrier frequency of 10 GHz (3 cm wavelength).

Figures 2.1062, 2.1067, and 2.1072 show histograms of the scattered signal amplitude (square root of the RCS) for the range of sounding azimuths −20° to +20° (sounding from the nose). The bold line shows the probability density functions of the distribution, which can be used to approximate the histogram of the signal amplitude.

extreme value distribution

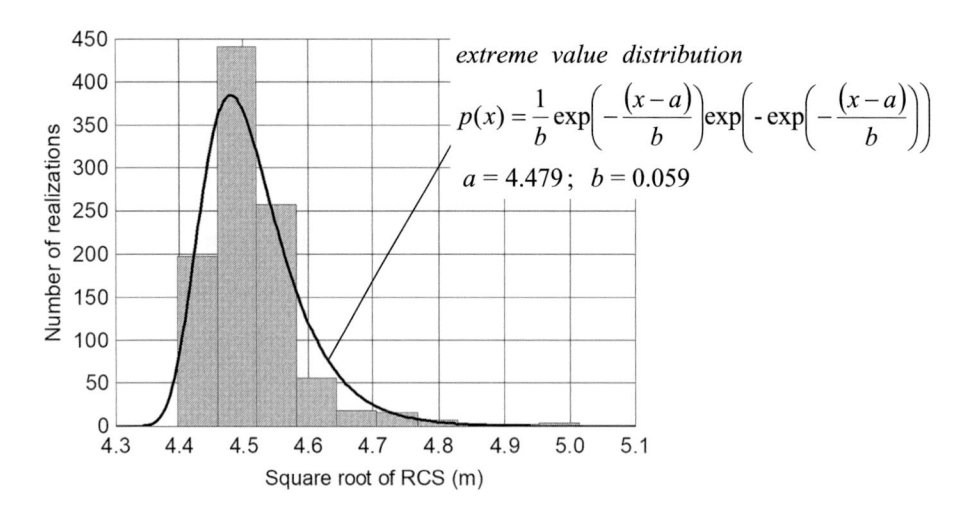

$$p(x) = \frac{1}{b} \exp\left(-\frac{(x-a)}{b}\right) \exp\left(-\exp\left(-\frac{(x-a)}{b}\right)\right)$$

$a = 4.479; \quad b = 0.059$

FIGURE 2.1062 Amplitude distribution of echo signal of ADM-141C model at a carrier frequency of 10 GHz given its horizontal polarization.

Scattering characteristics of ADM-141C model for sounding frequency 3 GHz (wavelength 10 cm).

Figure 2.1063 shows the RCS circular diagram of the ADM-141C. The noncoherent RCS circular diagram of the ADM-141C is shown in Figure 2.1064.

The circular mean RCS of the ADM-141C for horizontal polarization is 2.247 m². The circular median RCS for horizontal polarization is 0.701 m².

Figures 2.1065 and 2.1066 show the mean and median RCS for the main ranges of sounding azimuths (nose, side, and tail) and for ranges of 20°.

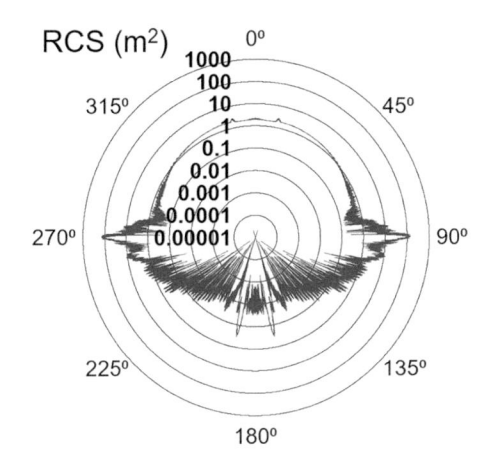

Scattering Characteristics of Aerial Objects

FIGURE 2.1063 Circular diagram of RCS given radar observation of ADM-141C model at a carrier frequency of 3 GHz (10 cm wavelength).

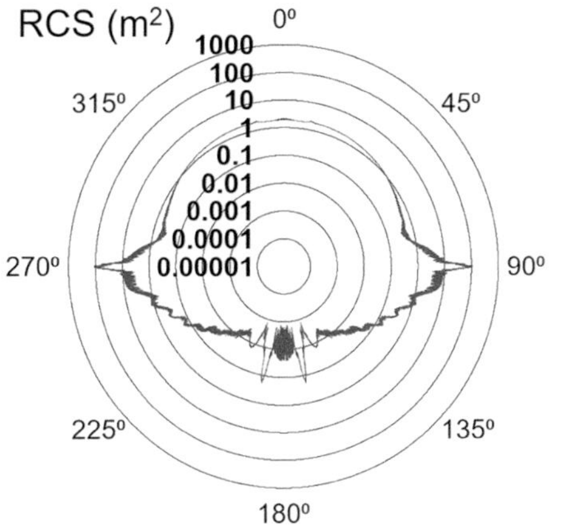

FIGURE 2.1064 Circular diagram of noncoherent RCS given radar observation of ADM-141C model at a carrier frequency of 3 GHz (10 cm wavelength).

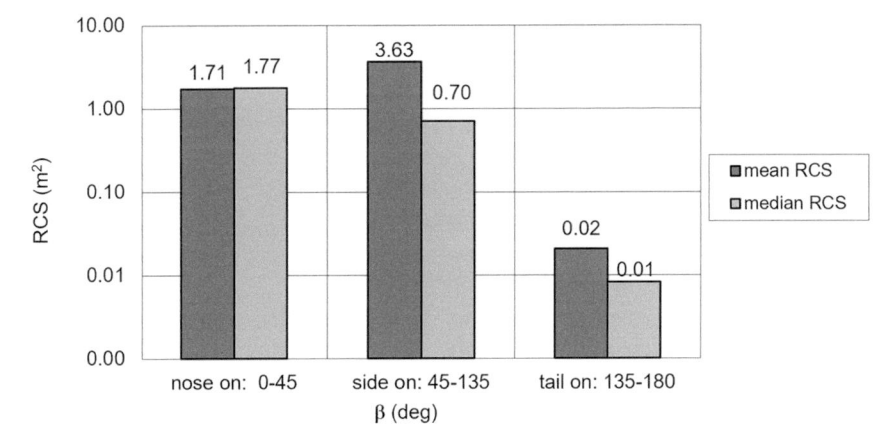

FIGURE 2.1065 Diagrams of mean and median RCS of ADM-141C model in three sectors of azimuth aspect given its radar observation at horizontal polarization and a carrier frequency of 3 GHz (10 cm wavelength).

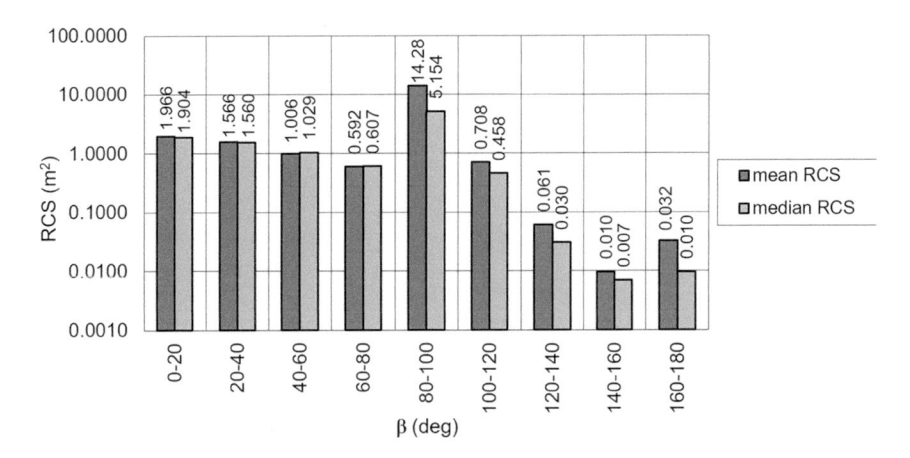

FIGURE 2.1066 Diagrams of mean and median RCS of ADM-141C model in 20-degree sectors of azimuth aspect given its radar observation at horizontal polarization and a carrier frequency of 3 GHz (10 cm wavelength).

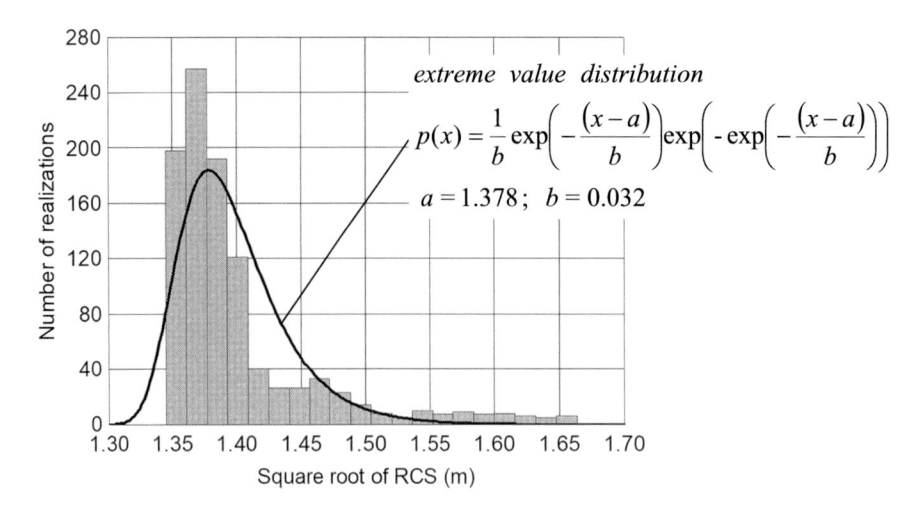

$$p(x) = \frac{1}{b} \exp\left(-\frac{(x-a)}{b}\right) \exp\left(-\exp\left(-\frac{(x-a)}{b}\right)\right)$$

extreme value distribution

$a = 1.378; \quad b = 0.032$

FIGURE 2.1067 Amplitude distribution of echo signal of ADM-141C model at a carrier frequency of 3 GHz given its horizontal polarization.

Scattering characteristics of ADM-141C model for sounding frequency 1 GHz (wavelength 30 cm).

Figure 2.1068 shows the RCS circular diagram of the ADM-141C. The noncoherent RCS circular diagram of the ADM-141C is shown in Figure 2.1069.

The circular mean RCS of the ADM-141C model for horizontal polarization is 1.910 m². The circular median RCS for horizontal polarization is 0.233 m².

Figures 2.1070 and 2.1071 show the mean and median RCS for the main ranges of sounding azimuths (nose, side, and tail) and for ranges of 20°.

The expressions and parameters of probability distributions that are most consistent with the empirical distributions of the square root of the RCS for different ranges of sounding azimuths and polarizations are given in the electronic appendix to this book. The expressions and parameters of the probability distributions that are most consistent with the empirical distributions of the RCS (energy characteristic) for different ranges of sounding azimuths and polarizations are also given there.

FIGURE 2.1069 Circular diagram of noncoherent RCS given radar observation of ADM-141C model at a carrier frequency of 1 GHz (30 cm wavelength).

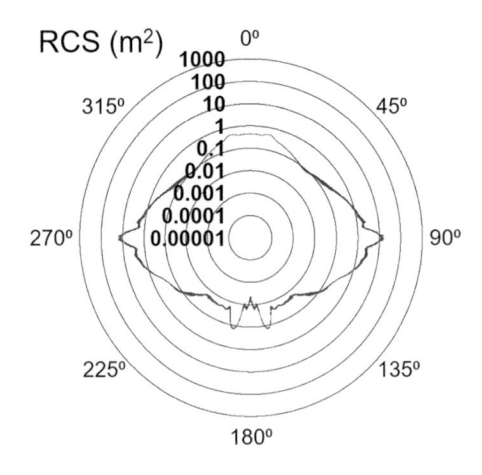

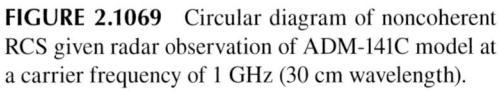

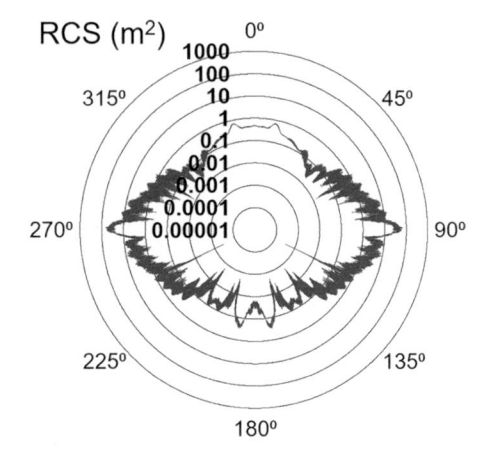

FIGURE 2.1068 Circular diagram of RCS given radar observation of ADM-141C model at a carrier frequency of 1 GHz (30 cm wavelength).

Scattering Characteristics of Aerial Objects

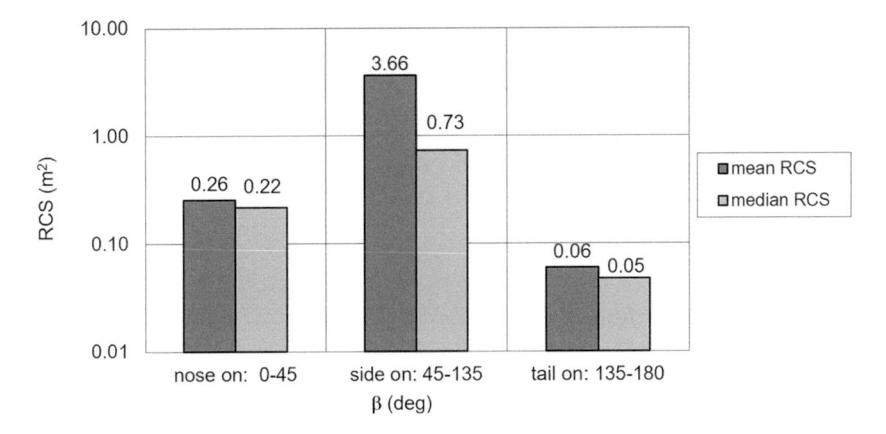

FIGURE 2.1070 Diagrams of mean and median RCS of ADM-141C model in three sectors of azimuth aspect given its radar observation at horizontal polarization and a carrier frequency of 1 GHz (30 cm wavelength).

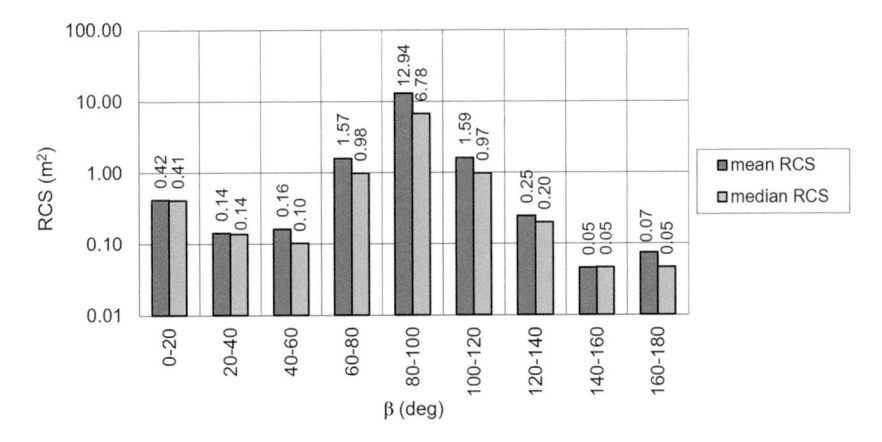

FIGURE 2.1071 Diagrams of mean and median RCS of ADM-141C model in 20-degree sectors of azimuth aspect given its radar observation at horizontal polarization and a carrier frequency of 1 GHz (30 cm wavelength).

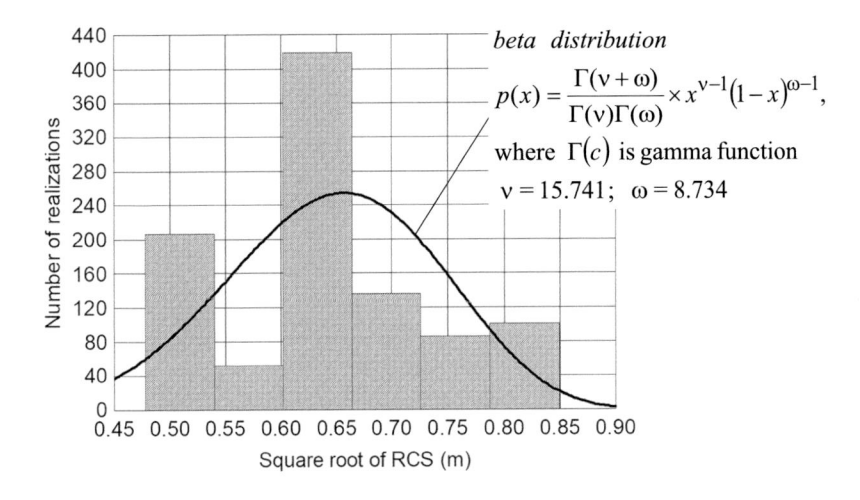

beta distribution

$$p(x) = \frac{\Gamma(\nu+\omega)}{\Gamma(\nu)\Gamma(\omega)} \times x^{\nu-1}(1-x)^{\omega-1},$$

where $\Gamma(c)$ is gamma function
$\nu = 15.741;\ \omega = 8.734$

FIGURE 2.1072 Amplitude distribution of echo signal of ADM-141C model at a carrier frequency of 1 GHz given its horizontal polarization.

3 Scattering Characteristics of Ground Objects

Based on the developed methods for calculating the radar characteristics of ground objects [1,2], mathematical modeling was performed, and data were obtained for 18 ground objects and one ship.

The scattering characteristics were obtained for the following main elevation sounding angles γ (Figure 3.1): 1° (sounding by ground radar systems); 10° and 30° (sounding by aerial radar systems). The step of change in the sounding azimuth was 1°, and the azimuth aspect angle β was calculated from the frontal view (0° corresponds to sounding the front of the object, 180° corresponds to sounding the tail of the object). The calculations have been carried out for three sounding frequencies: 10, 5, and 3 GHz (wavelengths of 3, 6, and 10 cm, respectively). Three variants of the underlying surface are considered: dry loam, dry soil, and wet soil (relative permittivity, respectively $\varepsilon' = 3 + j0,4$, $\varepsilon' = 7 + j0,1$, $\varepsilon' = 30 + j0,6$).

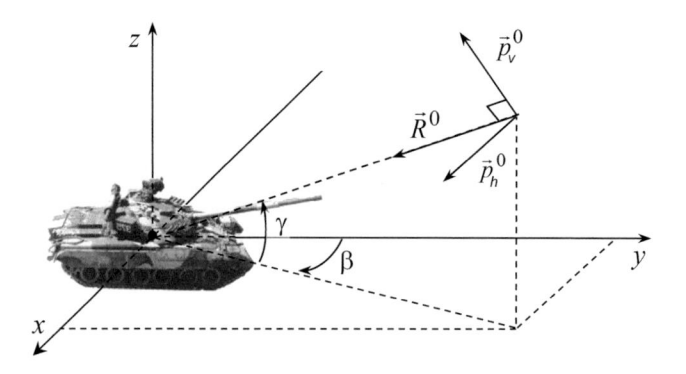

FIGURE 3.1 Scheme of sounding ground object.

The scattering characteristics are obtained for the case of monostatic reception in two orthogonal polarizations: horizontal – the electric field intensity vector of the incident wave $\vec{p}_h^0$ is parallel to underlying surface; and vertical – the electric field intensity vector of the incident wave $\vec{p}_v^0$ is orthogonal to $\vec{p}_h^0$ and lies in a plane that is perpendicular to the underlying surface and passes through the direction $\vec{R}^0$ of the incident plane wave (Figure 3.1).

Scattering Characteristics of Ground Objects

3.1 INFLUENCE OF THE UNDERLYING SURFACE ON THE SCATTERING CHARACTERISTICS OF A GROUND OBJECT

In addition to various factors that affect the scattering characteristics of a ground object, the underlying surface on which the object is placed plays an important role. In the book [1], it was shown that when a ground object is sounded by an electromagnetic wave with horizontal polarization, the highest level of its scattering will be observed for the underlying surface with wet soil parameters, and the lowest level is for the underlying surface with the parameters of dry loam. Approximately the average level of scattering between these types of underlying surface corresponds to the dry soil in the entire range of elevation angles of the sounding signal.

In connection with the above, it is interesting to study in more detail the scattering of electromagnetic waves for different types of the underlying surface under arbitrary probing conditions. This level can be described by the modulus of the reflection factor from the underlying surface |p|, which depends on the polarization of the electromagnetic wave, the elevation angle of the incidence probing signal, and the parameters of the underlying surface (Figs. 3.2 and 3.3).

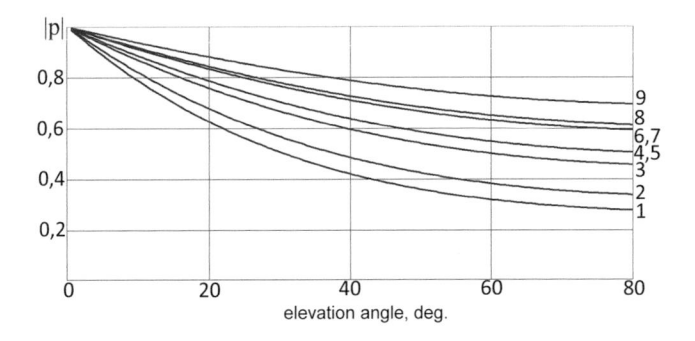

FIGURE 3.2 Modulus of reflection coefficient for horizontal polarization of illumination wave.

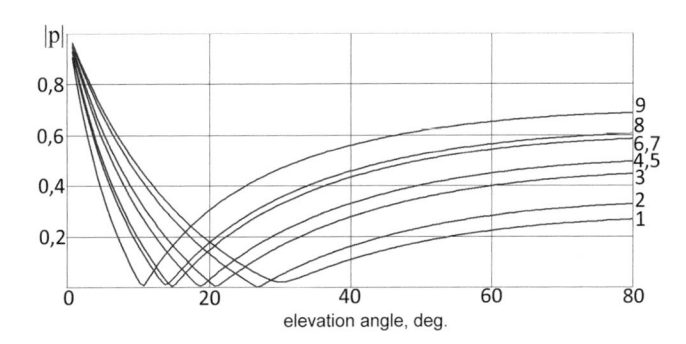

FIGURE 3.3 Modulus of reflection coefficient for vertical polarization of illumination wave.

Each of the lines in both figures corresponds to the |p| for one of the nine underlying surface types given as an example and marked with the corresponding numbers: 1 – dry loam, 2 – feldspar-quartz, 3 – dry soil, 4 – granite, 5 – sandstone, 6 – limestone-basalt, 7 – moist soil, 8 – moist 20% loam, and 9 – wet soil.

The analysis of Figure 3.2 shows that at the angles of the incidence probing signal close to 0°, |p| for different types of the underlying surface practically does not differ, and therefore the level of scattering of the object at these angles can be considered the same regardless of the parameters of the underlying surface on which it is placed. As the angle of incidence probing signal increases, there is an increasing mutual difference in the values of |p| between different types of the underlying surface.

In the case of vertical polarization of the probing signal (Figure 3.3), the behavior of |ρ| is significantly different from that observed with horizontal polarization of the signal. For example, |ρ| for wet soil has the smallest values compared to all other reflection coefficient modules in the range of site angles from 0° to approximately 12°. From the angle of 12°, the values for wet soil gradually exceed the values for other surface types, and starting from the angle of about 17°, the wet soil has the highest values, i.e., the level of scattering from an object placed on the wet soil will be the highest compared to other underlying surfaces.

3.2 MAIN BATTLE TANK T-90

Main battle tank T-90 (Figure 3.4) is the last modification of tank T-72. It has been in service since 1993. General characteristics of T-90 [75]: length with a gun – 9.53 m, width – 3.46 m, height – 2.23 m, and weight – 46.5 t.

In accordance with the design of the T-90, a model of its surface was created to obtain scattering characteristics (in particular, RCS). The model is shown in Figure 3.5. In the modeling, the smooth parts of the tank surface were approximated by parts of 89 triaxial ellipsoids. The surface breaks were modeled using 34 straight edge scattering parts.

FIGURE 3.4 Main battle tank T-90.

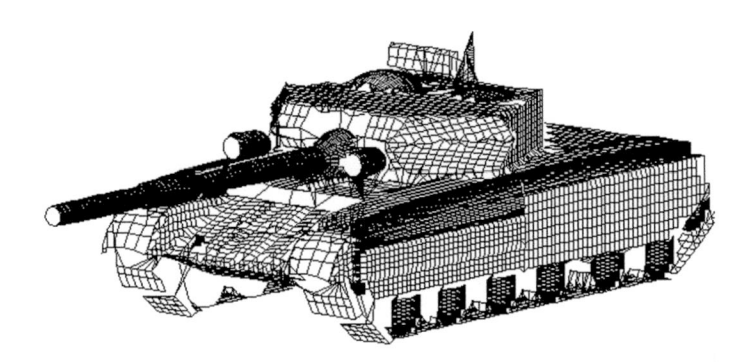

FIGURE 3.5 Surface model of main battle tank T-90.

Below are some scattering characteristics of the tank T-90 model at a sounding frequency of 10 GHz (wavelengths of 3 cm) for the following conditions: the elevation angle of sounding $\gamma = 10°$, wet soil, and the elevation angle of sounding $\gamma = 30°$, dry loam.

Scattering characteristics of the tank T-90 model for sounding frequency 10 GHz, elevation angle 10°, wet soil, and horizontal polarization.

Figure 3.6 shows the RCS circular diagram of the tank T-90 model. The noncoherent RCS circular diagram of the tank T-90 is presented in Figure 3.7. The circular mean RCS of the T-90 tank model is 6 057.88 m². The circular median RCS is 20.71 m².

Figures 3.8 and 3.9 show the mean and median RCS for the main ranges of sounding azimuths (head, side, and stern) and for ranges of 20°.

394   Scattering Characteristics of Aerial and Ground Radar Objects

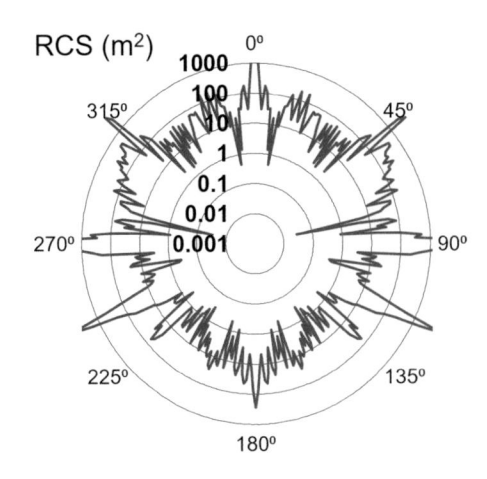

FIGURE 3.6 Circular diagrams of RCS given radar observation of T-90 tank model (sounding frequency 10 GHz, $\gamma = 10°$, underlying surface – wet soil, horizontal polarization).

FIGURE 3.7 Circular diagram of noncoherent RCS given radar observation of T-90 tank model (sounding frequency 10 GHz, $\gamma = 10°$, underlying surface – wet soil, horizontal polarization).

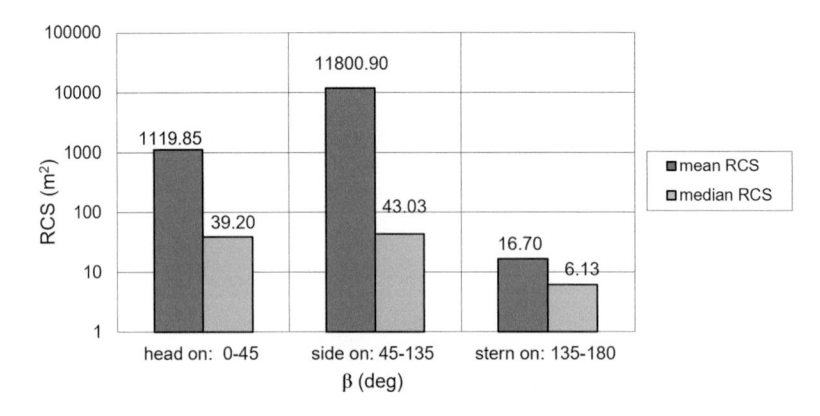

FIGURE 3.8 Diagrams of mean and median RCS of T-90 tank model in three sectors of azimuth aspect given its radar observation (sounding frequency 10 GHz, $\gamma = 10°$, underlying surface – wet soil, horizontal polarization).

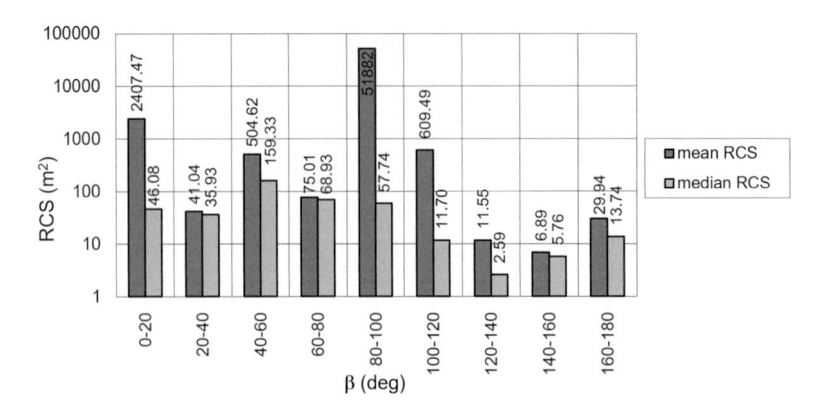

FIGURE 3.9 Diagrams of mean and median RCS of T-90 tank model in 20-degree sectors of azimuth aspect given its radar observation (sounding frequency 10 GHz, $\gamma = 10°$, underlying surface – wet soil, horizontal polarization).

Scattering characteristics of the tank T-90 model for sounding frequency 10 GHz, elevation angle 10°, wet soil, and vertical polarization.

Figure 3.10 shows the RCS circular diagram of the tank T-90 model. The noncoherent RCS circular diagram of the tank T-90 is presented in Figure 3.11.

The circular mean RCS of the T-90 tank model is $50.40 \, m^2$. The circular median RCS is $4.76 \, m^2$.

Figures 3.12 and 3.13 show the mean and median RCS for the main ranges of sounding azimuths (head, side, and stern) and for ranges of 20°.

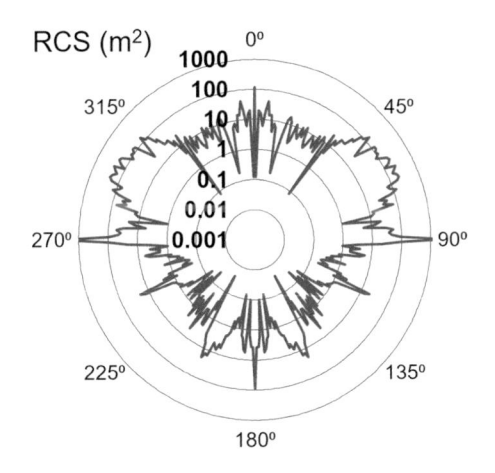

FIGURE 3.10 Circular diagrams of RCS given radar observation of T-90 tank model (sounding frequency 10 GHz, $\gamma = 10°$, underlying surface – wet soil, vertical polarization).

FIGURE 3.11 Circular diagram of noncoherent RCS given radar observation of T-90 tank model (sounding frequency 10 GHz, $\gamma = 10°$, underlying surface – wet soil, vertical polarization).

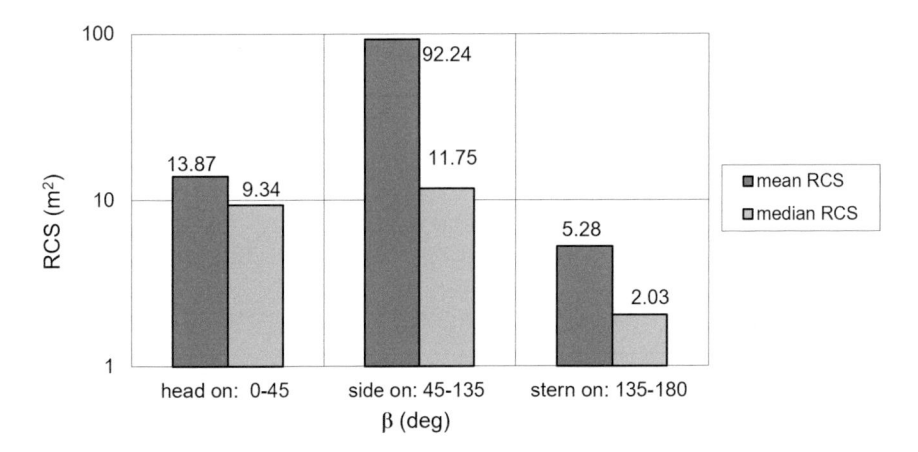

FIGURE 3.12 Diagrams of mean and median RCS of T-90 tank model in three sectors of azimuth aspect given its radar observation (sounding frequency 10 GHz, $\gamma = 10°$, underlying surface – wet soil, vertical polarization).

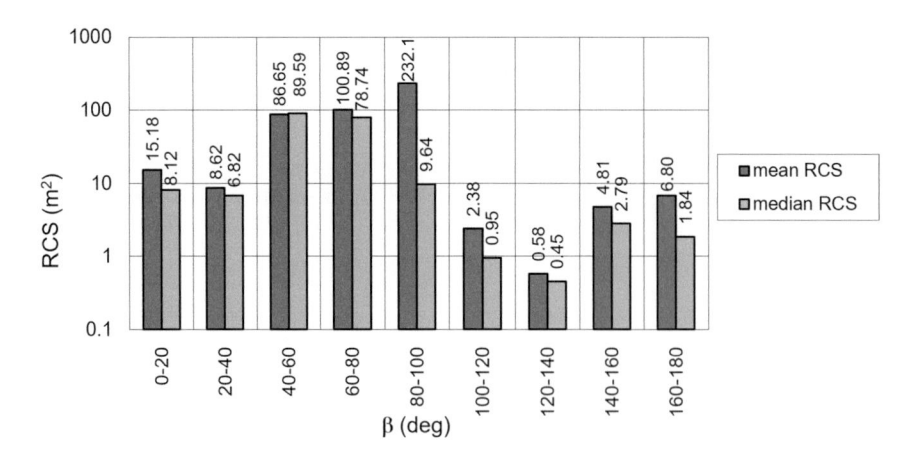

FIGURE 3.13 Diagrams of mean and median RCS of T-90 tank model in 20-degree sectors of azimuth aspect given its radar observation (sounding frequency 10 GHz, $\gamma = 10°$, underlying surface – wet soil, vertical polarization).

Scattering characteristics of the tank T-90 model for sounding frequency 10 GHz, elevation angle 30°, dry loam, and horizontal polarization.

Figure 3.14 shows the RCS circular diagram of the tank T-90 model. The noncoherent RCS circular diagram of the tank T-90 is presented in Figure 3.15. The circular mean RCS of the T-90 tank model is 1 261.44 m². The circular median RCS is 3.97 m².

Figures 3.16 and 3.17 show the mean and median RCS for the main ranges of sounding azimuths (head, side, and stern) and for ranges of 20°.

FIGURE 3.14 Circular diagrams of RCS given radar observation of T-90 tank model (sounding frequency 10 GHz, $\gamma = 30°$, underlying surface – dry loam, horizontal polarization).

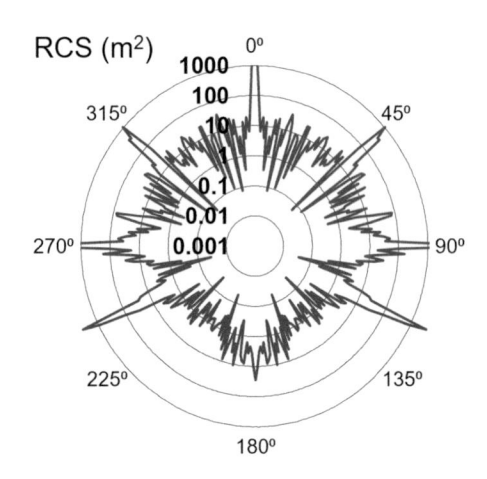

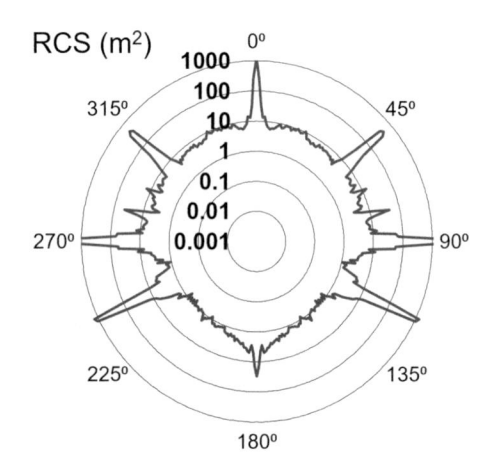

FIGURE 3.15 Circular diagram of noncoherent RCS given radar observation of T-90 tank model (sounding frequency 10 GHz, $\gamma = 30°$, underlying surface – dry loam, horizontal polarization).

Scattering Characteristics of Ground Objects 397

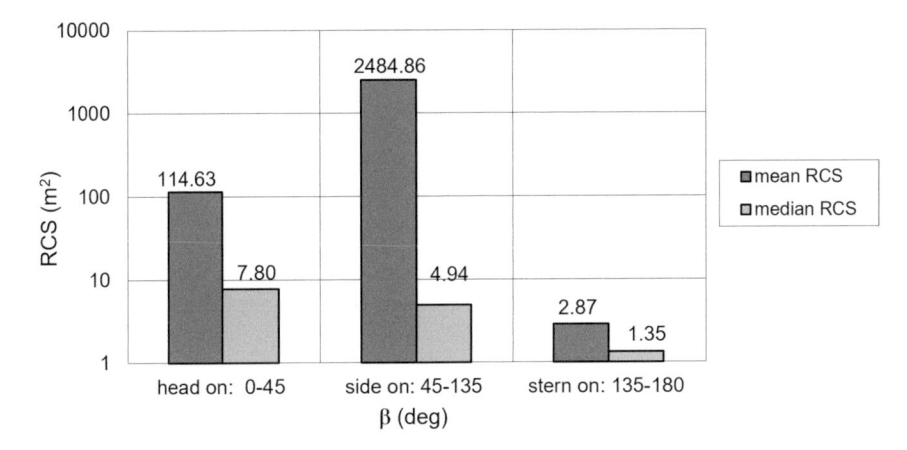

FIGURE 3.16 Diagrams of mean and median RCS of T-90 tank model in three sectors of azimuth aspect given its radar observation (sounding frequency 10 GHz, $\gamma=30°$, underlying surface – dry loam, horizontal polarization).

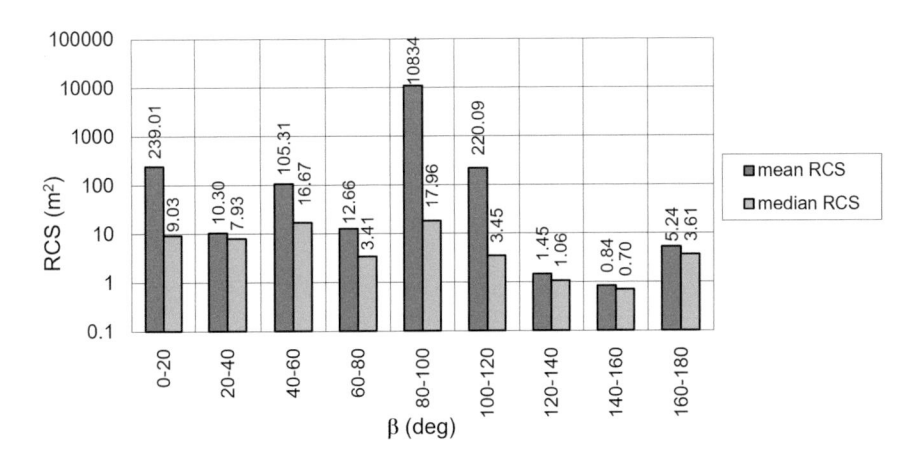

FIGURE 3.17 Diagrams of mean and median RCS of T-90 tank model in 20-degree sectors of azimuth aspect given its radar observation (sounding frequency 10 GHz, $\gamma=30°$, underlying surface – dry loam, horizontal polarization).

Scattering characteristics of the tank T-90 model for sounding frequency 10 GHz, elevation angle 30°, dry loam, and vertical polarization.

Figure 3.18 shows the RCS circular diagram of the tank T-90 model. The noncoherent RCS circular diagram of the tank T-90 is presented in Figure 3.19. The circular mean RCS of the T-90 tank model is 43.36 m². The circular median RCS is 2.76 m².

Figures 3.20 and 3.21 show the mean and median RCS for the main ranges of sounding azimuths (head, side, and stern) and for ranges of 20°.

The scattering characteristics for this model at two polarizations, sounding frequencies of 10, 5, and 3 GHz (wavelengths of 3, 6, and 10 cm, respectively), for three types of underlying surface (dry loam, dry soil, and wet soil) are given in the electronic appendix of this book.

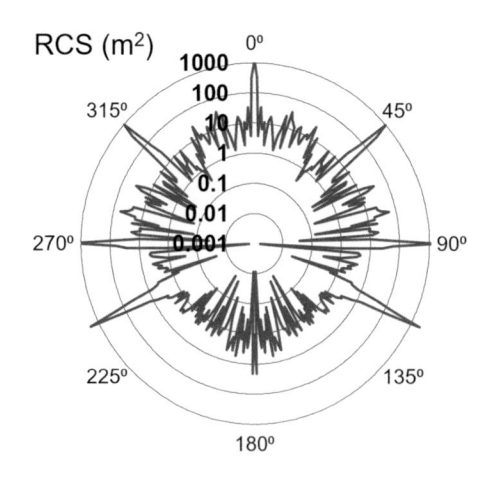

FIGURE 3.18 Circular diagrams of RCS given radar observation of T-90 tank model (sounding frequency 10 GHz, $\gamma=30°$, underlying surface – dry loam, vertical polarization).

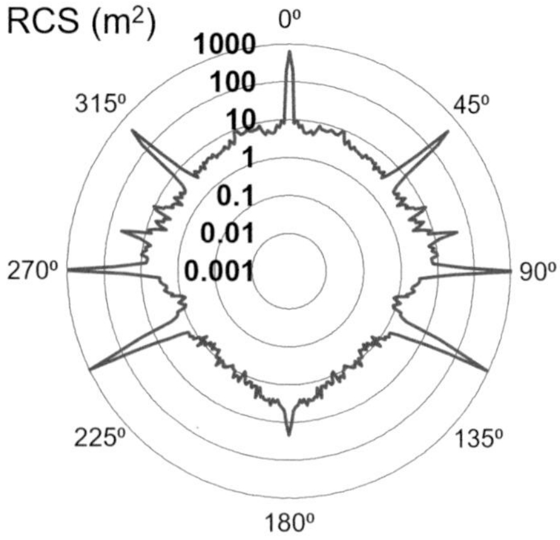

FIGURE 3.19 Circular diagram of noncoherent RCS given radar observation of T-90 tank model (sounding frequency 10 GHz, $\gamma=30°$, underlying surface – dry loam, vertical polarization).

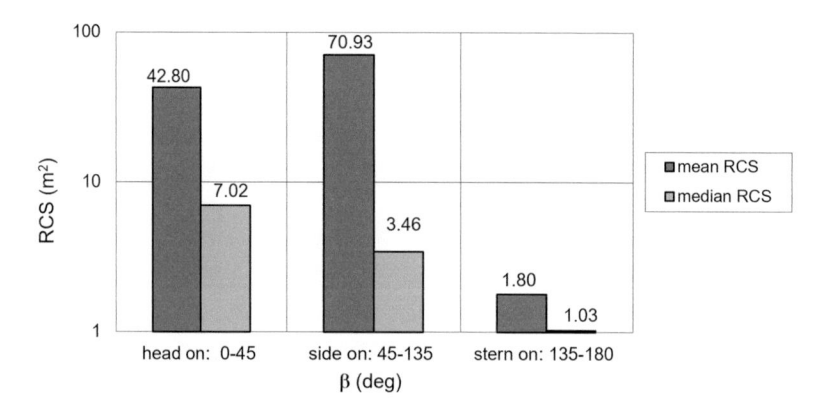

FIGURE 3.20 Diagrams of mean and median RCS of T-90 tank model in three sectors of azimuth aspect given its radar observation (sounding frequency 10 GHz, $\gamma=30°$, underlying surface – dry loam, vertical polarization).

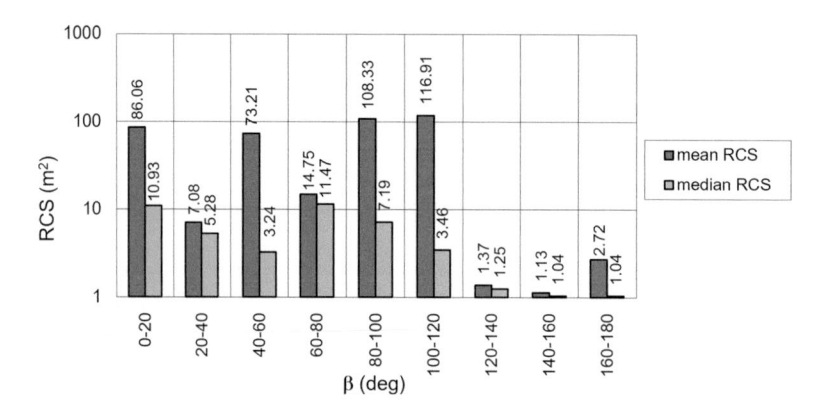

FIGURE 3.21 Diagrams of mean and median RCS of T-90 tank model in 20-degree sectors of azimuth aspect given its radar observation (sounding frequency 10 GHz, $\gamma=30°$, underlying surface – dry loam, vertical polarization).

Scattering Characteristics of Ground Objects

3.3 MAIN BATTLE TANK LEOPARD-2

Main battle tank Leopard-2 (Figure 3.22) is one of the most successful projects of the latest generation of main battle tanks. The total number of manufactured tanks is 3 200 units. It has been in service since 1979. General characteristics of Leopard-2 A4 [76]: length with a gun – 9.67 m, width – 3.70 m, height – 2.64 m, and weight – 55.15 t.

In accordance with the design of the Leopard-2 A4, a model of its surface was created to obtain scattering characteristics (in particular, RCS). The model is shown in Figure 3.23. In the modeling, the smooth parts of the tank surface were approximated by parts of 57 triaxial ellipsoids. The surface breaks were modeled using 24 straight edge scattering parts.

FIGURE 3.22 Main battle tank Leopard-2 A4.

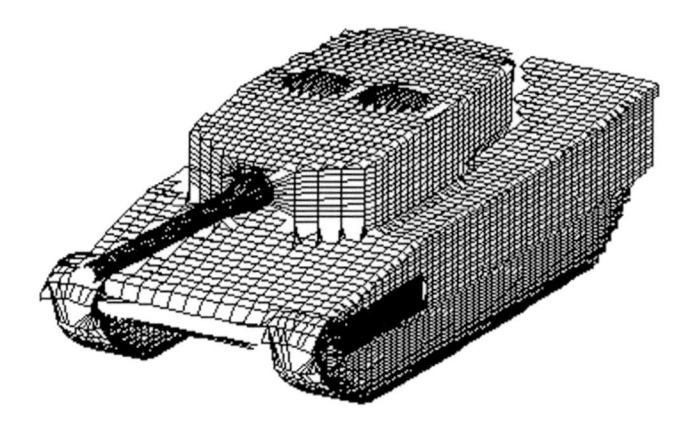

FIGURE 3.23 Surface model of main battle tank Leopard-2 A4.

Below are some scattering characteristics of the tank Leopard-2 model at a sounding frequency of 10 GHz (wavelengths of 3 cm) for the following conditions: the elevation angle of sounding $\gamma = 10°$, wet soil, and the elevation angle of sounding $\gamma = 30°$, dry loam.

Scattering characteristics of the tank Leopard-2 model for sounding frequency 10 GHz, elevation angle 10°, wet soil, and horizontal polarization.

Figure 3.24 shows the RCS circular diagram of the tank Leopard-2 model. The noncoherent RCS circular diagram of the tank Leopard-2 is presented in Figure 3.25. The circular mean RCS of the Leopard-2 tank model is 3 212 m². The circular median RCS is 7.38 m².

Figures 3.26 and 3.27 show the mean and median RCS for the main ranges of sounding azimuths (head, side, and stern) and for ranges of 20°.

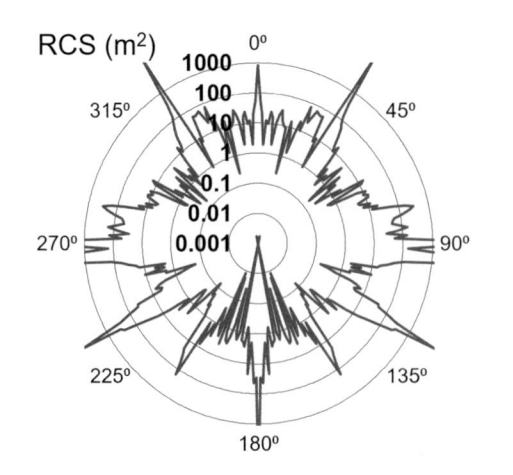

FIGURE 3.24 Circular diagrams of RCS given radar observation of Leopard-2 tank model (sounding frequency 10 GHz, $\gamma = 10°$, underlying surface – wet soil, horizontal polarization).

FIGURE 3.25 Circular diagram of noncoherent RCS given radar observation of Leopard-2 tank model (sounding frequency 10 GHz, $\gamma = 10°$, underlying surface – wet soil, horizontal polarization).

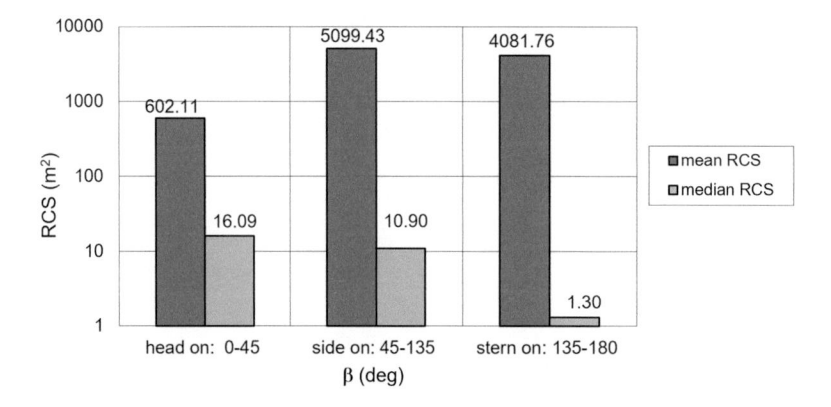

FIGURE 3.26 Diagrams of mean and median RCS of Leopard-2 tank model in three sectors of azimuth aspect given its radar observation (sounding frequency 10 GHz, $\gamma = 10°$, underlying surface – wet soil, horizontal polarization).

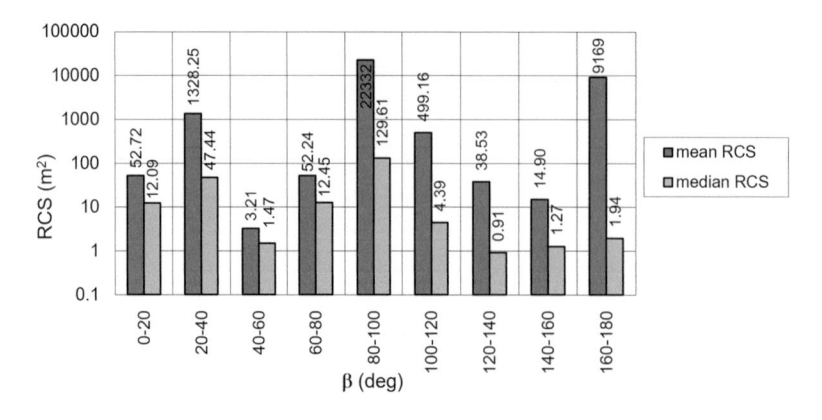

FIGURE 3.27 Diagrams of mean and median RCS of Leopard-2 tank model in 20-degree sectors of azimuth aspect given its radar observation (sounding frequency 10 GHz, $\gamma = 10°$, underlying surface – wet soil, horizontal polarization).

Scattering Characteristics of Ground Objects   401

Scattering characteristics of the tank Leopard-2 model for sounding frequency 10 GHz, elevation angle 10°, wet soil, and vertical polarization.

Figure 3.28 shows the RCS circular diagram of the tank Leopard-2 model. The noncoherent RCS circular diagram of the tank Leopard-2 is presented in Figure 3.29.

The circular mean RCS of the Leopard-2 tank model is 14.67 m². The circular median RCS is 0.61 m².

Figures 3.30 and 3.31 show the mean and median RCS for the main ranges of sounding azimuths (head, side, and stern) and for ranges of 20°.

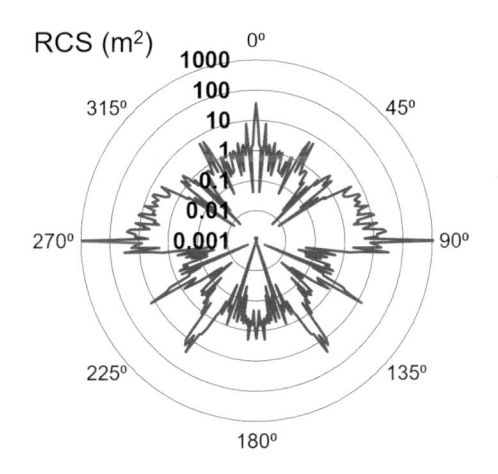

FIGURE 3.28 Circular diagrams of RCS given radar observation of Leopard-2 tank model (sounding frequency 10 GHz, $\gamma = 10°$, underlying surface – wet soil, vertical polarization).

FIGURE 3.29 Circular diagram of noncoherent RCS given radar observation of Leopard-2 tank model (sounding frequency 10 GHz, $\gamma = 10°$, underlying surface – wet soil, vertical polarization).

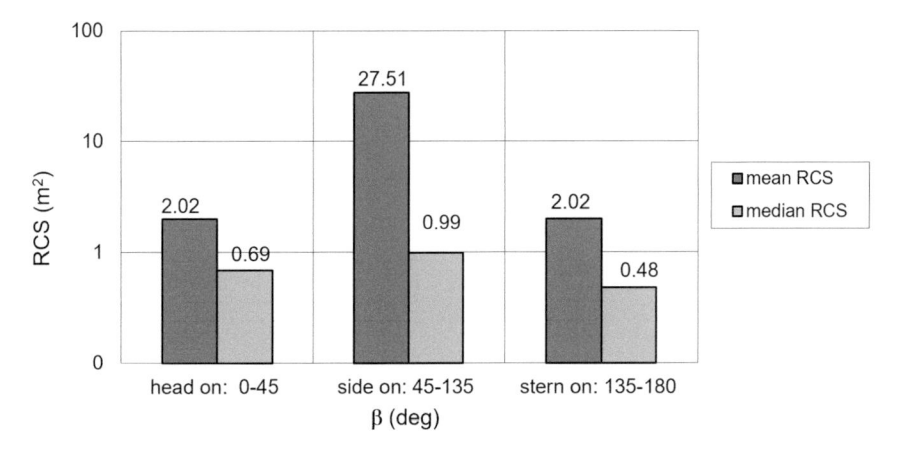

FIGURE 3.30 Diagrams of mean and median RCS of Leopard-2 tank model in three sectors of azimuth aspect given its radar observation (sounding frequency 10 GHz, $\gamma = 10°$, underlying surface – wet soil, vertical polarization).

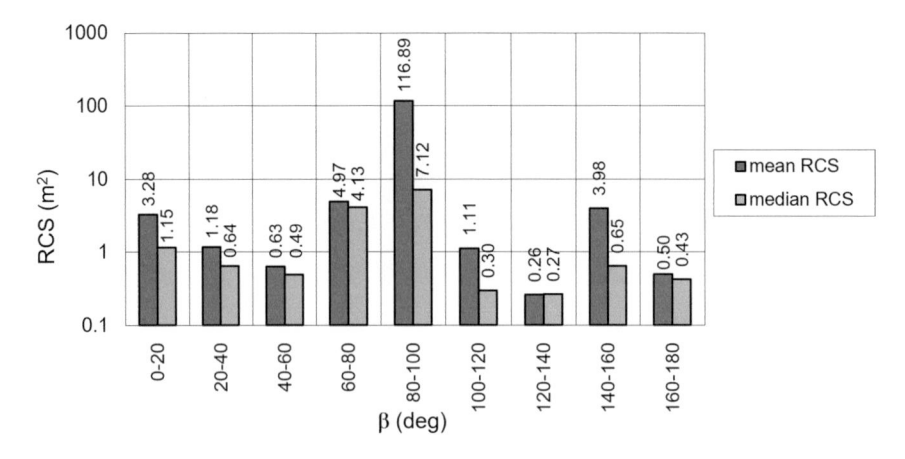

FIGURE 3.31 Diagrams of mean and median RCS of Leopard-2 tank model in 20-degree sectors of azimuth aspect given its radar observation (sounding frequency 10 GHz, $\gamma = 10°$, underlying surface – wet soil, vertical polarization).

Scattering characteristics of the tank Leopard-2 model for sounding frequency 10 GHz, elevation angle 30°, dry loam, and horizontal polarization.

Figure 3.32 shows the RCS circular diagram of the tank Leopard-2 model. The noncoherent RCS circular diagram of the tank Leopard-2 is presented in Figure 3.33. The circular mean RCS of the Leopard-2 tank model is 4 131 m². The circular median RCS is 1.64 m².

Figures 3.34 and 3.35 show the mean and median RCS for the main ranges of sounding azimuths (head, side, and stern) and for ranges of 20°.

FIGURE 3.32 Circular diagrams of RCS given radar observation of Leopard-2 tank model (sounding frequency 10 GHz, $\gamma = 30°$, underlying surface – dry loam, horizontal polarization).

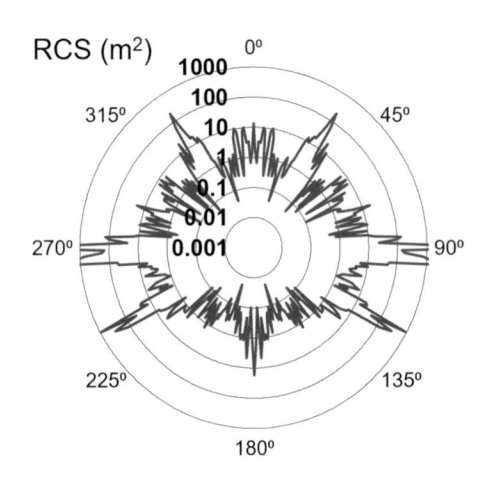

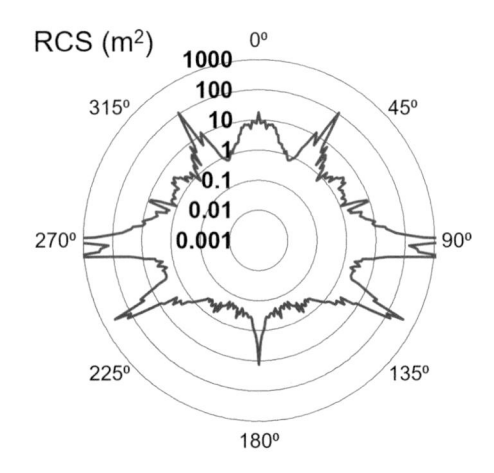

FIGURE 3.33 Circular diagram of noncoherent RCS given radar observation of Leopard-2 tank model (sounding frequency 10 GHz, $\gamma = 30°$, underlying surface – dry loam, horizontal polarization).

Scattering Characteristics of Ground Objects

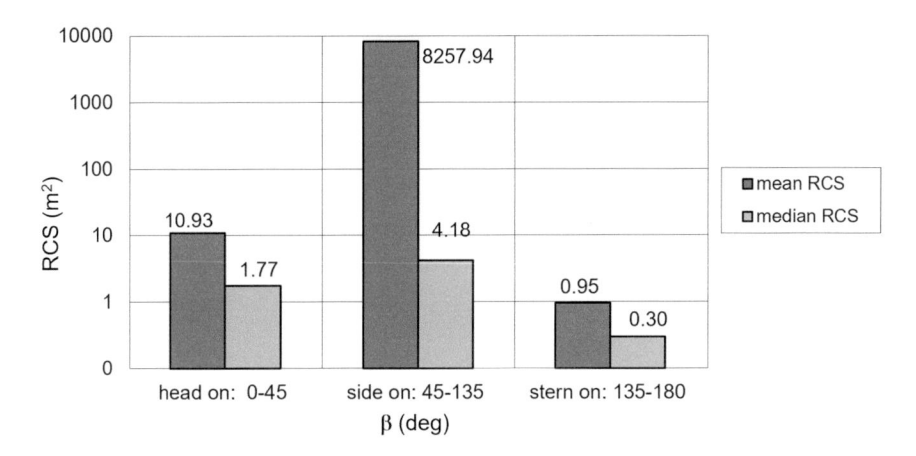

FIGURE 3.34 Diagrams of mean and median RCS of Leopard-2 tank model in three sectors of azimuth aspect given its radar observation (sounding frequency 10 GHz, $\gamma = 30°$, underlying surface – dry loam, horizontal polarization).

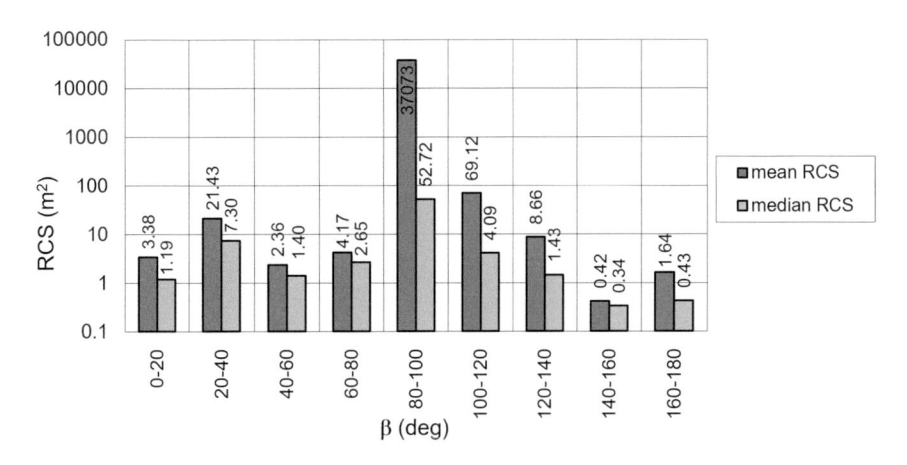

FIGURE 3.35 Diagrams of mean and median RCS of Leopard-2 tank model in 20-degree sectors of azimuth aspect given its radar observation (sounding frequency 10 GHz, $\gamma = 30°$, underlying surface – dry loam, horizontal polarization).

Scattering characteristics of the tank Leopard-2 model for sounding frequency 10 GHz, elevation angle 30°, dry loam, and vertical polarization.

Figure 3.36 shows the RCS circular diagram of the tank Leopard-2 model. The noncoherent RCS circular diagram of the tank Leopard-2 is presented in Figure 3.37. The circular mean RCS of the Leopard-2 tank model is 143.48 m². The circular median RCS is 0.80 m².

Figures 3.38 and 3.39 show the mean and median RCS for the main ranges of sounding azimuths (head, side, and stern) and for ranges of 20°.

The scattering characteristics for this model at two polarizations, sounding frequencies of 10, 5, and 3 GHz (wavelengths of 3, 6, and 10 cm, respectively), for three types of underlying surface (dry loam, dry soil, and wet soil) are given in the electronic appendix of this book.

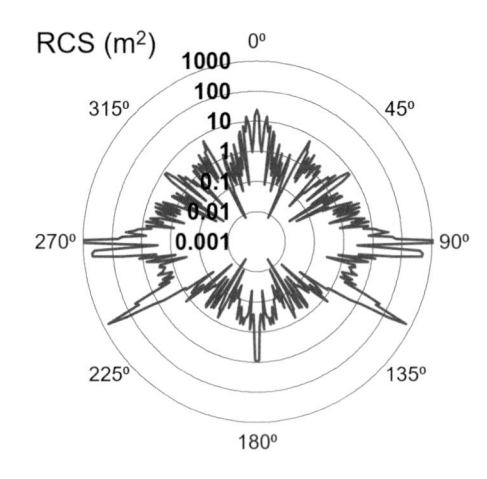

FIGURE 3.36 Circular diagrams of RCS given radar observation of Leopard-2 tank model (sounding frequency 10 GHz, $\gamma=30°$, underlying surface – dry loam, vertical polarization).

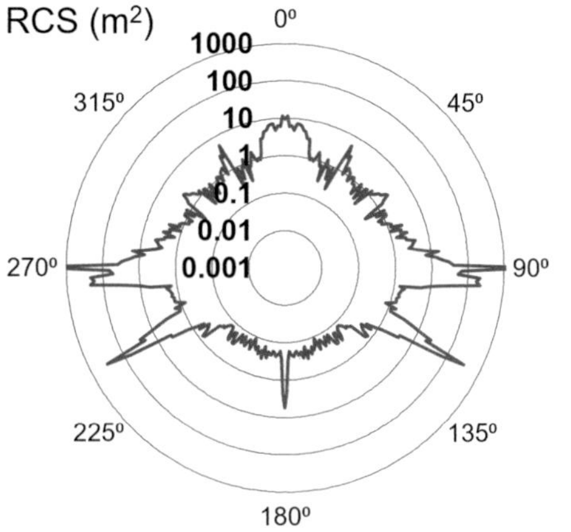

FIGURE 3.37 Circular diagram of noncoherent RCS given radar observation of Leopard-2 tank model (sounding frequency 10 GHz, $\gamma=30°$, underlying surface – dry loam, vertical polarization).

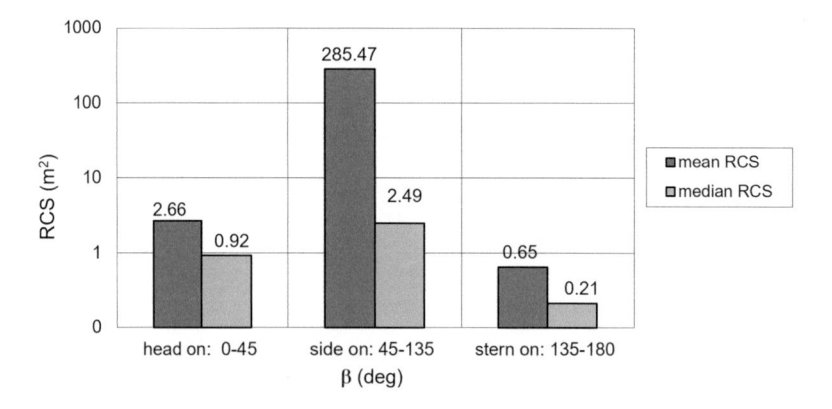

FIGURE 3.38 Diagrams of mean and median RCS of Leopard-2 tank model in three sectors of azimuth aspect given its radar observation (sounding frequency 10 GHz, $\gamma=30°$, underlying surface – dry loam, vertical polarization).

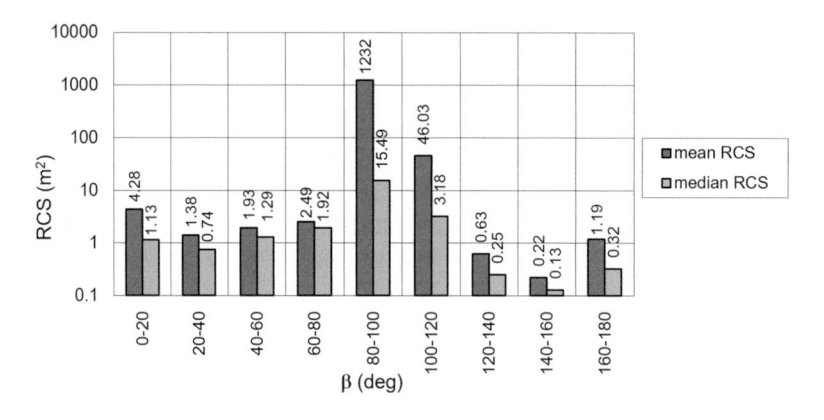

FIGURE 3.39 Diagrams of mean and median RCS of Leopard-2 tank model in 20-degree sectors of azimuth aspect given its radar observation (sounding frequency 10 GHz, $\gamma=30°$, underlying surface – dry loam, vertical polarization).

Scattering Characteristics of Ground Objects

3.4 MAIN BATTLE TANK M1 ABRAMS

M1 Abrams (Figure 3.40) is a third-generation American main battle tank. The total number of manufactured tanks is 10 000 units. It has been in service since 1981. General characteristics of M1 Abrams [77]: length with a gun – 9.83 m, width – 3.65 m, height – 2.44 m, weight (M1A1) – 57.15 t.

In accordance with the design of the M1 Abrams, a model of its surface was created to obtain scattering characteristics (in particular, RCS). The model is shown in Figure 3.41. In the modeling, the smooth parts of the tank surface were approximated by parts of 53 triaxial ellipsoids. The surface breaks were modeled using 22 straight edge scattering parts.

FIGURE 3.40 Main battle tank M1 Abrams.

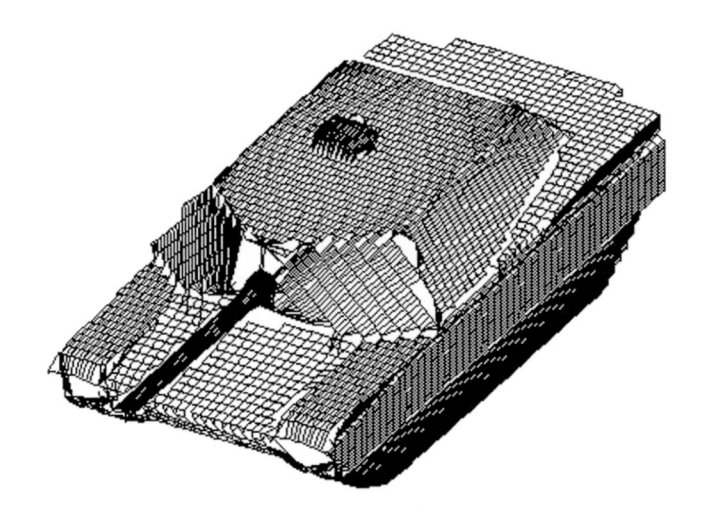

FIGURE 3.41 Surface model of main battle tank M1 Abrams.

Below are some scattering characteristics of the tank M1 Abrams model at a sounding frequency of 10 GHz (wavelengths of 3 cm) for the following conditions: the elevation angle of sounding $\gamma = 10°$, wet soil, and the elevation angle of sounding $\gamma = 30°$, dry loam.

Scattering characteristics of the tank M1 Abrams model for sounding frequency 10 GHz, elevation angle 10°, wet soil, and horizontal polarization.

Figure 3.42 shows the RCS circular diagram of the tank M1 Abrams model. The noncoherent RCS circular diagram of the tank M1 Abrams is presented in Figure 3.43. The circular mean RCS of the M1 Abrams tank model is 8 339 m². The circular median RCS is 9.31 m².

Figures 3.44 and 3.45 show the mean and median RCS for the main ranges of sounding azimuths (head, side, and stern) and for ranges of 20°.

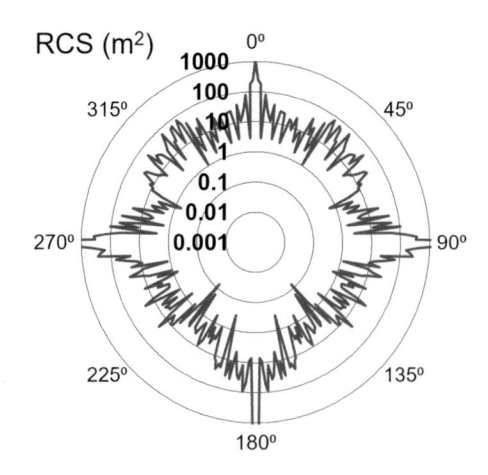

FIGURE 3.42 Circular diagrams of RCS given radar observation of M1 Abrams tank model (sounding frequency 10 GHz, $\gamma = 10°$, underlying surface – wet soil, horizontal polarization).

FIGURE 3.43 Circular diagram of noncoherent RCS given radar observation of M1 Abrams tank model (sounding frequency 10 GHz, $\gamma = 10°$, underlying surface – wet soil, horizontal polarization).

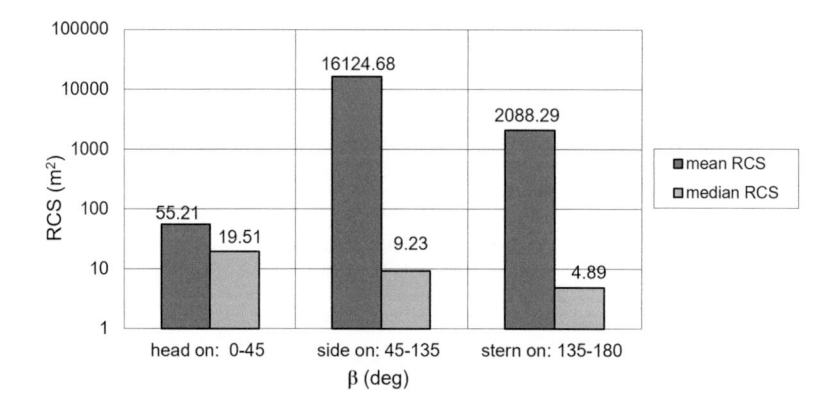

FIGURE 3.44 Diagrams of mean and median RCS of M1 Abrams tank model in three sectors of azimuth aspect given its radar observation (sounding frequency 10 GHz, $\gamma = 10°$, underlying surface – wet soil, horizontal polarization).

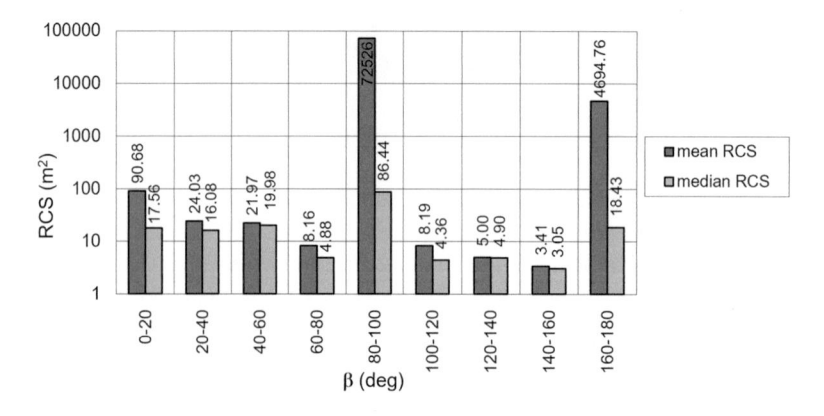

FIGURE 3.45 Diagrams of mean and median RCS of M1 Abrams tank model in 20-degree sectors of azimuth aspect given its radar observation (sounding frequency 10 GHz, $\gamma = 10°$, underlying surface – wet soil, horizontal polarization).

Scattering characteristics of the tank M1 Abrams model for sounding frequency 10 GHz, elevation angle 10°, wet soil, and vertical polarization.

Figure 3.46 shows the RCS circular diagram of the tank M1 Abrams model. The noncoherent RCS circular diagram of the tank M1 Abrams is presented in Figure 3.47.

The circular mean RCS of the M1 Abrams tank model is $10.92\,m^2$. The circular median RCS is $0.73\,m^2$.

Figures 3.48 and 3.49 show the mean and median RCS for the main ranges of sounding azimuths (head, side, and stern) and for ranges of 20°.

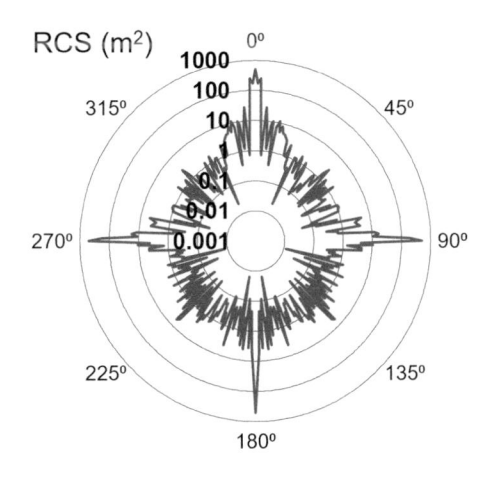

FIGURE 3.46 Circular diagrams of RCS given radar observation of M1 Abrams tank model (sounding frequency 10 GHz, $\gamma = 10°$, underlying surface – wet soil, vertical polarization).

FIGURE 3.47 Circular diagram of noncoherent RCS given radar observation of M1 Abrams tank model (sounding frequency 10 GHz, $\gamma = 10°$, underlying surface – wet soil, vertical polarization).

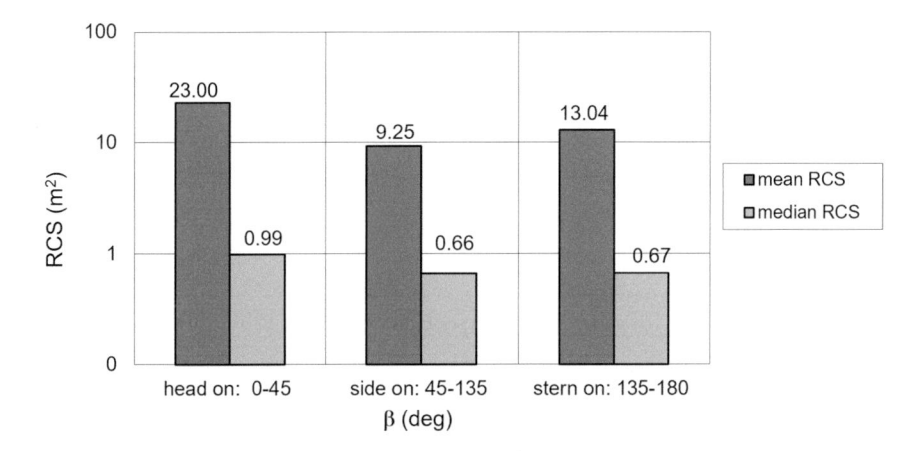

FIGURE 3.48 Diagrams of mean and median RCS of M1 Abrams tank model in three sectors of azimuth aspect given its radar observation (sounding frequency 10 GHz, $\gamma = 10°$, underlying surface – wet soil, vertical polarization).

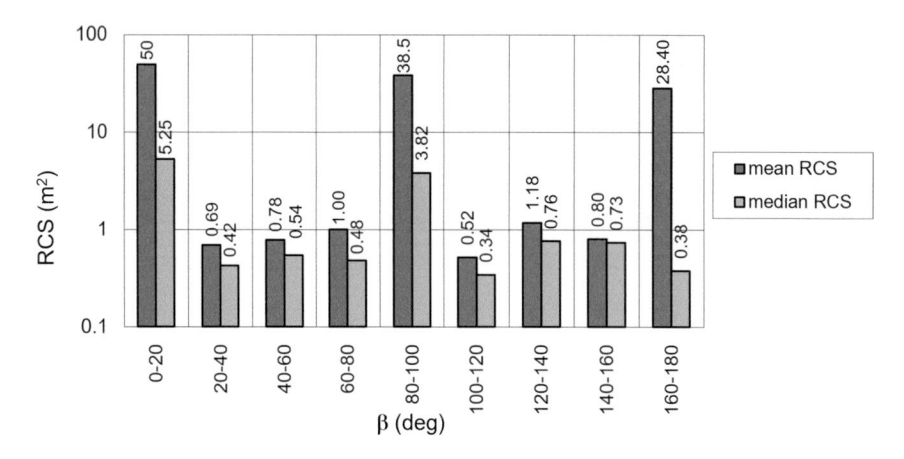

FIGURE 3.49 Diagrams of mean and median RCS of M1 Abrams tank model in 20-degree sectors of azimuth aspect given its radar observation (sounding frequency 10 GHz, $\gamma = 10°$, underlying surface – wet soil, vertical polarization).

Scattering characteristics of the tank M1 Abrams model for sounding frequency 10 GHz, elevation angle 30°, dry loam, and horizontal polarization.

Figure 3.50 shows the RCS circular diagram of the tank M1 Abrams model. The noncoherent RCS circular diagram of the tank M1 Abrams is presented in Figure 3.51. The circular mean RCS of the M1 Abrams tank model is 3 723 m². The circular median RCS is 2.30 m².

Figures 3.52 and 3.53 show the mean and median RCS for the main ranges of sounding azimuths (head, side, and stern) and for ranges of 20°.

FIGURE 3.50 Circular diagrams of RCS given radar observation of M1 Abrams tank model (sounding frequency 10 GHz, $\gamma = 30°$, underlying surface – dry loam, horizontal polarization).

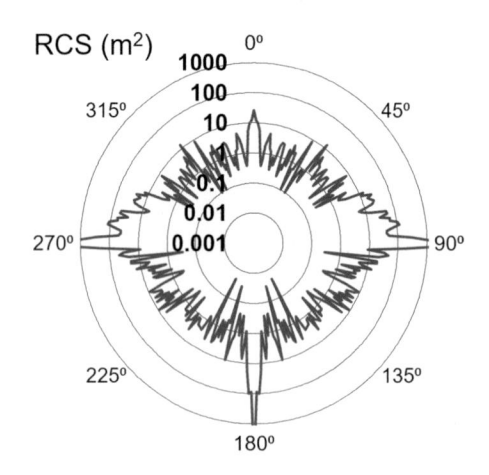

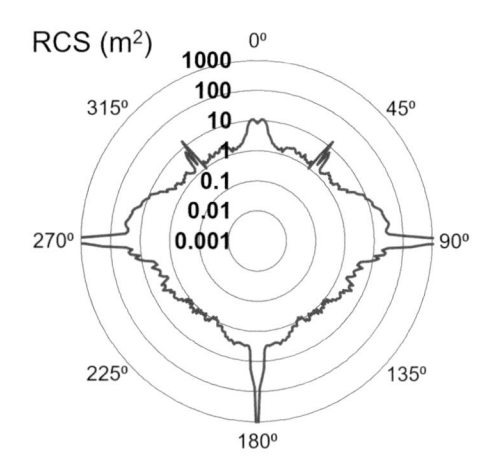

FIGURE 3.51 Circular diagram of noncoherent RCS given radar observation of M1 Abrams tank model (sounding frequency 10 GHz, $\gamma = 30°$, underlying surface – dry loam, horizontal polarization).

Scattering Characteristics of Ground Objects

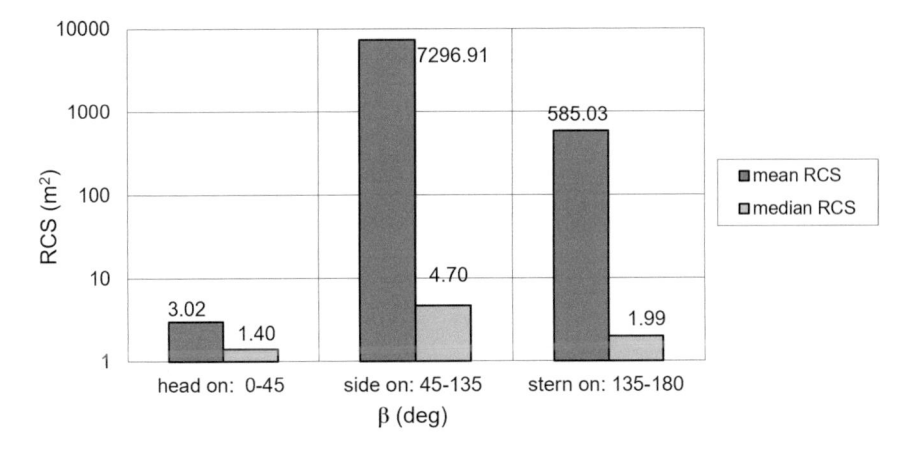

FIGURE 3.52 Diagrams of mean and median RCS of M1 Abrams tank model in three sectors of azimuth aspect given its radar observation (sounding frequency 10 GHz, $\gamma = 30°$, underlying surface – dry loam, horizontal polarization).

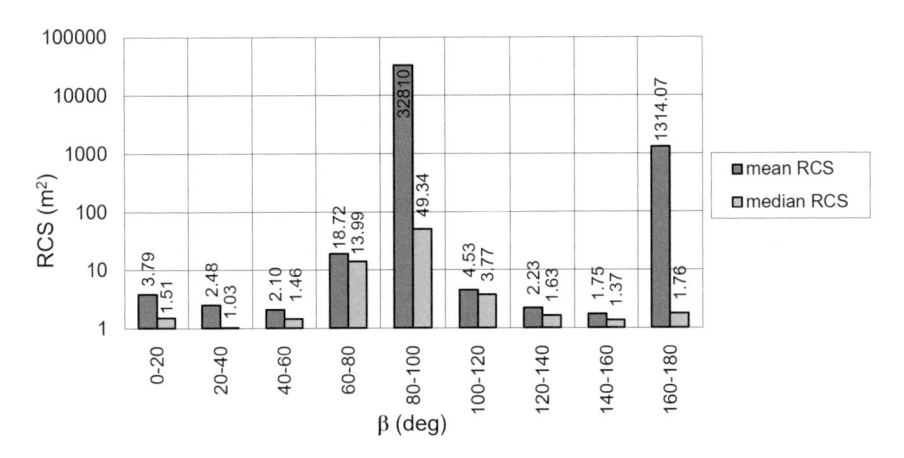

FIGURE 3.53 Diagrams of mean and median RCS of M1 Abrams tank model in 20-degree sectors of azimuth aspect given its radar observation (sounding frequency 10 GHz, $\gamma = 30°$, underlying surface – dry loam, horizontal polarization).

Scattering characteristics of the tank M1 Abrams model for sounding frequency 10 GHz, elevation angle 30°, dry loam, and vertical polarization.

Figure 3.54 shows the RCS circular diagram of the tank M1 Abrams model. The noncoherent RCS circular diagram of the tank M1 Abrams is presented in Figure 3.55. The circular mean RCS of the M1 Abrams tank model is 553.11 m². The circular median RCS is 0.93 m².

Figures 3.56 and 3.57 show the mean and median RCS for the main ranges of sounding azimuths (head, side, and stern) and for ranges of 20°.

The scattering characteristics for this model at two polarizations, sounding frequencies of 10, 5, and 3 GHz (wavelengths of 3, 6, and 10 cm, respectively), for three types of underlying surface (dry loam, dry soil, and wet soil) are given in the electronic appendix of this book.

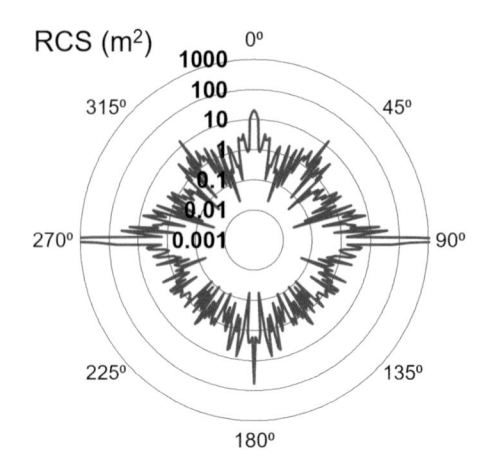

FIGURE 3.54 Circular diagrams of RCS given radar observation of M1 Abrams tank model (sounding frequency 10 GHz, $\gamma=30°$, underlying surface – dry loam, vertical polarization).

FIGURE 3.55 Circular diagram of noncoherent RCS given radar observation of M1 Abrams tank model (sounding frequency 10 GHz, $\gamma=30°$, underlying surface – dry loam, vertical polarization).

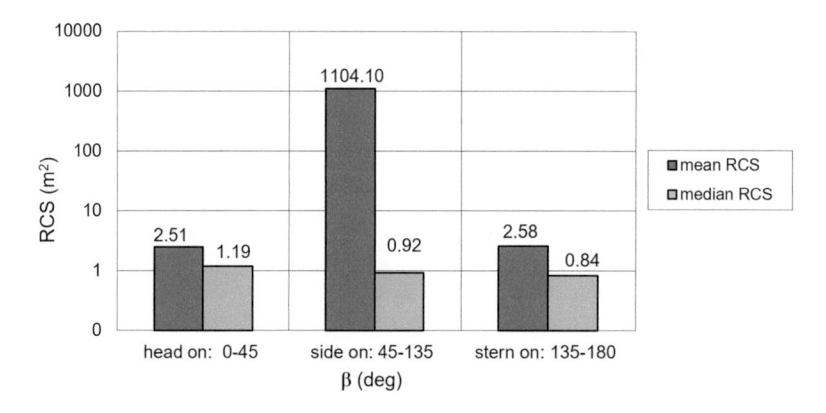

FIGURE 3.56 Diagrams of mean and median RCS of M1 Abrams tank model in three sectors of azimuth aspect given its radar observation (sounding frequency 10 GHz, $\gamma=30°$, underlying surface – dry loam, vertical polarization).

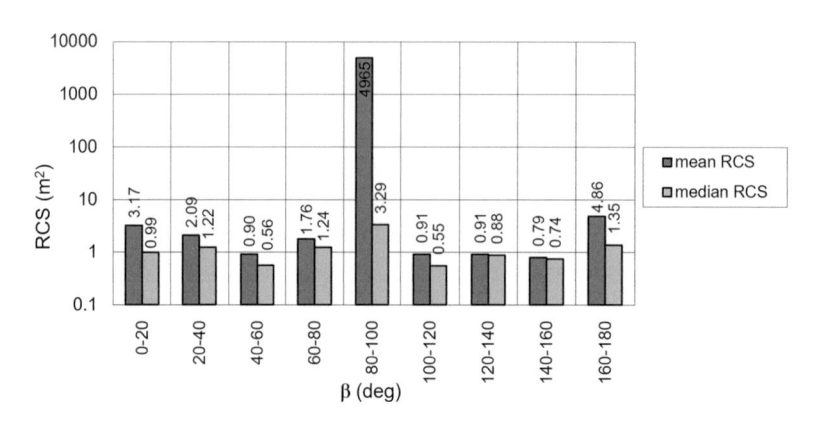

FIGURE 3.57 Diagrams of mean and median RCS of M1 Abrams tank model in 20-degree sectors of azimuth aspect given its radar observation (sounding frequency 10 GHz, $\gamma=30°$, underlying surface – dry loam, vertical polarization).

Scattering Characteristics of Ground Objects

3.5 BUK TARGET ACQUISITION RADAR 9S18M1

9S18M1- Snow Drift - (Figure 3.58) is a target acquisition radar for Buk missile system [78]. It has been in service since 1980.

In accordance with the design of the 9S18M1, a model of its surface was created to obtain scattering characteristics (in particular, RCS). The model is shown in Figure 3.59. In the modeling, the smooth parts of the object surface were approximated by parts of 140 triaxial ellipsoids. The surface breaks were modeled using 94 straight edge scattering parts.

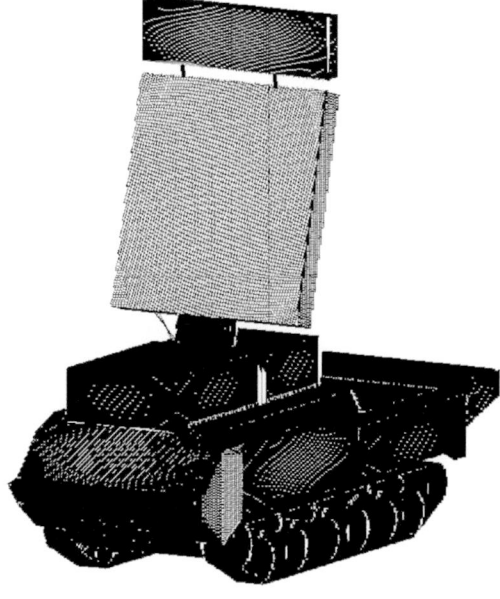

FIGURE 3.58 Buk target acquisition radar 9S18M1.

FIGURE 3.59 Surface model of Buk target acquisition radar 9S18M1.

Below are some scattering characteristics of the radar 9S18M1 model at a sounding frequency of 10 GHz (wavelengths of 3 cm) for the following conditions: the elevation angle of sounding $\gamma = 10°$, wet soil, and the elevation angle of sounding $\gamma = 30°$, dry loam.

Scattering characteristics of the radar 9S18M1 model for sounding frequency 10 GHz, elevation angle 10°, wet soil, and horizontal polarization.

Figure 3.60 shows the RCS circular diagram of the radar 9S18M1 model. The noncoherent RCS circular diagram of the radar 9S18M1 is presented in Figure 3.61. The circular mean RCS of the radar 9S18M1 model is 72 447 m². The circular median RCS is 5 513 m². Figures 3.62 and 3.63 show the mean and median RCS for the main ranges of sounding azimuths (head, side, and stern) and for ranges of 20°.

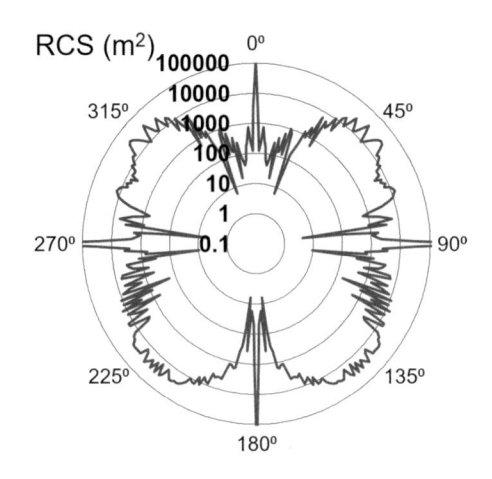

FIGURE 3.60 Circular diagrams of RCS given radar observation of radar 9S18M1 model (sounding frequency 10 GHz, $\gamma = 10°$, underlying surface – wet soil, horizontal polarization).

FIGURE 3.61 Circular diagram of noncoherent RCS given radar observation of radar 9S18M1 model (sounding frequency 10 GHz, $\gamma = 10°$, underlying surface – wet soil, horizontal polarization).

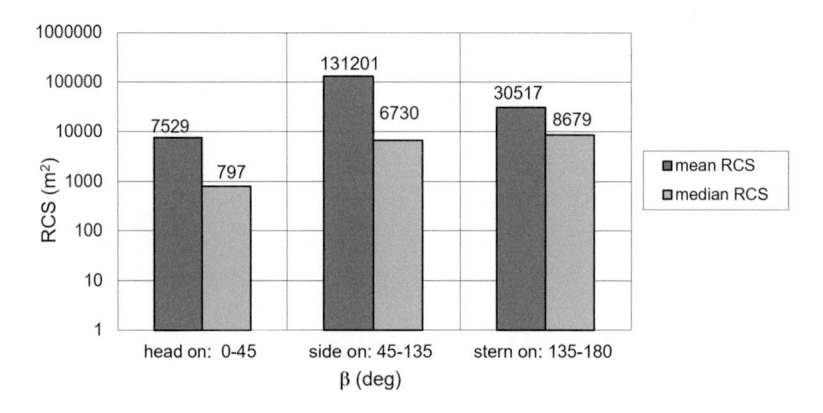

FIGURE 3.62 Diagrams of mean and median RCS of radar 9S18M1 model in three sectors of azimuth aspect given its radar observation (sounding frequency 10 GHz, $\gamma = 10°$, underlying surface – wet soil, horizontal polarization).

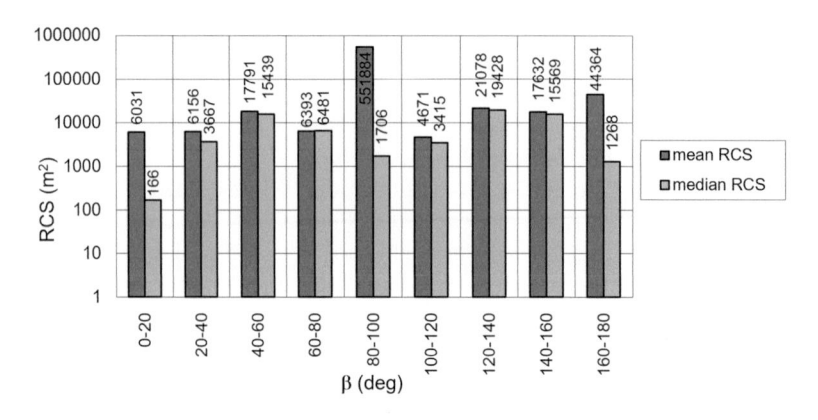

FIGURE 3.63 Diagrams of mean and median RCS of radar 9S18M1 model in 20-degree sectors of azimuth aspect given its radar observation (sounding frequency 10 GHz, $\gamma = 10°$, underlying surface – wet soil, horizontal polarization).

Scattering characteristics of the radar 9S18M1 model for sounding frequency 10 GHz, elevation angle 10°, wet soil, and vertical polarization.

Figure 3.64 shows the RCS circular diagram of the radar 9S18M1 model. The noncoherent RCS circular diagram of the radar 9S18M1 is presented in Figure 3.65.

The circular mean RCS of the radar 9S18M1 model is 7 117 m². The circular median RCS is 3 978 m².

Figures 3.66 and 3.67 show the mean and median RCS for the main ranges of sounding azimuths (head, side, and stern) and for ranges of 20°.

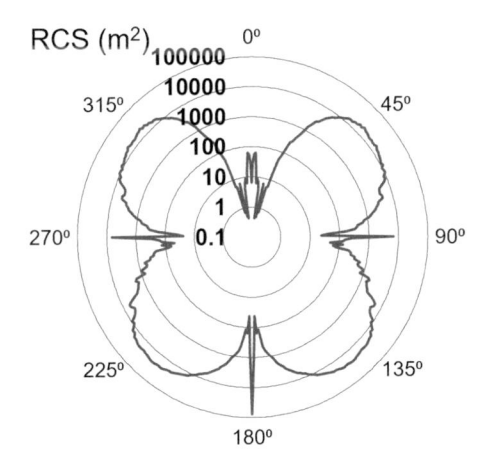

FIGURE 3.64 Circular diagrams of RCS given radar observation of radar 9S18M1 model (sounding frequency 10 GHz, $\gamma = 10°$, underlying surface – wet soil, vertical polarization).

FIGURE 3.65 Circular diagram of noncoherent RCS given radar observation of radar 9S18M1 model (sounding frequency 10 GHz, $\gamma = 10°$, underlying surface – wet soil, vertical polarization).

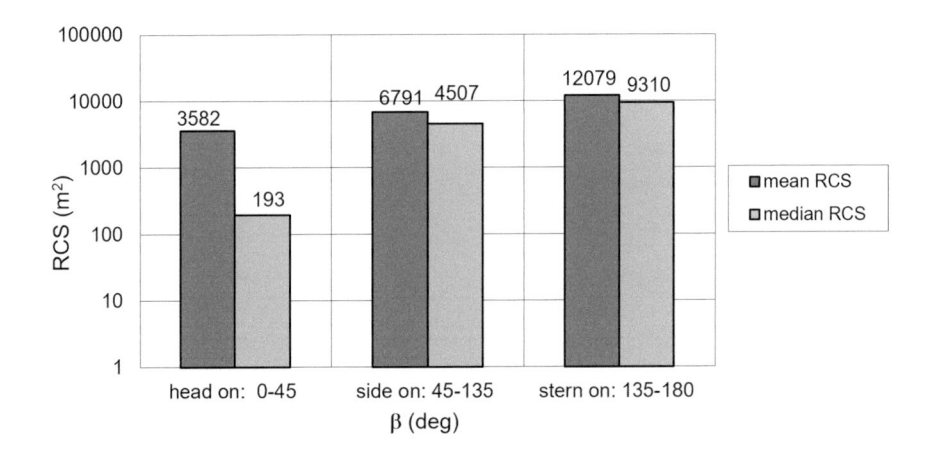

FIGURE 3.66 Diagrams of mean and median RCS of radar 9S18M1 model in three sectors of azimuth aspect given its radar observation (sounding frequency 10 GHz, $\gamma = 10°$, underlying surface – wet soil, vertical polarization).

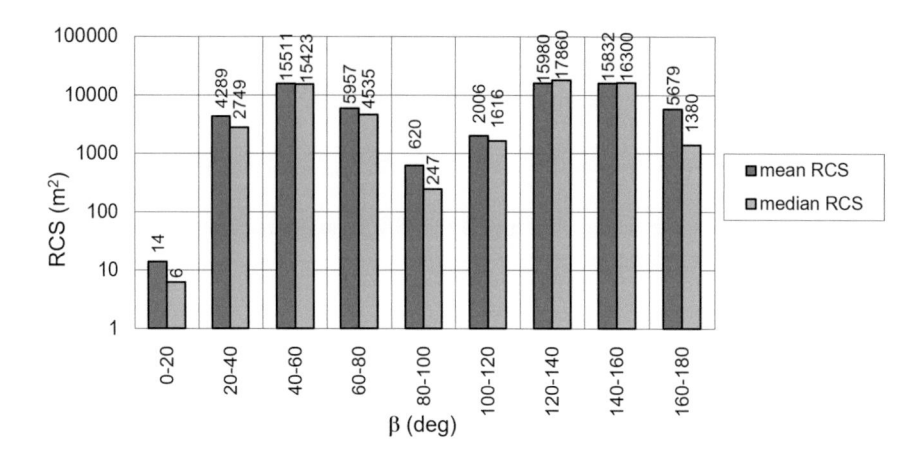

FIGURE 3.67 Diagrams of mean and median RCS of radar 9S18M1 model in 20-degree sectors of azimuth aspect given its radar observation (sounding frequency 10 GHz, $\gamma = 10°$, underlying surface – wet soil, vertical polarization).

Scattering characteristics of the radar 9S18M1 model for sounding frequency 10 GHz, elevation angle 30°, dry loam, and horizontal polarization.

Figure 3.68 shows the RCS circular diagram of the radar 9S18M1 model. The noncoherent RCS circular diagram of the radar 9S18M1 is presented in Figure 3.69. The circular mean RCS of the radar 9S18M1 model is 14 031 m². The circular median RCS is 1 503 m².

Figures 3.70 and 3.71 show the mean and median RCS for the main ranges of sounding azimuths (head, side, and stern) and for ranges of 20°.

FIGURE 3.68 Circular diagrams of RCS given radar observation of radar 9S18M1 model (sounding frequency 10 GHz, $\gamma = 30°$, underlying surface – dry loam, horizontal polarization).

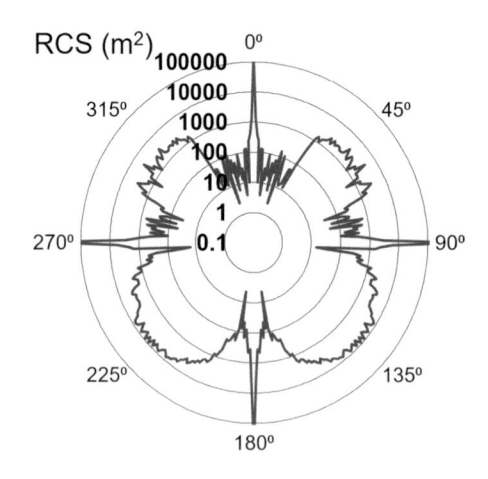

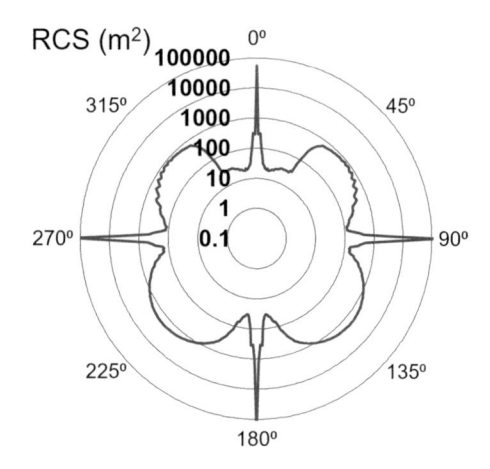

FIGURE 3.69 Circular diagram of noncoherent RCS given radar observation of radar 9S18M1 model (sounding frequency 10 GHz, $\gamma = 30°$, underlying surface – dry loam, horizontal polarization).

Scattering Characteristics of Ground Objects

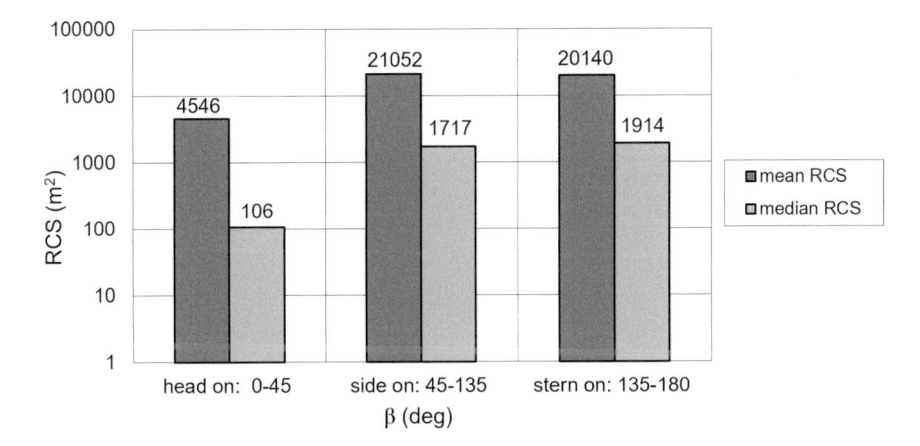

FIGURE 3.70 Diagrams of mean and median RCS of radar 9S18M1 model in three sectors of azimuth aspect given its radar observation (sounding frequency 10 GHz, $\gamma=30°$, underlying surface – dry loam, horizontal polarization).

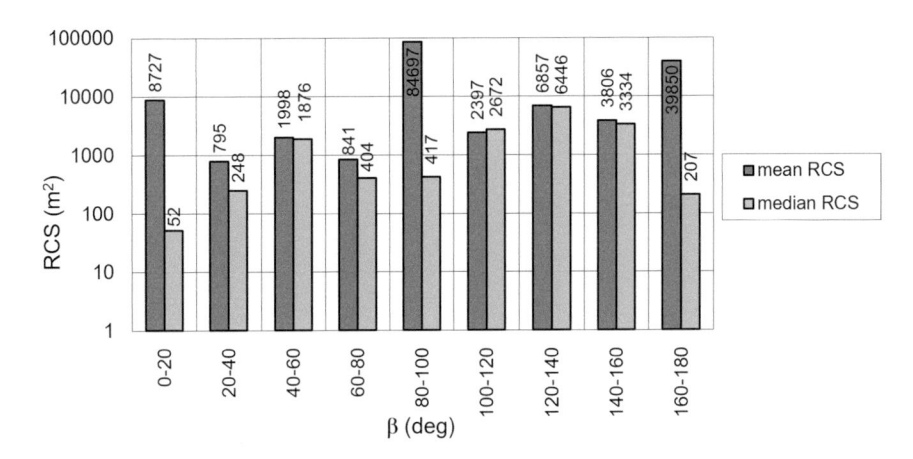

FIGURE 3.71 Diagrams of mean and median RCS of radar 9S18M1 model in 20-degree sectors of azimuth aspect given its radar observation (sounding frequency 10 GHz, $\gamma=30°$, underlying surface – dry loam, horizontal polarization).

Scattering characteristics of the radar 9S18M1 model for sounding frequency 10 GHz, elevation angle 30°, dry loam, and vertical polarization.

Figure 3.72 shows the RCS circular diagram of the radar 9S18M1 model. The noncoherent RCS circular diagram of the radar 9S18M1 is presented in Figure 3.73. The circular mean RCS of the radar 9S18M1 model is 4 661 m². The circular median RCS is 1 228 m².

Figures 3.74 and 3.75 show the mean and median RCS for the main ranges of sounding azimuths (head, side, and stern) and for ranges of 20°.

The scattering characteristics for this model at two polarizations, sounding frequencies of 10, 5, and 3 GHz (wavelengths of 3, 6, and 10 cm, respectively), for three types of underlying surface (dry loam, dry soil, and wet soil) are given in the electronic appendix of this book.

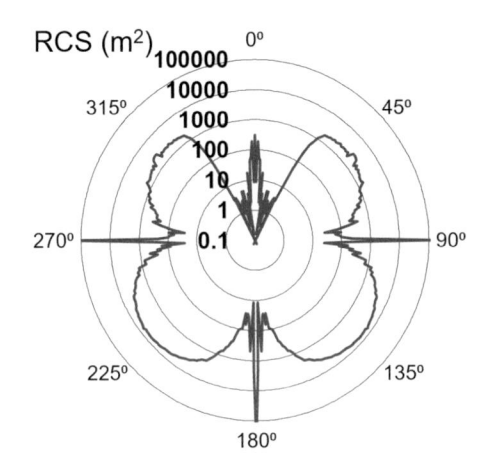

FIGURE 3.72 Circular diagrams of RCS given radar observation of radar 9S18M1 model (sounding frequency 10 GHz, $\gamma = 30°$, underlying surface – dry loam, vertical polarization).

FIGURE 3.73 Circular diagram of noncoherent RCS given radar observation of radar 9S18M1 model (sounding frequency 10 GHz, $\gamma = 30°$, underlying surface – dry loam, vertical polarization).

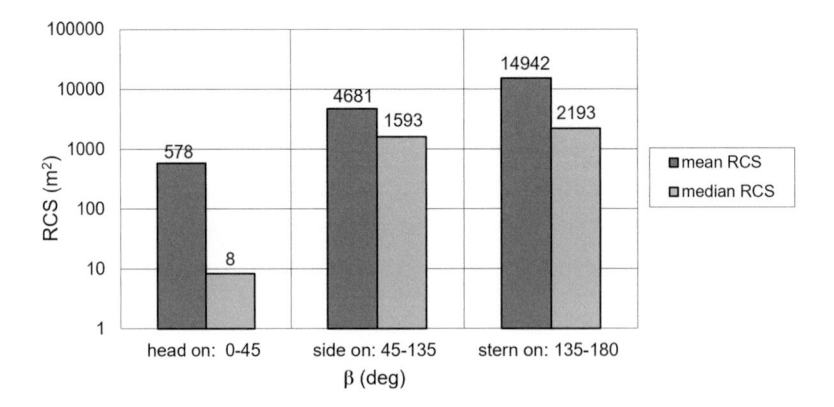

FIGURE 3.74 Diagrams of mean and median RCS of radar 9S18M1 model in three sectors of azimuth aspect given its radar observation (sounding frequency 10 GHz, $\gamma = 30°$, underlying surface – dry loam, vertical polarization).

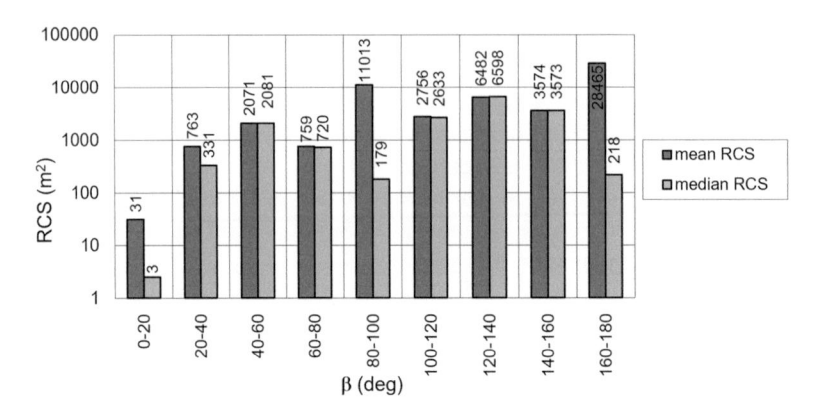

FIGURE 3.75 Diagrams of mean and median RCS of radar 9S18M1 model in 20-degree sectors of azimuth aspect given its radar observation (sounding frequency 10 GHz, $\gamma = 30°$, underlying surface – dry loam, vertical polarization).

3.6 BUK COMMAND POST VEHICLE 9S470

9S470 - (Figure 3.76) is a command post vehicle for Buk missile system [79]. It has been in service since 1980.

In accordance with the design of the 9S470, a model of its surface was created to obtain scattering characteristics (in particular, RCS). The model is shown in Figure 3.77. In the modeling, the smooth parts of the object surface were approximated by parts of 67 triaxial ellipsoids. The surface breaks were modeled using 36 straight edge scattering parts.

FIGURE 3.76 Buk command post vehicle 9S470.

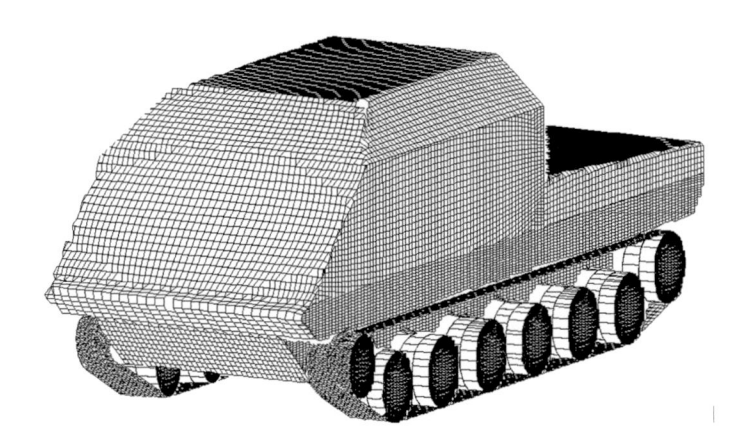

FIGURE 3.77 Surface model of Buk command post vehicle 9S470.

Below are some scattering characteristics of the vehicle 9S470 model at a sounding frequency of 10 GHz (wavelengths of 3 cm) for the following conditions: the elevation angle of sounding $\gamma = 10°$, wet soil, and the elevation angle of sounding $\gamma = 30°$, dry loam.

Scattering characteristics of the vehicle 9S470 model for sounding frequency 10 GHz, elevation angle 10°, wet soil, and horizontal polarization.

Figure 3.78 shows the RCS circular diagram of the vehicle 9S470 model. The noncoherent RCS circular diagram of the vehicle 9S470 is presented in Figure 3.79. The circular mean RCS of the vehicle 9S470 model is 20 236 m². The circular median RCS is 27.37 m².

Figures 3.80 and 3.81 show the mean and median RCS for the main ranges of sounding azimuths (head, side, and stern) and for ranges of 20°.

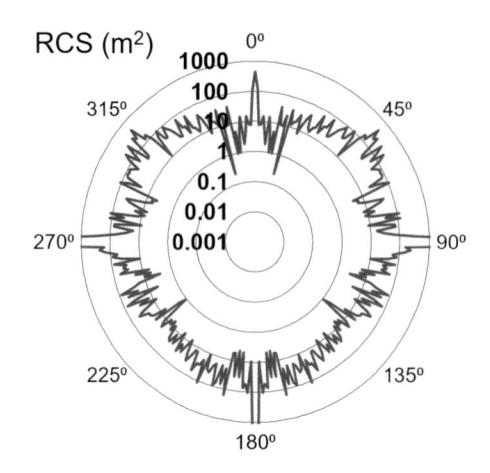

FIGURE 3.78 Circular diagrams of RCS given radar observation of vehicle 9S470 model (sounding frequency 10 GHz, $\gamma = 10°$, underlying surface – wet soil, horizontal polarization).

FIGURE 3.79 Circular diagram of noncoherent RCS given radar observation of vehicle 9S470 model (sounding frequency 10 GHz, $\gamma = 10°$, underlying surface – wet soil, horizontal polarization).

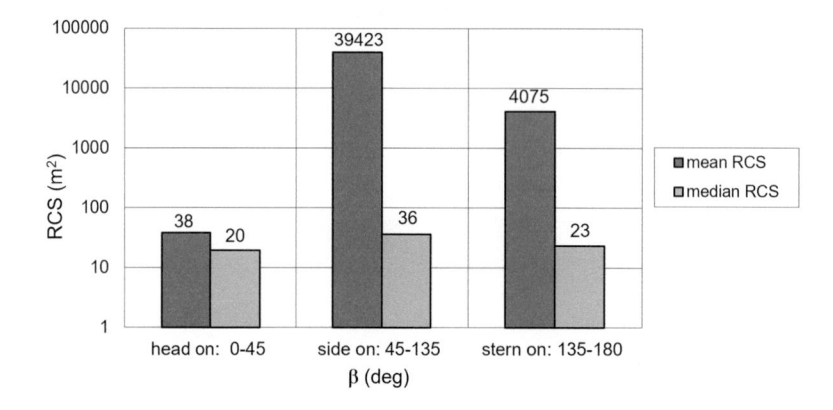

FIGURE 3.80 Diagrams of mean and median RCS of vehicle 9S470 model in three sectors of azimuth aspect given its radar observation (sounding frequency 10 GHz, $\gamma = 10°$, underlying surface – wet soil, horizontal polarization).

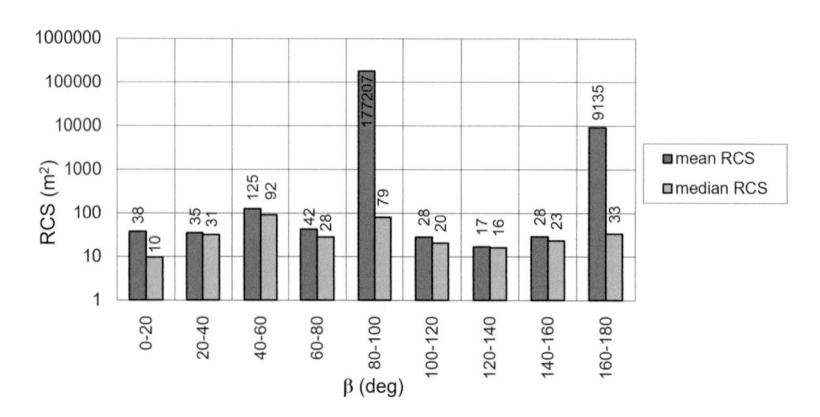

FIGURE 3.81 Diagrams of mean and median RCS of vehicle 9S470 model in 20-degree sectors of azimuth aspect given its radar observation (sounding frequency 10 GHz, $\gamma = 10°$, underlying surface – wet soil, horizontal polarization).

Scattering Characteristics of Ground Objects   419

Scattering characteristics of the vehicle 9S470 model for sounding frequency 10 GHz, elevation angle 10°, wet soil, and vertical polarization.

Figure 3.82 shows the RCS circular diagram of the vehicle 9S470 model. The noncoherent RCS circular diagram of the vehicle 9S470 is presented in Figure 3.83.

The circular mean RCS of the vehicle 9S470 model is 31.40 m². The circular median RCS is 6.08 m².

Figures 3.84 and 3.85 show the mean and median RCS for the main ranges of sounding azimuths (head, side, and stern) and for ranges of 20°.

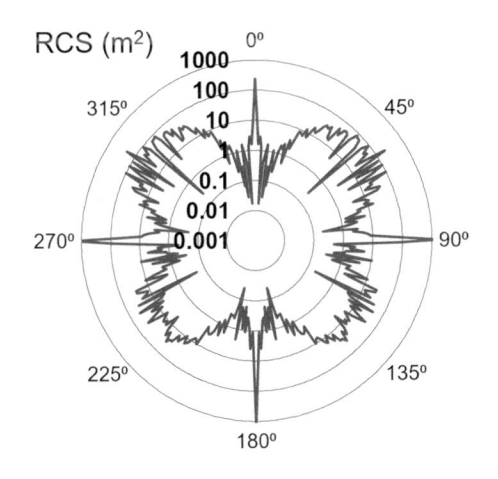

FIGURE 3.82 Circular diagrams of RCS given radar observation of vehicle 9S470 model (sounding frequency 10 GHz, $\gamma = 10°$, underlying surface – wet soil, vertical polarization).

FIGURE 3.83 Circular diagram of noncoherent RCS given radar observation of vehicle 9S470 model (sounding frequency 10 GHz, $\gamma = 10°$, underlying surface – wet soil, vertical polarization).

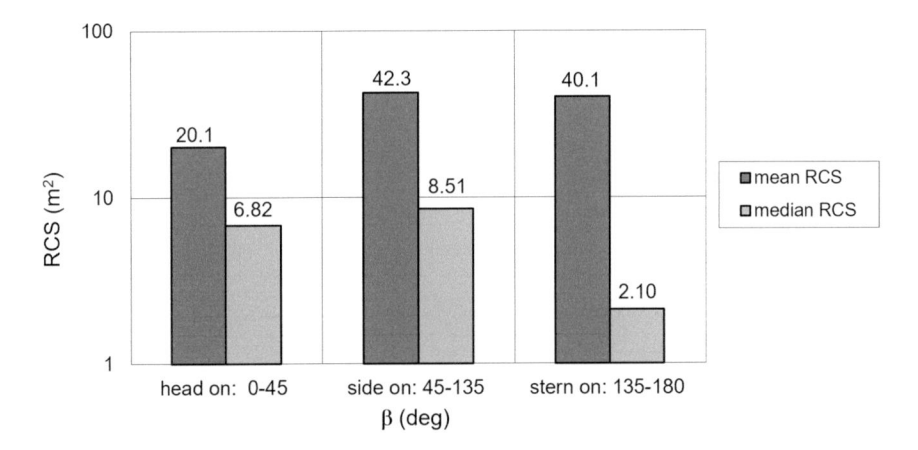

FIGURE 3.84 Diagrams of mean and median RCS of vehicle 9S470 model in three sectors of azimuth aspect given its radar observation (sounding frequency 10 GHz, $\gamma = 10°$, underlying surface – wet soil, vertical polarization).

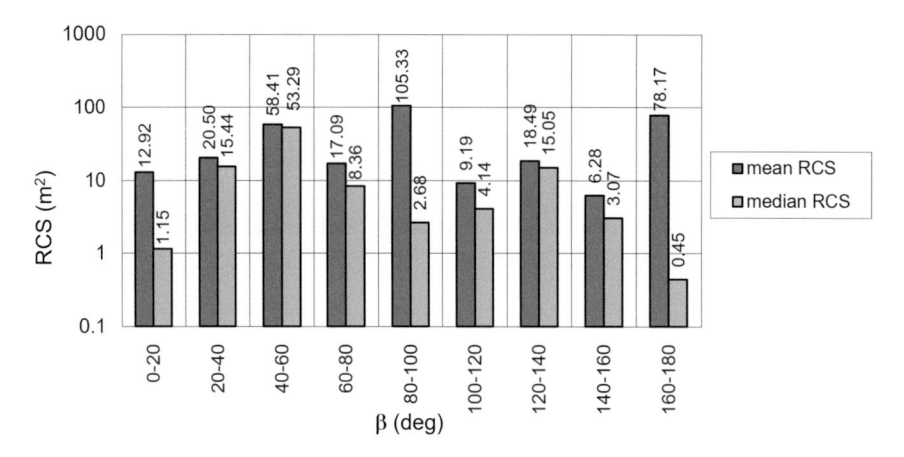

FIGURE 3.85 Diagrams of mean and median RCS of vehicle 9S470 model in 20-degree sectors of azimuth aspect given its radar observation (sounding frequency 10 GHz, $\gamma = 10°$, underlying surface – wet soil, vertical polarization).

Scattering characteristics of the vehicle 9S470 model for sounding frequency 10 GHz, elevation angle 30°, dry loam, and horizontal polarization.

Figure 3.86 shows the RCS circular diagram of the vehicle 9S470 model. The noncoherent RCS circular diagram of the vehicle 9S470 is presented in Figure 3.87. The circular mean RCS of the vehicle 9S470 model is 2 868 m². The circular median RCS is 5.10 m².

Figures 3.88 and 3.89 show the mean and median RCS for the main ranges of sounding azimuths (head, side, and stern) and for ranges of 20°.

FIGURE 3.86 Circular diagrams of RCS given radar observation of vehicle 9S470 model (sounding frequency 10 GHz, $\gamma = 30°$, underlying surface – dry loam, horizontal polarization).

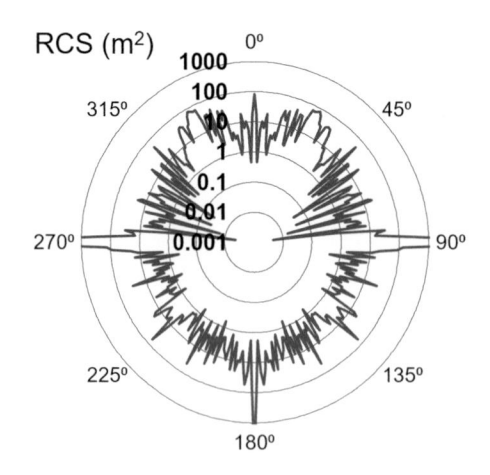

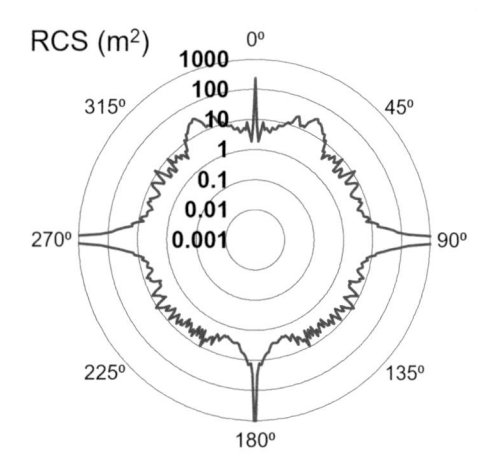

FIGURE 3.87 Circular diagram of noncoherent RCS given radar observation of vehicle 9S470 model (sounding frequency 10 GHz, $\gamma = 30°$, underlying surface – dry loam, horizontal polarization).

Scattering Characteristics of Ground Objects

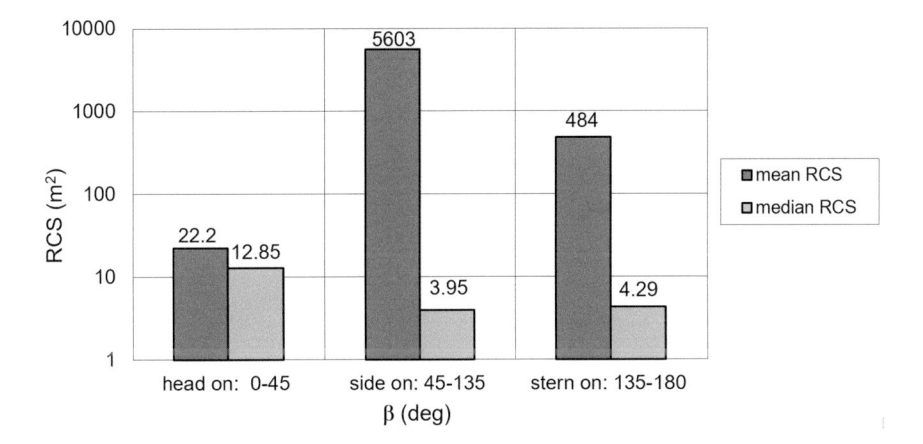

FIGURE 3.88 Diagrams of mean and median RCS of vehicle 9S470 model in three sectors of azimuth aspect given its radar observation (sounding frequency 10 GHz, $\gamma=30°$, underlying surface – dry loam, horizontal polarization).

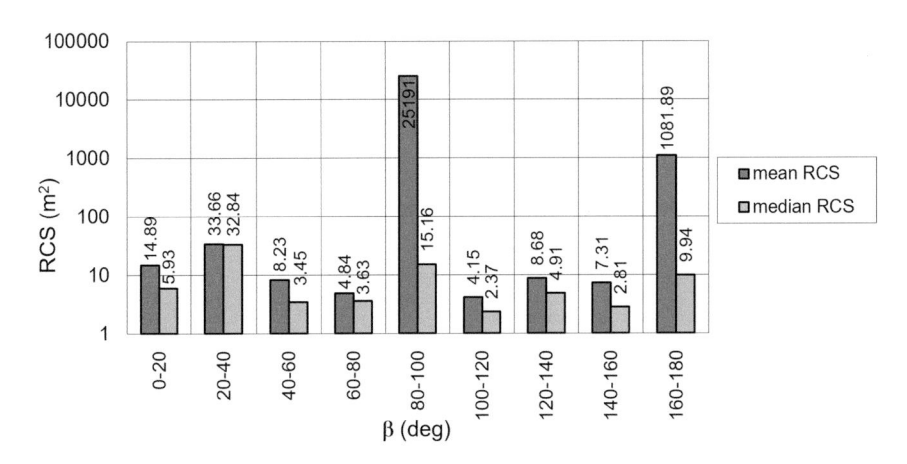

FIGURE 3.89 Diagrams of mean and median RCS of vehicle 9S470 model in 20-degree sectors of azimuth aspect given its radar observation (sounding frequency 10 GHz, $\gamma=30°$, underlying surface – dry loam, horizontal polarization).

Scattering characteristics of the vehicle 9S470 model for sounding frequency 10 GHz, elevation angle 30°, dry loam, and vertical polarization.

Figure 3.90 shows the RCS circular diagram of the vehicle 9S470 model. The noncoherent RCS circular diagram of the vehicle 9S470 is presented in Figure 3.91. The circular mean RCS of the vehicle 9S470 model is 12.68 m². The circular median RCS is 0.90 m².

Figures 3.92 and 3.93 show the mean and median RCS for the main ranges of sounding azimuths (head, side, and stern) and for ranges of 20°.

The scattering characteristics for this model at two polarizations, sounding frequencies of 10, 5, and 3 GHz (wavelengths of 3, 6, and 10 cm, respectively), for three types of underlying surface (dry loam, dry soil, and wet soil) are given in the electronic appendix of this book.

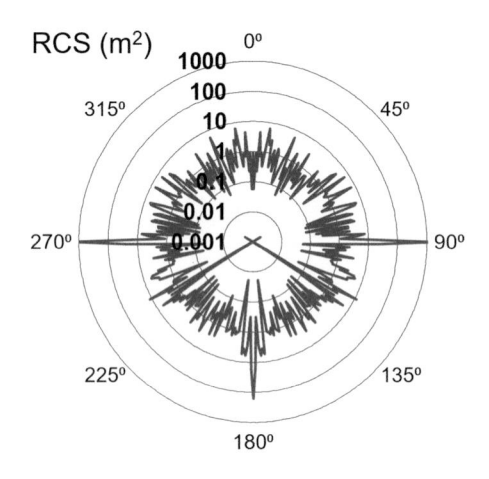

FIGURE 3.90 Circular diagrams of RCS given radar observation of vehicle 9S470 model (sounding frequency 10 GHz, $\gamma=30°$, underlying surface – dry loam, vertical polarization).

FIGURE 3.91 Circular diagram of noncoherent RCS given radar observation of vehicle 9S470 model (sounding frequency 10 GHz, $\gamma=30°$, underlying surface – dry loam, vertical polarization).

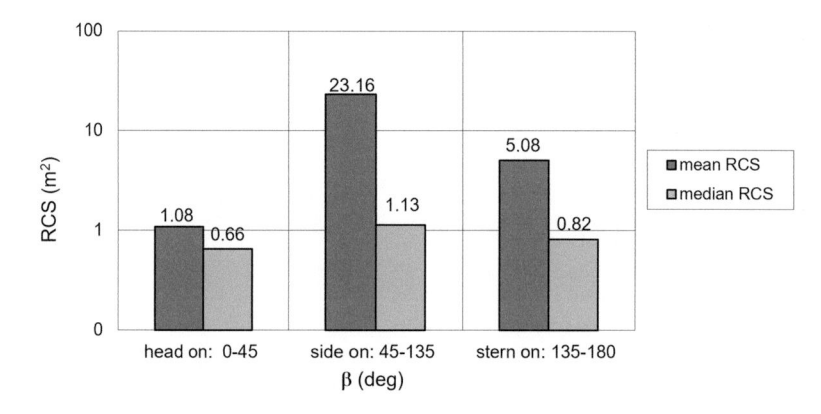

FIGURE 3.92 Diagrams of mean and median RCS of vehicle 9S470 model in three sectors of azimuth aspect given its radar observation (sounding frequency 10 GHz, $\gamma=30°$, underlying surface – dry loam, vertical polarization).

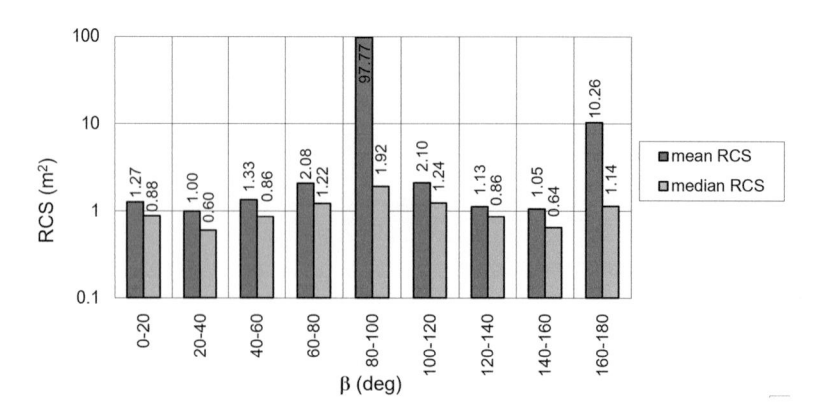

FIGURE 3.93 Diagrams of mean and median RCS of vehicle 9S470 model in 20-degree sectors of azimuth aspect given its radar observation (sounding frequency 10 GHz, $\gamma=30°$, underlying surface – dry loam, vertical polarization).

3.7 BUK TRANSPORTER ERECTOR LAUNCHER AND RADAR 9A310M1

9A310M1 - (Figure 3.94) is a transporter erector launcher and radar (TELAR) for Buk missile system [80]. It has been in service since 1980.

In accordance with the design of the TELAR 9A310M1, a model of its surface was created to obtain scattering characteristics (in particular, RCS). The model is shown in Figure 3.95. In the modeling, the smooth parts of the object surface were approximated by parts of 105 triaxial ellipsoids. The surface breaks were modeled using 63 straight edge scattering parts.

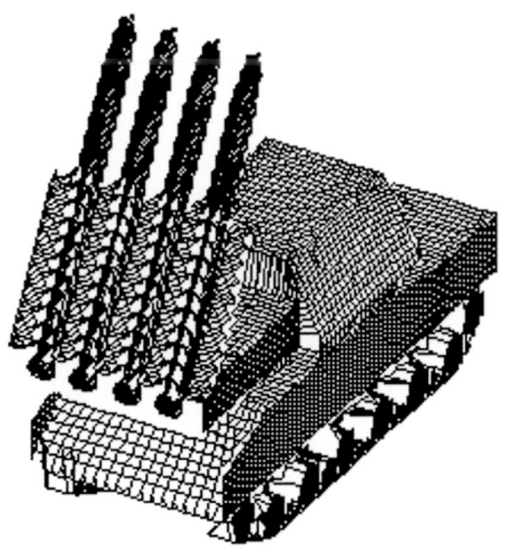

FIGURE 3.94 Buk transporter erector launcher and radar 9A310M1.

FIGURE 3.95 Surface model of Buk TELAR 9A310M1.

Below are some scattering characteristics of the TELAR 9A310M1 model at a sounding frequency of 10 GHz (wavelengths of 3 cm) for the following conditions: the elevation angle of sounding $\gamma = 10°$, wet soil, and the elevation angle of sounding $\gamma = 30°$, dry loam.

Scattering characteristics of the TELAR 9A310M1 model for sounding frequency 10 GHz, elevation angle 10°, wet soil, and horizontal polarization.

Figure 3.96 shows the RCS circular diagram of the TELAR 9A310M1 model. The noncoherent RCS circular diagram of the TELAR 9A310M1 is presented in Figure 3.97. The circular mean RCS of the TELAR 9A310M1 model is 49 970 m². The circular median RCS is 396.16 m².

Figures 3.98 and 3.99 show the mean and median RCS for the main ranges of sounding azimuths (head, side, and stern) and for ranges of 20°.

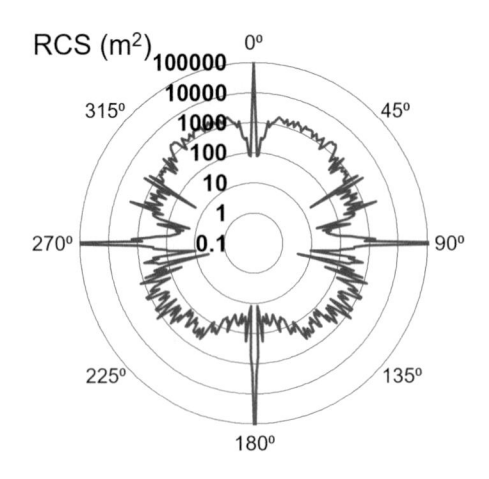

FIGURE 3.96 Circular diagrams of RCS given radar observation of TELAR 9A310M1 model (sounding frequency 10 GHz, $\gamma = 10°$, underlying surface – wet soil, horizontal polarization).

FIGURE 3.97 Circular diagram of noncoherent RCS given radar observation of TELAR 9A310M1 model (sounding frequency 10 GHz, $\gamma = 10°$, underlying surface – wet soil, horizontal polarization).

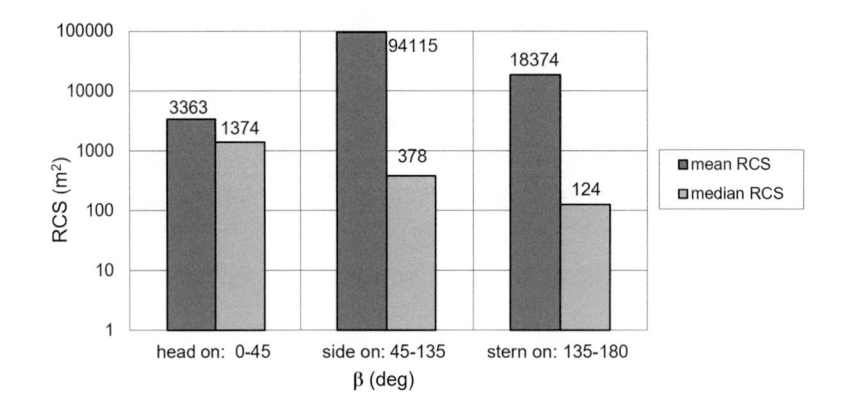

FIGURE 3.98 Diagrams of mean and median RCS of TELAR 9A310M1 model in three sectors of azimuth aspect given its radar observation (sounding frequency 10 GHz, $\gamma = 10°$, underlying surface – wet soil, horizontal polarization).

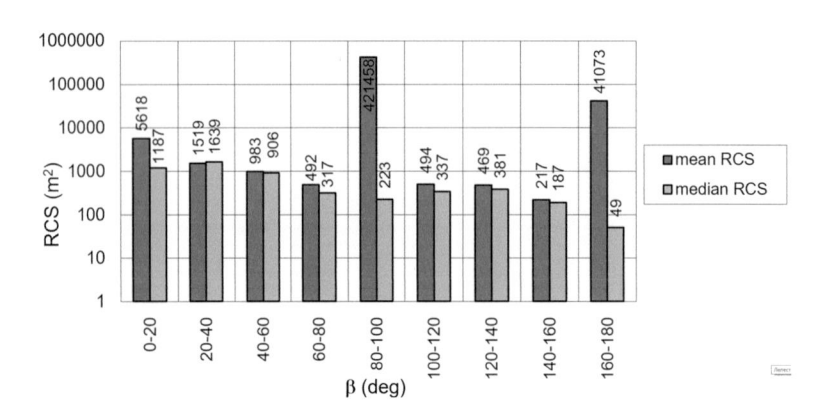

FIGURE 3.99 Diagrams of mean and median RCS of TELAR 9A310M1 model in 20-degree sectors of azimuth aspect given its radar observation (sounding frequency 10 GHz, $\gamma = 10°$, underlying surface – wet soil, horizontal polarization).

Scattering characteristics of the TELAR 9A310M1 model for sounding frequency 10 GHz, elevation angle 10°, wet soil, and vertical polarization.

Figure 3.100 shows the RCS circular diagram of the TELAR 9A310M1 model. The noncoherent RCS circular diagram of the TELAR 9A310M1 is presented in Figure 3.101.

The circular mean RCS of the TELAR 9A310M1 model is 56.25 m². The circular median RCS is 17.45 m².

Figures 3.102 and 3.103 show the mean and median RCS for the main ranges of sounding azimuths (head, side, and stern) and for ranges of 20°.

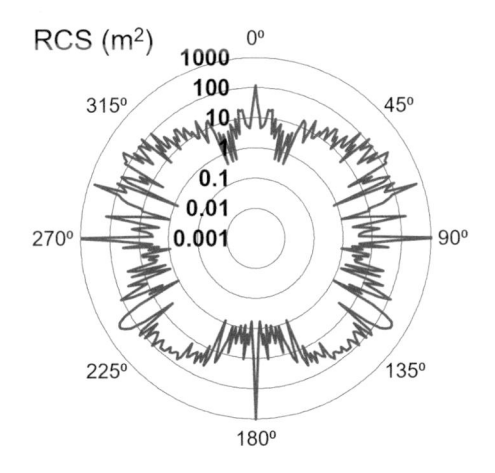

FIGURE 3.100 Circular diagrams of RCS given radar observation of TELAR 9A310M1 model (sounding frequency 10 GHz, $\gamma = 10°$, underlying surface – wet soil, vertical polarization).

FIGURE 3.101 Circular diagram of noncoherent RCS given radar observation of TELAR 9A310M1 model (sounding frequency 10 GHz, $\gamma = 10°$, underlying surface – wet soil, vertical polarization).

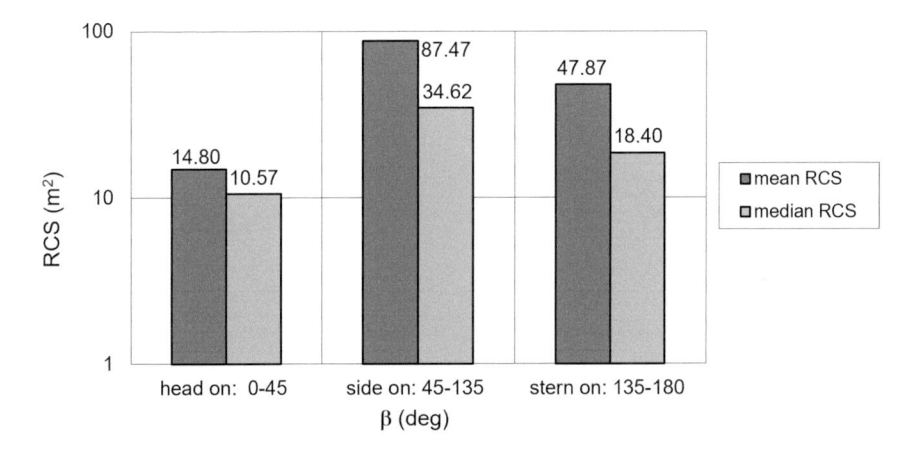

FIGURE 3.102 Diagrams of mean and median RCS of TELAR 9A310M1 model in three sectors of azimuth aspect given its radar observation (sounding frequency 10 GHz, $\gamma = 10°$, underlying surface – wet soil, vertical polarization).

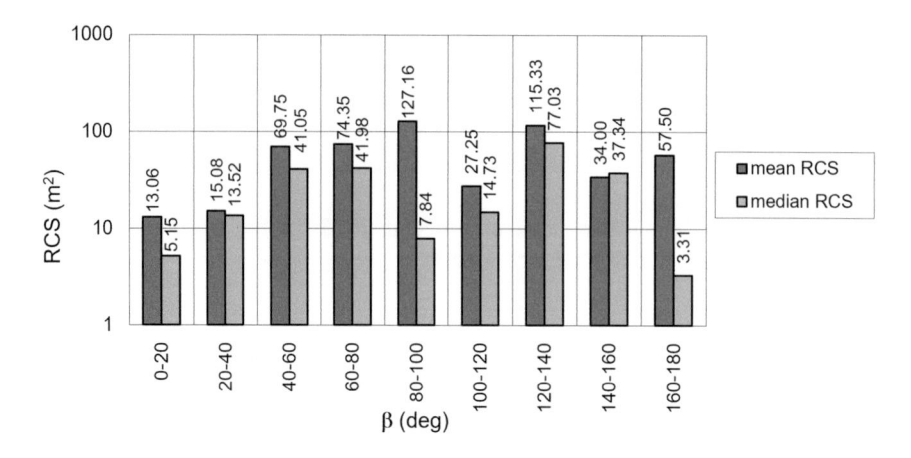

FIGURE 3.103 Diagrams of mean and median RCS of TELAR 9A310M1 model in 20-degree sectors of azimuth aspect given its radar observation (sounding frequency 10 GHz, $\gamma = 10°$, underlying surface – wet soil, vertical polarization).

Scattering characteristics of the TELAR 9A310M1 model for sounding frequency 10 GHz, elevation angle 30°, dry loam, and horizontal polarization.

Figure 3.104 shows the RCS circular diagram of the TELAR 9A310M1 model. The noncoherent RCS circular diagram of the TELAR 9A310M1 is presented in Figure 3.105. The circular mean RCS of the TELAR 9A310M1 model is 12 606 m². The circular median RCS is 101.31 m².

Figures 3.106 and 3.107 show the mean and median RCS for the main ranges of sounding azimuths (head, side, and stern) and for ranges of 20°.

FIGURE 3.104 Circular diagrams of RCS given radar observation of TELAR 9A310M1 model (sounding frequency 10 GHz, $\gamma = 30°$, underlying surface – dry loam, horizontal polarization).

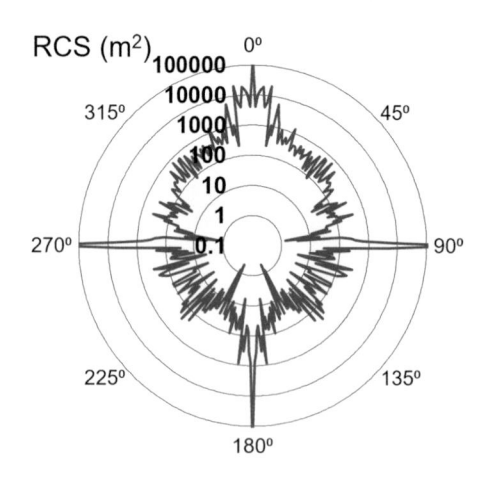

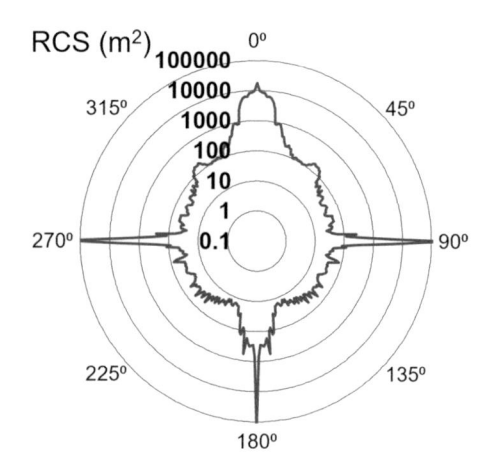

FIGURE 3.105 Circular diagram of noncoherent RCS given radar observation of TELAR 9A310M1 model (sounding frequency 10 GHz, $\gamma = 30°$, underlying surface – dry loam, horizontal polarization).

Scattering Characteristics of Ground Objects 427

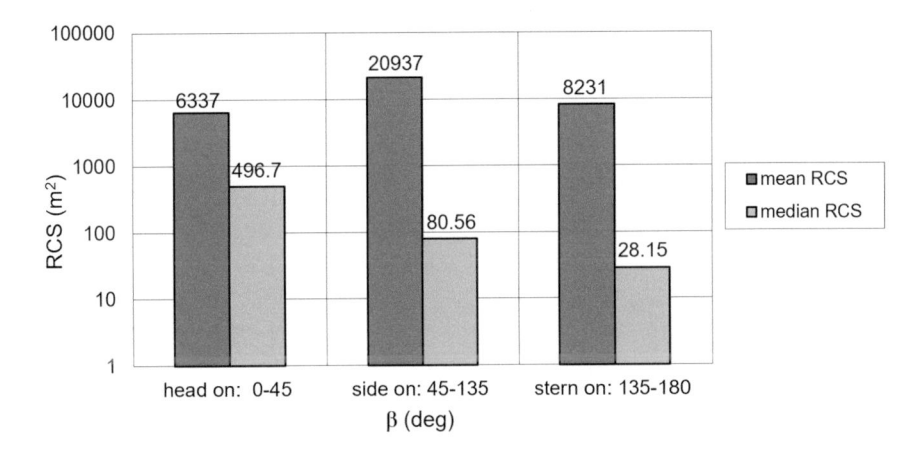

FIGURE 3.106 Diagrams of mean and median RCS of TELAR 9A310M1 model in three sectors of azimuth aspect given its radar observation (sounding frequency 10 GHz, $\gamma=30°$, underlying surface – dry loam, horizontal polarization).

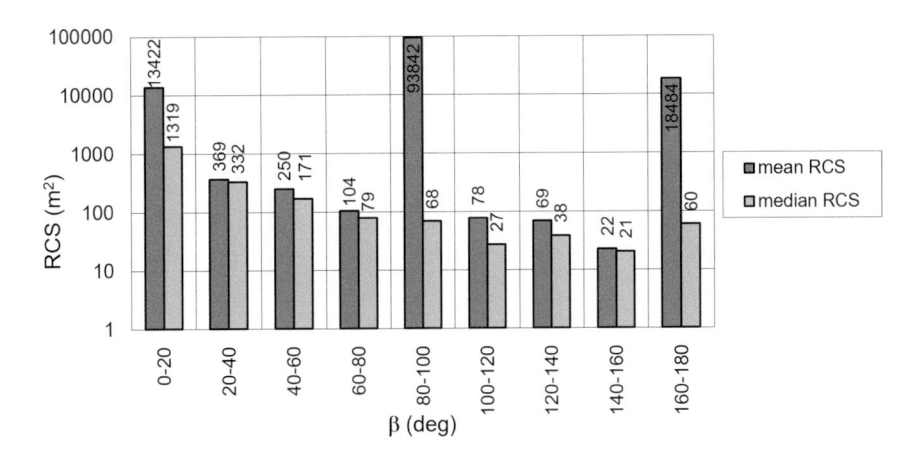

FIGURE 3.107 Diagrams of mean and median RCS of TELAR 9A310M1 model in 20-degree sectors of azimuth aspect given its radar observation (sounding frequency 10 GHz, $\gamma=30°$, underlying surface – dry loam, horizontal polarization).

Scattering characteristics of the TELAR 9A310M1 model for sounding frequency 10 GHz, elevation angle 30°, dry loam, and vertical polarization.

Figure 3.108 shows the RCS circular diagram of the TELAR 9A310M1 model. The noncoherent RCS circular diagram of the TELAR 9A310M1 is presented in Figure 3.109. The circular mean RCS of the TELAR 9A310M1 model is 1 485 m². The circular median RCS is 16.89 m².

Figures 3.110 and 3.111 show the mean and median RCS for the main ranges of sounding azimuths (head, side, and stern) and for ranges of 20°.

The scattering characteristics for this model at two polarizations, sounding frequencies of 10, 5, and 3 GHz (wavelengths of 3, 6, and 10 cm, respectively), for three types of underlying surface (dry loam, dry soil, and wet soil) are given in the electronic appendix of this book.

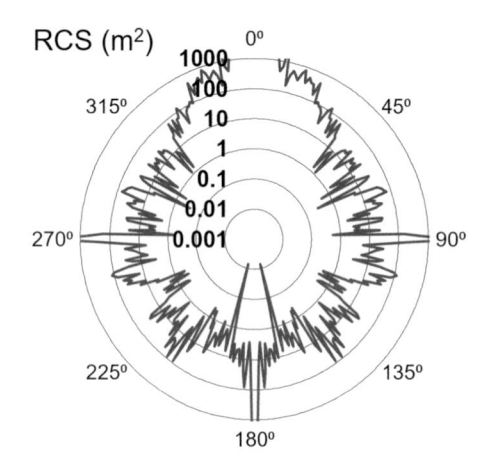

FIGURE 3.108 Circular diagrams of RCS given radar observation of TELAR 9A310M1 model (sounding frequency 10 GHz, $\gamma=30°$, underlying surface – dry loam, vertical polarization).

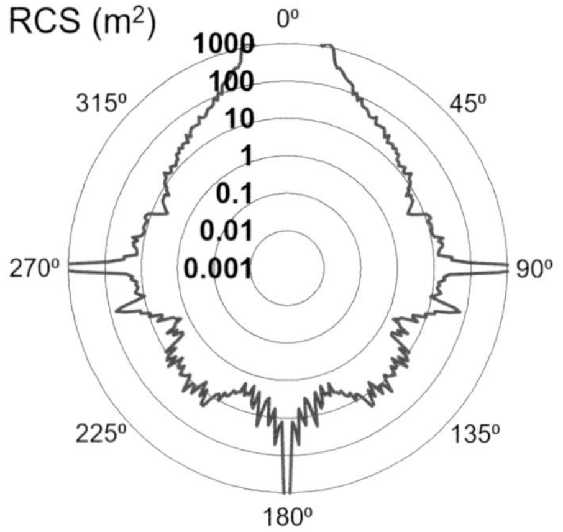

FIGURE 3.109 Circular diagram of noncoherent RCS given radar observation of TELAR 9A310M1 model (sounding frequency 10 GHz, $\gamma=30°$, underlying surface – dry loam, vertical polarization).

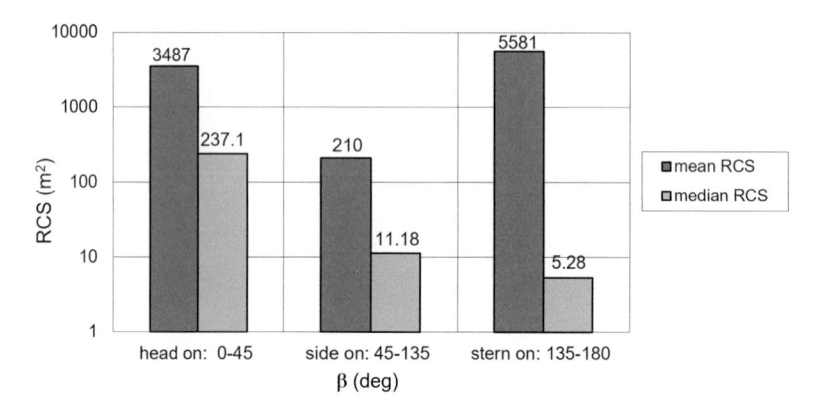

FIGURE 3.110 Diagrams of mean and median RCS of TELAR 9A310M1 model in three sectors of azimuth aspect given its radar observation (sounding frequency 10 GHz, $\gamma=30°$, underlying surface – dry loam, vertical polarization).

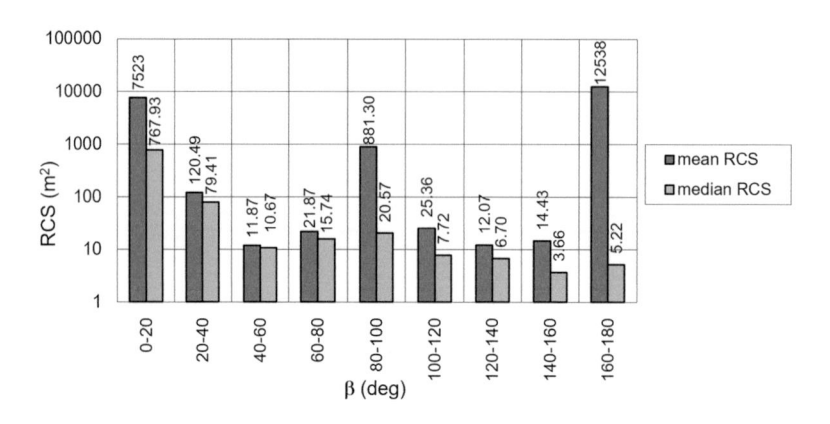

FIGURE 3.111 Diagrams of mean and median RCS of TELAR 9A310M1 model in 20-degree sectors of azimuth aspect given its radar observation (sounding frequency 10 GHz, $\gamma=30°$, underlying surface – dry loam, vertical polarization).

Scattering Characteristics of Ground Objects 429

3.8 S-300PS TRANSPORTER ERECTOR LAUNCHER 5P85D

5P85D - (Figure 3.112) is a transporter erector launcher (TEL) for S-300PS missile system (SA-10D Grumble) [81]. It has been in service since 1985.

In accordance with the design of the TEL 5P85D, a model of its surface was created to obtain scattering characteristics (in particular, RCS). The model is shown in Figure 3.113. In the modeling, the smooth parts of the object surface were approximated by parts of 156 triaxial ellipsoids. The surface breaks were modeled using 115 straight edge scattering parts.

FIGURE 3.112 S-300PS transporter erector launcher 5P85D.

FIGURE 3.113 Surface model of S-300PS TEL 5P85D.

Below are some scattering characteristics of the TEL 5P85D model at a sounding frequency of 10 GHz (wavelengths of 3 cm) for the following conditions: the elevation angle of sounding $\gamma = 10°$, wet soil, and the elevation angle of sounding $\gamma = 30°$, dry loam.

Scattering characteristics of the TEL 5P85D model for sounding frequency 10 GHz, elevation angle 10°, wet soil, and horizontal polarization.

Figure 3.114 shows the RCS circular diagram of the TEL 5P85D model. The noncoherent RCS circular diagram of the TEL 5P85D is presented in Figure 3.115. The circular mean RCS of the TEL 5P85D model is 77 429 m². The circular median RCS is 18 919 m². Figures 3.116 and 3.117 show the mean and median RCS for the main ranges of sounding azimuths (head, side, and stern) and for ranges of 20°.

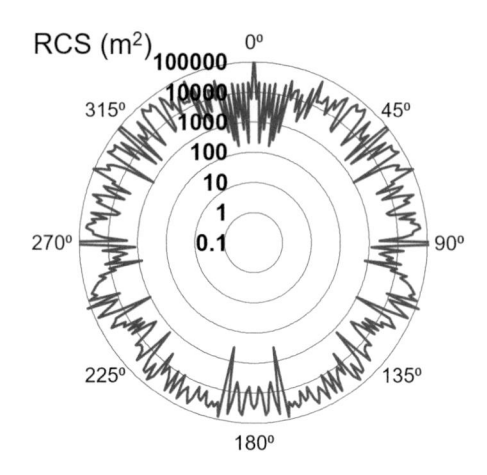

FIGURE 3.114 Circular diagrams of RCS given radar observation of TEL 5P85D model (sounding frequency 10 GHz, $\gamma = 10°$, underlying surface – wet soil, horizontal polarization).

FIGURE 3.115 Circular diagram of noncoherent RCS given radar observation of TEL 5P85D model (sounding frequency 10 GHz, $\gamma = 10°$, underlying surface – wet soil, horizontal polarization).

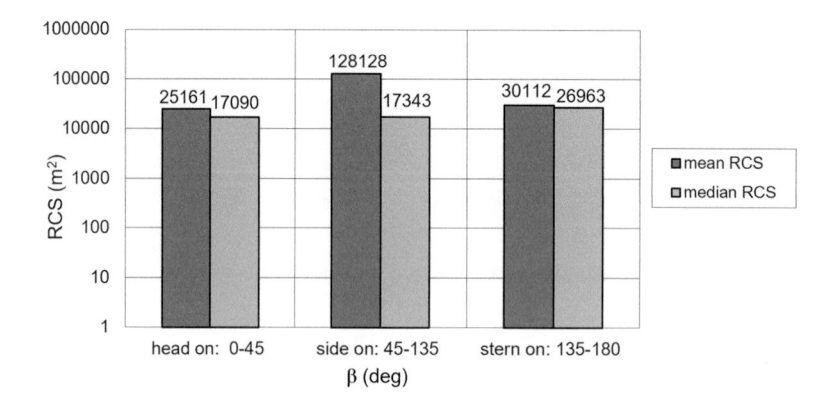

FIGURE 3.116 Diagrams of mean and median RCS of TEL 5P85D model in three sectors of azimuth aspect given its radar observation (sounding frequency 10 GHz, $\gamma = 10°$, underlying surface – wet soil, horizontal polarization).

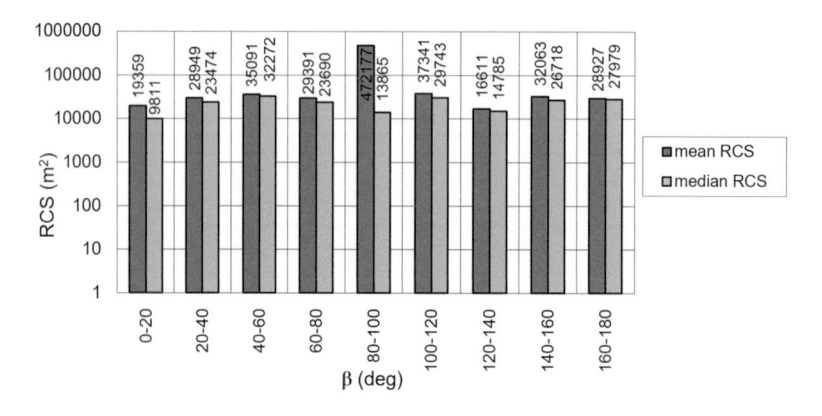

FIGURE 3.117 Diagrams of mean and median RCS of TEL 5P85D model in 20-degree sectors of azimuth aspect given its radar observation (sounding frequency 10 GHz, $\gamma = 10°$, underlying surface – wet soil, horizontal polarization).

Scattering characteristics of the TEL 5P85D model for sounding frequency 10 GHz, elevation angle 10°, wet soil, and vertical polarization.

Figure 3.118 shows the RCS circular diagram of the TEL 5P85D model. The noncoherent RCS circular diagram of the TEL 5P85D is presented in Figure 3.119.

The circular mean RCS of the TEL 5P85D model is 5 253 m². The circular median RCS is 277.73 m².

Figures 3.120 and 3.121 show the mean and median RCS for the main ranges of sounding azimuths (head, side, and stern) and for ranges of 20°.

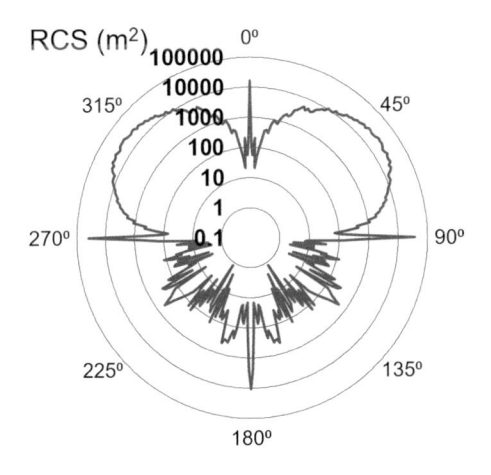

FIGURE 3.118 Circular diagrams of RCS given radar observation of TEL 5P85D model (sounding frequency 10 GHz, $\gamma = 10°$, underlying surface – wet soil, vertical polarization).

FIGURE 3.119 Circular diagram of noncoherent RCS given radar observation of TEL 5P85D model (sounding frequency 10 GHz, $\gamma = 10°$, underlying surface – wet soil, vertical polarization).

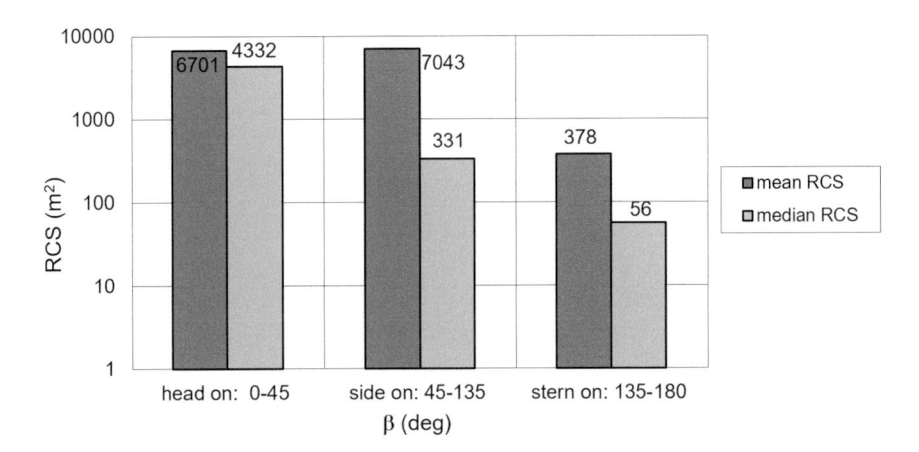

FIGURE 3.120 Diagrams of mean and median RCS of TEL 5P85D model in three sectors of azimuth aspect given its radar observation (sounding frequency 10 GHz, $\gamma = 10°$, underlying surface – wet soil, vertical polarization).

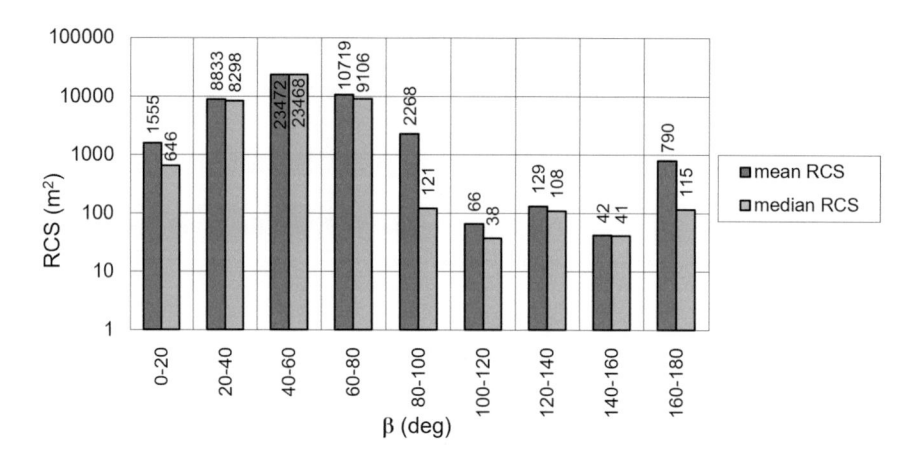

FIGURE 3.121 Diagrams of mean and median RCS of TEL 5P85D model in 20-degree sectors of azimuth aspect given its radar observation (sounding frequency 10 GHz, $\gamma = 10°$, underlying surface – wet soil, vertical polarization).

Scattering characteristics of the TEL 5P85D model for sounding frequency 10 GHz, elevation angle 30°, dry loam, and horizontal polarization.

Figure 3.122 shows the RCS circular diagram of the TEL 5P85D model. The noncoherent RCS circular diagram of the TEL 5P85D is presented in Figure 3.123. The circular mean RCS of the TEL 5P85D model is 21 914 m². The circular median RCS is 8 191 m².

Figures 3.124 and 3.125 show the mean and median RCS for the main ranges of sounding azimuths (head, side, and stern) and for ranges of 20°.

FIGURE 3.122 Circular diagrams of RCS given radar observation of TEL 5P85D model (sounding frequency 10 GHz, $\gamma = 30°$, underlying surface – dry loam, horizontal polarization).

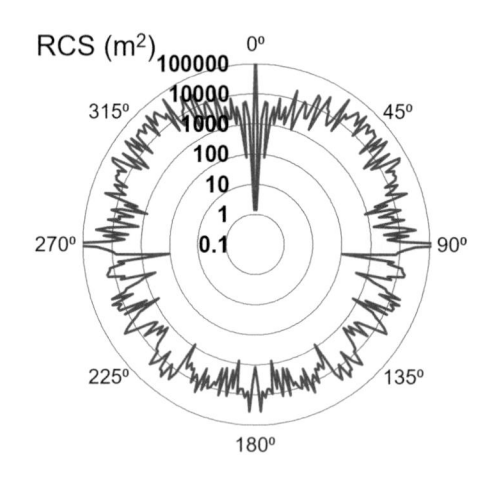

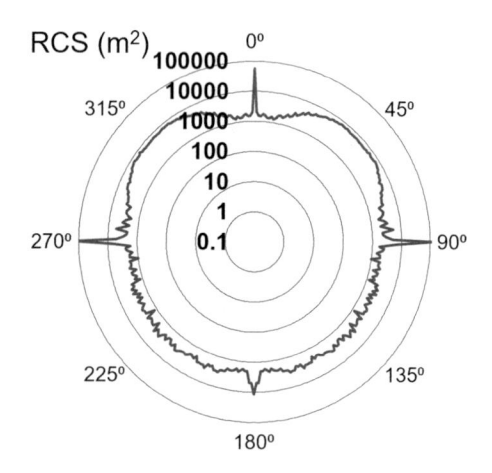

FIGURE 3.123 Circular diagram of noncoherent RCS given radar observation of TEL 5P85D model (sounding frequency 10 GHz, $\gamma = 30°$, underlying surface – dry loam, horizontal polarization).

Scattering Characteristics of Ground Objects 433

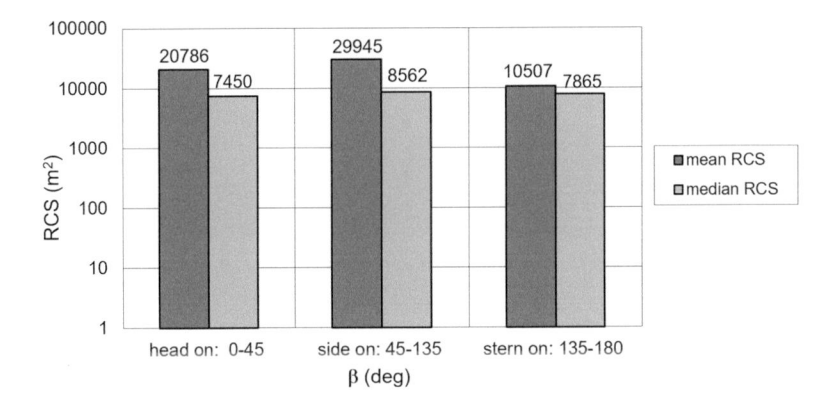

FIGURE 3.124 Diagrams of mean and median RCS of TEL 5P85D model in three sectors of azimuth aspect given its radar observation (sounding frequency 10 GHz, $\gamma = 30°$, underlying surface – dry loam, horizontal polarization).

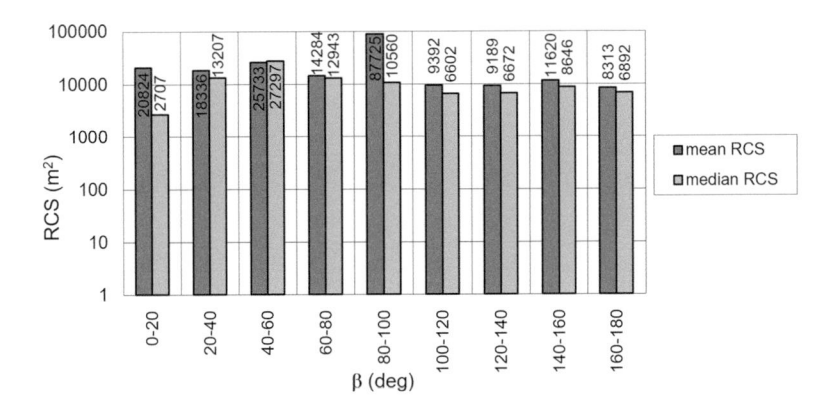

FIGURE 3.125 Diagrams of mean and median RCS of TEL 5P85D model in 20-degree sectors of azimuth aspect given its radar observation (sounding frequency 10 GHz, $\gamma = 30°$, underlying surface – dry loam, horizontal polarization).

Scattering characteristics of the TEL 5P85D model for sounding frequency 10 GHz, elevation angle 30°, dry loam, and vertical polarization.

Figure 3.126 shows the RCS circular diagram of the TEL 5P85D model. The noncoherent RCS circular diagram of the TEL 5P85D is presented in Figure 3.127. The circular mean RCS of the TEL 5P85D model is 5 768 m². The circular median RCS is 372.21 m².

Figures 3.128 and 3.129 show the mean and median RCS for the main ranges of sounding azimuths (head, side, and stern) and for ranges of 20°.

The scattering characteristics for this model at two polarizations, sounding frequencies of 10, 5, and 3 GHz (wavelengths of 3, 6, and 10 cm, respectively), for three types of underlying surface (dry loam, dry soil, and wet soil) are given in the electronic appendix of this book.

434 Scattering Characteristics of Aerial and Ground Radar Objects

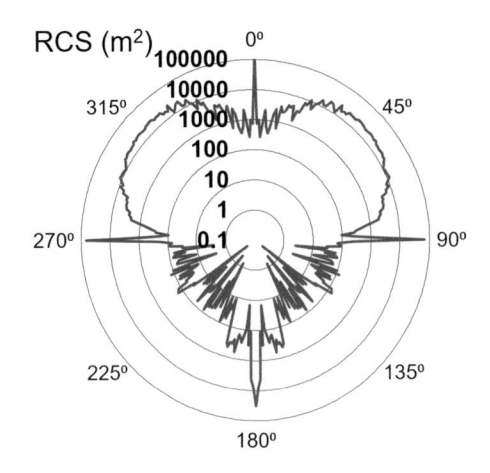

FIGURE 3.126 Circular diagrams of RCS given radar observation of TEL 5P85D model (sounding frequency 10 GHz, $\gamma = 30°$, underlying surface – dry loam, vertical polarization).

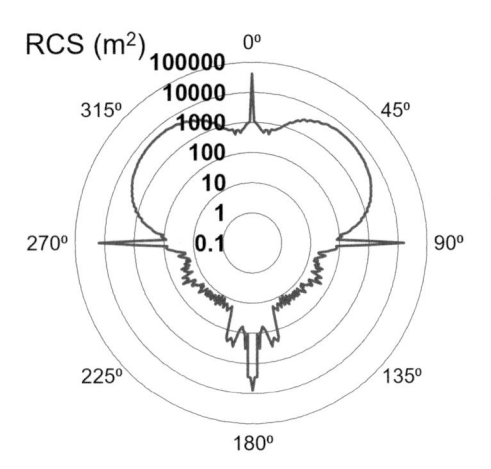

FIGURE 3.127 Circular diagram of noncoherent RCS given radar observation of TEL 5P85D model (sounding frequency 10 GHz, $\gamma = 30°$, underlying surface – dry loam, vertical polarization).

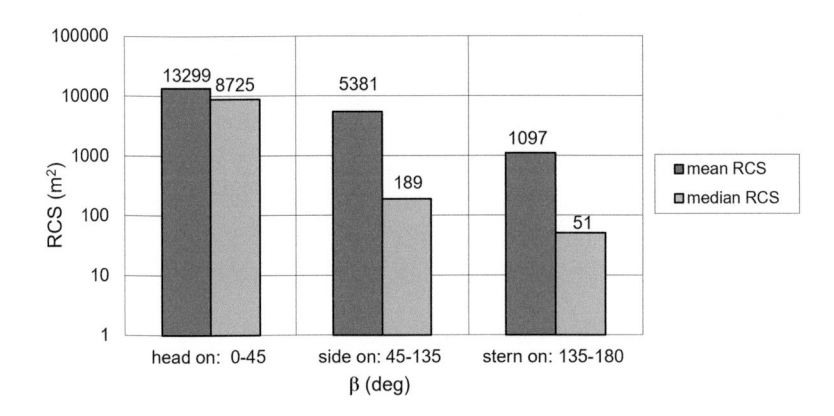

FIGURE 3.128 Diagrams of mean and median RCS of TEL 5P85D model in three sectors of azimuth aspect given its radar observation (sounding frequency 10 GHz, $\gamma = 30°$, underlying surface – dry loam, vertical polarization).

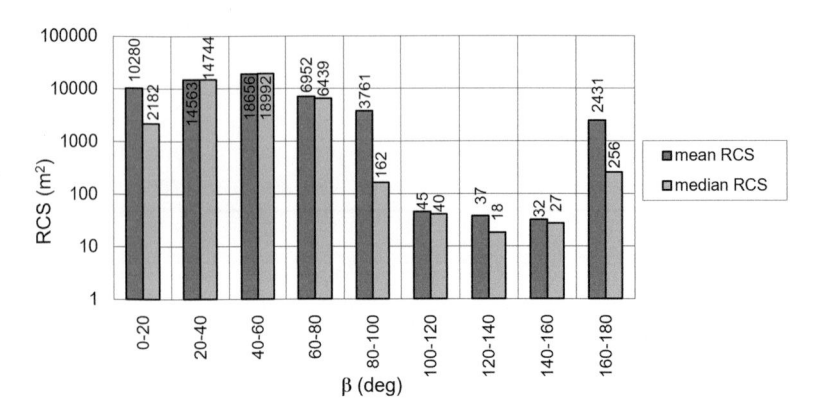

FIGURE 3.129 Diagrams of mean and median RCS of TEL 5P85D model in 20-degree sectors of azimuth aspect given its radar observation (sounding frequency 10 GHz, $\gamma = 30°$, underlying surface – dry loam, vertical polarization).

3.9 S-300PS TRANSPORTER ERECTOR LAUNCHER 5P85S

5P85S - (Figure 3.130) is a "master" TEL for S-300PS missile system (SA-10D Grumble) [81]. It has been in service since 1985.

In accordance with the design of the TEL 5P85S, a model of its surface was created to obtain scattering characteristics (in particular, RCS). The model is shown in Figure 3.131. In the modeling, the smooth parts of the object surface were approximated by parts of 136 triaxial ellipsoids. The surface breaks were modeled using 116 straight edge scattering parts.

FIGURE 3.130 S-300PS transporter erector launcher 5P85S.

FIGURE 3.131 Surface model of S-300PS TEL 5P85S.

Below are some scattering characteristics of the TEL 5P85S model at a sounding frequency of 10 GHz (wavelengths of 3 cm) for the following conditions: the elevation angle of sounding $\gamma = 10°$, wet soil, and the elevation angle of sounding $\gamma = 30°$, dry loam.

Scattering characteristics of the TEL 5P85S model for sounding frequency 10 GHz, elevation angle 10°, wet soil, and horizontal polarization.

Figure 3.132 shows the RCS circular diagram of the TEL 5P85S model. The noncoherent RCS circular diagram of the TEL 5P85S is presented in Figure 3.133. The circular mean RCS of the TEL 5P85S model is 101 493 m². The circular median RCS is 19 757 m². Figures 3.134 and 3.135 show the mean and median RCS for the main ranges of sounding azimuths (head, side, and stern) and for ranges of 20°.

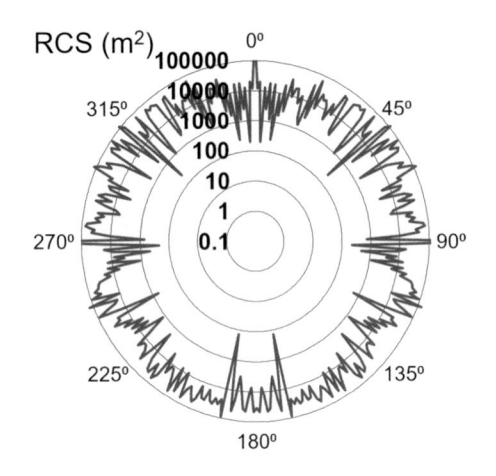

FIGURE 3.132 Circular diagrams of RCS given radar observation of TEL 5P85S model (sounding frequency 10 GHz, $\gamma = 10°$, underlying surface – wet soil, horizontal polarization).

FIGURE 3.133 Circular diagram of noncoherent RCS given radar observation of TEL 5P85S model (sounding frequency 10 GHz, $\gamma = 10°$, underlying surface – wet soil, horizontal polarization).

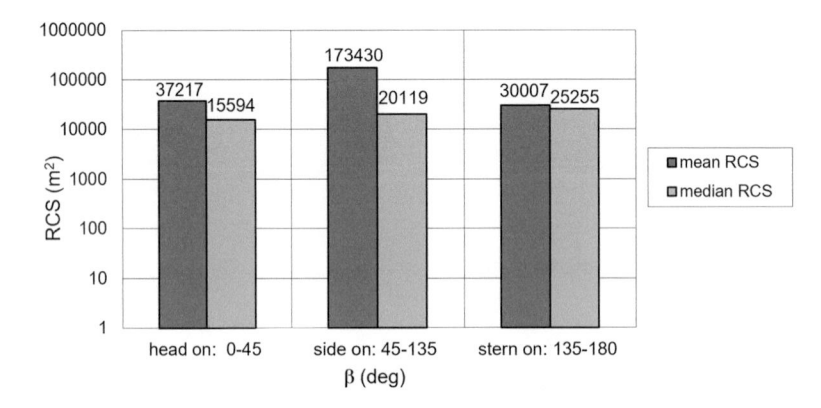

FIGURE 3.134 Diagrams of mean and median RCS of TEL 5P85S model in three sectors of azimuth aspect given its radar observation (sounding frequency 10 GHz, $\gamma = 10°$, underlying surface – wet soil, horizontal polarization).

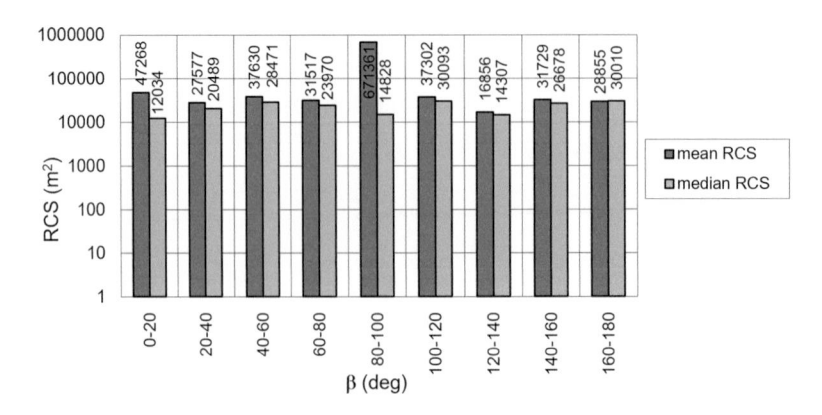

FIGURE 3.135 Diagrams of mean and median RCS of TEL 5P85S model in 20-degree sectors of azimuth aspect given its radar observation (sounding frequency 10 GHz, $\gamma = 10°$, underlying surface – wet soil, horizontal polarization).

Scattering Characteristics of Ground Objects

Scattering characteristics of the TEL 5P85S model for sounding frequency 10 GHz, elevation angle 10°, wet soil, and vertical polarization.

Figure 3.136 shows the RCS circular diagram of the TEL 5P85S model. The noncoherent RCS circular diagram of the TEL 5P85S is presented in Figure 3.137.

The circular mean RCS of the TEL 5P85S model is 5 320 m². The circular median RCS is 253.39 m².

Figures 3.138 and 3.139 show the mean and median RCS for the main ranges of sounding azimuths (head, side, and stern) and for ranges of 20°.

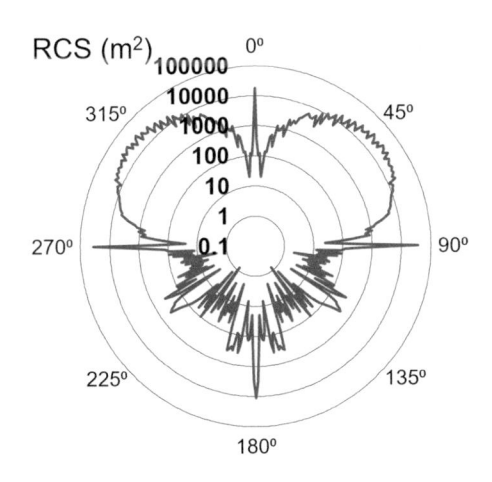

FIGURE 3.136 Circular diagrams of RCS given radar observation of TEL 5P85S model (sounding frequency 10 GHz, $\gamma = 10°$, underlying surface – wet soil, vertical polarization).

FIGURE 3.137 Circular diagram of noncoherent RCS given radar observation of TEL 5P85S model (sounding frequency 10 GHz, $\gamma = 10°$, underlying surface – wet soil, vertical polarization).

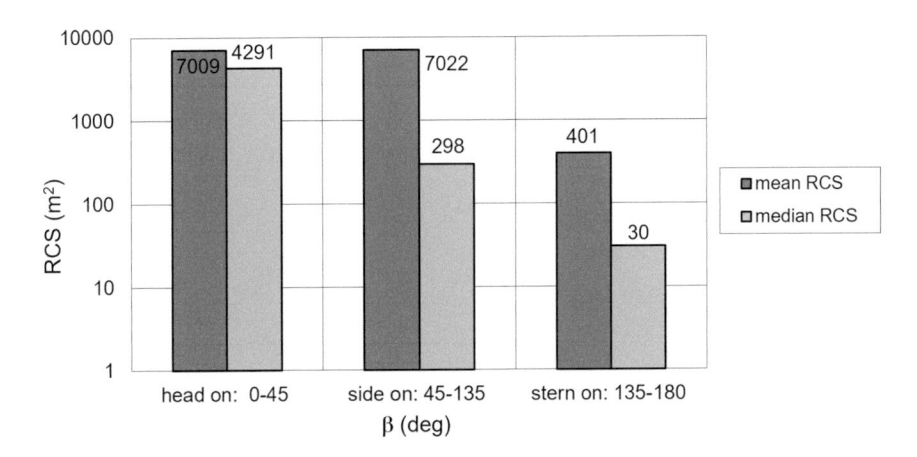

FIGURE 3.138 Diagrams of mean and median RCS of TEL 5P85S model in three sectors of azimuth aspect given its radar observation (sounding frequency 10 GHz, $\gamma = 10°$, underlying surface – wet soil, vertical polarization).

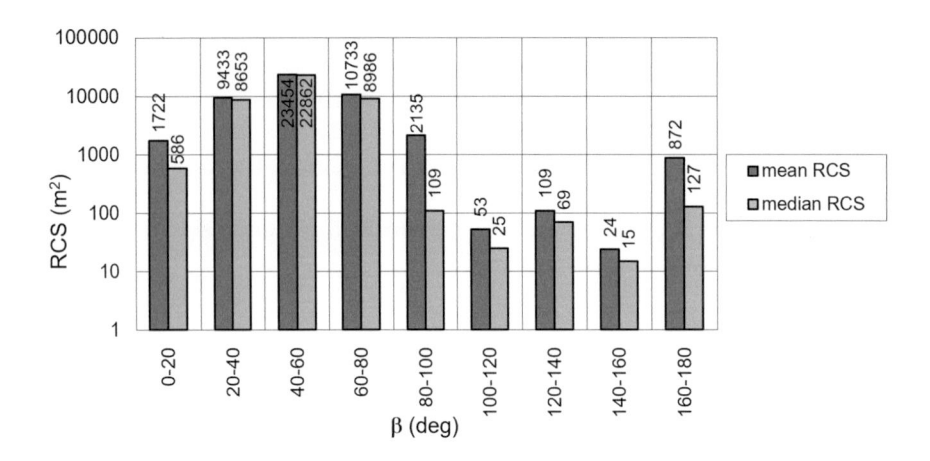

FIGURE 3.139 Diagrams of mean and median RCS of TEL 5P85S model in 20-degree sectors of azimuth aspect given its radar observation (sounding frequency 10 GHz, $\gamma = 10°$, underlying surface – wet soil, vertical polarization).

Scattering characteristics of the TEL 5P85S model for sounding frequency 10 GHz, elevation angle 30°, dry loam, and horizontal polarization.

Figure 3.140 shows the RCS circular diagram of the TEL 5P85S model. The noncoherent RCS circular diagram of the TEL 5P85S is presented in Figure 3.141. The circular mean RCS of the TEL 5P85S model is 26 807 m². The circular median RCS is 7 917 m².

Figures 3.142 and 3.143 show the mean and median RCS for the main ranges of sounding azimuths (head, side, and stern) and for ranges of 20°.

FIGURE 3.140 Circular diagrams of RCS given radar observation of TEL 5P85S model (sounding frequency 10 GHz, $\gamma = 30°$, underlying surface – dry loam, horizontal polarization).

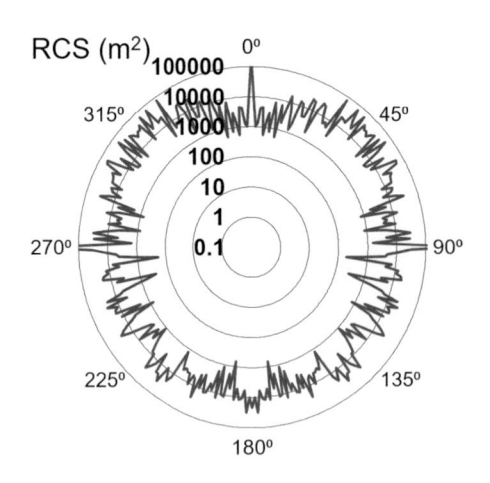

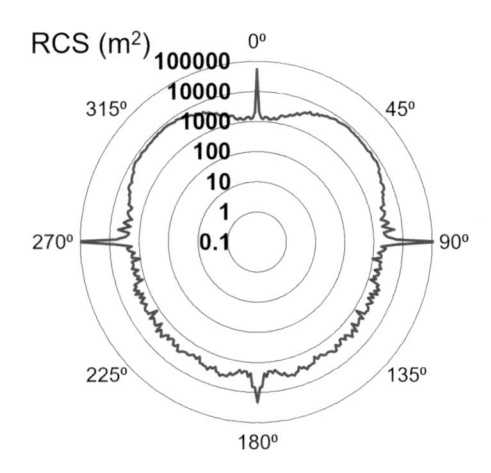

FIGURE 3.141 Circular diagram of noncoherent RCS given radar observation of TEL 5P85S model (sounding frequency 10 GHz, $\gamma = 30°$, underlying surface – dry loam, horizontal polarization).

Scattering Characteristics of Ground Objects

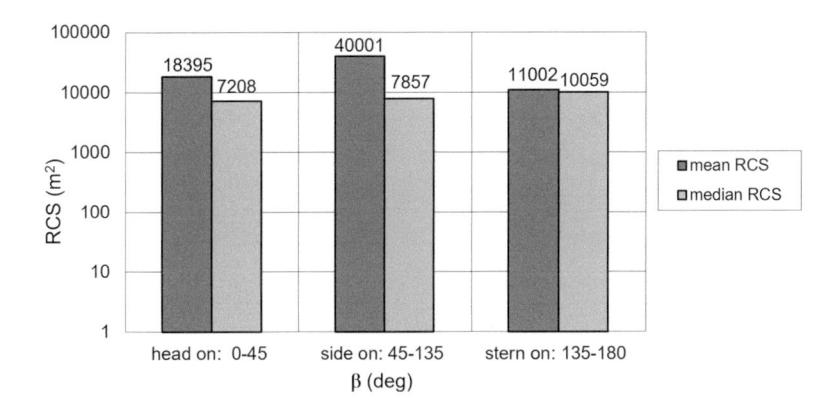

FIGURE 3.142 Diagrams of mean and median RCS of TEL 5P85S model in three sectors of azimuth aspect given its radar observation (sounding frequency 10 GHz, $\gamma=30°$, underlying surface – dry loam, horizontal polarization).

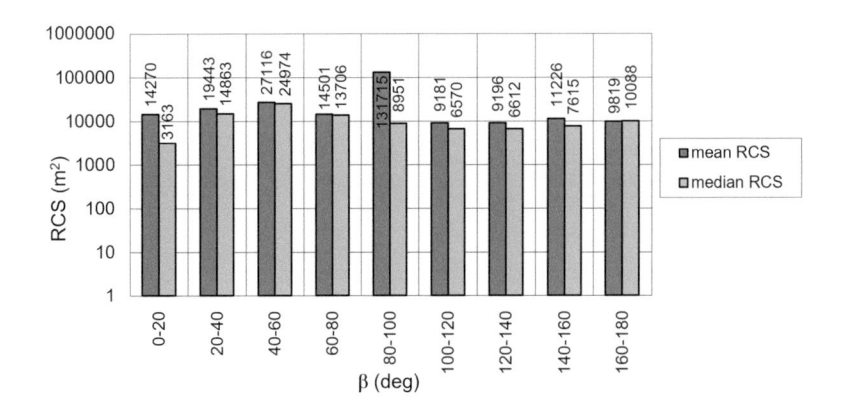

FIGURE 3.143 Diagrams of mean and median RCS of TEL 5P85S model in 20-degree sectors of azimuth aspect given its radar observation (sounding frequency 10 GHz, $\gamma=30°$, underlying surface – dry loam, horizontal polarization).

Scattering characteristics of the TEL 5P85S model for sounding frequency 10 GHz, elevation angle 30°, dry loam, and vertical polarization.

Figure 3.144 shows the RCS circular diagram of the TEL 5P85S model. The noncoherent RCS circular diagram of the TEL 5P85S is presented in Figure 3.145. The circular mean RCS of the TEL 5P85S model is 5 957 m². The circular median RCS is 532.99 m².

Figures 3.146 and 3.147 show the mean and median RCS for the main ranges of sounding azimuths (head, side, and stern) and for ranges of 20°.

The scattering characteristics for this model at two polarizations, sounding frequencies of 10, 5, and 3 GHz (wavelengths of 3, 6, and 10 cm, respectively), for three types of underlying surface (dry loam, dry soil, and wet soil) are given in the electronic appendix of this book.

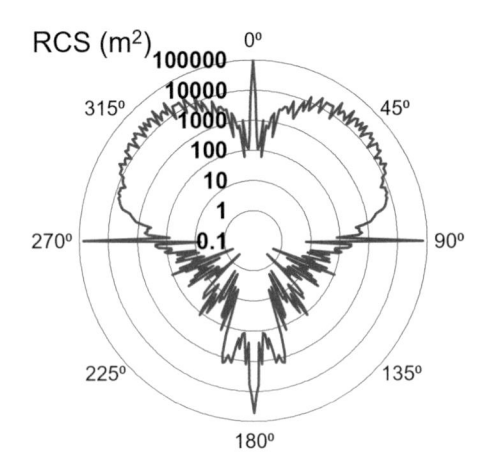

FIGURE 3.144 Circular diagrams of RCS given radar observation of TEL 5P85S model (sounding frequency 10 GHz, $\gamma=30°$, underlying surface – dry loam, vertical polarization).

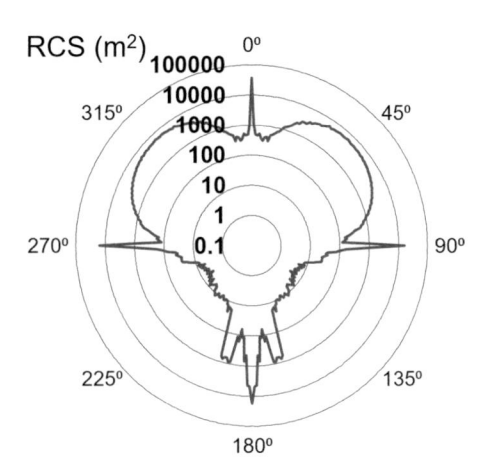

FIGURE 3.145 Circular diagram of noncoherent RCS given radar observation of TEL 5P85S model (sounding frequency 10 GHz, $\gamma=30°$, underlying surface – dry loam, vertical polarization).

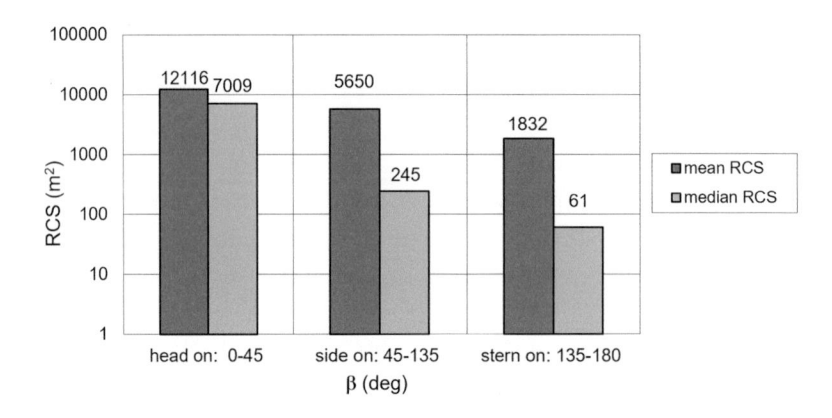

FIGURE 3.146 Diagrams of mean and median RCS of TEL 5P85S model in three sectors of azimuth aspect given its radar observation (sounding frequency 10 GHz, $\gamma=30°$, underlying surface – dry loam, vertical polarization).

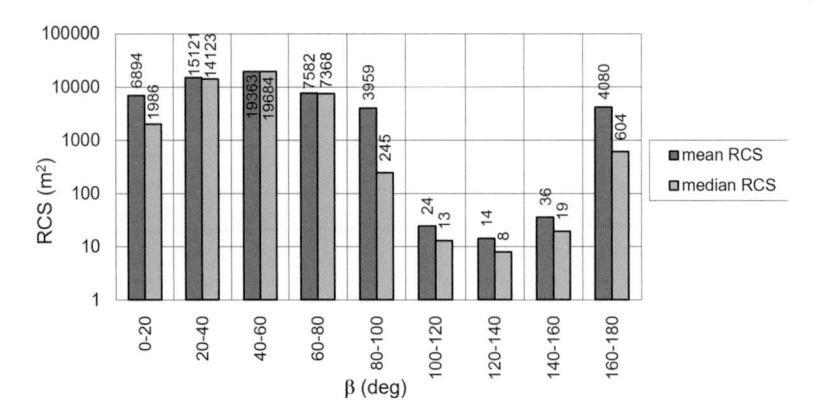

FIGURE 3.147 Diagrams of mean and median RCS of TEL 5P85S model in 20-degree sectors of azimuth aspect given its radar observation (sounding frequency 10 GHz, $\gamma=30°$, underlying surface – dry loam, vertical polarization).

3.10 S-300PS FIRE CONTROL SYSTEM 30N6

30N6- Flap Lid - (Figure 3.148) is a fire control system with the target engagement radar for the S300PS missile system (SA-10D Grumble) [82,83]. It has been in service since 1978.

In accordance with the design of the vehicle 30N6, a model of its surface was created to obtain scattering characteristics (in particular, RCS). The model is shown in Figure 3.149. In the modeling, the smooth parts of the object surface were approximated by parts of 98 triaxial ellipsoids. The surface breaks were modeled using 87 straight edge scattering parts.

FIGURE 3.148 S-300PS fire control system 30N6.

FIGURE 3.149 Surface model of S-300PS fire control system 30N6.

Below are some scattering characteristics of the vehicle 30N6 model at a sounding frequency of 10 GHz (wavelengths of 3 cm) for the following conditions: the elevation angle of sounding $\gamma = 10°$, wet soil, and the elevation angle of sounding $\gamma = 30°$, dry loam.

Scattering characteristics of the vehicle 30N6 model for sounding frequency 10 GHz, elevation angle 10°, wet soil, and horizontal polarization.

Figure 3.150 shows the RCS circular diagram of the vehicle 30N6 model. The noncoherent RCS circular diagram of the vehicle 30N6 is presented in Figure 3.151. The circular mean RCS of the vehicle 30N6 model is 360 007 m². The circular median RCS is 4 516 m². Figures 3.152 and 3.153 show the mean and median RCS for the main ranges of sounding azimuths (head, side, and stern) and for ranges of 20°.

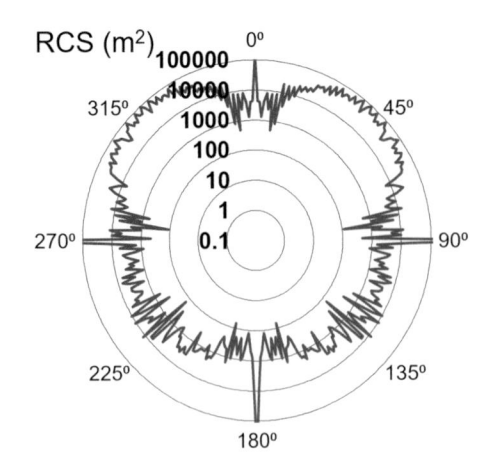

FIGURE 3.150 Circular diagrams of RCS given radar observation of vehicle 30N6 model (sounding frequency 10 GHz, $\gamma=10°$, underlying surface – wet soil, horizontal polarization).

FIGURE 3.151 Circular diagram of noncoherent RCS given radar observation of vehicle 30N6 model (sounding frequency 10 GHz, $\gamma=10°$, underlying surface – wet soil, horizontal polarization).

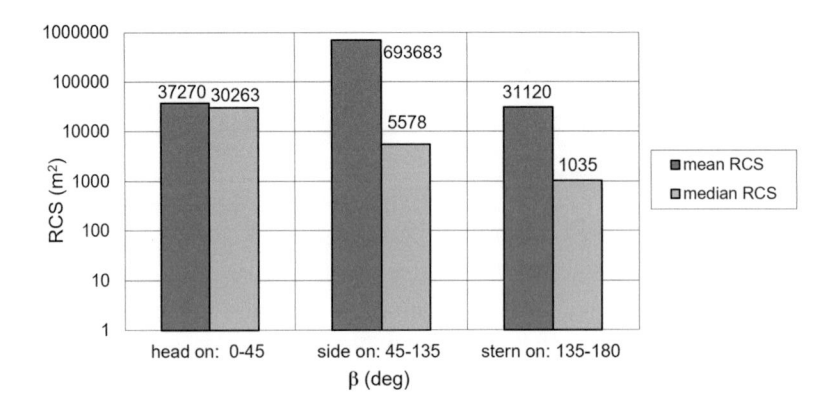

FIGURE 3.152 Diagrams of mean and median RCS of vehicle 30N6 model in three sectors of azimuth aspect given its radar observation (sounding frequency 10 GHz, $\gamma=10°$, underlying surface – wet soil, horizontal polarization).

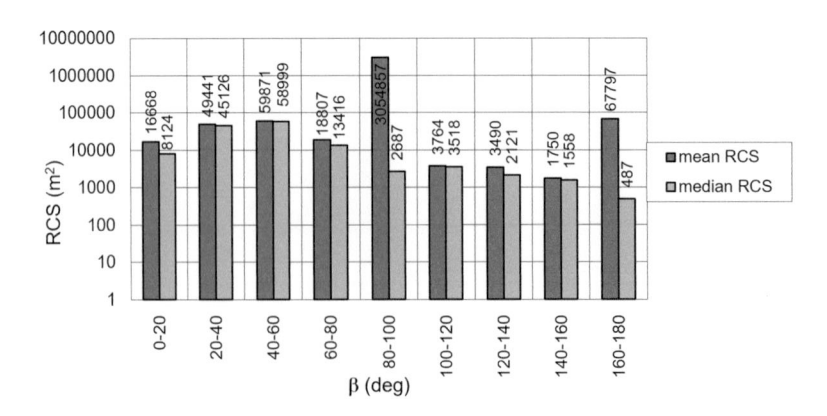

FIGURE 3.153 Diagrams of mean and median RCS of vehicle 30N6 model in 20-degree sectors of azimuth aspect given its radar observation (sounding frequency 10 GHz, $\gamma=10°$, underlying surface – wet soil, horizontal polarization).

Scattering Characteristics of Ground Objects 443

Scattering characteristics of the vehicle 30N6 model for sounding frequency 10 GHz, elevation angle 10°, wet soil, and vertical polarization.

Figure 3.154 shows the RCS circular diagram of the vehicle 30N6 model. The noncoherent RCS circular diagram of the vehicle 30N6 is presented in Figure 3.155.

The circular mean RCS of the vehicle 30N6 model is 10 424 m². The circular median RCS is 490.08 m².

Figures 3.156 and 3.157 show the mean and median RCS for the main ranges of sounding azimuths (head, side, and stern) and for ranges of 20°.

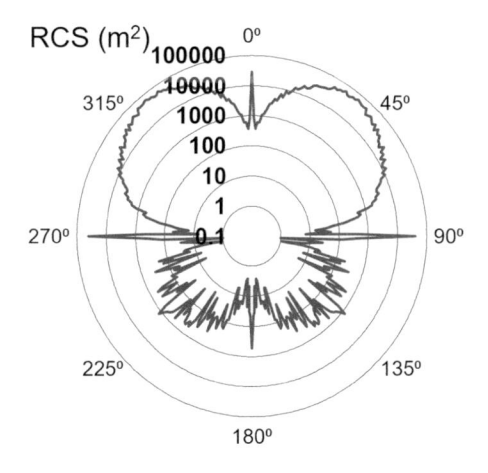

FIGURE 3.154 Circular diagrams of RCS given radar observation of vehicle 30N6 model (sounding frequency 10 GHz, $\gamma = 10°$, underlying surface – wet soil, vertical polarization).

FIGURE 3.155 Circular diagram of noncoherent RCS given radar observation of vehicle 30N6 model (sounding frequency 10 GHz, $\gamma = 10°$, underlying surface – wet soil, vertical polarization).

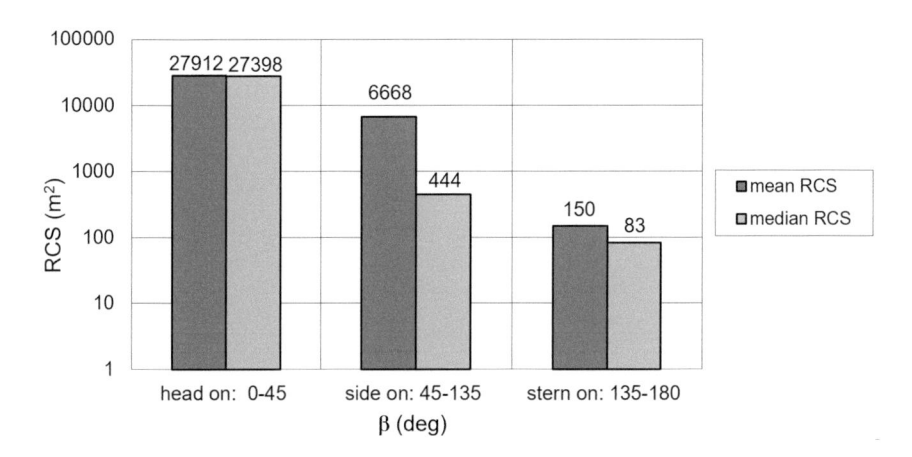

FIGURE 3.156 Diagrams of mean and median RCS of vehicle 30N6 model in three sectors of azimuth aspect given its radar observation (sounding frequency 10 GHz, $\gamma = 10°$, underlying surface – wet soil, vertical polarization).

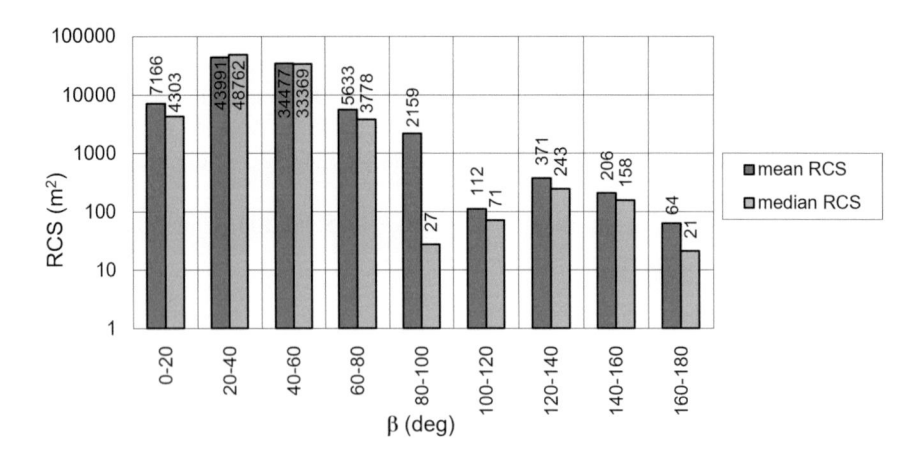

FIGURE 3.157 Diagrams of mean and median RCS of vehicle 30N6 model in 20-degree sectors of azimuth aspect given its radar observation (sounding frequency 10 GHz, $\gamma = 10°$, underlying surface – wet soil, vertical polarization).

Scattering characteristics of the vehicle 30N6 model for sounding frequency 10 GHz, elevation angle 30°, dry loam, and horizontal polarization.

Figure 3.158 shows the RCS circular diagram of the vehicle 30N6 model. The noncoherent RCS circular diagram of the vehicle 30N6 is presented in Figure 3.159. The circular mean RCS of the vehicle 30N6 model is 73 409 m². The circular median RCS is 783.62 m².

Figures 3.160 and 3.161 show the mean and median RCS for the main ranges of sounding azimuths (head, side, and stern) and for ranges of 20°.

FIGURE 3.158 Circular diagrams of RCS given radar observation of vehicle 30N6 model (sounding frequency 10 GHz, $\gamma = 30°$, underlying surface – dry loam, horizontal polarization).

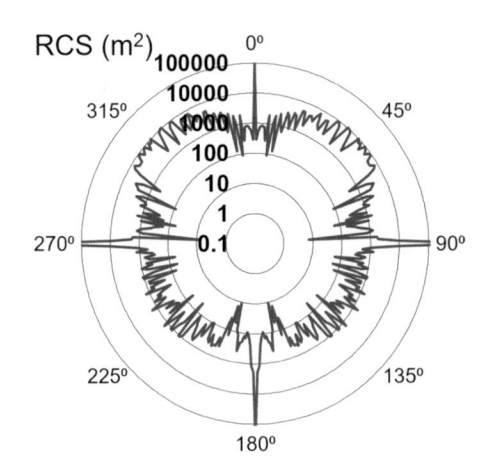

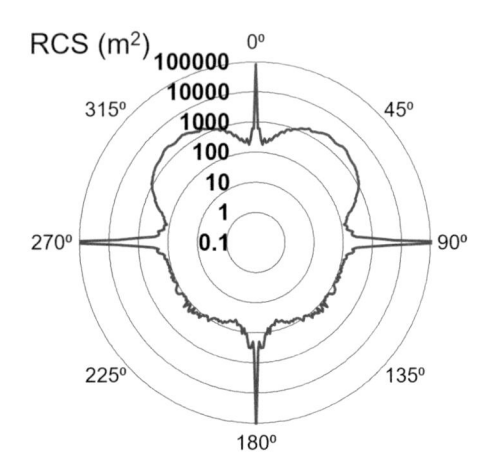

FIGURE 3.159 Circular diagram of noncoherent RCS given radar observation of vehicle 30N6 model (sounding frequency 10 GHz, $\gamma = 30°$, underlying surface – dry loam, horizontal polarization).

Scattering Characteristics of Ground Objects

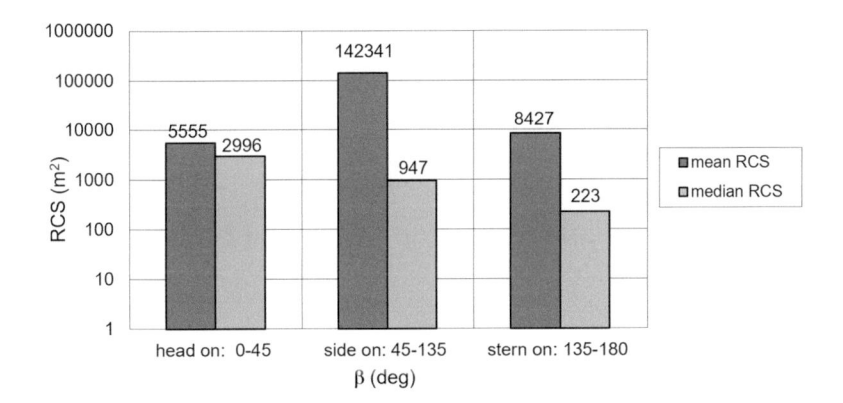

FIGURE 3.160 Diagrams of mean and median RCS of vehicle 30N6 model in three sectors of azimuth aspect given its radar observation (sounding frequency 10 GHz, $\gamma=30°$, underlying surface – dry loam, horizontal polarization).

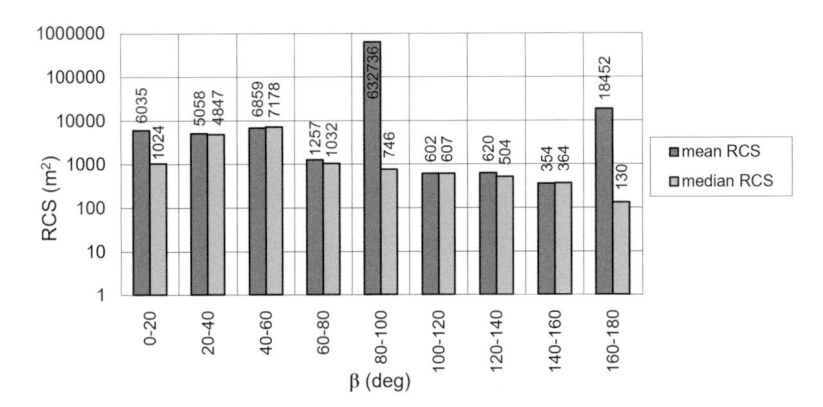

FIGURE 3.161 Diagrams of mean and median RCS of vehicle 30N6 model in 20-degree sectors of azimuth aspect given its radar observation (sounding frequency 10 GHz, $\gamma=30°$, underlying surface – dry loam, horizontal polarization).

Scattering characteristics of the vehicle 30N6 model for sounding frequency 10 GHz, elevation angle 30°, dry loam, and vertical polarization.

Figure 3.162 shows the RCS circular diagram of the vehicle 30N6 model. The noncoherent RCS circular diagram of the vehicle 30N6 is presented in Figure 3.163. The circular mean RCS of the vehicle 30N6 model is 1 849 m². The circular median RCS is 63.30 m².

Figures 3.164 and 3.165 show the mean and median RCS for the main ranges of sounding azimuths (head, side, and stern) and for ranges of 20°.

The scattering characteristics for this model at two polarizations, sounding frequencies of 10, 5, and 3 GHz (wavelengths of 3, 6, and 10 cm, respectively), for three types of underlying surface (dry loam, dry soil, and wet soil) are given in the electronic appendix of this book.

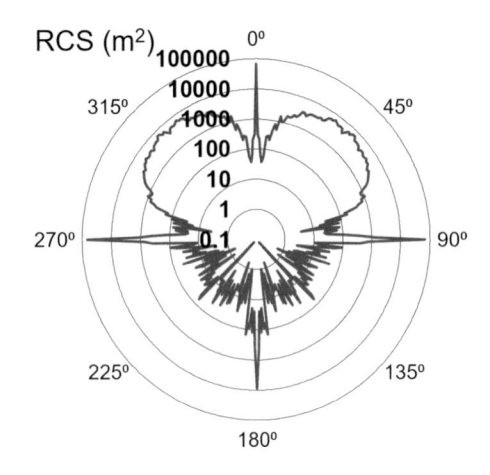

FIGURE 3.162 Circular diagrams of RCS given radar observation of vehicle 30N6 model (sounding frequency 10 GHz, $\gamma=30°$, underlying surface – dry loam, vertical polarization).

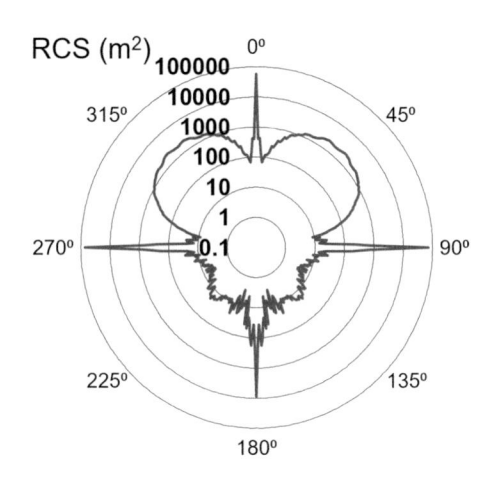

FIGURE 3.163 Circular diagram of noncoherent RCS given radar observation of vehicle 30N6 model (sounding frequency 10 GHz, $\gamma=30°$, underlying surface – dry loam, vertical polarization).

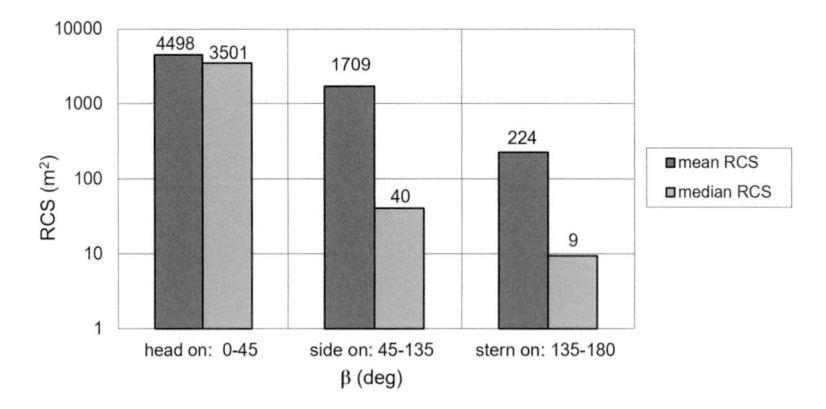

FIGURE 3.164 Diagrams of mean and median RCS of vehicle 30N6 model in three sectors of azimuth aspect given its radar observation (sounding frequency 10 GHz, $\gamma=30°$, underlying surface – dry loam, vertical polarization).

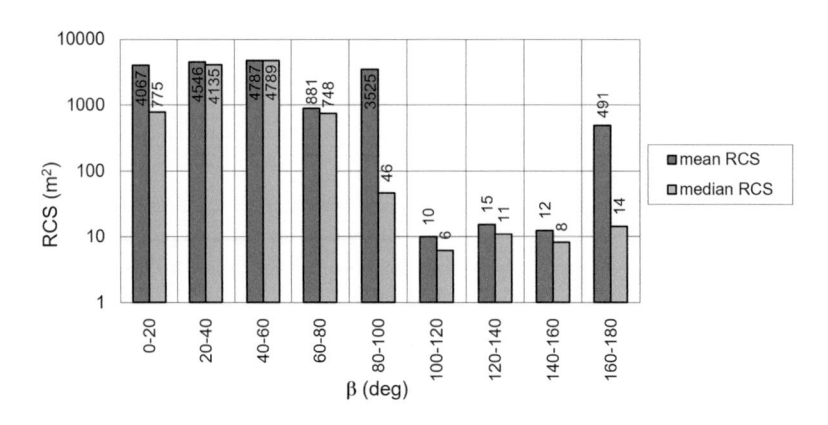

FIGURE 3.165 Diagrams of mean and median RCS of vehicle 30N6 model in 20-degree sectors of azimuth aspect given its radar observation (sounding frequency 10 GHz, $\gamma=30°$, underlying surface – dry loam, vertical polarization).

Scattering Characteristics of Ground Objects   447

3.11 S-300PS LOW ALTITUDE ACQUISITION RADAR 76N6

76N6- Clam Shell - (Figure 3.166) is a low altitude acquisition radar for S300PS missile system (SA-10D Grumble) [84]. It has been in service since 1978.

In accordance with the design of the radar 76N6, a model of its surface was created to obtain scattering characteristics (in particular, RCS). The model is shown in Figure 3.167. In the modeling, the smooth parts of the object surface were approximated by parts of 111 triaxial ellipsoids. The surface breaks were modeled using 86 straight edge scattering parts.

FIGURE 3.166 S-300PS low altitude acquisition radar 76N6.

FIGURE 3.167 Surface model of S-300PS low altitude acquisition radar 76N6.

Below are some scattering characteristics of the radar 76N6 model at a sounding frequency of 10 GHz (wavelengths of 3 cm) for the following conditions: the elevation angle of sounding $\gamma = 10°$, wet soil, and the elevation angle of sounding $\gamma = 30°$, dry loam.

Scattering characteristics of the radar 76N6 model for sounding frequency 10 GHz, elevation angle 10°, wet soil, and horizontal polarization.

Figure 3.168 shows the RCS circular diagram of the radar 76N6 model. The noncoherent RCS circular diagram of the radar 76N6 is presented in Figure 3.169. The circular mean RCS of the radar 76N6 model is 440 228 m². The circular median RCS is 87 064 m². Figures 3.170 and 3.171 show the mean and median RCS for the main ranges of sounding azimuths (head, side, and stern) and for ranges of 20°.

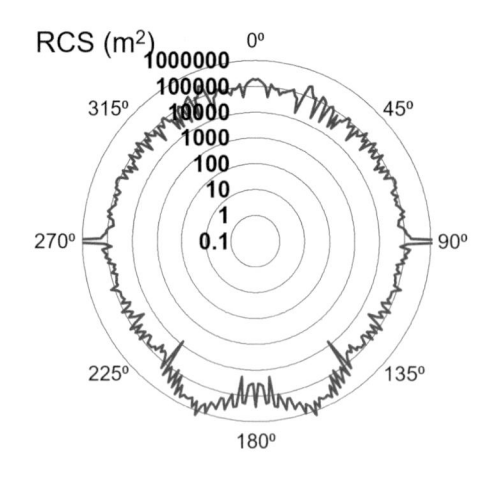

FIGURE 3.168 Circular diagrams of RCS given radar observation of radar 76N6 model (sounding frequency 10 GHz, $\gamma = 10°$, underlying surface – wet soil, horizontal polarization).

FIGURE 3.169 Circular diagram of noncoherent RCS given radar observation of radar 76N6 model (sounding frequency 10 GHz, $\gamma = 10°$, underlying surface – wet soil, horizontal polarization).

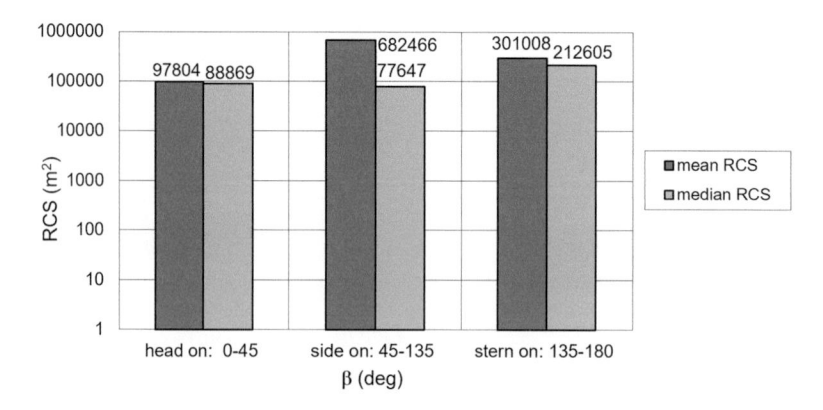

FIGURE 3.170 Diagrams of mean and median RCS of radar 76N6 model in three sectors of azimuth aspect given its radar observation (sounding frequency 10 GHz, $\gamma = 10°$, underlying surface – wet soil, horizontal polarization).

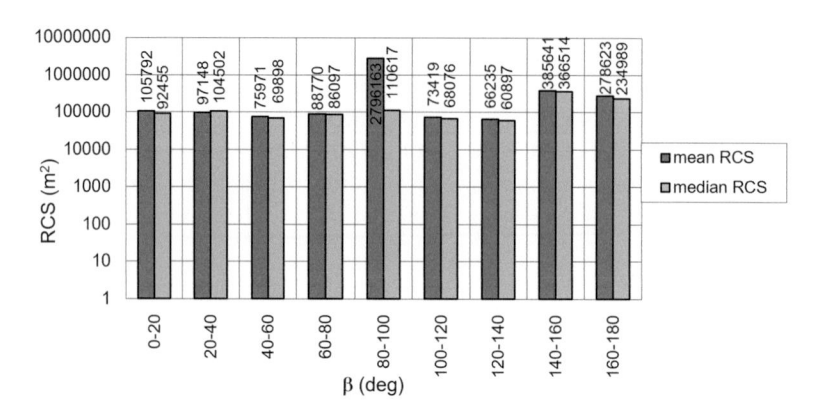

FIGURE 3.171 Diagrams of mean and median RCS of radar 76N6 model in 20-degree sectors of azimuth aspect given its radar observation (sounding frequency 10 GHz, $\gamma = 10°$, underlying surface – wet soil, horizontal polarization).

Scattering Characteristics of Ground Objects 449

Scattering characteristics of the radar 76N6 model for sounding frequency 10 GHz, elevation angle 10°, wet soil, and vertical polarization.

Figure 3.172 shows the RCS circular diagram of the radar 76N6 model. The noncoherent RCS circular diagram of the radar 76N6 is presented in Figure 3.173.

The circular mean RCS of the radar 76N6 model is 1 210 m². The circular median RCS is 193.55 m².

Figures 3.174 and 3.175 show the mean and median RCS for the main ranges of sounding azimuths (head, side, and stern) and for ranges of 20°.

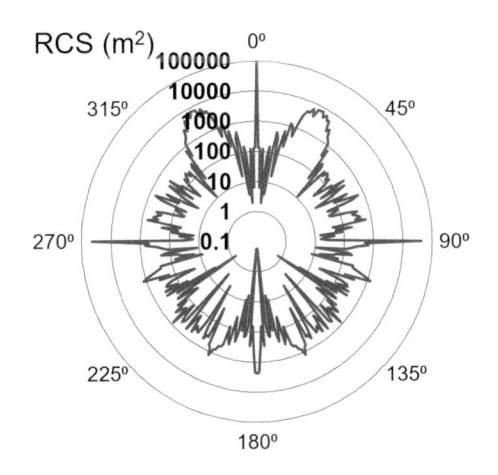

FIGURE 3.172 Circular diagrams of RCS given radar observation of radar 76N6 model (sounding frequency 10 GHz, $\gamma = 10°$, underlying surface – wet soil, vertical polarization).

FIGURE 3.173 Circular diagram of noncoherent RCS given radar observation of radar 76N6 model (sounding frequency 10 GHz, $\gamma = 10°$, underlying surface – wet soil, vertical polarization).

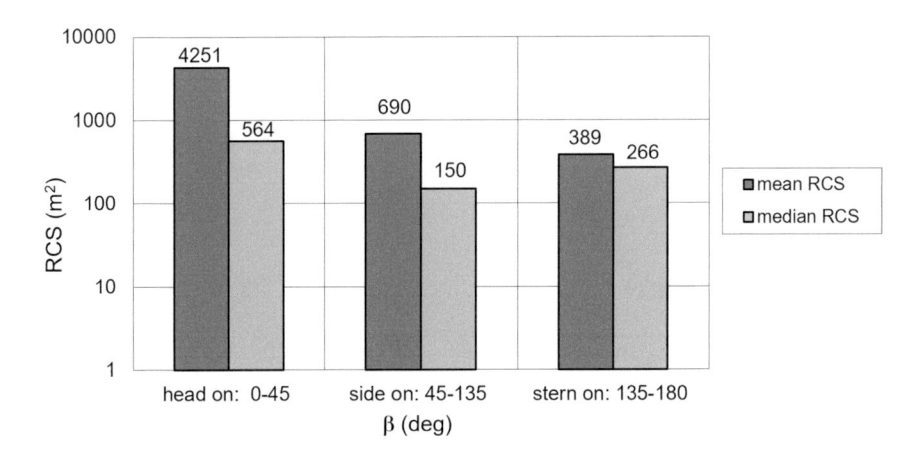

FIGURE 3.174 Diagrams of mean and median RCS of radar 76N6 model in three sectors of azimuth aspect given its radar observation (sounding frequency 10 GHz, $\gamma = 10°$, underlying surface – wet soil, vertical polarization).

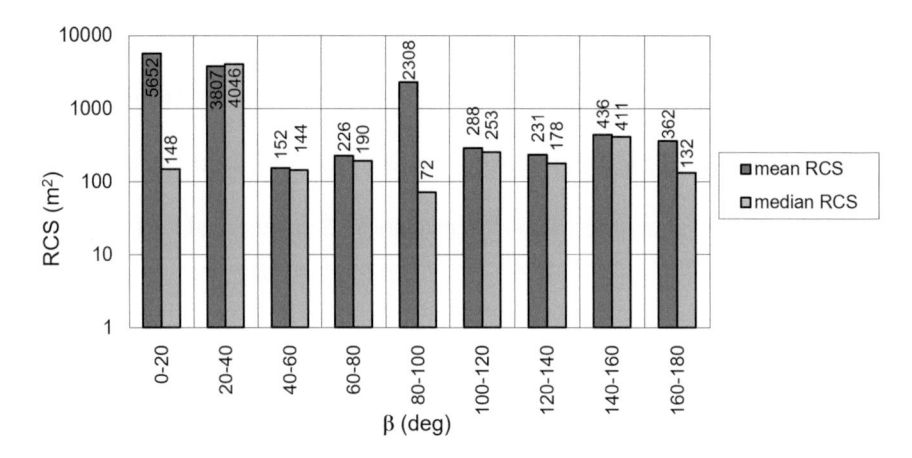

FIGURE 3.175 Diagrams of mean and median RCS of radar 76N6 model in 20-degree sectors of azimuth aspect given its radar observation (sounding frequency 10 GHz, $\gamma = 10°$, underlying surface – wet soil, vertical polarization).

Scattering characteristics of the radar 76N6 model for sounding frequency 10 GHz, elevation angle 30°, dry loam, and horizontal polarization.

Figure 3.176 shows the RCS circular diagram of the radar 76N6 model. The noncoherent RCS circular diagram of the radar 76N6 is presented in Figure 3.177. The circular mean RCS of the radar 76N6 model is 129 888 m². The circular median RCS is 35 052 m².

Figures 3.178 and 3.179 show the mean and median RCS for the main ranges of sounding azimuths (head, side, and stern) and for ranges of 20°.

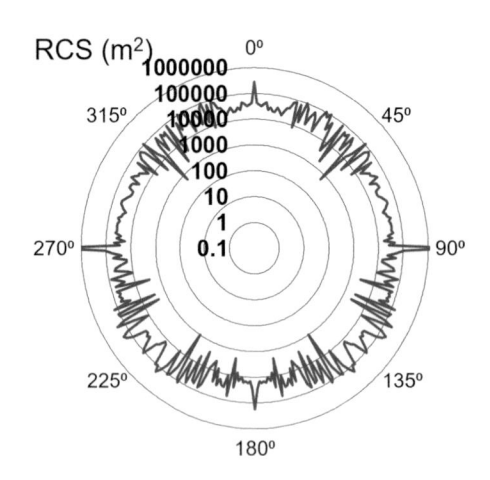

FIGURE 3.176 Circular diagrams of RCS given radar observation of radar 76N6 model (sounding frequency 10 GHz, $\gamma = 30°$, underlying surface – dry loam, horizontal polarization).

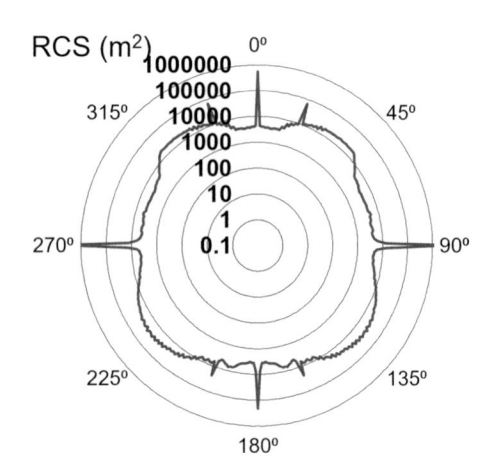

FIGURE 3.177 Circular diagram of noncoherent RCS given radar observation of radar 76N6 model (sounding frequency 10 GHz, $\gamma = 30°$, underlying surface – dry loam, horizontal polarization).

Scattering Characteristics of Ground Objects

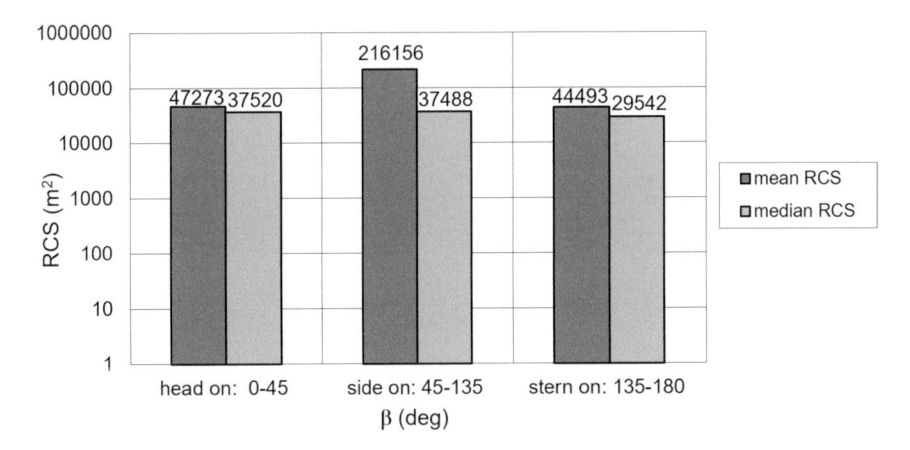

FIGURE 3.178 Diagrams of mean and median RCS of radar 76N6 model in three sectors of azimuth aspect given its radar observation (sounding frequency 10 GHz, $\gamma=30°$, underlying surface – dry loam, horizontal polarization).

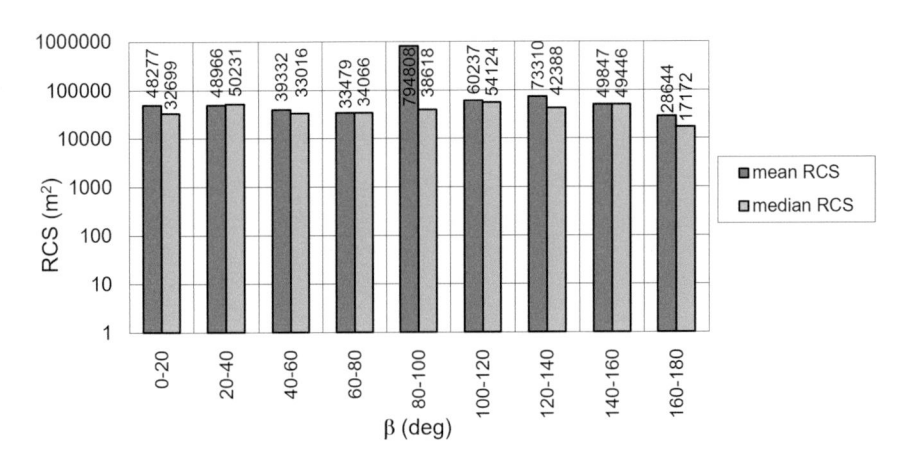

FIGURE 3.179 Diagrams of mean and median RCS of radar 76N6 model in 20-degree sectors of azimuth aspect given its radar observation (sounding frequency 10 GHz, $\gamma=30°$, underlying surface – dry loam, horizontal polarization).

Scattering characteristics of the radar 76N6 model for sounding frequency 10 GHz, elevation angle 30°, dry loam, and vertical polarization.

Figure 3.180 shows the RCS circular diagram of the radar 76N6 model. The noncoherent RCS circular diagram of the radar 76N6 is presented in Figure 3.181. The circular mean RCS of the radar 76N6 model is 6 943 m². The circular median RCS is 548.78 m².

Figures 3.182 and 3.183 show the mean and median RCS for the main ranges of sounding azimuths (head, side, and stern) and for ranges of 20°.

The scattering characteristics for this model at two polarizations, sounding frequencies of 10, 5, and 3 GHz (wavelengths of 3, 6, and 10 cm, respectively), for three types of underlying surface (dry loam, dry soil, and wet soil) are given in the electronic appendix of this book.

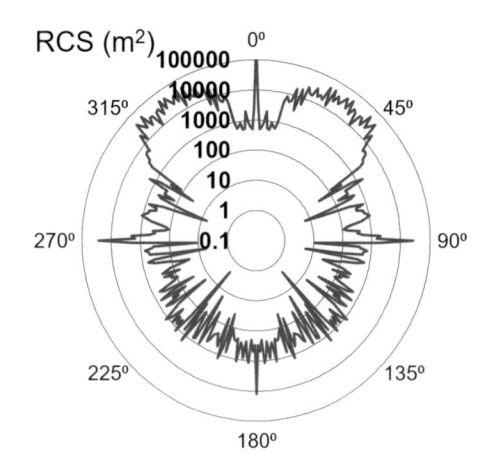

FIGURE 3.180 Circular diagrams of RCS given radar observation of radar 76N6 model (sounding frequency 10 GHz, $\gamma = 30°$, underlying surface – dry loam, vertical polarization).

FIGURE 3.181 Circular diagram of noncoherent RCS given radar observation of radar 76N6 model (sounding frequency 10 GHz, $\gamma = 30°$, underlying surface – dry loam, vertical polarization).

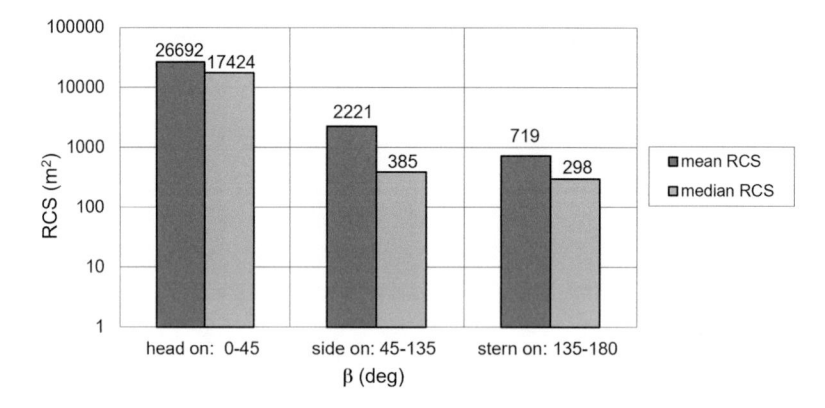

FIGURE 3.182 Diagrams of mean and median RCS of radar 76N6 model in three sectors of azimuth aspect given its radar observation (sounding frequency 10 GHz, $\gamma = 30°$, underlying surface – dry loam, vertical polarization).

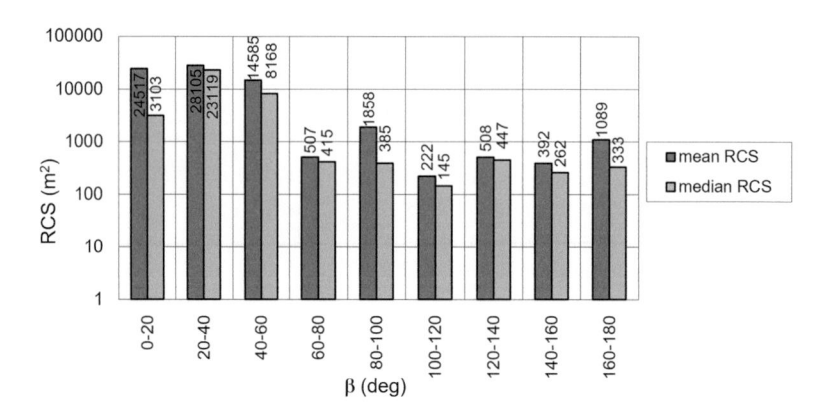

FIGURE 3.183 Diagrams of mean and median RCS of radar 76N6 model in 20-degree sectors of azimuth aspect given its radar observation (sounding frequency 10 GHz, $\gamma = 30°$, underlying surface – dry loam, vertical polarization).

Scattering Characteristics of Ground Objects 453

3.12 S-300PS COMMAND POST VEHICLE 54K6

54K6 (Figure 3.184) is a command post used to control several batteries of the S300PS missile system (SA-10D Grumble) [85]. It has been in service since 1985.

In accordance with the design of the vehicle 54K6, a model of its surface was created to obtain scattering characteristics (in particular, RCS). The model is shown in Figure 3.185. In the modeling, the smooth parts of the object surface were approximated by parts of 81 triaxial ellipsoids. The surface breaks were modeled using 66 straight edge scattering parts.

FIGURE 3.184 S-300PS command post vehicle 54K6.

FIGURE 3.185 Surface model of S-300PS command post vehicle 54K6.

Below are some scattering characteristics of the vehicle 54K6 model at a sounding frequency of 10 GHz (wavelengths of 3 cm) for the following conditions: the elevation angle of sounding $\gamma = 10°$, wet soil, and the elevation angle of sounding $\gamma = 30°$, dry loam.

Scattering characteristics of the vehicle 54K6 model for sounding frequency 10 GHz, elevation angle 10°, wet soil, and horizontal polarization.

Figure 3.186 shows the RCS circular diagram of the vehicle 54K6 model. The noncoherent RCS circular diagram of the vehicle 54K6 is presented in Figure 3.187. The circular mean RCS of the vehicle 54K6 model is 199 411 m². The circular median RCS is 2 992 m². Figures 3.188 and 3.189 show the mean and median RCS for the main ranges of sounding azimuths (head, side, and stern) and for ranges of 20°.

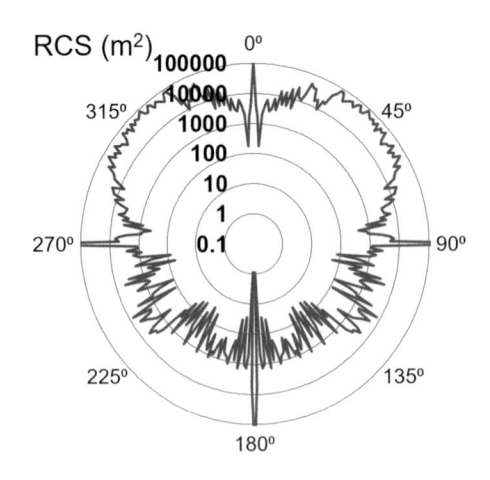

FIGURE 3.186 Circular diagrams of RCS given radar observation of vehicle 54K6 model (sounding frequency 10 GHz, $\gamma = 10°$, underlying surface – wet soil, horizontal polarization).

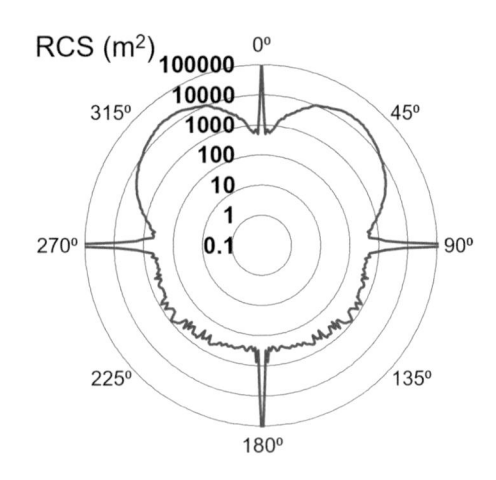

FIGURE 3.187 Circular diagram of noncoherent RCS given radar observation of vehicle 54K6 model (sounding frequency 10 GHz, $\gamma = 10°$, underlying surface – wet soil, horizontal polarization).

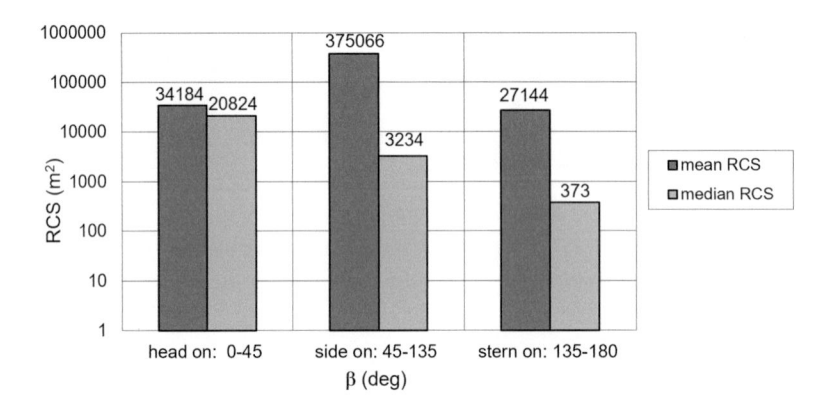

FIGURE 3.188 Diagrams of mean and median RCS of vehicle 54K6 model in three sectors of azimuth aspect given its radar observation (sounding frequency 10 GHz, $\gamma = 10°$, underlying surface – wet soil, horizontal polarization).

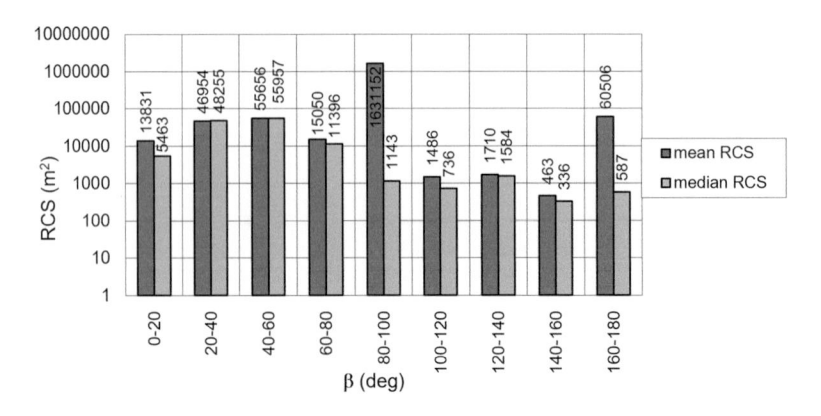

FIGURE 3.189 Diagrams of mean and median RCS of vehicle 54K6 model in 20-degree sectors of azimuth aspect given its radar observation (sounding frequency 10 GHz, $\gamma = 10°$, underlying surface – wet soil, horizontal polarization).

Scattering characteristics of the vehicle 54K6 model for sounding frequency 10 GHz, elevation angle 10°, wet soil, and vertical polarization.

Figure 3.190 shows the RCS circular diagram of the vehicle 54K6 model. The noncoherent RCS circular diagram of the vehicle 54K6 is presented in Figure 3.191.

The circular mean RCS of the vehicle 54K6 model is 10 415 m². The circular median RCS is 465.02 m².

Figures 3.192 and 3.193 show the mean and median RCS for the main ranges of sounding azimuths (head, side, and stern) and for ranges of 20°.

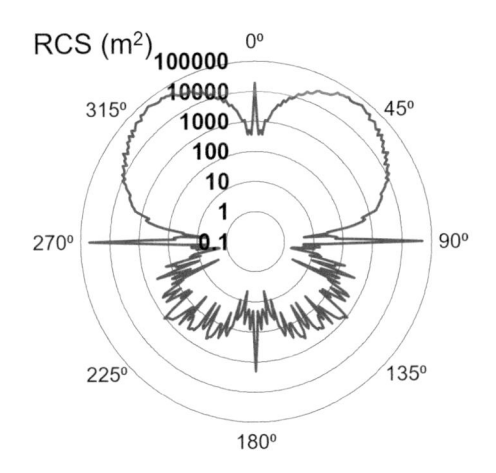

FIGURE 3.190 Circular diagrams of RCS given radar observation of vehicle 54K6 model (sounding frequency 10 GHz, $\gamma = 10°$, underlying surface – wet soil, vertical polarization).

FIGURE 3.191 Circular diagram of noncoherent RCS given radar observation of vehicle 54K6 model (sounding frequency 10 GHz, $\gamma = 10°$, underlying surface – wet soil, vertical polarization).

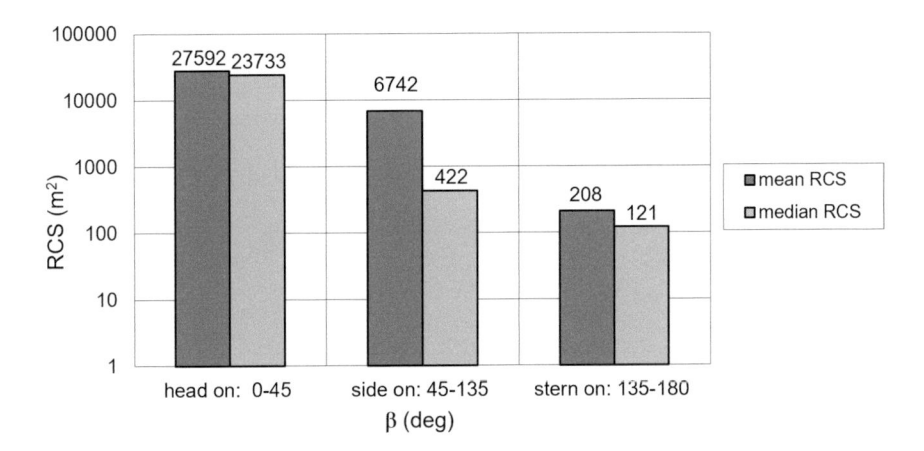

FIGURE 3.192 Diagrams of mean and median RCS of vehicle 54K6 model in three sectors of azimuth aspect given its radar observation (sounding frequency 10 GHz, $\gamma = 10°$, underlying surface – wet soil, vertical polarization).

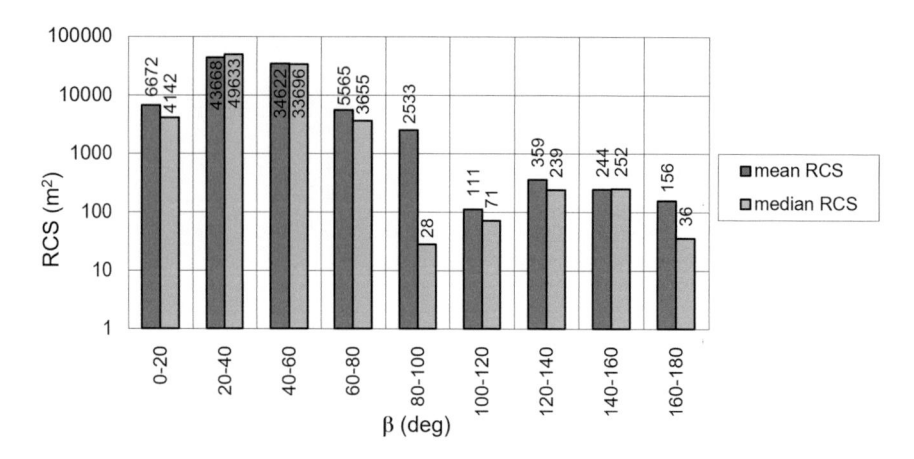

FIGURE 3.193 Diagrams of mean and median RCS of vehicle 54K6 model in 20-degree sectors of azimuth aspect given its radar observation (sounding frequency 10 GHz, $\gamma = 10°$, underlying surface – wet soil, vertical polarization).

Scattering characteristics of the vehicle 54K6 model for sounding frequency 10 GHz, elevation angle 30°, dry loam, and horizontal polarization.

Figure 3.194 shows the RCS circular diagram of the vehicle 54K6 model. The noncoherent RCS circular diagram of the vehicle 54K6 is presented in Figure 3.195. The circular mean RCS of the vehicle 54K6 model is 46 012 m². The circular median RCS is 376.05 m².

Figures 3.196 and 3.197 show the mean and median RCS for the main ranges of sounding azimuths (head, side, and stern) and for ranges of 20°.

FIGURE 3.194 Circular diagrams of RCS given radar observation of vehicle 54K6 model (sounding frequency 10 GHz, $\gamma = 30°$, underlying surface – dry loam, horizontal polarization).

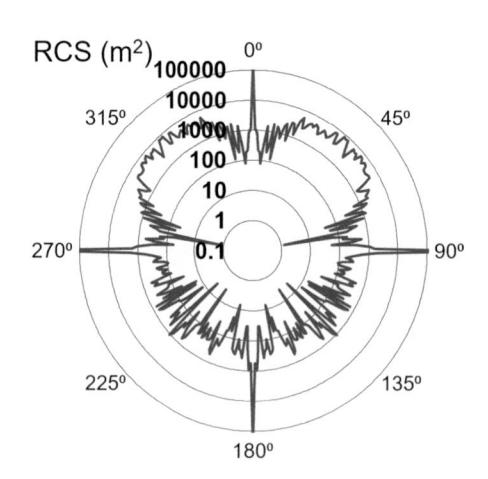

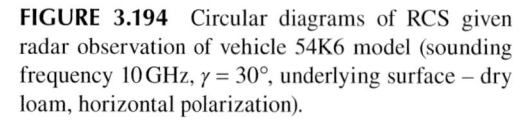

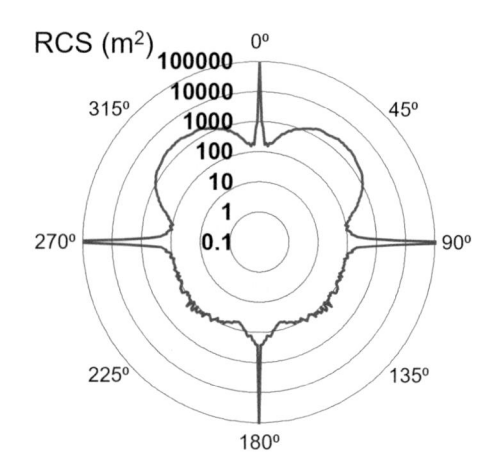

FIGURE 3.195 Circular diagram of noncoherent RCS given radar observation of vehicle 54K6 model (sounding frequency 10 GHz, $\gamma = 30°$, underlying surface – dry loam, horizontal polarization).

Scattering Characteristics of Ground Objects 457

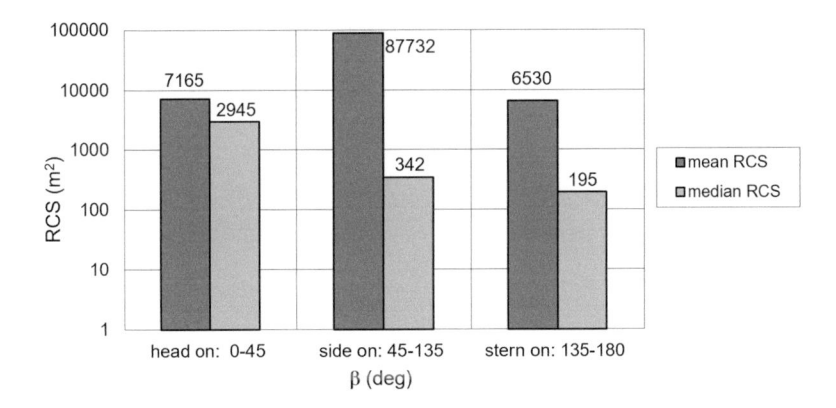

FIGURE 3.196 Diagrams of mean and median RCS of vehicle 54K6 model in three sectors of azimuth aspect given its radar observation (sounding frequency 10 GHz, $\gamma=30°$, underlying surface – dry loam, horizontal polarization).

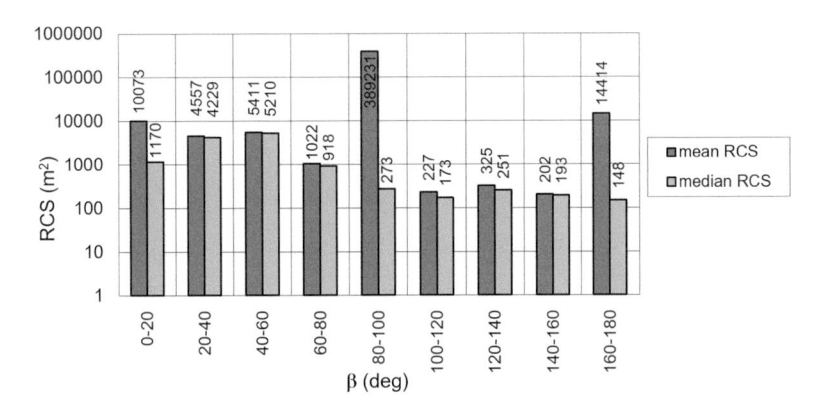

FIGURE 3.197 Diagrams of mean and median RCS of vehicle 54K6 model in 20-degree sectors of azimuth aspect given its radar observation (sounding frequency 10 GHz, $\gamma=30°$, underlying surface – dry loam, horizontal polarization).

Scattering characteristics of the vehicle 54K6 model for sounding frequency 10 GHz, elevation angle 30°, dry loam, and vertical polarization.

Figure 3.198 shows the RCS circular diagram of the vehicle 54K6 model. The noncoherent RCS circular diagram of the vehicle 54K6 is presented in Figure 3.199. The circular mean RCS of the vehicle 54K6 model is 3 018 m². The circular median RCS is 74.31 m².

Figures 3.200 and 3.201 show the mean and median RCS for the main ranges of sounding azimuths (head, side, and stern) and for ranges of 20°.

The scattering characteristics for this model at two polarizations, sounding frequencies of 10, 5, and 3 GHz (wavelengths of 3, 6, and 10 cm, respectively), for three types of underlying surface (dry loam, dry soil, and wet soil) are given in the electronic appendix of this book.

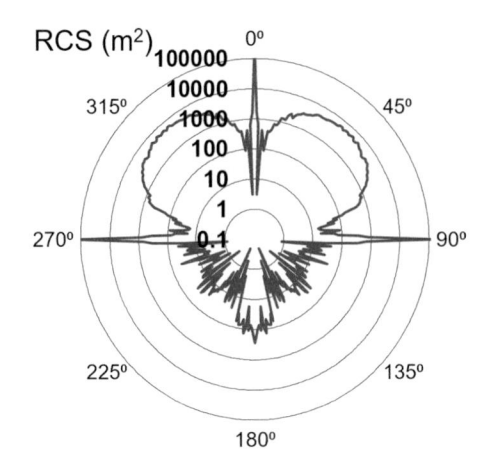

FIGURE 3.198 Circular diagrams of RCS given radar observation of vehicle 54K6 model (sounding frequency 10 GHz, $\gamma = 30°$, underlying surface – dry loam, vertical polarization).

FIGURE 3.199 Circular diagram of noncoherent RCS given radar observation of vehicle 54K6 model (sounding frequency 10 GHz, $\gamma = 30°$, underlying surface – dry loam, vertical polarization).

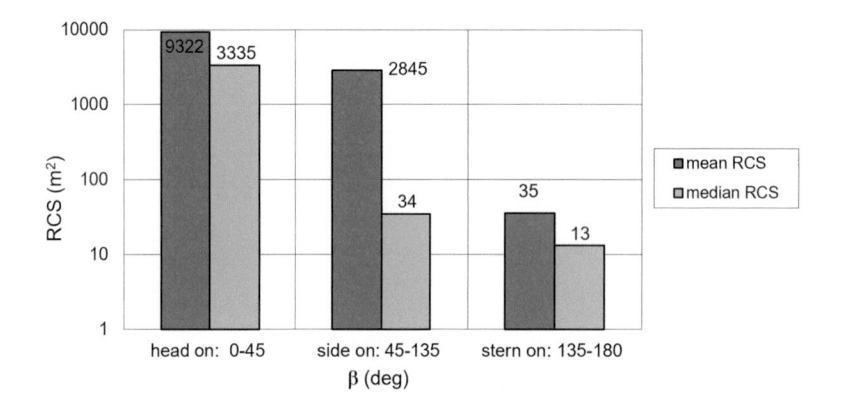

FIGURE 3.200 Diagrams of mean and median RCS of vehicle 54K6 model in three sectors of azimuth aspect given its radar observation (sounding frequency 10 GHz, $\gamma = 30°$, underlying surface – dry loam, vertical polarization).

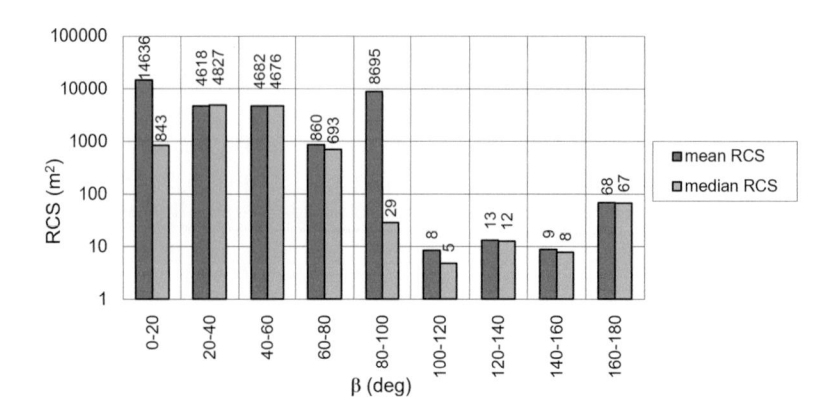

FIGURE 3.201 Diagrams of mean and median RCS of vehicle 54K6 model in 20-degree sectors of azimuth aspect given its radar observation (sounding frequency 10 GHz, $\gamma = 30°$, underlying surface – dry loam, vertical polarization).

Scattering Characteristics of Ground Objects   459

3.13 S-300V TRANSPORTER ERECTOR LAUNCHER AND RADAR 9A83

9A83 (Figure 3.202) is a transporter erector and radar (TELAR) of the S300V missile system (SA-12 Gladiator) [86]. It has been in service since 1988.

In accordance with the design of the TELAR 9A83, a model of its surface was created to obtain scattering characteristics (in particular, RCS). The model is shown in Figure 3.203. In the modeling, the smooth parts of the object surface were approximated by parts of 350 triaxial ellipsoids. The surface breaks were modeled using 166 straight edge scattering parts.

FIGURE 3.202   S-300V transporter erector launcher and radar 9A83.

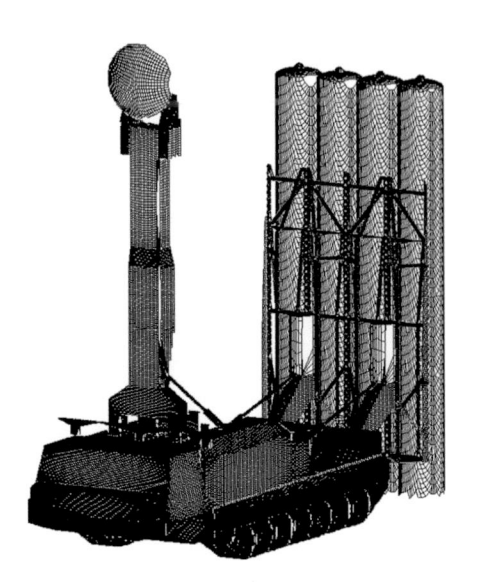

FIGURE 3.203   Surface model of S-300V TELAR 9A83.

Below are some scattering characteristics of the TELAR 9A83 model at a sounding frequency of 10 GHz (wavelengths of 3 cm) for the following conditions: the elevation angle of sounding $\gamma = 10°$, wet soil, and the elevation angle of sounding $\gamma = 30°$, dry loam.

Scattering characteristics of the TELAR 9A83 model for sounding frequency 10 GHz, elevation angle 10°, wet soil, and horizontal polarization.

Figure 3.204 shows the RCS circular diagram of the TELAR 9A83 model. The noncoherent RCS circular diagram of the TELAR 9A83 is presented in Figure 3.205. The circular mean RCS of the TELAR 9A83 model is 66 026 m². The circular median RCS is 18 575 m². Figures 3.206 and 3.207 show the mean and median RCS for the main ranges of sounding azimuths (head, side, and stern) and for ranges of 20°.

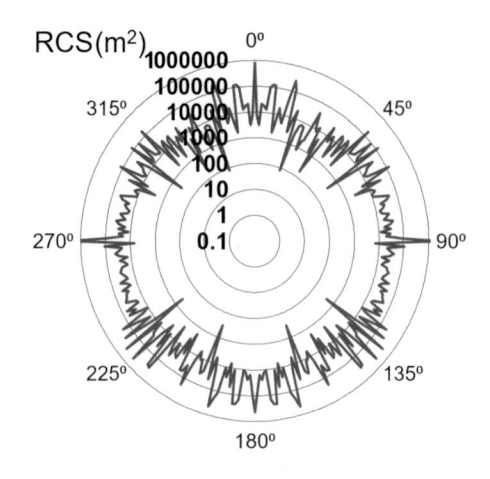

FIGURE 3.204 Circular diagrams of RCS given radar observation of TELAR 9A83 model (sounding frequency 10 GHz, $\gamma = 10°$, underlying surface – wet soil, horizontal polarization).

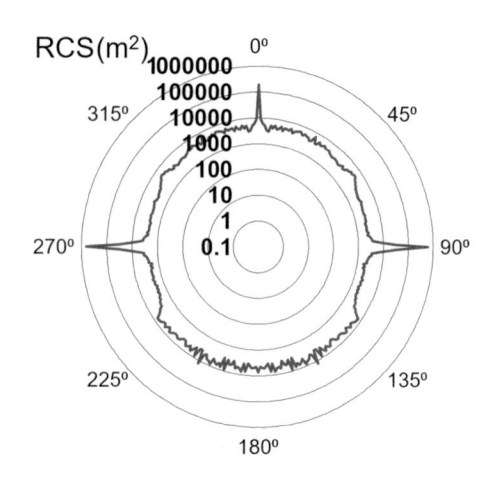

FIGURE 3.205 Circular diagram of noncoherent RCS given radar observation of TELAR 9A83 model (sounding frequency 10 GHz, $\gamma = 10°$, underlying surface – wet soil, horizontal polarization).

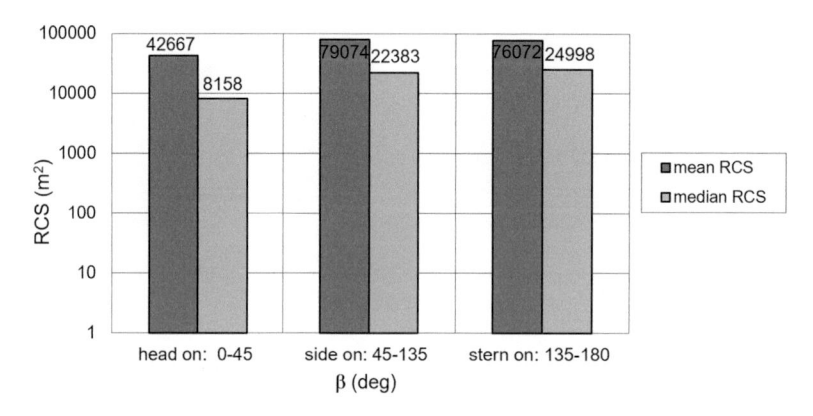

FIGURE 3.206 Diagrams of mean and median RCS of TELAR 9A83 model in three sectors of azimuth aspect given its radar observation (sounding frequency 10 GHz, $\gamma = 10°$, underlying surface – wet soil, horizontal polarization).

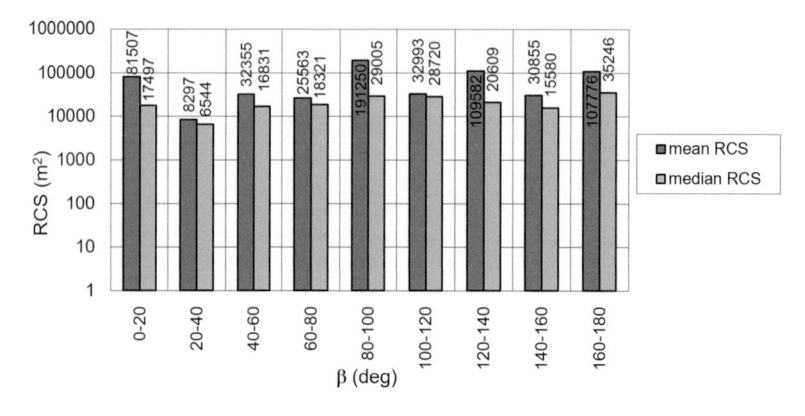

FIGURE 3.207 Diagrams of mean and median RCS of TELAR 9A83 model in 20-degree sectors of azimuth aspect given its radar observation (sounding frequency 10 GHz, $\gamma = 10°$, underlying surface – wet soil, horizontal polarization).

Scattering Characteristics of Ground Objects

Scattering characteristics of the TELAR 9A83 model for sounding frequency 10 GHz, elevation angle 10°, wet soil, and vertical polarization.

Figure 3.208 shows the RCS circular diagram of the TELAR 9A83 model. The noncoherent RCS circular diagram of the TELAR 9A83 is presented in Figure 3.209.

The circular mean RCS of the TELAR 9A83 model is 122.67 m². The circular median RCS is 51.96 m².

Figures 3.210 and 3.211 show the mean and median RCS for the main ranges of sounding azimuths (head, side, and stern) and for ranges of 20°.

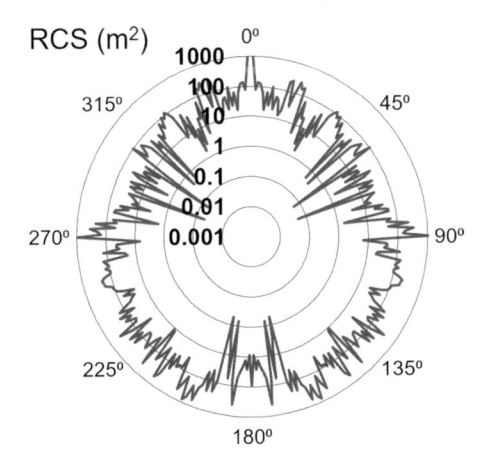

FIGURE 3.208 Circular diagrams of RCS given radar observation of TELAR 9A83 model (sounding frequency 10 GHz, $\gamma = 10°$, underlying surface – wet soil, vertical polarization).

FIGURE 3.209 Circular diagram of noncoherent RCS given radar observation of TELAR 9A83 model (sounding frequency 10 GHz, $\gamma = 10°$, underlying surface – wet soil, vertical polarization).

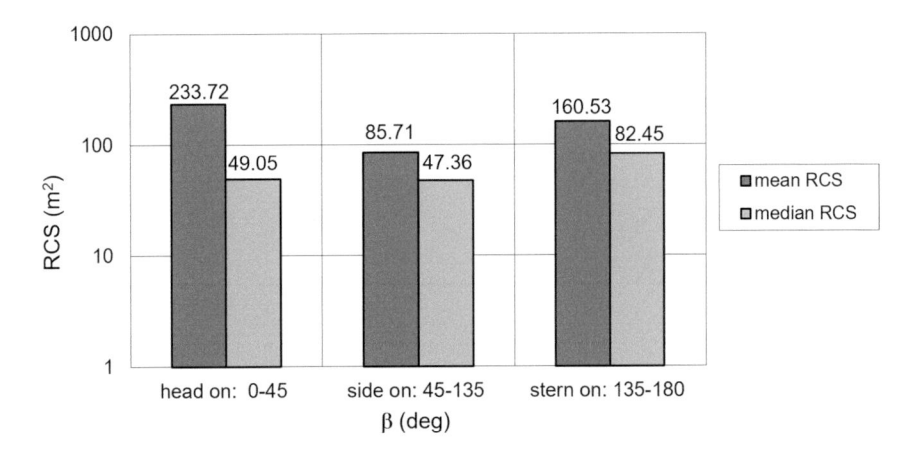

FIGURE 3.210 Diagrams of mean and median RCS of TELAR 9A83 model in three sectors of azimuth aspect given its radar observation (sounding frequency 10 GHz, $\gamma = 10°$, underlying surface – wet soil, vertical polarization).

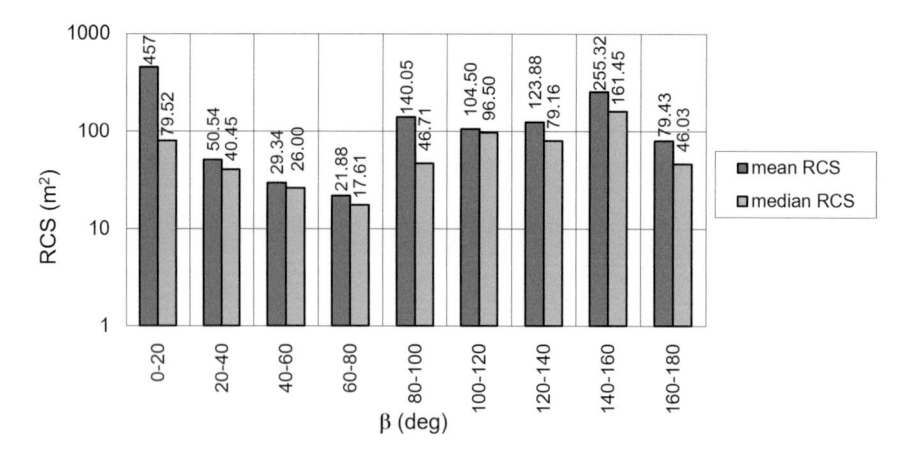

FIGURE 3.211 Diagrams of mean and median RCS of TELAR 9A83 model in 20-degree sectors of azimuth aspect given its radar observation (sounding frequency 10 GHz, $\gamma = 10°$, underlying surface – wet soil, vertical polarization).

Scattering characteristics of the TELAR 9A83 model for sounding frequency 10 GHz, elevation angle 30°, dry loam, and horizontal polarization.

Figure 3.212 shows the RCS circular diagram of the TELAR 9A83 model. The noncoherent RCS circular diagram of the TELAR 9A83 is presented in Figure 3.213. The circular mean RCS of the TELAR 9A83 model is 27 264 m². The circular median RCS is 7 020 m².

Figures 3.214 and 3.215 show the mean and median RCS for the main ranges of sounding azimuths (head, side, and stern) and for ranges of 20°.

FIGURE 3.212 Circular diagrams of RCS given radar observation of TELAR 9A83 model (sounding frequency 10 GHz, $\gamma = 30°$, underlying surface – dry loam, horizontal polarization).

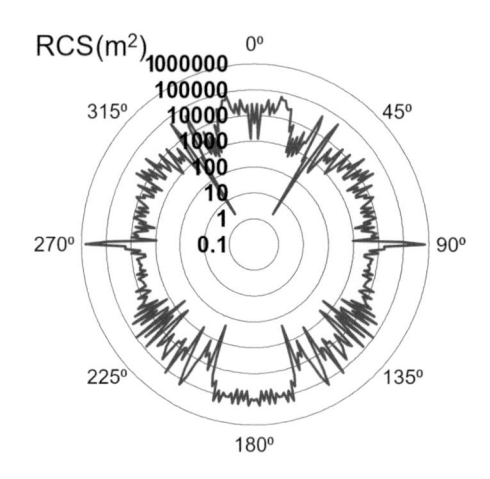

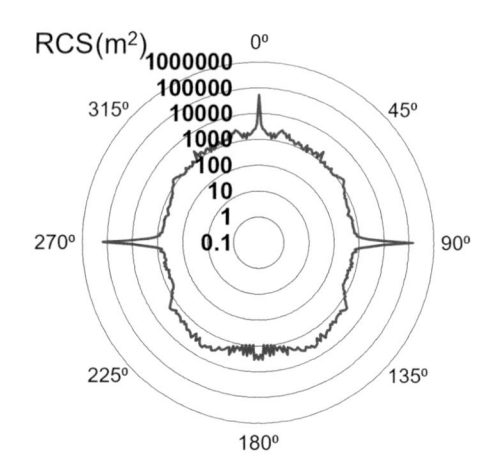

FIGURE 3.213 Circular diagram of noncoherent RCS given radar observation of TELAR 9A83 model (sounding frequency 10 GHz, $\gamma = 30°$, underlying surface – dry loam, horizontal polarization).

Scattering Characteristics of Ground Objects

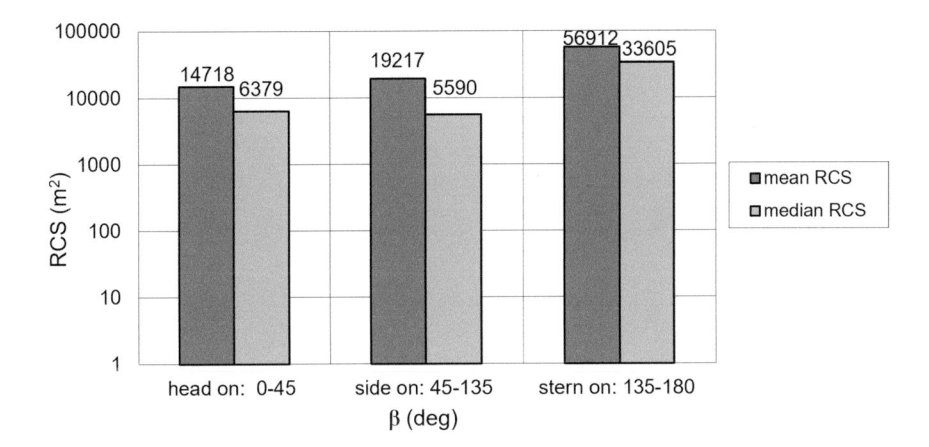

FIGURE 3.214 Diagrams of mean and median RCS of TELAR 9A83 model in three sectors of azimuth aspect given its radar observation (sounding frequency 10 GHz, $\gamma = 30°$, underlying surface – dry loam, horizontal polarization).

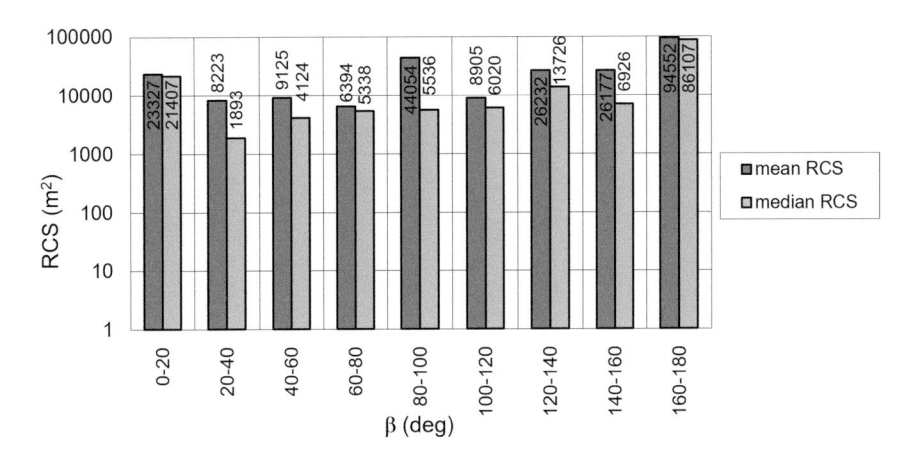

FIGURE 3.215 Diagrams of mean and median RCS of TELAR 9A83 model in 20-degree sectors of azimuth aspect given its radar observation (sounding frequency 10 GHz, $\gamma = 30°$, underlying surface – dry loam, horizontal polarization).

Scattering characteristics of the TELAR 9A83 model for sounding frequency 10 GHz, elevation angle 30°, dry loam, and vertical polarization.

Figure 3.216 shows the RCS circular diagram of the TELAR 9A83 model. The noncoherent RCS circular diagram of the TELAR 9A83 is presented in Figure 3.217. The circular mean RCS of the TELAR 9A83 model is 1 132 m². The circular median RCS is 186.77 m².

Figures 3.218 and 3.219 show the mean and median RCS for the main ranges of sounding azimuths (head, side, and stern) and for ranges of 20°.

The scattering characteristics for this model at two polarizations, sounding frequencies of 10, 5, and 3 GHz (wavelengths of 3, 6, and 10 cm, respectively), for three types of underlying surface (dry loam, dry soil, and wet soil) are given in the electronic appendix of this book.

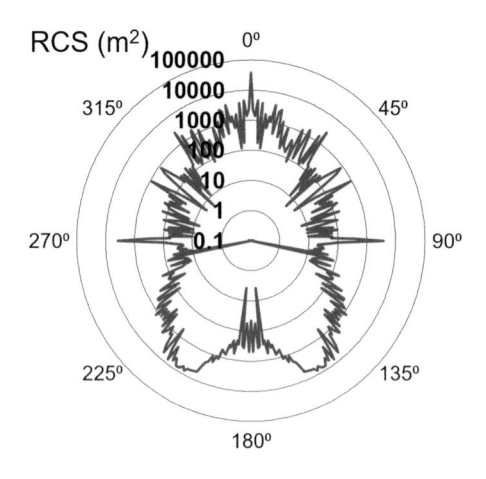

FIGURE 3.216 Circular diagrams of RCS given radar observation of TELAR 9A83 model (sounding frequency 10 GHz, $\gamma=30°$, underlying surface – dry loam, vertical polarization).

FIGURE 3.217 Circular diagram of noncoherent RCS given radar observation of TELAR 9A83 model (sounding frequency 10 GHz, $\gamma=30°$, underlying surface – dry loam, vertical polarization).

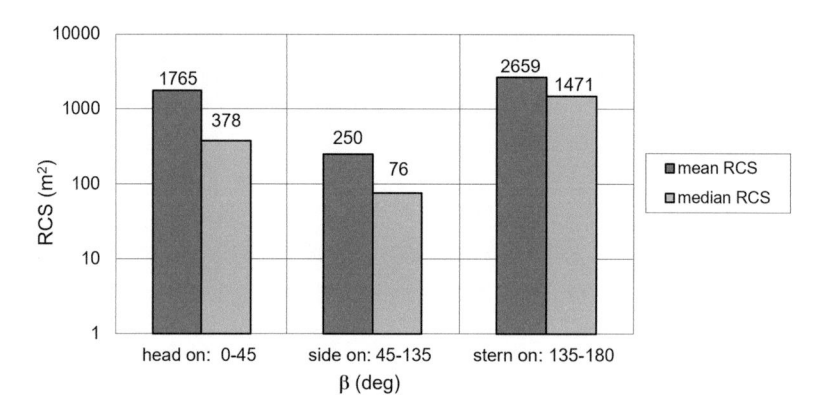

FIGURE 3.218 Diagrams of mean and median RCS of TELAR 9A83 model in three sectors of azimuth aspect given its radar observation (sounding frequency 10 GHz, $\gamma=30°$, underlying surface – dry loam, vertical polarization).

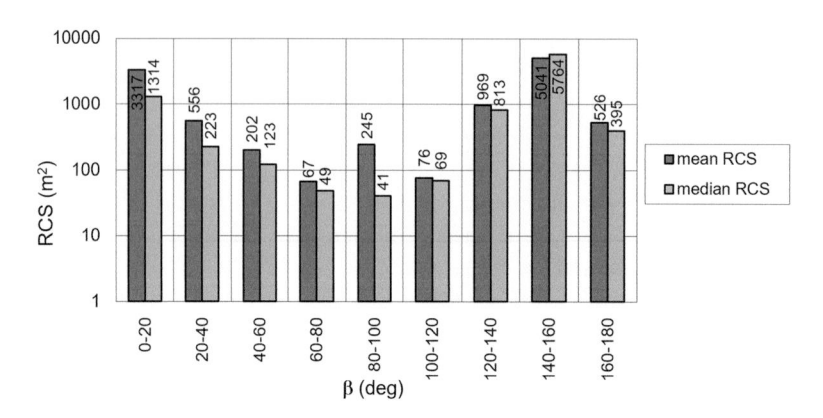

FIGURE 3.219 Diagrams of mean and median RCS of TELAR 9A83 model in 20-degree sectors of azimuth aspect given its radar observation (sounding frequency 10 GHz, $\gamma=30°$, underlying surface – dry loam, vertical polarization).

Scattering Characteristics of Ground Objects

3.14 S-300V COMMAND POST VEHICLE 9S457

9S457 (Figure 3.220) is a command post vehicle (CPV) of the S300V missile system (SA-12 Gladiator) [85]. It has been in service since 1988.

In accordance with the design of the CPV 9S457, a model of its surface was created to obtain scattering characteristics (in particular, RCS). The model is shown in Figure 3.221. In the modeling, the smooth parts of the object surface were approximated by parts of 192 triaxial ellipsoids. The surface breaks were modeled using 184 straight edge scattering parts.

FIGURE 3.220 S-300V command post vehicle 9S457.

FIGURE 3.221 Surface model of S-300V CPV 9S457.

Below are some scattering characteristics of the CPV 9S457 model at a sounding frequency of 10 GHz (wavelengths of 3 cm) for the following conditions: the elevation angle of sounding $\gamma = 10°$, wet soil, and the elevation angle of sounding $\gamma = 30°$, dry loam.

Scattering characteristics of the CPV 9S457 model for sounding frequency 10 GHz, elevation angle 10°, wet soil, and horizontal polarization.

Figure 3.222 shows the RCS circular diagram of the CPV 9S457 model. The noncoherent RCS circular diagram of the CPV 9S457 is presented in Figure 3.223. The circular mean RCS of the CPV 9S457 model is 32 306 m². The circular median RCS is 983.38 m². Figures 3.224 and 3.225 show the mean and median RCS for the main ranges of sounding azimuths (head, side, and stern) and for ranges of 20°.

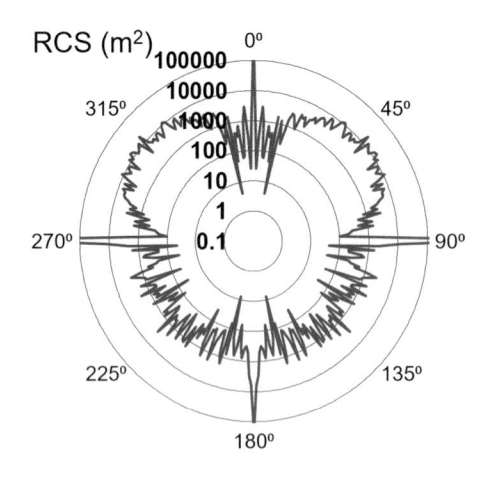

FIGURE 3.222 Circular diagrams of RCS given radar observation of CPV 9S457 model (sounding frequency 10 GHz, $\gamma = 10°$, underlying surface – wet soil, horizontal polarization).

FIGURE 3.223 Circular diagram of noncoherent RCS given radar observation of CPV 9S457 model (sounding frequency 10 GHz, $\gamma = 10°$, underlying surface – wet soil, horizontal polarization).

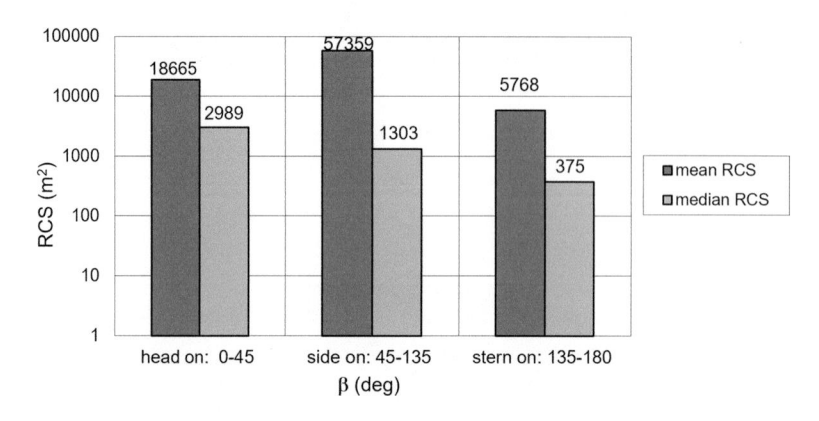

FIGURE 3.224 Diagrams of mean and median RCS of CPV 9S457 model in three sectors of azimuth aspect given its radar observation (sounding frequency 10 GHz, $\gamma = 10°$, underlying surface – wet soil, horizontal polarization).

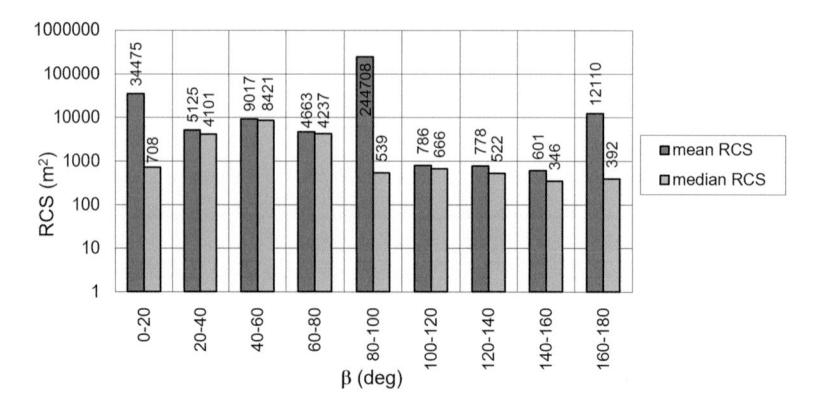

FIGURE 3.225 Diagrams of mean and median RCS of CPV 9S457 model in 20-degree sectors of azimuth aspect given its radar observation (sounding frequency 10 GHz, $\gamma = 10°$, underlying surface – wet soil, horizontal polarization).

Scattering characteristics of the CPV 9S457 model for sounding frequency 10 GHz, elevation angle 10°, wet soil, and vertical polarization.

Figure 3.226 shows the RCS circular diagram of the CPV 9S457 model. The noncoherent RCS circular diagram of the CPV 9S457 is presented in Figure 3.227.

The circular mean RCS of the CPV 9S457 model is 531.83 m². The circular median RCS is 194.96 m².

Figures 3.228 and 3.229 show the mean and median RCS for the main ranges of sounding azimuths (head, side, and stern) and for ranges of 20°.

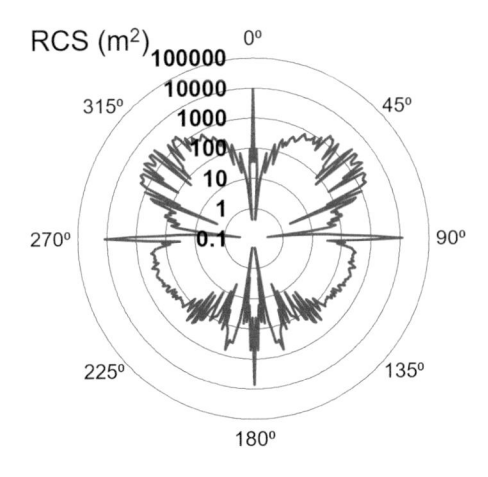

FIGURE 3.226 Circular diagrams of RCS given radar observation of CPV 9S457 model (sounding frequency 10 GHz, $\gamma = 10°$, underlying surface – wet soil, vertical polarization).

FIGURE 3.227 Circular diagram of noncoherent RCS given radar observation of CPV 9S457 model (sounding frequency 10 GHz, $\gamma = 10°$, underlying surface – wet soil, vertical polarization).

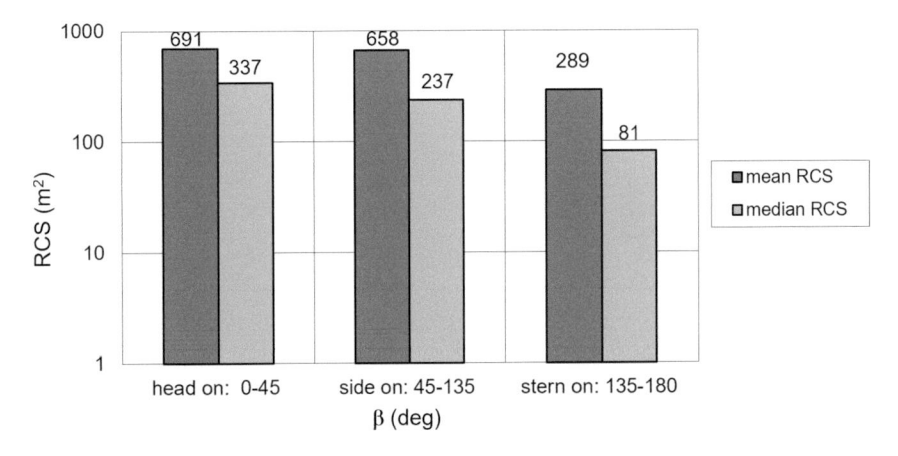

FIGURE 3.228 Diagrams of mean and median RCS of CPV 9S457 model in three sectors of azimuth aspect given its radar observation (sounding frequency 10 GHz, $\gamma = 10°$, underlying surface – wet soil, vertical polarization).

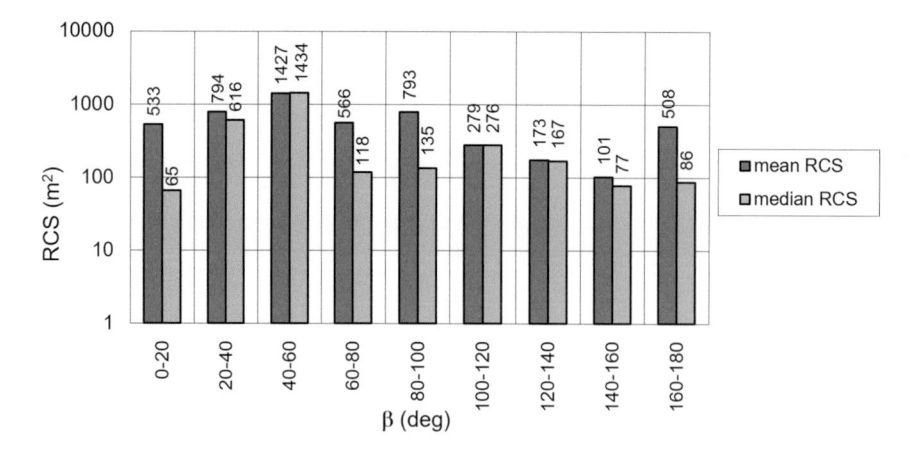

FIGURE 3.229 Diagrams of mean and median RCS of CPV 9S457 model in 20-degree sectors of azimuth aspect given its radar observation (sounding frequency 10 GHz, $\gamma = 10°$, underlying surface – wet soil, vertical polarization).

Scattering characteristics of the CPV 9S457 model for sounding frequency 10 GHz, elevation angle 30°, dry loam, and horizontal polarization.

Figure 3.230 shows the RCS circular diagram of the CPV 9S457 model. The noncoherent RCS circular diagram of the CPV 9S457 is presented in Figure 3.231. The circular mean RCS of the CPV 9S457 model is 4 720 m². The circular median RCS is 423.86 m².

Figures 3.232 and 3.233 show the mean and median RCS for the main ranges of sounding azimuths (head, side, and stern) and for ranges of 20°.

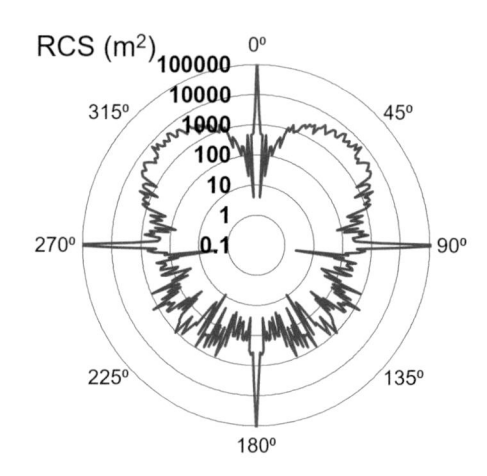

FIGURE 3.230 Circular diagrams of RCS given radar observation of CPV 9S457 model (sounding frequency 10 GHz, $\gamma = 30°$, underlying surface – dry loam, horizontal polarization).

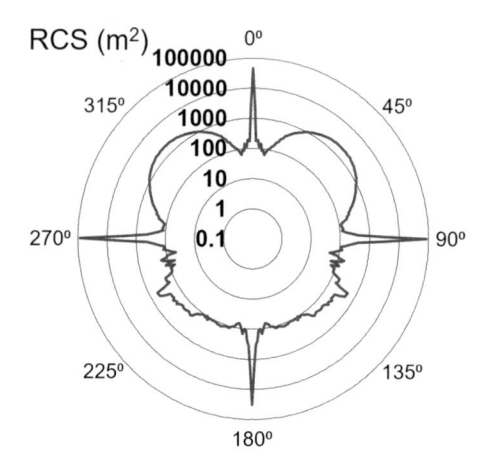

FIGURE 3.231 Circular diagram of noncoherent RCS given radar observation of CPV 9S457 model (sounding frequency 10 GHz, $\gamma = 30°$, underlying surface – dry loam, horizontal polarization).

Scattering Characteristics of Ground Objects

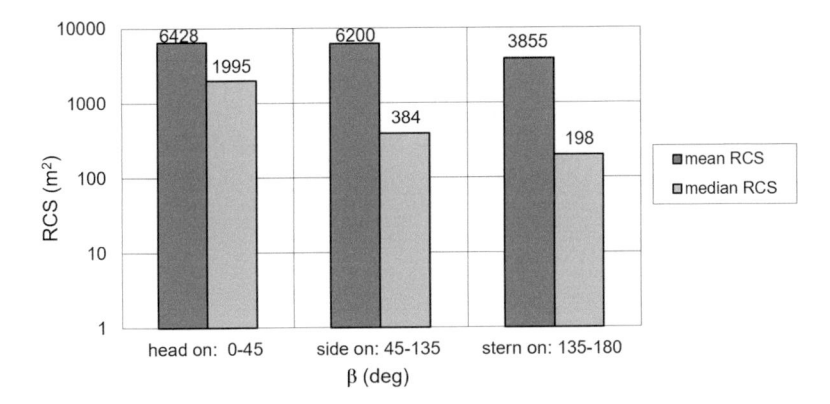

FIGURE 3.232 Diagrams of mean and median RCS of CPV 9S457 model in three sectors of azimuth aspect given its radar observation (sounding frequency 10 GHz, $\gamma=30°$, underlying surface – dry loam, horizontal polarization).

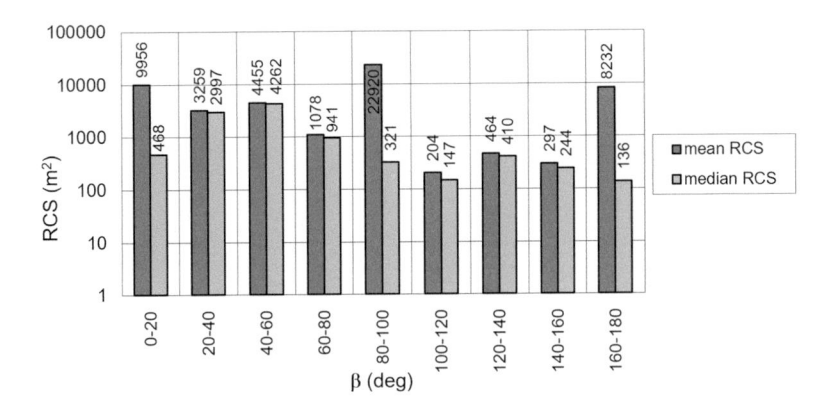

FIGURE 3.233 Diagrams of mean and median RCS of CPV 9S457 model in 20-degree sectors of azimuth aspect given its radar observation (sounding frequency 10 GHz, $\gamma=30°$, underlying surface – dry loam, horizontal polarization).

Scattering characteristics of the CPV 9S457 model for sounding frequency 10 GHz, elevation angle 30°, dry loam, and vertical polarization.

Figure 3.234 shows the RCS circular diagram of the CPV 9S457 model. The noncoherent RCS circular diagram of the CPV 9S457 is presented in Figure 3.235. The circular mean RCS of the CPV 9S457 model is 1 498 m². The circular median RCS is 280.88 m².

Figures 3.236 and 3.237 show the mean and median RCS for the main ranges of sounding azimuths (head, side, and stern) and for ranges of 20°.

The scattering characteristics for this model at two polarizations, sounding frequencies of 10, 5, and 3 GHz (wavelengths of 3, 6, and 10 cm, respectively), for three types of underlying surface (dry loam, dry soil, and wet soil) are given in the electronic appendix of this book.

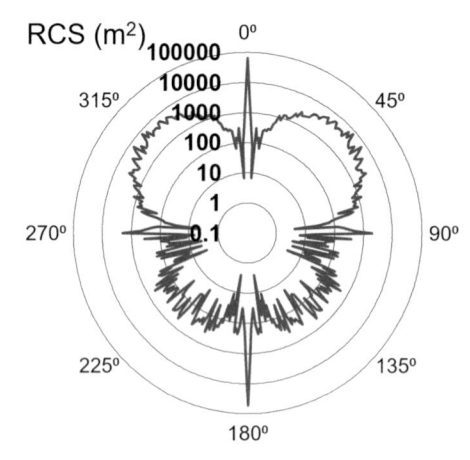

FIGURE 3.234 Circular diagrams of RCS given radar observation of CPV 9S457 model (sounding frequency 10 GHz, $\gamma=30°$, underlying surface – dry loam, vertical polarization).

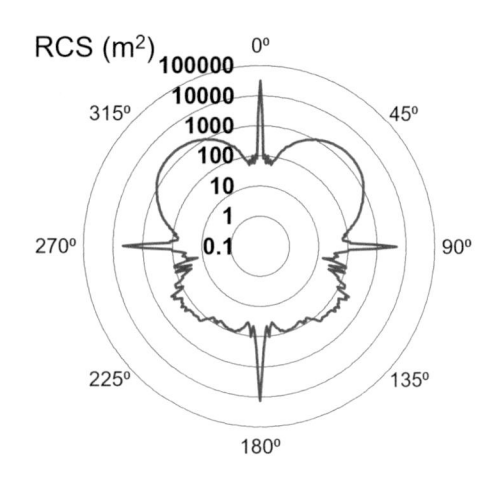

FIGURE 3.235 Circular diagram of noncoherent RCS given radar observation of CPV 9S457 model (sounding frequency 10 GHz, $\gamma=30°$, underlying surface – dry loam, vertical polarization).

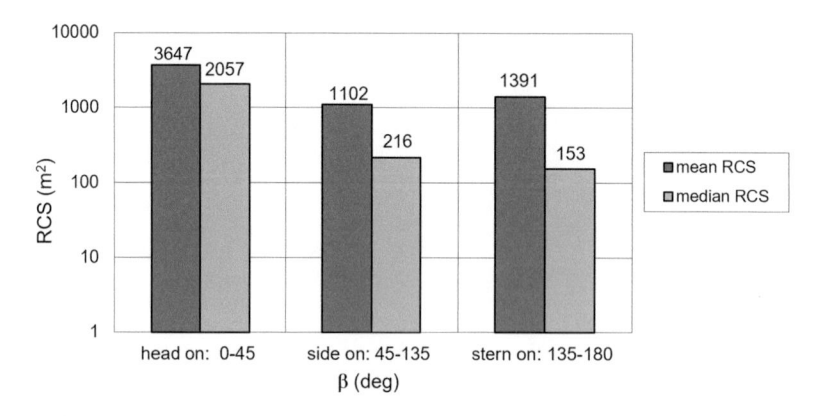

FIGURE 3.236 Diagrams of mean and median RCS of CPV 9S457 model in three sectors of azimuth aspect given its radar observation (sounding frequency 10 GHz, $\gamma=30°$, underlying surface – dry loam, vertical polarization).

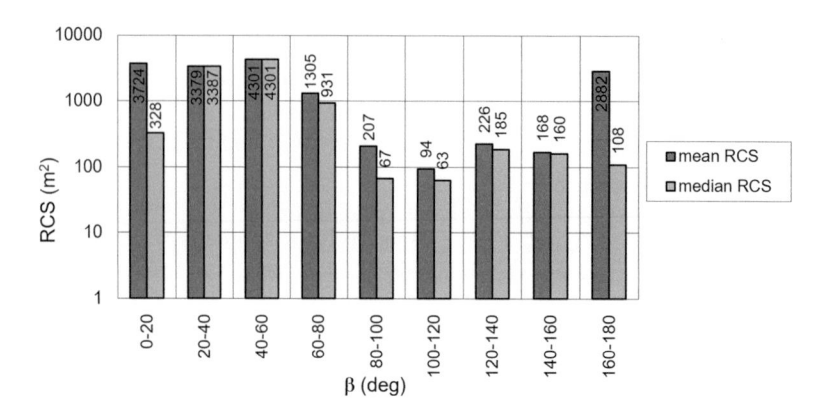

FIGURE 3.237 Diagrams of mean and median RCS of CPV 9S457 model in 20-degree sectors of azimuth aspect given its radar observation (sounding frequency 10 GHz, $\gamma=30°$, underlying surface – dry loam, vertical polarization).

3.15 S-200 LAUNCHER 5P72

5P72 (Figure 3.238) is a launcher of the S200 long-range, high-altitude missile system (SA-5 Gamon) [87]. It has been in service since 1967.

In accordance with the design of the launcher 5P72, a model of its surface was created to obtain scattering characteristics (in particular, RCS). The model is shown in Figure 3.239. In the modeling, the smooth parts of the object surface were approximated by parts of 97 triaxial ellipsoids. The surface breaks were modeled using 66 straight edge scattering parts.

FIGURE 3.238 S-200 launcher 5P72.

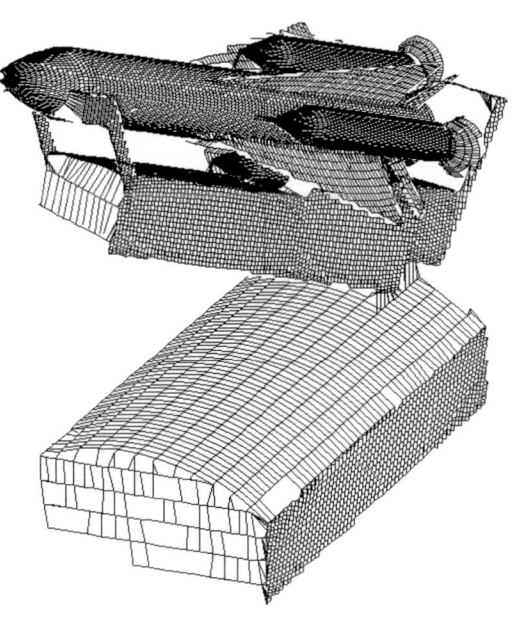

FIGURE 3.239 Surface model of S-200 launcher 5P72.

Below are some scattering characteristics of the launcher 5P72 model at a sounding frequency of 10 GHz (wavelengths of 3 cm) for the following conditions: the elevation angle of sounding $\gamma = 10°$, wet soil, and the elevation angle of sounding $\gamma = 30°$, dry loam.

Scattering characteristics of the launcher 5P72 model for sounding frequency 10 GHz, elevation angle 10°, wet soil, and horizontal polarization.

Figure 3.240 shows the RCS circular diagram of the launcher 5P72 model. The noncoherent RCS circular diagram of the launcher 5P72 is presented in Figure 3.241. The circular mean RCS of the launcher 5P72 model is 53 598 m². The circular median RCS is 370.30 m². Figures 3.242 and 3.243 show the mean and median RCS for the main ranges of sounding azimuths (head, side, and stern) and for ranges of 20°.

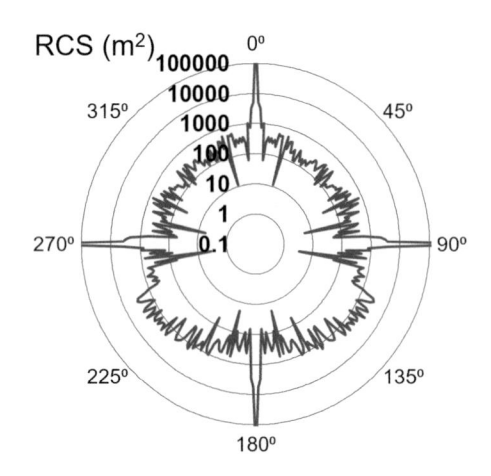

FIGURE 3.240 Circular diagrams of RCS given radar observation of launcher 5P72 model (sounding frequency 10 GHz, $\gamma = 10°$, underlying surface – wet soil, horizontal polarization).

FIGURE 3.241 Circular diagram of noncoherent RCS given radar observation of launcher 5P72 model (sounding frequency 10 GHz, $\gamma = 10°$, underlying surface – wet soil, horizontal polarization).

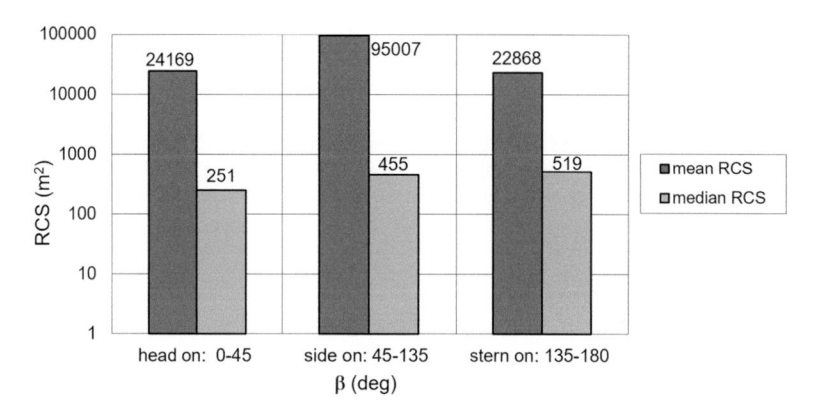

FIGURE 3.242 Diagrams of mean and median RCS of launcher 5P72 model in three sectors of azimuth aspect given its radar observation (sounding frequency 10 GHz, $\gamma = 10°$, underlying surface – wet soil, horizontal polarization).

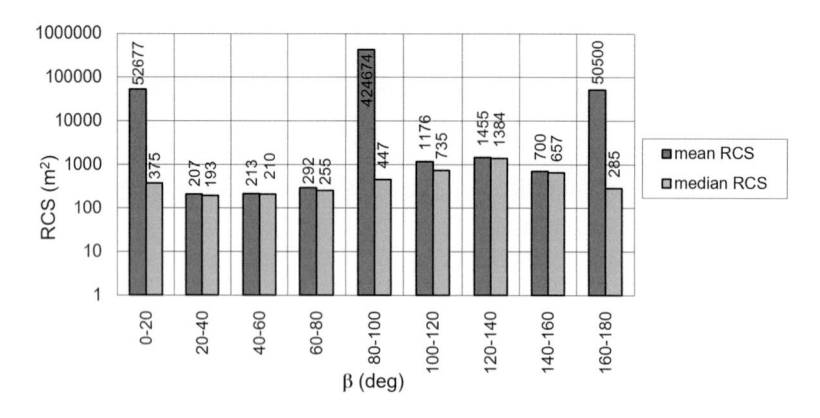

FIGURE 3.243 Diagrams of mean and median RCS of launcher 5P72 model in 20-degree sectors of azimuth aspect given its radar observation (sounding frequency 10 GHz, $\gamma = 10°$, underlying surface – wet soil, horizontal polarization).

Scattering Characteristics of Ground Objects

Scattering characteristics of the launcher 5P72 model for sounding frequency 10 GHz, elevation angle 10°, wet soil, and vertical polarization.

Figure 3.244 shows the RCS circular diagram of the launcher 5P72 model. The noncoherent RCS circular diagram of the launcher 5P72 is presented in Figure 3.245.

The circular mean RCS of the launcher 5P72 model is 14.98 m². The circular median RCS is 1.08 m².

Figures 3.246 and 3.247 show the mean and median RCS for the main ranges of sounding azimuths (head, side, and stern) and for ranges of 20°.

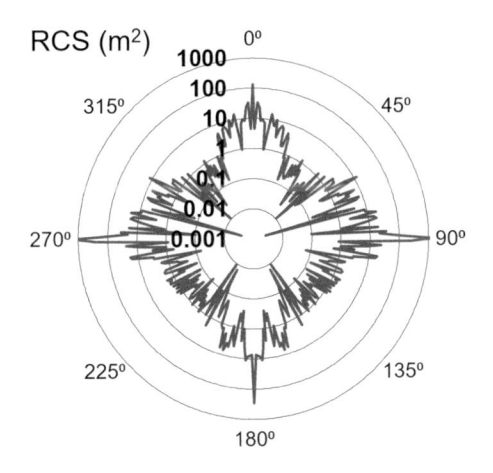

FIGURE 3.244 Circular diagrams of RCS given radar observation of launcher 5P72 model (sounding frequency 10 GHz, $\gamma = 10°$, underlying surface – wet soil, vertical polarization).

FIGURE 3.245 Circular diagram of noncoherent RCS given radar observation of launcher 5P72 model (sounding frequency 10 GHz, $\gamma = 10°$, underlying surface – wet soil, vertical polarization).

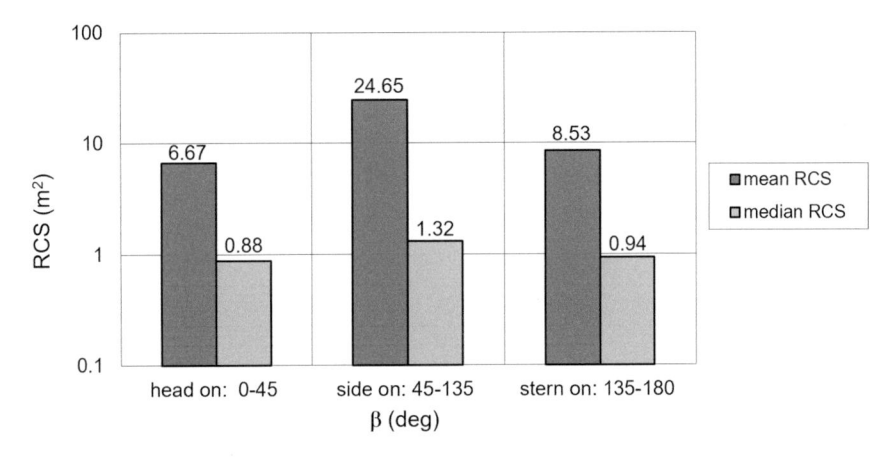

FIGURE 3.246 Diagrams of mean and median RCS of launcher 5P72 model in three sectors of azimuth aspect given its radar observation (sounding frequency 10 GHz, $\gamma = 10°$, underlying surface – wet soil, vertical polarization).

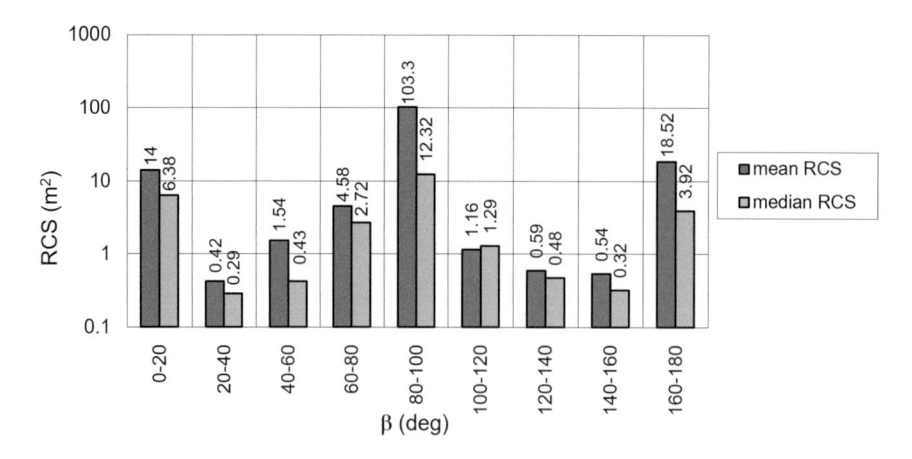

FIGURE 3.247 Diagrams of mean and median RCS of launcher 5P72 model in 20-degree sectors of azimuth aspect given its radar observation (sounding frequency 10 GHz, $\gamma = 10°$, underlying surface – wet soil, vertical polarization).

Scattering characteristics of the launcher 5P72 model for sounding frequency 10 GHz, elevation angle 30°, dry loam, and horizontal polarization.

Figure 3.248 shows the RCS circular diagram of the launcher 5P72 model. The noncoherent RCS circular diagram of the launcher 5P72 is presented in Figure 3.249. The circular mean RCS of the launcher 5P72 model is 12 552 m². The circular median RCS is 175.90 m².

Figures 3.250 and 3.251 show the mean and median RCS for the main ranges of sounding azimuths (head, side, and stern) and for ranges of 20°.

FIGURE 3.248 Circular diagrams of RCS given radar observation of launcher 5P72 model (sounding frequency 10 GHz, $\gamma = 30°$, underlying surface – dry loam, horizontal polarization).

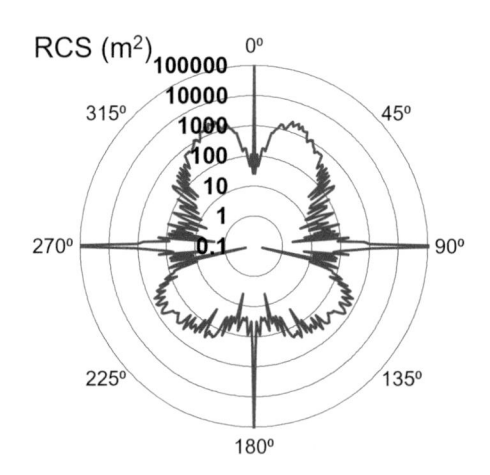

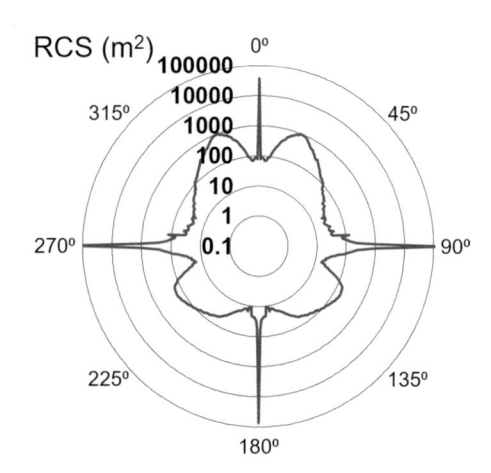

FIGURE 3.249 Circular diagram of noncoherent RCS given radar observation of launcher 5P72 model (sounding frequency 10 GHz, $\gamma = 30°$, underlying surface – dry loam, horizontal polarization).

Scattering Characteristics of Ground Objects

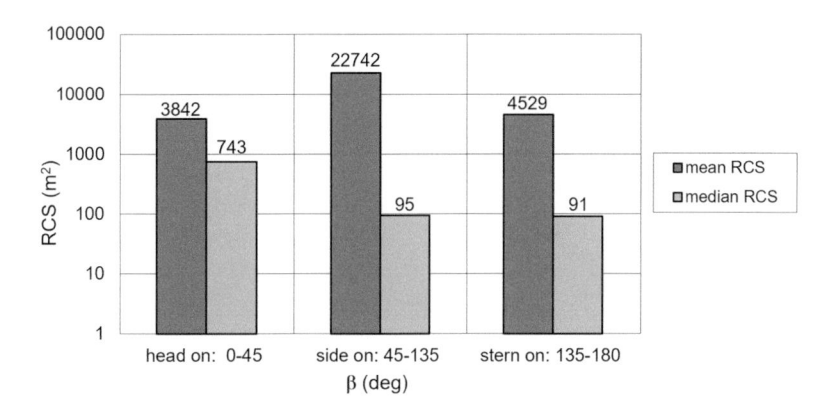

FIGURE 3.250 Diagrams of mean and median RCS of launcher 5P72 model in three sectors of azimuth aspect given its radar observation (sounding frequency 10 GHz, $\gamma=30°$, underlying surface – dry loam, horizontal polarization).

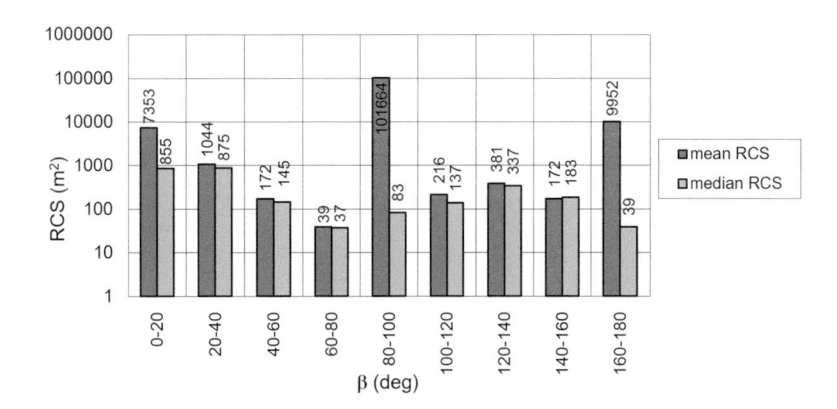

FIGURE 3.251 Diagrams of mean and median RCS of launcher 5P72 model in 20-degree sectors of azimuth aspect given its radar observation (sounding frequency 10 GHz, $\gamma=30°$, underlying surface – dry loam, horizontal polarization).

Scattering characteristics of the launcher 5P72 model for sounding frequency 10 GHz, elevation angle 30°, dry loam, and vertical polarization.

Figure 3.252 shows the RCS circular diagram of the launcher 5P72 model. The noncoherent RCS circular diagram of the launcher 5P72 is presented in Figure 3.253. The circular mean RCS of the launcher 5P72 model is 117.80 m². The circular median RCS is 1.98 m².

Figures 3.254 and 3.255 show the mean and median RCS for the main ranges of sounding azimuths (head, side, and stern) and for ranges of 20°.

The scattering characteristics for this model at two polarizations, sounding frequencies of 10, 5, and 3 GHz (wavelengths of 3, 6, and 10 cm, respectively), for three types of underlying surface (dry loam, dry soil, and wet soil) are given in the electronic appendix of this book.

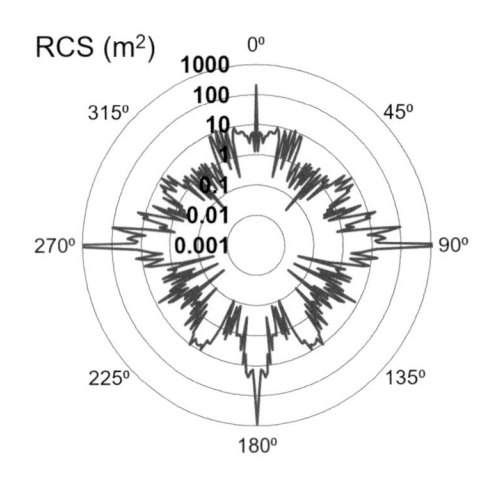

FIGURE 3.252 Circular diagrams of RCS given radar observation of launcher 5P72 model (sounding frequency 10 GHz, $\gamma=30°$, underlying surface – dry loam, vertical polarization).

FIGURE 3.253 Circular diagram of noncoherent RCS given radar observation of launcher 5P72 model (sounding frequency 10 GHz, $\gamma=30°$, underlying surface – dry loam, vertical polarization).

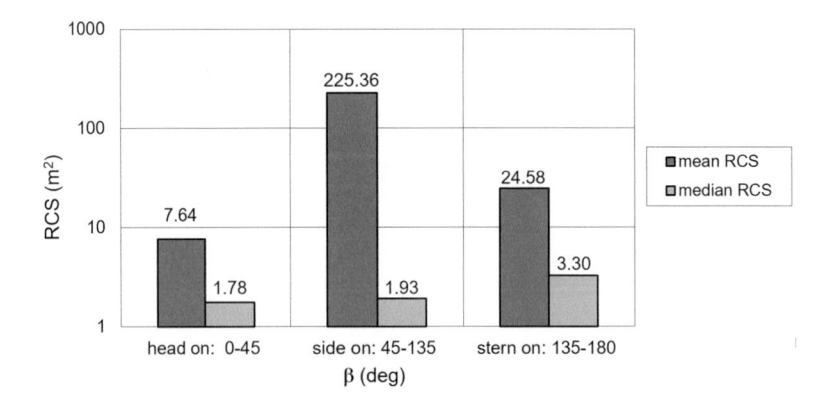

FIGURE 3.254 Diagrams of mean and median RCS of launcher 5P72 model in three sectors of azimuth aspect given its radar observation (sounding frequency 10 GHz, $\gamma=30°$, underlying surface – dry loam, vertical polarization).

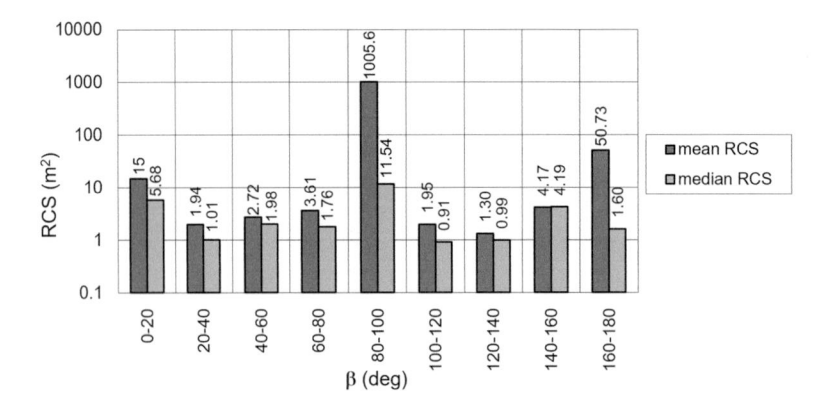

FIGURE 3.255 Diagrams of mean and median RCS of launcher 5P72 model in 20-degree sectors of azimuth aspect given its radar observation (sounding frequency 10 GHz, $\gamma=30°$, underlying surface – dry loam, vertical polarization).

Scattering Characteristics of Ground Objects

3.16 COMBAT VEHICLE 2S6 OF ANTI-AIRCRAFT GUN-MISSILE SYSTEM 2K22 TUNGUSKA

2S6- SA-19 Grison - (Figure 3.256) is an anti-aircraft self-propelled unit of anti-aircraft gun-missile system 2K22 Tunguska [88]. It has been in service since 1982.

In accordance with the design of the vehicle 2S6, a model of its surface was created to obtain scattering characteristics (in particular, RCS). The model is shown in Figure 3.257. In the modeling, the smooth parts of the object surface were approximated by parts of 268 triaxial ellipsoids. The surface breaks were modeled using 157 straight edge scattering parts.

FIGURE 3.256 Combat vehicle 2S6.

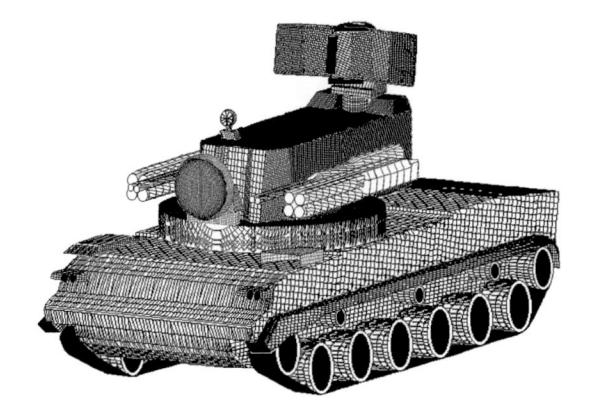

FIGURE 3.257 Surface model of combat vehicle 2S6.

Below are some scattering characteristics of the vehicle 2S6 model at a sounding frequency of 10 GHz (wavelengths of 3 cm) for the following conditions: the elevation angle of sounding $\gamma = 10°$, wet soil, and the elevation angle of sounding $\gamma = 30°$, dry loam.

Scattering characteristics of the vehicle 2S6 model for sounding frequency 10 GHz, elevation angle 10°, wet soil, and horizontal polarization.

Figure 3.258 shows the RCS circular diagram of the vehicle 2S6 model. The noncoherent RCS circular diagram of the vehicle 2S6 is presented in Figure 3.259. The circular mean RCS of the vehicle 2S6 model is 4 201 m². The circular median RCS is 326.09 m². Figures 3.260 and 3.261 show the mean and median RCS for the main ranges of sounding azimuths (head, side, and stern) and for ranges of 20°.

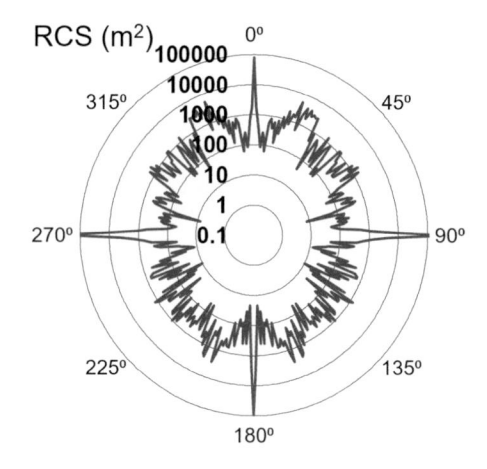

FIGURE 3.258 Circular diagrams of RCS given radar observation of vehicle 2S6 model (sounding frequency 10 GHz, $\gamma = 10°$, underlying surface – wet soil, horizontal polarization).

FIGURE 3.259 Circular diagram of noncoherent RCS given radar observation of vehicle 2S6 model (sounding frequency 10 GHz, $\gamma = 10°$, underlying surface – wet soil, horizontal polarization).

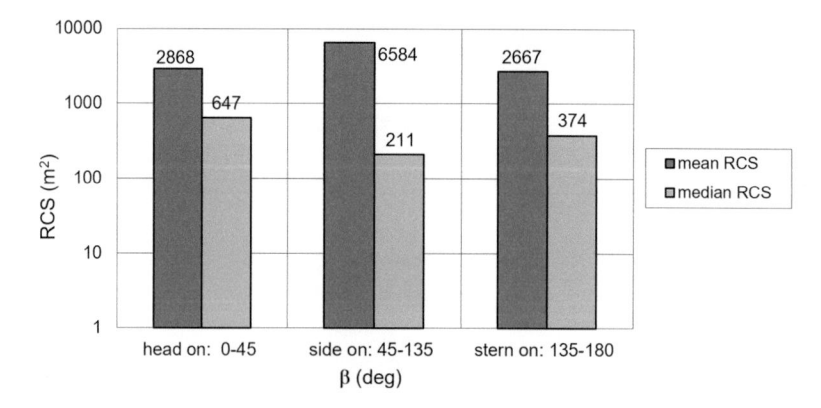

FIGURE 3.260 Diagrams of mean and median RCS of vehicle 2S6 model in three sectors of azimuth aspect given its radar observation (sounding frequency 10 GHz, $\gamma = 10°$, underlying surface – wet soil, horizontal polarization).

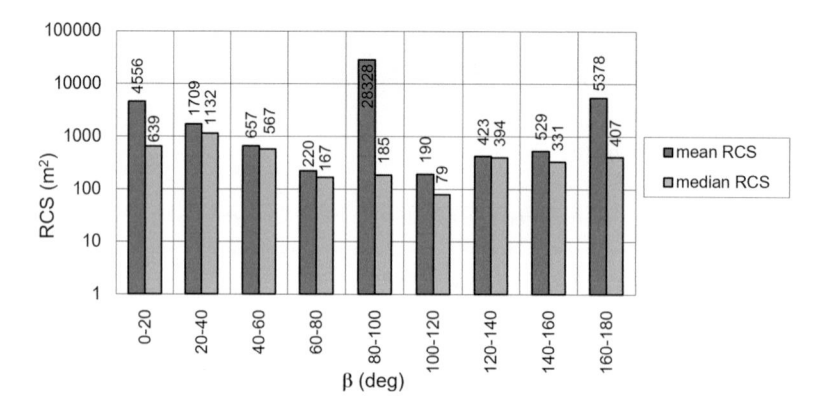

FIGURE 3.261 Diagrams of mean and median RCS of vehicle 2S6 model in 20-degree sectors of azimuth aspect given its radar observation (sounding frequency 10 GHz, $\gamma = 10°$, underlying surface – wet soil, horizontal polarization).

Scattering Characteristics of Ground Objects

Scattering characteristics of the vehicle 2S6 model for sounding frequency 10 GHz, elevation angle 10°, wet soil, and vertical polarization.

Figure 3.262 shows the RCS circular diagram of the vehicle 2S6 model. The noncoherent RCS circular diagram of the vehicle 2S6 is presented in Figure 3.263.

The circular mean RCS of the vehicle 2S6 model is $43.59\,m^2$. The circular median RCS is $15.91\,m^2$.

Figures 3.264 and 3.265 show the mean and median RCS for the main ranges of sounding azimuths (head, side, and stern) and for ranges of 20°.

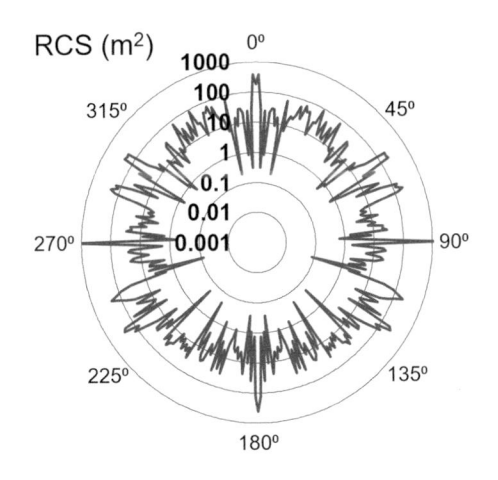

FIGURE 3.262 Circular diagrams of RCS given radar observation of vehicle 2S6 model (sounding frequency 10 GHz, $\gamma = 10°$, underlying surface – wet soil, vertical polarization).

FIGURE 3.263 Circular diagram of noncoherent RCS given radar observation of vehicle 2S6 model (sounding frequency 10 GHz, $\gamma = 10°$, underlying surface – wet soil, vertical polarization).

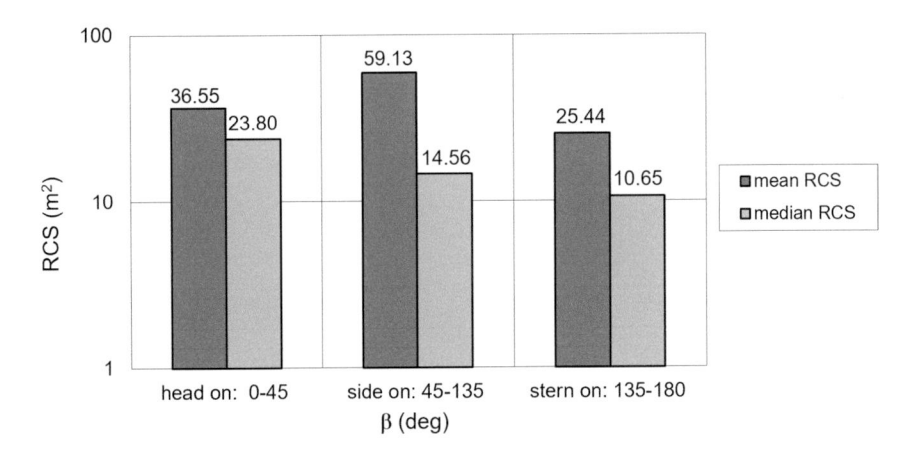

FIGURE 3.264 Diagrams of mean and median RCS of vehicle 2S6 model in three sectors of azimuth aspect given its radar observation (sounding frequency 10 GHz, $\gamma = 10°$, underlying surface – wet soil, vertical polarization).

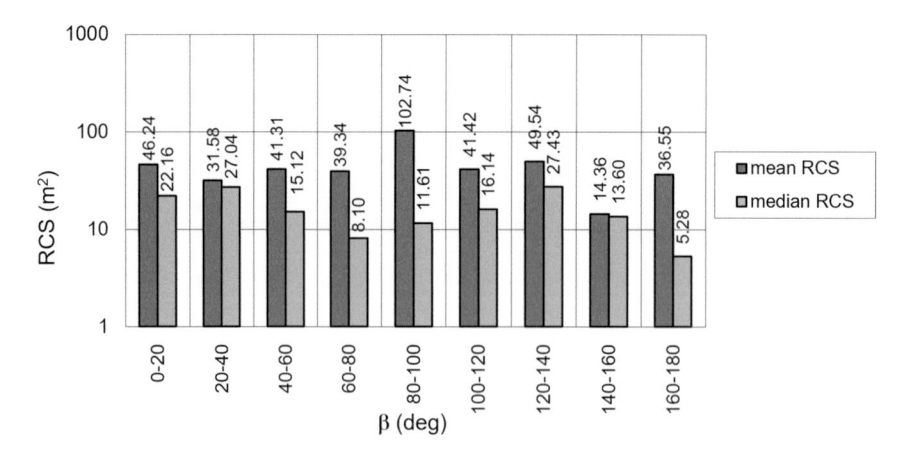

FIGURE 3.265 Diagrams of mean and median RCS of vehicle 2S6 model in 20-degree sectors of azimuth aspect given its radar observation (sounding frequency 10 GHz, $\gamma = 10°$, underlying surface – wet soil, vertical polarization).

Scattering characteristics of the vehicle 2S6 model for sounding frequency 10 GHz, elevation angle 30°, dry loam, and horizontal polarization.

Figure 3.266 shows the RCS circular diagram of the vehicle 2S6 model. The noncoherent RCS circular diagram of the vehicle 2S6 is presented in Figure 3.267. The circular mean RCS of the vehicle 2S6 model is 1 788 m². The circular median RCS is 96.74 m².

Figures 3.268 and 3.269 show the mean and median RCS for the main ranges of sounding azimuths (head, side, and stern) and for ranges of 20°.

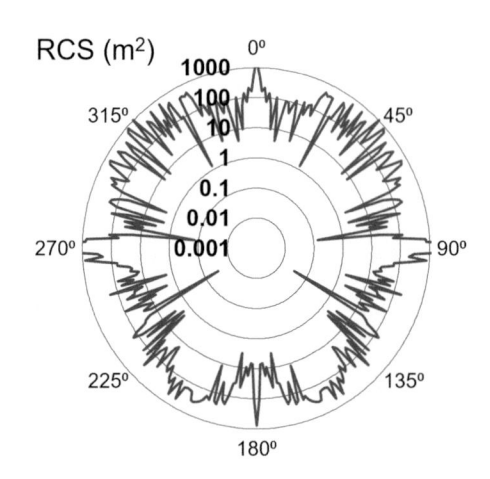

FIGURE 3.266 Circular diagrams of RCS given radar observation of vehicle 2S6 model (sounding frequency 10 GHz, $\gamma = 30°$, underlying surface – dry loam, horizontal polarization).

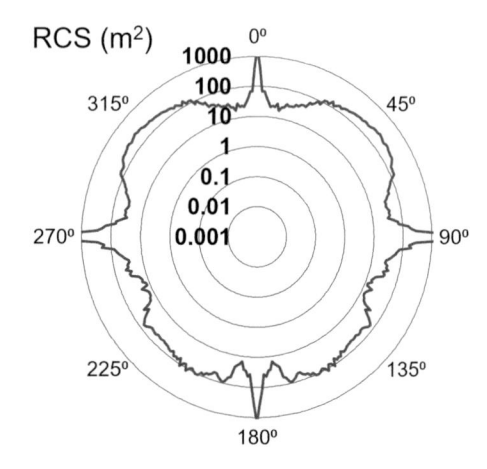

FIGURE 3.267 Circular diagram of noncoherent RCS given radar observation of vehicle 2S6 model (sounding frequency 10 GHz, $\gamma = 30°$, underlying surface – dry loam, horizontal polarization).

Scattering Characteristics of Ground Objects

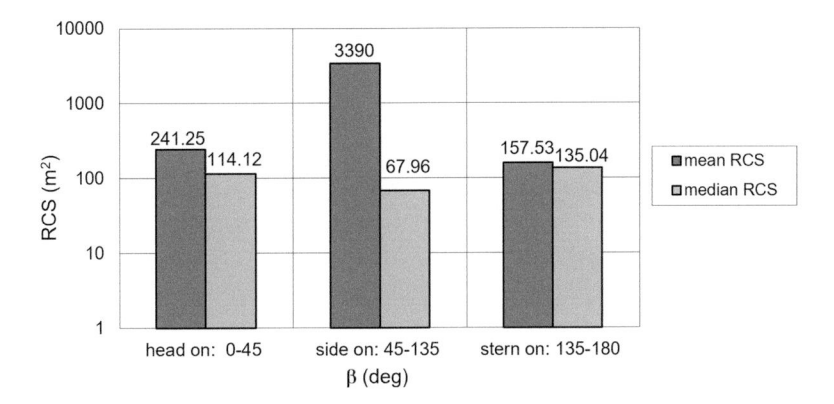

FIGURE 3.268 Diagrams of mean and median RCS of vehicle 2S6 model in three sectors of azimuth aspect given its radar observation (sounding frequency 10 GHz, $\gamma=30°$, underlying surface – dry loam, horizontal polarization).

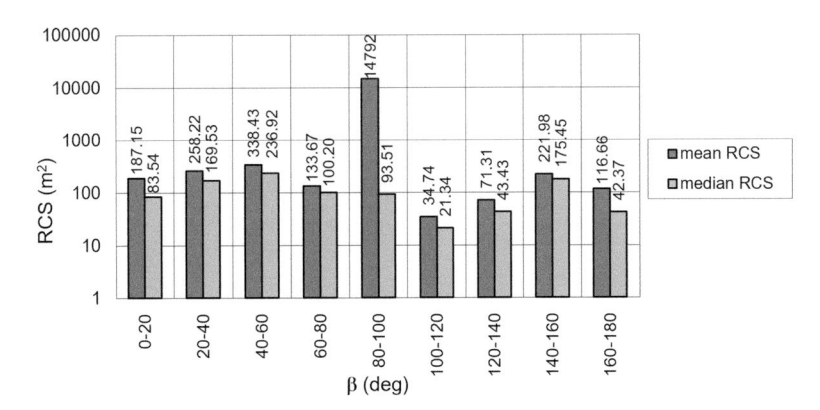

FIGURE 3.269 Diagrams of mean and median RCS of vehicle 2S6 model in 20-degree sectors of azimuth aspect given its radar observation (sounding frequency 10 GHz, $\gamma=30°$, underlying surface – dry loam, horizontal polarization).

Scattering characteristics of the vehicle 2S6 model for sounding frequency 10 GHz, elevation angle 30°, dry loam, and vertical polarization.

Figure 3.270 shows the RCS circular diagram of the vehicle 2S6 model. The noncoherent RCS circular diagram of the vehicle 2S6 is presented in Figure 3.271. The circular mean RCS of the vehicle 2S6 model is 89.68 m². The circular median RCS is 30.79 m².

Figures 3.272 and 3.273 show the mean and median RCS for the main ranges of sounding azimuths (head, side, and stern) and for ranges of 20°.

The scattering characteristics for this model at two polarizations, sounding frequencies of 10, 5, and 3 GHz (wavelengths of 3, 6, and 10 cm, respectively), for three types of underlying surface (dry loam, dry soil, and wet soil) are given in the electronic appendix of this book.

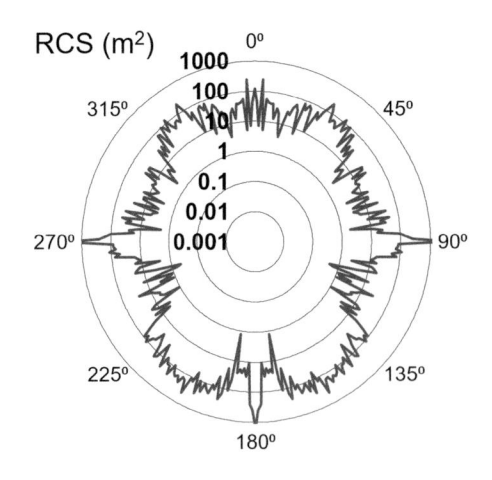

FIGURE 3.270 Circular diagrams of RCS given radar observation of vehicle 2S6 model (sounding frequency 10 GHz, $\gamma=30°$, underlying surface – dry loam, vertical polarization).

FIGURE 3.271 Circular diagram of noncoherent RCS given radar observation of vehicle 2S6 model (sounding frequency 10 GHz, $\gamma=30°$, underlying surface – dry loam, vertical polarization).

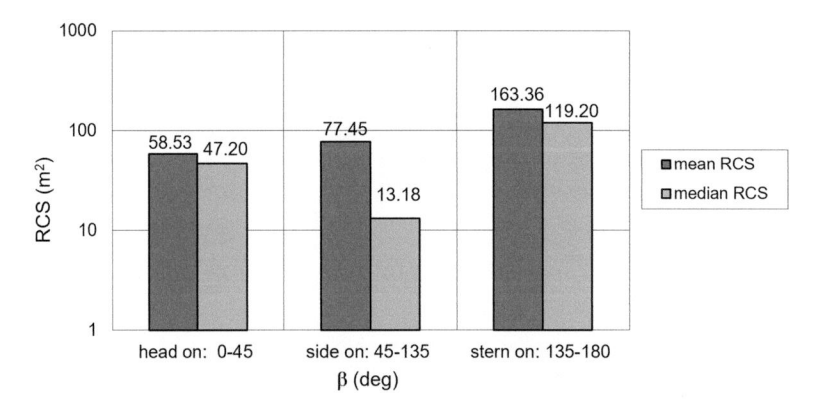

FIGURE 3.272 Diagrams of mean and median RCS of vehicle 2S6 model in three sectors of azimuth aspect given its radar observation (sounding frequency 10 GHz, $\gamma=30°$, underlying surface – dry loam, vertical polarization).

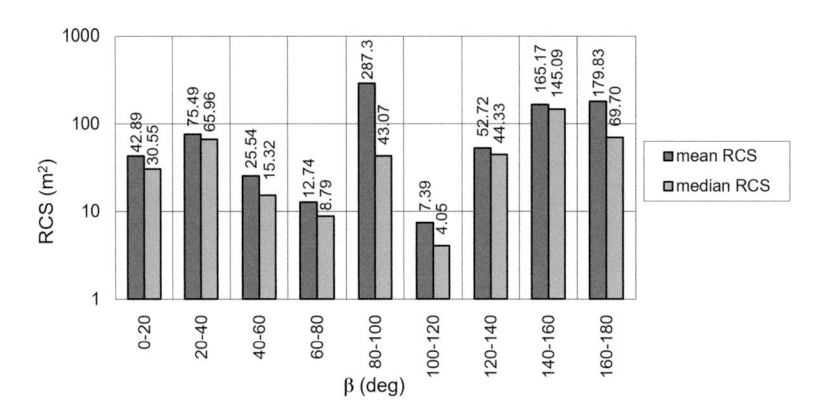

FIGURE 3.273 Diagrams of mean and median RCS of vehicle 2S6 model in 20-degree sectors of azimuth aspect given its radar observation (sounding frequency 10 GHz, $\gamma=30°$, underlying surface – dry loam, vertical polarization).

Scattering Characteristics of Ground Objects

3.17 ISKANDER LAUNCHER 9P78-1

9P78-1 (Figure 3.274) is a TEL of the mobile short-range ballistic missile system 9K720 Iskander (SS-26 Stone) [89]. It has been in service since 2006.

In accordance with the design of the launcher 9P78-1, a model of its surface was created to obtain scattering characteristics (in particular, RCS). The model is shown in Figure 3.275. In the modeling, the smooth parts of the object surface were approximated by parts of 216 triaxial ellipsoids. The surface breaks were modeled using 181 straight edge scattering parts.

FIGURE 3.274 Iskander launcher 9P78-1

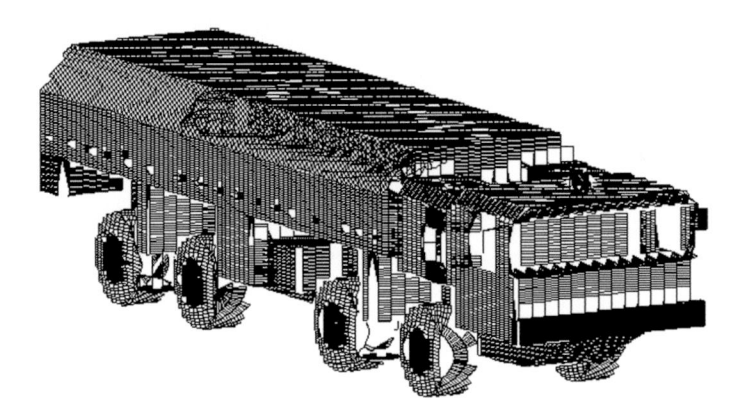

FIGURE 3.275 Surface model of Iskander launcher 9P78-1.

Below are some scattering characteristics of the launcher 9P78-1 model at a sounding frequency of 10 GHz (wavelengths of 3 cm) for the following conditions: the elevation angle of sounding $\gamma = 10°$, wet soil, and the elevation angle of sounding $\gamma = 30°$, dry loam.

Scattering characteristics of the launcher 9P78-1 model for sounding frequency 10 GHz, elevation angle 10°, wet soil, and horizontal polarization.

Figure 3.276 shows the RCS circular diagram of the launcher 9P78-1 model. The noncoherent RCS circular diagram of the launcher 9P78-1 is presented in Figure 3.277. The circular mean RCS of the launcher 9P78-1 model is 89 024 m². The circular median RCS is 7 663 m². Figure 3.278 and 3.279 show the mean and median RCS for the main ranges of sounding azimuths (head, side, and stern) and for ranges of 20°.

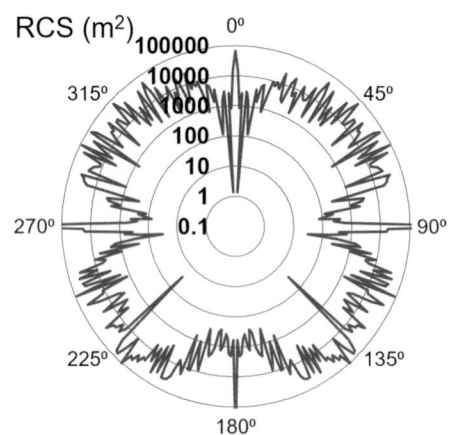

FIGURE 3.276 Circular diagrams of RCS given radar observation of launcher 9P78-1 model (sounding frequency 10 GHz, $\gamma = 10°$, underlying surface – wet soil, horizontal polarization).

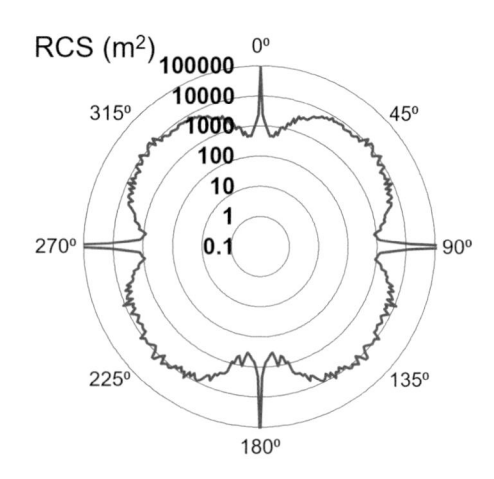

FIGURE 3.277 Circular diagram of noncoherent RCS given radar observation of launcher 9P78-1 model (sounding frequency 10 GHz, $\gamma = 10°$, underlying surface – wet soil, horizontal polarization).

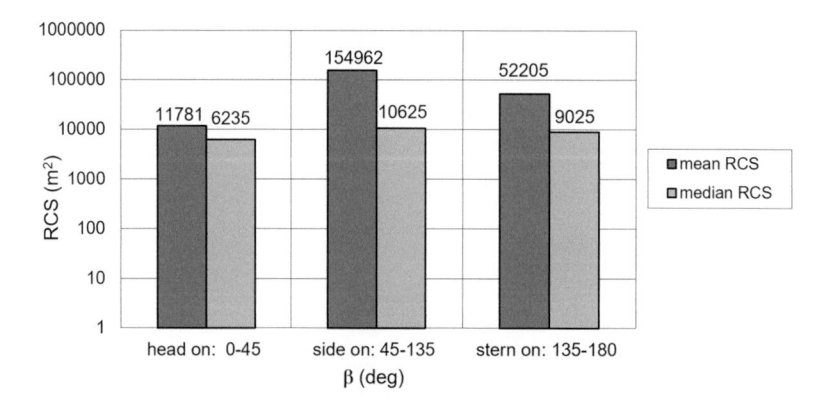

FIGURE 3.278 Diagrams of mean and median RCS of launcher 9P78-1 model in three sectors of azimuth aspect given its radar observation (sounding frequency 10 GHz, $\gamma = 10°$, underlying surface – wet soil, horizontal polarization).

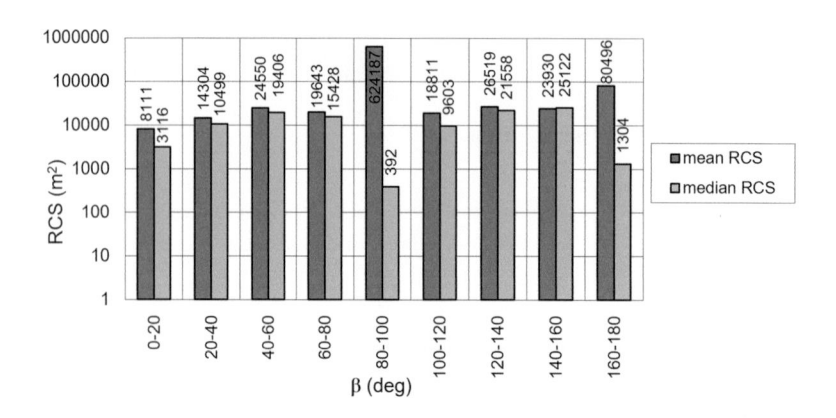

FIGURE 3.279 Diagrams of mean and median RCS of launcher 9P78-1 model in 20-degree sectors of azimuth aspect given its radar observation (sounding frequency 10 GHz, $\gamma = 10°$, underlying surface – wet soil, horizontal polarization).

Scattering Characteristics of Ground Objects

Scattering characteristics of the launcher 9P78-1 model for sounding frequency 10 GHz, elevation angle 10°, wet soil, and vertical polarization.

Figure 3.280 shows the RCS circular diagram of the launcher 9P78-1 model. The noncoherent RCS circular diagram of the launcher 9P78-1 is presented in Figure 3.281.

The circular mean RCS of the launcher 9P78-1 model is 775.94 m². The circular median RCS is 364.75 m².

Figures 3.282 and 3.283 show the mean and median RCS for the main ranges of sounding azimuths (head, side, and stern) and for ranges of 20°.

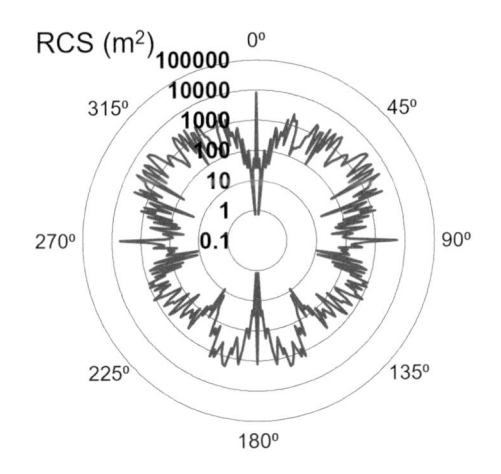

FIGURE 3.280 Circular diagrams of RCS given radar observation of launcher 9P78-1 model (sounding frequency 10 GHz, $\gamma = 10°$, underlying surface – wet soil, vertical polarization).

FIGURE 3.281 Circular diagram of noncoherent RCS given radar observation of launcher 9P78-1 model (sounding frequency 10 GHz, $\gamma = 10°$, underlying surface – wet soil, vertical polarization).

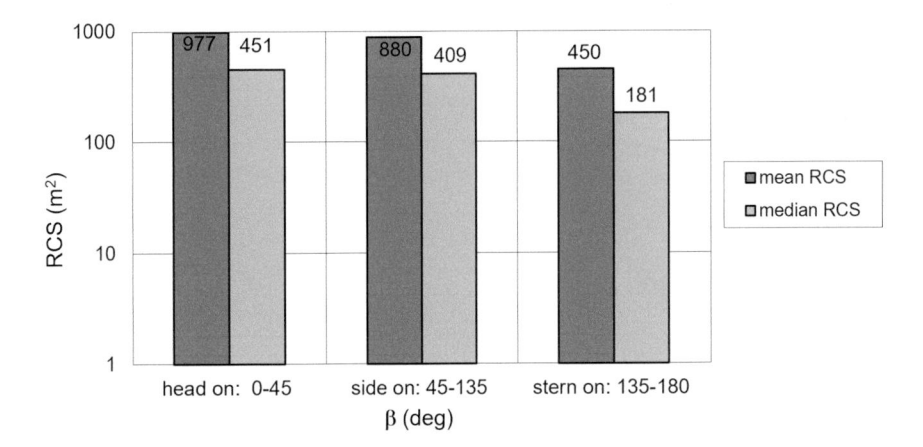

FIGURE 3.282 Diagrams of mean and median RCS of launcher 9P78-1 model in three sectors of azimuth aspect given its radar observation (sounding frequency 10 GHz, $\gamma = 10°$, underlying surface – wet soil, vertical polarization).

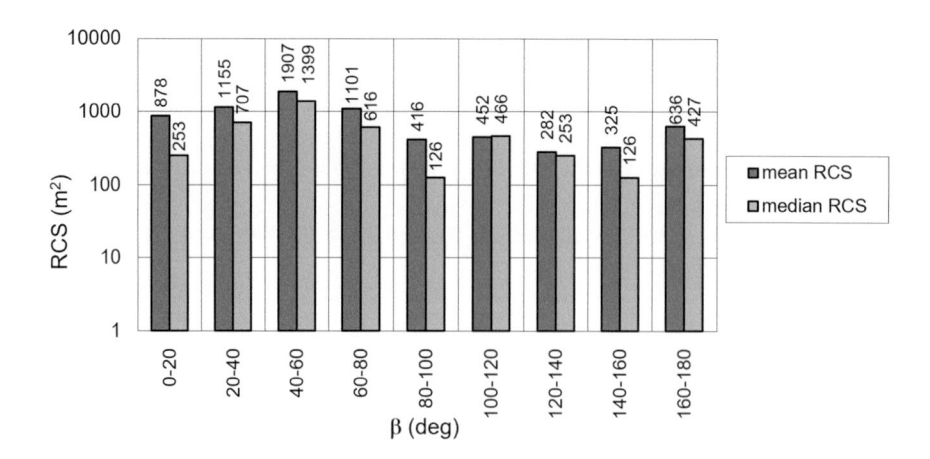

FIGURE 3.283 Diagrams of mean and median RCS of launcher 9P78-1 model in 20-degree sectors of azimuth aspect given its radar observation (sounding frequency 10 GHz, $\gamma = 10°$, underlying surface – wet soil, vertical polarization).

Scattering characteristics of the launcher 9P78-1 model for sounding frequency 10 GHz, elevation angle 30°, dry loam, and horizontal polarization.

Figure 3.284 shows the RCS circular diagram of the launcher 9P78-1 model. The noncoherent RCS circular diagram of the launcher 9P78-1 is presented in Figure 3.285. The circular mean RCS of the launcher 9P78-1 model is 5 444 m². The circular median RCS is 1 519 m².

Figures 3.286 and 3.287 show the mean and median RCS for the main ranges of sounding azimuths (head, side, and stern) and for ranges of 20°.

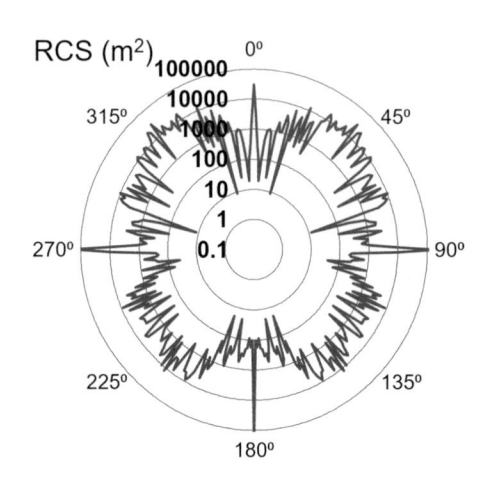

FIGURE 3.284 Circular diagrams of RCS given radar observation of launcher 9P78-1 model (sounding frequency 10 GHz, $\gamma = 30°$, underlying surface – dry loam, horizontal polarization).

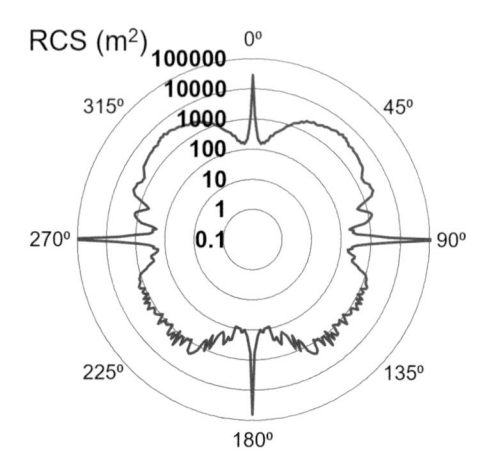

FIGURE 3.285 Circular diagram of noncoherent RCS given radar observation of launcher 9P78-1 model (sounding frequency 10 GHz, $\gamma = 30°$, underlying surface – dry loam, horizontal polarization).

Scattering Characteristics of Ground Objects

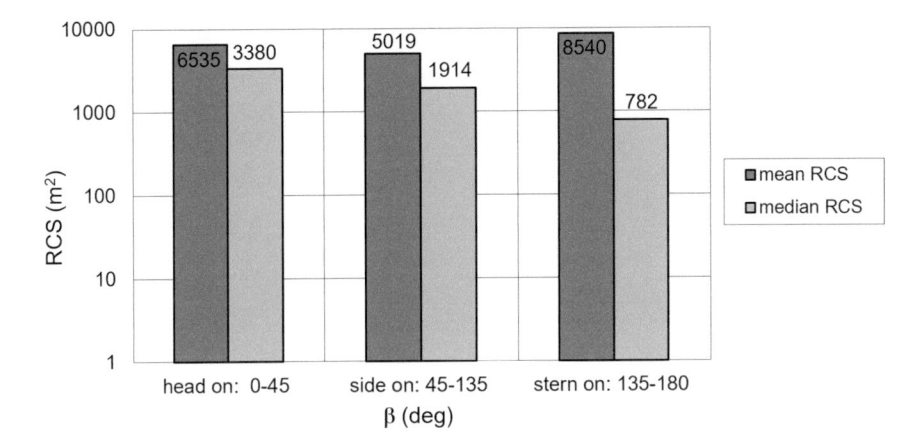

FIGURE 3.286 Diagrams of mean and median RCS of launcher 9P78-1 model in three sectors of azimuth aspect given its radar observation (sounding frequency 10 GHz, $\gamma=30°$, underlying surface – dry loam, horizontal polarization).

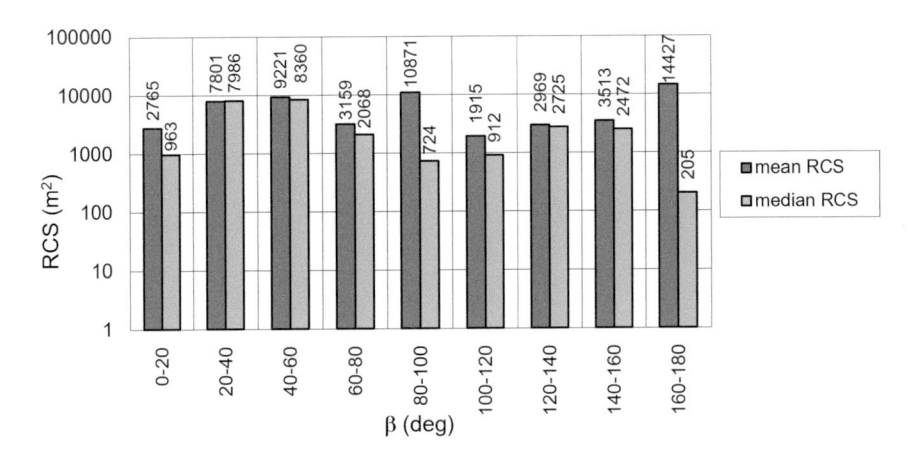

FIGURE 3.287 Diagrams of mean and median RCS of launcher 9P78-1 model in 20-degree sectors of azimuth aspect given its radar observation (sounding frequency 10 GHz, $\gamma=30°$, underlying surface – dry loam, horizontal polarization).

Scattering characteristics of the launcher 9P78-1 model for sounding frequency 10 GHz, elevation angle 30°, dry loam, and vertical polarization.

Figure 3.288 shows the RCS circular diagram of the launcher 9P78-1 model. The noncoherent RCS circular diagram of the launcher 9P78-1 is presented in Figure 3.289. The circular mean RCS of the launcher 9P78-1 model is 2 307 m². The circular median RCS is 582.81 m².

Figures 3.290 and 3.291 show the mean and median RCS for the main ranges of sounding azimuths (head, side, and stern) and for ranges of 20°.

The scattering characteristics for this model at two polarizations, sounding frequencies of 10, 5, and 3 GHz (wavelengths of 3, 6, and 10 cm, respectively), for three types of underlying surface (dry loam, dry soil, and wet soil) are given in the electronic appendix of this book.

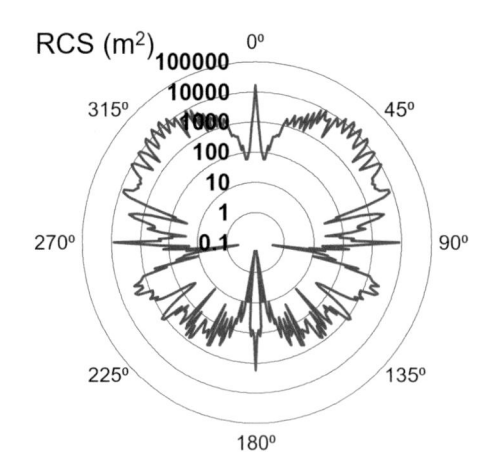

FIGURE 3.288 Circular diagrams of RCS given radar observation of launcher 9P78-1 model (sounding frequency 10 GHz, $\gamma=30°$, underlying surface – dry loam, vertical polarization).

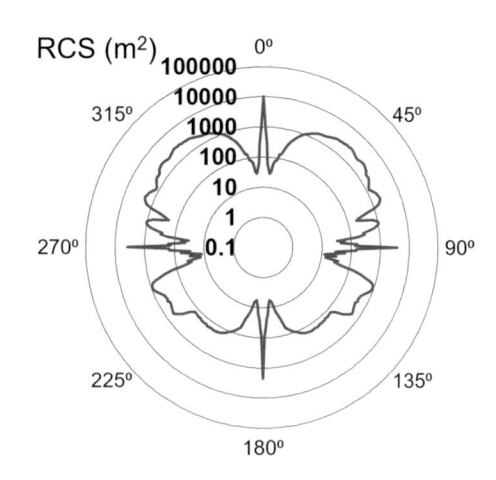

FIGURE 3.289 Circular diagram of noncoherent RCS given radar observation of launcher 9P78-1 model (sounding frequency 10 GHz, $\gamma=30°$, underlying surface – dry loam, vertical polarization).

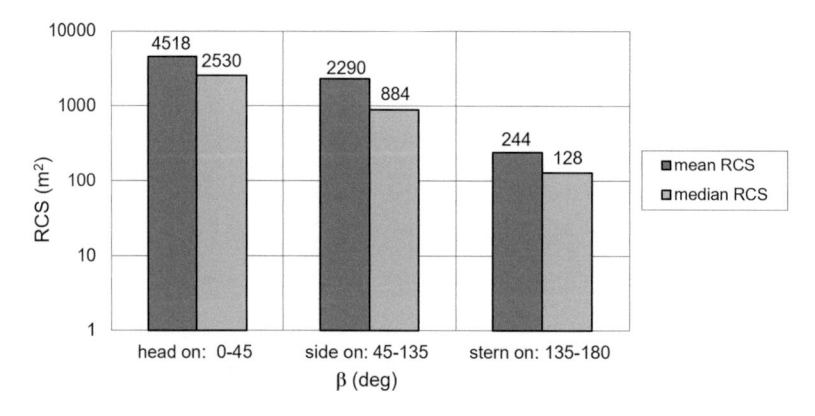

FIGURE 3.290 Diagrams of mean and median RCS of launcher 9P78-1 model in three sectors of azimuth aspect given its radar observation (sounding frequency 10 GHz, $\gamma=30°$, underlying surface – dry loam, vertical polarization).

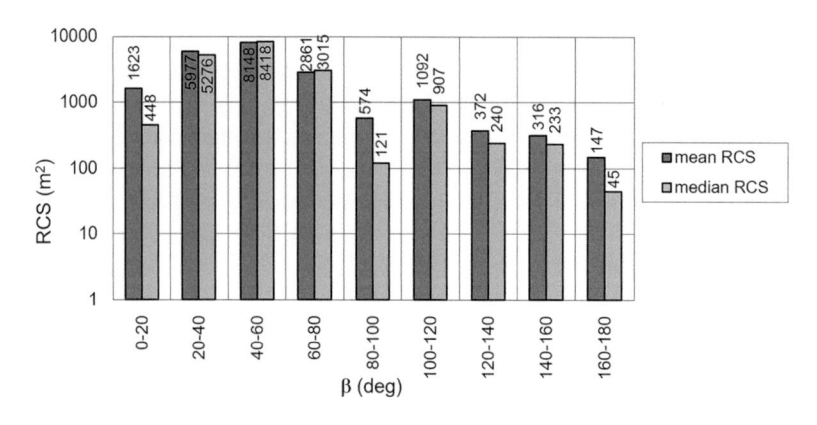

FIGURE 3.291 Diagrams of mean and median RCS of launcher 9P78-1 model in 20-degree sectors of azimuth aspect given its radar observation (sounding frequency 10 GHz, $\gamma=30°$, underlying surface – dry loam, vertical polarization).

Scattering Characteristics of Ground Objects

3.18 AIR SURVEILLANCE RADAR ST-68U

ST-68U - Tin Shield, 19Zh6- (Figure 3.292) is a medium-range 3D air defense radar [90]. It has been in service since 1981.

In accordance with the design of the radar ST-68U, a model of its surface was created to obtain scattering characteristics (in particular, RCS). The model is shown in Figure 3.293. In the modeling, the smooth parts of the object surface were approximated by parts of 60 triaxial ellipsoids. The surface breaks were modeled using 73 straight edge scattering parts.

FIGURE 3.292 Air surveillance radar ST68U.

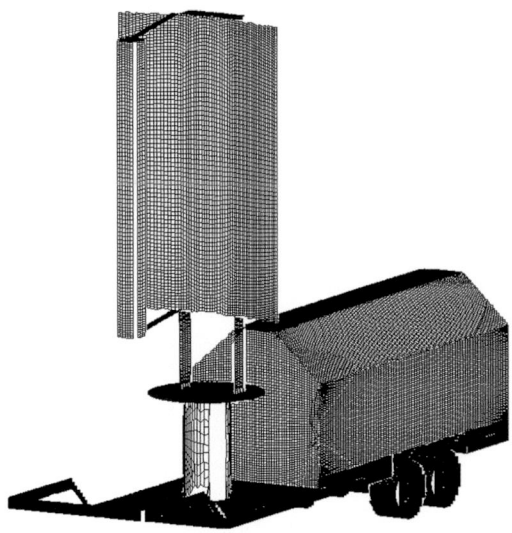

FIGURE 3.293 Surface model of air surveillance radar ST-68U.

Below are some scattering characteristics of the radar ST-68U model at a sounding frequency of 10 GHz (wavelengths of 3 cm) for the following conditions: the elevation angle of sounding $\gamma = 10°$, wet soil, and the elevation angle of sounding $\gamma = 30°$, dry loam.

Scattering characteristics of the radar ST-68U model for sounding frequency 10 GHz, elevation angle 10°, wet soil, and horizontal polarization.

Figure 3.294 shows the RCS circular diagram of the radar ST-68U model. The noncoherent RCS circular diagram of the radar ST-68U is presented in Figure 3.295. The circular mean RCS of the radar ST-68U model is 65 967 m². The circular median RCS is 2 315 m². Figures 3.296 and 3.297 show the mean and median RCS for the main ranges of sounding azimuths (head, side, and stern) and for ranges of 20°.

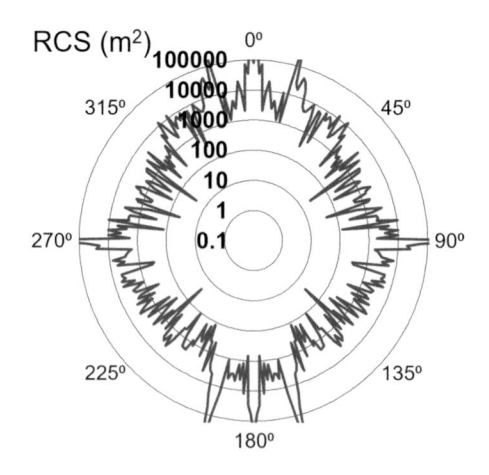

FIGURE 3.294 Circular diagrams of RCS given radar observation of radar ST-68U model (sounding frequency 10 GHz, $\gamma=10°$, underlying surface – wet soil, horizontal polarization).

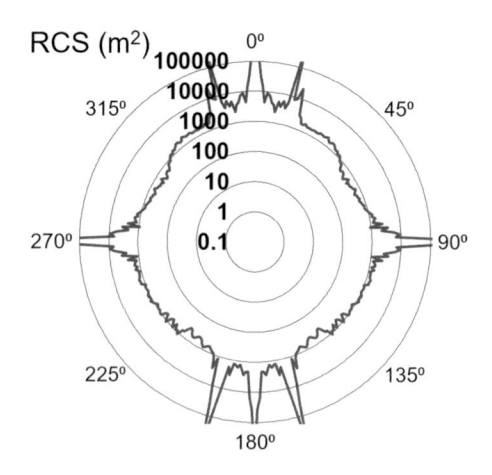

FIGURE 3.295 Circular diagram of noncoherent RCS given radar observation of radar ST-68U model (sounding frequency 10 GHz, $\gamma=10°$, underlying surface – wet soil, horizontal polarization).

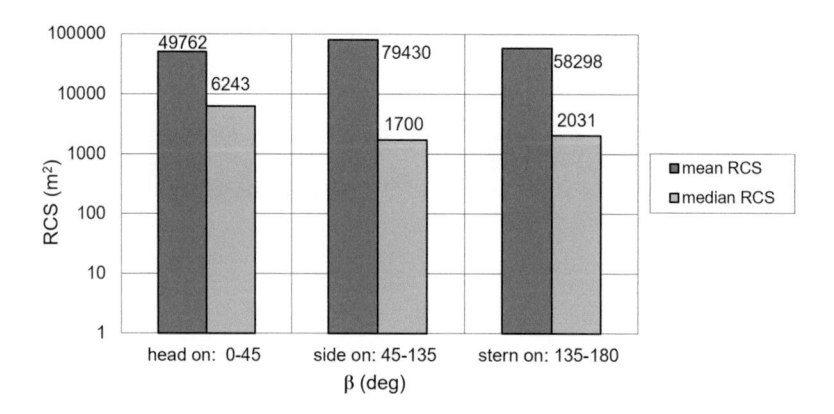

FIGURE 3.296 Diagrams of mean and median RCS of radar ST-68U model in three sectors of azimuth aspect given its radar observation (sounding frequency 10 GHz, $\gamma=10°$, underlying surface – wet soil, horizontal polarization).

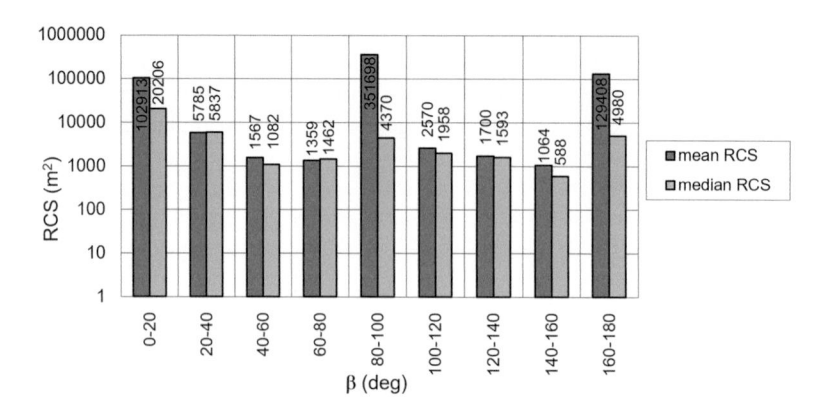

FIGURE 3.297 Diagrams of mean and median RCS of radar ST-68U model in 20-degree sectors of azimuth aspect given its radar observation (sounding frequency 10 GHz, $\gamma=10°$, underlying surface – wet soil, horizontal polarization).

Scattering characteristics of the radar ST-68U model for sounding frequency 10 GHz, elevation angle 10°, wet soil, and vertical polarization.

Figure 3.298 shows the RCS circular diagram of the radar ST-68U model. The noncoherent RCS circular diagram of the radar ST-68U is presented in Figure 3.299.

The circular mean RCS of the radar ST-68U model is 201.32 m². The circular median RCS is 11.36 m².

Figures 3.300 and 3.301 show the mean and median RCS for the main ranges of sounding azimuths (head, side, and stern) and for ranges of 20°.

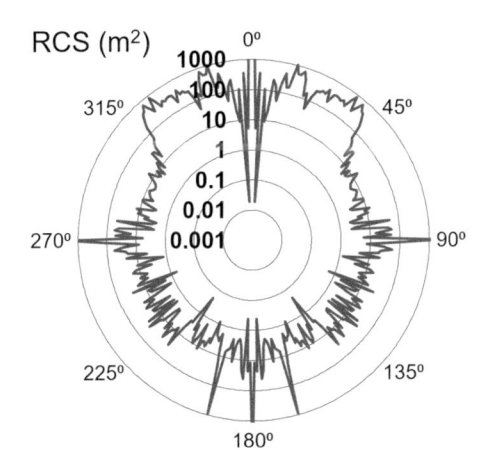

FIGURE 3.298 Circular diagrams of RCS given radar observation of radar ST-68U model (sounding frequency 10 GHz, $\gamma = 10°$, underlying surface – wet soil, vertical polarization).

FIGURE 3.299 Circular diagram of noncoherent RCS given radar observation of radar ST-68U model (sounding frequency 10 GHz, $\gamma = 10°$, underlying surface – wet soil, vertical polarization).

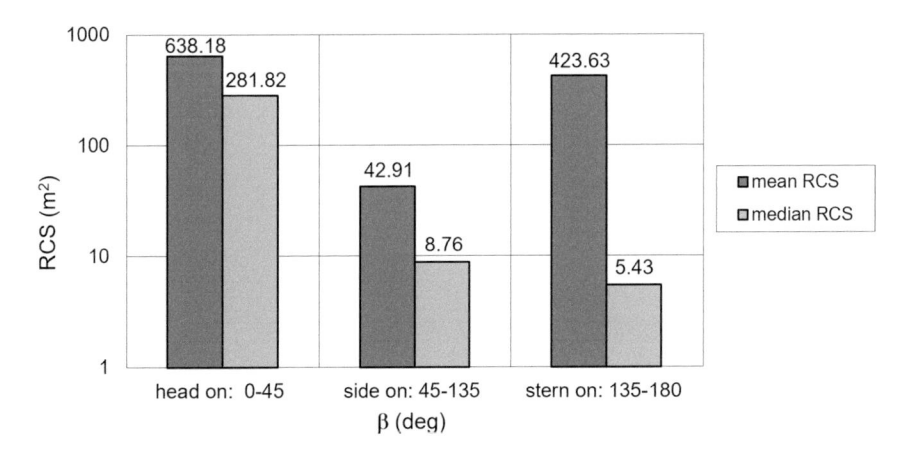

FIGURE 3.300 Diagrams of mean and median RCS of radar ST-68U model in three sectors of azimuth aspect given its radar observation (sounding frequency 10 GHz, $\gamma = 10°$, underlying surface – wet soil, vertical polarization).

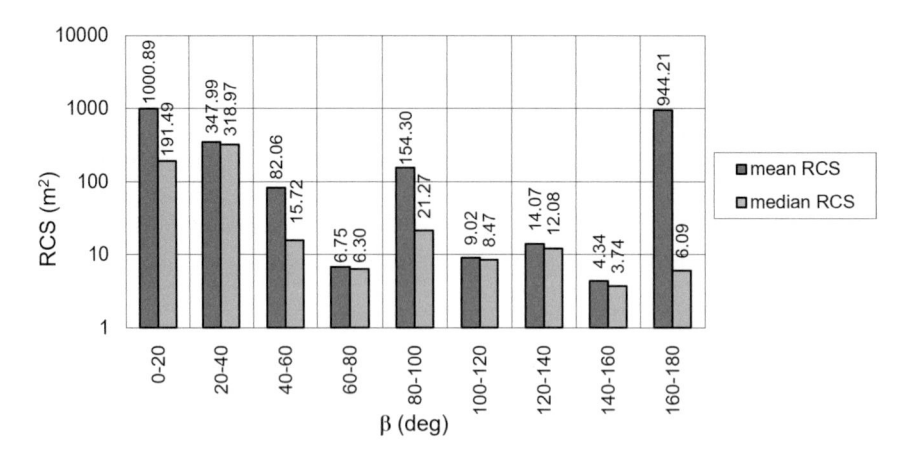

FIGURE 3.301 Diagrams of mean and median RCS of radar ST-68U model in 20-degree sectors of azimuth aspect given its radar observation (sounding frequency 10 GHz, $\gamma = 10°$, underlying surface – wet soil, vertical polarization).

Scattering characteristics of the radar ST-68U model for sounding frequency 10 GHz, elevation angle 30°, dry loam, and horizontal polarization.

Figure 3.302 shows the RCS circular diagram of the radar ST-68U model. The noncoherent RCS circular diagram of the radar ST-68U is presented in Figure 3.303. The circular mean RCS of the radar ST-68U model is 28 135 m². The circular median RCS is 431.70 m².

Figures 3.304 and 3.305 show the mean and median RCS for the main ranges of sounding azimuths (head, side, and stern) and for ranges of 20°.

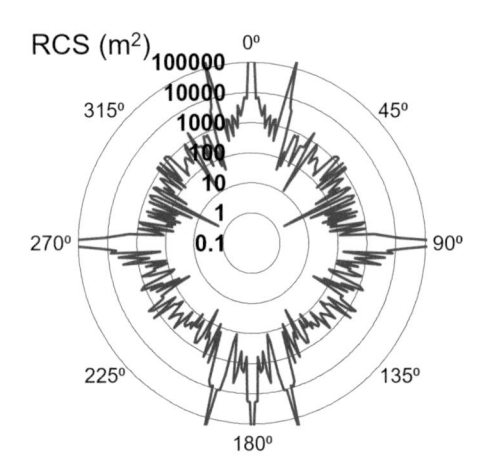

FIGURE 3.302 Circular diagrams of RCS given radar observation of radar ST-68U model (sounding frequency 10 GHz, $\gamma = 30°$, underlying surface – dry loam, horizontal polarization).

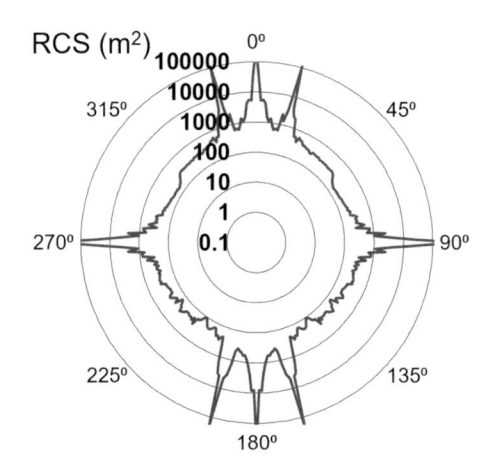

FIGURE 3.303 Circular diagram of noncoherent RCS given radar observation of radar ST-68U model (sounding frequency 10 GHz, $\gamma = 30°$, underlying surface – dry loam, horizontal polarization).

Scattering Characteristics of Ground Objects

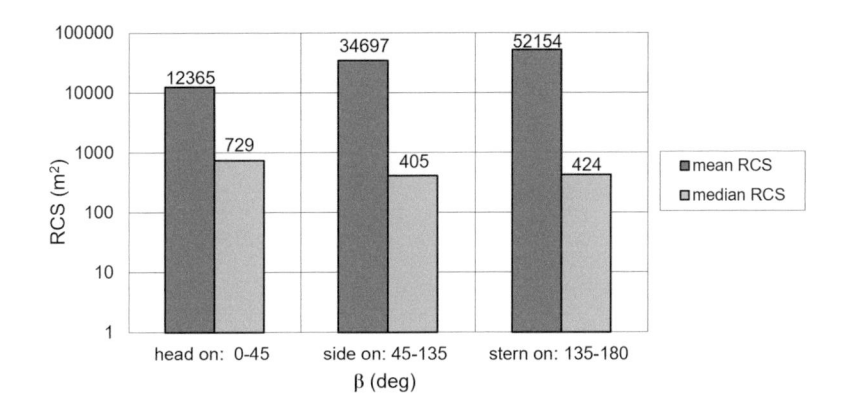

FIGURE 3.304 Diagrams of mean and median RCS of radar ST-68U model in three sectors of azimuth aspect given its radar observation (sounding frequency 10 GHz, $\gamma=30°$, underlying surface – dry loam, horizontal polarization).

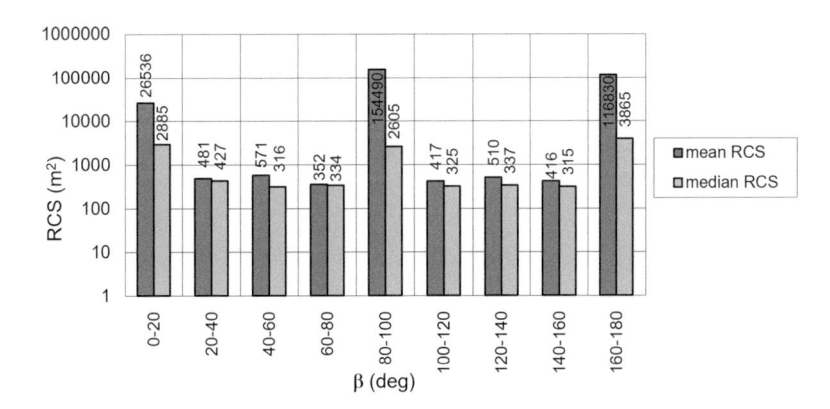

FIGURE 3.305 Diagrams of mean and median RCS of radar ST-68U model in 20-degree sectors of azimuth aspect given its radar observation (sounding frequency 10 GHz, $\gamma=30°$, underlying surface – dry loam, horizontal polarization).

Scattering characteristics of the radar ST-68U model for sounding frequency 10 GHz, elevation angle 30°, dry loam, and vertical polarization.

Figure 3.306 shows the RCS circular diagram of the radar ST-68U model. The noncoherent RCS circular diagram of the radar ST-68U is presented in Figure 3.307. The circular mean RCS of the radar ST-68U model is 2 182 m². The circular median RCS is 54.41 m².

Figures 3.308 and 3.309 show the mean and median RCS for the main ranges of sounding azimuths (head, side, and stern) and for ranges of 20°.

The scattering characteristics for this model at two polarizations, sounding frequencies of 10, 5, and 3 GHz (wavelengths of 3, 6, and 10 cm, respectively), for three types of underlying surface (dry loam, dry soil, and wet soil) are given in the electronic appendix of this book.

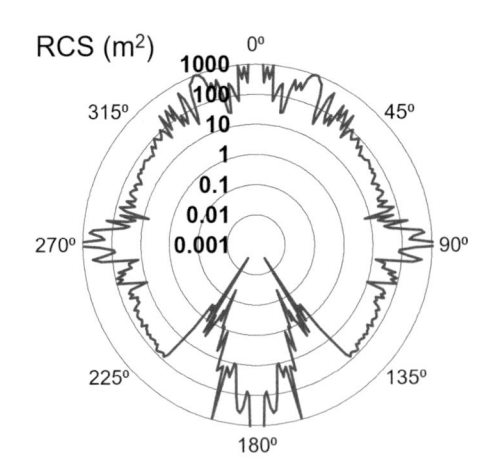

FIGURE 3.306 Circular diagrams of RCS given radar observation of radar ST-68U model (sounding frequency 10 GHz, $\gamma=30°$, underlying surface – dry loam, vertical polarization).

FIGURE 3.307 Circular diagram of noncoherent RCS given radar observation of radar ST-68U model (sounding frequency 10 GHz, $\gamma=30°$, underlying surface – dry loam, vertical polarization).

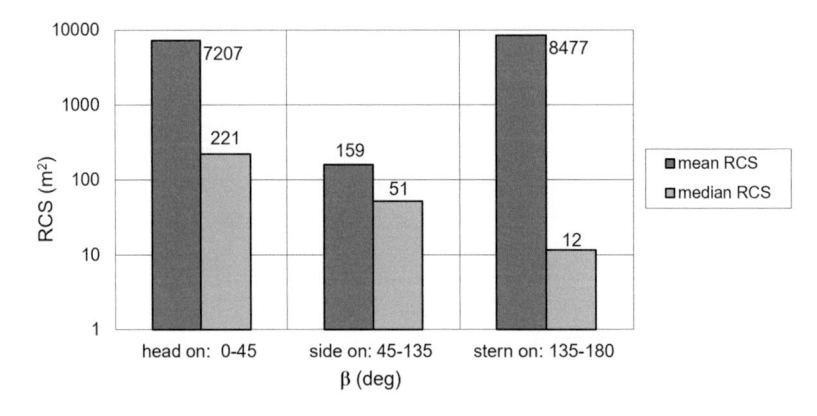

FIGURE 3.308 Diagrams of mean and median RCS of radar ST-68U model in three sectors of azimuth aspect given its radar observation (sounding frequency 10 GHz, $\gamma=30°$, underlying surface – dry loam, vertical polarization).

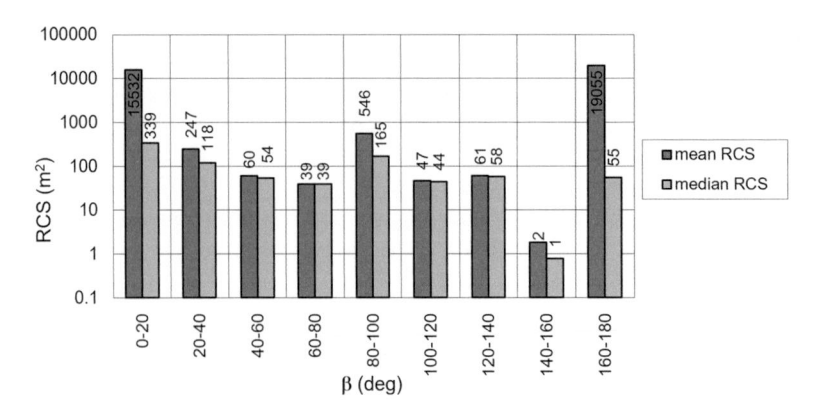

FIGURE 3.309 Diagrams of mean and median RCS of radar ST-68U model in 20-degree sectors of azimuth aspect given its radar observation (sounding frequency 10 GHz, $\gamma=30°$, underlying surface – dry loam, vertical polarization).

Scattering Characteristics of Ground Objects

3.19 RADAR P-18

P-18- Spoon Rest D - (Figure 3.310) is a general-purpose early warning radar operating in the VHF band [91]. It has been in service since 1971.

In accordance with the design of the radar P-18, a model of its surface was created to obtain scattering characteristics (in particular, RCS). Radar P-18 is a truck Ural-375D with an antenna system. The truck model is shown in Figure 3.311. In the modeling, the smooth parts of the truck surface were approximated by parts of 88 triaxial ellipsoids. The surface breaks were modeled using 106 straight edge scattering parts. To calculate the scattering characteristics of antenna system elements (Figs. 3.312 and 3.313), we used the calculation method obtaining the field scattered by a single cylinder [92].

FIGURE 3.310 Radar P-18.

FIGURE 3.311 Surface model of truck Ural-375D.

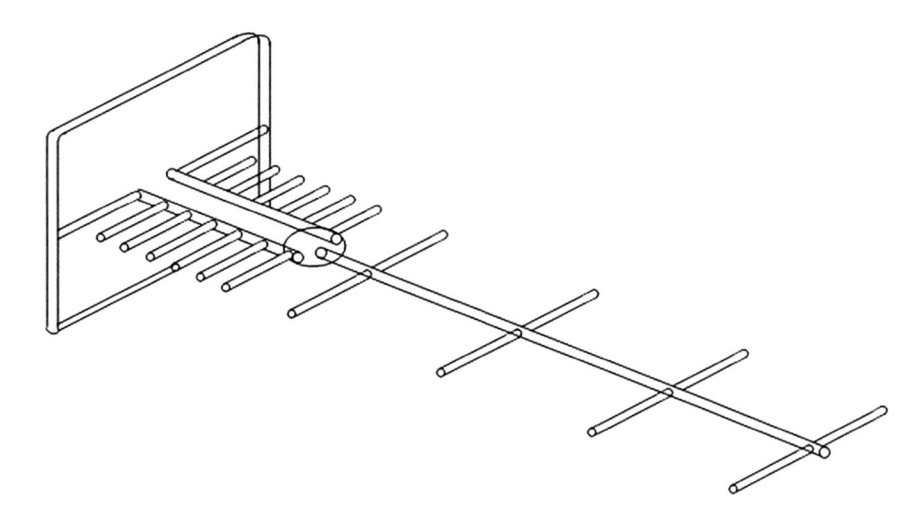

FIGURE 3.312 Antenna system element of the radar P-18- Yagi–Uda antenna.

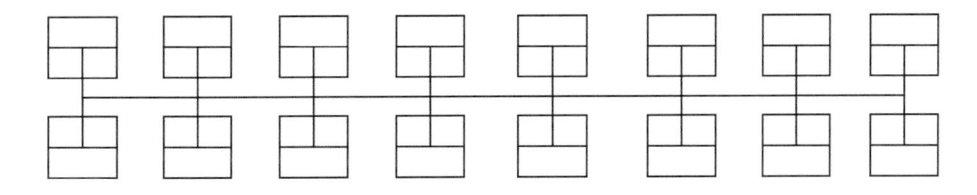

FIGURE 3.313 Antenna system scheme of the radar P-18.

Below are some scattering characteristics of the radar P-18 model at a sounding frequency of 10 GHz (wavelengths of 3 cm) for the following conditions: the elevation angle of sounding $\gamma = 10°$, wet soil, and the elevation angle of sounding $\gamma = 30°$, dry loam.

Scattering characteristics of the radar P-18 model for sounding frequency 10 GHz, elevation angle 10°, wet soil, and horizontal polarization.

Figure 3.314 shows the RCS circular diagram of the radar P-18 model. The noncoherent RCS circular diagram of the radar P-18 is presented in Figure 3.315. The circular mean RCS of the radar P-18 model is 31 588 m². The circular median RCS is 574.43 m². Figures 3.316 and 3.317 show the mean and median RCS for the main ranges of sounding azimuths (head, side, and stern) and for ranges of 20°.

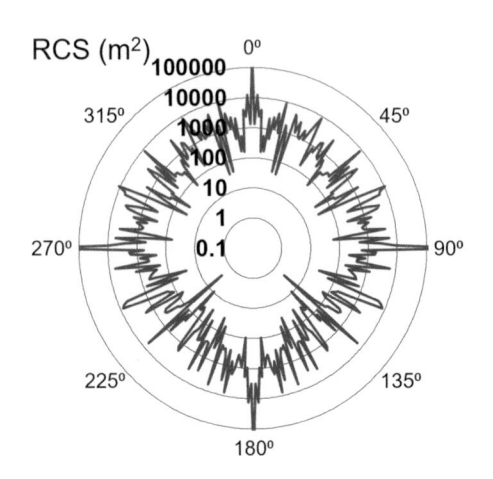

FIGURE 3.314 Circular diagrams of RCS given radar observation of radar P-18 model (sounding frequency 10 GHz, $\gamma = 10°$, underlying surface – wet soil, horizontal polarization).

FIGURE 3.315 Circular diagram of noncoherent RCS given radar observation of radar P-18 model (sounding frequency 10 GHz, $\gamma = 10°$, underlying surface – wet soil, horizontal polarization).

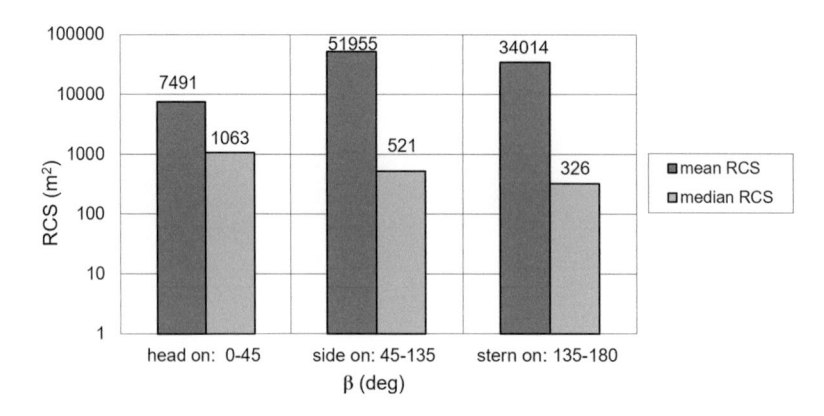

FIGURE 3.316 Diagrams of mean and median RCS of radar P-18 model in three sectors of azimuth aspect given its radar observation (sounding frequency 10 GHz, $\gamma = 10°$, underlying surface – wet soil, horizontal polarization).

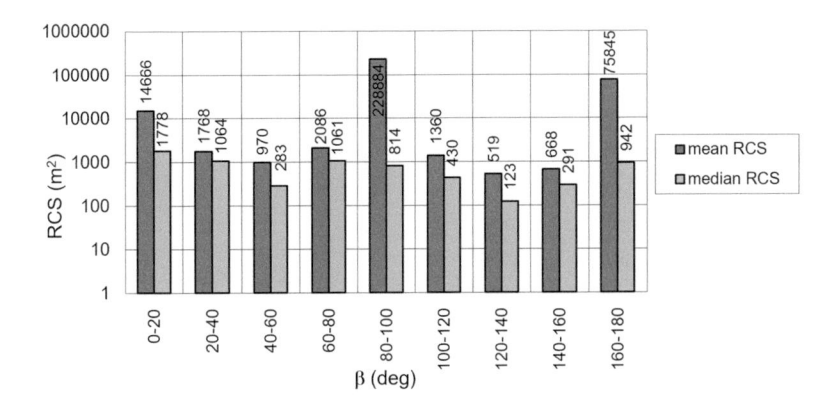

FIGURE 3.317 Diagrams of mean and median RCS of radar P-18 model in 20-degree sectors of azimuth aspect given its radar observation (sounding frequency 10 GHz, $\gamma = 10°$, underlying surface – wet soil, horizontal polarization).

Scattering characteristics of the radar P-18 model for sounding frequency 10 GHz, elevation angle 10°, wet soil, and vertical polarization.

Figure 3.318 shows the RCS circular diagram of the radar P-18 model. The noncoherent RCS circular diagram of the radar P-18 is presented in Figure 3.319. The circular mean RCS of the radar P-18 model is 31.35 m². The circular median RCS is 4.92 m².

Figures 3.320 and 3.321 show the mean and median RCS for the main ranges of sounding azimuths (head, side, and stern) and for ranges of 20°.

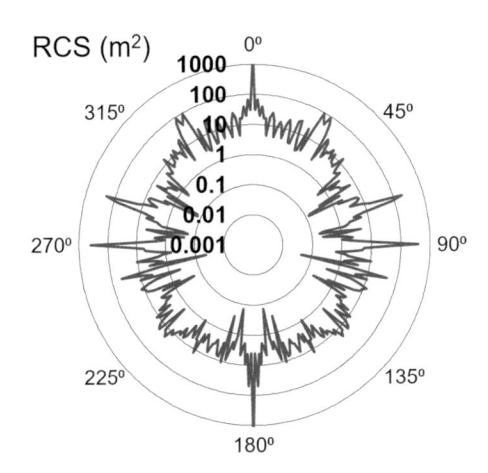

FIGURE 3.318 Circular diagrams of RCS given radar observation of radar P-18 model (sounding frequency 10 GHz, $\gamma = 10°$, underlying surface – wet soil, vertical polarization).

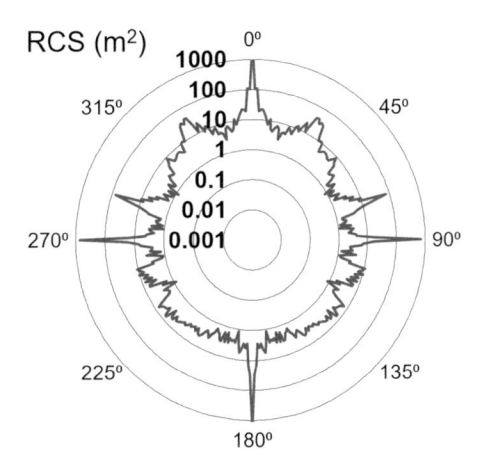

FIGURE 3.319 Circular diagram of noncoherent RCS given radar observation of radar P-18 model (sounding frequency 10 GHz, $\gamma = 10°$, underlying surface – wet soil, vertical polarization).

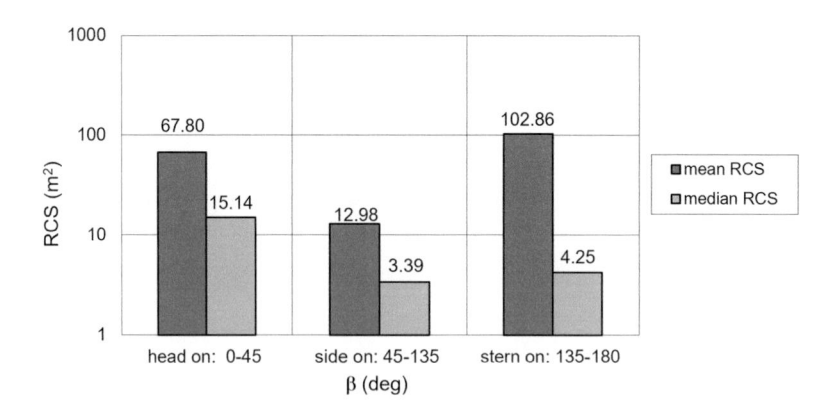

FIGURE 3.320 Diagrams of mean and median RCS of radar P-18 model in three sectors of azimuth aspect given its radar observation (sounding frequency 10 GHz, $\gamma = 10°$, underlying surface – wet soil, vertical polarization).

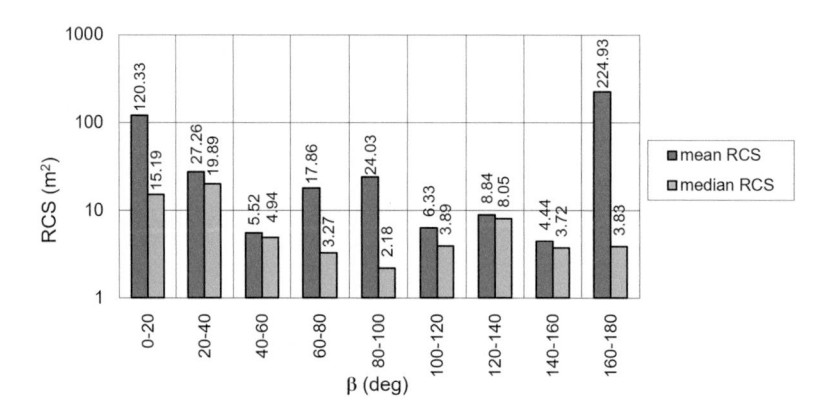

FIGURE 3.321 Diagrams of mean and median RCS of radar P-18 model in 20-degree sectors of azimuth aspect given its radar observation (sounding frequency 10 GHz, $\gamma = 10°$, underlying surface – wet soil, vertical polarization).

Scattering characteristics of the radar P-18 model for sounding frequency 10 GHz, elevation angle 30°, dry loam, and horizontal polarization.

Figure 3.322 shows the RCS circular diagram of the radar P-18 model. The noncoherent RCS circular diagram of the radar P-18 is presented in Figure 3.323. The circular mean RCS of the radar P-18 model is 6 520 m². The circular median RCS is 202.28 m².

Figures 3.324 and 3.325 show the mean and median RCS for the main ranges of sounding azimuths (head, side, and stern) and for ranges of 20°.

Scattering characteristics of the radar P-18 model for sounding frequency 10 GHz, elevation angle 30°, dry loam, and vertical polarization.

Figure 3.326 shows the RCS circular diagram of the radar P-18 model. The noncoherent RCS circular diagram of the radar P-18 is presented in Figure 3.327. The circular mean RCS of the radar P-18 model is 135.25 m². The circular median RCS is 6.83 m².

Figures 3.328 and 3.329 show the mean and median RCS for the main ranges of sounding azimuths (head, side, and stern) and for ranges of 20°.

The scattering characteristics for this model at two polarizations, sounding frequencies of 10, 5, and 3 GHz (wavelengths of 3, 6, and 10 cm, respectively), for three types of underlying surface (dry loam, dry soil, and wet soil) are given in the electronic appendix of this book.

Scattering Characteristics of Ground Objects

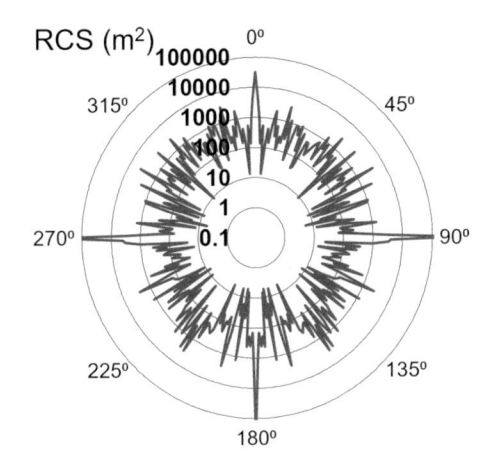

FIGURE 3.322 Circular diagrams of RCS given radar observation of radar P-18 model (sounding frequency 10 GHz, $\gamma=30°$, underlying surface – dry loam, horizontal polarization).

FIGURE 3.323 Circular diagram of noncoherent RCS given radar observation of radar P-18 model (sounding frequency 10 GHz, $\gamma=30°$, underlying surface – dry loam, horizontal polarization).

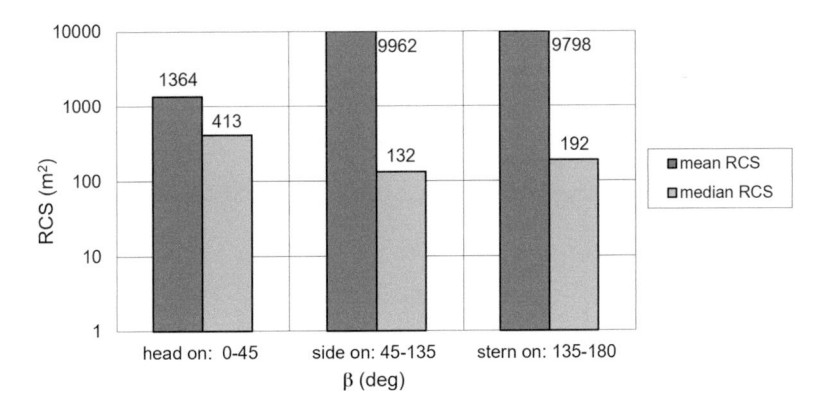

FIGURE 3.324 Diagrams of mean and median RCS of radar P-18 model in three sectors of azimuth aspect given its radar observation (sounding frequency 10 GHz, $\gamma=30°$, underlying surface – dry loam, horizontal polarization).

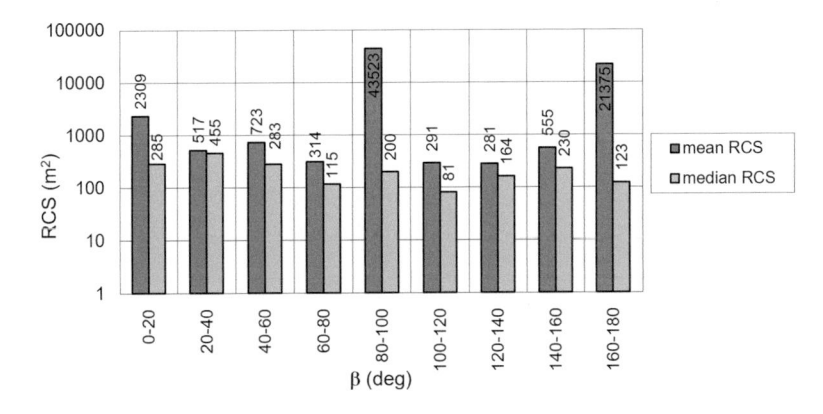

FIGURE 3.325 Diagrams of mean and median RCS of radar P-18 model in 20-degree sectors of azimuth aspect given its radar observation (sounding frequency 10 GHz, $\gamma=30°$, underlying surface – dry loam, horizontal polarization).

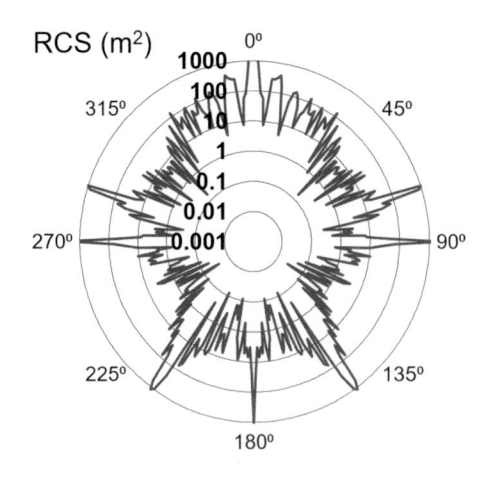

FIGURE 3.326 Circular diagrams of RCS given radar observation of radar P-18 model (sounding frequency 10 GHz, $\gamma=30°$, underlying surface – dry loam, vertical polarization).

FIGURE 3.327 Circular diagram of noncoherent RCS given radar observation of radar P-18 model (sounding frequency 10 GHz, $\gamma=30°$, underlying surface – dry loam, vertical polarization).

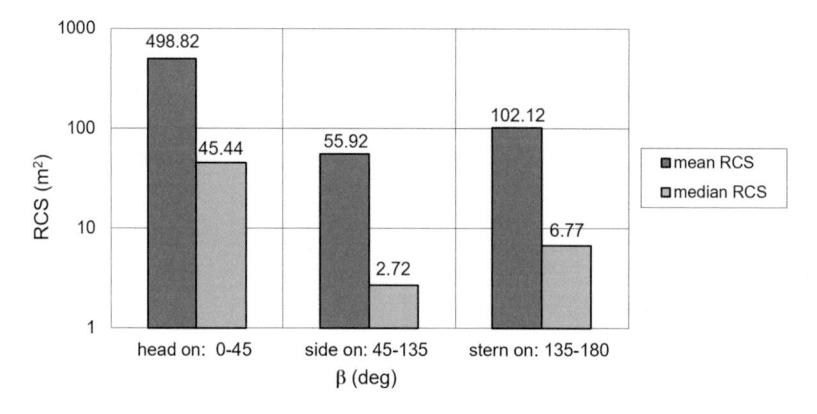

FIGURE 3.328 Diagrams of mean and median RCS of radar P-18 model in three sectors of azimuth aspect given its radar observation (sounding frequency 10 GHz, $\gamma=30°$, underlying surface – dry loam, vertical polarization).

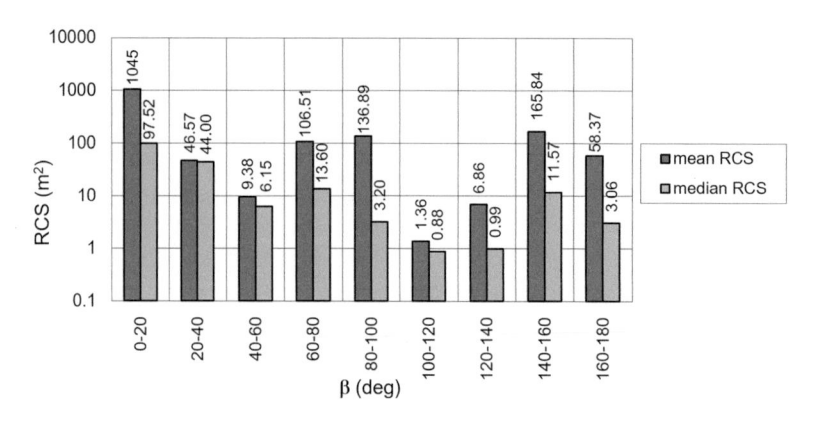

FIGURE 3.329 Diagrams of mean and median RCS of radar P-18 model in 20-degree sectors of azimuth aspect given its radar observation (sounding frequency 10 GHz, $\gamma=30°$, underlying surface – dry loam, vertical polarization).

Scattering Characteristics of Ground Objects

3.20 ROPUCHA-CLASS LANDING SHIP

The Ropucha (toad), or Project 775 class landing ships (Figure 3.330) are classified as "large landing craft". They were built in Poland in the Stocznia Północna shipyards, in Gdansk. They are designed for beach landings and can carry a 450-ton cargo. The ships have both bow and stern doors for loading and unloading vehicles, and the 630 m² of vehicle deck stretches the length of the hull. Up to 25 armored personnel carriers can be embarked. They have been in service since 1975.

The general characteristics of the Ropucha-class landing ship [93]: length – 112.5 m, beam – 15 m, draft: 3.7 m, height under water – 24.2 m, speed - 18 knots (33 km/h), range is 6 100 km at 15 knots (28 km/h).

The scattering characteristics for this object were obtained as an example of how calculation methods proposed in [1] for ground vehicles are applicable to surface ships.

In accordance with the design of the ship, a model of its surface was created to obtain scattering characteristics (in particular, RCS). The model is shown in Figure 3.331. In the modeling, the smooth parts of the object surface were approximated by parts of 160 triaxial ellipsoids. The surface breaks were modeled using 140 straight edge scattering parts.

FIGURE 3.330 Ropucha-class landing ship.

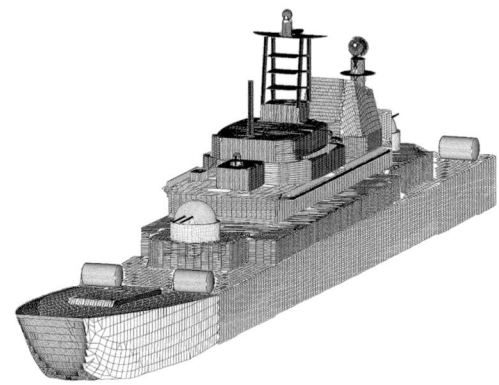

FIGURE 3.331 Surface model of the Ropucha-class landing ship.

Below are some scattering characteristics of the ship model at a sounding frequency of 10 GHz (wavelengths of 3 cm) for the following conditions: the elevation angle of sounding $\gamma = 10°$, rough sea ($\varepsilon' = 80 + j3.4$), and the elevation angle of sounding $\gamma = 30°$, smooth sea. For both conditions, the sea water has relative permeability $\varepsilon' = 80 + j3.4$. But the reflection factor of rough sea is assumed to be zero.

Scattering characteristics of the ship model for sounding frequency 10 GHz, elevation angle 10°, rough sea, and horizontal polarization.

Figure 3.332 shows the RCS circular diagram of the ship model. The noncoherent RCS circular diagram of the ship is presented in Figure 3.333. The circular mean RCS of the ship model is 433 545 m². The circular median RCS is 171 670 m².

Figures 3.334 and 3.335 show the mean and median RCS for the main ranges of sounding azimuths (head, side, and stern) and for ranges of 20°.

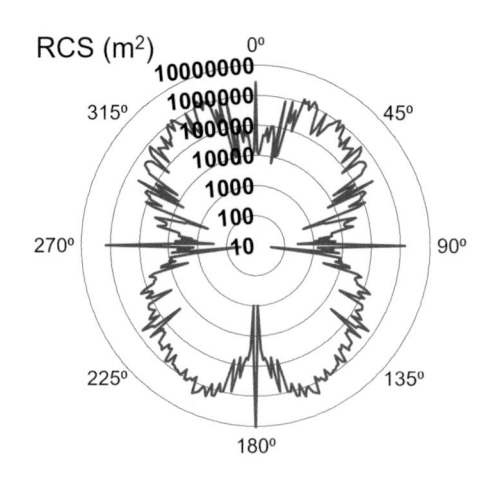

FIGURE 3.332 Circular diagrams of RCS given radar observation of ship model (sounding frequency 10 GHz, $\gamma = 10°$, underlying surface – rough sea, horizontal polarization).

FIGURE 3.333 Circular diagram of noncoherent RCS given radar observation of ship model (sounding frequency 10 GHz, $\gamma = 10°$, underlying surface – rough sea, horizontal polarization).

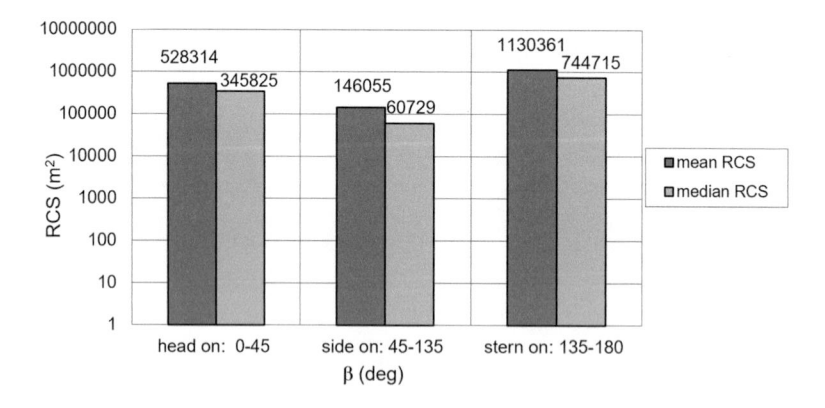

FIGURE 3.334 Diagrams of mean and median RCS of ship model in three sectors of azimuth aspect given its radar observation (sounding frequency 10 GHz, $\gamma = 10°$, underlying surface – rough sea, horizontal polarization).

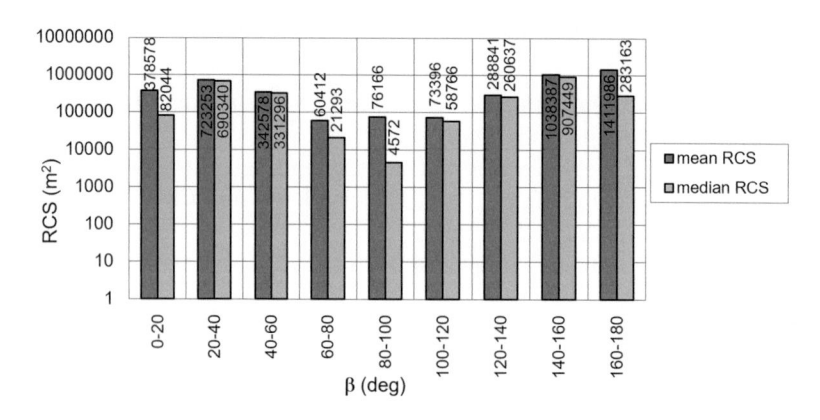

FIGURE 3.335 Diagrams of mean and median RCS of ship model in 20-degree sectors of azimuth aspect given its radar observation (sounding frequency 10 GHz, $\gamma = 10°$, underlying surface – rough sea, horizontal polarization).

Scattering Characteristics of Ground Objects   503

Scattering characteristics of the ship model for sounding frequency 10 GHz, elevation angle 10°, rough sea, and vertical polarization.

Figure 3.336 shows the RCS circular diagram of the ship model. The noncoherent RCS circular diagram of the ship is presented in Figure 3.337. The circular mean RCS of the ship model is 429 023 m². The circular median RCS is 156 756 m².

Figures 3.338 and 3.339 show the mean and median RCS for the main ranges of sounding azimuths (head, side, and stern) and for ranges of 20°.

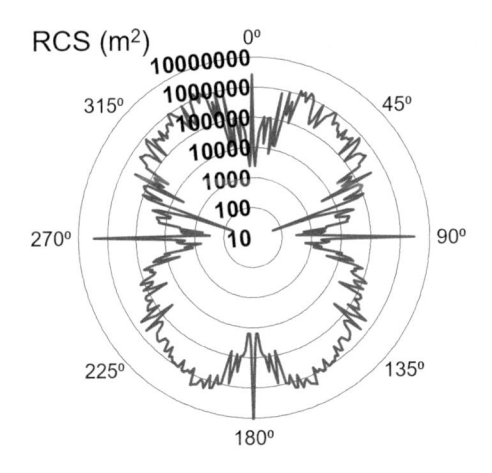

FIGURE 3.336 Circular diagrams of RCS given radar observation of ship model (sounding frequency 10 GHz, $\gamma = 10°$, underlying surface – rough sea, vertical polarization).

FIGURE 3.337 Circular diagram of noncoherent RCS given radar observation of ship model (sounding frequency 10 GHz, $\gamma = 10°$, underlying surface – rough sea, vertical polarization).

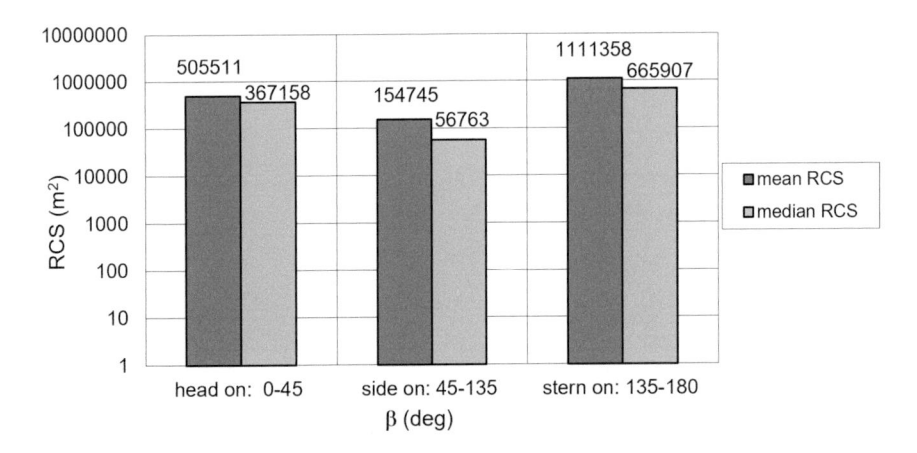

FIGURE 3.338 Diagrams of mean and median RCS of ship model in three sectors of azimuth aspect given its radar observation (sounding frequency 10 GHz, $\gamma = 10°$, underlying surface – rough sea, vertical polarization).

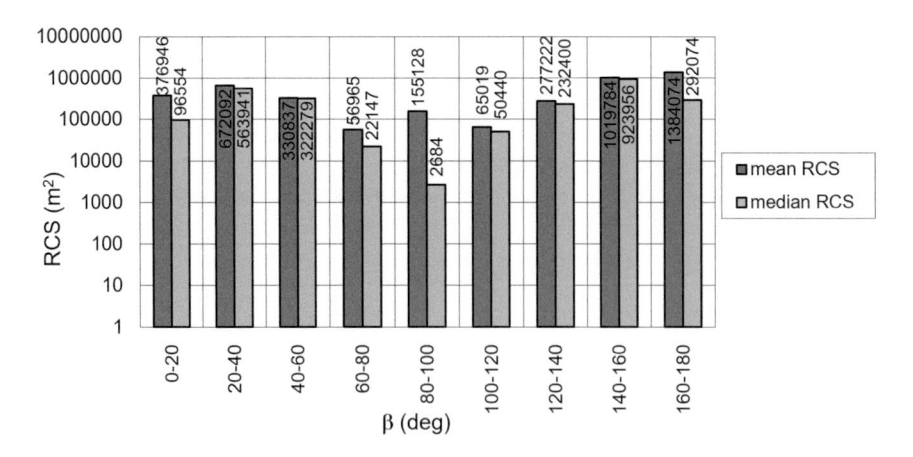

FIGURE 3.339 Diagrams of mean and median RCS of ship model in 20-degree sectors of azimuth aspect given its radar observation (sounding frequency 10 GHz, $\gamma=10°$, underlying surface – rough sea, vertical polarization).

Scattering characteristics of the ship model for sounding frequency 10 GHz, elevation angle 30°, smooth sea, and horizontal polarization.

Figure 3.340 shows the RCS circular diagram of the ship model. The noncoherent RCS circular diagram of the ship is presented in Figure 3.341. The circular mean RCS of the ship model is 11 653 888 m². The circular median RCS is 802 534 m².

Figures 3.342 and 3.343 show the mean and median RCS for the main ranges of sounding azimuths (head, side, and stern) and for ranges of 20°.

FIGURE 3.340 Circular diagrams of RCS given radar observation of ship model (sounding frequency 10 GHz, $\gamma=30°$, underlying surface – smooth sea, horizontal polarization).

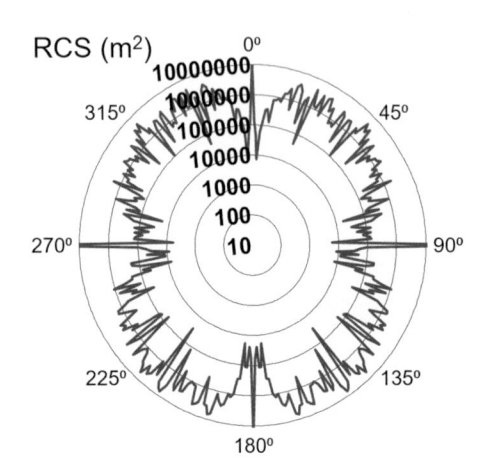

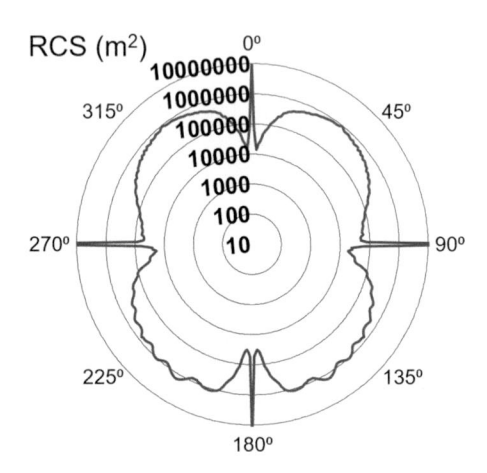

FIGURE 3.341 Circular diagram of noncoherent RCS given radar observation of ship model (sounding frequency 10 GHz, $\gamma=30°$, underlying surface – smooth sea, horizontal polarization).

Scattering Characteristics of Ground Objects

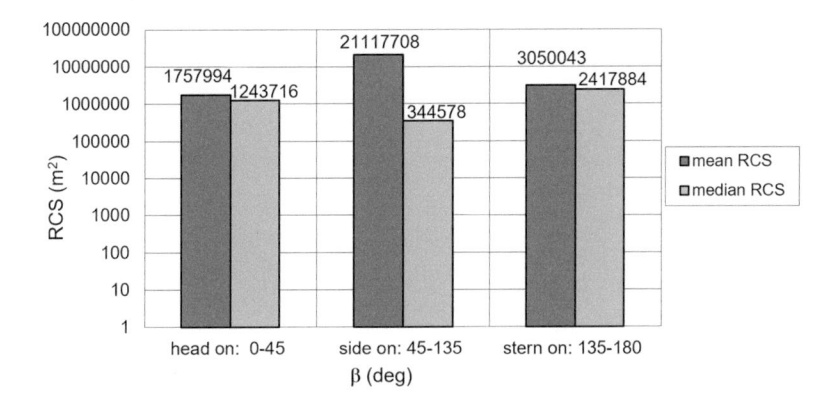

FIGURE 3.342 Diagrams of mean and median RCS of ship model in three sectors of azimuth aspect given its radar observation (sounding frequency 10 GHz, $\gamma = 30°$, underlying surface – smooth sea, horizontal polarization).

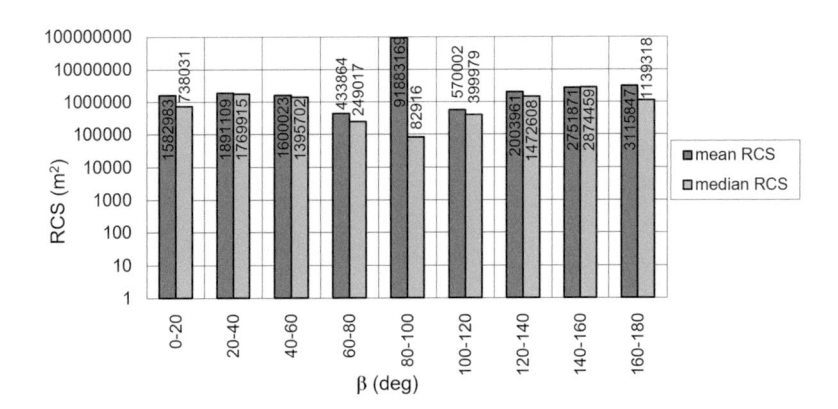

FIGURE 3.343 Diagrams of mean and median RCS of ship model in 20-degree sectors of azimuth aspect given its radar observation (sounding frequency 10 GHz, $\gamma = 30°$, underlying surface – smooth sea, horizontal polarization).

Scattering characteristics of the ship model for sounding frequency 10 GHz, elevation angle 30°, smooth sea, and vertical polarization.

Figure 3.344 shows the RCS circular diagram of the ship model. The noncoherent RCS circular diagram of the ship is presented in Figure 3.345. The circular mean RCS of the ship model is 12 933 627 m². The circular median RCS is 751 522 m².

Figures 3.346 and 3.347 show the mean and median RCS for the main ranges of sounding azimuths (head, side, and stern) and for ranges of 20°.

The scattering characteristics for this model at two polarizations, sounding frequencies of 10, 5, and 3 GHz (wavelengths of 3, 6, and 10 cm, respectively), for two types of underlying surface (smooth sea and rough sea) are given in the electronic appendix of this book.

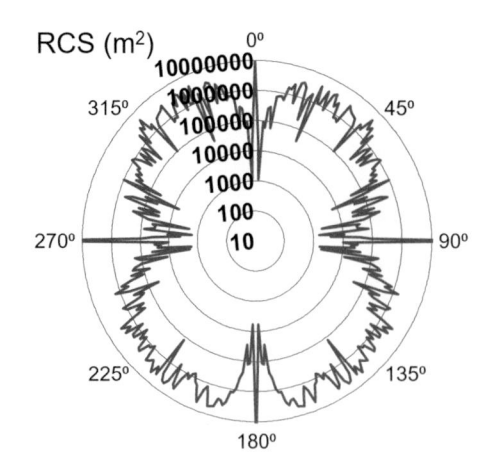

FIGURE 3.344 Circular diagrams of RCS given radar observation of ship model (sounding frequency 10 GHz, $\gamma=30°$, underlying surface – smooth sea, vertical polarization).

FIGURE 3.345 Circular diagram of noncoherent RCS given radar observation of ship model (sounding frequency 10 GHz, $\gamma=30°$, underlying surface – smooth sea, vertical polarization).

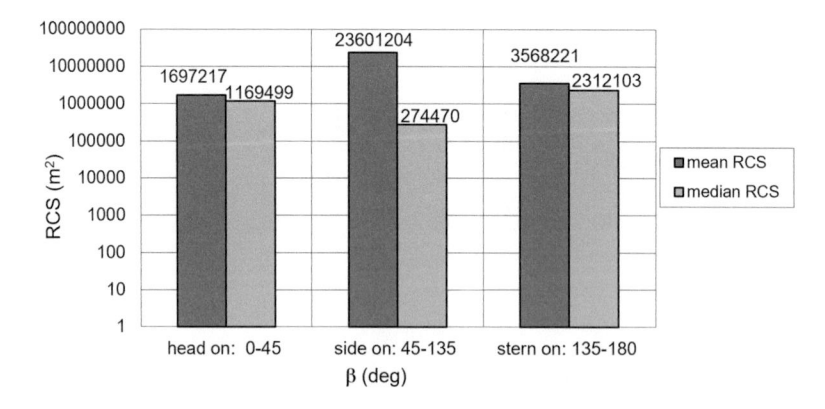

FIGURE 3.346 Diagrams of mean and median RCS of ship model in three sectors of azimuth aspect given its radar observation (sounding frequency 10 GHz, $\gamma=30°$, underlying surface – smooth sea, vertical polarization).

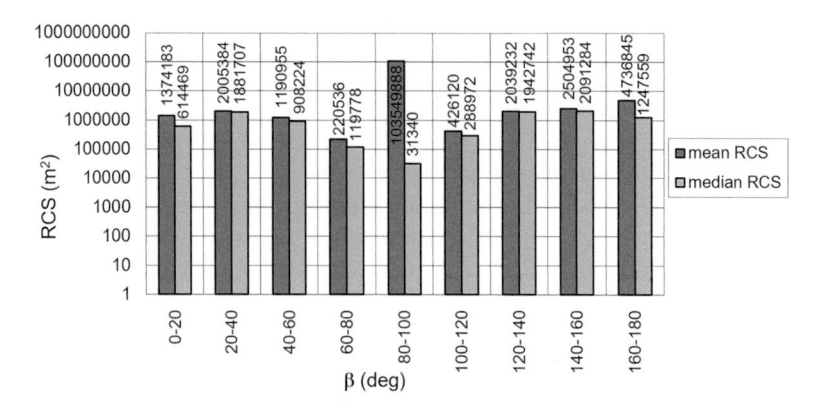

FIGURE 3.347 Diagrams of mean and median RCS of ship model in 20-degree sectors of azimuth aspect given its radar observation (sounding frequency 10 GHz, $\gamma=30°$, underlying surface – smooth sea, vertical polarization).

Conclusion

Dear readers. We, the authors, are glad that our book has interested you and that you have found time to study it. In conclusion, we would like to make a few small but, in our opinion, essential remarks.

First, it should be noted that this book presents the scattering characteristics not of the aerial and ground objects themselves, but of their models. We have tried to make the models as close as possible to the originals in size, construction, and used materials. However, many of the objects we have never seen or had access to, so we used only the information we could find. We are a bit like the artists at Madame Tussauds. However, we believe that the results obtained are consistent with the electrodynamic processes that occur on real objects and will be useful to you.

This book contains only a small part of the obtained scattering characteristics. More information, especially numerical and statistical, can be found in the electronic appendix of this book. The electronic appendix is available on the publisher's website and can be accessed at the beginning of this book. The electronic appendix is also available on our website at http://radar.dinos.net. The data provided in the electronic appendix allows the reader to analyze the scattering characteristics of the considered objects. The numerical data can also be used in the simulation modeling of complex systems with interactions between different radar objects.

In addition, our website http://radar.dinos.net provides access to our previous Open Access books. Since we refer to many of these books here, it will be useful for readers of this book to familiarize themselves with these studies.

This book will be suited for scientists who specialize in the area of electromagnetic wave scattering by radar objects and for engineers who work with radar detection and recognition algorithms for aerial and ground radar targets.

References

1. Sukharevsky, O.I., Vasilets, V.A., Nechitaylo, S.V., and Orlenko, V.M., eds. (2014). *Electromagnetic wave scattering by aerial and ground radar objects*. Boca Raton, FL, London, New York: CRC Press Taylor & Francis Group, 334 p. https://doi.org/10.1201/9781315214511.
2. Sukharevsky, O.I., Vasylets, V.O., and Nechitaylo, S.V. (2015). Scattering characteristics computation method for corner reflectors in arbitrary illumination conditions. In *X Anniversary International Conference on Antenna Theory and Technique (ICATT' 2015)*, April 21–24, 2015, Kharkiv, Ukraine, pp. 219–221. https://doi.org/10.1109/ICATT.2015.7136836.
3. Sukharevsky, O.I., Vasilets, V.A., Orlenko, V.M., and Ryapolov, I. (2020, April). Radar scattering characteristics of a UAV model in X-band. *IET Radar, Sonar & Navigation*, 14(4), pp. 532–537. https://doi.org/10.1049/iet-rsn.2019.0243.
4. Sukharevsky, O.I., Nechitaylo, S.V., Orlenko, V.M., Vasilets, V.A., and Zalevsky, G.S., eds. (2022). *Applied problems in the theory of electromagnetic wave scattering*. Bristol: IOP Publishing, 282 p. https://doi.org/10.1088/978-0-7503-3979-7.
5. Electromagnetic simulation software Altair FEKO – https://altairhyperworks.com/product/FEKO.
6. Sukharevsky, O.I., Vasilets, V.A., and Nechitaylo, S.V. (2017). Scattering and radiation characteristics of antenna systems under nose dielectric radomes. *Progress in Electromagnetics Research B*, 76, pp. 141–157. https://doi.org/10.2528/PIERB17032208.
7. Sukharevsky, O.I., Vasilets, V.A., Sazonov, A.Z., and Tkachuk, K.I. (2000). Calculation of electromagnetic wave scattering on perfectly conducting object partly coated by radar absorbing material with using triangulation cubature formula. *Radio Physics and Radio Astronomy*, 5(1), pp. 47–54. http://rpra-journal.org.ua/index.php/ra/article/view/957.
8. Brekhovskikh, L.M. (1976). *Waves in layered media*. New York: Academic.
9. Sukharevsky, O.I., and Dobrodnyak, A.F. (1988). Three-dimensional problem of diffraction by an ideally conducting wedge with a radio-absorbing cylinder on the edge. *Radiophysics and Quantum Electronics*, 31(9), pp. 763–769. https://doi.org/10.1007/BF01039335.
10. Ufimtsev, P.Y. (2007). *Fundamentals of the physical theory of diffraction*. Hoboken, NJ: Wiley, 329 p.
11. Sukharevsky, O.I., and Vasilets, V.A. (2008). Scattering of reflector antenna with conic dielectric radome. *Progress in Electromagnetic Research B*, 4, pp. 159–169. https://doi.org/10.2528/PIERB08011404.
12. Gordon, Y., and Davison, P. (2006). *Sukhoi Su-27 Flanker - Warbird Tech Vol. 42*. North Branch, MN: Speciality Press. ISBN 978-1-58007-091-1.
13. MiG-29 Fulcrum Fighter – https://www.airforce-technology.com/projects/mig29/.
14. Sukhoi Su-57 – https://aerocorner.com/aircraft/sukhoi-su-57-felon.
15. F-22 Raptor – https://www.af.mil/About-Us/Fact-Sheets/Display/Article/104506/f-22-raptor/.
16. F-35A Lightning II – https://www.af.mil/About-Us/Fact-Sheets/Display/Article/478441/f-35a-lightning-ii/.
17. F-16 Fighting Falcon – https://www.af.mil/About-Us/Fact-Sheets/Display/Article/104505/f-16-fighting-falcon/.
18. F-15 Eagle – https://www.af.mil/About-Us/Fact-Sheets/Display/Article/104501/f-15-eagle/.
19. F/A-18A-D Hornet and F/A-18E/F Super Hornet Strike Fighter – https://www.navy.mil/Resources/Fact-Files/Display-FactFiles/Article/2383479/fa-18a-d-hornet-and-fa-18ef-super-hornet-strike-fighter/.
20. Mehrzweckkampfflugzeug PA-200 Tornado – https://www.bundeswehr.de/de/ausruestung-technik-bundeswehr/luftsysteme-bundeswehr/pa-200-tornado.
21. About the Typhoon FGR4 – https://www.raf.mod.uk/aircraft/typhoon-fgr4/.
22. Jackson, P., ed. (2003). *Jane's all the world's aircraft 2003–2004*. Coulsdon, Surrey: Jane's Information Group.
23. A-10C Thunderbolt II – https://www.af.mil/About-Us/Fact-Sheets/Display/Article/104490/a-10c-thunderbolt-ii/.
24. Su-24M Fencer Bomber – https://www.airforce-technology.com/projects/su24/.
25. Frawley, G. (2002). The international directory of military aircraft 2002/03. Motorbooks Intl, Osceola, WI, 200 p.
26. Wilson, S. (2000). Combat aircraft since 1945. Australian Aviation, Sydney, 192 p.
27. Su-34 (Su-32) Fullback Fighter Bomber – https://www.airforce-technology.com/projects/su34/.
28. B-2 Spirit – https://www.af.mil/About-Us/Fact-Sheets/Display/Article/104482/b-2-spirit/.

References

29. Zakhariev, L.N., and Lemansky, A.A. (1972). *Wave scattering by black bodies*. Moscow (in Russian): Soviet Radio.
30. B-52H Stratofortress – https://www.af.mil/About-Us/Fact-Sheets/Display/Article/104465/b-52h-stratofortress/.
31. B-1B Lancer – https://www.af.mil/About-Us/Fact-Sheets/Display/Article/104500/b-1b-lancer/.
32. Beriev A-50 – https://www.militarytoday.com/aircraft/a50.htm.
33. E-3 Sentry (AWACS) – https://www.globalsecurity.org/military/systems/aircraft/e-3-specs.htm.
34. E-2 Hawkeye Airborne Command and Control Aircraft – https://www.navy.mil/Resources/Fact-Files/Display-FactFiles/Article/2382134/e-2-hawkeye-airborne-command-and-control-aircraft/.
35. L-39 Jet Fighter – https://fighterpilot.com.au/aircraft/l-39-albatros.
36. Taylor, J.W.R., ed. (1988). *Jane's all the world's aircraft 1988–89*. Coulsdon, Surrey: Jane's Defence Data.
37. Ilyushin Il-76 – https://www.maximus-air.com/fleet/ilyushin-il-76.
38. The 737-300/-400/-500 Offers Flexibility to Meet Market Demands – https://www.boeing.com/resources/boeingdotcom/company/about_bca/startup/pdf/historical/737-classic-passenger.pdf.
39. Lambert, M., Munson, K., and Taylor, M.J.H., eds. (1992). *Jane's all the world's aircraft 1988–89* (83rd ed.). Coulson, Surrey: Jane's Information Group.
40. Mil Mi-24 – NATO code: HIND – https://www.army.cz/en/armed-forces/equipment/air-force/helicopters/mil-mi-24—nato-code:-hind-38157/.
41. Tupolev Tu-143 (Reys/Flight) – https://www.militaryfactory.com/aircraft/detail.php?aircraft_id=2470.
42. Orlan-10 Uncrewed Aerial Vehicle (UAV) – https://www.airforce-technology.com/projects/orlan-10-unmanned-aerial-vehicle-uav/.
43. Predator RQ-1/MQ-1/MQ-9 Reaper UAV – https://www.airforce-technology.com/projects/predator-uav/.
44. RQ-4 Global Hawk – https://www.af.mil/About-Us/Fact-Sheets/Display/Article/104516/rq-4-global-hawk/.
45. Shadow 200 RQ-7 Tactical Unmanned Aircraft System – https://www.army-technology.com/projects/shadow-200-uav/.
46. Bayraktar TB2 – https://www.baykartech.com/en/uav/bayraktar-tb2/.
47. Mohajer-6 UAV – https://cmano-db.com/aircraft/5022/.
48. Searcher II – https://www.israeli-weapons.com/weapons/aircraft/uav/searcher2/Searcher2.html.
49. Dozor-600/Dozor-3 – https://www.globalsecurity.org/military/world/russia/dozor.htm.
50. Kronstadt Orion (Inokhodets) – https://www.militaryfactory.com/aircraft/detail.php?aircraft_id=1848.
51. Shahed 136 – https://www.militarytoday.com/aircraft/shahed_136.htm.
52. Kh-555 – https://www.armyrecognition.com/russia_russian_missile_system_vehicle_uk/kh-555_air-launched_subsonic_cruise_missile.technical_data.html.
53. Kh-101/Kh-102 – https://missilethreat.csis.org/missile/kh-101-kh-102/.
54. Anti-Ship Cruise Missile P-700 Granite (3M-45) – https://en.missilery.info/missile/granit.
55. P-800 Oniks/Yakhont/Bastion (SS-N-26 Strobile) – https://missilethreat.csis.org/missile/ss-n-26/.
56. 3M-14 Kalibr (SS-N-30A) – https://missilethreat.csis.org/missile/ss-n-30a/.
57. 3M-54 Kalibr/Club (SS-N-27) – https://missilethreat.csis.org/missile/ss-n-27-sizzler/.
58. AGM-86C/D Conventional Air Launched Cruise Missile – https://nuke.fas.org/guide/usa/bomber/calcm.htm.
59. Kh-47M2 Kinzhal – https://www.militarytoday.com/missiles/kh_47m2_kinzhal.htm.
60. Zvezda Kh-25 MP – https://weaponsystems.net/system/175-Zvezda+Kh-25MP.
61. Kh-29T – https://wiki.warthunder.com/Kh-29T.
62. Antimissile Missile X-31P (X-31PD) – https://en.missilery.info/missile/x31p.
63. Raduga Kh-32 – https://www.globalsecurity.org/wmd/world/russia/kh-32.htm.
64. Kh-35 Uran – https://weaponsystems.net/system/1188-Kh-35+Uran.
65. Kh-38/X-38 Air-to-Surface Missile – https://www.globalsecurity.org/military/world/russia/kh-38.htm.
66. Soviet/Russian Tactical Air to Surface Missiles – https://www.ausairpower.net/APA-Rus-ASM.html#mozTocId919852.
67. X-59M "Ovod-M" Rocket – https://en.missilery.info/missile/x59m.
68. AGM-65 Maverick – https://www.af.mil/About-Us/Fact-Sheets/Display/Article/104577/agm-65-maverick/.
69. AGM-114 Hellfire - https://www.military.com/equipment/agm-114-hellfire.
70. AGM-88 HARM – https://www.af.mil/About-Us/Fact-Sheets/Display/Article/104574/agm-88-harm//.
71. Lenkflugkörper Taurus KEPD-350 – https://www.bundeswehr.de/de/ausruestung-technik-bundeswehr/ausruestung-bewaffnung/marschflugkoerper-taurus-kepd-350.
72. SCALP EG/Storm Shadow/SCALP Naval/Black Shaheen/APACHE AP – https://missilethreat.csis.org/missile/apache-ap/.

73. S-300 – https://missilethreat.csis.org/defsys/s-300/.
74. IMI (Brunswick) ADM-141 TALD – https://www.designation-systems.net/dusrm/m-141.html.
75. T-90 Main Battle Tank – https://www.militarytoday.com/tanks/t90.htm.
76. Die Leistungsmerkmale des LEOPARD 2 A4 – https://www.knds.de/systeme-produkte/kettenfahrzeuge/kampfpanzer/leopard-2-a4/.
77. M1 Abrams. Main Battle Tank – https://www.militarytoday.com/tanks/m1_abrams.htm.
78. 9S18 "Kupol" TUBE ARM – https://www.globalsecurity.org/military/world/russia/9s18.htm.
79. Führungspunkt 9S470 – https://www.rwd-mb3.de/technik_g/pages/9s470.htm.
80. Anti-Aircraft Missile System 9K37 Buk-M1 – https://en.missilery.info/missile/bukm1.
81. 5P85 TEL (S-300 PMU) – https://truck-encyclopedia.com/coldwar/ussr/5P85.php.
82. Anti-Aircraft Missile System S-300PS (S-300PMU) – https://en.missilery.info/missile/c300ps.
83. 30N6E Flap Lid B – https://www.armyrecognition.com/russia_russian_missile_system_vehicle_uk/30n6_30n6e_5n63s_flap_lid_b_tracking_and_missile_guidance_radar_sa-10_grumble_technical_data_sheet.html.
84. 76N6 Clam Shell Russian Low Altitude Acquisition Radar – https://odin.tradoc.army.mil/WEG/Asset/76N6_Clam_Shell_Russian_Low_Altitude_Acquisition_Radar.
85. Warsaw Pact/Russian Air Defence Command Posts – https://www.ausairpower.net/APA-Rus-ADCP-CP.html#mozTocId207381.
86. 9A83/9A83M Transporter Erector Launcher and Radar – https://www.ausairpower.net/APA-Giant-Gladiator.html#mozTocId336111.
87. S-200 (SA-5 Gammon) – https://missilethreat.csis.org/defsys/s-200-sa-5-gammon/.
88. 2S6 Tunguska – https://weaponsystems.net/system/60-2S6+Tunguska.
89. SS-26 (Stone)/9K720 Iskander – https://www.militaryfactory.com/armor/detail.php?armor_id=791.
90. ST-68U – https://www.radartutorial.eu/19.kartei/11.ancient/karte059.en.html.
91. P-18 – https://www.radartutorial.eu/19.kartei/11.ancient/karte049.en.html.
92. Sukharevsky, O., Belevshchuk, Y., Vasilets, V., and Nechitaylo, S. (2010). Radar cross-section calculation method for antenna of P-18 radar station. In *2010 International Conference on Mathematical Methods in Electromagnetic Theory*, Kyiv, Ukraine, pp. 101–104. https://doi.org/10.1109/MMET.2010.5611345.
93. Ropucha (Class)/Project 775 – https://www.militaryfactory.com/ships/detail.php?ship_id=ropucha-class-landing-ship-russia.

Index

2S6, anti-aircraft gun-missile system 2K22
    Tunguska 477
    computer model 477
    mean and median RCS 477, 479, 480, 481
    noncoherent RCS 478, 479, 480, 482
    RCS 478, 479, 480, 482
    scattering characteristics 477, 479, 480, 481
3M-14 Kalibr, cruise missile 276
    amplitude distribution of echo signal 278, 280, 281
    computer model 276
    geometrical characteristics 276
    mean and median RCS 277, 278, 280
    noncoherent RCS 277, 279, 280
    parameters of probability distributions 278, 280, 281
    RCS 277, 279, 280
    scattering characteristics 277, 278, 280
3M-54 Kalibr, cruise missile 282
    amplitude distribution of echo signal 284, 286, 287
    computer model 282
    geometrical characteristics 282
    mean and median RCS 283, 284, 286
    noncoherent RCS 283, 285, 286
    parameters of probability distributions 284, 286, 287
    RCS 283, 285, 286
    scattering characteristics 283, 284, 286
30N6, S-300PS fire control system 441
    computer model 441
    mean and median RCS 441, 443, 444, 445
    noncoherent RCS 442, 443, 444, 446
    RCS 442, 443, 444, 446
    scattering characteristics 441, 443, 444, 445
5P72, S-200 launcher 471
    computer model 471
    mean and median RCS 471, 473, 474, 475
    noncoherent RCS 472, 473, 474, 476
    RCS 472, 473, 474, 476
    scattering characteristics 471, 473, 474, 475
5P85D, S-300PS transporter erector launcher 429
    computer model 429
    mean and median RCS 429, 431, 432, 433
    noncoherent RCS 430, 431, 432, 434
    RCS 430, 431, 432, 434
    scattering characteristics 429, 431, 432, 433
5P85S, S-300PS transporter erector launcher 435
    computer model 435
    mean and median RCS 435, 437, 438, 439
    noncoherent RCS 436, 437, 438, 440
    RCS 436, 437, 438, 440
    scattering characteristics 435, 437, 438, 439
5V55R, anti-aircraft missile 378
    amplitude distribution of echo signal 380, 382, 383
    computer model 378
    geometrical characteristics 378
    mean and median RCS 379, 380, 382
    noncoherent RCS 379, 381, 382
    parameters of probability distributions 380, 382, 383
    RCS 379, 381, 382
    scattering characteristics 379, 380, 382

54K6, S-300PS command post vehicle 453
    computer model 453
    mean and median RCS 453, 455, 456, 457
    noncoherent RCS 454, 455, 456, 458
    RCS 454, 455, 456, 458
    scattering characteristics 453, 455, 456, 457
76N6, S-300PS low altitude acquisition radar 447
    computer model 447
    mean and median RCS 447, 449, 450, 451
    noncoherent RCS 448, 449, 450, 452
    RCS 448, 449, 450, 452
    scattering characteristics 447, 449, 450, 451
9A310M1, Buk transporter erector launcher and radar 423
    computer model 423
    mean and median RCS 423, 425, 426, 427
    noncoherent RCS 424, 425, 426, 428
    RCS 424, 425, 426, 428
    scattering characteristics 423, 425, 426, 427
9A83, S-300V transporter erector launcher and radar 459
    computer model 459
    mean and median RCS 459, 461, 462, 463
    noncoherent RCS 460, 461, 462, 464
    RCS 460, 461, 462, 464
    scattering characteristics 459, 461, 462, 463
9P78-1, Iskander launcher 483
    computer model 483
    mean and median RCS 483, 485, 486, 487
    noncoherent RCS 484, 485, 486, 488
    RCS 484, 485, 486, 488
    scattering characteristics 483, 485, 486, 487
9S18M1, Buk target acquisition radar 411
    computer model 411
    mean and median RCS 411, 413, 414, 415
    noncoherent RCS 412, 413, 414, 416
    RCS 412, 413, 414, 416
    scattering characteristics 411, 413, 414, 415
9S457, S-300V command post vehicle 465
    computer model 465
    mean and median RCS 465, 467, 468, 469
    noncoherent RCS 466, 467, 468, 470
    RCS 466, 467, 468, 470
    scattering characteristics 465, 467, 468, 469
9S470, Buk command post vehicle 417
    computer model 417
    mean and median RCS 417, 419, 420, 421
    noncoherent RCS 418, 419, 420, 422
    RCS 418, 419, 420, 422
    scattering characteristics 417, 419, 420, 421

A-10 Thunderbolt II, attack aircraft 78
    amplitude distribution of echo signal 80, 82, 83
    computer model 78
    geometrical characteristics 78
    mean and median RCS 79, 80, 82
    noncoherent RCS 79, 81, 82
    parameters of probability distributions 80, 82, 83
    RCS 79, 81, 82
    scattering characteristics 79, 80, 82

A-50, airborne early warning and control aircraft 132
    amplitude distribution of echo signal 134, 136, 137
    computer model 132
    geometrical characteristics 132
    mean and median RCS 133, 134, 136
    noncoherent RCS 133, 135, 136
    parameters of probability distributions 134, 136, 137
    RCS 133, 135, 136
    scattering characteristics 133, 134, 136
ADM-141C iTALD, decoy missile 384
    amplitude distribution of echo signal 386, 388, 389
    computer model 384
    geometrical characteristics 384
    mean and median RCS 385, 386, 388
    noncoherent RCS 385, 387, 388
    parameters of probability distributions 386, 388, 389
    RCS 385, 387, 388
    scattering characteristics 385, 386, 388
aerial objects 1, 10
    asymptotic method 1
    cubature formula 3
    local edge scatterer 4
    optimal distribution 11, 14
    radar scattering 2
    scattering characteristics 10
AGM-65 Maverick, air-to-surface missile 348
    amplitude distribution of echo signal 350, 352, 353
    computer model 348
    geometrical characteristics 348
    mean and median RCS 349, 350, 352
    noncoherent RCS 349, 351, 352
    parameters of probability distributions 350, 352, 353
    RCS 349, 351, 352
    scattering characteristics 349, 350, 352
AGM-86C CALCM, cruise missile 288
    amplitude distribution of echo signal 290, 292, 293
    computer model 288
    geometrical characteristics 288
    mean and median RCS 289, 290, 292
    noncoherent RCS 289, 291, 292
    parameters of probability distributions 290, 292, 293
    RCS 289, 291, 292
    scattering characteristics 289, 290, 292
AGM-88 HARM, air-to-surface missile 360
    amplitude distribution of echo signal 362, 364, 365
    computer model 360
    geometrical characteristics 360
    mean and median RCS 361, 363, 364
    noncoherent RCS 361, 363, 364
    parameters of probability distributions 362, 364, 365
    RCS 361, 363, 364
    scattering characteristics 361, 362, 364
AGM-114 Hellfire, air-to-surface missile 354
    amplitude distribution of echo signal 356, 358, 359
    computer model 354
    geometrical characteristics 354
    mean and median RCS 355, 356, 358
    noncoherent RCS 355, 357, 358
    parameters of probability distributions 356, 358, 359
    RCS 355, 357, 358
    scattering characteristics 355, 356, 358
Airborne Warning and Control System (AWACS) 7
aircraft model 6, 12
An-26, multipurpose transport aircraft 156
    amplitude distribution of echo signal 158, 160, 161
    computer model 156
    geometrical characteristics 156
    mean and median RCS 157, 158, 160
    noncoherent RCS 157, 159, 160
    parameters of probability distributions 158, 160, 161
    RCS 157, 159, 160
    scattering characteristics 157, 158, 160
antenna system 7
    scattering 7
    calculation method 7, 8
    with radome 8
AWACS see Airborne Warning and Control System (AWACS)

B-1B, strategic bomber 126
    amplitude distribution of echo signal 128, 130, 131
    computer model 126
    geometrical characteristics 126
    mean and median RCS 127, 128, 130
    noncoherent RCS 127, 129, 130
    parameters of probability distributions 128, 130, 131
    RCS 127, 129, 130
    scattering characteristics 127, 128, 130
B-2, strategic bomber 114
    amplitude distribution of echo signal 116, 118, 119
    computer model 114
    geometrical characteristics 114
    mean and median RCS 115, 116, 118
    noncoherent RCS 115, 117, 118
    parameters of probability distributions 116, 118, 119
    RCS 115, 117, 118
    scattering characteristics 115, 116, 118
B-52, strategic bomber 120
    amplitude distribution of echo signal 122, 124, 125
    computer model 120
    geometrical characteristics 120
    mean and median RCS 121, 122, 124
    noncoherent RCS 121, 123, 124
    parameters of probability distributions 122, 124, 125
    RCS 121, 123, 124
    scattering characteristics 121, 122, 124
Bayraktar TB2, unmanned aerial vehicle 216
    amplitude distribution of echo signal 218, 220, 221
    computer model 216
    geometrical characteristics 216
    mean and median RCS 217, 218, 220
    noncoherent RCS 217, 219, 220
    parameters of probability distributions 218, 220, 221
    RCS 217, 219, 220
    scattering characteristics 217, 218, 220
Boeing 737–400 medium-range airliner 168
    amplitude distribution of echo signal 170, 171, 173
    computer model 168
    geometrical characteristics 168
    mean and median RCS 168, 170, 172
    noncoherent RCS 169, 170, 172
    parameters of probability distributions 170, 171, 173
    RCS 169, 170, 172
    scattering characteristics 168, 170, 172
Buk, missile system 411
    9A310M1 launcher and radar 423
    9S18M1 radar 411
    9S470 vehicle 417

Index

command post vehicle 417, 453, 465
    Buk 417
    S-300PS 453
    S-300V 465
complex-shaped objects 1, 10
cruise missile 252
    3M-14 Kalibr 276
    3M-54 Kalibr 282
    AGM-86C CALCM 288
    Kh-32 318
    Kh-35 324
    Kh-101 258
    Kh-555 252
    P-700 Granit 264
    P-800 Oniks 270
    Storm Shadow 372
    Taurus KEPD-350 366
cubature formula 3

Dozor-600, unmanned aerial vehicle 234
    amplitude distribution of echo signal 236, 238, 239
    computer model 234
    geometrical characteristics 234
    mean and median RCS 235, 236, 238
    noncoherent RCS 235, 237, 238
    parameters of probability distributions 236, 238, 239
    RCS 235, 237, 238
    scattering characteristics 235, 236, 238

E-2C, airborne early warning and control aircraft 144
    amplitude distribution of echo signal 146, 148, 149
    computer model 144
    geometrical characteristics 144
    mean and median RCS 145, 146, 148
    noncoherent RCS 145, 147, 148
    parameters of probability distributions 146, 148, 149
    RCS 145, 147, 148
    scattering characteristics 145, 146, 148
E-3A, airborne early warning and control aircraft 138
    amplitude distribution of echo signal 140, 142, 143
    computer model 138
    geometrical characteristics 138
    mean and median RCS 139, 140, 142
    noncoherent RCS 139, 141, 142
    parameters of probability distributions 140, 142, 143
    RCS 139, 141, 142
    scattering characteristics 139, 140, 142
ellipsoid 7
    parameters 8
    three-axial 7

F-15, air superiority fighter 48
    amplitude distribution of echo signal 50, 52, 53
    computer model 48
    geometrical characteristics 48
    mean and median RCS 49, 50, 52
    noncoherent RCS 49, 51, 52
    parameters of probability distributions 50, 52, 53
    RCS 49, 51, 52
    scattering characteristics 49, 50, 52
F-16, multirole fighter 42
    amplitude distribution of echo signal 44, 46, 47
    computer model 42
    geometrical characteristics 42
    mean and median RCS 43, 44, 46
    noncoherent RCS 43, 45, 46
    parameters of probability distributions 44, 46, 47
    RCS 43, 45, 46
    scattering characteristics 43, 44, 46
F/A-18, multirole attack and fighter aircraft 54
    amplitude distribution of echo signal 56, 58, 59
    computer model 54
    geometrical characteristics 54
    mean and median RCS 55, 56, 58
    noncoherent RCS 55, 57, 58
    parameters of probability distributions 56, 58, 59
    RCS 55, 57, 58
    scattering characteristics 55, 56, 58
F-22, tactical fighter 30
    amplitude distribution of echo signal 32, 34, 35
    computer model 30
    geometrical characteristics 30
    mean and median RCS 31, 32, 34
    noncoherent RCS 31, 33, 34
    parameters of probability distributions 32, 34, 35
    RCS 31, 33, 34
    scattering characteristics 31, 32, 34
F-35, multirole fighter 36
    amplitude distribution of echo signal 38, 40, 41
    computer model 36
    geometrical characteristics 36
    mean and median RCS 37, 38, 40
    noncoherent RCS 37, 39, 40
    parameters of probability distributions 38, 40, 41
    RCS 37, 39, 40
    scattering characteristics 37, 38, 40
FEKO software x

ground complex-shaped objects 6
    electromagnetic wave propagation 6
    nonperfectly reflecting surface 6
    perfectly conducting model 6
    plane electromagnetic wave scattering 390
    scattering characteristic computation method 6
ground objects 5
    computation method 6
    method for computing RCS 6
    model 6
    paths of electromagnetic wave propagation 6
    scattering characteristics 390
    underlying surface 390, 391

IAI Searcher II (Forpost), unmanned aerial vehicle 228
    amplitude distribution of echo signal 230, 232, 233
    computer model 228
    geometrical characteristics 228
    mean and median RCS 229, 230, 232
    noncoherent RCS 229, 231, 232
    parameters of probability distributions 230, 232, 233
    RCS 229, 231, 232
    scattering characteristics 229, 230, 232
Il-76, srategic airlifter 162
    amplitude distribution of echo signal 164, 166, 167
    computer model 162
    geometrical characteristics 162
    mean and median RCS 163, 164, 166

Il-76, srategic airlifter (*cont.*)
   noncoherent RCS 163, 165, 166
   parameters of probability distributions 164, 166, 167
   RCS 163, 165, 166
   scattering characteristics 163, 164, 166

Kh-25MPU, anti-radiation missile 300
   amplitude distribution of echo signal 302, 304, 305
   computer model 300
   geometrical characteristics 300
   mean and median RCS 301, 302, 304
   noncoherent RCS 301, 303, 304
   parameters of probability distributions 302, 304, 305
   RCS 301, 303, 304
   scattering characteristics 301, 302, 304

Kh-29T, air-to-surface missile 306
   amplitude distribution of echo signal 308, 310, 311
   computer model 306
   geometrical characteristics 306
   mean and median RCS 307, 308, 310
   noncoherent RCS 307, 309, 310
   parameters of probability distributions 308, 310, 311
   RCS 307, 309, 310
   scattering characteristics 307, 308, 310

Kh-31PD, anti-radiation missile 312
   amplitude distribution of echo signal 314, 316, 317
   computer model 312
   geometrical characteristics 312
   mean and median RCS 313, 314, 316
   noncoherent RCS 313, 315, 316
   parameters of probability distributions 314, 316, 317
   RCS 313, 315, 316
   scattering characteristics 313, 314, 316

Kh-32, cruise missile 318
   amplitude distribution of echo signal 320, 322, 323
   computer model 318
   geometrical characteristics 318
   mean and median RCS 319, 320, 322
   noncoherent RCS 319, 321, 322
   parameters of probability distributions 320, 322, 323
   RCS 319, 321, 322
   scattering characteristics 319, 320, 322

Kh-35, cruise missile 324
   amplitude distribution of echo signal 326, 328, 329
   computer model 324
   geometrical characteristics 324
   mean and median RCS 325, 326, 328
   noncoherent RCS 325, 327, 328
   parameters of probability distributions 326, 328, 329
   RCS 325, 327, 328
   scattering characteristics 325, 326, 328

Kh-38ML, anti-radiation missile 330
   amplitude distribution of echo signal 332, 334, 335
   computer model 330
   geometrical characteristics 330
   mean and median RCS 331, 332, 334
   noncoherent RCS 331, 333, 334
   parameters of probability distributions 332, 334, 335
   RCS 331, 333, 334
   scattering characteristics 331, 332, 334

Kh-47M2 Kinzhal, hypersonic missile 294
   amplitude distribution of echo signal 296, 298, 299
   computer model 294
   geometrical characteristics 294
   mean and median RCS 295, 296, 298
   noncoherent RCS 295, 297, 298
   parameters of probability distributions 296, 298, 299
   RCS 295, 297, 298
   scattering characteristics 295, 296, 298

Kh-58UShKE, anti-radiation missile 336
   amplitude distribution of echo signal 338, 340, 341
   computer model 336
   geometrical characteristics 336
   mean and median RCS 337, 338, 340
   noncoherent RCS 337, 339, 340
   parameters of probability distributions 338, 340, 341
   RCS 337, 339, 340
   scattering characteristics 337, 338, 340

Kh-59M, air-to-surface missile 342
   amplitude distribution of echo signal 344, 346, 347
   computer model 342
   geometrical characteristics 342
   mean and median RCS 343, 344, 346
   noncoherent RCS 343, 345, 346
   parameters of probability distributions 344, 346, 347
   RCS 343, 345, 346
   scattering characteristics 343, 344, 346

Kh-101, cruise missile 258
   amplitude distribution of echo signal 260, 262, 263
   computer model 258
   geometrical characteristics 258
   mean and median RCS 259, 260, 262
   noncoherent RCS 259, 261, 262
   parameters of probability distributions 260, 262, 263
   RCS 259, 261, 262
   scattering characteristics 259, 260, 262

Kh-555, cruise missile 252
   amplitude distribution of echo signal 254, 256, 257
   computer model 252
   geometrical characteristics 252
   mean and median RCS 253, 254, 256
   noncoherent RCS 253, 255, 256
   parameters of probability distributions 254, 256, 257
   RCS 253, 255, 256
   scattering characteristics 253, 254, 256

Kronshtadt Orion, unmanned aerial vehicle 240
   amplitude distribution of echo signal 242, 244, 245
   computer model 240
   geometrical characteristics 240
   mean and median RCS 241, 242, 244
   noncoherent RCS 241, 243, 244
   parameters of probability distributions 242, 244, 245
   RCS 241, 243, 244
   scattering characteristics 241, 242, 244

L-39, jet trainer aircraft 150
   amplitude distribution of echo signal 152, 154, 155
   computer model 150
   geometrical characteristics 150
   mean and median RCS 151, 152, 154
   noncoherent RCS 151, 153, 154
   parameters of probability distributions 152, 154, 155
   RCS 151, 153, 154
   scattering characteristics 151, 152, 154

Leopard-2, main battle tank 399
   computer model 399
   geometrical characteristics 399
   mean and median RCS 399, 401, 402, 403

Index

noncoherent RCS 400, 401, 402, 404
RCS 400, 401, 402, 404
scattering characteristics 399, 401, 402, 403
loitering munition 246

M1A1 Abrams main battle tank 405
   computer model 405
   geometrical characteristics 405
   mean and median RCS 405, 407, 408, 409
   noncoherent RCS 406, 407, 408, 410
   RCS 406, 407, 408, 410
   scattering characteristics 405, 407, 408, 409
Mi-8, multi-purpose helicopter 174
   amplitude distribution of echo signal 176, 178, 179
   computer model 174
   geometrical characteristics 174
   mean and median RCS 175, 176, 178
   noncoherent RCS 175, 177, 178
   parameters of probability distributions 176, 178, 179
   RCS 175, 177, 178
   scattering characteristics 175, 176, 178
Mi-24, multi-purpose combat helicopter 180
   amplitude distribution of echo signal 182, 184, 185
   computer model 180
   geometrical characteristics 180
   mean and median RCS 181, 182, 184
   noncoherent RCS 181, 183, 184
   parameters of probability distributions 182, 184, 185
   RCS 181, 183, 184
   scattering characteristics 181, 182, 184
MiG-29 front-line fighter 18
   amplitude distribution of echo signal 20, 22, 23
   computer model 18
   geometrical characteristics 18
   mean and median RCS 19, 20, 22
   noncoherent RCS 19, 21, 22
   parameters of probability distributions 20, 22, 23
   RCS 19, 21, 22
   scattering characteristics 19, 20, 22
Mohajer-6, unmanned aerial vehicle 222
   amplitude distribution of echo signal 224, 226, 227
   computer model 222
   geometrical characteristics 222
   mean and median RCS 223, 224, 226
   noncoherent RCS 223, 225, 226
   parameters of probability distributions 224, 226, 227
   RCS 223, 225, 226
   scattering characteristics 223, 224, 226

noncoherent RCS 1
nonperfectly reflecting surface 3

Orlan-10, unmanned aerial vehicle 192
   amplitude distribution of echo signal 194, 196, 197
   computer model 192
   geometrical characteristics 192
   mean and median RCS 193, 194, 196
   noncoherent RCS 193, 195, 196
   parameters of probability distributions 194, 196, 197
   RCS 193, 195, 196
   scattering characteristics 193, 194, 196

P-18, radar 495
   computer model 495

mean and median RCS 496, 497, 498
noncoherent RCS 496, 497, 499, 500
RCS 496, 497, 499, 500
scattering characteristics 496, 497, 498
P-700 Granit, cruise missile 264
   amplitude distribution of echo signal 266, 268, 269
   computer model 264
   geometrical characteristics 264
   mean and median RCS 265, 266, 268
   noncoherent RCS 265, 267, 268
   parameters of probability distributions 266, 268, 269
   RCS 265, 267, 268
   scattering characteristics 265, 266, 268
P-800 Oniks, cruise missile 270
   amplitude distribution of echo signal 272, 274, 275
   computer model 270
   geometrical characteristics 270
   mean and median RCS 271, 272, 274
   noncoherent RCS 271, 273, 274
   parameters of probability distributions 272, 274, 275
   RCS 271, 273, 274
   scattering characteristics 271, 272, 274
plane electromagnetic wave scattering 1
   "object–ground" system 6
   physical optics approximation 2

radar absorbing material (RAM) 2
   coating x, 24, 30, 36, 114
   edge RAM toroidal coating 4
radar cross-section (RCS) x
   Mean x, 13
   Median x, 13
   noncoherent 1
radio transparent antenna radome 8
radome 8
RAM *see* Radar absorbent material (RAM)
Ropucha-class landing ship 501
   computer model 501
   mean and median RCS 501, 503, 504, 505
   noncoherent RCS 502, 503, 504, 506
   RCS 502, 503, 504, 506
   scattering characteristics 501, 503, 504, 505
RQ-1 Predator, unmanned aerial vehicle 198
   amplitude distribution of echo signal 200, 202, 203
   computer model 198
   geometrical characteristics 198
   mean and median RCS 199, 200, 202
   noncoherent RCS 199, 201, 202
   parameters of probability distributions 200, 202, 203
   RCS 200, 202, 203
   scattering characteristics 199, 200, 202
RQ-4 Global Hawk, unmanned aerial vehicle 204
   amplitude distribution of echo signal 206, 208, 209
   computer model 204
   geometrical characteristics 204
   mean and median RCS 205, 206, 208
   noncoherent RCS 205, 207, 208
   parameters of probability distributions 206, 208, 209
   RCS 205, 207, 208
   scattering characteristics 205, 206, 208
RQ-7 Shadow, unmanned aerial vehicle 210
   amplitude distribution of echo signal 212, 214, 215
   computer model 210
   geometrical characteristics 210

RQ-7 Shadow, unmanned aerial vehicle (*cont.*)
    mean and median RCS 211, 212, 214
    noncoherent RCS 211, 213, 214
    parameters of probability distributions 212, 214, 215
    RCS 211, 213, 214
    scattering characteristics 211, 212, 214

S-300PS, missile system 378, 429
    5P85D TEL 429
    5P85S TEL 435
    5V55R missile 378
    30N6 system 441
    54K6 vehicle 453
    76N6 radar 447
S-300V, missile system 459
    9A83 TELAR 459
    9S457 vehicle 465
Shahed 136, loitering munition 246
    amplitude distribution of echo signal 248, 250, 251
    computer model 246
    geometrical characteristics 246
    mean and median RCS 247, 248, 250
    noncoherent RCS 247, 249, 250
    parameters of probability distributions 248, 250, 251
    RCS 247, 249, 250
    scattering characteristics 247, 248, 250
Sommerfeld-type absorbent 114
ST-68U, radar 489
    computer model 489
    mean and median RCS 489, 491, 492, 493
    noncoherent RCS 490, 491, 492, 494
    RCS 490, 491, 492, 494
    scattering characteristics 489, 491, 492, 493
Storm Shadow, cruise missile 372
    amplitude distribution of echo signal 374, 376, 377
    computer model 372
    geometrical characteristics 372
    mean and median RCS 373, 374, 376
    noncoherent RCS 373, 375, 376
    parameters of probability distributions 374, 376, 377
    RCS 373, 375, 376
    scattering characteristics 373, 374, 376
Su-24, tactical bomber 84
    amplitude distribution of echo signal 86, 88, 89
    computer model 84
    geometrical characteristics 84
    mean and median RCS 85, 86, 88
    noncoherent RCS 85, 87, 88
    parameters of probability distributions 86, 88, 89
    RCS 85, 87, 88
    scattering characteristics 85, 86, 88
Su-25, attack aircraft 72
    amplitude distribution of echo signal 74, 76, 77
    computer model 72
    geometrical characteristics 72
    mean and median RCS 73, 74, 76
    noncoherent RCS 73, 75, 76
    parameters of probability distributions 74, 76, 77
    RCS 73, 75, 76
    scattering characteristics 73, 74, 76
Su-27, multirole fighter 12
    amplitude distribution of echo signal 14, 16, 17
    computer model 12
    geometrical characteristics 12
    mean and median RCS 13, 14, 16
    noncoherent RCS 13, 15, 16
    parameters of probability distributions 14, 16, 17
    RCS 13, 15, 16
    scattering characteristics 13, 14, 16
Su-34, tactical bomber 108
    amplitude distribution of echo signal 110, 112, 113
    computer model 108
    geometrical characteristics 108
    mean and median RCS 109, 110, 112
    noncoherent RCS 109, 111, 112
    parameters of probability distributions 110, 112, 113
    RCS 109, 111, 112
    scattering characteristics 109, 110, 112
Su-57, multirole fighter 24
    amplitude distribution of echo signal 26, 28, 29
    computer model 24
    geometrical characteristics 24
    mean and median RCS 25, 26, 28
    noncoherent RCS 25, 27, 28
    parameters of probability distributions 26, 28, 29
    RCS 25, 27, 28
    scattering characteristics 25, 26, 28
surface geometry modeling 6, 7

T-90 main battle tank 393
    computer model 393
    geometrical characteristics 393
    mean and median RCS 393, 395, 396, 397
    noncoherent RCS 394, 395, 396, 398
    RCS 394, 395, 396, 398
    scattering characteristics 393, 395, 396, 397
Taurus KEPD 350, cruise missile 366
    amplitude distribution of echo signal 368, 370, 371
    computer model 366
    geometrical characteristics 366
    mean and median RCS 367, 368, 370
    noncoherent RCS 367, 369, 370
    parameters of probability distributions 368, 370, 371
    RCS 367, 369, 370
    scattering characteristics 367, 368, 370
TEL *see* transporter erector launcher
TELAR *see* transporter erector launcher and radar
Tornado IDS, multirole combat aircraft 60
    amplitude distribution of echo signal 62, 64, 65
    computer model 60
    geometrical characteristics 60
    mean and median RCS 61, 62, 64
    noncoherent RCS 61, 63, 64
    parameters of probability distributions 62, 64, 65
    RCS 61, 63, 64
    scattering characteristics 61, 62, 64
transporter erector launcher 429
    S-300PS 5P85D 429
    S-300PS 5P85S 435
transporter erector launcher and radar 423, 459
    Buk 9A310M1 423
    S-300V 9A83 459
Tu-22M3 long-range strategic bomber 90
    amplitude distribution of echo signal 92, 93, 95
    computer model 90
    geometrical characteristics 90
    mean and median RCS 90, 92, 94
    noncoherent RCS 91, 92, 94

Index

parameters of probability distributions 92, 93, 95
RCS 91, 92, 94
scattering characteristics 90, 92, 94
Tu-95, strategic bomber 96
    amplitude distribution of echo signal 98, 100, 101
    computer model 96
    geometrical characteristics 96
    mean and median RCS 97, 98, 100
    noncoherent RCS 97, 99, 100
    parameters of probability distributions 98, 100, 101
    RCS 97, 99, 100
    scattering characteristics 97, 98, 100
Tu-143 Reys, unmanned aerial vehicle 186
    amplitude distribution of echo signal 188, 190, 191
    computer model 186
    geometrical characteristics 186
    mean and median RCS 187, 188, 190
    noncoherent RCS 187, 189, 190
    parameters of probability distributions 188, 190, 191
    RCS 187, 189, 190
    scattering characteristics 187, 188, 190
Tu-160, strategic bomber 102
    amplitude distribution of echo signal 104, 106, 107
    computer model 102
    geometrical characteristics 102
    mean and median RCS 103, 104, 106
    noncoherent RCS 103, 105, 106
    parameters of probability distributions 104, 106, 107
    RCS 103, 105, 106
    scattering characteristics 103, 104, 106

UAV *see* Unmanned aerial vehicle
underlying surface 6, 390
unmanned aerial vehicle 8
    asymptotic method 9
    Bayraktar TB2 216
    Dozor-600 234
    IAI Searcher II (Forpost) 228
    Kronshtadt Orion 240
    Mohajer-6 222
    Orlan-10 192
    RQ-1 Predator 198
    RQ-4 Global Hawk 204
    RQ-7 Shadow 210
    scattering characteristics 9
    Tu-143 Reys 186

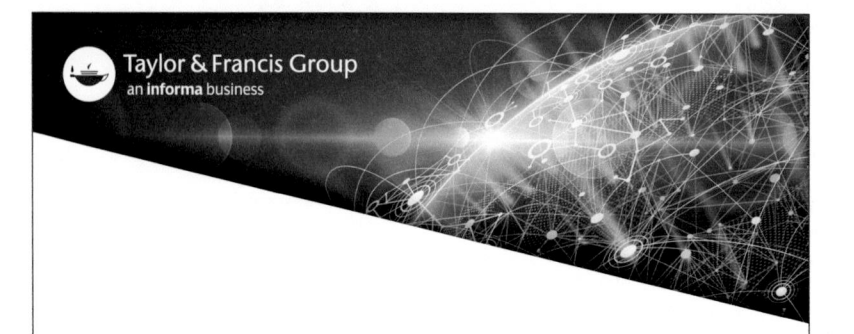